Mass Spectrometry in Food Analysis

Food Analysis & Properties

Series Editor
Leo M.L. Nollet
University College Ghent, Belgium

This CRC series **Food Analysis and Properties** is designed to provide a state-of-the-art coverage on topics to the understanding of physical, chemical, and functional properties of foods: including (1) recent analysis techniques of a choice of food components; (2) developments and evolutions in analysis techniques related to food; and (3) recent trends in analysis techniques of specific food components and/or a group of related food components.

Fingerprinting Techniques in Food Authenticity and Traceability
Edited by K.S. Siddiqi and Leo M.L. Nollet

Hyperspectral Imaging Analysis and Applications for Food Quality
Edited by Nrusingha Charan Basantia, Leo M.L. Nollet, and Mohammed Kamruzzaman

Ambient Mass Spectroscopy Techniques in Food and the Environment
Edited by Leo M.L. Nollet and Basil K. Munjanja

Food Aroma Evolution
During Food Processing, Cooking and Aging
Edited by Matteo Bordiga and Leo M.L. Nollet

Mass Spectrometry Imaging in Food Analysis
Edited by Leo M.L. Nollet

Proteomics for Food Authentication
Edited by Leo M.L. Nollet, Otles, and Semih

Analysis of Nanoplastics and Microplastics in Food
Edited by Leo M.L. Nollet and Khwaja Salahuddin Siddiqi

Chiral Organic Pollutants
Monitoring and Characterization in Food and the Environment
Edited by Edmond Sanganyado, Basil Munjanja, and Leo M.L. Nollet

Sequencing Technologies in Microbial Food Safety and Quality
Edited by Devarajan Thangadurai, Leo M.L. Nollet, Saher Islam, and Jeyabalan Sangeetha

Nanoemulsions in Food Technology
Development, Characterization, and Applications
Edited by Javed Ahmad and Leo M.L. Nollet

Mass Spectrometry in Food Analysis
Edited by Leo M.L. Nollet and Robert Winkler

For more information, please visit the Series Page: https://www.crcpress.com/Food-Analysis--Properties/book-series/CRCFOODANPRO

Mass Spectrometry in Food Analysis

Edited by
Leo M.L. Nollet
Robert Winkler

CRC Press
Taylor & Francis Group
Boca Raton London New York

CRC Press is an imprint of the
Taylor & Francis Group, an **informa** business

First edition published 2022
by CRC Press
6000 Broken Sound Parkway NW, Suite 300, Boca Raton, FL 33487-2742

and by CRC Press
2 Park Square, Milton Park, Abingdon, Oxon, OX14 4RN

CRC Press is an imprint of Taylor & Francis Group, LLC

Library of Congress Cataloging-in-Publication Data
Names: Nollet, Leo M. L., 1948- editor. | Winkler, Robert, editor.
Title: Mass spectrometry in food analysis / edited by Leo M.L. Nollet, Robert Winkler.
Description: First edition. | Boca Raton : Taylor and Francis, 2022. |
Includes bibliographical references and index.
Identifiers: LCCN 2021038581 (print) | LCCN 2021038582 (ebook) |
ISBN 9780367548797 (hardback) | ISBN 9780367549367 (paperback) |
ISBN 9781003091226 (ebook)
Subjects: LCSH: Mass spectrometry. | Food analysis.
Classification: LCC QD96.M3 M344 2022 (print) | LCC QD96.M3 (ebook) | DDC 543/.65—dc23/eng/20211021
LC record available at https://lccn.loc.gov/2021038581
LC ebook record available at https://lccn.loc.gov/2021038582

ISBN: 978-0-367-54879-7 (hbk)
ISBN: 978-0-367-54936-7 (pbk)
ISBN: 978-1-003-09122-6 (ebk)

DOI: 10.1201/9781003091226

Typeset in Times
by codeMantra

Contents

Series Preface ix
Preface xi
Editors xiii
Contributors xv

SECTION 1 Mass Spectrometry – Theoretics

Chapter 1 Basic Principles and Fundamental Aspects of Mass Spectrometry 3

Javed Ahamad, Faraat Ali, Manjoor Ahmad Sayed, Javed Ahmad, and Leo M.L. Nollet

Chapter 2 Targeted and Untargeted Analyses 19

Dev Kant Shandilya

SECTION 2 MS Analysis of Food Components

Chapter 3 Mass Spectrometry of Lipids 31

Aynur Gunenc

Chapter 4 Proteomic Strategies to Evaluate the Impact of Farming Conditions on Food Quality and Safety in Aquaculture Products 43

Mónica Carrera, Carmen Piñeiro, and Iciar Martinez

Chapter 5 Recent Advances in Vitamin Analysis by Mass Spectrometry 61

Maria Katsa and Charalampos Proestos

Chapter 6 Carbohydrate Analysis by Mass Spectrometry 85

Alexander O. Chizhov

Chapter 7 Mass Spectrometry Analysis of Allergens 109

Behnam Keshavarz

Chapter 8 Mass Spectrometry of Food Pigments 119

Laurent Dufossé, Fabio Gosetti, Daria Maria Monti, Antonello Merlino, and Sameer A. S. Mapari

Chapter 9 Flavour and Mass Spectrometry 137

Jean-Luc Le Quéré and Géraldine Lucchi

Chapter 10 Evaluation of Nutraceutical Value 181

Mariana Martínez-Ávila, Janet A. Gutiérrez-Uribe, and Marilena Antunes-Ricardo

SECTION 3 MS Analysis of Residues

Chapter 11 Pesticide Analysis in Food Samples by GC-MS, LC-MS, and Tandem Mass Spectrometry 211

Jeyabalan Sangeetha, Mahesh Pattabhiramaiah, Shanthala Mallikarjunaiah, Devarajan Thangadurai, Suraj Shashikant Dabire, Ravichandra Hospet, Muniswamy David, Saher Islam, M. Supriya, Akhil Silla, R. Preetha, Inamul Hasan Madar, Ghazala Sultan, Ramachandran Chelliah, Deog-Hwan Oh, Anand Torvi, and Zaira Zaman Chowdhury

Chapter 12 Mass Spectrometry of Food Contact Materials 231

Deepthi Eswar and Selvi Chellamuthu

Chapter 13 Mass Analysis in Veterinary Drugs 247

Semih Ötleş and Vasfiye Hazal Özyurt

Chapter 14 Multi-Target Analysis and Suspect Screening of Xenobiotics in Milk by UHPLC-HRMS/MS 255

Mikel Musatadi, Belén González-Gaya, Mireia Irazola, Ailette Prieto, Nestor Etxebarria, Maitane Olivares, and Olatz Zuloaga

Chapter 15 Analytical Determination of Persistent Organic Pollutants from Food Sources Using High-Resolution Mass Spectrometry 279

Jeyabalan Sangeetha, Mahesh Pattabhiramaiah, Shanthala Mallikarjunaiah, Devarajan Thangadurai, Lokeshkumar Prakash, Ravichandra Hospet, Muniswamy David, Inamul Hasan Madar, Saher Islam, Akhil Silla, M. Supriya, R. Preetha, Ghazala Sultan, Ramachandran Chelliah, Deog-Hwan Oh, Anand Torvi, and Zaira Zaman Chowdhury

Chapter 16 MS and Food Forensics 291

Leo M.L. Nollet

Chapter 17 Detecting Food Pathogens through Mass Spectrometry Approaches 329

Saher Islam, Devarajan Thangadurai, Ravichandra Hospet, Zabin Khoje, Lokeshkumar Prakash, Muniswamy David, Namita Bedi, Neeta Bhagat, Umar Farooq, Khizar Hayat, Jeyabalan Sangeetha, Adil Hussain, and Zaira Zaman Chowdhury

Chapter 18 Biogenic Amines Analysis by Mass Spectrometry 347

Sadaf Jamal Gilani, Chandra Kala, Mohammad Taleuzzaman, Syed Sarim Imam, Sultan Alsheri, and Mohammad Asif

Chapter 19 Mass Spectrometry of Food Preservatives 357

Emmanouil D. Tsochatzis

Chapter 20 Mass Spectrometry for Mycotoxin Detection in Food and Food Byproducts 379

Devarajan Thangadurai, Saher Islam, Afshan Shafi, Namita Bedi, Jeyabalan Sangeetha, Adil Hussain, Ravichandra Hospet, Jarnain Naik, Muhammad Zaki Khan, Anand Torvi, Muniswamy David, Zaira Zaman Chowdhury, and Amjad Iqbal

SECTION 4 MS Analysis in Food Authentication

Chapter 21 Mass Spectrometry in Food Authentication 403

Javed Ahamad, Subasini Uthirapathy, and Javed Ahmad

SECTION 5 Emerging Fields

Chapter 22 Challenges and Trends of Mass Spectrometry for Food Analysis 419

Robert Winkler

Index 431

Series Preface

There will always be a need to analyze food compounds and their properties. Current trends in analyzing methods include automation, increasing the speed of analyses, and miniaturization. Over the years, the unit of detection has evolved from micrograms to pictograms.

A classical pathway of analysis is sampling, sample preparation, cleanup, derivatization, separation, and detection. At every step, researchers are working and developing new methodologies. A large number of papers are published every year on all facets of analysis. So, there is a need for books that gather information on one kind of analysis technique or on the analysis methods for a specific group of food components.

The scope of the CRC Series on Food Analysis & Properties aims to present a range of books edited by distinguished scientists and researchers who have significant experience in scientific pursuits and critical analysis. This series is designed to provide state-of-the-art coverage on topics such as:

1. Recent analysis techniques on a range of food components.
2. Developments and evolution in analysis techniques related to food.
3. Recent trends in analysis techniques for specific food components and/or a group of related food components.
4. The understanding of physical, chemical, and functional properties of foods.

The book *Mass Spectrometry in Food Analysis* is volume number 18 of this series.

I am happy to be a series editor of such books for the following reasons:

- I am able to pass on my experience in editing high-quality books related to food.
- I get to know colleagues from all over the world more personally.
- I continue to learn about interesting developments in food analysis.

Much work is involved in the preparation of a book. I have been assisted and supported by a number of people, all of whom I would like to thank. I would especially like to thank the team at CRC Press/Taylor & Francis, with a special word of thanks to Steve Zollo, senior editor.

Many, many thanks to all the editors and authors of this volume and future volumes. I very much appreciate all their effort, time, and willingness to do a great job.

I dedicate this series to:

- My wife, for her patience with me (and all the time I spend on my computer).
- All patients suffering from prostate cancer; knowing what this means, I am hoping they will have some relief.

Preface

The quality and safety of food depend to a large extent on its chemical composition. For the analyst, food is a complex mixture of compounds embedded in a matrix. Thus, efficient food analysis needs suitable extraction protocols and methods with a high analytical resolution.

Gas and liquid chromatography, coupled to mass spectrometry, enable the comprehensive study of substance mixtures on a molecular level. Besides the determination of known materials, the discovery and structure elucidation of new compounds is possible. Therefore, mass spectrometry techniques are ideal for monitoring food quality and discovering novel compounds relevant to the consumer.

For supporting the implementation of mass spectrometry methods in routine and research food analysis labs, we edited this book, which contains five sections:

1. The first section explains the basics of mass spectrometry and its application for targeted and untargeted analyses.
2. Section two compiles chapters dealing with the analysis of food components, such as the mass spectrometry of the macronutrient lipids, proteins, and carbohydrates, evaluating vitamins and nutraceuticals, pigments, flavors, and allergens.
3. Residuals analyses are treated in section three, including examining pesticides, food contact materials, veterinary drugs, xenobiotics in milk, food preservatives, biogenic amines, and organic pollutants. Further, food forensics and the detection of pathogens and mycotoxins with mass spectrometry are presented.
4. The fourth section explains the use of mass spectrometry for food authentication.
5. The closing chapter presents challenges and trends of food analysis with mass spectrometry.

We hope that this book will be useful for practitioners and scientists in food research.

We cordially thank all authors for contributing to this book and for the editorial support of CRC.

Leo M.L. Nollet and Robert Winkler

Editors

Leo M.L. Nollet earned an MS (1973) and PhD (1978) in biology from the Katholieke Universiteit Leuven, Belgium. He is an editor and associate editor of numerous books. He edited for M. Dekker, New York—now CRC Press of Taylor & Francis Publishing Group—the first, second, and third editions of *Food Analysis by HPLC* and *Handbook of Food Analysis*. The last edition is a two-volume book. Dr. Nollet also edited the *Handbook of Water Analysis* (first, second, and third editions) and *Chromatographic Analysis of the Environment*, third and fourth editions (CRC Press). With F. Toldrá, he coedited two books published in 2006, 2007, and 2017: *Advanced Technologies for Meat Processing* (CRC Press) and *Advances in Food Diagnostics* (Blackwell Publishing—now Wiley). With M. Poschl, he coedited the book *Radionuclide Concentrations in Foods and the Environment*, also published in 2006 (CRC Press). Dr. Nollet has also coedited several books with Y. H. Hui and other colleagues: *Handbook of Food Product Manufacturing* (Wiley, 2007), *Handbook of Food Science, Technology, and Engineering* (CRC Press, 2005), *Food Biochemistry and Food Processing* (first and second editions; Blackwell Publishing—now Wiley—2006 and 2012), and the *Handbook of Fruits and Vegetable Flavors* (Wiley, 2010). In addition, he edited the *Handbook of Meat, Poultry, and Seafood Quality*, first and second editions (Blackwell Publishing—now Wiley—2007 and 2012). From 2008 to 2011, he published five volumes on animal product-related books with F. Toldrá: *Handbook of Muscle Foods Analysis*, *Handbook of Processed Meats and Poultry Analysis*, *Handbook of Seafood and Seafood Products Analysis*, *Handbook of Dairy Foods Analysis* (second edition in 2021), and *Handbook of Analysis of Edible Animal By-Products*. Also, in 2011, with F. Toldrá, he coedited two volumes for CRC Press: *Safety Analysis of Foods of Animal Origin* and *Sensory Analysis of Foods of Animal Origin*. In 2012, they published the *Handbook of Analysis of Active Compounds in Functional Foods*. In a coedition with Hamir Rathore, *Handbook of Pesticides: Methods of Pesticides Residues Analysis* was marketed in 2009; *Pesticides: Evaluation of Environmental Pollution* in 2012; *Biopesticides Handbook* in 2015; and *Green Pesticides Handbook: Essential Oils for Pest Control* in 2017. Other finished book projects include *Food Allergens: Analysis, Instrumentation, and Methods* (with A. van Hengel; CRC Press, 2011) and *Analysis of Endocrine Compounds in Food* (Wiley-Blackwell, 2011). Dr. Nollet's recent projects include *Proteomics in Foods* with F. Toldrá (Springer, 2013) and *Transformation Products of Emerging Contaminants in the Environment: Analysis, Processes, Occurrence, Effects, and Risks* with D. Lambropoulou (Wiley, 2014). In the series Food Analysis & Properties, he edited (with C. Ruiz-Capillas) *Flow Injection Analysis of Food Additives* (CRC Press, 2015) and *Marine Microorganisms: Extraction and Analysis of Bioactive Compounds* (CRC Press, 2016). With A.S. Franca, he coedited *Spectroscopic Methods in Food Analysis* (CRC Press, 2017), and with Horacio Heinzen and Amadeo R. Fernandez-Alba, he coedited *Multiresidue Methods for the Analysis of Pesticide Residues in Food* (CRC Press, 2017). Further volumes in the series Food Analysis & Properties are *Phenolic Compounds in Food: Characterization and Analysis* (with Janet Alejandra Gutierrez-Uribe, 2018), *Testing and Analysis of GMO-containing Foods and Feed* (with Salah E. O. Mahgoub, 2018), *Fingerprinting Techniques in Food Authentication and Traceability* (with K. S. Siddiqi, 2018), *Hyperspectral Imaging Analysis and Applications for Food Quality* (with N.C. Basantia and Mohammed Kamruzzaman, 2018), *Ambient Mass Spectroscopy Techniques in Food and the Environment* (with Basil K. Munjanja, 2019), *Food Aroma Evolution: During Food Processing, Cooking, and Aging* (with M. Bordiga, 2019), *Mass Spectrometry Imaging in Food Analysis* (2020), *Proteomics in Food Authentication* (with S. Ötleş, 2020), *Analysis of Nanoplastics and Microplastics in Food* (with K.S. Siddiqi, 2020), *Chiral Organic Pollutants, Monitoring and Characterization in Food and the Environment* (with Edmond Sanganyado and Basil K. Munjanja, 2020), *Sequencing Technologies in Microbial Food Safety and Quality* (with

Devarajan Thangardurai, Saher Islam, and Jeyabalan Sangeetha, 2021), and *Nanoemulsions in Food Technology: Development, Characterization, and Applications* (with Javed Ahmad, 2021).

Robert Winkler studied biotechnology (Fachhochschule Jena, Germany) and biochemical engineering (University of Birmingham, UK). In 2004, he started a Ph.D. in natural products chemistry at the Leibniz Institute for Natural Product Research and Infection Biology, Hans Knöll Institute (HKI) Jena, Germany. During this time, Robert Winkler used mass spectrometry for structure elucidation. After graduating in 2007, he stayed at the HKI as PostDoc and head of the mass spectrometry and proteomics unit.

In 2010, Robert Winkler founded the biochemical and instrumental analysis laboratory (labABI) at the Center for Research and Advanced Studies of the National Polytechnic Institute (CINVESTAV), Unidad Irapuato, Mexico. He is a faculty member of two postgraduate programs: Plant Biotechnology and Integrative Biology. The labABI develops novel analytical platforms to study biological systems and food products, such as ambient ionization methods, mass spectrometry imaging tools, and mass spectrometry software for metabolomics and proteomics. Robert Winkler published about 100 articles in international magazines, including high-impact journals such as *Nature*, *Analytical Chemistry*, *Angewandte Chemie*, and *Talanta*. Besides this, he holds patents on the technology developed in his group in Mexico, the USA, and Europe. In 2018/2019, he did a sabbatical stay at the Max Planck Institute of Ecological Chemistry, Jena, Germany.

In 2020, Robert Winkler published the first book he edited: *Processing Metabolomics and Proteomics Data with Open Software: A Practical Guide*, 1st edition, and *New Developments in Mass Spectrometry Series*, Royal Society of Chemistry, Cambridge, UK.

Contributors

Javed Ahamad
Faculty of Pharmacy, Department of Pharmacognosy
Tishk International University
Erbil, Kurdistan Region, Iraq

Javed Ahmad
Department of Pharmaceutics
College of Pharmacy, Najran University
Najran, Kingdom of Saudi Arabia

Faraat Ali
Department of Inspection and Enforcement, Laboratory Services
Botswana Medicines Regulatory Authority
Gaborone, Botswana

Sultan Alsheri
Department of Pharmaceutics
King Saud University
Riyadh, Kingdom of Saudi Arabia

Marilena Antunes-Ricardo
Tecnológico de Monterrey, Centro de Biotecnología FEMSA
Escuela de Ingeniería y Ciencias
Monterrey, Mexico

Mohammad Asif
Faculty of Pharmacy, Department of Pharmacognosy
Lachoo Memorial College of Science and Technology
Jodhpur, Rajasthan, India

Namita Bedi
Amity Institute of Biotechnology
Amity University
Noida, Uttar Pradesh, India

Neeta Bhagat
Amity Institute of Biotechnology
Amity University
Noida, Uttar Pradesh, India

Mónica Carrera
Food Technology Department
Institute of Marine Research (IIM), Spanish National Research Council (CSIC)
Pontevedra, Spain

Selvi Chellamuthu
Tamil Nadu Agricultural University
Coimbatore, Tamil Nadu, India

Ramachandran Chelliah
Department of Food Science and Biotechnology
College of Agriculture and Life Sciences, Kangwon National University
Chuncheon, Gangwon, Korea

Alexander O. Chizhov
Department of Structural Studies
N.D. Zelinsky Institute of Organic Chemistry, Russian Academy of Sciences
Moscow, Russia

Zaira Zaman Chowdhury
Nanotechnology and Catalysis Research Center (NANOCAT)
Institute of Advanced Studies (IAS), University of Malaya
Kuala Lumpur, Malaysia

Suraj Shashikant Dabire
Department of Zoology
Karnatak University
Dharwad, Karnataka, India

Muniswamy David
Department of Zoology
Karnatak University
Dharwad, Karnataka, India

Laurent Dufossé
CHEMBIOPRO Chimie et Biotechnologie des Produits Naturels, ESIROI Département Agroalimentaire
Université de La Réunion
Ile de La Réunion, Indian Ocean, France

Deepthi Eswar
Department of Agriculture
Government of Tamil Nadu
Tamil Nadu, India

Nestor Etxebarria
Department of Analytical Chemistry
University of the Basque Country (UPV/EHU)
Leioa, Basque Country, Spain
and
Research Centre for Experimental Marine Biology and Biotechnology
University of the Basque Country (PiE-UPV/EHU)
Plentzia, Basque Country, Spain

Umar Farooq
Department of Food Science and Technology
MNS University of Agriculture
Multan, Pakistan

Sadaf Jamal Gilani
Department of Basic Health Sciences
Princess Noura Bint Abdulrahman University
Riyadh, Kingdom of Saudi Arabia

Belén González-Gaya
Department of Analytical Chemistry
University of the Basque Country (UPV/EHU)
Leioa, Basque Country, Spain
and
Research Centre for Experimental Marine Biology and Biotechnology
University of the Basque Country (PiE-UPV/EHU)
Plentzia, Basque Country, Spain

Fabio Gosetti
Department of Earth and Environmental Sciences (DISAT)
University of Milano-Bicocca
Milano, Italy

Aynur Gunenc
Chemistry Department (Food Science and Nutrition)
Carleton University
Ottawa, ON, Canada

Janet A. Gutiérrez-Uribe
Tecnológico de Monterrey, Centro de Biotecnología FEMSA
Escuela de Ingeniería y Ciencias
Monterrey, Mexico
and Tecnologico de Monterrey, Campus Puebla
Puebla, México

Khizar Hayat
Department of Food Science and Technology
MNS University of Agriculture
Multan, Pakistan

Ravichandra Hospet
Department of Botany
Karnatak University
Dharwad, Karnataka, India

Adil Hussain
Department of Agriculture
Abdul Wali Khan University
Mardan, Khyber Pakhtunkhwa, Pakistan

Syed Sarim Imam
Department of Pharmaceutics
King Saud University
Riyadh, Kingdom of Saudi Arabia

Amjad Iqbal
Department of Food Science and Technology
Abdul Wali Khan University Mardan
Mardan, Pakistan

Mireia Irazola
Department of Analytical Chemistry
University of the Basque Country (UPV/EHU)
Leioa, Basque Country, Spain
and
Research Centre for Experimental Marine Biology and Biotechnology
University of the Basque Country (PiE-UPV/EHU)
Plentzia, Basque Country, Spain

Saher Islam
Department of Biotechnology
Lahore College for Women University
Lahore, Pakistan

Chandra Kala
Faculty of Pharmacy, Department of Pharmaceutical Chemistry
Maulana Azad University
Jodhpur, Rajasthan, India

Maria Katsa
Laboratory of Food Chemistry, Department of Chemistry
National and Kapodistrian University of Athens
Athens, Greece

Behnam Keshavarz
Division of Allergy and Clinical Immunology, Department of Medicine
University of Virginia
Charlottesville, Virginia

Muhammad Zaki Khan
Department of Food Science and Technology
MNS University of Agriculture
Multan, Pakistan

Zabin Khoje
Department of Zoology
Karnatak University
Dharwad, Karnataka, India

Géraldine Lucchi
Centre for Taste, Smell and Feeding Behaviour
French National Institute for Agricultural Food and Environment Research (INRAE)
Dijon, France

Inamul Hasan Madar
Department of Biomedical Science and Environmental Biology
Kaohsiung Medical University
Kaohsiung, Taiwan

Shanthala Mallikarjunaiah
Department of Zoology
Centre for Applied Genetics, Bangalore University
Bangalore, Karnataka, India

Sameer A. S. Mapari
Department of Biotechnology
VIVA College [Affiliated to University of Mumbai]
Palghar, Maharashtra, India

Iciar Martinez
Research Centre for Experimental Marine Biology and Biotechnology—Plentzia Marine Station (PiE)
University of the Basque Country (UPV/EHU)
Plentzia, Spain
and
IKERBASQUE Basque Foundation for Science
Bilbao, Spain

Mariana Martínez-Ávila
Tecnológico de Monterrey, Centro de Biotecnología FEMSA
Escuela de Ingeniería y Ciencias
Monterrey, Mexico

Antonello Merlino
Department of Chemical Sciences
University of Naples Federico II
Napoli, Italy

Daria Maria Monti
Department of Chemical Sciences
University of Naples Federico II
Napoli, Italy

Mikel Musatadi
Department of Analytical Chemistry
University of the Basque Country (UPV/EHU)
Leioa, Basque Country, Spain

Jarnain Naik
Department of Zoology
Karnatak University
Dharwad, Karnataka, India

Leo M.L. Nollet
University College Ghent
Ghent, Belgium

Deog-Hwan Oh
Department of Food Science and Biotechnology
College of Agriculture and Life Sciences, Kangwon National University
Chuncheon, Gangwon-do, Korea

Maitane Olivares
Department of Analytical Chemistry
University of the Basque Country (UPV/EHU)
Leioa, Basque Country, Spain
and
Research Centre for Experimental Marine Biology and Biotechnology
University of the Basque Country (PiE-UPV/EHU)
Plentzia, Basque Country, Spain

Semih Ötleş
Faculty of Engineering, Department of Food Engineering
Ege University
Bornova, Izmir, Turkey

Vasfiye Hazal Özyurt
Faculty of Engineering, Department of Food Engineering
Near East University
Nicosia, Turkey

Mahesh Pattabhiramaiah
Department of Zoology
Centre for Applied Genetics, Bangalore University
Bangalore, Karnataka, India

Carmen Piñeiro
Scientific Instrumentation and Quality Service (SICIM)
Institute of Marine Research (IIM), Spanish National Research Council (CSIC)
Pontevedra, Spain

Lokeshkumar Prakash
Department of Zoology
Karnatak University
Dharwad, Karnataka, India

R. Preetha
Department of Food Process Engineering
School of Bioengineering, The College of Engineering and Technology, SRM Institute of Science and Technology
Kattankulathur, Tamil Nadu, India

Ailette Prieto
Department of Analytical Chemistry
University of the Basque Country (UPV/EHU)
Leioa, Basque Country, Spain
and
Research Centre for Experimental Marine Biology and Biotechnology
University of the Basque Country (PiE-UPV/EHU)
Plentzia, Basque Country, Spain

Charalampos Proestos
Laboratory of Food Chemistry, Department of Chemistry
National and Kapodistrian University of Athens
Athens, Greece

Jean-Luc Le Quéré
Centre for Taste, Smell and Feeding Behaviour
French National Institute for Agricultural Food and Environment Research (INRAE)
Dijon, France

Jeyabalan Sangeetha
Department of Environmental Science
Central University of Kerala
Kasaragod, Kerala, India

Manjoor Ahmad Sayed
Department of Pharmacy
College of Public Health and Medical Sciences, Mettu University
Mettu, Ethiopia

Afshan Shafi
Department of Food Science and Technology
MNS University of Agriculture
Multan, Pakistan

Dev Kant Shandilya
Department of Research
Bhagwant University
Ajmer, India

Akhil Silla
Department of Food Process Engineering
School of Bioengineering, The College of Engineering and Technology, SRM Institute of Science and Technology
Kattankulathur, Tamil Nadu, India

Ghazala Sultan
Faculty of Science, Department of Computer Science
Aligarh Muslim University
Aligarh, Uttar Pradesh, India

M. Supriya
Department of Food Process Engineering, School of Bioengineering
The College of Engineering and Technology, SRM Institute of Science and Technology
Kattankulathur, Tamil Nadu, India

Mohammad Taleuzzaman
Faculty of Pharmacy, Department of Pharmaceutical Chemistry
Maulana Azad University
Jodhpur, Rajasthan, India

Devarajan Thangadurai
Department of Botany
Karnatak University
Dharwad, Karnataka, India

Anand Torvi
Centre of Nano and Material Science
Jain University
Bangalore, Karnataka, India

Emmanouil D. Tsochatzis
Department of Food Science
Aarhus University
Aarhus, Denmark

Subasini Uthirapathy
Faculty of Pharmacy, Department of Pharmacology
Tishk International University
Erbil, Kurdistan Region, Iraq

Robert Winkler
Department of Biotechnology and Biochemistry
Center for Research and Advanced Studies of the National Polytechnic Institute (CINVESTAV), Unidad Irapuato
Irapuato, Mexico

Olatz Zuloaga
Department of Analytical Chemistry
University of the Basque Country (UPV/EHU)
Leioa, Basque Country, Spain
and
Research Centre for Experimental Marine Biology and Biotechnology
University of the Basque Country (PiE-UPV/EHU)
Plentzia, Basque Country, Spain

Section 1

Mass Spectrometry – Theoretics

1 Basic Principles and Fundamental Aspects of Mass Spectrometry

Javed Ahamad
Tishk International University

Faraat Ali
Botswana Medicines Regulatory Authority

Manjoor Ahmad Sayed
Mettu University

Javed Ahmad
Najran University

Leo M.L. Nollet
University College Ghent

CONTENTS

1.1 Introduction 4
1.2 Principle of Mass Spectrometry 4
1.3 Instrumentation of Mass Spectrometry 5
1.3.1 Ionization Techniques 5
1.3.1.1 Gas Phase 6
1.3.1.2 Desorption Phase 6
1.3.2 Mass Analyzers 7
1.3.2.1 Time-of-Flight 7
1.3.2.2 Quadrupole 7
1.3.2.3 Orbitrap Mass Analyzer 8
1.3.2.4 Fourier Transform–Ion Cyclotron Resistance 8
1.3.2.5 Tandem Mass Analyzer 8
1.3.3 Mass Detectors 8
1.3.3.1 Electron Multiplier 8
1.3.3.2 Faraday Cup 8
1.3.3.3 Photomultipliers 8
1.4 Interpretation of Mass Spectra 9
1.4.1 Types of Peaks in Mass Spectroscopy 9
1.4.2 Fragmentation Rules 9
1.5 Application of Mass Spectrometry 9
1.5.1 Qualitative Applications 10
1.5.1.1 Determination of Molecular Weight 10

DOI: 10.1201/9781003091226-2

1.5.1.2 Determination of the Molecular Formula 10
1.5.1.3 Determination of the Partial Molecular Formula 10
1.5.1.4 Determination of the Compound Structure 11
1.5.2 Quantitative Applications 12
1.5.2.1 Isotope Abundance Assessment 12
1.5.2.2 Determination of Isotope Ratio 13
1.5.2.3 Differentiation between *cis*-Isomers and *trans*-Isomers 13
1.5.2.4 Mass Spectrometry in Thermodynamics 13
1.5.2.5 Measurement of Ionization Potential 13
1.5.2.6 Determination of Ion–Molecule Reactions 13
1.5.2.7 Detection and Identification of Impurity 13
1.5.2.8 Identification of Unknown Compounds 14
1.5.3 Both Qualitative and Quantitative Applications 14
1.5.3.1 Phytochemical Analysis 14
1.5.3.2 Structural Elucidation of Unknown Phytochemicals 14
1.5.3.3 Drug Metabolism Studies 14
1.5.3.4 Clinical Studies 14
1.5.3.5 Forensic Applications 14
1.6 Conclusion 15
References 15

1.1 INTRODUCTION

Over the past decades, mass spectrometry (MS) has developed exponentially, with its applications ranging from each discipline across the health and life disciplines (Henderson and McIndoe, 2005; Banoub et al., 2005). Depending on their mass-to-charge (*m/z*) ratio, MS relies on the creation of gas-phase ions (negatively or positively charged) that can be electrically (or magnetically) separated (Korfmacher, 2005). The *x*-axis reflects *m/z* values in an MS spectrum, while the *y*-axis shows complete ion counts. Mass spectrometers, technically speaking, should indeed be called *m/z* spectrometers (Grayson, 2002). The mass spectrometric study, including their composition, potency, and purity, may provide considerable details about the analytes (Kitson et al., 1996; Herbert and Johnstone, 2003).

The use of advanced mass spectrometric methods has been restricted to volatile compounds with a low-weight spectrum of molecules (>1,000 Da) (Beckey, 1969; Torgerson et al., 1974). Furthermore, a traditional ionization technique, namely electron impact (EI), is extreme and can contribute to the degradation of complex organic compounds (e.g., enzymes, nucleic acids, and carbohydrates). In particular, MS was restricted to the gas chromatography (GC)-MS instrument separating volatile compounds within biology laboratories (Barber et al., 1981). Under such circumstances, without substantial destruction and deterioration, highly complex compounds, such as proteins, cannot be transmitted to the gas phase (Tanaka et al., 1988). Only after soft ionization techniques were introduced, MS was used for genomics studies (Hillenkamp and Karas, 1990). Many separation methods, such as high-pressure liquid chromatography (HPLC-MS) and capillary electrophoresis (CE), can now be combined with MS (Pramanik et al., 2002). MS has emerged as a powerful technique for ensuring the quality and safety of food products such as honey, cereals, meat products, wine, and milk and dairy products (Ahamad et al., 2018; 2020a,b). MS analysis of purified materials or synthetic conjugates without using a separation technique is also popular (Baldwin, 2005). This chapter comprehensively summarizes the principles, recent advancements in ionization and detectors, and MS applications.

1.2 PRINCIPLE OF MASS SPECTROMETRY

MS measures the specific fragments or ions that occur from organic molecules' breakdown (Griffiths et al., 2001). The fundamental concept requires the bombardment of organic matter. A compound is attacked with a beam of electrons to generate positively charged ions (Emmett and Caprioli, 1994;

Gale and Smith, 1993). A mass spectrum represents each ion in the form of peak strength. Ion deflection is based on charge, mass, and velocity; ion separation is based on the *m/z* ratio; and detection is proportional to the abundance of ions (Karas et al., 2000; Banoub et al., 2003; Banoub et al., 2004).

1.3 INSTRUMENTATION OF MASS SPECTROMETRY

In 1912, the first mass spectroscope was developed and used by J.J. Thompson to analyze the atomic weight of elements and to track the abundance of elemental isotopes. With the advent of techniques to vaporize organic compounds in the 1950s, it became possible to study bioactive compounds via a mass spectroscope. MS consists of an ionization source, analyzer, detector, and data processor. The analyzer and detector are kept in a vacuum so that air molecules do not collide with the ions created, and the trajectories of the ions are held at a certain speed. The sample is ionized in the ionization chamber after it is pumped into the instrument through the inlet. In the analyzer, ionized species (cations/anions) are then separated, resolving the ions based on their *m/z* ratio. Finally, these ions are detected by detectors, and the relative abundance is recorded in the form of mass spectra for each resolved ionic species (Silverstein and Webster, 1998). Figure 1.1 shows the schematic diagram of different parts of MS.

1.3.1 Ionization Techniques

Depending on the types of organic compounds, different types of ionization techniques are used. Volatile samples are subjected to either electron or chemical ionization (CI), and fast-atom bombardment (FAB), matrix-assisted laser desorption ionization (MALDI), and electrospray ionization (ESI) methods are used to ionize non-volatile samples (Silverstein and Webster, 1998). In Table 1.1, the characteristics of various methods of ionization used in MS are summarized.

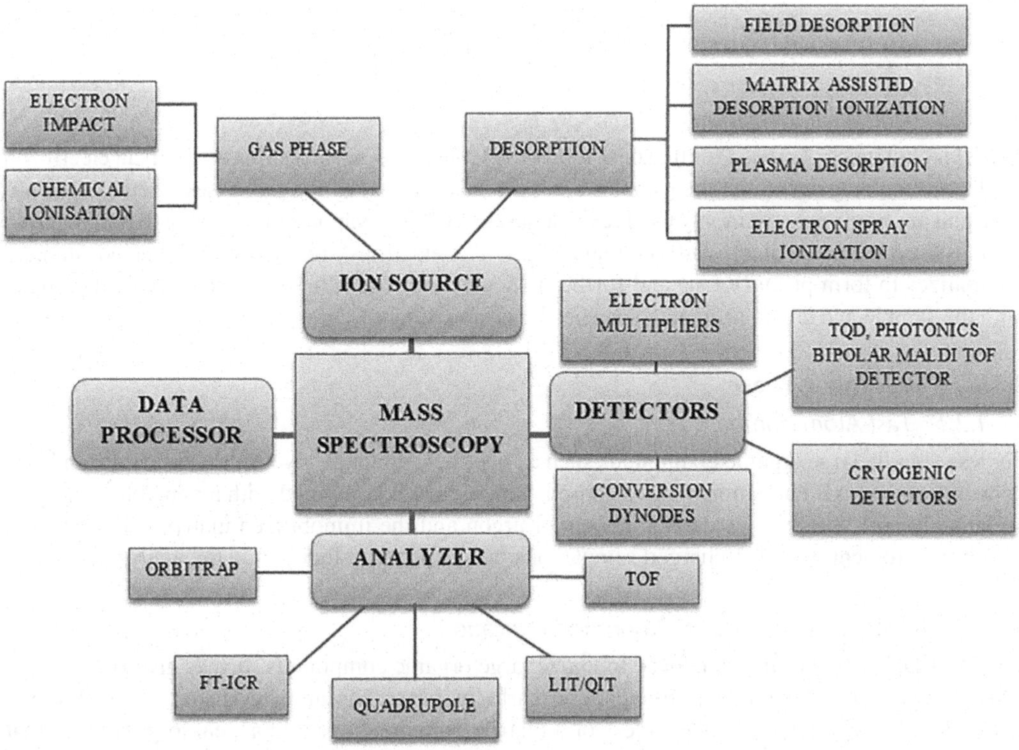

FIGURE 1.1 Schematic diagram representing different parts of mass spectrometry.

TABLE 1.1
Summary of the Features of Different Ionization Methods Used in Mass Spectrometry

Ionization Mode	Nature	Sample	Range of Mass	Description
Electron impact (EI)	Volatile: thermally stable	GC solid or liquid probe	<1,100 Da	Hard method; mainly fragmented ions
Chemical ionization (CI)	Volatile: thermally stable	GC solid or liquid probe	<1,000 Da	Soft method; molecular ion
Fast-atom bombardment (FAB)	Organometallic compounds	LC direct injection	<5,200 Da; optimal range 250–2,000	Soft method; require matrix
Matrix-assisted laser desorption ionization (MALDI)	Proteomics, genomics	Sample is crystallized with a matrix	Can go beyond 510,000 Da	Very soft method singly charged ions
Electrospray ionization (ESI)	Organic and inorganic compounds	Sample in solution	From very low masses to extraordinarily high	Very soft method; multiply charged ions

1.3.1.1 Gas Phase

1.3.1.1.1 Electron Impact

EI ionization is extensively applied for ionization of volatile organic compounds such as essential oils, fixed oils, and hydrocarbons. The ionizing agent is a high-energy electron emitted from a heated filament and accelerated by a potential difference of approximately 70eV. By removing an electron and producing positively charged ion species, the sample is ionized (cation). There is further fragmentation of the charged radical cation into positive ions.

$$M + e^- = M^+ + 2e^-$$

1.3.1.1.2 Chemical Ionization

The CI technique is also used for the ionization of volatile compounds. In this method, an electron is used as an ionizing agent, but the samples are indirectly ionized by reagent gases. Methane, isobutene, and ammonia are commonly used reagent gases in CI. The vaporized samples are pumped into the mass spectrometer, along with the reagent gas. The electrons are created by a heated filament that ionizes to form primary ions and further to secondary ions. These secondary ions then react with the sample and create protonated species by protonating it.

1.3.1.2 Desorption Phase

1.3.1.2.1 Fast-Atom Bombardment

For less volatile large organic compounds such as peptides, proteins, glycosides, alkaloids, and carbohydrates, the FAB ionization method is used. The sample is combined with a non-volatile matrix such as glycerol, nitrobenzyl alcohol, xenon, or argon and the immobilized matrix is bombarded. This bombardment produces charged sample ions, which are then focused on the analyzer.

1.3.1.2.2 Matrix-Assisted Laser Desorption Ionization

The MALDI method is also employed to ionize large organic compounds such as glycosides, alkaloids, resins, tannins, peptides, proteins, and carbohydrates. The analyte is first co-crystallized in this process with a large molar excess of a matrix compound, usually a weak organic acid that absorbs ultraviolet (UV). Irradiation by a laser of this analyte–matrix mixture results in vaporization of the matrix. The samples' co-crystallized molecules often vaporize but without having to absorb energy directly from the laser (Hillenkamp et al., 1991; Canas et al., 2006).

1.3.1.2.3 Electrospray Ionization

The ESI method is used in low-resolution spectrometer molecular mass determination for peptide, protein, and hydrocarbon analysis. ESI is the softest ionization technique, where the movement interactions of a macromolecule are preserved in its gas phase. The sample is passed through a small electrical capillary of high potential in this procedure, resulting in an electrostatic spray consisting of many charged droplets. These droplets move under a heated capillary where a solvent is evaporated. Micro-droplets explode, and charged ionized analytes are isolated from the micro-drops, moving into the analyzer (El-Aneed et al., 2009; Yamashita and Fenn, 1984).

1.3.2 Mass Analyzers

Mass spectrometer analyzers, based on *m/z* ratio, resolve and separate the ions, and this separation is driven by an electrical or magnetic field. All ions resolved by mass are concentrated on a single focal point. The desirable functions for any successful mass analyzer are as follows (Rajawat and Jhingan, 2019; Silverstein and Webster, 1998):

- The maximum permissible *m/z* ratio is the mass range.
- The mass's precision is expressed as *parts per million* (ppm) and specifies how similar the measured weight is to the exact mass.
- The ability to resolve molecular species with comparable yet distinct masses is the resolution of the mass spectrometer. The resolution is determined by dividing the peak value of *m/z* at half the maximum intensity by its distance.
- Efficiency is the transmission multiplied by the duty cycle.
- Sensitivity is the concentration that the mass spectrometer detects.
- Mass spectrometer analyzer should have a linear dynamic range.

Based on the above criteria, the following mass analyzers are currently used for the analysis of organic compounds.

1.3.2.1 Time-of-Flight

Time-of-flight (ToF) mass analyzer is suitable for analysis of small organic molecules such as proteins, peptides, polynucleotides, lipids, and carbohydrates. ToF is the simplest mass analyzer consisting of a high vacuum flight tube. Depending on their mass, ionized molecules fly through the flight tube at various velocities, where the mass is inversely proportional to the velocity. The ToF to the detector is determined in a ToF analyzer where ions with a lower mass enter the detector faster than ions with a higher mass. The flight time can vary from 1 to 50 μs in ToF. The ToF analyzer's key benefits are that, depending on the mass, all ions can eventually enter the detector, and this type of analyzer can distinguish ions with a very high mass range (Canas et al., 2006).

1.3.2.2 Quadrupole

A quadrupole mass analyzer consists of four conducting rods arranged in parallel, with a space in the middle; the opposing pairs of rods are electrically connected. The field is generated when a radio frequency (RF) voltage is applied between opposing rods within the quadrupole, creating an electrically oscillating field. As ions are pulsed into the quadrupole from the ionization chamber, the positive ions travel toward the negative rod. Still, the ions take up a complicated trajectory due to changing polarity. The trajectory of ions is therefore controlled by the variance of the direct current/RF voltage. Ultimately, the ions within a small *m/z* range will take up a stable trajectory in the quadrupole. The ions with an incorrect trajectory interfere with the rods and do not enter the detector.

The ions with different *m/z* ratios can be passed to the detector by controlling the voltage and frequency (Rajawat and Jhingan, 2019).

1.3.2.3 Orbitrap Mass Analyzer

Orbitrap mass analyzers are made up of three electrodes, where the two outer electrodes face each other in the shape of cups and are separated by a central dielectric ring. The central electrode is spindle-shaped. It retains the trap and aligns it with the central dielectric end electrode, forcing the ion to take up harmonic axial oscillations with the centrifugal force. Outer electrodes act as receiver plates that detect axial oscillations to detect image current and then convert additional signals into mass spectra (Zubarev and Makarov, 2013).

1.3.2.4 Fourier Transform–Ion Cyclotron Resistance

The Fourier transform (FT)-ion cyclotron resistance mass analyzer consists of four electrodes placed in a magnetic field to create a penning trap where the electric field is perpendicular to the magnetic field. With cyclotron frequency, the ions get stuck in the penning trap and oscillate. The frequency of the cyclotron is inversely proportional to the *m/z* ratio and directly proportional to the magnetic field's power. The signal is observed on the detector plates in the form of a picture current, and this signal is referred to as a free induction decay (FID) signal with a frequency equal to the frequency of ion induction. Using a mathematical equation known as the FT, the data are derived from this signal, thereby generating mass spectra (Rajawat and Jhingan, 2019).

1.3.2.5 Tandem Mass Analyzer

Tandem mass analyzers are arising due to a combination of two or more mass analyzers. Suppose two or more mass analyzers are linked in mass spectroscopy in series. In that case, this instrument is referred to as a tandem mass spectrometry (MS/MS) instrument. Analyte mass analysis is conducted in two consecutive steps, in which the first analyzer performs ion isolation, then fragmentation is caused by collision-induced dissociation in the collision cell, and finally separation of fragmented ions based on the *m/z* ratio in the final analyzer (Silverstein and Webster, 1998).

1.3.3 Mass Detectors

A mass spectrometer has the following four types of detectors (Rajawat and Jhingan, 2019).

1.3.3.1 Electron Multiplier

The electron multiplier consists of dynodes made up of copper–beryllium, which are particularly useful in detecting both positive and negative ions. The theory requires the electron emission from the first dynode to the transduction of the initial ion current, concentrated on the next dynode, etc. This present cascade is eventually amplified and documented a million times.

1.3.3.2 Faraday Cup

In a Faraday cup mass detector, the dynode consists of secondary emitting materials such as gallium phosphide (GaP) and cesium antimony (CsSb). In this process, electrons are emitted that cause current, amplified, and reported in the form of mass spectra by the ion beam striking the dynode.

1.3.3.3 Photomultipliers

In this type of detector, the dynode consists of a photon-emitting scintillator. Light is emitted from the dynode and is converted by a photomultiplier tube into an electrical current, and the current is reported in the form of mass spectra.

1.4 INTERPRETATION OF MASS SPECTRA

1.4.1 Types of Peaks in Mass Spectroscopy

Different types of peaks present in mass spectra are discussed below:

Molecular ion peak: it represents the molecular ion and is produced by bombarding 9–15 eV energy electrons.

Fragment ion peak: it is produced when molecular ions are further bombarded with high-energy electrons of up to 70 eV. The resultant ions are known as fragment ions, and peaks represent fragment ion peaks.

Base peak: in the mass spectrum, the highest and most extreme peak is the base peak.

Rearrangement ion peak: it is the reflection of fragment ion recombination.

Metastable ion peak: it is a broad peak arising and corresponding to metastable ions that are formed after decomposition of analyte between source and magnetic analyzer.

Multicharged ion peak: with more than one charge, some ions are formed and are represented as multicharge ion peaks.

Negative ion peak: electron capture by a molecule contributes to negative ion formation during electron bombardment, which forms a negative ion peak in the spectrum.

1.4.2 Fragmentation Rules

The general rules for predicting the fragmentation pattern of organic compounds are as follows:

i. Increasing the degree of branching results in a decrease in the maximum height of M^+.
ii. Molecular weight increase reduces M^+ peak height.
iii. Aromatic rings, cyclic structures, and double bonds are stabilized by the M^+ peak.
iv. Double bond-favored allylic cleavage results in resonance-stabilized cations.
v. Cleavage occurs with carbon substituted for alkyl, resulting in the formation of a carbocation.
vi. The alkyl side chain is removed in the saturated ring, while the retro Diels–Alder reaction occurs in unsaturated rings.
vii. Preferably, aromatic compounds are split at β-bond, resulting in resonance-stabilized benzyl ion.
viii. C–C bonds are often broken next to heteroatoms (O, NH, and S), creating charged heteroatoms stabilized by resonance.
ix. Small neutral molecules have been removed during bond cleavage and atom rearrangements, which is known as the McLafferty rearrangement.

Based on the base peak, fragment peaks, fragmentation pattern, and molecular weight, known organic compounds are determined. For unknown compounds, nuclear magnetic resonance (NMR), Fourier transform infrared (FTIR), UV–visible, and mass spectroscopic data are combined to predict the chemical structure of natural or organic compounds (Silverstein and Webster, 1998).

1.5 APPLICATION OF MASS SPECTROMETRY

MS has extensive application in different fields, including pharmaceuticals/nutraceuticals and food science technology. The various applications of MS are illustrated in Figure 1.2.

The main applications of MS are categorized as follows:

- Qualitative
- Quantitative
- Both Qualitative and Quantitative.

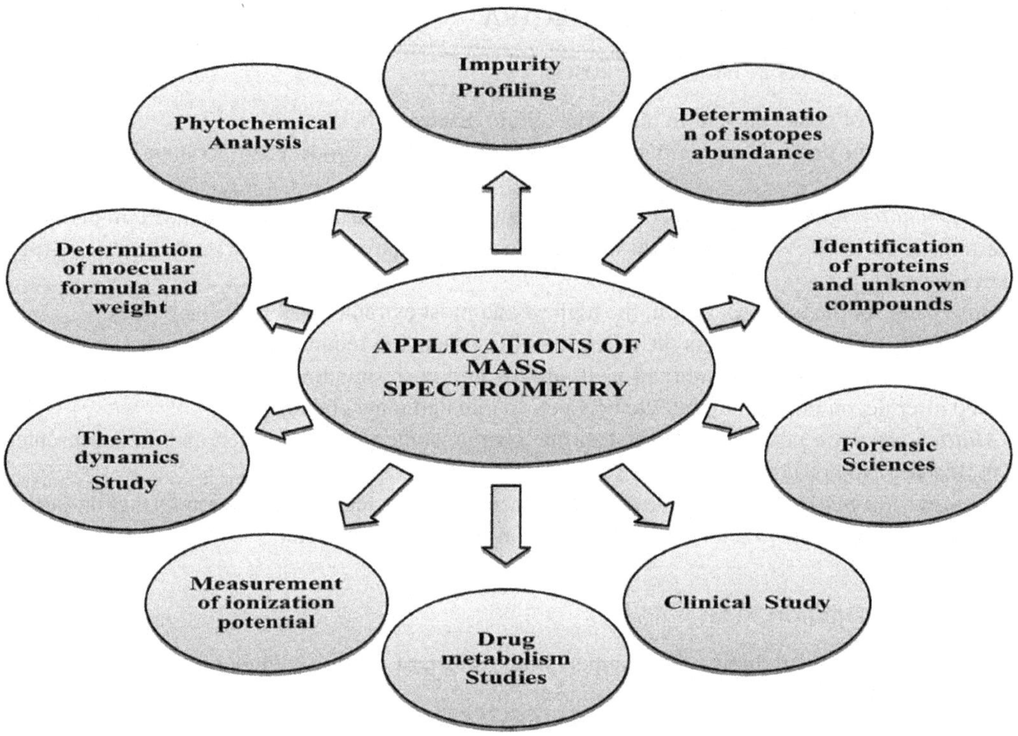

FIGURE 1.2 Application of mass spectrometry.

1.5.1 Qualitative Applications

1.5.1.1 Determination of Molecular Weight

MS is the only technique used for the determination of molecular weight by which the molecular formula can be predicted. It is based on the sample's conversion into an ionized state, with or without fragmentation, which is then characterized by its *m/z* ratio. One of the methods used by organic chemists to determine organic molecules' structure is electron ionization-MS. In most first texts on organic chemistry, the approach is introduced and is commonly represented by idealized spectra, where the molecular ion is labeled (William and Price, 2002). The example of ethanol is shown in Figure 1.3.

1.5.1.2 Determination of the Molecular Formula

The compound's molecular mass is needed to determine the molecular formula of a compound. For example, benzene (C_6H_6) and acetylene (C_2H_2) have both the empirical formula CH but with different molecular masses and molecular formulas (McLafferty and Turecek, 1993; Silverstein and Webster, 1998). The molecular formula's determination requires employing the molecular ion, isotope abundance, MS fragment, exact mass, combustion analysis, ^{13}C NMR, and FTIR analysis.

1.5.1.3 Determination of the Partial Molecular Formula

The partial molecular formula serves to identify it. Generally, atoms are polyisotopic, and the ions are chosen according to their real mass in a mass spectrometer. The mass distribution of molecular ions provides exact knowledge about the chosen ions' atomic composition (Rockwood et al., 2018; Bucknall et al., 2002).

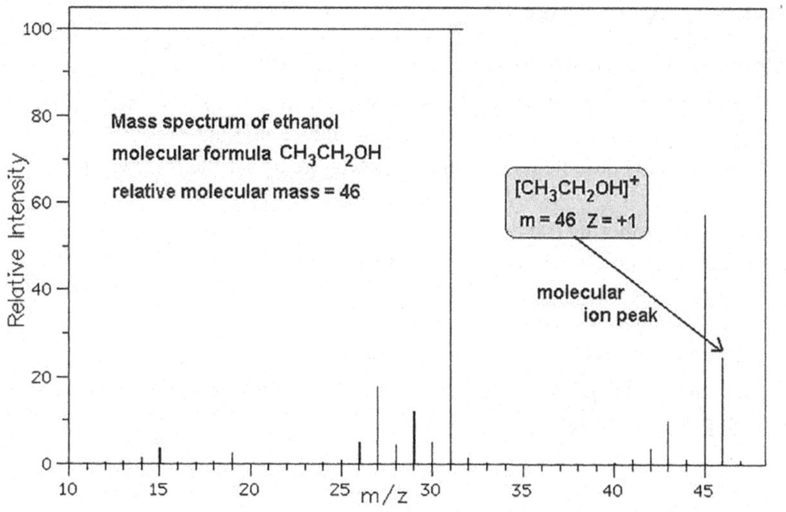

FIGURE 1.3 Mass spectrum of ethanol where the maximum molecular ion peak has an *m/z* value of 46 (M^+).

1.5.1.4 Determination of the Compound Structure

A molecule is bombarded with a high-energy electron beam in a mass spectrometer and produces smaller ions with different molecular masses. This molecular mass is represented by the *m/z* ratio, which helps separate ionized atoms and molecules from each other to determine the compound structure. The set of fragment ions formed by the decomposition of the starting molecule gives information about the structure of the compound. To identify and determine the compound structure, both computer libraries of mass spectra and manual interpretation of the spectra by using fragmentation law are needed (Albert and Zaikin, 2008). An example of acetone is shown in Figure 1.4.

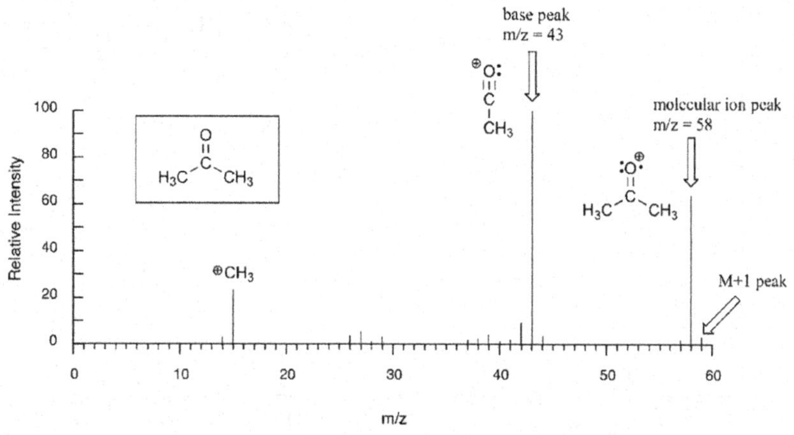

FIGURE 1.4 Mass spectra of acetone.

1.5.2 Quantitative Applications

1.5.2.1 Isotope Abundance Assessment

The variations in the masses of isotopes of an element are very small. With MS, the isotope abundance, that is, the isotopic composition of molecules can be calculated within an easily vaporizable sample (Rozett, 1974). This data helps to collect knowledge about the following:

- Experiments of isotopes as tracers
- Determination of atomic weights of compounds
- Determination of age rocks and minerals
- Study of origin and the nature of the solar system.

The chemical structure of bromobenzene is shown in Figure 1.5. Its mass spectra shows that bromine has two isotopes (Figure 1.6).

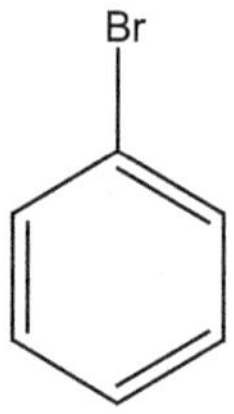

FIGURE 1.5 Chemical structure of bromobenzene.

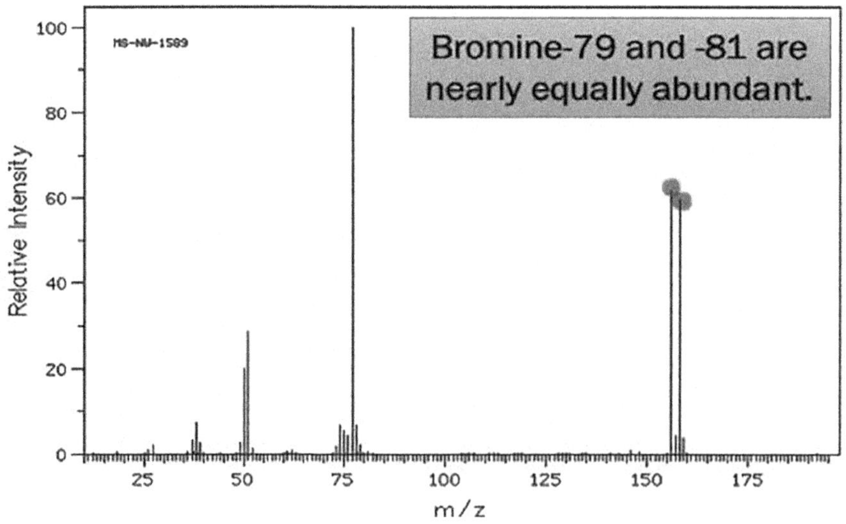

FIGURE 1.6 Mass spectra of bromobenzene.

1.5.2.2 Determination of Isotope Ratio

Mass spectroscopy is used to determine the isotope ratio. It helps determine the concentration of individual components present in a complex mixture that cannot be quantitatively isolated. Isotope ratio MS is a sophisticated technique used to provide information on substances' geographical, chemical, and biological sources. The ability to identify the source of an organic substance derives from the relative isotopic abundances of the material-containing elements. The isotope ratios measurement can be used to distinguish between samples that otherwise share similar chemical compositions. Non-volatile substances such as foods, drugs, amino acids, and fatty acids can be measured most easily with this technique (Zeland and Glen, 2009).

1.5.2.3 Differentiation between *cis*-Isomers and *trans*-Isomers

FAB-MS in the negative ion mode can be used to unambiguously distinguish between *cis*- and *trans*-isomers. The molecular ion peak of the *trans*-isomer is more intense than the *cis*-isomer (Hong et al., 2007). An example is the determination of isomers of monounsaturated fatty acids by the relative signal strengths of an intense pair of ion signals.

1.5.2.4 Mass Spectrometry in Thermodynamics

Determination of heat vaporization: The intensity of the mass spectrum's peak height and temperature determine the heat of vaporization of analyte. The data collected from the spectrum is plotted on the graph by taking intensities on y-axis and temperature on x-axis (Babeliowsky et al., 1962).

For example, as a function of the temperature of a Knudsen cell within the ion source, mass spectrometric determinations have been made of the rate of vaporization of CaO. Log (IT) gives the heat of vaporization at 2,210°K directly as a function of $1/T$.

Determination of heat sublimation: The sublimated solids' vapors are moved to the mass spectrometer's ionization chamber. In this spectrum, the peak intensities are directly proportional to the vapor pressure (VP) of the sample calculated from a Knudsen cell in the ionization chamber (Pattoret et al., 1971). For example, the VP of uranium was determined between 1,720°K and 2,340°K.

1.5.2.5 Measurement of Ionization Potential

The electron bombardment of a gaseous molecule by the mass spectrometer is often used to measure the appearance potential of ions. These two gases are introduced into instruments simultaneously. One gas is the gas under investigation, while the other one is a standard gas (usually argon or krypton), and the ionization potential is accurately measured from spectroscopic data (Warren, 1950).

1.5.2.6 Determination of Ion–Molecule Reactions

In the ion–molecule response study, which is the reaction between the fragment ion and the unionized molecules, MS finds its application. The rate of these responses depends directly on the operating pressure (Lindinger et al., 1993).

1.5.2.7 Detection and Identification of Impurity

A small number of impurities present in the sample, even in low concentrations (ppm), can be detected by MS and provide the impurities' molecular weights that differ from the original component. The impurity structure was identified by LC-MS using a quadrupole ionizer with different ionization techniques such as ESI and CI.

A chromatogram of carmidole solution under light at a 10 mg/mL concentration under preparative LC for 2 hours shows an impurity. The impurity has a molecular weight of 108 amu lower than that of carmidole. The protonated ion of impurity loses two neutrals (m/z 59 and 44 amu), and the protonated ion of carmidole loses three neutrals (m/z 59, 44, and 108 amu) during the collision-induced decomposition process (Ning et al., 2007).

1.5.2.8 Identification of Unknown Compounds

A mass spectrometer can be used to identify an unknown compound via molecular weight determination to determine the structure and chemical properties by comparing with the available data of known compounds from a public reference library such as the Chemical Abstract Service (CAS) registry, which contains over 54 million entries. Accurate mass measurements can be used to query the CAS registry by either molecular formulae or average molecular weights (James et al., 2011; McLafferty et al., 1999).

MS serves as a versatile tool for the study of the structure and functions of proteins. For this purpose, ESI and MALDI are the most widely used ionization methods to produce a good result for the determination of proteins. MS in proteomics mainly deals with the analysis of protein digested by a protease such as trypsin (Bruno, 2006).

1.5.3 Both Qualitative and Quantitative Applications

1.5.3.1 Phytochemical Analysis

This kind of technique is widely used in phytochemical analysis because of its ability to identify and measure metabolites having a very low molecular weight at shallow concentration ranges below nanogram per milliliter (ng/mL) with gas chromatography, capillary electrophoresis, and HPLC (GC-MS, CE-MS, and HPLC-MS). Analyses of fatty acids, sucrose, diterpenes, and sesquiterpenes with the help of HPLC-MS and vitamin E by using GC-MS are examples (Safer et al., 2011; Shellie and Marriot, 2003).

1.5.3.2 Structural Elucidation of Unknown Phytochemicals

One of the major applications of MS is structural elucidation by taking into consideration the following peaks: base peak, molecular ion peak, isotopic peak, fragment ion peak, and metastable ion peak. The structure of flavonoid monoglycosides is elucidated using liquid secondary-ion mass spectrometry (LSI-MS) and electron impact mass spectrometry (EI-MS) (Franski et al., 1999). Mass spectroscopic data, along with NMR, FTIR, and UV–visible data, are extensively used to elucidate unknown natural products (Ahamad et al., 2013; 2014a,b).

1.5.3.3 Drug Metabolism Studies

There are two types of metabolic reactions, i.e., Phase-I and Phase-II reactions. During a metabolic reaction, the transformation of the metabolic product corresponds with the molecular weight. These changes can be analyzed by a mass spectrometer. Examples are drug metabolism studies of denopamine and promethazine, in which the hydroxylated metabolites are easily separated from N- or S-oxide. A three-stage quadrupole MS equipped with ionization techniques for EI, FAB, atmospheric pressure chemical ionization, ESI, and thermospray (TSP) was used (Ohashi et al., 1998).

1.5.3.4 Clinical Studies

In any disease condition, the chemistry of the body changes to produce various chemical products in body liquid and excretion products that can be detected for diagnosis with the help of GC-MS (Becket and Stenlake, 2007).

Nowadays, it is a trend to directly analyze and image pharmaceutical compounds in intact tissue by MALDI-MS (Reyzer et al., 2003).

1.5.3.5 Forensic Applications

The sample in drug abuse and murder cases or death due to poisoning is mainly urine, hair, and blood. Samples may be analyzed using MS (Folz, 1992).

1.6 CONCLUSION

Mass spectroscopy has emerged as a powerful analytical technique for the correct identification of organic compounds. Mass spectroscopy is extensively applied in conjugation with other analytical methods such as liquid chromatography and GC to enhance online identification, quantification, and simultaneous structural characterization. In conjugation with NMR, FTIR, and UV–visible spectroscopy, it is used for the structure elucidation of unknown natural and synthetic compounds. Recent advancements in ionization techniques and upgrading of detectors make MS more suitable for food analysis and drug discovery.

REFERENCES

Ahamad, J., Ahmad, J., Ameen, M.S.M., Anwer, E.T., Mir, and S.R., Ameeduzzafa. (2020b). Chapter 12: Proteomics in authentication of honey. *Proteomics for Food Authentication* (Edited by Leo M.L. Nollet, Semih Ötleş). CRS Press, Taylor & Francis Group: Boca Raton, FL.

Ahamad, J., Ahmad, J., Shahzad, N., and Mohsin, N. (2018). *Authentication and Traceability of Wine. Fingerprinting Techniques in Food Authentication and Traceability*, vol. 1. CRS Press, Taylor & Francis Group: Boca Raton, FL.

Ahamad, J., Mir, S.R., Kaskoos, R.A., and Ahmad, J. (2020a). Chapter 11: Proteomics in authentication of wine. *Proteomics for Food Authentication* (Edited by Leo M.L. Nollet, Semih Ötleş). CRS Press, Taylor & Francis Group: Boca Raton, FL.

Ahamad, J., Naquvi, K.J., Ali, M., and Mir, S.R. (2013). New Glycoside Esters from the Aerial Parts of *Artemisia absinthium* Linn. *Nat. Prod. J.*, 3(4): 260–267.

Ahamad, J., Naquvi, K.J., Ali, M., and Mir, S.R. (2014a). Isoflavone glycosides from aerial parts of *Artemisia absinthium* Linn. *Chem. Nat. Comp.*, 49(6): 696–700.

Ahamad, J., Naquvi, K.J., Ali, M., and Mir, S.R. (2014b). New isoquinoline alkaloids from the stem bark of *Berberis aristata. Ind. J. Chem.-B*, 53: 1237–1241.

Albert, T. and Zaikin, V. (2008). Organic mass spectrometry at the beginning of the 21st century. *J. Anal. Chem.*, 63: 1236–1264.

Babeliowsky, T.P.J.H., Boerboom, A.J.H., and Kistemaker, J. (1962). Mass spectrometric determination of the heat of vaporization of CaO. *Physica*, 28: 1155–1159.

Baldwin, M.A. (2005) Mass spectrometers for the analysis of biomolecules. *Meth. Enzymol.*, 402: 3–48.

Banoub, J., Cohen, A., Mansour, A., and Thibault, P. (2004). Characterization and de novo sequencing of Atlantic salmon vitellogenin protein by electrospray tandem and matrix-assisted laser desorption/ionization mass spectrometry. *Eur. J. Mass Spectrom.*, 10: 121–134.

Banoub, J., Thibault, P., Mansour, A., Cohen, A., Heeley, D.H., and Jackman, D. (2003). Characterisation of the intact rainbow trout vitellogenin protein and analysis of its derived tryptic and cyanogen bromide peptides by matrix-assisted laser desorption/ionisation time-of-flight-mass spectrometry and electrospray ionisation quadrupole/time-of-flight mass spectrometry. *Eur. J. Mass Spectrom. (Chichester, Eng).*, 9: 509–524.

Banoub, J.H., Newton, R.P., Esmans, E., Ewing, D.F., and Mackenzie, G. (2005). Recent developments in mass spectrometry for the characterization of nucleosides, nucleotides, oligonucleotides, and nucleic acids. *Chem. Rev.*, 105: 1869–1915.

Barber, M., Bordoli, R.S., Sedgwick, R.D., and Tyler, A. (1981). Fast atom bombardment of solids as an ion source in mass spectrometry. *Nature*, 293: 270–275.

Becket, A.H. and Stenlake, J.B. (2007). *Practical Pharmaceutical Chemistry*, 4th ed. CBS publishers and distributors: India, p. 493.

Beckey, H.D. (1969). Field desorption mass spectrometry: A technique for the study of thermally unstable substances of low volatility. *Int. J. Mass Spectrom. Ion Phys.*, 2: 500–503.

Bruno, D. (2006). Mass spectrometry and protein analysis. *Science*, 312: 212–217.

Bucknall, M., Fung, K.Y., and Duncan, M.W. (2002). Practical quantitative biomedical applications of MALDI-TOF mass spectrometry. *J. Am. Soc. Mass Spectrom.*, 13: 1015–1027.

Canas, B., Lopez-Ferrer, D., Ramos-Fernandez, A., Camafeita, E., and Calvo, E. (2006). Mass spectrometry technologies for proteomics. *Brief Funct. Genom.*, 4(4): 295–320.

El-Aneed, A., Cohen, A., and Banoub, J. (2009). Mass spectrometry, review of the basics: Electrospray, MALDI, and commonly used mass analyzers. *Appl. Spectroscopy Rev.*, 44: 210–230.

Emmett, M.R. and Caprioli, R.M. (1994). Micro electrospray mass spectrometry: Ultra-high sensitivity analysis of peptides and proteins. *J. Am. Chem. Soc. Mass Spectrom.*, 5: 605–613.

Folz, R.L. (1992). Recent applications of mass spectrometry in forensic toxicology. *Int. J. Mass Spectrom. Ion Processes*, 118/119: 237–263.

Franski, R., Bednarek, P., Siatkowska, D., Wojtaszek, P., and Stobiecki, M. (1999). Application of mass spectrometry to structural identiication of lavonoid monoglycosides isolated from shoot of lupin (Lupinus LuteusA cLt.a). *Biochimica Polonica.*, 46: 459–473.

Gale, D.C. and Smith, R.D. (1993). Small volume and low flow-rate electrospray ionization mass spectrometry of aqueous samples. *Rapid Comm. Mass Spectrom.*, 7: 1017–1021.

Grayson, M.A., Ed. (2002). *Measuring Mass: From Positive Rays to Proteins.* Chemical Heritage Press: Philadelphia, PA.

Griffiths, W.J., Jonsson, A.P., Liu, S., Rai, D.K., and Wang, Y. (2001). Electrospray and tandem mass spectrometry in biochemistry. *Biochem. J.*, 355: 545–561.

Henderson, W. and McIndoe, J.S. (2005). *Mass Spectrometry of Inorganic and Organometallic Compounds.* John Wiley & Sons: Chichester.

Herbert, C.G. and Johnstone, R.A.W. (2003). *Mass Spectrometry Basics.* CRC Press: Boca Raton, FL.

Hillenkamp, F. and Karas, M. (1990) Mass spectrometry of peptides and proteins by matrix-assisted ultraviolet laser desorption/ionization. *Meth. Enzymol.*, 193: 280–295.

Hillenkamp, F., Karas, M., Beavis, R.C., and Chait, B.T. (1991). Matrix-assisted laser desorption/ionization mass spectrometry of biopolymers. *Anal. Chem.*, 63: 1193A–1203A.

Hong, Ji., Valery, G.V., Max, L.D., and Douglas, F.B. (2007). Distinguishing between Cis/Trans isomers of monounsaturated fatty acids by FAB MS. *Anal. Chem.*, 79: 1519–1522.

James, L.L., Curtis, D.C., and Stacy, D.B. (2011). Identification of "Known Unknowns" utilizing accurate mass data and chemical abstracts service databases. *J. Am. Soc. Mass Spectrum*, 22: 348–359.

Karas, M., Bahr, U., and Dulcks, T. (2000). Nano-electrospray ionization mass spectrometry: Addressing analytical problems beyond routine. *Fresen. J. Anal. Chem.*, 366: 669–676.

Kitson, F.G., Larsen, B.S., and McEwen, C.N. (1996). *Gas Chromatography and Mass Spectrometry: A Practical Guide.* Academic Press: San Diego, CA.

Korfmacher, W.A., Ed. (2005). *Using Mass Spectrometry for Drug Metabolism Studies.* CRC Press: Boca Raton, FL.

Lindinger, W., Hirber, J., and Paretzke, H. (1993). An ion/molecule-reaction mass spectrometer used for online trace gas analysis. *Int. J. Mass Spectrom. Ion Processes*, 129: 79–88.

McLafferty, F. and Turecek, F. (1993). *Interpretation of Mass Spectra*, 4th ed. University Science: Mull Valey, CA, pp. 355–359.

Mclafferty, F.W., Douglas, A.S., Stanton, Y.L., and Chrystomos, W. (1999). Unknown identification using reference mass spectra. Quality evaluation of databases. *J. Am. Soc. Mass Spectrom.*, 10: 1229–1240.

Ning, Li., Jing, Y., Fang, Q., Famie, Li., and Ping, G. (2007). Isolation and identification of a major impurity in a new bulk drug candidate by preparative LC, ESI-MS, LC-MS, and NMR. *J. Chromatogr. Sci.*, 45: 45–49.

Ohashi, N., Furuuchi, S., and Yoshikawa, M. (1998). Usefulness of the hydrogen-deuterium exchange method in the study of drug metabolism using liquid chromatography-tandem mass spectrometry. *J. Pharma. Biomed. Anal.*, 18: 325–334.

Pattoret, A., Drowart, J., and Smoes, S. (1971). Mass spectrometric determination of heat of sublimation of uranium. *Trans. Faraday Soc.*, 67; 001–002.

Pramanik, B.N., Ganguly, A.K., and Gross, M.L. (Eds.). (2002). *Applied Electrospray Mass Spectrometry.* Marcel Dekker: New York.

Rajawat, J. and Jhingan, G. (2019). Chapter 1: Mass spectroscopy. *Data Processing Handbook for Complex Biological Data Sources.* DOI: 10.1016/B978-0-12-816548-5.00001-0.

Reyzer, M.L., Hsieh, Y., Ng, K., Korfmacher, W.A., and Caprioli, R.M. (2003). Direct analysis of drug candidates in tissue by matrix-assisted laser desorption/ionization mass spectrometry. *J. Mass Spectrom.*, 38: 1081–1092.

Rockwood, A.L., Kushnir, M.M., and Clarke, N.J. (2018). Mass spectrometry. Principles and Applications of Clinical Mass Spectrometry, 33–65.

Rozett, R.W. (1974). Isotope abundance from Mass spectra. *Anal. Chem.*, 46: 2085–2089.

Safer, S., Cicek, S.S., Pieri, V., Schwaiger, S., Schneider, P., Wissemann, V., and Stuppner, H. (2011). Metabolic fingerprinting of *Leontopodium* species (Asteraceae) using ^{1}H NMR and HPLCESI- MS. *Phytochemistry*, 72: 1379–1389.

Shellie, R.A. and Marriot, P.J. (2003). Comprehensive two-dimensional gas chromatography-mass spectrometry analysis of Pelargonium Graveolens essential oil using rapid scanning quadrupole mass spectrometry. *The Analyst.*, 128: 879–883.

Silverstein, R. and Webster, F. (1998). *Spectrometric Identification of Organic Compounds.* Wiley: New York, pp. 45–65.

Tanaka, K., Waki, Y., Ido, Y., Akita, S., Yoshida, Y., and Yoshida, T. (1988). Protein and polymer analyses up to m/z 100,000 by laser ionization time-of-flight mass spectrometry. *Rapid Comm. Mass Spectrom.*, 2: 151–153.

Torgerson, D.F., Skowronski, R.P., and Macfarlane, R.D. (1974). New approach to the mass spectroscopy of non-volatile compounds. *Biochem. Biophys. Res. Comm.*, 60: 616–621.

Warren, J. (1950). Measurement of Appearance Potentials of Ions Produced by Electron Impact, using a Mass Spectrometer. *Nature*, 165: 810–811.

William, F.W. and Price, D. (2002). Mass spectrometry finding the molecular ion and what it can tell you: An undergraduate organic laboratory experiment. *Chem. Educator*, 7(4): 226–232.

Yamashita, M. and Fenn, J.B. (1984). Electrospray ion source. Another variation on the free-jet theme. *J. Phys. Chem.*, 88: 4451–4459.

Zeland, M. and Glen, P.J. (2009). Isotope ratio mass spectrometry. *Analyst*, 134: 213–222.

Zubarev, R.A. and Makarov, A. (2013). Orbitrap mass spectrometry. *Anal. Chem.*, 85: 5288–5296.

2 Targeted and Untargeted Analyses

Dev Kant Shandilya
Bhagwant University

CONTENTS

2.1 Overview of Mass Spectrometry Analysis 19
2.1.1 Targeted Analysis 20
2.1.1.1 Targeted Acquisition Using MS 20
2.1.1.2 Targeted Analysis Using MS/MS 21
2.1.2 Untargeted Analysis 21
2.1.2.1 Data-Dependent Analysis 22
2.1.2.2 Data-Independent Acquisition 22
2.1.3 Extracted Ion Chromatogram (XIC) 23
2.2 Targeted and Untargeted Analyses – Workflows 24
2.3 Conclusion 25
References 25

2.1 OVERVIEW OF MASS SPECTROMETRY ANALYSIS

This chapter depicts the details of mass spectrometry (MS) analysis: instrumentation and basic principles that are already discussed in Chapter 1. MS analysis can be covered under two clouds: targeted analysis and untargeted analysis. With the advancement in the technology, targeted and untargeted MS analyses can be performed using a wide variety of tandem mass spectrometry (MS/MS) analyzers – analyzer 1 (Quadrupole)–analyzer 2 (Quadrupole), analyzer 1 (Quadrupole)–analyzer 2 (time-of-flight [ToF]), analyzer 1 (Quadrupole)–analyzer 2 (Orbitrap), and others. These analyzers along with targeted and untargeted analyses scan types facilitate several quantitative and qualitative applications of food analysis [1–4]. These applications are typically developed with hyphenation of MS with chromatography, i.e., high-pressure liquid chromatography, micro-liquid chromatography, gas chromatography (GC), capillary electrophoresis (CE), and others.

During targeted analysis, the molecule of interest or mass-to-charge (*m/z*) values (for MS1 and MS2) are predefined, and it is one of the most popular analyses for quantitation of analytes present in a complex food matrix, with high sensitivity, accuracy, and specificity [5,6]; a detailed view of target analysis is discussed in Section 2.2.1. The untargeted analysis (using MS1 and MS2) is the analysis to identify the unknowns in a sample, using parent ion and product or fragment ion analysis (MS/MS) [7,8]. The features of targeted and untargeted analyses are summarized in Figure 2.1.

During targeted and untargeted analyses, the ionization source [9] is a vital module of the mass spectrometer, which significantly impacts the quality of mass spectra. Electron impact, electrospray ionization, atmospheric pressure chemical ionization, atmospheric pressure photoionization, and matrix-assisted laser desorption analyzer are the most used ionization sources for MS analysis (GC-MS, GC-MS/MS, LC-MS, and LC-MS/MS); however, three recent ionization sources added advantage for fast screening of food authentication samples with limited or no sample preparation: (i) direct analysis in real-time [10–13], (ii) atmospheric pressure solids analysis probe [14,15], and (iii) desorption electrospray ionization [16].

DOI: 10.1201/9781003091226-3

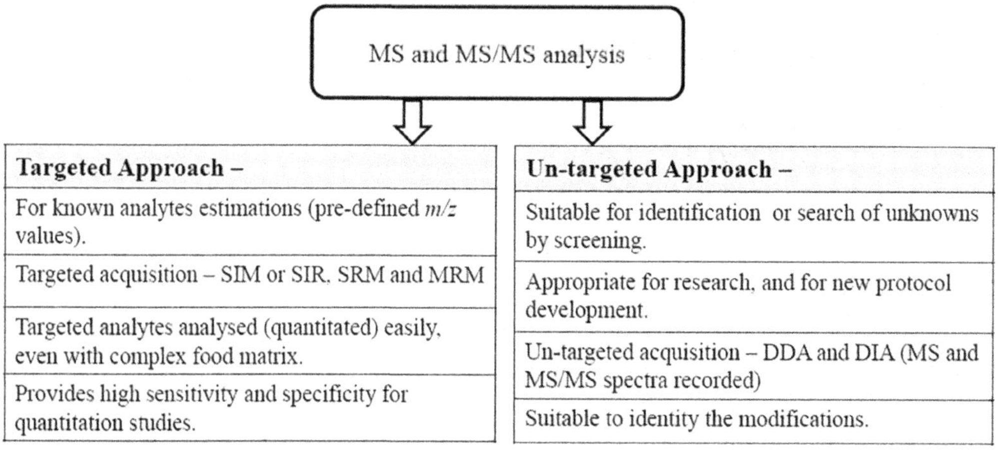
MS and MS/MS analysis
Targeted Approach –
For known analytes estimations (pre-defined m/z values).
Targeted acquisition – SIM or SIR, SRM and MRM
Targeted analytes analysed (quantitated) easily, even with complex food matrix.
Provides high sensitivity and specificity for quantitation studies.
Un-targeted Approach –
Suitable for identification or search of unknowns by screening.
Appropriate for research, and for new protocol development.
Un-targeted acquisition – DDA and DIA (MS and MS/MS spectra recorded)
Suitable to identity the modifications.

FIGURE 2.1 Key features of targeted and untargeted MS analysis. SIM, Selected ion monitoring; SIR, Selected ion recording; SRM, Selected reaction monitoring; MRM, Multiple reaction monitoring; DDA, Data-dependent analysis; DIA, Data-independent analysis.

2.1.1 Targeted Analysis

Targeted analysis approach serves numerous applications of food analysis by the use of various types of mass analyzers. During the target analysis approach, the *m/z* ratio of concerned ions (parent and product ions of analyte) needs to be predefined. Target approaches can be classified into two acquisition methods: (i) using MS (protonated $[M+H]^+$ or molecular ion peak M^+ or in-source fragment ion) and (ii) using MS and MS/MS (parent and product ions based).

2.1.1.1 Targeted Acquisition Using MS

During the analysis, only an analyzer (MS1) is used to filter/separate the selected ions. MS-based targeted acquisition mode is also known as selected ion monitoring (SIM). The schematic representation of SIM acquisition mode is presented in Figure 2.2. In SIM mode, pre-selected *m/z* values (concerned ion or ions of targeted analyte) are defined in the acquisition method. The analyzer separates/filters the selected *m/z* ions, which are then constantly monitored by the detector throughout the chromatography run time [17–19]. Detector outcome is used for the estimation of analytes concentration in a wide variety of matrixes. During SIM mode, additional confirmation about the identity of analyte or targeted molecule can be obtained using a tandem mass analyzer (MS2; product or fragmentation ions). MS2 analyzer is used under full-scan function (appropriate mass range

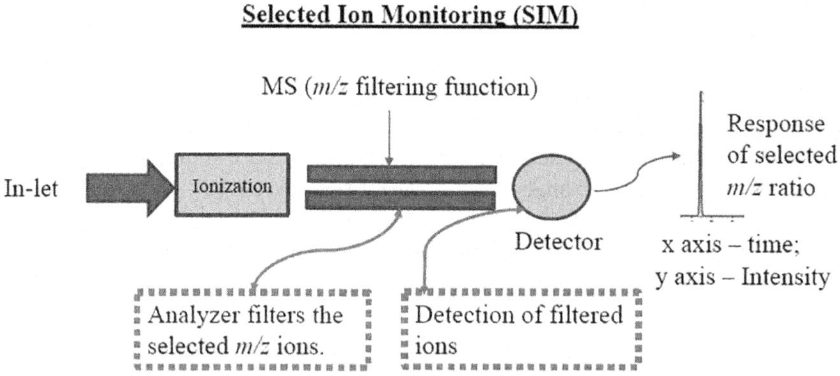

FIGURE 2.2 Schematic representation of selected ion monitoring/recording (SIM or SIR).

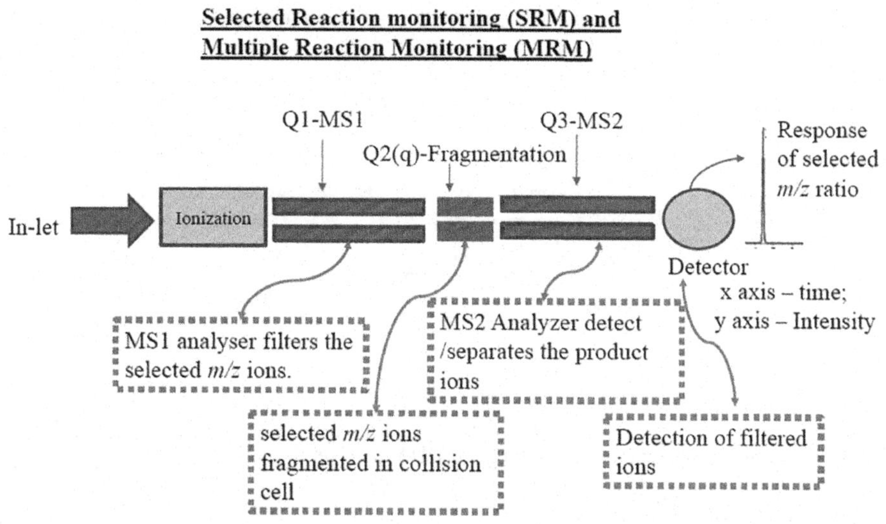

FIGURE 2.3 Schematic representation of selected reaction monitoring (SRM) and multiple reaction monitoring (MRM) acquisition.

to cover all fragments of targeted analyte); it facilitates product ions spectra (MS/MS). SIM acquisition mode is also known as selected ion recording (SIR).

2.1.1.2 Targeted Analysis Using MS/MS

During the analysis, MS1, MS2 analyzers, and collision cell are used (tandem mass analyzers); by stating the *m/z* values of precursor and its fragment, a pair of precursor and fragment is known as transition. MS1 analyzer filters the predefined *m/z* values (precursor ions). These are then fragmented in collision cell and MS2 analyzer detects the predefined product ions, which are constantly monitored by a detector. The detector response is used for estimation purposes. Targeted MS/MS-based analysis with single transition is known as selected reaction monitoring (SRM). Multiple reaction monitoring (MRM) [20] acquisition contains multiple SRM transitions in same experiment [21]. The schematic representation of targeted MS/MS analysis, SRM, and MRM is presented in Figure 2.3.

Targeted analysis facilitates highly specific, accurate, and reproducible analysis results with high sensitivity [22,23]. MS-based targeted analysis or SIM or SIR acquisition facilitates to acquire the selected ion or ions of selected analyte, and MS/MS-based targeted analysis or SRM and MRM acquisitions enable estimation of the fragment ions produced from the selected precursor ion. During targeted analysis, estimation is done by peak area response comparison (known versus unknown) and by peak area response versus concentration linearity curve.

2.1.2 Untargeted Analysis

Unlike targeted analysis, untargeted analysis is a full-scan analysis. MS1 analyzer separates the ions based on *m/z* ratio. These are then fragmented in collision cell and MS2 analyzer separates the product ions, which are constantly monitored by detector. Untargeted analysis can be classed into two types of acquisition modes: data-dependent analysis (DDA) and data-independent acquisition (DIA) scans. During both analysis modes, the purpose is to acquire MS and MS/MS spectra [24–26]. Collected information is used for structural confirmation, and quantitation by using extracted ion chromatograms (XICs) and high-resolution extracted ion chromatograms (HR-XICs). XIC is further explained in Section 2.2.3 of this chapter.

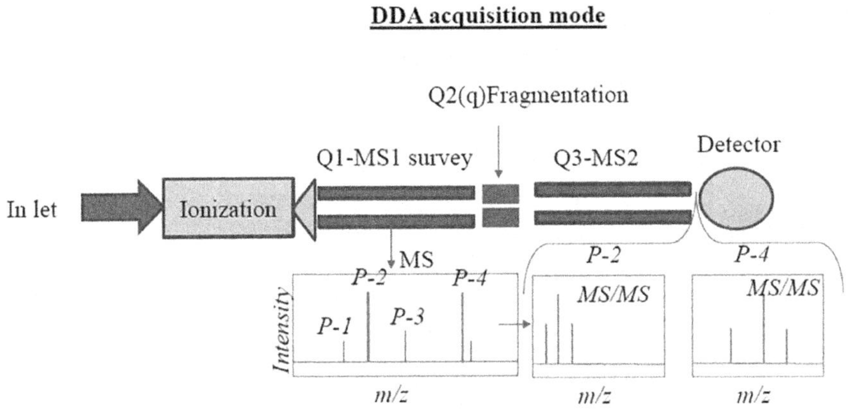

FIGURE 2.4 Schematic representation of untargeted DDA analysis.

2.1.2.1 Data-Dependent Analysis

A DDA is the most used MS/MS analysis. During DDA, precursor selection for fragmentation (MS/MS) analysis is done based on information gathered during the MS1 scan. This analysis allows acquiring the MS and MS/MS spectral data in a single run, which can be used for both qualitative and quantitative analyses.

In DDA, MS and MS/MS data are acquired for analytes separated or eluted as a function of time during the chromatography run (GC/LC/CE). MS/MS data of the analytes are acquired based on the *m/z* values obtained throughout the MS analysis or survey scan. The MS acquisition method parameters are set to decide the selection criteria of *m/z* values for product ion (MS/MS) analysis. i.e., every cycle, the analyzer scans a short precursor ion scan (MS1) or survey scan for the analytes eluting from the chromatography run. This enables to monitor ionized analytes *m/z* values and intensities and to identify *m/z* values to fragment, followed by sequential product ion (MS/MS) scans. During the individual scan, a parent ion is selected and fragmented, and its product ions are detected. Parent ions are fragmented in order of decreasing intensity.

To exclude the fragment of the same parent ion (*m/z*) again and again dynamic exclusion criteria are set based on time and *m/z* value to ensure that the parent ion which was selected recently is not re-selected for fragmentation or MS/MS analysis. Analogous criteria can also apply for the exclusion of background ions. The quality of the MS/MS spectra also depends on the chromatography run time covered by *m/z* value during the survey scan and intensity during the survey scan. The schematic representation of the DDA experiment is presented in Figure 2.4.

Selection of scan parameters is crucial for the success of the DDA experiment: Q1 MS mass range, MS/MS mass range, and cycle time for MS and MS data collection. DDA experiments have also some limitations, including less precursor selection and low reproducibility. To overcome these limitations and to cover all precursor ions for MS/MS, the additional data acquisition type is developed – data-independent acquisition (DIA) [27], also known as MS/MS-All. A detailed explanation of DIA is given in Section 2.2.2.2.

2.1.2.2 Data-Independent Acquisition

DIA analysis has a more substantial number of MS2 spectra in comparison with DDA. However, the noise level increases during DIA ; the level of noise can be controlled by applying a wide mass range (instead of full scan) to the Q1MS spectra, and the noise level can be further reduced by applying background ions exclusion function. The flow of the DIA is presented in Figure 2.5.

Throughout the survey scan, a dynamic *m/z* window [28] is used for the selection of parent ions during DIA (to cover the full scan range). On the other hand, during the DDA experiment, a full spectrum is scanned (broader mass range), and precursor selection is based on the intensities of

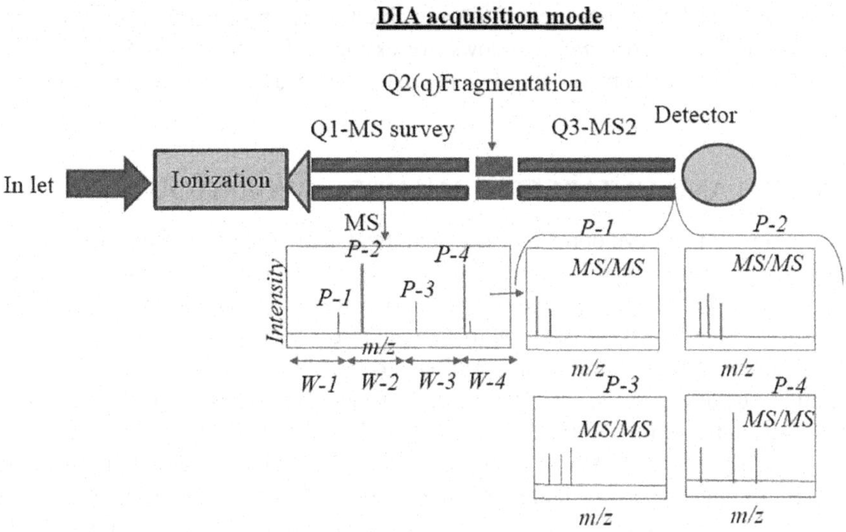

FIGURE 2.5 Schematic representation of untargeted DIA analysis.

m/z values. During a DIA experiment, we get MS/MS information of almost all precursor ions or Q1MS *m/z* values. The noise level in DIA spectra is higher in comparison with DDA spectral data. The parameters for DIA can be rationally optimized to achieve a high quality of data. Numerous automated DIA scans are available, i.e., sequential windowed acquisition of all theoretical fragment ion mass spectra (SWATH) and alternating low-energy collision-induced dissociation and high-energy collision-induced dissociation (MSE). DIA scan modes are also used for quantitative information [29].

2.1.3 Extracted Ion Chromatogram (XIC)

XIC [30] provides additional assurance to the data, i.e., the presence of a specific *m/z* value at a specific retention time. XIC is a data processing function of ion intensity of a specific *m/z* value or range as a function of chromatographic retention time. XIC function is applied to scan data, i.e., to total ion chromatogram. The XIC data processing function can be used for quantitative studies and delivers with high sensitivity and specificity. The pictorial representation of XIC is presented in Figure 2.6. XIC of ToF and Orbitrap data (high-resolution analyzers) is known as HR-XIC.

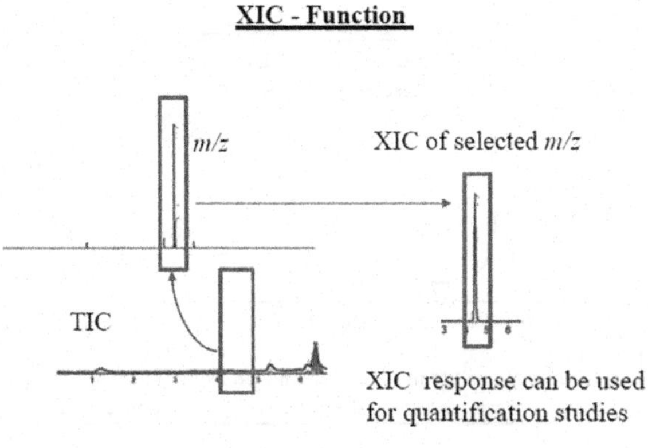

FIGURE 2.6 Representation of XIC processing function.

HR-XIC of high-resolution MS spectral data enable highly specific quantification using untargeted data. Ions of very narrow *m/z* windows are extracted to obtain HR-XIC (chromatogram). Quantitation can be done by comparison of peak areas or by peak area versus concentration linearity curve.

2.2 TARGETED AND UNTARGETED ANALYSES – WORKFLOWS

Workflows are essential for developing the right analytical methodology. In this section, step-by-step analysis workflows are explained for both types of analysis: targeted and untargeted. The targeted analysis workflows explained in Figures 2.7 and 2.8 depict the workflow for untargeted analysis.

MS analysis workflows (targeted and untargeted analyses) are paved in all aspects of food analysis; ensure the absence of contaminations in food – pesticides, packaging material traces, pathogens, estimation of preservatives, flavoring agents, sweeteners and confirm the nutritional value, composition, and source of foodstuff.

Targeted analysis using MS, MS/MS, and high-resolution tandem mass analyzer facilitates all quantitative applications, whereas untargeted analysis can be used for the qualitative and quantitative applications and facilitates the identification of the unknowns. Ion mobility-equipped mass analyzer enables improved sensitivity and more certain MS/MS spectra. The software tools and databases facilitate the data interpretation. Latest advancements in instrumentation and software with workflows support researchers to apply MS for food analysis. Detailed applications of targeted and untargeted analyses are explained in Chapters 3–22 of this book.

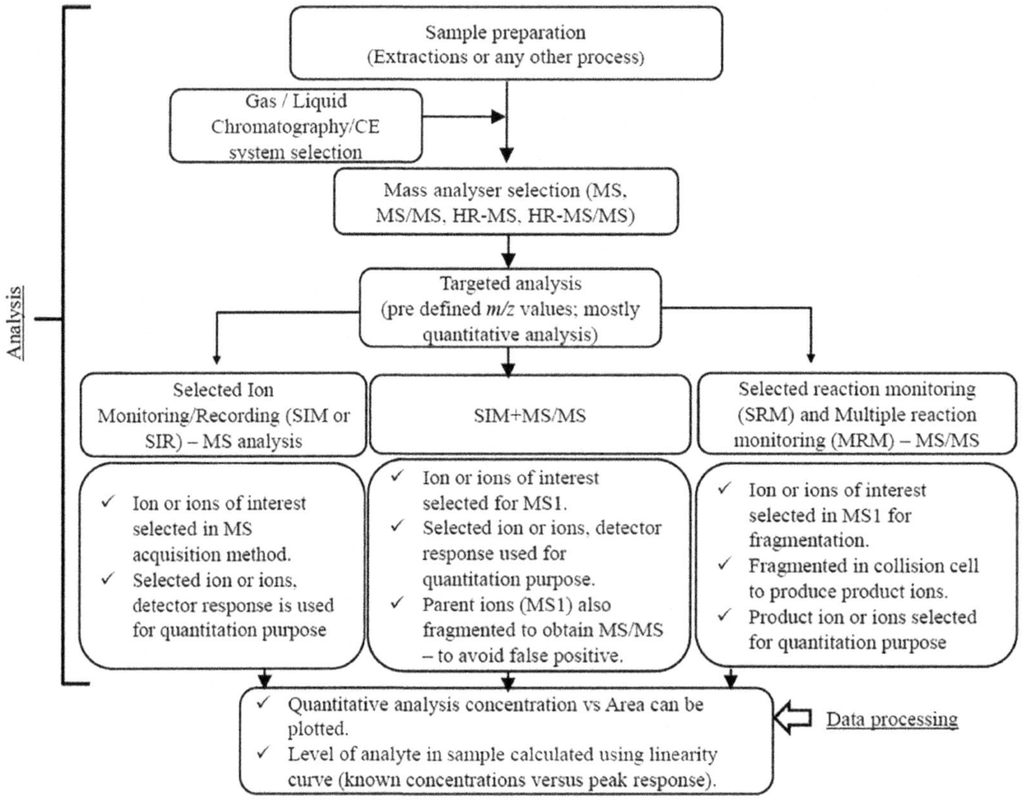

FIGURE 2.7 Workflow for targeted analysis.

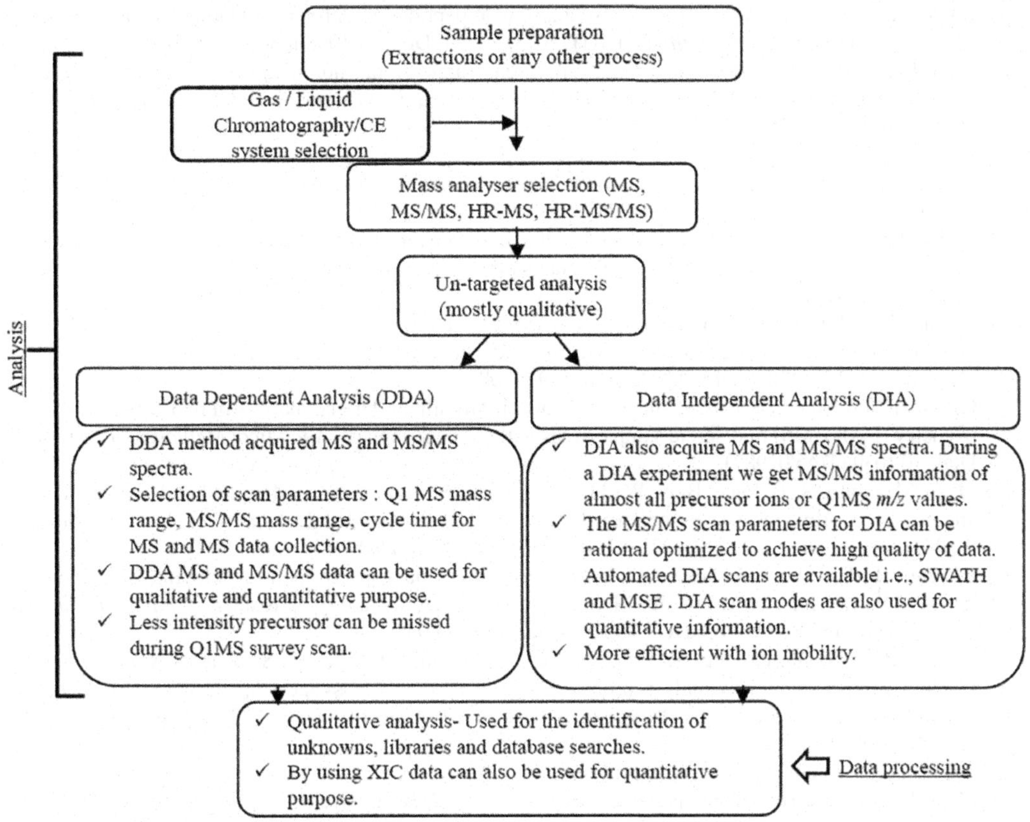

FIGURE 2.8 Workflow for untargeted analysis.

2.3 CONCLUSION

This chapter provides an outline and details of targeted and untargeted MS analyses, its applications and workflows for food analysis, and how new advancements of mass spectrometers helped a lot to ensure the quality of foodstuff and its safety for the consumers. This chapter also explained the selection criteria, advantages, and disadvantages related to targeted and untargeted MS analyses. This chapter will enable us to build MS analysis skills in the food analysis arena.

REFERENCES

1. M. E. Dasenaki, S. K. Drakopoulou, R. Aalizadeh, and N. S. Thomaidis, Targeted and untargeted metabolomics as an enhanced tool for the detection of pomegranate juice adulteration. *Foods*, 8 (6) (2019) 212. DOI: 10.3390/foods8060212.
2. M. Kunzelmann, M. Winter, M Åberg, K. -E. Hellenäs, and J. Rosén, Non-targeted analysis of unexpected food contaminants using LC-HRMS. *Analytical and Bioanalytical Chemistry*, 410 (2018) 5593–5602. DOI: 10.1007/s00216-018-1028-4.
3. G. Cao, K. Li, J. Guo, M. Lu, Y. Hong, and Z. Cai, Mass spectrometry for analysis of changes during food storage and processing. *Journal of Agricultural and Food Chemistry*, 68 (26) (2020) 6956–6966. DOI: 10.1021/acs.jafc.0c02587.
4. M. Careri, F. Bianchi, and C. Corradini, Recent advances in the application of mass spectrometry in food-related analysis. *Journal of Chromatography A*, 970 (1–2) (2002) 3–64. DOI: 10.1016/s0021-9673(02)00903-2.
5. I. Domínguez, A. G. Frenich, and R. Romero-González, Mass spectrometry approaches to ensure food safety. *Analytical Methods*, 9 (2020). DOI: 10.1039/C9AY02681A.

6. M. A. Gillette and S. A Carr, Quantitative analysis of peptides and proteins in biomedicine by targeted mass spectrometry. *Nature Methods*, 10 (1) (2013) 28–34. DOI: 10.1038/nmeth.2309.
7. T. Wang, L. Duedahl-Olesen, and H. L. Frandsen, Targeted and non-targeted unexpected food contaminants analysis by LC/HRMS: Feasibility study on rice. *Food Chemistry*, 338 (2021) 127957. DOI: 10.1016/j.foodchem.2020.127957.
8. N. Z. Ballin and K. H. Laursen, To target or not to target? Definitions and nomenclature for targeted versus non-targeted analytical food authentication. *Trends in Food Science & Technology*, 86 (2019) 537–543. DOI: 10.1016/j.tifs.2018.09.025.
9. G. Kaklamanos, E. Aprea, and G. Theodoridis, *Mass Spectrometry: Principles and Instrumentation, Encyclopedia of Food and Health*, Academic Press, (2016) pp. 661–668, DOI: 10.1016/B978-0-12-384947-2.00447-5.
10. B. Gao, S. E. Holroyd, J. C. Moore, K. Laurvick, S. M. Gendel, and Z. Xie, Opportunities and challenges using non-targeted methods for food fraud detection. *Journal of Agricultural and Food Chemistry*, 67 (31) (2019) 8425–8430. DOI: 10.1021/acs.jafc.9b03085.
11. G. Asher Newsome, I. Kayama, and S. A. Brogdon-Grantham, Direct analysis in real-time mass spectrometry (DART-MS) of discrete sample areas without heat damage. *Analytical Methods*. 10 (9) (2018). DOI: 10.1039/C7AY02987J.
12. M. G. Mazzotta, R. B. Pace, B. N. Wallgren, S. A. 3rd Morton, K. M. Miller, and D. L. Smith, Direct analysis in real-time mass spectrometry (DART-MS) of ionic liquids. *Journal of the American Society for Mass Spectrometry*, 24 (10) (2013) 1616–1619. DOI: 10.1007/s13361-013-0696-8.
13. L. Vaclavik, T. Cajka, V. Hrbek, and J. Hajslova, Ambient mass spectrometry employing direct analysis in real-time (DART) ion source for olive oil quality and authenticity assessment. *Analytica Chimica Acta*, 645 (1–2) (2009) 56–63. DOI: 10.1016/j.aca.2009.04.043.
14. S. B. J. McCullough, K. Patel, R. Francis, P. Cain, D. Douce, K. Whyatt, S. Bajic, N. Lumley, and C. Hopley, Atmospheric solids analysis probe coupled to a portable mass spectrometer for rapid identification of bulk drug. *Journal of the American Society for Mass Spectrometry*, 31 (2) (2020) 386–393. DOI: 10.1021/jasms.9b00020.
15. https://ionsense.com/products/asap_probe/en.
16. M. W. F. Nielen, H. Hooijerink, P. Zomer, and J. G. J. Mol, Desorption electrospray ionization mass spectrometry in the analysis of chemical food contaminants in food. *TrAC Trends in Analytical Chemistry*, 30 (2) (2011) 165–180. DOI: 10.1016/j.trac.2010.11.006.
17. A. Stachniuk and E. Fornal, Liquid chromatography-mass spectrometry in the analysis of pesticide residues in food. *Food Analytical Methods*, 9 (6) (2016) 1654–1665. DOI: 10.1007/s12161-015-0342-0.
18. C. Ruiz-Samblás, A. González-Casado, L. Cuadros-Rodríguez, and F. P. García, Application of selected ion monitoring to the analysis of triacylglycerols in olive oil by high temperature-gas chromatography mass spectrometry. *Talanta*, 82 (1) (2010) 255–260. DOI: 10.1016/j.talanta.2010.04.030.
19. S. Thurnhofer and W. Vetter, A gas chromatography/electron ionization–mass spectrometry–selected ion monitoring method for determining the fatty acid pattern in food after formation of fatty acid methyl esters. *Journal of Agricultural and Food Chemistry*, 53 (23) (2005) 8896–8903. DOI: 10.1021/jf051468u.
20. https://www.ionsource.com/tutorial/msquan/intro.htm.
21. http://www.mrmatlas.org/glossary2.php.
22. R. W. Kondrat, G. A. McClusky, and R. G. Cooks, Multiple reaction monitoring in mass spectrometry/mass spectrometry for direct analysis of complex mixtures. *Analytical Chemistry*, 50 (14) (1978) 2017–2021. DOI: 10.1021/ac50036a020.
23. A. D. Watson, Y. Gunning, N. M. Rigby, M. Philo, and E. K. Kemsley, Meat authentication via multiple reaction monitoring mass spectrometry of myoglobin peptides. *Analytical Chemistry*, 87 (20) (2015) 10315–10322. DOI: 10.1021/acs.analchem.5b02318.
24. https://www.creative-proteomics.com/blog/index.php/data-dependent-acquisition-and-data-independent-acquisition-mass-spectrometry/.
25. P. Barbier Saint Hilaire, K. Rousseau, A. Seyer, S. Dechaumet, A. Damont, C. Junot, and F. Fenaille, Comparative evaluation of data dependent and data independent acquisition workflows implemented on an orbitrap fusion for untargeted metabolomics. *Metabolites*, 10 (4) (2020) 158. DOI: 10.3390/metabo10040158.
26. J. Wang, W. Chow, J. W Wong, D. Leung, J. Chang, and M. Li, Non-target data acquisition for target analysis (nDATA) of 845 pesticide residues in fruits and vegetables using UHPLC/ESI Q-Orbitrap. *Analytical and Bioanalytical Chemistry*, 411 (7) (2019) 1421–1431. DOI: 10.1007/s00216-019-01581-z.

27. D. Martins-de-Souza, V. M. Faça, and F. C. Gozzo, DIA is not a new mass spectrometry acquisition method. *Proteomics*, 17 (7) (2017) 1700017. DOI: 10.1002/pmic.201700017.
28. D. Amodei, J. Egertson, B. X. MacLean, R. Johnson, G. E. Merrihew, A. Keller, D. Marsh, O. Vitek, P. Mallick, and M. J. MacCoss, Improving precursor selectivity in data-independent acquisition using overlapping windows. *Journal of the American Society for Mass Spectrometry*, 30 (4) (2019) 669–684. DOI: 10.1007/s13361-018-2122-8.
29. C. Ludwig, L. Gillet, G. Rosenberger, S. Amon, B. C Collins, and R. Aebersold, Data-independent acquisition-based SWATH-MS for quantitative proteomics: a tutorial. *Molecular Systems Biology*, 14 (8) (2018) e8126. DOI: 10.15252/msb.20178126.
30. M. Katajamma and M. Oresic, Processing methods for differential analysis of LC/MS profile data. *BMC Bioinformatics*, 6 (1) (2005) 179. DOI: 10.1186/1471-2105-6-179.

Section 2

MS Analysis of Food Components

3 Mass Spectrometry of Lipids

Aynur Gunenc
Carleton University

CONTENTS

3.1 Introduction 31
3.2 Food Lipid Classifications 32
3.3 Lipid Extraction 33
3.4 Bulk Oil 33
3.5 Dairy Products 37
3.6 Conclusions 40
References 40

3.1 INTRODUCTION

Lipids are one of the key structural elements of foods like proteins and carbohydrates. Generally lipids are amphiphilic organic molecules, soluble in organic solvents (e.g., hexane, ether, and chloroform) but poorly in water [1]. They can take the role of a cell membrane component, and be active in metabolic and signaling pathways in cells [2]. Food materials contain mainly triacylglycerols (TAGs) and phospholipids (PL). According to the US Food and Drug Administration, the sum of fatty acids from C4 to C24 is calculated as total triglycerides [3]. Simple lipids are esters of fatty acids with alcohol; fats are esters of fatty acids with glycerol; and waxes are esters of fatty acids with long-chain alcohols (vitamin A, D esters, and myricyl palmitate). At room temperature, liquid triacylglycerides (TAGs) largely from plants are named as oils, whereas solid TAGs are named as fats.

Edible lipid analysis could be grouped as bulk oil (for instance, soybean, corn, coconut, and olive oil) and foodstuff oil analyses. For the determination of foodstuff oil, extraction is necessary as a first step, and then similar techniques used for bulk oil analysis can be applied to quantify and characterize the lipids in the foodstuff. Widely accepted methods are developed by the American Oil Chemists' Society (AOCS), Association of Official Analytical Collaboration (AOAC) International, or the International Union of Pure and Applied Chemists. Some of the instrumental methods for the characterization of those mentioned lipids are gas chromatography (GC), high-performance liquid chromatography (HPLC), nuclear magnetic resonance (NMR), and Fourier transform infrared [4]. Mass spectrometry (MS) combined with GC or HPLC plays a crucial analytical role in lipid analysis, and has advantages in terms of sensitivity and capability to analyze highly polar compounds. An MS has three basic compartments are as follows: an ion source for producing ions from samples; a mass analyzer that separates ions depending on their mass-to-charge (*m/z*); and a detector for the identification of ions as a mass spectrum graph [5]. Some other MS-based techniques used to identify lipid composition (e.g., free fatty acids [FFAs] and TAGs) of food products are summarized in this chapter. Therefore, the main focus is to give an overview of MS analysis of lipids in food products including bulk oil/edible oils and dairy food products.

DOI: 10.1201/9781003091226-5

3.2 FOOD LIPID CLASSIFICATIONS

Lipids can be found mainly as TAG or PL [6]. Based on chemical functionalities and structural diversity, lipids can be categorized into the following eight groups (Figure 3.1): glycerolipids, sphingolipids, glycerophospholipids, sterol lipids, fatty acyls (FA), prenol lipids, polyketides (PK), and saccharolipids (SL) [7–9].

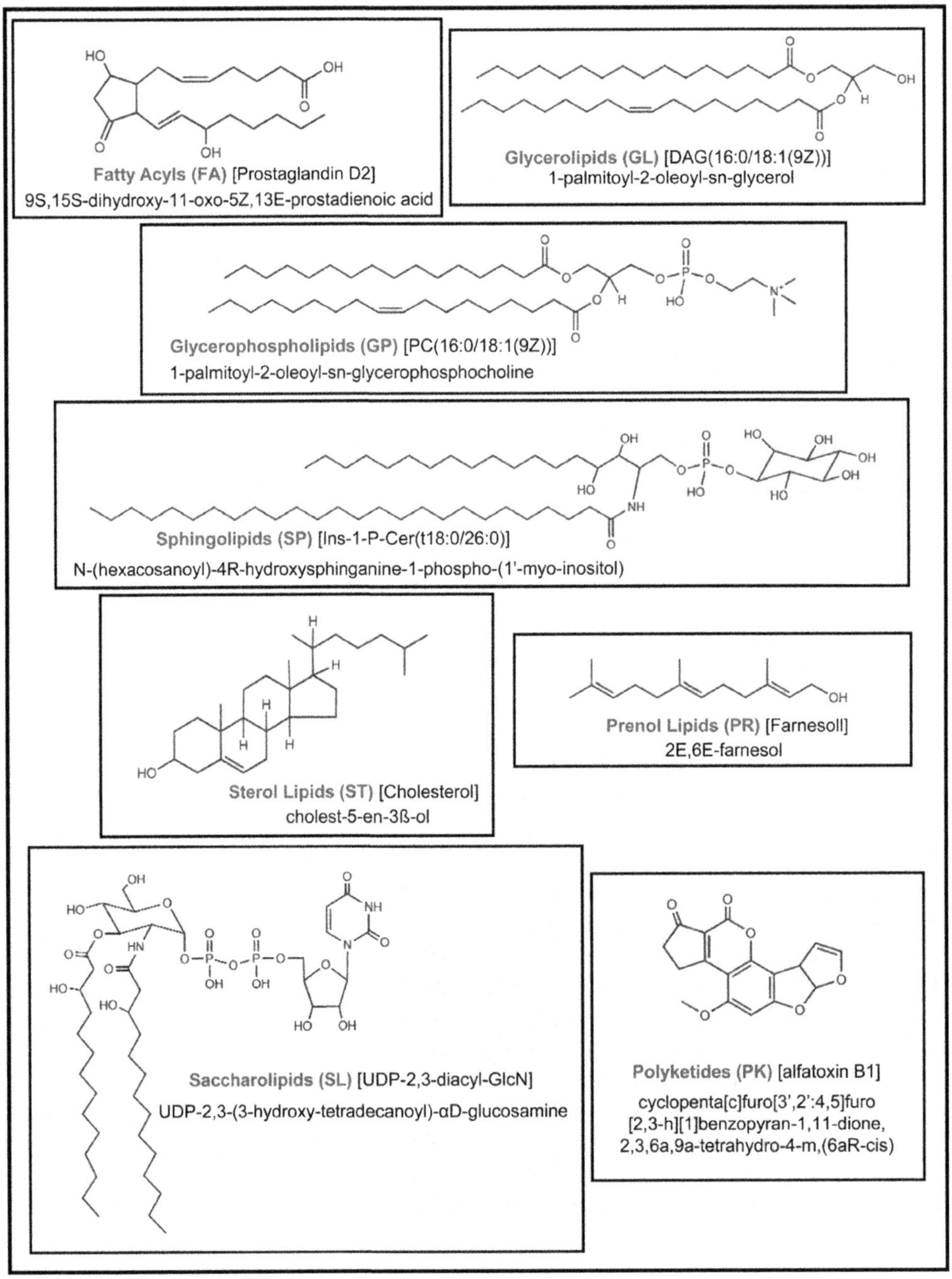

FIGURE 3.1 Examples of each of the eight categories of lipids as defined by the Lipid Metabolites and Pathways Strategy (LIPID MAPS) Consortium [10].

Food lipids are mainly fatty acids and their glycerides (mono-, di-, and triacylglycerides), waxes, and lipid-soluble pigments and vitamins [4]. For nutritional studies, and food labeling, mostly GC of fatty acid methyl esters (FAME), saturated fatty acids, polyunsaturated fatty acids (PUFA), *trans*-fatty acids, and conjugated linoleic acid (CLA) are some important terminologies used to define the lipids in the food materials. Health concerns and food labeling requirements make it prerequisite to characterize such parameters and measure the total lipid content of a foodstuff [11] whether it matches with required standard specifications. Dairy (butter, milk, cheese, and yogurt), salad spreads, margarine, shortening, cooking and salad oils, mayonnaise, peanut butter, meat, poultry, fish, and many others are examples of food products containing high amounts of fats and oils [12,13]. Detailed information (definition and specifications) related to food fatty acid constituents and bulk fats and oils can be found in Section I of the AOCS Official Methods [14], in the Merck Index [15], and in Fats and Oils [16].

3.3 LIPID EXTRACTION

As mentioned earlier, lipids are soluble in organic solvents, and this specific character (being insoluble in water) gives an opportunity to separate them from proteins, carbohydrates, and water in foods. At the same time, lipids in the food matrix can exist as a complex of either proteins (lipoproteins) or carbohydrates (liposaccharides). Therefore, based on their wide structural diversity (relative hydrophobicity) and different properties of each food, one solvent system for all lipids and/or fats is not possible. At the same time, because of the wide structural diversity, sample extraction stands very critical for the MS analysis [1]. For an accurate solvent extraction (ethyl ether, petroleum ether, hexane, pentane, or combination of one or two solvents) to determine the total lipid content of a food product, there are some general considerations or pretreatments suggested to be followed. They are pre-drying (vacuum oven drying or lyophilization to let solvent penetrate) [17], particle size reduction (grinding, milling to increase surface area), and acid or alkaline hydrolysis (to break down the bond with proteins or carbohydrates) depending on food matrix. Also, all steps in sample preparations need to be carried out under an inert atmosphere to minimize or prevent lipid oxidation. Extraction methods could be continuous solvent (Goldfish), semi-continuous solvent (Soxhlet), discontinuous solvent (Mojonnier) as well as non-solvent (Babcock, Gerber) and instrumental (NMR, specific gravity, infrared, and colorimetric) [3]. A detailed review on lipid extraction techniques for MS analysis in complex biological matrices has been published by Aldana et al. [1].

3.4 BULK OIL

PUFAs mainly in bulk oils (vegetable fats and olive oils) are inversely correlated with the risk of upper digestive, stomach, and urinary tract cancer development [18]. Olive and olive oil that represent essential components of the Mediterranean diet are also well-known for their potentials in the prevention of coronary heart disease, plus its high unsaturated fatty acids and phenolic compounds [19]. Therefore, the assessments of its authenticity and quality are crucial and considerable works are done in these aspects [20]. Extra virgin olive oil (EVOO) is cold pressed and does not contain any trace of refined oil according to EU Regulation EC 1531/2001 same as virgin olive oil (VOO) has extra value due to its mechanical extraction with no chemicals compared with most edible oils. Vegetable oils, on the other hand, are complex mixtures and have their unique composition of FFAs and TAGs with different concentrations. These fingerprint compositional characteristics make it possible to distinguish different vegetable oils and detect adulterations [21]. Mostly VOO is subjected to adulteration with lower-priced seed oils and refined olive oil [22]. For VOO authentication, there are many official analytical methods by the International Olive Council [23] relying on for instance GC-flame ionization detector, and HPLC – refraction index). Additionally, for investigation of adulteration or contamination of EVOO or VOO, other analytical methods could be used such as chromatographic, differential scanning calorimetry, and NMR [24].

MS-based analysis with specific features such as speed, specificity, and sensitivity is one of the best unique analytical tools for authentication and quality control of vegetable oils [25]. There are different MS methods that could be used for direct oil analysis. They can be classified as (i) by means of atmospheric pressure ionization methods (electrospray ionization [ESI], atmospheric pressure chemical ionization [APCI] or atmospheric pressure photoionization); (ii) ambient desorption ionization MS using atmospheric pressure sampling (ESI-based desorption electrospray ionization [DESI], APCI-based direct analysis on real time [DART]); and (iii) under vacuum such as headspace-MS sampling for volatile analysis (electron impact or chemical ionization) or matrix-assisted laser desorption/ionization (MALDI) for TAGs profiling [22]. Beneito-Cambra et al. [22] critically reviewed the potentials of those mentioned MS-based methods for direct analysis of VOO and other vegetable oils.

Vegetable oils can be distinguished based on their FFAs and TAGs composition, and mostly they are analyzed similarly to olive oil by GC-MS [26,27] and HPLC with MS or ultraviolet detectors such as photo diode array (PDA) [28–30], respectively. Those methods need a derivatization procedure and time-consuming sample preparations, whereas other MS-based methods such as ESI [31], APCI [32], DART [33], and MALDI [34], as reviewed in the study by Beneito-Cambra et al. [22], have allowed direct analysis of FFAs and TGAs in real time. Another technique, Fourier transform ion cyclotron resonance mass spectrometry (FT-ICR-MS) coupled with ESI is used to characterize the molecular composition of FFAs and TAGs in different vegetable oils (soybean, rapeseed, corn, sunflower, peanut, linseed, and olive oil) [21]. Moreover, in the same study by Li et al. [21], commercial genetically modified (GM) and non-GM vegetable oils were distinguished by ESI-FT-ICR MS technique. This technique unravels the complexity of different vegetable oils; enhanced selectivity provides profiling of fatty acids as well as minor components such as tocopherols.

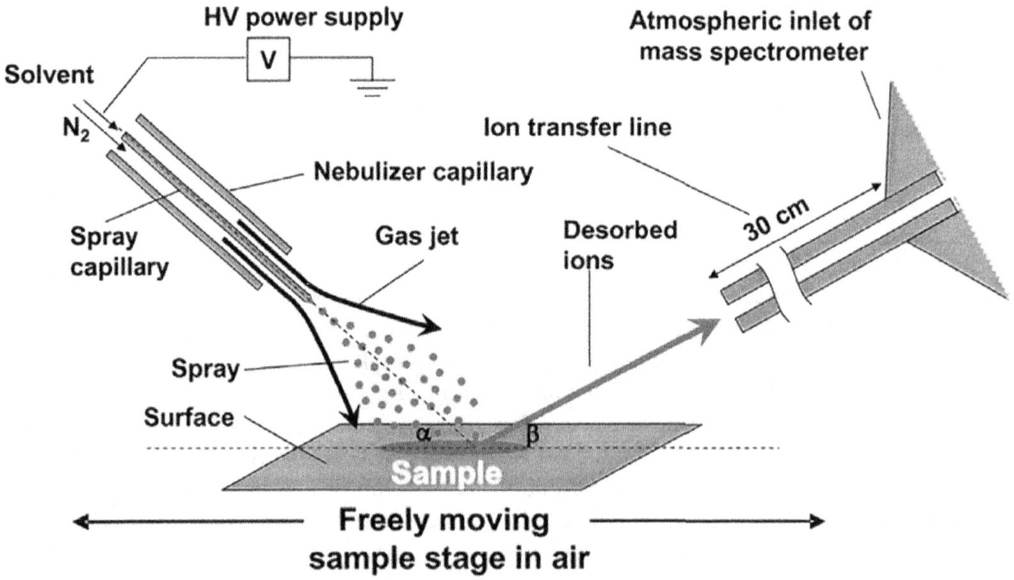

FIGURE 3.2 Schematic of typical DESI experiment. The sample solution is deposited from the solution and dried onto a PTFE surface and an appropriate solvent is sprayed at a flow rate of 3–15 mL/min under the influence of a high voltage (4.0 kV). (Adapted from Ref. [22] with permission.) The nominal linear velocity of the nebulizing gas was set to 350 m/s.

One of the most used ambient MS methods is DESI-MS as shown in Figure 3.2 [35]. This is a rapid analysis with no sample preparation steps; mainly, the sample solution is transferred on a polytetrafluoroethylene (PTFE) surface and dried, then a solvent is sprayed at optimum conditions (3–15 mL/min, 4.0 kV, and nebulizing gas at 350 m/s) and charged particles on the surface produces gaseous ions and mass analyzed [36–38].

Matrix-assisted laser desorption/ionization Fourier transform ion cyclotron resonance mass spectrometry (MALDI-FT-ICR-MS) is also a suitable MS-based method to identify and quantify TAGs of edible oils and determine the adulterants [34]. MALDI-MS of TAGs profile identification has been used for many other oils including canola oil, [39] castor oil [40], and pomegranate oil [41]. Some selected studies for MS-based lipid analysis have been summarized in Table 3.1.

Another type of ion source such as DART-based ambient MS method (Figure 3.3) has been used in vegetable oil either for fingerprint analysis or for target analysis [33,42]. DART principle is based on the interactions of long-lived electronic excited-state atoms with the sample and atmospheric gases as shown in Figure 3.3. In a typical DART source, gas will flow through a chamber (electrical discharge produces ions, electrons, and excited-metastables); during the gas passing the perforated chamber only metastables and neutral gas molecules remain, and the rest (charged particles) will be removed [43]. DART analysis can be completed in seconds and no solvent is required for lipid characterizations (fatty acids, mono-, di-, and triglycerides) in cooking oils and detects adulterated olive oils [33, 44].

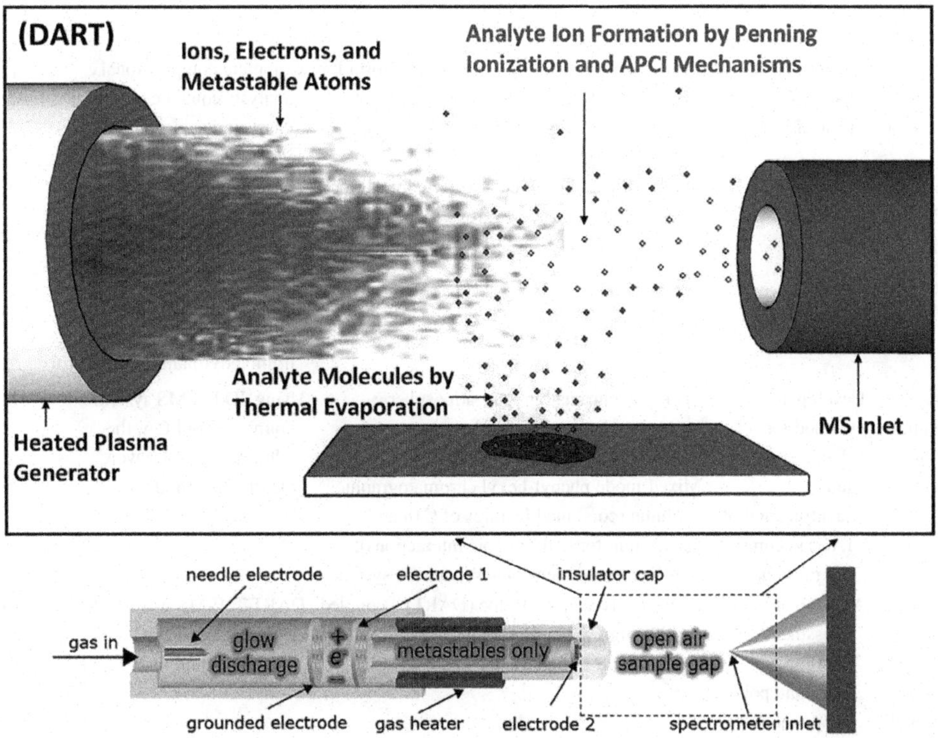

FIGURE 3.3 Schematic representation of a DART source. (Adapted from Ref. [22] with permission.)

TABLE 3.1
MS-Based Food Lipid Analysis in Edible/Bulk Oils

Sample	Focus	Instrumentation	Results	Source
Virgin olive oil and other vegetable oils	Authentication and quality control of VOO	Atmospheric pressure ionization methods (ESI and APCI); under vacuum (MALDI); headspace (HS-MS) or chemical ionization (PTR-MS); selected ion flow tube (SIFT-MS)	A critical overview of all these methods and their potential use for vegetable oil characterization.	[22]
Extra virgin olive oil (EVOO)	Determine the adulteration of EVOO with corn oil	Matrix-Assisted Laser Desorption/Ionization Mass Spectrometry (MALDI-ToF-MS); laser frequency, 20 Hz; ion source 1 voltage, 20 kV; ion source 2 voltage, 18.4 kV; lens voltage, 9.1 kV; mass-to-charge (*m/z*) 300–1,300 Da; FlexControl v.3.0	MALDI-ToF-MS coupled with unsupervised hierarchical clustering (UHC), principal component (PCA) and Pearson's correlation analysis was able to detect corn oil adulterations at very low levels (0.5%).	[24]
Virgin olive oil (VOO)	Investigation of the oxidative status of VOO	Atmospheric pressure chemical ionization-direct infusion mass spectrometry (APCI-DIMS); with the application of linear discriminant analysis (LDA)	Very quick and simple procedure has been followed; can be automated for the future; DIMS can predict the oxidative status of VOO during storage.	[45]
Soybean, rapeseed, corn, sunflower, peanut, linseed, and olive oil	Free fatty acids Triacylglycerols Structural analysis	Electrospray Ionization Fourier Transform Ion Cyclotron Resonance Mass Spectrometry (ESI FT-ICR MS); in both positive and negative mode	Shown that each sample has a unique composition of FFAs and TAGs. Genetically modified samples were also detected.	[21]
Canola, olive, and soybean oil	Compositional-based fingerprint analysis	Electrospray Ionization Fourier Transform Ion Cyclotron Resonance Mass Spectrometry (ESI FT-ICR MS); in both positive and negative mode	Complete compositional analysis could be used to characterize, detect, and identify adulterants.	[31]
Grape seed oil (32 hybrids)	Triglyceride profile of seed oil	Direct infusion in the electrospray ionization (ESI)-tandem mass spectrometry and matrix-assisted laser desorption ionization and time of flight	Among 32 hybrid grape seed oil: six were the principal TAGs; trilinolein (LLL) was the most abundant (43%) followed by dilinoleoyl-oleoylglycerol (LOL) and dilinoleoyl-palmitoylglycerol	[46]
Peanut, soybean, and lord oils	Develop a method for characterization and quantification of TAGs in complex natural samples	2D liquid chromatography-atmospheric pressure chemical ionization mass spectrometry (LC-APCI-MS) Mixed-mode phenyl-hexyl chromatographic column (combined features of C18 and silver-ion; for hydrophobic interactions); off-line two-dimensional separation system	Using 2D LC-MS system coupled with PCA, the adultered samples were clearly identified	[32]
EVOO, olive pomace oil (OPO), olive oil (OO), hazelnut oil (HO)	Authenticity assessment/ Profiling of TAGs and polar compounds	Direct analysis in real time (DART) – coupled with a high-resolution time-of-flight mass spectrometry (ToF-MS)	DART-ToF-MS was able to differentiate EVOO, OPO, OO, and adultered EVOO with HO	[33]

(*Continued*)

TABLE 3.1 (*Continued*)
MS-Based Food Lipid Analysis in Edible/Bulk Oils

Sample	Focus	Instrumentation	Results	Source
Lard, corn, peanut, colz, soybean, rice, sunflower, sesame oils, mutton fat,	A rapid quantitative method to determine TAGs profile and adulterations	Matrix-assisted laser desorption/ionization Fourier transform ion cyclotron resonance mass spectrometry (MALDI-FT-ICR-MS) Using pencil graphite combined with 2,5-dihydroxybenzoic acid (DHB) as matrix	MALDI-FT-ICR-MS enables to identify TAGs of edible oils, and adulterants (10% lard fat) in corn oil	[34]
Castor oil, canola oil, olive oil	A fast-drying sample prep method with α-cyano-4-hydroxycinnamic acid as a matrix	Matrix-assisted laser desorption/ionization (MALDI) time-of-flight (ToF) mass spectra	It has been proven that sample preparation with α-cyano-4-hydroxycinnamic acid as a matrix in acetonitrile/tetrahydrofuran solvent system could be used for MALDI-ToF-MS of vegetable oils	[39]
Castor oil	Characterization of castor oil	A non-aqueous reverse-phase HPLC based on two columns in series used to separate TAGs; on-line negative ion atmospheric pressure chemical ionization (APCI) or off-line positive ion matrix-assisted laser desorption ionization (MALDI)/MS; MALDI time of flight (ToF)/MS; seamless post-source decay (PSD) fragment ion analysis	PSD with ToF-MS analysis allowed to determine the fatty acid composition of TAGs; Castrol oil has eight different TAGs-determined by both APCI and MALDI	[40]
Pomegranate oil (PO)	Detailed examination of PO profile	Fatty acid composition by GC-MS; triacylglycerol by MALDI-ToF/MS	The most abundant FA is linolenic acid (64–83%); MALDI-ToF/MS showed fingerprinting results based on different geometric isomers of conjugated linolenic acid, punicic acid as a major one	[41]

3.5 DAIRY PRODUCTS

Lipids as one of the main building blocks of bovine milk affect physicochemical characteristics of dairy products [47] and some lipid species can act as biomarkers of the physiological status of animals [48]. Investigation of milk lipids increased interest from academia and milk producers since milk lipids as a biomarker gives the opportunity to detect diseases earlier and improve milk quality [49]. Most commonly GC-MS is used to detect milk lipids and fatty acids as FAME [50]. Approximately 400 different fatty acids have been reported in milk fat and more than 98% of milk fat is triglycerides [51]. Markiewicz-Kęszycka et al. studied the fatty acid profile of different milk types (goat, sheep, cow) and reported palmitic acid is the most abundant fatty acid in all samples as well as C6:0, C8:0 and C10:0 are the highest in goat milk. Sheep milk has the highest amount of CLA and α-linolenic acid [52]. Table 3.2 is the summary of some selected studies carried on MS-based food lipid analysis in the dairy industry. A recent study based on the accurate mass of the parent ion and MS2 information of bovine milk provides inventory information about the most abundant tTAGs groups (C34:0, C36:0 C38:1, and C40:1). The determination of TAGs was done at the group level since isomeric species cannot be identified chromatographically. The outcomes of

TABLE 3.2
MS-Based Food Lipid Analysis in Dairy Food Products

Sample	Focus	Instrumentation	Results	Source
Cheese (cow, buffalo, sheep, goat)	A bottom-up proteomic approach to assess the milk species cheese production	HPLC-Tandem MS; a triple quadruple-linear ion trap MS with a Turbo V electrospray ionization; positive ion mode; Mascot Software	Multianalyte multiple reaction monitoring methods were showing high specificity in the detection of species	[57]
Cheese	Animal species identification of dairy products	MALDI-ToF-MS; Microflex LT mass spectrometer controlled by FlexControl (version3.4); MALDI Biotyper software (MBT v 3.1), positive linear mode, laser frequency 60 Hz ion source1:2 kV, ion source 2:18 kV; Bruker's MBT_FC and MBT_AutoX; mass range 2,000–20,000 Da	This study showed a rapid, easy and robust technique to identify the origin of cheese samples. The animal origin of dairy products was confirmed with the fatty acid profile. The exchange of reference spectra is at http://maldi-up.ua-bw.bw.de	[59]
Cheese	Investigate the effect of ruminant diets on the nutritional value of sheep cheese fat	GC-MS; FAME	Eighty-four individual fatty acids were identified; hay forages may be an alternative low-cost for cheese production with a fatty acid profile suitable for human health.	[60]
Cheese	Determination of free FAs, especially conjugated linoleic acids (CLAs)	Matrix Solid-Phase Dispersion (MSPD) – UHPLC-MS/MS	It provides a rapid and selective sample pretreatment; selective elution of the FFAs. The method's correlation coefficients were greater than 0.99 and recovery ranges between 75% and 105% as well as a good reproducibility (≤12%)	[61]
Bovine milk	Survey on triglycerides (TAGs)	LC-MS/MS; Q Exactive Plus MS, electrospray ionization; positive ion mode (4.2 kV); a full scan (120–1,800 *m/z*); a resolution of 70,000 followed by 5 data-dependent MS2 scant at a resolution of 17,500 and 25 eV collision energy	Total of 220 groups (having the same chemical formula and mass) and 3,454 molecular species of TAGs were identified; a complete lipid inventory	[53]
Milk (goat, cow)	Free FAs determination	LC-MS/MS; ESI in negative mode; ToF-MS full scan *m/z* 50–850; information-dependent acquisition (IDA)-ToF-MS/MS 40 V collision energy (CE) with 15 V collision energy spread (CES)	Major FFAs for all milk samples are C16:0, C18:0, and C18:1. The quantities of C6:0, C8:0, and C10:0 were found higher in goat milk than in cow milk.	[62]
Ewe milk	Qualitative and quantitative method development of fatty acids	GC-MS; GC; the software (MassLynx v 4.0); MS; *m/z* 45–700; electron ionization ion source producing 70 eV; NIST 08 as a reference library	38 polyunsaturated FAs, mono-unsaturated FAs (C6:0-C24:1) were used; FAME was analyzed; LOD and LOQ in the ng/mL.	[63]

(*Continued*)

TABLE 3.2 (*Continued*)
MS-Based Food Lipid Analysis in Dairy Food Products

Sample	Focus	Instrumentation	Results	Source
Bovine milk	Quantitative determination of 2- and 3-OH-FAs	GC-ECNI-MS-SIM; electron-capture negative ion in the selected ion mode	Ten OH-FAs methyl esters were quantified at the range of 0.02±0.00 to 4.49±0.29 mg/100 mg of milk fat.	[64]
Milk, yogurt and concentrated yogurt from Peranakan Etawah goat milk	Characterize the fatty acid profiles of goat milk and its fermented products	GC-MS and FAME	Twenty-six FAs were determined in milk, yogurt, and concentrated yogurt; oleic, stearic, and palmitic acids were the major FAs in all examined samples.	[65]
Bovine milk	To detect adulteration (e.g., soya or coconut) in bovine milk	MALDI-ToF-MS analyses in the positive linear ion mode; *m/z* 400 to 1,000; Flex Analysis v.3.4 software	Method is diluted samples in water (1:4) and followed by MALDI-ToF-MS. The optimized method has the potential for use in food authenticity applications.	[66]
Butter and butter oils	Investigate the potential of PTR-MS to classify milk fats	Proton Transfer Reaction Mass Spectrometry (PTR-MS); head space of the samples were analyzed; *m/z* 20–120	Milk fats were grouped by both PTR-MS and PLS-DA (partial least square-discriminant analysis) mode; the outcomes of this study have the potential for use in quality control.	[67]
Milk powder	Method development to quantify lipid oxidation	Headspace Solid-Phase Micro Extraction-Gas Chromatography-Mass Spectrometry (HS-SPME-GC-MS)	Using response surface methodology, the method was validated; important model terms are extraction T and the interaction between time and T.	[68]

this comprehensive survey could be a reference for studying milk lipids [53]. Milk fatty acids identification is usually through GC and ~25 fatty acids (C4:0–C28:0) have been reported [54, 55]. Fatty acid composition could be affected by environmental factors. A study by O'Donnell-Megaro et al. examined seasonal and geographical impacts on US milk fat and its fatty acid composition (FAME analysis by GC) by following up 56 milk processing plants regularly (every 3 months) during a year. This work showed that environmental factors had some minor differences for some fatty acids, and the majority of fatty acids profile is consistent from the human dietary perspective [55].

A study by Verma and Ambatipudi gives an overview of milk fat analysis by MS-based techniques and discusses about the challenges and they suggest a multi-pronged MS strategy to better explore different milk components for future studies [56].

Bernardi et al. examined the four most common milk-producing species (cow, buffalo, sheep, and goat) in the Italian dairy industry, and they applied a bottom-up proteomic approach to differentiate those species in all investigated cheese samples by using HPLC-MS/MS with a multianalyte multiple reaction monitoring methods [57].

Mass spectrometry imaging (MSI) is a novel technology that provides a two-dimensional ionization to detect food components in tissue with no sample preparation such as extraction, separation, and purification. Yoshimura and Zaima reviewed applications of MSI in food science and related fields [58]. This MSI technology could be one of the top analytical tools to visualize food components distribution and better improve end products quality attributes during food processing.

3.6 CONCLUSIONS

MS-based analytical instrumentation is one of the fastest-growing fields in food science. Most of the lipid analyses for food products are based on official methods (GC or HPLC with non-specific detectors). However, new emerging techniques, more specifically direct MS methods such as atmospheric pressure ionization (ESI, APCI, and ambient MS) or under vacuum conditions, offer more specific molecular identifications of lipid components in food products. Also, MSI is another novel technology, has the potential to detect food components with no sample preparation, and it could be one of the top analytical instruments to visualize food components to better improve end-product quality during processing.

REFERENCES

1. Aldana, J., A. Romero-Otero, and M.P. Cala, Exploring the lipidome: Current lipid extraction techniques for mass spectrometry analysis. *Metabolites*, 2020. **10**(6): p. 231.
2. Harayama, T. and H. Riezman, Understanding the diversity of membrane lipid composition. *Nature Reviews. Molecular Cell Biology*, 2018. **19**(5): pp. 281–296.
3. Min, D.B. and W.C. Ellefson, Fat analysis, in *Food Analysis*, S.S. Nielsen, Editor. 2010, Springer US: Boston, MA. pp. 117–132.
4. Pike, O.A. and S. O'Keefe, Fat characterization, in *Food Analysis*, S.S. Nielsen, Editor. 2017, Springer International Publishing: Cham. pp. 407–429.
5. Mellmann, A. and J. Müthing, MALDI-TOF mass spectrometry-based microbial identification, in *Advanced Techniques in Diagnostic Microbiology*, Y.-W. Tang and C.W. Stratton, Editors. 2013, Springer US: Boston, MA. pp. 187–201.
6. Jacobsen, C., Oxidative rancidity, in *Encyclopedia of Food Chemistry*, L. Melton, F. Shahidi, and P. Varelis, Editors. 2019, Academic Press: Oxford. pp. 261–269.
7. Fahy, E., et al., Lipid classification, structures and tools. *Biochimica et Biophysica Acta (BBA) - Molecular and Cell Biology of Lipids*, 2011. **1811**(11): pp. 637–647.
8. Fahy, E., et al., A comprehensive classification system for lipids. *Journal of Lipid Research*, 2005. **46**(5): pp. 839–862.
9. Han, X. and R.W. Gross, Shotgun lipidomics: Electrospray ionization mass spectrometric analysis and quantitation of cellular lipidomes directly from crude extracts of biological samples. *Mass Spectrometry Reviews*, 2005. **24**(3): pp. 367–412.
10. Harkewicz, R. and E.A. Dennis, Applications of mass spectrometry to lipids and membranes. *Annual Review of Biochemistry*, 2011. **80**(1): pp. 301–325.
11. Khanal, R.C. and T.R. Dhiman, Biosynthesis of conjugated linoleic acid (CLA): A review. *Pakistan Journal of Nutrition*, 2004. **3**: pp. 72–81.
12. Hui, Y.H., *Bailey's Industrial Oil and Fat Products*. 5th ed. 1996, Wiley: New York.
13. Nawar, W.W., Lipids, in *Food Chemistry*, O.R. Fennema, Editor. 1996, Marcel Dekker: New York.
14. AOCS, *Official Methods and Recommended Practices of the AOCS*. 2009, American Oil Chemists' Society: Champaign, IL.
15. Budavari, S., *The Merck Index. An Encyclopedia of Chemicals, Drugs, and Biologicals*. 12th ed. 1996, Merck: Whitehouse Station, NJ.
16. Stauffer, C.E., Fats and oils, in *Eagan Press Handbook Series*. 1996, American Association of Cereal Chemists: St.Paul, MN.
17. Pomeranz, Y. and C.F. Meloan, Editors. *Food Analysis: Theory and Practice*. 3rd ed. 1994, Van Nostrand Reinhold: New York.
18. Vecchia, C. and S. Franceschi, Nutrition and gastric cancer. *Canadian Journal of Gastroenterology = Journal Canadien de gastroenterologie*, 2000. **14 Suppl D**: pp. 51D–54D.
19. Uylaşer, V. and G. Yildiz, The historical development and nutritional importance of olive and olive oil constituted an important part of the Mediterranean diet. *Critical Reviews in Food Science and Nutrition*, 2014. **54**(8): pp. 1092–1101.

20. Reboredo-Rodríguez, P., et al., Quality of extra virgin olive oils produced in an emerging olive growing area in north-western Spain. *Food Chemistry*, 2014. **164**: pp. 418–426.
21. Li, Y., et al., Molecular characterization of edible vegetable oils via free fatty acid and triacylglycerol fingerprints by electrospray ionization Fourier transform ion cyclotron resonance mass spectrometry. *International Journal of Food Science & Technology*, 2020. **55**(1): pp. 165–174.
22. Beneito-Cambra, M., et al., Direct analysis of olive oil and other vegetable oils by mass spectrometry: A review. *TrAC Trends in Analytical Chemistry*, 2020. **132**: p. 116046.
23. Council, I.O.O., Trade Standard Applying to Olive Oils and Olive Pomace Oils. COI/T.15/NC N. 3/rev. 14, in International Olive Oil Council. 2020.
24. Di Girolamo, F., et al., A simple and effective mass spectrometric approach to identify the adulteration of the Mediterranean diet component extra-virgin olive oil with corn oil. *International Journal of Molecular Sciences*, 2015. **16**(9): pp. 20896–20912.
25. Aparicio, R., et al., Authenticity of olive oil: Mapping and comparing official methods and promising alternatives. *Food Research International*, 2013. **54**(2): pp. 2025–2038.
26. Steenbergen, H., et al., Direct analysis of intact glycidyl fatty acid esters in edible oils using gas chromatography–mass spectrometry. *Journal of Chromatography A*, 2013. **1313**: pp. 202–211.
27. Zhang, M., et al., A quick method for routine analysis of C18 trans fatty acids in non-hydrogenated edible vegetable oils by gas chromatography–mass spectrometry. *Food Control*, 2015. **57**: pp. 293–301.
28. Dubois, M., et al., Determination of seven glycidyl esters in edible oils by gel permeation chromatography extraction and liquid chromatography coupled to mass spectrometry detection. *Journal of Agricultural and Food Chemistry*, 2011. **59**(23): pp. 12291–12301.
29. Zeb, A. and M. Murkovic, Analysis of triacylglycerols in refined edible oils by isocratic HPLC-ESI-MS. *European Journal of Lipid Science and Technology*, 2010. **112**(8): pp. 844–851.
30. Zhang, S.-D., et al., Separation of triacylglycerols from edible oil using a liquid chromatography–mass spectrometry system with a porous graphitic carbon column and a toluene–isopropanol gradient mobile phase. *Journal of the American Oil Chemists' Society*, 2018. **95**(10): pp. 1253–1266.
31. Wu, Z., R.P. Rodgers, and A.G. Marshall, Characterization of vegetable oils: Detailed compositional fingerprints derived from electrospray ionization Fourier transform ion cyclotron resonance mass spectrometry. *Journal of Agricultural and Food Chemistry*, 2004. **52**(17): pp. 5322–5328.
32. Wei, F., et al., Quantitation of triacylglycerols in edible oils by off-line comprehensive two-dimensional liquid chromatography–atmospheric pressure chemical ionization mass spectrometry using a single column. *Journal of Chromatography A*, 2015. **1404**: pp. 60–71.
33. Vaclavik, L., et al., Ambient mass spectrometry employing direct analysis in real time (DART) ion source for olive oil quality and authenticity assessment. *Analytica Chimica Acta*, 2009. **645**(1): pp. 56–63.
34. Zhang, Y.-L., et al., Rapid quantitative determination of triglycerides in edible oils by matrix-assisted laser desorption/ionisation Fourier transform ion cyclotron resonance mass spectrometry using pencil graphite combined with 2,5-dihydroxybenzoic acid as matrix. *International Journal of Mass Spectrometry*, 2018. **431**: pp. 56–62.
35. Takáts, Z., et al., Mass spectrometry sampling under ambient conditions with desorption electrospray ionization. *Science*, 2004. **306**(5695): p. 471.
36. Monge, M.E., et al., Mass spectrometry: Recent advances in direct open air surface sampling/ionization. *Chemical Reviews*, 2013. **113**(4): pp. 2269–2308.
37. Venter, A.R., et al., Mechanisms of real-time, proximal sample processing during ambient ionization mass spectrometry. *Analytical Chemistry*, 2014. **86**(1): pp. 233–249.
38. Weston, D.J., Ambient ionization mass spectrometry: Current understanding of mechanistic theory; analytical performance and application areas. *Analyst*, 2010. **135**(4): pp. 661–668.
39. Ayorinde, F.O., E. Elhilo, and C. Hlongwane, Matrix-assisted laser desorption/ionization time-of-flight mass spectrometry of canola, castor and olive oils. *Rapid Communications in Mass Spectrometry*, 1999. **13**(8): pp. 737–739.
40. Stübiger, G., E. Pittenauer, and G. Allmaier, Characterisation of castor oil by on-line and off-line non-aqueous reverse-phase high-performance liquid chromatography–mass spectrometry (APCI and UV/MALDI). *Phytochemical Analysis*, 2003. **14**(6): pp. 337–346.
41. Kaufman, M. and Z. Wiesman, Pomegranate oil analysis with emphasis on MALDI-TOF/MS triacylglycerol fingerprinting. *Journal of Agricultural and Food Chemistry*, 2007. **55**(25): pp. 10405–10413.
42. Vaclavik, L., et al., Rapid monitoring of heat-accelerated reactions in vegetable oils using direct analysis in real time ionization coupled with high resolution mass spectrometry. *Food Chemistry*, 2013. **138**(4): pp. 2312–2320.

43. Cody, R.B., et al., *DART (Direct Analysis in Real Time) Applications Notebook*, L. Joel, Editor. 2016, Jeol Ltd.: Tokyo.
44. Moravcova, E., et al., Novel approaches to analysis of 3-chloropropane-1,2-diol esters in vegetable oils. *Analytical and Bioanalytical Chemistry*, 2012. **402**(9): pp. 2871–2883.
45. Lerma-García, M., et al., Evaluation of the oxidative status of virgin olive oils with different phenolic content by direct infusion atmospheric pressure chemical ionization mass spectrometry. *Analytical and Bioanalytical Chemistry*, 2009. **395**(5): pp. 1543–1550.
46. De Marchi, F., et al., Seed oil triglyceride profiling of thirty-two hybrid grape varieties. *Journal of Mass Spectrometry*, 2012. **47**(9): pp. 1113–1119.
47. Smiddy, M.A., T. Huppertz, and S.M. van Ruth, Triacylglycerol and melting profiles of milk fat from several species. *International Dairy Journal*, 2012. **24**(2): pp. 64–69.
48. Liu, Z., et al., Heat stress in dairy cattle alters lipid composition of milk. *Scientific Reports*, 2017. **7**(1): p. 961.
49. Benbrook, C.M., et al., Organic production enhances milk nutritional quality by shifting fatty acid composition: A United States-wide, 18-month study. *PLoS One*, 2013. **8**(12): pp. e82429–e82429.
50. Stefanov, I., B. Vlaeminck, and V. Fievez, A novel procedure for routine milk fat extraction based on dichloromethane. *Journal of Food Composition and Analysis*, 2010. **23**(8): pp. 852–855.
51. Lindmark Månsson, H., Fatty acids in bovine milk fat. *Food & Nutrition Research*, 2008. **52**(1): p. 1821.
52. Markiewicz-Kęszycka, M., et al., Fatty acid profile of milk - A review. *Bulletin of the Veterinary Institute in Pulawy*, 2013. **57**(2): pp. 135–139.
53. Liu, Z., et al., Comprehensive characterization of bovine milk lipids: Triglycerides. *ACS Omega*, 2020. **5**(21): pp. 12573–12582.
54. Moate, P.J., et al., Grape marc reduces methane emissions when fed to dairy cows. *Journal of Dairy Science*, 2014. **97**(8): pp. 5073–5087.
55. O'Donnell-Megaro, A.M., D.M. Barbano, and D.E. Bauman, Survey of the fatty acid composition of retail milk in the United States including regional and seasonal variations. *Journal of Dairy Science*, 2011. **94**(1): pp. 59–65.
56. Verma, A. and K. Ambatipudi, Challenges and opportunities of bovine milk analysis by mass spectrometry. *Clinical Proteomics*, 2016. **13**(1): p. 8.
57. Bernardi, N., et al., A rapid high-performance liquid chromatography-tandem mass spectrometry assay for unambiguous detection of different milk species employed in cheese manufacturing. *Journal of Dairy Science*, 2015. **98**(12): pp. 8405–8413.
58. Yoshimura, Y. and N. Zaima, Application of mass spectrometry imaging for visualizing food components. *Foods*, 2020. **9**(5): p. 575.
59. Rau, J., et al., Rapid animal species identification of feta and mozzarella cheese using MALDI-TOF mass-spectrometry. *Food Control*, 2020. **117**: p. 107349.
60. Renes, E., et al., Effect of forage type in the ovine diet on the nutritional profile of sheep milk cheese fat. *Journal of Dairy Science*, 2020. **103**(1): pp. 63–71.
61. Simeoni, M.C., et al., Determination of free fatty acids in cheese by means of matrix solid-phase dispersion followed by ultra-high performance liquid chromatography and tandem mass spectrometry analysis. *Food Analytical Methods*, 2018. **11**(10): pp. 2961–2968.
62. Kokotou, M.G., C. Mantzourani, and G. Kokotos, Development of a liquid chromatography–high resolution mass spectrometry method for the determination of free fatty acids in milk. *Molecules (Basel, Switzerland)*, 2020. **25**(7): p. 1548.
63. Devle, H., et al., A GC – magnetic sector MS method for identification and quantification of fatty acids in ewe milk by different acquisition modes. *Journal of Separation Science*, 2009. **32**(21): pp. 3738–3745.
64. Jenske, R. and W. Vetter, Gas chromatography/electron-capture negative ion mass spectrometry for the quantitative determination of 2- and 3-Hydroxy fatty acids in bovine milk fat. *Journal of Agricultural and Food Chemistry*, 2008. **56**(14): pp. 5500–5505.
65. Sumarmono, J., M. Sulistyowati, and Soenarto, Fatty acids profiles of fresh milk, yogurt and concentrated yogurt from *Peranakan etawah* goat milk. *Procedia Food Science*, 2015. **3**: pp. 216–222.
66. Larrouy-Maumus, G., et al., Discrimination of bovine milk from non-dairy milk by lipids fingerprinting using routine matrix-assisted laser desorption ionization mass spectrometry. 2020.
67. van Ruth, S.M., et al., Butter and butter oil classification by PTR-MS. *European Food Research and Technology*, 2008. **227**(1): pp. 307–317.
68. Clarke, H.J., et al., Development of a headspace solid-phase microextraction gas chromatography mass spectrometry method for the quantification of volatiles associated with lipid oxidation in whole milk powder using response surface methodology. *Food Chemistry*, 2019. **292**: pp. 75–80.

4 Proteomic Strategies to Evaluate the Impact of Farming Conditions on Food Quality and Safety in Aquaculture Products

Mónica Carrera and Carmen Piñeiro
Spanish National Research Council (CSIC)

Iciar Martinez
University of the Basque Country UPV/EHU
IKERBASQUE Basque Foundation for Science

CONTENTS

4.1 Introduction to Proteomics in Aquaculture 43
4.2 Workflow of Proteomics: Discovery and Targeted Proteomics 44
4.3 Application of Proteomics to Evaluate the Farming Conditions on Food Quality and Safety in Aquaculture Products 46
4.3.1 Dietary Management in Aquaculture 47
4.3.2 Fish Welfare and Stress Response in Aquaculture 47
4.3.3 Food Safety in Aquaculture 49
4.3.3.1 Food Safety in Aquaculture: Biotic Hazards 49
4.3.3.2 Food Safety in Aquaculture: Abiotic Hazards 50
4.3.4 Antibiotic Resistance in Aquaculture 51
4.4 Concluding Remarks and Future Directions 53
References 54

4.1 INTRODUCTION TO PROTEOMICS IN AQUACULTURE

Aquaculture is the breeding of aquatic organisms under controlled conditions, involving both marine and freshwater fish along with algae, crustaceans, and mollusks. According to the Food and Agriculture Organization of the United Nations (FAO), this food sector represents a significant source of nutrients for the human diet and produces approximately 97.2 million tons of fish annually, which represents 47% of the global fish production [1]. With 9 billion people expected to be living on the planet by 2050, maintaining the current level of fish consumption (9.0–20.2 kg annually per capita) is a challenging task [1]. Aquaculture, the most rapidly growing food-producing sector in

Open Access
Carrera, M.; Piñeiro, C.; Martinez, I. Proteomic Strategies to Evaluate the Impact of Farming Conditions on Food Quality and Safety in Aquaculture Products. Foods 2020, 9, 1050.

DOI: 10.3390/foods9081050-6

the world, offers an excellent source of high-value food and is expected to significantly contribute to meeting this demand for fish products.

The globalization of aquaculture markets not only presents important nutritional and economic benefits but also poses potential risks for food safety, such as the fraudulent substitution of fish species and the presence of food microorganisms, viruses, parasites, and vectors of their corresponding foodborne diseases [2]. Moreover, fish and seafood are easily spoiled, resulting in a fast loss of food quality due to the presence of fish microbiota, the elevated amount of unsaturated fatty acids, and the abundance of proteases. Improving aquaculture practices to offer products of optimal quality and to reduce chronic stress throughout improved farming conditions to maintain fish welfare are two major questions in aquaculture research. Consumer awareness and rising demand for aquaculture products have motivated scientists to develop procedures to enhance productivity and to improve the quality and safety of these foodstuffs. In this context, proteomics has been established as a powerful methodology for the evaluation of quality and safety in aquaculture products [3–5].

Proteomics is the high-throughput analysis of the proteins of a specific biological sample [6]. Proteomics involves the identification, localization, and quantification of proteins as well as the analysis of protein modifications and the elucidation of protein–protein networks [7]. Among proteomic analytical techniques, mass spectrometry (MS) is recognized as an indispensable instrument to precisely analyze a large number of proteins from complex samples in the majority of food proteomics studies [8,9]. In addition, the computational analysis of MS data has improved the discriminatory power of proteomics techniques, making them effective methodologies for the global analysis of proteins and peptides [10]. Thus, the latest advances in proteomics and bioinformatics approaches have turned them into useful tools to develop promising strategies for food science investigations [11,12]. Within that framework, this chapter summarizes some highly relevant applications of proteomics to evaluate the impact of farming conditions on fish wellbeing as well as the quality and safety of aquaculture products.

4.2 WORKFLOW OF PROTEOMICS: DISCOVERY AND TARGETED PROTEOMICS

Proteomics has the potential to provide information useful to improve the production, welfare, health, nutritional value, and wholesomeness of farmed fish. Figure 4.1 shows the classical proteomics approaches, that is, discovery and targeted proteomics, with their corresponding workflows.

Discovery proteomics aims at identifying biological markers in a given proteome, frequently employing a bottom-up approach, in which the proteins of the sample are separated, proteolyzed with enzymes such as trypsin or Glu-C, and the peptides obtained are subsequently analyzed by tandem mass spectrometry (MS/MS). Two-dimensional gel electrophoresis (2-DE) has traditionally been the technique selected for the separation of proteins samples [13]. This gel-based procedure is the most suitable approach for species whose protein sequences are not yet known, which includes many fish. In these cases, identification is performed by comparison of the MS/MS spectra of the peptides obtained with orthologous protein sequences from related species or by *de novo* MS/MS sequencing [14]. Programs such as Progenesis and PDQuest can analyze the 2-DE gels themselves.

In gel-free approaches, also known as shotgun proteomics, the proteins are directly digested in the extract with a selected enzyme, and the obtained mixture of peptides is subsequently analyzed by liquid chromatography coupled with tandem mass spectrometry (LC-MS/MS) [15,16]. It is possible to perform multidimensional LC separations, combining, for example, strong anion/cation exchange chromatography and reverse-phase chromatography [17]. Database searching programs, such as SEQUEST, X! Tandem, or Mascot [18,19], allow the tentative identification of presumed peptide sequences based on the obtained fragmentation spectra, and additional software programs, such as Percolator are used to validate the identification [20]. When the protein is not present in the database, then the peptides must be sequenced *de novo* [21], either manually or using programs such as PEAKS and DeNovoX [22,23]. This approach has been successfully used in the *de novo* sequencing of some fish allergens, such as parvalbumins and shrimp arginine kinases [14,24,25]. When protein quantification is deemed necessary, the methods of choice include metabolic stable-isotope

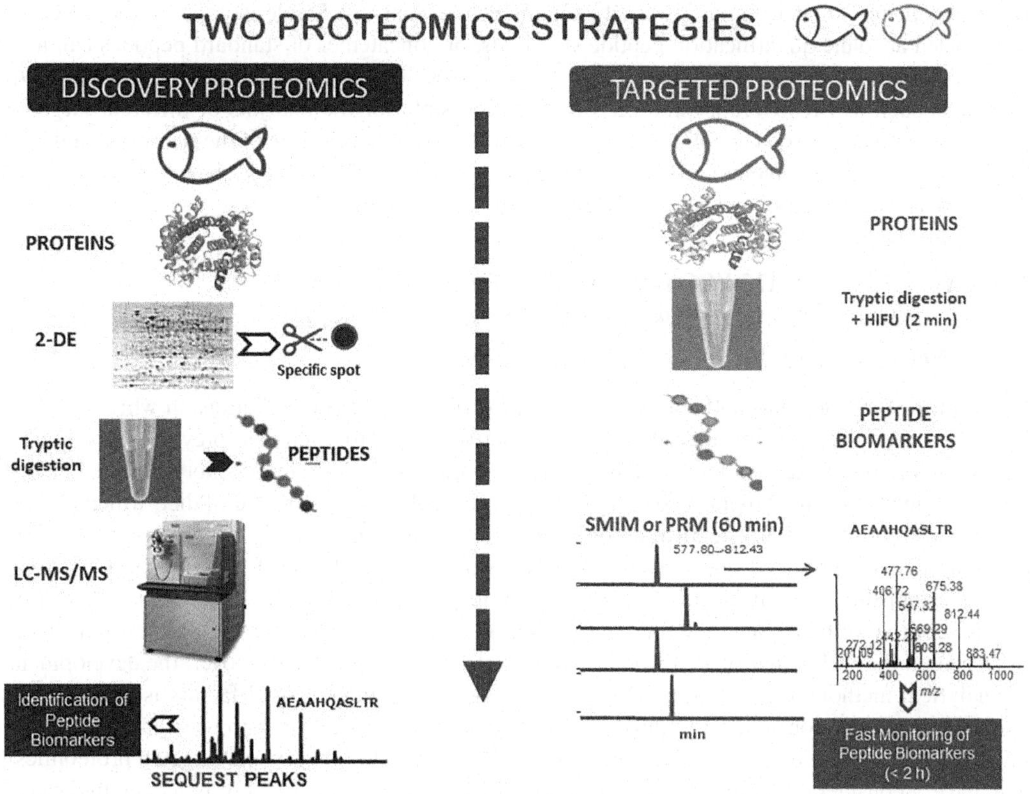

FIGURE 4.1 Workflow of proteomics: discovery and targeted proteomics.

labeling (such as stable-isotope labeling by/with amino acids in cell culture) [26]; isotope tagging by chemical reaction, such as isobaric tags for relative and absolute quantitation (iTRAQ), tandem mass tag (TMT), and difference gel electrophoresis [27–29]; stable-isotope incorporation via enzyme reaction (i.e., ^{18}O) [30]; and label-free quantification (i.e., measuring the intensity of the peptides at the MS level) [31]. After matching the obtained peptides and proteins by alignment software programs such as BLAST (https://blast.ncbi.nlm.nih.gov/), it is possible to select relevant peptide biomarkers to be used in the subsequent phase namely, targeted proteomics.

Targeted proteomics refers to the monitoring of the relevant peptide biomarkers and it has become a recognized methodology to detect selected proteins with significant accuracy, reproducibility, and sensitivity [32]. In targeted proteomics, the MS analyzer is focused on detecting only the peptide/s chosen by selected/multiple-reaction monitoring (SRM/MRM) [33]. Monitoring appropriate transitions (events of precursor and fragment ions of a mass-to-charge ratio), represents a common analysis for detecting and identifying peptide biomarkers. These techniques are selective, sensitive, highly reproducible, with a high dynamic range and an excellent signal-to-noise ratio [34]. SRM/MRM modes are usually performed on triple quadrupole instruments. This method possesses a highly sensitive scanning procedure but its optimization for a final SRM/MRM analysis is very time-consuming and, most importantly, this scanning mode does not produce entire MS/MS spectra. Because the spectrum of a peptide is critical to verify its sequence, new procedures are being used to obtain entire structural information; for instance, SRM-triggered MS/MS using hybrid quadrupole-ion trap mass spectrometers, selected MS/MS ion monitoring (SMIM), parallel reaction monitoring (PRM) in ion trap or high-resolution Q-Orbitrap instruments are alternative targeted modes that enable the monitoring of precise peptides [35–37]. The development of targeted data-independent analysis, conducted on a sequential windowed acquisition of all theoretical fragment ion mass spectra (SWATH-MS) [38], can identify and quantify thousands of proteins without the prerequisite of specifying a group of proteins before analysis. Stable-isotope dilution, ^{13}C- or ^{15}N-labeled absolute quantification peptide standards, or concatemer of standard peptides can also be introduced to the sample as internal standards for absolute quantification of the proteins [39]. Programs such as SRMCollider and Skyline are accessible for the analysis of different targeted proteomic modes [40,41]. The following sections will show the application of the scanning mode for the follow-up of peptide biomarkers identified in the discovery phase to assess the impact of farming conditions on food quality and safety of farmed fish.

4.3 APPLICATION OF PROTEOMICS TO EVALUATE THE FARMING CONDITIONS ON FOOD QUALITY AND SAFETY IN AQUACULTURE PRODUCTS

Fish farming environments and conditions are very different from the conditions in which fish live in nature. Farmed fish, for instance, do not need to actively swim to catch their prey or escape predators; therefore, they exercise less, which affects their muscle growth and phenotype. The heavily processed feed consumed by farmed fish differs considerably from their natural diet, which affects muscle metabolism and biochemical composition. In addition, the farming conditions in aquaculture may not be optimized regarding stocking densities, the incidence of parasites and diseases, and the establishment of hierarchies due to competition for space or feed, all of which have consequences for the wellbeing and development of abnormal behavior. All these variables exert a strong influence on the yield, quality, and wholesomeness of farmed seafood. Moreover, the development of analytical methods to ensure that fish was farmed minimizing stressful factors is also of high relevance to satisfy consumer demands and labeling on the welfare of fish to be used for food. To investigate all these topics, powerful proteomic methodologies (discovery and targeted proteomics) may have a considerable impact on the understanding of current aquaculture practices in several major areas are as follows: (i) dietary management, (ii) fish welfare and response to stress, (iii) food safety, and (iv) antibiotic resistance (Figure 4.2).

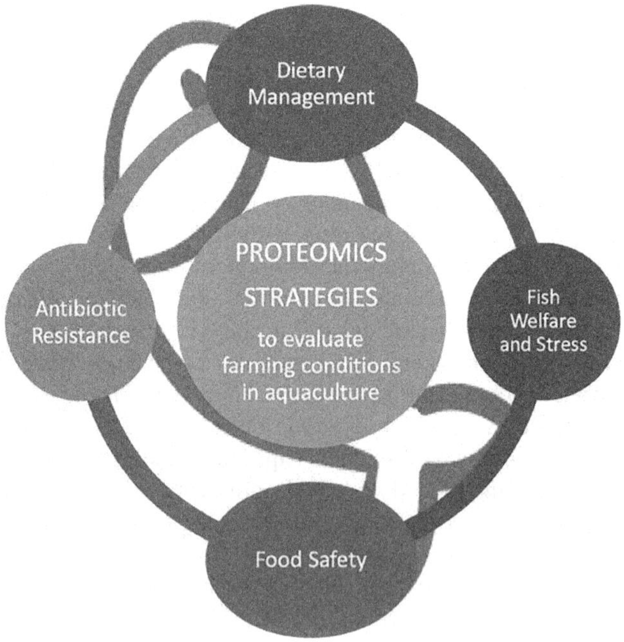

FIGURE 4.2 Summary of the main applications of proteomics techniques to evaluate the farming conditions in aquaculture reviewed in this publication.

4.3.1 Dietary Management in Aquaculture

Dietary management in aquaculture attempts to improve growth performance, health, and immune status in living aquaculture organisms. Numerous works have shown that the composition of the feed influences the composition of fish fillets. Initial works using 2-DE revealed differences between the proteomes of skeletal muscle samples from wild and farmed fish. Carpene et al. [42], found differences by 2-DE in the level of abundance of the fast skeletal myosin light chain type 3 which seemed to be more abundant in the wild than in farmed fish and, surprisingly, was also present in the red muscle of farmed, but not of wild, fish [42]. The 2-DE protein pattern of skeletal muscle of cod excised within 5 hours of death revealed the presence of spots in the ranges of molecular weight between 35 and 45 kDa and between 50 and 100 kDa in the muscles of the farmed fish that were not present in the wild cod [43]. The authors attributed the differences in the proteome to differences during cultivation that may have affected not only the make-up of the muscle *in vivo*, but also the postmortem muscle conditions (e.g., pH) and the abundance and regulation of proteases relevant in postmortem muscle tenderization. A large-scale study by Chiozzi and coworkers [44] comparing the proteome of wild and farmed European seabass with that of wild specimens from the same area in the Mediterranean by label-free multidimensional shotgun proteomics to identify relationships between farming conditions and quality and safety of the fish, confirmed muscle atrophy in farmed fish [44]. The most abundant upregulated proteins in farmed sea bass were some structural proteins and proteins involved in binding and catalytic activities, while the main downregulated proteins also involved catalytic activities and binding.

Optimization of fish diets has been a priority in the aquaculture sector for many years [45]. One ingredient whose incorporation in feed seems to improve growth performance and the humoral immune response of some fish species is β-glucan [46]. Feeding β-glucan to rainbow trout induced an increase in the amounts of tropomyosin isoforms and it lowered those of myosin light and heavy chain isoforms in the proteome of the fillet in treated trout [47]. Evaluation of the effects of partial substitution of fishmeal by plant proteins on the fish proteome in different tissues has also been the target of several studies. Thus, partial substitution of fishmeal with soybean meal caused an increase in the amount of enzymes involved in protein catabolism and turnover in the liver of rainbow trout [48] and it affected the proteome of gut mucosa in gilthead bream [49].

Reduction of the use of fishmeal and oil in aquaculture is a priority [45], which has led to investigating the effects of novel diets on fish physiology where some marine ingredients were substituted by vegetable protein and oils [50]. The inclusion of vegetable oil feed induced a specific response in the intestinal proteome in salmonids, indicating a defense against oxidative cellular stress [51]. Significantly, downregulated proteins were those related to oxidative stress and motility, including the myosin light chains, peroxiredoxin-1, and hemopexin-like protein.

Using analytical techniques based on microfluidic electrophoresis and sequencing, some authors have consistently confirmed differences in the protein abundance and/or regulation in fish muscles depending on the production method. Monti et al. [52], using sodium dodecyl sulphate-polyacrylamide gel electrophoresis (SDS-PAGE), matrix-assisted laser desorption ionization (MALDI) and electrospray ionization (ESI) MS/MS, and capillary electrophoresis (CE) for protein identification and relative quantification, showed that the enzymes involved in the metabolism of carbohydrates were upregulated in farmed sea bass muscle (i.e., glyceraldehyde-3-phosphate dehydrogenase and aldolase), while creatine kinase, nuclease diphosphate kinase B, and parvalbumin were downregulated, displaying the expected proteome pattern of muscle in farmed fish [52]. In addition, new protein sources, such as insect meal, have been characterized for aquafeeds by direct comparison through LC-MS/MS analysis [53].

4.3.2 Fish Welfare and Stress Response in Aquaculture

The effects of stress on growth have been studied extensively in animal production and aquaculture [54]. Different chronic stress conditions, such as confinement, overcrowding, repetitive handling, deficient water and diet, and hypoxia, affect the welfare and stress response of aquaculture organisms and several proteomic studies have identified robust protein signatures for chronic stress in fish [55]. Elevated cortisol due to long-term stress conditions has a strong impact on the entire organism and is directly linked to the inhibition of muscle growth by inhibiting protein synthesis and increasing protein catabolism to obtain energy from amino acids [56]. The proteome of fish farmed under these stressful conditions displays an increase in the amount of enzymes related to protein catabolism, and a decrease in the amount of the structural proteins, with the latter being degraded to provide energy. Stress-related depletion of the required energy for muscle growth has been shown to lead to muscle atrophy [57].

Farming itself affects the levels of stress the fish suffer, their growth, and the biochemical composition of different tissues and organs, including the liver, brain, and muscle. For instance, exercising in salmonids lowers the levels of aggression and the building up of hierarchies, leading to increased fish welfare and growth [58–60]. It is reasonable to assume that these observations would apply to any species with similar behavioral characteristics, that is, active swimmers with shoaling behavior and schooling responses [61]. Interestingly, while muscle atrophy is provoked by not using the muscle, its use induces both muscle growth and a type of muscle damage due to the need to develop and grow both the activated muscle satellite cells (to regenerate the lesion) and the existing myofibrils that need to increase their volume, that is, inducing both muscle hyperplasia and hypertrophy [62]. Thus, while normal muscle growth in adult fish will be accomplished by hyperplasia and, mostly, by hypertrophy [63–66], muscle regeneration and subsequent growth, as observed in exercising fish [62], recapitulates embryonic myogenesis through the activation of satellite cells and the consequent larger contribution from hyperplasia followed by hypertrophy of the muscle fibers [64,67]. This process involves alterations in protein abundance and muscle metabolism to achieve muscle growth [68] and should ultimately lead to an increase in the yield of fillet and to improved welfare of the fish.

One study on sea bream, however, was not able to show consistent differences between the 2-DE patterns of muscle from two natural repopulation lagoons and those of fish from four offshore mariculture plants in Italy by 2-DE, MALDI-MS, and liquid chromatography coupled with electrospray ionization–quadrupole-time of flight–mass spectrometry (LC-ESI-Q-ToF-MS) [69]. The similarity between the proteomes of farmed and wild fish would indicate the suitability of the farming conditions and locations. The authors did, however, find significant individual differences in the relative expression of parvalbumin isoforms and of spots corresponding to the myosin-binding protein H isoelectric series with no apparent relationship to the length of the fish, its production method, or geographical location [69]. Muscle protein patterns obtained by 2-DE analysis showed variations attributed to different factors; for example, acclimation to higher water temperature significantly increased the amount of the warm temperature acclimation-related protein-65 isoforms, and the ratio of structural proteins vs. glycolytic enzymes increased as fish grew larger [69]. This work is particularly interesting because it shows that it is possible to achieve offshore-farmed gilthead sea breams of commercial size whose protein expression profile is comparable to that of wild fish [69].

Discovery proteomics has been applied to improve our understanding of the mechanisms implicated in skeletal deformities [70]. Analysis of how preslaughter stress affects the postmortem processes in gilthead seabream muscle was performed by 2-DE and MALDI-ToF-ToF-MS [70]. Moreover, 2-DE followed by LC-MS/MS of grass carp gills uncovered alterations in the metabolic pathways after hypoxic stress [71], some of which were involved in energy generation, metabolic, immunity and oxidative processes, and proteolytic activities. It must be emphasized that the improvement of farming practices, leading to minimizing chronic stress and preserving fish welfare not only is one of the primary challenges for fish farmers, but it is also a demand by European consumers. The Pacific geoduck clam is one of the species whose farming is seeing a successful bloom and, consequently, a species under study to improve its production [72]. Spencer et al. [73] performed proteomic studies on how geoduck production may be impacted by conditions susceptible to being modified by ocean acidification, such as pH and temperature. The results showed that the amounts of heat shock protein 90-α, puromycin-sensitive aminopeptidase, and tri-functional-enzyme β-subunit as well as shell growth, kept a negative correlation with the average temperature and a positive one with the amount of dissolved oxygen. That indicates that geoducks may be more resistant to acidification under natural conditions and more susceptible to variations in the concentration of dissolved oxygen and the temperature of the water.

4.3.3 Food Safety in Aquaculture

Aquaculture has the capacity to provide food for millions of people over the world; however, inappropriate facility management may severely damage aquatic ecosystems and affect health risks to consumers through contamination with environmental or human-made hazards. Moreover, several pathogenic microorganisms can be found in the aquatic environment with the potential to negatively affect, not only aquatic life, but also human health. In the last decade, the risk of dissemination of infectious or toxic agents and the occurrence of disease outbreaks has risen mainly due to increases in the following factors: (i) intake of raw or scarcely processed seafood; (ii) international trade of aquaculture products; (iii) suboptimal monitoring methodologies; (iv) alterations in ecological stability; and (v) contamination and climatic change [74]. Food safety in the aquaculture sector is of crucial relevance to avoid health hazards that can be biotic (bacteria, allergies, parasites, virus, or harmful algae blooms) and abiotic (aromatic hydrocarbons, dioxins, heavy metals, and plastics) [75]. To control and minimize the presence of hazards, the FAO implemented a code of practice for aquaculture products [1], where the handling of fish is presented according to the requirements of Hazard Analysis Critical Control Points (HACCP).

To make the reading of this chapter more comprehensible, the authors have decided to divide the hazards to which consumers of aquaculture products are exposed into two general groups: biotic hazards and abiotic hazards.

4.3.3.1 Food Safety in Aquaculture: Biotic Hazards

Foodborne poisonings are a relevant cause of mortality and morbidity, which result from drinking water or eating food contaminated with such pathogens as viruses, bacteria and parasites, and their toxins.

Regarding biotic hazards, there are two main groups of bacteria that affect food products of aquaculture are as follows: those naturally present in its habitat (*Aeromonas* spp., *Clostribuim botulinum*, *Listeria monocytogenes*, *Vibrio cholerae*, and *Vibrio parahaemolyticus*) and those derived from environmental contamination (*Enterobacteriaceae*, *Escherichia coli*, and *Salmonella* spp.) [76]. Additionally, *Staphylococcus aureus* can infect aquaculture species during management due to inadequate hygiene conduct of operators in the processing factories [77,78].

Proteomics has been applied to the detection and identification of bacterial species from aquaculture products both after their direct detection in fish products, or after their isolation and growing in different culture media. For instance, MALDI-ToF-MS techniques enabled us to achieve mass spectral fingerprints of *Vibrio* spp., a Gram-negative bacteria that cause gastrointestinal diseases in humans after ingestion of poorly cooked infected seafood, such as seabream and mollusks [79,80]. A study by Li and coworkers reported that the exhaustive proteome and transcriptome data analysis obtained by iTRAQ coupled with MRM provided some critical protein signatures for the study of the regulatory mechanisms of the intestinal mucosal immunity in grass carp (*Ctenopharyngodon idella*) against *Vibrio mimicus* [81]. Vaccination altered the regulation of 5,339 genes and of 1,173 proteins in the grass carp intestines. The conclusions of the study suggest that the integration of the five activated immune-related pathways is relevant to the improved immune response of the intestinal mucosal in immunized carp. MALDI-ToF-MS analyses have been performed to obtain available reference spectral libraries for diverse bacterial strains isolated from seafood [82]. Recently, the open MALDI Biotyper Library (Bruker MALDI Biotyper) allowed for the precise identification of 75 pathogenic bacterial isolates [83].

Targeted proteomics applications in the field of pathogenic bacteria have increased substantially in recent years. For example, both SRM and PRM have produced sensitive quantitative results about the proteins associated with bacterial infection, particularly in the fields of clinical diagnosis and antibiotic resistance [84]. Thus, iTRAQ-based quantitative proteomics followed by MRM studies were used to contrast the differentially regulated proteins of *Aeromonas veronii*. This bacterium, a Gram-negative virulent pathogen associated with infections in freshwater fish species and mammals, is capable of adhering to biotic and abiotic surfaces surrounded by the extracellular matrix produced by the resident microorganisms. The study, which used an *in vitro* biofilm model [85], showed that the upregulated TonB protein increased the nutrient absorption capacity, and the enolase gene was involved in the regulation of multiple pathways, leading to enhancement of the bacteria's ability to undergo invasion and metastasis. These changes may be the principal cause for the capability of *A. veronii* to create biofilms and its increased dissemination.

Protozoan parasites, such as *Ichthyophthirius multifiliis*, cause important economic losses to the aquaculture sector. Upon exposure to the parasite, LC-ESI-MS/MS revealed the differential regulation of some immune-related signal transduction proteins in the skin mucus of common carp [86]. Multiple lectins and several serpins with protease inhibitor activity were likely implicated in lectin pathway activation and regulation of proteolysis, indicating that these proteins support the carp innate immune system and the preventive characteristics of the skin mucus.

Virus infections can decimate the production in fish and shrimp farms. Among the latter, the white spot syndrome virus is currently one of the most serious global hazards. A protein interactomics map for the white spot syndrome virus has been produced by means of a co-immunoprecipitation assay from a yeast two-hybrid approach [87].

Harmful algal blooms (HABs) generate shellfish poisoning toxins that affect aquaculture, particularly mussel farming. Gel-based proteomics approaches were used to distinguish and identify nontoxic dinoflagellates from the toxic dinoflagellate *Alexandrium tamarense* [88]. An alternative approach consists of the generation and application of monoclonal antibodies directed against intracellular antigens of the toxic dinoflagellate *Alexandrium minutum* as described by Carrera et al. [89]. Recent proteomic studies have contributed to enlarging the volume of data in sequence databases suitable to identify how the proteomes of HABs are modulated by physiological parameters and in response to changes in the environment, such as climate change [90]. In that regard, Piñeiro et al. [91] reviewed the application of proteomics methods to study the effects of climate change on the quality and safety of wild and cultivated seafood products.

Proteomic studies and systems biology analysis of allergenic proteins have also been critical determinants for the evaluation of the quality and safety of wild and cultivated fish and crustacean food products [92]. The major allergen identified in fish is β-parvalbumin. A rapid strategy for the detection of fish β-parvalbumin in fish products was performed by targeted proteomics using SMIM [93]. On the other hand, tropomyosin is the major allergen in shrimp and mollusks. Proteomic profiling of the allergen tropomyosin was performed to obtain the full amino acid sequence in a Q-ToF instrument [94]. Recently, the impact of ethylenediaminetetraacetic acid (EDTA)-enriched diets on farmed fish allergenicity was studied using 2-DE [95].

4.3.3.2 Food Safety in Aquaculture: Abiotic Hazards

Abiotic hazards in aquaculture have been extensively studied in mollusks exposed to contaminants in polluted areas. Discovery proteomics studies have been mainly performed for environmental assessment and marine pollution monitoring in the digestive glands of mussels *Mytilus galloprovincialis* using 2-DE and MS [96]. Bottom-up proteomics approaches on mussels exposed to fresh fuel and weathered fuel in a laboratory experiment that attempted to mimic the effects of the Prestige's oil spill were performed using 2-DE and MS [97]. Moreover, 2-DE and MALDI-ToF/ToF-MS analyses have also been applied to the identification of differentially regulated proteins in the gonads of the oyster *Crassostrea angulata* after $HgCl_2$ contamination [98]. The first shotgun proteomics analysis of mussels after exposure to pharmaceutical environmental contaminants, such as propranolol, was performed by Campos et al. [99].

To assess complex field contamination, targeted proteomics using SRM methodologies was applied to the quantification of dozens of protein biomarkers in caged amphipods (*Gammarus fossarum*) after *in situ* exposure to several aquatic environments [100]. The work detected some of the previously identified and currently well-established protein biomarkers for amphipod crustaceans, such as the detoxification/antioxidant enzymes glutathione S-transferase, acetylcholinesterase, catalase, superoxide dismutase, and some digestive enzymes [100].

Nanoparticle pollution is a recent issue of concern that has also been addressed by proteomic techniques. Thus, targeted proteomics has shown that the ionic form of silver (Ag) affected the growth of *Pseudomonas* spp. more strongly than did the nanoparticulate form of Ag in a bacterium isolated from waters in a region where fish is farmed for human consumption [101]. Possessing broad-spectrum antimicrobial properties, silver nanoparticles (AgNPs) are widely used in textiles and medical drugs. Approximately 20–130 tons of ionic silver (Ag^+) have been predicted to reach EU freshwaters annually, mostly due to the leaching of ionic AgNPs from biocidal plastics and textiles. Proteomic analysis using SWATH-MS allowed the identification of 166 proteins affected by exposure to the nanoparticulate form of Ag, which also affected the growth of *Pseudomonas* spp. The form of Ag induced different adaptive responses in the metabolic, stress, and energetic pathways in *Pseudomonas* spp., and proteins affected were transmembrane transporters, chaperones, and proteins related to the metabolism of carbohydrates and proteins, indicating their potential value as biomarkers of the stress induced by Ag^+ and/or AgNPs. Among all the modified proteins, 59 had their content significantly changed by one or both forms of silver. In view of the pieces of evidence obtained in these studies, we believe that nanoparticle pollution should be considered an emergent hazard in waters with aquaculture production.

4.3.4 Antibiotic Resistance in Aquaculture

Antibiotics are natural and synthetic compounds that kill bacteria and have been heavily used and abused in aquaculture for over 50 years [102] not only to treat and prevent infections, but also to promote growth. Fortunately, success in the development of vaccines for the most relevant infections and the implementation of vaccination programs has greatly reduced their use in some countries (e.g., Norway) although it is still a very serious problem in other countries and the breeding of some species.

The abuse in antibiotic treatments has provoked the development and spreading of bacterial resistance and the appearance and expansion of multidrug-resistant (MDR) strains in such a way that the aquatic environment has become an important reservoir of antibiotic-resistance genes/proteins (ARG/Ps) and a route for their dissemination and potential transmission to human pathogens. Until now, five main mechanisms of antibiotic resistance have been accurately recorded due to the related development of resistance to drugs that (a) deteriorate enzymes, (b) bypass target pathways, (c) change antibiotic focus sites, (d) alter the penetrability of porins, and/or (e) trigger flow systems [103].

Proteomics techniques have the potential to significantly contribute to increase the knowledge about molecular mechanisms related to antibiotic resistance [104]. In the last 5 years, metagenomics and metaproteomics have been applied to identify correlations between the "resistome" (the antibiotic immunity genes/proteins) and the transmission of ARG/Ps from natural microflora to human pathogenic microorganisms, which could become a serious health issue. Conventional methodologies to evaluate water quality had been used to analyze marine sediments close to aquaculture farms, evidencing that the native "resistome" had been enriched by the use of antibiotics at the farming sites, although the findings were restricted to only a group of genes/proteins [5].

Proteomic studies on antibiotic resistance of fish- and shellfish-borne bacteria have largely been performed on *Aeromonas* spp., because this pathogen is responsible for hemorrhagic septicemia and hemolytic diseases in aquaculture, which causes large financial losses to farmers. Some of these studies are described below.

Tetracyclines are commonly used antibiotics comprising the monocyclines group, doxycycline, and chlortetracycline (CTC), and they are very efficient against both Gram-positive and Gram-negative microbes. In aquaculture, tetracycline-resistant *Aeromonas hydrophila* (a notorious pathogen causing infections in many relevant wild and farmed species including carp, shellfish, grass carp, and shrimp) has been confirmed by different proteomics analyses. Comparison between the fitness and acquired resistance to CTC in an *A. hydrophila* biofilm by TMT-labeling-based quantitative proteomics indicated an increase in translation-related ribosomal proteins in both cases and an increase in proteins involved in fatty acid biosynthesis only in biofilm fitness, while proteins involved in other pathways were less abundant in acquired resistance biofilm. Targeting the upregulation of fatty acid biosynthesis, the authors found that a mixture of CTC and triclosan (a fatty acid biosynthesis inhibitor) had a more powerful antimicrobial effect than either one of them alone. This information is highly relevant in the fight against this pathogen when forming biofilms, which are always a challenge in seafood farming and processing [105].

Two different quantitative proteomic studies, using dimethyl labeling and label-free methods, performed in the same year as the previous work, were conducted to examine the differential regulation of proteins in response to several doses of oxytetracycline (OXY) in *A. hydrophila* [103].

The results showed an increase in translation-related proteins, although the amount of many central metabolic-related proteins decreased upon OXY treatment and, also, antibiotic sensitivity seemed to be significantly inhibited by numerous external metabolites when they were compounded with OXY antibiotics.

In 2018, Li et al. published a quantitative proteomics experiment based on iTRAQ methodology, to compare proteins differentially regulated in CTC-resistant *A. hydrophila* and in control strains [106]. The majority of the detected differentially regulated proteins were involved in key energy biosynthesis pathways, such as metabolic and catabolic processes, transportation, and signal transduction. Chemotaxis-related proteins were downregulated in CTC-resistant strains, but exogenous metabolite addition increased bacterial susceptibility in *A. hydrophila*. In addition, Elbehiry and colleagues described the application of MALDI-ToF-MS for the discrimination of the *Aeromonas* genus from meat and water samples, with a spotlight on the antimicrobial resistance of *A. hydrophila* [107].

Recently, another proteomic study using 2-DE and MALDI-ToF/ToF analysis was conducted by Zhu and coworkers with MDR and sensitive *A. hydrophila* strains to find differences in the regulation of proteins [108]. The work showed that, in the sensitive strains, proteins engaged in glycolysis/gluconeogenesis and antibiotic biosynthesis were upregulated in the MDR strain, while those involved in the biosynthesis of secondary metabolites, cationic antimicrobial peptide resistance, metabolic processes related to carbon regulation, and bacterial metabolism were downregulated. Other proteomics approaches have been used to obtain knowledge about antibiotic resistance in other pathogenic genera, such as *Edwardsiella*, a Gram-negative microorganism that also generates hemorrhagic septicemia in a broad number of cultivated fish species, including yellowtail carp and eels. As in the case of *Aeromonas*, the antibiotic-resistance status of *Edwardsiella tarda* is of high relevance for seafood safety, particularly when the bacterium is forming biofilms. Sun and coworkers, used iTRAQ-based quantitative proteomics and high-resolution LC-MS/MS to analyze the differential protein regulation of *E. tarda* in response to OXY stress in biofilms [109]. Their work showed a total of 281 modified proteins, 193 of which were downregulated and 88 upregulated. As *A. hydrophila* in biofilms, many ribosomal proteins were upregulated in response to the stress in *E. tarda*, while treatment with OXY increased the amount of Uvr C, a member of the UvrABC system that plays an important role in multiple antibiotic-resistance processes.

iTRAQ and LC-MS/MS were used in conjunction to evidence the differential proteome of the ampicillin-resistant LTB4 (LTB4-RAMP) strain of the Gram-negative facultative aerobic bacteria *Edwardsiella piscicida* and showed that a depressed P cycle seemed to be a characteristic of the differential proteome in the LTB4-RAMP [110] strain, leading the authors to conclude that the depressed P cycle caused the ampicillin resistance in *E. piscicida*.

The above research works in aquatic organisms, using iTRAQ technologies, seem to indicate that the acquisition of antibiotic resistance involves chemotaxis, energy metabolism, biofilm characteristics, and external membrane proteins, as well as networks of proteins associated with antibiotic resistance.

Finally, "reprogramming proteomics" needs to be developed in a general manner to revert an antibiotic-resistance proteome to an antibiotic-sensitive proteome for the control of antibiotic-resistant pathogens [104].

4.4 CONCLUDING REMARKS AND FUTURE DIRECTIONS

As presented in this chapter, proteomic approaches help to characterize some of the principal issues associated with farming conditions and to address some of the main challenges in aquaculture, such as dietary management, fish welfare, stress responses, food safety, and antibiotic resistance.

Proteomics helps to elucidate how dietary management in the aquaculture sector influences the production, growth, immunity, and wellness/welfare of living aquaculture organisms and assists in the selection of optimal diets. A large number of publications have shown that the composition of the feed influences the fish muscle's nutritional value and its proteome. Efforts have been made to discover protein markers for such quality traits. In addition, various proteomics investigations have been published on the identification of robust protein signatures for fish chronic stress. From this perspective, the amelioration of aquaculture conditions to reduce chronic stress during farming and maintain fish welfare is one of the principal issues that can be addressed with discovery proteomics. Additionally, innovative fast targeted proteomics workflows have demonstrated the rapid detection of fish allergens, parasites, and microorganisms in aquaculture. The characterization of species-specific peptides by MS/MS-based proteomics and their monitoring by targeted proteomics demonstrated the adequacy of these approaches for food safety control, enabling the differential detection of several hazards in the aquaculture sector. In this way, the utilization of rapid sample preparation methods, combined with sensitive and accurate MS for both the discovery and targeting of fish quality and safety biomarkers, may enhance quality control and safety in aquaculture. Moreover, proteomics offers a more holistic point of view on the molecular mechanisms of antibiotic resistance in the aquaculture sector and it can be directly linked to the metagenomic/metaproteomic approaches that are being applied to the study of a new concept known as the resistome, a current challenge of high relevance that needs an effective and rapid response and that may be elucidated through proteomics techniques.

As the proteins are considered the principal functional macromolecules in all biological systems, we consider that proteomics strategies and their associated techniques can offer several advantages compared with other methodologies for the study of the impact of farming conditions on food quality and safety in aquaculture products. This is the case primarily because, with those methodologies, it is possible to identify and directly quantify protein/peptide signatures without the necessity of inferring conclusions based on other approaches such as genomics tools. Second, the benefits of proteomic analysis may be adapted for fish products with a short shelf-life. Finally, the current advances in proteomic methodologies allow for the implementation of precise methods that may be useful for routine control tests with a potentially lower cost and in a relatively short estimated time (<30 minutes).

Lastly, the development and practical implementation of new advances based on protein arrays, microfluidics, and biosensors to the aquaculture sector offers a promising research area in which the results of proteomic studies can be established for the routine control test and diagnosis of fish products. We also assume that the digitalization of these new devices may be relevant to the aquaculture industry and control authorities in the next several years and may supply rapid monitoring information to effectively drive decision enforcement by the industry and authorities.

Author Contributions: All authors listed have made a substantial, direct, and intellectual contribution to the work, and have approved it for publication. All authors have read and agreed to the published version of the manuscript.

Funding: This research was funded by GAIN-Xunta de Galicia Project (IN607D 2017/01) and the Agencia Estatal de Investigación (AEI) of Spain and the European Regional Development Fund through project CTM2017-84763-C3-1-R.

M.C. is supported by the Ramón y Cajal Contract (RYC-2016-20419, Ministry of Science, Innovation, and Universities of Spain).

Conflicts of Interest: The authors declare no conflicts of interest.

REFERENCES

1. Food and Agriculture Organization of the United Nations (FAO). *The State of World Fisheries and Aquaculture 2018: Meeting the Sustainable Development Goals*; Food and Agriculture Organization of the United Nations: Rome, Italy, 2008.
2. Chintagari, S.; Hazard, N.; Edwards, H.G.; Jadeja, R.; Janes, M. Risks associated with fish and seafood. *Microbiol. Spectr.* **2017**, *5*, 1–16.
3. D'Alessandro, A.; Zolla, L. We are what we eat: Food safety and proteomics. *J. Proteome Res.* **2012**, *11*, 26–36.
4. Rodrigues, P.M.; Silva, T.S.; Dias, J.; Jessen, F. Proteomics in aquaculture: Applications and trends. *J. Proteom.* **2012**, *75*, 4325–4345.
5. Rodrigues, P.M.; Campos, A.; Kuruvilla, J.; Schrama, D.; Cristobal, S. Proteomics in aquaculture: Quality and safety. In *Proteomics in Food Science, from Farm to Fork*; Colgrave, M.L., Ed.; Elsevier: London, UK, 2017; pp. 279–290.
6. Pandey, A.; Mann, M. Proteomics to study genes and genomes. *Nature* **2000**, *405*, 837–846.
7. Aebersold, R.; Mann, M. Mass-spectrometric exploration of proteome structure and function. *Nature* **2016**, *537*, 347–355.
8. Carrera, M.; Cañas, B.; Gallardo, J.M. Proteomics for the assessment of quality and safety of fishery products. *Food Res. Int.* **2013**, *54*, 972–979.
9. Piñeiro, C.; Carrera, M.; Cañas, B.; Lekube, X.; Martínez, I. Proteomics and food analysis: Principles, techniques and applications. In *Handbook of Food Analysis-Two Volume Set*; Leo, M.L.; Nollet, F.T., Eds.; CRC Press: Boca Raton, FL, 2015; pp. 393–416.
10. Holton, T.A.; Vijayakumar, V.; Khaldi, N. Bioinformatics: Current perspectives and future directions for food and nutritional research facilitated by a Food-Wiki database. *Trends Food Sci. Technol.* **2013**, *34*, 5–17.
11. Gallardo, J.M.; Carrera, M.; Ortea, I. Proteomics in food science. In *Foodomics: Advanced Mass Spectrometry in Modern Food Science and Nutrition*; Cifuentes, A., Ed.; JohnWiley & Sons Inc.: Hoboken, NJ, 2013; pp. 125–165.
12. Carrera, M.; Mateos, J.; Gallardo, J.M. Data treatment in food proteomics. In *Reference Module in Food Science*; Smithers, S., Ed.; Elsevier: London, UK, 2019.
13. Rabilloud, T.; Lelong, C. Two-dimensional gel electrophoresis in proteomics: A tutorial. *J. Proteom.* **2011**, *74*, 1829–1841.
14. Carrera, M.; Cañas, B.; Piñeiro, C.; Vázquez, J.; Gallardo, J.M. *De novo* mass spectrometry sequencing and characterization of species-specific peptides from nucleoside diphosphate kinase B for the classification of commercial fish species belonging to the family Merlucciidae. *J. Proteome Res.* **2007**, *6*, 3070–3080.
15. Wolters, D.A.; Washburn, M.P.; Yates, J.R., 3rd. An automated multidimensional protein identification technology for shotgun proteomics. *Anal. Chem.* **2001**, *73*, 5683–5690.
16. Carrera, M.; Ezquerra-Brauer, J.M.; Aubourg, S.P. Characterization of the jumbo squid (*Dosidicus gigas*) skin by-product by shotgun proteomics and protein-based bioinformatics. *Mar. Drugs* **2019**, *18*, 31.
17. Zhang, Y.; Fonslow, B.R.; Shan, B.; Baek, M.C.; Yates, J.R., 3rd. Protein analysis by shotgun/bottom-up proteomics. *Chem. Rev.* **2013**, *113*, 2343–2394.
18. Perkins, D.N.; Pappin, D.J.C.; Creasy, D.M.; Cottrell, J.S. Probability-based protein identification by searching sequence databases using mass spectrometry data. *Electrophoresis* **1999**, *20*, 3551–3567.
19. Eng, J.K.; McCormack, A.L.; Yates, J.R.I.I.I. An approach to correlate tandem mass spectral data of peptides with amino acid sequences in a protein database. *J. Am. Soc. Mass Spectrom.* **1994**, *5*, 976–989.
20. Kall, L.; Canterbury, J.D.; Weston, J.; Noble, W.S.; MacCoss, M.J. Semi-supervised learning for peptide identification from shotgun proteomics datasets. *Nat. Methods* 2007, *4*, 923–925.
21. Shevchenko, A.; Wilm, M.; Mann, M. Peptide sequencing by mass spectrometry for homology searches and cloning of genes. *J. Protein Chem.* **1997**, *16*, 481–490.

22. Ma, B.; Zhang, K.; Hendrie, C.; Liang, C.; Li, M.; Doherty-Kirby, A.; Lajoie, G. PEAKS: Powerful software for peptide *de novo* sequencing by tandem mass spectrometry. *Rapid Commun. Mass Spectrom.* **2003**, *17*, 2337–2342.
23. Scigelova, M.; Maroto, F.; Dufresne, C.; Vázquez, J. High Throughput *de novo* Sequencing. 2007. Available online: http://www.thermo.com/ (accessed on 23 July 2020).
24. Carrera, M.; Cañas, B.; Vázquez, J.; Gallardo, J.M. Extensive *de novo* sequencing of new parvalbumin isoforms using a novel combination of bottom-up proteomics, accurate molecular mass measurement by FTICR-MS, and selected MS/MS ion monitoring. *J. Proteome Res.* **2010**, *9*, 4393–4406.
25. Ortea, I.; Cañas, B.; Gallardo, J.M. Mass spectrometry characterization of species-specific peptides from arginine kinase for the identification of commercially relevant shrimp species. *J. Proteome Res.* **2009**, *8*, 5356–5362.
26. Ong, S.E.; Blagoev, B.; Kratchmarova, I.; Kristensen, D.B.; Steen, H.; Pandey, A.; Mann, M. Stable isotope labelling by amino acids in cell culture, SILAC, as a simple and accurate approach to expression proteomics. *Mol. Cell. Proteom.* **2002**, *1*, 376–386.
27. Mateos, J.; Landeira-Abia, A.; Fafián-Labora, J.A.; Fernández-Pernas, P.; Lesende-Rodríguez, I.; Fernández-Puente, P.; Fernández-Moreno, M.; Delmiro, A.; Martín, M.A.; Blanco, F.J.; et al. iTRAQ-based analysis of progerin expression reveals mitocondrial dysfunction, reactive oxygen species accumulation and altered proteostasis. *Stem Cell Res. Ther.* **2015**, *6*, 119.
28. Robotti, E.; Marengo, E. 2D-DIGE and fluorescence image analysis. *Methods Mol. Biol.* **2018**, *1664*, 25–39.
29. Stryiński, R.; Mateos, J.; Pascual, S.; González, A.F.; Gallardo, J.M.; Łopieńska-Biernat, E.; Medina, I.; Carrera, M. Proteome profiling of L3 and L4 *Anisakis simplex* development stages by TMT-based quantitative proteomics. *J. Proteom.* **2019**, 201, 1–11.
30. López-Ferrer, D.; Ramos-Fernández, A.; Martínez-Bartolomé, S.; García-Ruiz, P.; Vázquez, J. Quantitative proteomics using 16O/18O labeling and linear ion trap mass spectrometry. *Proteomics* **2006**, *6* (Suppl. S1), S4–S11.
31. Mueller, L.N.; Rinner, O.; Schmidt, A.; Letarte, S.; Bodenmiller, B.; Brusniak, M.Y.; Vitek, O.; Aebersold, R.; Müller, M. SuperHirn—A novel tool for high resolution LC-MS-based peptide/protein profiling. *Proteomics* **2007**, *7*, 3470–3480.
32. Borràs, E.; Sabidó, E. What is targeted proteomics? A concise revision of targeted acquisition and targeted data analysis in mass spectrometry. *Proteomics* 2017, 17, 17–18.
33. Aebersold, R.; Bensimon, A.; Collins, B.C.; Ludwig, C.; Sabido, E. Applications and developments in targeted proteomics: From SRM to DIA/SWATH. *Proteomics* 2016, 16, 2065–2067.
34. Lange, V.; Picotti, P.; Domon, B.; Aebersold, R. Selected reaction monitoring for quantitative proteomics: A tutorial. *Mol. Syst. Biol.* **2008**, *4*, 1–14.
35. Jorge, I.; Casas, E.M.; Villar, M.; Ortega-Pérez, I.; López-Ferrer, D.; Martínez-Ruiz, A.; Carrera, M.; Marina, A.; Martínez, P.; Serrano, H.; et al. High-sensitivity analysis of specific peptides in complex samples by selected MS/MS ion monitoring and linear ion trap mass spectrometry: Application to biological studies. *J. Mass Spectrom.* **2007**, *42*, 1391–1403.
36. Carrera, M.; Cañas, B.; López-Ferrer, D.; Piñeiro, C.; Vázquez, J.; Gallardo, J.M. Fast monitoring of species-specific peptide biomarkers using high-intensity-focused-ultrasound-assisted tryptic digestion and selected MS/MS ion monitoring. *Anal. Chem.* **2011**, *83*, 5688–5695.
37. Carrera, M.; Gallardo, J.M.; Pascual, S.; González, A.F.; Medina, I. Protein biomarker discovery and fast monitoring for the identification and detection of Anisakids by parallel reaction monitoring (PRM) mass spectrometry. *J. Proteom.* **2016**, *142*, 130–137.
38. Gillet, L.C.; Navarro, P.; Tate, S.; Röst, H.; Selevsek, N.; Reiter, L.; Bonner, R.; Aebersold, R. Targeted data extraction of the MS/MS spectra generated by data-independent acquisition: A new concept for consistent and accurate proteome analysis. *Mol. Cell. Proteom.* **2012**, *11*, 016717.
39. Beynon, R.J.; Doherty, M.K.; Pratt, J.M.; Gaskell, S.J. Multiplexed absolute quantification in proteomics using artificial QCAT proteins of concatenated signature peptides. *Nat. Methods* 2005, *2*, 587–589.
40. Röst, H.; Malmström, L.; Aebersold, R. A computational tool to detect and avoid redundancy in selected reaction monitoring. *Mol. Cell. Proteom.* **2012**, *11*, 540–549.
41. Bereman, M.S.; MacLean, B.; Tomazela, D.M.; Liebler, D.C.; MacCoss, M.J. The development of selected reaction monitoring methods for targeted proteomics via empirical refinement. *Proteomics* 2012, *12*, 1134–1141.
42. Carpene, E.; Martin, B.; Dalla Libera, L. Biochemical differences in lateral muscle of wild and farmed gilthead sea bream (series *Sparus aurata* L.). *Fish Physiol. Biochem.* **1998**, *19*, 229–238.

43. Martinez, I.; Standal, I.B.; Aursand, M.; Yamashita, Y.; Yamashita, M. Analytical Methods to differentiate farmed from wild seafood. In *Handbook of Seafood and Seafood Products Analysis*; Nollet, L., Toldrá, F., Eds.; CRC Press: Boca Raton, FL, 2010; pp. 215–232.
44. Chiozzi, R.Z.; Capriotti, A.L.; Cavalieri, C.; La Barbera, G.; Montone, C.M.; Piovesana, S.; Lagana, A. Label-Free shotgun proteomics approach to characterize muscle tissue from farmed and wild European sea bass (*Dicentrarchus labrax*). *Food Anal. Method.* **2018**, *11*, 292–301.
45. Torstensen, B.E.; Espe, M.; Sanden, M.; Stubhaug, I.; Waagbø, R.; Hemre, G.I.; Fontanillas, F.; Nordgarden, U.; Hevrøy, E.M.; Olsvik, P.; et al. Novel production of Atlantic salmon (*Salmo salar*) protein based on combined replacement of fish meal and fish oil with plant meal and vegetable oil blends. *Aquaculture* **2008**, *285*, 193–200.
46. Dalmo, R.A.; Bøgwald, J. ß-glucans as conductors of immune symphonies. *Fish Shellfish Immunol.* **2008**, *25*, 384–396.
47. Ghaedi, G.; Keyvanshokooh, S.; Mohammadi Azarm, H.; Akhlaghi, M. Proteomic analysis of muscle tissue from rainbow trout (*Oncorhynchus mykiss*) fed dietary β-glucan. *Iran. J. Vet. Res.* **2016**, *17*, 184–189.
48. Martin, S.A.M.; Vilhelmsson, O.; Médale, F.; Watt, P.; Kaushik, S.; Houlihan, D.F. Proteomic sensitivity to dietary manipulations in rainbow trout. *Biochim. Biophys. Acta* **2003**, 1651, 17–29.
49. Estruch, G.; Martínez-Llorens, S.; Tomás-Vidal, A.; Monge-Ortiz, R.; Jover-Cerdá, M.; Brown, P.B.; Peñaranda, D.S. Impact of high dietary plant protein with or without marine ingredients in gut musosa proteome of gilthead seabream (*Sparus aurata*, L.). *J. Proteom.* **2020**, 216, 103672.
50. Nasopoulou, C.; Zabetakis, I. Benefits of fish oil replacement by plant originated oils in compounded fish feeds. A review. *LWT* **2012**, *47*, 217–224.
51. Morais, S.; Silva, T.; Cordeiro, O.; Rodrigues, P.; Guy, D.R.; Bron, J.E.; Taggart, J.B.; Bell, J.G.; Tocher, D.R. Effects of genotype and dietary fish oil replacement with vegetable oil on the intestinal transcriptome and proteome of Atlantic salmon (*Salmo salar*). *BMC Genom.* **2012**, *13*, 448.
52. Monti, G.; De Napoli, L.; Mainolfi, P.; Barone, R.; Guida, M.; Marino, G.; Amoresano, A. Monitoring food quality by microfluidic electrophoresis, gas chromatography, and mass spectrometry techniques: Effects of aquaculture on the sea bass (*Dicentrarchus labrax*). *Anal. Chem.* **2005**, *77*, 2587–2594.
53. Belghit, I.; Lock, E.J.; Fumière, O.; Lecrenier, M.C.; Renard, P.; Dieu, M.; Berntssen, M.H.G.; Palmblad, M.; Rasinger, J.D. Species-specific discrimination of insect meals for aquafeeds by direct comparison of tandem mass spectra. *Animals* 2019, *9*, 222.
54. Marco-Ramell, A.; de Almeida, A.M.; Cristobal, S.; Rodrigues, P.; Roncada, P.; Bassols, A. Proteomics and the search for welfare and stress biomarkers in animal production in the one-health context. *Mol. Biosyst.* **2016**, *12*, 2024–2035.
55. Raposo de Magalhães, C.; Schrama, D.; Farinha, A.P.; Revets, D.; Kuehn, A.; Planchon, S.; Rodrigues, P.M.; Marco Cerqueira, M.A. Protein changes as robust signatures of fish chronic stress: A proteomics approach in fish welfare research. *BMC Genom.* **2020**, *21*, 309.
56. Mommsen, T.P.; Vijayan, M.M.; Moon, T.W. Cortisol in teleosts: Dynamics, mechanisms of action, and metabolic regulation. *Rev. Fish Biol. Fish.* **1999**, *9*, 211–268.
57. Torres-Velarde, J.; Llera-Herrera, R.; García-Gasca, T.; García-Gasca, A. Mechanisms of stress-related muscle atrophy in fish: An ex vivo approach. *Mech. Dev.* **2018**, *154*, 162–169.
58. Christiansen, J.S.; Ringø, E.; Jobling, M. Effects of sustained exercise on growth and body composition of first-feeding fry of Arctic charr, *Salvelinus alpinus* (L.). *Aquaculture* **1989**, *79*, 329–335.
59. Christiansen, J.S.; Jobling, M. The behaviour and the relationship between food intake and growth of juvenile Arctic charr, *Salvelinus alpinus* L. subjected to sustained exercise. *Can. J. Zool.* **1990**, *68*, 2185–2191.
60. Jobling, M.; Baardvik, B.M.; Christiansen, J.S.; Jørgensen, E.H. The effects of prolonged exercise training on growth performance and production parameters in fish. *Aquac. Int.* **1993**, *1*, 95–111.
61. Eguiraun, H.; Casquero, O.; Sørensen, A.J.; Martinez, I. Reducing the number of individuals to monitor shoaling fish systems—Application of the Shannon entropy to construct a biological warning system model. *Front. Physiol.* **2018**, *9*, 493.
62. Christiansen, J.S.; Martinez, I.; Jobling, M.; Amin, A. Rapid somatic growth and muscle damage in a salmonid fish. *Basic Appl. Myol.* **1992**, *2*, 235–239.
63. Stickland, N.C. Growth and development of muscle fibres in the rainbow trout (*Salmo gairdneri*). *J. Anat.* **1983**, *137*, 323–333.
64. Rossi, G.; Messina, G. Comparative myogenesis in teleosts and mammals. *Cell. Mol. Life Sci.* **2014**, *71*, 3081–3099.
65. Nemova, N.N.; Lysenko, L.A.; Kantserova, N.P. Degradation of skeletal muscle protein during growth and development of salmonid fish. *Russ. J. Dev. Biol.* **2016**, *47*, 161–172.

66. Vélez, E.J.; Lutfi, E.; Azizi, S.; Perelló, M.; Salmerón, C.; Riera-Codina, M.; Ibarz, A.; Fernández-Borràs, J.; Blasco, J.; Capilla, E.; et al. Understanding fish muscle growth regulation to optimize aquaculture production. *Aquaculture* **2017**, *467*, 28–40.
67. Stockdale, F.E. Myogenic cell lineages. *Dev. Biol.* **1992**, *154*, 284–298.
68. Bigard, A.X.; Janmot, C.; Sanchez, H.; Serrurier, B.; Pollet, S.; d'Albis, A. Changes in myosin heavy chain profile of mature regenerated muscle with endurance training in rat. *Acta Physiol. Scand.* **1999**, *165*, 185–192.
69. Addis, M.F.; Cappuccinelli, R.; Tedde, V.; Pagnozzi, D.; Porcu, M.C.; Bonaglini, E.; Roggio, T.; Uzzau, S. Proteomic analysis of muscle tissue from gilthead sea bream (*Sparus aurata*, L.) farmed in offshore floating cages. *Aquaculture* 2010, *309*, 245–252.
70. Silva, T.S.; Cordeiro, O.D.; Matos, E.D.; Wulff, T.; Dias, J.P.; Jessen, F.; Rodrigues, P.M. Effects of preslaughter stress levels on the post-mortem sarcoplasmic proteomic profile of gilthead seabream muscle. *J. Agric. Food Chem.* **2012**, *60*, 9443–9453.
71. Xu, Z.N.; Zheng, G.D.; Wu, C.B.; Jiang, X.Y.; Zou, S.M. Identification of proteins differentially expressed in the gills of grass carp (*Ctenopharyngodon idella*) after hypoxic stress by two-dimensional gel electrophoresis analysis. *Fish Physiol. Biochem.* **2019**, *45*, 743–752.
72. Timmins-Schiffman, E.B.; Crandall, G.A.; Vadopalas, B.; Riffle, M.E.; Nunn, B.L.; Roberts, S.B. Integrating discovery-driven proteomics and selected reaction monitoring to develop a noninvasive assay for geoduck reproductive maturation. *J. Proteome Res.* **2017**, *16*, 3298–3309.
73. Spencer, L.H.; Horwith, M.; Lowe, A.T.; Venkataraman, Y.R.; Timmins-Schiffman, E.; Nunn, B.L.; Roberts, S.B. Pacific geoduck (*Panopea generosa*) resilience to natural pH variation. *Comp. Biochem. Physiol. Part D Genomics Proteom.* **2019**, *30*, 91–101.
74. Freitas, J.; Vaz-Pires, P.; Câmara, J.S. From aquaculture production to consumption: Freshness, safety, traceability and authentication, the four pillars of quality. *Aquaculture* 2020, *518*, 734857.
75. Teklemariam, A.D.; Tessema, F.; Abayneh, T. Review on evaluation of safety of fish and fish products. *Int. J. Fish Aquat. Stud.* **2015**, 3, 111–117.
76. Huss, H.H. Assurance of Seafood Quality. In *FAO Fishery Technical Paper No. 334*; FAO: Rome, Italy, 1994; p. 169.
77. Jahncke, M.L.; Schwarz, M.H. Public, animal and environmental aquaculture health issues in industrialized countries. In *Public, Animal and Environmental Aquaculture Health Issues*; Jahncke, M., Garrett, E.S., Reilly, A., Martin, R.E., Cole, E., Eds.; John Wiley & Sons, Inc.: New York, 2002; pp. 67–102.
78. Carrera, M.; Böhme, K.; Gallardo, J.M.; Barros-Velázquez, J.; Cañas, B.; Calo-Mata, P. Characterization of foodborne strains of *Staphylococcus aureus* by shotgun proteomics: Functional networks, virulence factors and species-specific peptide biomarkers. *Front. Microbiol.* **2017**, 8, 2458.
79. Hazen, T.H.; Martinez, R.J.; Chen, Y.F.; Lafon, P.C.; Garrett, N.M.; Parsons, M.B.; Bopp, C.A.; Sullards, M.C.; Sobecky, P.A. Rapid identification of *Vibrio parahaemolyticus* by whole-cell matrix-assisted laser desorption ionization-time of flight mass spectrometry. *Appl. Environ. Microbiol.* **2009**, 75, 6745–6756.
80. Kazazić, S.P.; Topić Popović, N.; Strunjak-Perović, I.; Babić, S.; Florio, D.; Fioravanti, M.; Bojanić, K.; Čož-Rakovac, R. Matrix-assisted laser desorption/ionization time of flight mass spectrometry identification of *Vibrio* (Listonella) *anguillarum* isolated from sea bass and sea bream. *PLoS One* **2019**, *14*, e0225343.
81. Li, J.N.; Zhao, Y.T.; Cao, S.L.; Wang, H.; Zhang, J.J. Integrated transcriptomic and proteomic analyses of grass carp intestines after vaccination with a double-targeted DNA vaccine of *Vibrio mimicus. Fish Shellfish Immunol.* **2020**, *98*, 641–652.
82. Böhme, K.; Fernández-No, I.C.; Barros-Velázquez, J.; Gallardo, J.M.; Calo-Mata, P.; Cañas, B. Species differentiation of seafood spoilage and pathogenic gram-negative bacteria by MALDI-TOF mass fingerprinting. *J. Proteome Res.* **2010**, *9*, 3169–3183.
83. Piamsomboon, P.; Jaresitthikunchai, J.; Hung, T.Q.; Roytrakul, S.; Wongtavatchai, J. Identification of bacterial pathogens in cultured fish with a custom peptide database constructed by matrix-assisted laser desorption/ionization time-of-flight mass spectrometry (MALDI-TOF MS). *BMC Vet. Res.* **2020**, *16*, 52.
84. Saleh, S.; Staes, A.; Deborggraeve, S.; Gevaert, K. Targeted proteomics for studying pathogenic bacteria. *Proteomics* 2019, *19*, 1–10.
85. Li, Y.; Yang, B.; Tian, J.; Sun, W.; Wang, G.; Qian, A.; Wang, C.; Shan, X.; Kang, Y. An iTRAQ-based comparative proteomics analysis of the biofilm and planktonic states of *Aeromonas veronii* TH0426. *Int. J. Mol. Sci.* **2020**, *21*, 1450.
86. Saleh, M.; Kumar, G.; Abdel-Baki, A.S.; Dkhil, M.A.; El-Matbouli, M.; Al-Quraishy, S. Quantitative proteomic profiling of immune responses to *Ichthyophthirius multifiliis* in common carp skin mucus. *Fish Shellfish Immunol.* **2019**, *84*, 834–842.

87. Sangsuriya, P.; Huang, J.Y.; Chu, Y.F.; Phiwsaiya, K.; Leekitcharoenphon, P.; Meemetta, W.; Senapin, S.; Huang, W.P.; Withyachumnarnkul, B.; Flegel, T.W.; et al. Construction and application of a protein interaction map for white spot syndrome virus (WSSV). *Mol. Cell. Proteom.* **2014**, *13*, 269–282.
88. Chan, L.L.; Sit, W.H.; Lam, P.K.; Hsieh, D.P.; Hodgkiss, I.J.; Wan, J.M.; Ho, A.Y.; Choi, N.M.; Wang, D.Z.; Dudgeon, D. Identification and characterization of a "biomarker of toxicity" from the proteome of the paralytic shellfish toxin-producing dinoflagellate *Alexandrium tamarense* (Dinophyceae). *Proteomics* **2006**, *6*, 654–666.
89. Carrera, M.; Garet, E.; Barreiro, A.; Garcés, E.; Pérez, D.; Guisande, C.; González-Fernández, A. Generation of monoclonal antibodies for the specific immunodetection of the toxic dinoflagellate *Alexandrium minutum* Halim from Spanish waters. *Harmful Algae* **2010**, *9*, 272–280.
90. Hennon, G.M.M.; Dyhrman, S.T. Progress and promise of omics for predicting the impacts of climate change on harmful algal blooms. *Harmful Algae* **2020**, *91*, 101587.
91. Piñeiro, C.; Cañas, B.; Carrera, M. The role of proteomics in the study of the influence of climate change on seafood products. *Food Res. Int.* **2010**, *43*, 1791–1802.
92. Carrera, M.; Cañas, B.; Gallardo, J.M. Advanced proteomics and systems biology applied to study food allergy. *Curr. Opin. Food Sci.* **2018**, *22*, 9–16.
93. Carrera, M.; Cañas, B.; Gallardo, J.M. Rapid direct detection of the major fish allergen, parvalbumin, by selected MS/MS ion monitoring mass spectrometry. *J. Proteom.* **2012**, *75*, 3211–3220.
94. Abdel Rahman, A.M.; Kamath, S.; Lopata, A.L.; Helleur, R.J. Analysis of the allergenic proteins in black tiger prawn (*Penaeus monodon*) and characterization of the major allergen tropomyosin using mass spectrometry. *Rapid Commun. Mass Spectrom.* **2010**, *24*, 2462–2470.
95. De Magalhães, C.R.; Schrama, D.; Fonseca, F.; Kuehn, A.; Morisset, M.; Ferreira, S.R.; Gonçalves, A.; Rodrigues, P.M. Effect of EDTA enriched diets on farmed fish allergenicity and muscle quality; a proteomics approach. *Food Chem.* **2020**, *305*, 125508.
96. Mi, J.; Orbea, A.; Syme, N.; Ahmed, M.; Cajaraville, M.P.; Cristobal, S. Peroxisomal proteomics, a new tool for risk assessment of peroxisome proliferating pollutants in the marine environment. *Proteomics* 2005, *5*, 3954–3965.
97. Apraiz, I.; Leoni, G.; Lindenstrand, D.; Persson, J.O.; Cristobal, S. Proteomic analysis of mussels exposed to fresh and weathered Prestige's oil. *J. Proteom. Bioinf.* **2009**, *2*, 255–261.
98. Zhang, Q.H.; Huang, L.; Zhang, Y.; Ke, C.H.; Huang, H.Q. Proteomic approach for identifying gonad differential proteins in the oyster (*Crassostrea angulata*) following food-chain contamination with HgCl2. *J. Proteom.* **2003**, *94*, 37–53.
99. Campos, A.; Danielsson, G.; Farinha, A.P.; Kuruvilla, J.; Warholm, P.; Cristobal, S. Shotgun proteomics to unravel marine mussel (*Mytilus edulis*) response to long-term exposure to low salinity and propranolol in a Baltic Sea microcosm. *J. Proteom.* **2016**, *137*, 97–106.
100. Gouveia, D.; Chaumot, A.; Charnot, A.; Almunia, C.; François, A.; Navarro, L.; Armengaud, J.; Salvador, A.; Geffard, O. Ecotoxico-proteomics for aquatic environmental monitoring: First in situ application of a new proteomics-based multibiomarker assay using caged amphipods. *Environ. Sci. Technol.* **2017**, *51*, 13417–13426.
101. Barros, D.; Pradhan, A.; Mendes, V.M.; Manadas, B.; Santos, P.M.; Pascoal, C.; Cássio, F. Proteomics and antioxidant enzymes reveal different mechanisms of toxicity induced by ionic and nanoparticulate silver in bacteria. *Environ. Sci. Nano* **2019**, *6*, 1207–1218.
102. Lulijwa, R.; Rupia, E.J.; Alfaro, A.C. Antibiotic use in aquaculture, policies and regulation, health and environmental risks: A review of the top 15 major producers. *Rev. Aquac.* **2020**, *12*, 640–663.
103. Yao, Z.; Li, W.; Lin, Y.; Wu, Q.; Yu, F.; Lin, W.; Lin, X. Proteomic analysis reveals that metabolic flows affect the susceptibility of *Aeromonas hydrophila* to antibiotics. *Sci. Rep.* **2016**, *6*, 39413.
104. Peng, B.; Li, H.; Peng, X. Proteomics approach to understand bacterial antibiotic resistance strategies. *Expert Rev. Proteom.* **2019**, *16*, 829–839.
105. Li, W.; Yao, Z.; Sun, L.; Hu, W.; Cao, J.; Lin, W.; Lin, X. Proteomics analysis reveals a potential antibiotic cocktail therapy strategy for *Aeromonas hydrophila* infection in biofilm. *J. Proteome Res.* **2016**, *15*, 1810–1820.
106. Li, W.; Ali, F.; Cai, Q.; Yao, Z.; Sun, L.; Lin, W.; Lin, X. Reprint of: Quantitative proteomic analysis reveals that chemotaxis is involved in chlortetracycline resistance of *Aeromonas hydrophila*. *J. Proteom.* **2018**, *180*, 138–146.
107. Elbehiry, A.; Marzouk, E.; Abdeen, E.; Al-Dubaib, M.; Alsayeqh, A.; Ibrahem, M.; Hamada, M.; Alenzi, A.; Moussa, I.; Hemeg, H.A. Proteomic characterization and discrimination of *Aeromonas* species recovered from meat and water samples with a spotlight on the antimicrobial resistance of *Aeromonas hydrophila*. *Microbiologyopen* **2019**, *8*, e782.

108. Zhu, W.; Zhou, S.; Chu, W. Comparative proteomic analysis of sensitive and multi-drug resistant *Aeromonas hydrophila* isolated from diseased fish. *Microb. Pathog.* **2019**, *139*, 103930.
109. Sun, L.; Chen, H.; Lin, W.; Lin, X. Quantitative proteomic analysis of *Edwardsiella tarda* in response to oxytetracycline stress in biofilm. *J. Proteom.* **2017**, *150*, 141–148.
110. Su, Y.B.; Kuang, S.F.; Peng, X.X.; Li, H. The depressed P cycle contributes to the acquisition of ampicillin resistance in *Edwardsiella piscicida*. *J. Proteom.* **2020**, *212*, 103562.

5 Recent Advances in Vitamin Analysis by Mass Spectrometry

Maria Katsa and Charalampos Proestos
National and Kapodistrian University of Athens

CONTENTS

5.1 Introduction 61
5.2 Mass Spectrometry 66
5.3 Extraction Techniques 67
 5.3.1 Fat-Soluble Vitamins 67
 5.3.2 Water-Soluble Vitamins 68
5.4 Liquid Chromatography 70
 5.4.1 Fat-Soluble Vitamins 70
 5.4.2 Water-Soluble Vitamins 71
 5.4.3 Simultaneous Determination of FSVs and WSVs 75
5.5 Gas Chromatography 77
5.6 Conclusions 79
Acknowledgments 79
References 79

5.1 INTRODUCTION

In the past decades, the consumers' concern about the content and the safety of the foods has increased significantly. As food products contain a variety of ingredients, the insurance of their quality and safety is a complicated task but necessary not only for the raw materials but also for the final products. Thus, food researchers and food companies have to overcome the challenges to give appropriate answers to consumers' questions. To achieve this goal, laboratories have to replace the classic analytical techniques with modern approaches in order to develop fit-for-purpose methods (Cifuentes 2013).

Vitamins are organic compounds that are essential for several biochemical and physiological functions in the human body such as normal growth, cellular differentiation, and self-maintenance. Vitamins vary in structure and polarity but are classified into two groups based on their solubility: water-soluble vitamins (WSVs) and fat-soluble vitamins (FSVs). WSV group includes vitamin C and the B-complex group consisting of vitamins B1, B2, B3, B5, B6, B7, B9, and B12, while FSV group consists of vitamins A, D, E, and K. Each vitamin occurs with various homologs, known as vitamers, and metabolites with a different biological role (Fanali et al. 2017). An equilibrate diet should provide all the necessary amounts of vitamins as they cannot be synthesized in the human body. The dietary guidelines of the United States Department of Agriculture recommend dietary allowances of vitamins according to sex and age (US Department of Health and Human Services and US Department of Agriculture 2020). However, health problems can be caused by their deficiency and excessive consumption (Herrero, Cifuentes, and Ibáñez 2012; Melfi et al. 2018; Katsa, Proestos, and Komaitis 2016). Table 5.1 shows some characteristic information about their structure, sources, and functions.

DOI: 10.1201/9781003091226-7

TABLE 5.1
List of Vitamins with their Chemical Name, Chemical Structure, Sources and Functions

Vitamin	Chemical Names	Chemical Structure	Food Sources	Daily Intake (mg/day), Age 19–70	Function
B1	Thiamine, thiamine chloride, thiamine hydrochloride, thiamine mononitrate, thiamine monophosphate(TMP), thiamine pyrophosphate (TDP), and thiamine triphosphate	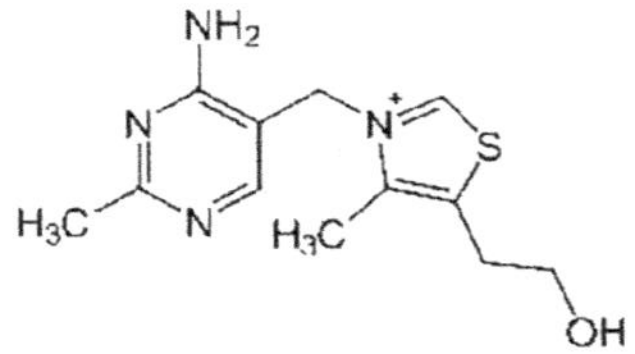	Whole grains, enriched grain products, yeast, nuts, seeds, vegetables, potatoes, liver, pork, and eggs	1.20 (males) 1.10 (females)	Antioxidant, erythropoietic, mood modulating, and glucose-regulating activities
B2	Riboflavin (RF), flavin mononucleotide (FMN), flavin adenine dinucleotide (FAD), riboflavin hydrochloride, and riboflavin 5′-phosphate	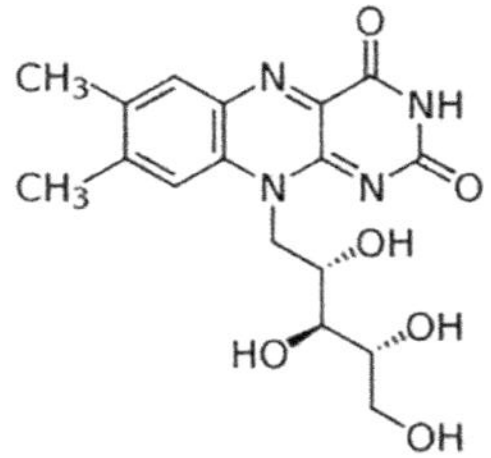	Enriched grain products, yeast, milk, cheese, eggs, mushrooms, spinach, bananas, meat, kidney, fish, and seafood	1.30 (males) 1.10 (females)	Precursor of the coenzymes FMN and FAD, role in the conversion of food into energy
B3	Nicotinic acid (NA), nicotinamide (NM), nicotinamide riboside (NAR), nicotinuric acid (NUA), nicotinamide adenine dinucleotide phosphate (NADPH), and nicotinamide adenine dinucleotide (NAD)	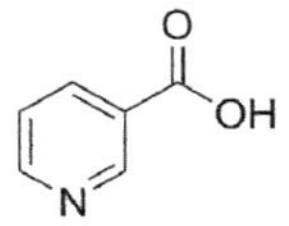	Whole grains, enriched grain products, vegetables, mushrooms, eggs, peanut butter, pork, liver, poultry, fish, and seafood	16.0 (males) 14.0 (females)	Component of the coenzymes NAD and its phosphate form NADP, role in the production of cholesterol, the digestion and the conversion of food into energy
B5	D-pantothenic acid, panthenol, dexpanthenol, pantethine, and calcium pantothenate	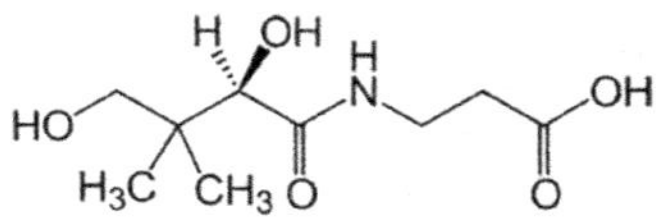	Yeast, beans, eggs, broccoli, mushrooms, avocado, kidney, liver, and poultry	5.00	Component of coenzyme A (CoA), growth factor, essential for the synthesis of cholesterol, lipids, neurotransmitters, steroid hormones, and hemoglobin

(Continued)

TABLE 5.1 ***(Continued)***
List of Vitamins with their Chemical Name, Chemical Structure, Sources and Functions

Vitamin	Chemical Names	Chemical Structure	Food Sources	Daily Intake (mg/day), Age 19–70	Function
B6	Pyridoxine (PN), pyridoxal (PL), pyridoxamine (PM), pyridoxine 5′-phosphate, pyridoxal 5′-phosphate(PLP), pyridoxamine 5′-phosphate (PMP), pyridoxamine phosphate, pyridoxine phosphate, pyridoxic 4-acid (Pyr)	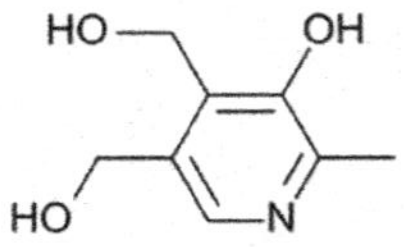	Whole grains, yeast, nuts, vegetables, potatoes, fruits, liver, salmon, tuna and meat	1.30–1.70 (males) 1.30–1.50 (females)	Coenzyme for the synthesis of amino acids, neurotransmitters, sphingolipids,aminolevulinic acid, and various enzymes, important for the metabolism of proteins and sugars
B7	D-(+)-biotin, and D-biocytin (ε-N-biotinyl-L-lysine)	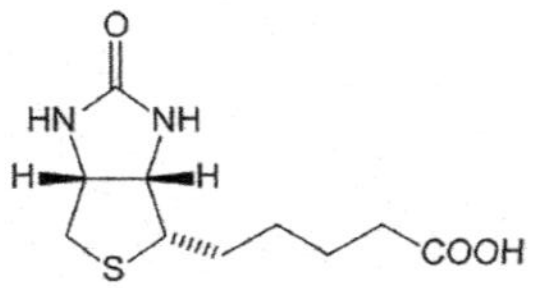	Whole grains, peanuts, soybeans, eggs, vegetables, fruits, liver, and salmon	0.03	Cofactor for 4 carboxylases, involved in the synthesis of fats, glycogen, and amino acids
B9	Folic acid (FA), folinic acid, 5-methyltetrahydrofolate (5-MTHF), 5-formyltetrahydrofolate (5-HCO-THF), 10-formylfolic acid (10-HCO-FA), 10-formyldihydrofolate, tetrahydrofolic acid (THF), 5,10-methylene-tetrahydrofolate (5,10-CH2-THF), 5,10-methenyl-tetrahydrofolate, 7,8-dihydrofolic acid (DHF), formyltetrahydrofolate (5-CHO-THF)	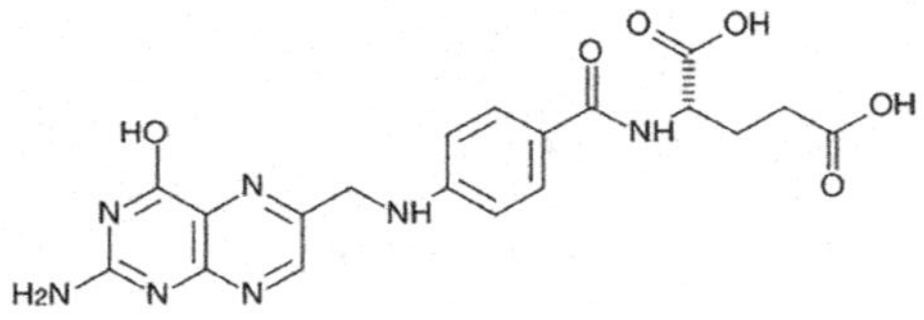	Enriched grain products, green leafy vegetables, asparagus, orange, avocado, citrus fruits, and liver	0.40	Involved in the production of DNA, protein metabolism, and heart health

(Continued)

TABLE 5.1 (*Continued*)
List of Vitamins with their Chemical Name, Chemical Structure, Sources and Functions

Vitamin	Chemical Names	Chemical Structure	Food Sources	Daily Intake (mg/day), Age 19–70	Function
B12	Cyanocobalamin, hydroxocobalamin, hydroxocobalamin acetate, methylcobalamin, 5′-deoxyadenosylcobalamin (coenzyme B12), and pseudo-cyanocobalamin	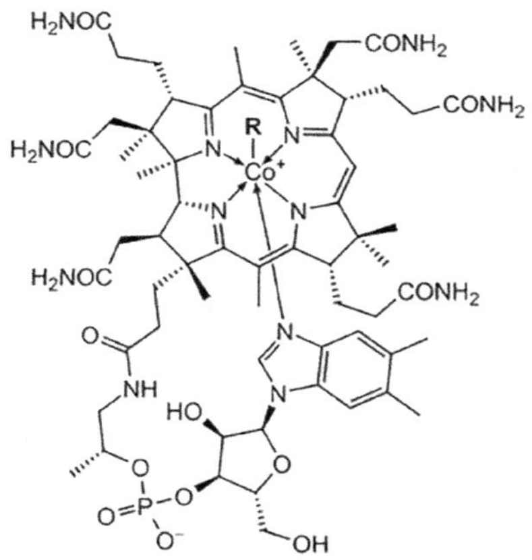 R = 5′-deoxyadenosyl, CH_3, OH, CN	Fortified cereals, milk, and dairy products, eggs, meat, poultry, fish, and seafood	0.0024	Necessary for hematopoiesis, neural metabolism, DNA and RNA production, and carbohydrate, fat, and protein metabolism improves iron functions in the metabolic cycle and assists folic acid in choline synthesis
C	L-ascorbic acid (AA), L-dehydroascorbic acid (DHAA), D-isoascorbic acid (D-IAA), ascorbic phosphate, ascorbyl palmitate/phosphate/stearate, and calcium/potassium/sodium ascorbate	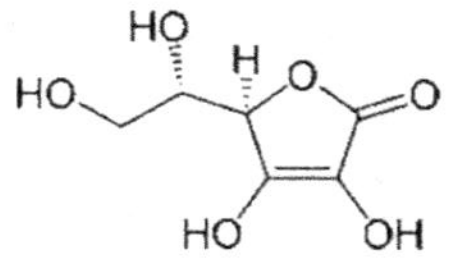	Fresh fruits (orange, strawberries, kiwifruit, citrus fruits and juices, papaya) and vegetables (peppers, broccoli, tomato, lettuce)	90.0 (males) 75.0 (females)	Antioxidant, boosts the immune system, forms the collagen
A	Retinol, retinal, retinyl acetate, retinyl palmitate, retinoic acid, retinaldehyde and carotenoids (α-,β- and γ-carotene, xanthophyll)	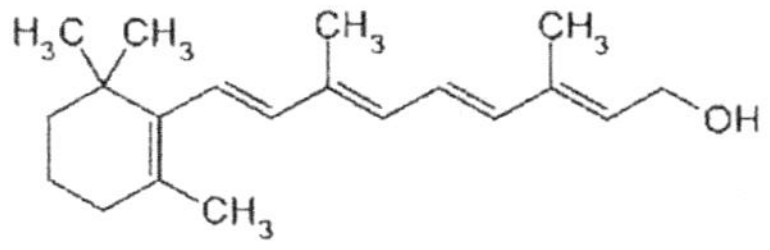	Fortified cereals, dairy products, eggs, butter, fish oils, carrots, sweet potatoes, red peppers, leafy vegetables, orange, ripe yellow fruits, liver and meat	0.900 (males) 0.700 (females)	Necessary for normal vision, immune function and gene expression in the embryonic development

(*Continued*)

TABLE 5.1 (*Continued*)
List of Vitamins with their Chemical Name, Chemical Structure, Sources and Functions

Vitamin	Chemical Names	Chemical Structure	Food Sources	Daily Intake (mg/day), Age 19–70	Function
D	Cholecalciferol (D3), ergocalciferol (D2), Calcitriol, 25-OH-D3, and 25-OH-D2	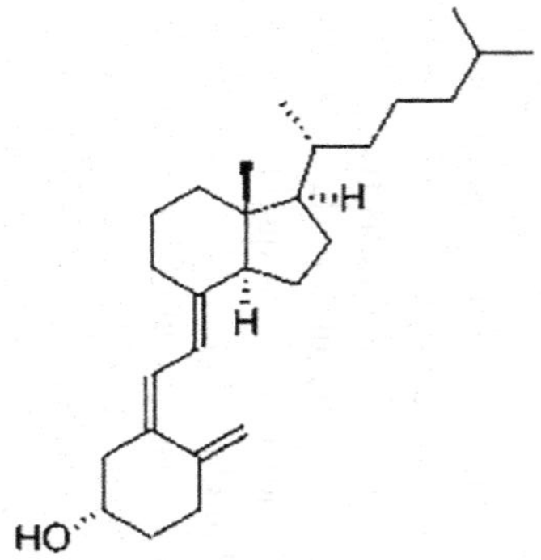	Fortified cereals, fortified dairy products and soy beverages, fortified margarines, orange juice, soy beverages, fortified juices, mushrooms, egg yolks, liver, and fatty fishes	0.02	important for calcium metabolism, mobilization in bones, and absorption
E	Tocopherols (α-, β-, γ- and δ-), tocotrienols (α-,β-, γ- and δ-), and a-tocopheryl acetate	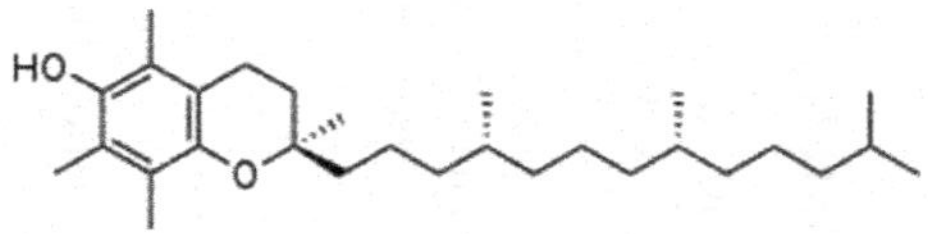	Fortified cereals, nuts, beans, seeds, peanuts and peanut butter, egg yolks, vegetables oils, vegetables, fruits, and fortified juices	15.0.0	Antioxidant protecting cells from damage and lipids from peroxidation
K	Phylloquinone (K1), menaquinones (K2, MK-n), menadiones (K3)	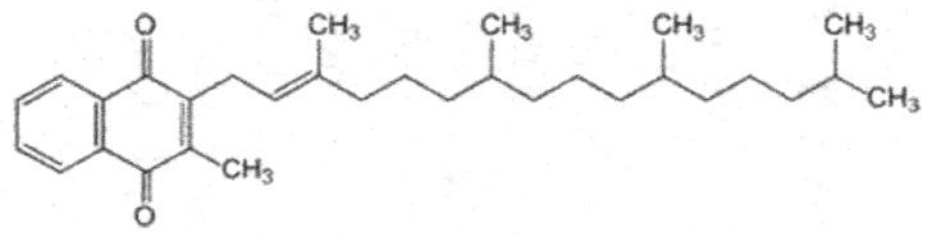	Milk, green leafy vegetables and vegetables oils	0.090 (males) 0.120 (females)	Antihemorrhagic, essential for blood clotting and bone health

Source: Aqel, Yusuf, and Al-Rifai (2014), Herrero, Cifuentes, and Ibáñez (2012), FDA (2016), Ross, Taylor, and Yaktine et al. (2020)

Vitamins are very susceptible to degradation because of their instability to oxygen, temperature, pH, and light. As a result, losses during technological processes and storage are possible. To cover these losses, food fortification strategies were realized, although it is an expensive technique. Food fortification with vitamins has a scope to enrich the foodstuffs with adequate quantities of vitamins in order to enhance the human body and prevent all vitamin-related diseases such as scurvy, pellagra, and neglected tropical diseases (NTDs) (Katsa and Proestos 2019).

Their instability in combination with their heterogeneity and their low concentrations in matrices make their determination a difficult task (Vazquez et al. 2009; Dwyer et al. 2015). For this reason, the precise labeling of vitamins in foodstuffs is necessary and requires reliable analytical methods for their identification and their determination (Nimalaratne et al. 2014). Until now, a wide range of analytical methods has been reported in the literature in various matrices including food, biological samples, and pharmaceuticals as well as animal feed.

Liquid chromatography (LC) and gas chromatography (GC) techniques have been developed for the determination of vitamins in various matrices. LC is the most common technique coupled with various detectors such as ultraviolet (UV), diode array detection (DAD), fluorescence (FL), and mass spectrometry (MS). GC analysis coupled with both flame ionization detector and MS was performed. Nowadays, MS is mainly used for the quantitative and qualitative determination of vitamins because of its higher accuracy, selectivity, and sensitivity (Demirkaya and Kadioglu 2007; Bartosinska Ewa, Buszewska-Forajta, and Siluk 2016; Wang et al. 2018).

5.2 MASS SPECTROMETRY

In recent years, the use of MS techniques is increased significantly and constitute an important tool. MS combined with chromatography is the best analytical methodology for the molecular characterization of the compounds of interest even in pictograms (pg). Improvements in the extraction techniques and in the data processing tools contributed to the evolution of the analytical methods and consequently all the analytes can be identified and quantified properly (Careri, Bianchi, and Corradini 2002; Eichhorn, Pérez, and Barceló 2012).

The detection by MS is based on the measurement of the mass-to-charge (*m/z*) of ionized molecules and the selection of the precursor ion. Usually, the protonated ions $[M+H]^+$ are used for evaluation in positive mode, and the deprotonated ions $[M-H]^-$ in negative mode. The advantage of the MS technique is that the complete chromatographic separation of the analytes of interest is not necessary for their detection. However, it will help the analysis by decreasing significantly the matrix effect.

In the MS methods, the matrix effect should be examined because of the existence of interference compounds such as lipids, proteins, etc. that compete with the analytes of interest and affects their ionization. It can provoke either suppression or enhancement of the signals detected. For this reason, the appropriate sample extraction and purification are two important steps in order to minimize this effect. In addition, the use of stable isotope-labeled standards and matrix-matched calibration curves improves the quantitation of the analytes (Di Stefano et al. 2012; Y. Zhang et al. 2018).

There are various types of mass spectrometer that are used in combination with chromatography. The most common are as follows:

1. **Single Quadrupole** is the most simple and cheap low-resolution instrument for quantitative analysis. It measures the *m/z* ratio of an ion to the nearest integer value using low voltages (Ardrey 2003).
2. **Triple Quadrupole (QQQ)** is the most widespread instrument used for low- and high-molecular-weight compounds. High sensitivity and selectivity are its advantages, although the analytes of interest coelute. Two transitions in selected reaction monitoring (SRM) mode are necessary for the identification of the target analyte. The two most abundant fragments from the precursor ion are selected as quantifier and qualifier ions for the

identification and quantification of the analytes. However, the SRM mode cannot provide data in full scan and cannot detect unknown substances. This technique also permits the simultaneous monitoring of a lot of compounds (Herebian et al. 2009; Y. Zhang et al. 2018).

3. **Quadrupole Time-of-Flight (QToF)** constitutes a high-resolution instrument with the tandem mass spectrometric capability of 10,000–12,000 full width at half maximum (FWHM). It is used to identify the target, suspect, and non-target compounds. Its disadvantage is the inaccurate mass measurements that might occur to the matrix interferences in complex food matrices.
4. **Orbitrap** systems can achieve mass resolving power of up to 100,000 FWHM by maintaining excellent mass accuracy (<5 ppm) without the use of internal mass correction (Rubert, Zachariasova, and Hajslova 2015).
5. **Ion Trap (IT)** is the most sensitive instrument when coupled with MS. Simple ITs are cheap but not high-resolution instruments. When ITs are combined with Orbitrap or ion cyclotron resonance analyzers (Fourier transform-ion cyclotron resonance instruments), they are converted to high-resolution instruments that can provide the characterization of unknown compounds.
6. **Hybrid Triple Quadrupole-linear Ion Trap** is a powerful instrument for large-scale screenings of unknown or suspect compounds that combines the advantages of the classical QQQ and IT scanning. Its third quadrupole (Q3) can be used either as a conventional quadrupole mass filter or a linear IT (Y. Zhang et al. 2018).

The ionization methods that can be used are electron ionization (EI), chemical ionization (CI), fast-atom bombardment, thermospray, electrospray ionization (ESI), atmospheric pressure chemical ionization (APCI), and atmospheric pressure photoionization (APPI). Each source has a different ion formation mechanism.

The most widely used interfaces are ESI, APCI, and APPI in positive and negative modes. They are considered soft ionization sources because the ionized molecules in positive and negative modes are formatted without fragmentation. ESI was usually preferred to identify polar, ionized, and ionizable molecules such as WSVs, whereas APCI and APPI are suitable for low polarity substances such as FSVs and carotenoids. Depending on the selected ionization source, the matrix effect also differs. APCI is less susceptible to matrix effect than ESI source due to the source design and the fact that the ionization is realized in the gas phase (Di Stefano et al. 2012; Souverain, Rudaz, and Veuthey 2004; Ardrey 2003).

5.3 EXTRACTION TECHNIQUES

For the determination of the analyte of interest, the selection of the extraction technique is crucial as well as the chromatographic technique. Conventional extraction techniques usually demand a large amount of organic solvents and are time-consuming as the procedures are laborious for the analysts. The most common classic techniques are liquid–liquid extraction (LLE), Soxhlet extraction and solid-phase extraction (SPE).

Nowadays, these techniques have been replaced by new, advanced, automatic, and environmental friendly techniques such as supercritical fluid extraction (SFE), pressurized liquid extraction (PLE), microwave-assisted extraction, ultrasound-assisted extraction, dispersive liquid–liquid microextraction, and liquid-phase microextraction (LPME) (Herrero, Cifuentes, and Ibáñez 2012).

5.3.1 Fat-Soluble Vitamins

FSVs are a family of groups of vitamins A, D, E, and K. Each group consists of various homologs and metabolites (vitamers) with different biological and physiological functions. Their extraction is a difficult task due to their chemical heterogeneity and their chemical instability to light, oxygen, temperature, and pH value. Moreover, their low concentrations in the samples in combination with the matrix complexity play an important role in the selection of the adequate extraction procedure.

FSVs and carotenoids can be found in the lipid fraction of the samples, which is composed of triglycerides, sterols, and phospholipids. These compounds have similar solubility to FSVs and as a result, they are coextracted from the matrix provoking interferences during the analysis (Fanali et al. 2017).

For the determination of FSVs, conventional extraction techniques are used. Due to its solubility, solvents of low polarity are preferred such as hexane, diethyl ether, petroleum ether, dichloromethane, acetone, etc. However, a saponification step is necessary before their extraction in order to remove the unwanted lipids, chlorophylls, and all the macromolecular interferences of hydrophobic nature, to release them from the matrix, and to convert their esters to their free forms. In the saponification, an ethanolic or methanolic solution of sodium hydroxide is added to lipid-rich samples to remove the neutral fat and incubated to 70–100°C for 30 minutes. In some cases, overnight saponification is conducted under mild conditions to extract also vitamin K that is sensitive to alkaline conditions. The optimization of the saponification conditions is necessary to maximize the efficiency and the recoveries of the determined analytes. The addition of antioxidants (e.g., ascorbic acid, butylated hydroxytoluene, and pyrogallol) prevents the degradation of the vitamins during the saponification and the extraction step (Bates 2005; Herrero, Cifuentes, and Ibáñez 2012; Kienen et al. 2008).

After the extraction, an SPE step is followed in order to clean up the sample from interferences and pre-concentrate the vitamins, especially vitamins D2 and D3, whose concentrations are extremely low in the majority of the food matrices. The most common stationary phases for the SPE cartridges are silica, C18, C8, C2, and CH (non-polar) as well as CN and NH_2 (polar) (GB 5009.82-2016 2016; Jakobsen et al. 2004; Iwase 2000).

Enzymatic digestion constitutes an alternative procedure instead of saponification. It can remove the triglycerides and all the unwanted lipids allowing the dephosphorylation of the vitamins. For products with starch, α-amylase was used in order to catalyze the α-1,4-glycosidic linkages to glucose, maltose, and other oligosaccharides. Lipases were selected for the extraction of FSVs and the samples were digested at 37°C for 1–3 hours depending on the matrix. An example could be the Lee et al. (2017) method for the determination of *cis–trans* isomers of vitamin K1 in infant formula using LC-APCI-MS/MS. The separation of the isomers was conducted by a C30 column and the use of isotope-labeled internal standards to check the accuracy and the precision of the method. The experimental procedure was based on the enzymatic hydrolysis with lipase at 37°C for 3 hours, the precipitation of the proteins with a mixture of ethanol:methanol and the extraction of the vitamins with *n*-hexane (Nollet 2013; Lee et al. 2017; Sundarram and Murthy 2014).

SFE is an advanced technique used for the extraction of FSVs. The most appropriate supercritical solvent due to the low polarity of the FSVs is CO_2 containing 5% methanol or 3% ethanol. It can be followed by a saponification step and enzymatic hydrolysis as occurring in traditional methods. Membranes can be used. This technique is more automatic, faster, and less time-consuming (Turner, King, and Mathiasson 2001; Turner, Mattias et al. 2001).

Finally, LPME constitutes a technique based on a microdroplet of an organic solvent that is introduced at the end of a microsyringe needle. The final volume of the sample is larger than a microdroplet and for this reason, the extract can be analyzed with LC. In combination with an orthogonal array experimental design, it can be used for the optimization of the experimental parameters for the determination of FSVs in aqueous samples (Sobhi et al. 2008).

5.3.2 Water-Soluble Vitamins

WSVs are a group of B-complex vitamins and vitamin C that can be found in various food matrices but in small concentrations. They are usually either found bounded to proteins and carbohydrates or free in foods and feeds. To release them, their extraction is based on acid and enzymatic hydrolysis methods.

For vitamins B1 (thiamine), B2 (riboflavin), and B6 (pyridoxine), acid hydrolysis is usually conducted by autoclaving the sample at 60°C–150°C for 10–30 minutes with HCl. Then, the pH is adjusted around 4.0–4.5. An enzymatic treatment can be followed with the use of various enzymes such as takadiastase, acid phosphatase, and β-glucosidase depending on the nature of the matrix. Finally, the sample is incubated for 18 hours in order to release the phosphorylated forms of vitamins. In case the acid hydrolysis is omitted, only the free forms of vitamins will be detected (Herrero, Cifuentes, and Ibáñez 2012; Blake 2007; Rodriguez et al. 2012).

Vitamin B3 (niacin) can be found in foods as nicotinic acid or nicotinamide in free and bounded forms. Enzyme hydrolysis and acid hydrolysis are the treatments that can be used for the determination of niacin in foodstuffs. Enzyme hydrolysis with NADase can convert nicotinamide adenine dinucleotide and nicotinamide adenine dinucleotide phosphate to nicotinamide that can be determined simultaneously with nicotinic acid. Acid hydrolysis can cause the degradation of nicotinamide in some food matrices but in the meantime, it can release the bounded forms of the vitamins that are not bioavailable (Ndaw et al. 2002; Lebiedzińska and Szefer 2006).

Vitamin B5 exists in the foods either in its free form or bounded in coenzyme A (CoA) and acyl carrier protein (ACP). Thus, the determination of pantothenic acid requires its release from its bounded forms. Acid and alkaline hydrolysis is not appropriate for pantothenic acid as it can provoke important losses. However, enzymatic procedures with pepsin, pantotheinase, and alkaline phosphatase are the most adequate for its liberation from its bounded forms except from the forms bounded to ACP (Engel et al. 2010; Pakin et al. 2004).

Vitamin B7 (biotin) is an important cofactor for five biotin-dependent carboxylases are as follows: acetyl-CoA carboxylase-a, acetyl-CoA carboxylase-b, propionyl-CoA carboxylase, pyruvate carboxylase, and b-methylcrotonyl-CoA carboxylase. Acid hydrolysis can release the protein-bounded forms of vitamins and can be combined with enzymatic hydrolysis with papain (digestion at 37°C for 16 hours). In case that biotin does not naturally exist and is added in fortified samples, alkaline hydrolysis with ammonium hydroxide can be conducted (Staggs et al. 2004; Herrero, Cifuentes, and Ibáñez 2012).

The determination of vitamin B9 (folic acid) and the total folate content requires enzymatic hydrolysis. More specifically, a tri-enzyme procedure with protease, a-amylase, and folate conjugase enzymes is conducted to release the folates of their mono- and diglutamyl derivatives. Then, a purification step is followed. C18 cartridges as well as strong anion exchange (SAX) cartridges can be used for their isolation (Camara, Lowenthal, and Phinney 2013; Freisleben, Schieberle, and Rychlik 2003).

Vitamin B12 (cyanocobalamin) is one of the most important vitamins in foods due to its carbon–metal bond. The determination of this vitamin is a challenging task because of its low concentration in food (3–20 ng/g). It can be found in various forms as Adl-Cbl and OH-Cbl in animal products and as Me-Cbl and OH-Cbl in dairy products. Before the quantitative determination of vitamin B12, these coenzyme forms should be released from the proteins that are bounded. A typical extraction procedure includes extraction with sodium acetate (pH 4.0) in combination with the addition of the enzyme a-amylase and sodium cyanide solution to convert all the naturally occurring forms to cyanocobalamin. The sample is incubated at 100°C for 30 minutes and finally is followed a cleanup SPE step with immunoaffinity columns EASY-EXTRACT VITAMIN B12 LGE (R-Biopharm AG) as well as C18, C8, and hydrophilic-lipophilic balance (HLB) columns (Campos-Giménez et al. 2008; Li et al. 2019; Kumar, Chouhan, and Thakur 2009).

Vitamin C (ascorbic acid) is also one of the most important vitamins as is involved in several physiological functions. It is easily oxidated to dehydroascorbic acid under increased temperatures and the presence of some cations of metals and enzymes. For its determination as ascorbic and dehydroascorbic acids, a protein precipitation treatment in combination with the use of an acid solution is used. Metaphosphoric acid (MPA), trichloroacetic acid (TCA), o-phosphoric acid, oxalic acid, and EDTA are some of the acids that can be used for the stabilization of the analyte (Nováková et al. 2008; Engel et al. 2010).

5.4 LIQUID CHROMATOGRAPHY

LC-MS method is one of the most promising approaches for the molecular characterization of analytes at scientific and industrial levels. As vitamins are sensible and thermally unstable compounds, high-performance liquid chromatography (HPLC), ultra-high-performance liquid chromatography (UHPLC), and nano-HPLC coupled with MS and tandem mass spectrometry (MS/MS) are the most appropriate techniques for their determination providing better chromatographic resolution.

UHPLC offers the possibility of the use of small particle size columns (<5 μm) as it can resist higher pressures than the traditional HPLC systems. Using smaller particles (<2.5 μm) improves the sensitivity, the resolution, and the efficiency of the separation. Thus, lower limits of detection (LODs) can be achieved. In addition, UHPLC gives the possibility of smaller flow rates than 1 mL/min and accelerates the analysis reducing the solvent consumption and the total cost of the analysis.

In LC, normal-phase (NP) and reverse-phase (RP) systems can be used for the chromatographic determination of vitamins. NP is ideal for matrices rich in lipids as the NP columns do not absorb the lipids and can be easily cleaned by non-polar solvents. It is more suitable to be combined with fluorescence detector (FLD) than with UV and MS detectors due to the lipids interferences that can be coextracted and the nature of the mobile phases that do not favor the ionization either in ESI or in APCI. FSVs, due to its lipophilic nature, can be analyzed either by NP or RP-HPLC chromatographic systems. When NP is selected, the geometric and positional isomers of vitamins can be separated such as carotenoids, *cis–trans* isomers of retinol, α-, β-, γ-, and δ-isomers of tocopherols and tocotrienols.

On the other hand, RP is compatible with MS detection and the chromatographic characteristics such as the peak shape and the reproducibility of the retention times are improved. WSVs and FSVS can be determined by RP using mainly C18 columns. In this case, the structural isomers cannot be separated and the use of triacontyl (C30) and pentafluorophenyl (PFP) columns is necessary in order to achieve it. C30 columns are less efficient than C18 columns with higher LOD, while PFP columns improve the selectivity of analytes and favor the retention of polar and ionic compounds. Moreover, fused-core and core-shell technology columns are ideal for conventional HPLC systems as their efficiency is similar to a sub-2 mm particles column with the half backpressure for the same column length. C18 and PFP columns with 2.6 mm core-shell particles have been tested for the determination of vitamins (Di Stefano et al. 2012; Ardrey 2003; Aqel, Yusuf, and Al-Rifai 2014; Fanali et al. 2017).

In ESI, the composition and the pH of the mobile phase affect the output of the analysis. Volatile buffers of low concentration (2–50 mM) in combination with acids such as acetic acid and formic acid were added to water, methanol, and acetonitrile. Strong acids and high concentration buffers were avoided as the ionization efficiency was affected and ion suppression or enhancement can be caused. APCI is more tolerable to a range of mobile phases and especially to the presence of buffers. The optimum flow rates for ESI ranged between 0.05 and 0.2 mL/min because the source ionization efficiency is increased in lower flow rates. On the contrary, APCI requires higher flow rates (up to 2 mL/min).

Several methods have been developed for the determination of FSVs and WSVs in various matrices using HPLC coupled with conventional UV and FL detectors. In the past decades, LC-MS methods have replaced the HPLC-based methods due to their sensitivity and selectivity permitting the separation of analytes. However, there is a need to develop high-resolution methods in order to detect and quantify all the possible vitamers (Careri, Bianchi, and Corradini 2002).

5.4.1 FAT-SOLUBLE VITAMINS

For the determination of FSVs, the most recent trend constitutes the characterization of vitamins as naturally occurring in the samples. This trend requires advanced techniques not only for their extraction but also for their detection. High-resolution MS (HRMS) techniques provide the possibility of identification of all the vitamers. Ion mobility MS is a promising HRMS technique that permits the separation of structural and geometric isomers as well as isobaric compounds and interferences from the analytes of interest (Fanali et al. 2017).

Dugo et al. (2008) developed a novel two-dimensional LC×LC-DAD-APCI-MS method for the determination of 41 carotenoids (esters and free forms) in red-orange essential oil. This technique uses two independent and orthogonal separation mechanisms and is appropriate for the analysis of complex matrices. In the first NP-LC system, a Discovery Cyano column (1.0 mm×10 mm, 5 μm) and a mixture of n-hexane-butyl acetate-acetone as mobile phase were used. In the second RP-LC system, a Chromolith Performance C18 (4.6 mm×100 mm, 5 μm) and a mixture of 2-propanol and acetonitrile as mobile phase were selected. As a result, 16 carotenoid monoesters, 21 carotenoid diesters, and 4 free carotenoids were identified (Dugo et al. 2008).

Jumaah et al. (2016) studied the determination of nine metabolites of vitamin D using supercritical fluid chromatography (SFC) and more specifically SFC-QToF-MS in less than 8 minutes. In this novel method, vitamins D2, D3, 25OHD2, 25OHD3, 1OHD3, 1OHD2, 24,25(OH)2D3, 1,25(OH)2D2, and 1,25(OH)2D3 were precipitated and extracted with acetonitrile. Six different columns were examined, and a 1-aminoanthracene (1-AA) column (3.0 mm×100 mm, 1.7 μm) was selected as the best column for the separation of metabolites of vitamins D2 and D3. After SFC separation, ESI in positive mode was used for their detection due to its higher sensitivity (six times) compared with APCI. This SFC method constitutes an alternative to NP and RP-LC and the LOD were satisfactory for all the analytes.

Apart from the abovementioned novel methods, several LC-MS/MS methods have been developed for the determination of FSVs in various matrices such as milk, infant formulas, cereals, vegetables, fruits, animal feed, and biological samples. Table 5.2 presents some characteristic examples of the extraction procedures that were used for the isolation and the determination of FSVs through LC-MS/MS.

5.4.2 Water-Soluble Vitamins

The determination of WSVs with LC methods has been studied a lot in the last decades in a wide range of matrices. The majority of the methods are based on LLE extraction and a purification step with SPE is added for vitamin B9 (folates) and vitamin B12 because their concentrations are in traces (μg/kg). Some typical experimental procedures for the isolation of WSVs are presented in Table 5.3. However, only a few HRMS methods have been developed until now.

Halvin et al. have developed two novel HRMS methods and more specifically two LC-ESI-QToF methods for the simultaneous determination of B1 (thiamine), B2 (riboflavin), B3 (nicotinic acid and nicotinamide), B5 (pantothenic acid), and B6 (pyridoxal and pyridoxine) in two different matrices: nutritional yeast and quinoa. Isotope-labeled internal standards were used in order to increase the accuracy and the precision of the methods. As shown in Figure 5.1, various treatments were tested for each matrix in order to maximize the extraction and the liberation of all the vitamins (Hälvin, Paalme, and Nisamedtinov 2013, 2014).

Various methods have been developed for the determination of vitamin B12 by LC-MS/MS in various matrices such as dairy products, infant formulae, dietary supplements, beef, and plant samples (Nakos et al. 2017; Vyas and O'Kane 2011; Lee et al. 2015; Zironi et al. 2014; Szterk et al. 2012).

Eldenmann et al. (2019) and Chamlagain et al. (2018) developed a UHPLC method for the determination of B12 by LC-ESI-QToF-MS using a Waters Acquity HSS T3 C18 column (2.1×100 mm; 1.8 μm) in two microalgae powders and cereal matrices, respectively. The two microalgae species were *Arthospira* sp. (known as Spirulina) and *Chlorella* sp. considered as food in European Union. Vitamin B12 was extracted with an acid buffer (pH 4.5) and 100 μL of 1% sodium cyanide. Then, α-amylase was added and incubated at 37°C for 40 minutes. After the extraction, the sample was purified on an immunoaffinity column. Two ions were detected in positive ion mode: *m/z* 678.2882 ($[M+2\ H]^{2+}$ of cyanocobalamine) and *m/z* 672.7752 ($[M+2\ H]^{2+}$ of pseudovitamin B12).

TABLE 5.2
LC Methods for the Simultaneous Determination of Fat-Soluble Vitamins

Vitamin	Matrix	Sample Preparation	Chromatographic Conditions		Instrumentation	References
			Column	Mobile Phase		
A, D, E, K (9 vitamers)	Infant formulae, infant cereals, and ready-to-feed adult nutritional products	Enzymatic extraction of a-amylase and papain/extraction with acidified methanol and isooctane containing BHT	Acquity UPC2 Torus 1-aminoanthracene (3.0 mm×100 mm, 1.8 µm)	(A) Carbon dioxide and (B) 10 mM ammonium formate in methanol:water (98:2)	SFC-APCI-MS/MS	Oberson et al. (2018)
A, D, E, -β-carotenoids	Bovine milk	Saponification with ethanol and potassium hydroxide/extraction with hexane	Polaris C18 (2.1 mm×150 mm, 5 µm)	(A) Water and (B) Methanol	LC-Ion-trap-APCI-MS^n	Plozza, Craige Trenerry, and Caridi (2012)
A, D, E, K, and β-carotenoids	Avocado	Saponification with ethanol and potassium hydroxide/extraction with hexane	YMC Carotenoid C30 (4.6 mm×150 mm, 3 µm)	(A) Methanol (B) Acetonitrile and (C) 2-Propanol	LC-ESI-MS	Cortés-Herrera et al. (2019)
D2, D3, K1, K2, K3	Vegetables and infant formulae	Extraction with acetonitrile/DLLME procedure with carbon tetrachloride and water	Zorbax Eclipse ODS non–end-capped (4.6×250 mm, 5 µm)	(A) Water and (B) Acetonitrile	LC-DAD-APCI-MS	Viñas et al. (2013)
A (retinyl esters)	Milk samples from different animal species	Deproteinization with ethanol in an ultrasound bath for 6 minutes/extraction, with hexane	TSKgel Super-ODS (4.6 mm×100 mm, 2 µm)	(A) methanol and (B) Isopropanol:hexane (50:50)	LC-APCI-MS/MS	Rocchi et al. (2016)
Carotenoids	Vegetables	Homogenization with tetrahydrofuran:methanol (1:1, *v/v*)/extraction with petroleum ether and sodium chloride	RP Si C18 (2.0×250 mm, 5 µm)	Mixture of 50% acetonitrile, 40% methanol, 5% chloroform, and 5% n-heptane	LC-APCI-MS	Huck, Popp, and Scherz (2000)
D (D2 and D3)	Cereals, cheese, infant formulae, pet food	Overnight saponification with ethanol and potassium hydroxide/extraction with hexane	YMC Carotenoid S-3 (2.0×150 mm, 3 µm)	(A) acetonitrile – 0.1 acetic acid and (B) Methanol	LC-APCI-MS/MS	Huang et al. (2009)
D3	Skim milk, orange juice, breakfast cereal, salmon, and cheese	Saponification with ethanol and potassium hydroxide/extraction with ethyl ether and petroleum ether	Inertsil ODS-2 (4.6 mm×250 mm, 5 µm)	(A) Methanol and (B) Acetonitrile	LC-APCI-MS/MS	Byrdwell (2009)
E (tocopherols and tocotrienols)	Cereals	Extraction by PLE with methanol at 50°C, and pressurized at 110 bar	XTerra MS C18 (2.1×100 mm 3.5 µm)	(A) 6 mM Ammonia in methanol and (B) 6 mM Ammonia in water	LC-PDA-ESI-MS	Bustamante-Rangel et al. (2007)
K (phylloquinone and menaquinones)	Feces, serum, and food	Protein denaturation with, mix of 2-propanol:hexane (3:2)/purification with silica SPE	Kinetex C18 (3.0×150 mm, 2.6 µm)	(A) Methanol and (B) Methylene chloride	LC-APCI-MS	Karl et al. (2014)

TABLE 5.3
LC Methods for the Simultaneous Determination of Water-Soluble Vitamins

Vitamin	Matrix	Sample Preparation	Chromatographic Conditions		Instrumentation	References
			Column	Mobile Phase		
B5, B7, B9, and B12	Fortified infant foods	Extraction with ammonium acetate solution/shaker 5 minutes/ ultrasonic bath for 15 min	ACQUITY UPLC BEH C18 (2.1 mm×100 mm, 1.7 µm)	(A) Acetonitrile –0.1% formic acid and (B) Water – 0.1% formic acid	UPLC-ESI-MS/MS	Lu et al. (2008)
B1, B2, B3, B6, and B9 (18 vitamers)	Breast milk	Incubation with a-amylase for 30 minutes/extraction with methanol containing 1% acetic acid	ACE-3 C18-pentafluorophenyl (2.1 mm×150 mm, 3 µm)	(A) Water – 5% acetic acid and 0.2% HFBA (*v/v*) and (B) Acetonitrile	LC-ESI-MS/MS	Redeuil et al. (2017)
B1, B2, B3, B5, and B6	Standard reference materials (milk, IFs, chocolate, cereal, soy flour, etc.)	Extraction with 1% acetic acid in water/ultrasonic bath for 30–120 minutes without heating	Cadenza CD-C18 (4.6 mm×250 mm, 3 µm)	(A) Ammonium formate (20 mM, pH 4.00) and (B) Methanol	LC-ESI-MS/MS	Phillips (2015)
B1, B2, B3, B5, B6, and B9	Italian pasta	B1, B2, B3, B6: extraction with HCl in a boiling bath for 30 minutes B5: extraction acetate buffer (pH 5.6) at 121°C for 15 minutes B9: extraction with phosphate–ascorbate buffer, pH 8.0 in a boiling bath for 10 minutes/addition of a-amylase and cleanup with SAX cartridges	Supelcosil C18 (4.6 mm×250 mm, 5 µm)	(A) Ammonium formate (20 mM, pH 3.75) and (B) Methanol	LC-ESI-MS/MS LC-APCI-MS/MS	Leporati et al. (2005)
B1, B2, B3, B5, and B6	Wheat flour products	Extraction with enzymatic hydrolysis (phosphatase, papain, taka-diastase, and β-glucosidase)	Atlantis dC18 (2.1 mm×150 mm, 5 µm)	(A) Water – 0.1% formic acid and (B) Methanol – 0.1% formic acid	LC-ESI-MS/MS	Nurit et al. (2015)
B1, B2, B3, B5, B6, B7, B9, B12, and C	Maize flour, green and golden kiwi, and tomato pulp	Extraction with ethanol: water (50:50, *v/v*)/SPE with C18-bonded silica and glass cartridges	Alltima C18 (4.6 mm×250 mm, 5 µm)	(A) Acetonitrile – 5 mM formic acid and (B) Water – 5 mM formic acid	LC-ESI-MS/MS	Gentili et al. (2008)
B1, B2,B3, B5,B6, B7, B9, B12, and C	Infant formula	Extraction with 10 mM ammonium acetate and chloroform/ ultrasonic bath for 15 minutes	ACQUITY UPLC BEH Shield RP18 (2.1 mm×100 mm, 1.7 µm)	(A) Ammonium acetate (10 mM) and (B) Methanol	UPLC-ESI-MS/MS	Zhang, Chen, and Ren (2009)

(*Continued*)

TABLE 5.3 (*Continued*)
LC Methods for the Simultaneous Determination of Water-Soluble Vitamins

			Chromatographic Conditions			
Vitamin	**Matrix**	**Sample Preparation**	**Column**	**Mobile Phase**	**Instrumentation**	**References**
B1, B2, B3, B5, B6, B7, B9, B12, and C	Beverages and dietary supplements	Beverages: ultrasonically degassed for 10 minutes and diluted with water supplements: extraction with water: acetonitrile (95:5 *v*/*v*) with acetic acid 1% (*v*/*v*) in a water bath at 65°C for 10 minutes	Scherzo SM-C18 (2.0 mm × 150 mm, 3 μm)	(A) Ammonium formate (5 mM)-0.05% formic acid and (B) Acetonitrile – 0.3% formic acid	UPLC-ESI-MS/MS	Kakitani et al. (2014)
B1, B2, B3, B5, and B6	Nutritional supplements	Extraction with methanol: aq. hydrochloric acid (5mM, pH 2.85)	Cadenza CD-C18 (4.6 mm × 250 mm, 3 μm)	Mixture of methanol: water:FA (50:50:0.1)	Flow-injection ESI–MS/MS	Bhandari and Van Berkel (2012)
B2, B3, and B5	Energy drinks	Ultrasonically degassed for 20 minutes and diluted with methanol: water	N/A	(A) Ammonium formate (10 mM, pH 4.00) and (B) Methanol	LC-ESI-MS	Aranda and Morlock (2006)

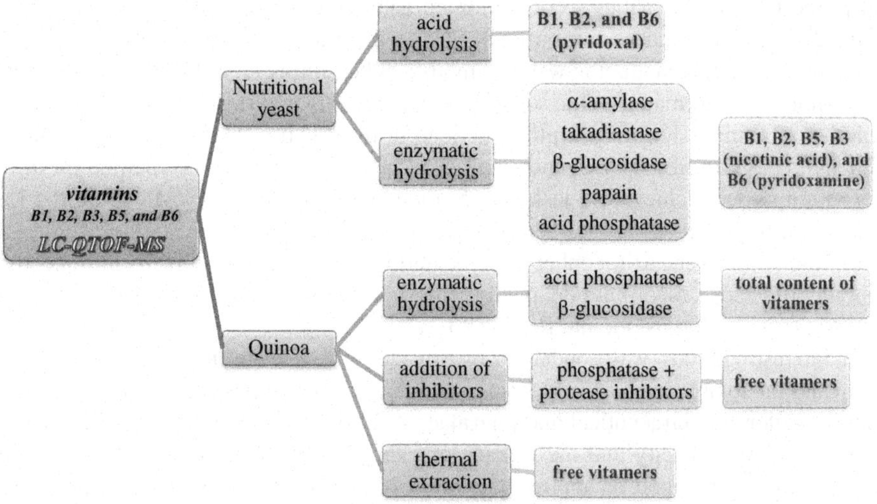

FIGURE 5.1 Experimental procedures for the determination of water-soluble vitamins in nutritional yeast and quinoa using LC-QToF-MS.

In Spirulina, pseudocyanocobalamin was the main active form of the vitamin (Chamlagain et al. 2018; Edelmann et al. 2019).

Vitamin B9 is a B group of vitamers known as folates. Folic acid is the main form of vitamin and is fully oxidized mono-glutamate form. The folates differ in the oxidation state of the pteridine ring and substitution at the 5- and 10-carbon positions. Some folates are 5-methyl-tetrahydrofolate (5-CH_3-H_4 folate), 5-formyl-tetrahydrofolate (5-HCO-H_4 folate), 5,10-methylenetetrahydrofolate (5,10-CH_2-H_4 folate), tetrahydrofolate (H_4 folate), 7,8-dihydrofolic acid, and 10-formylfolic acid (10-HCO-FA). Their separation was conducted on a C18 column by LC-ESI-MS/MS in a variety of foods such as vegetables, fruits, fruit juices, cereals, beer, meat, and seed crops.

Their extraction differs depending on the matrix, but there are a lot of methodologies with similarities. A common extraction of folates in foods includes extraction with 2-(*N*-morpholino) ethanesulfonic acid (MES) buffer or 4-(2-hydroxyethyl)-1-piperazineethanesulfonic acid (HEPES)-*N*-cyclohexyl-2-aminoethanesulfonic acid (CHES) buffer, and a tri-enzyme treatment: protease solution, α-amylase, and rat serum. The protease was added only for foods rich in protein, α-amylase for foods rich in starch, and rat serum for the deconjugation of the folates. After the addition of each enzyme, the samples were incubated at 37°C for a few hours depending on the matrix and the enzyme. Then, the samples were subjected to cleanup by SPE. For this step SAX and Oasis HLB cartridges can be used (Ringling and Rychlik 2017; Zou et al. 2019; Bertuzzi et al. 2019; H. Zhang et al. 2018; Freisleben, Schieberle, and Rychlik 2003).

5.4.3 Simultaneous Determination of FSVs and WSVs

The simultaneous extraction and determination of all the vitamins is a difficult task because of their different physical and chemical properties. For their simultaneous determination, the extraction should be optimized in order to maximize the concentration of each vitamin. To achieve it, a compromise should be done to the experimental procedure.

In the literature, various LC methods coupled with UV and FL detectors but only a few coupled with MS are reported. The main advantage of the MS methods is their sensitivity and selectivity in comparison with the abovementioned detectors.

Taguchi et al. (2014) developed a novel method for the simultaneous determination of FSVs and WSVs combining two chromatographic techniques: supercritical fluid and LC. In this way, various chemical properties and polarities were combined achieving a wide range of log *P* values from –2.11 to 10.12. As a mobile phase, a mixture of $SCCO_2$ and methanol was selected with 2-propanol, acetonitrile, methanol, and methanol: water as modifiers beginning from 100% CO_2 and finishing to 100% methanol. The chromatographic separation was investigated in six different Waters Acquity UCP^2 columns (3.0 mm × 100 mm, 2 μm); ethylene-bridged hybrid (BEH), 2-Ethylpyridine, CSH Fluoro-Phenyl, HSS C18 SB, HSS Cyano, and CSH Phenyl-Hexyl. The best method for the separation of 17 vitamins was achieved in less than 5 minutes by SFC-ESI-MS/MS using the HSS C18 SB column with methanol as a modifier. The determined vitamins were retinol, retinyl acetate, retinyl palmitate, β-carotene, α-tocopherol, ergocalciferol, phylloquinone, menaquinone, thiamine, riboflavin, nicotinic acid, nicotinamide, pantothenic acid, pyridoxine, biotin, cyanocobalamin, and ascorbic acid (Taguchi, Fukusaki, and Bamba 2014).

Another interesting study was conducted by Tayade et al. (2013). In this method, 15 vitamins both water- and fat-soluble in their free forms were determined in *Rhodiola imbricata* root that is used as an ingredient in nutraceutical and pharmaceutical products. *R. imbricata* root is a complex matrix due to its heterogeneity and the linkage of vitamins with other matrix components. Thus, this extraction procedure could be applied also to other complicated matrices such as food supplements, fortified beverages, plant samples, etc. The determination of vitamins was achieved by rapid resolution liquid chromatography-tandem mass spectrometry (RRLC-ESI-MS/MS). The separation was conducted within 30 minutes using an Agilent poroshell 120 EC-C18 narrow bore column (2.1 mm × 100 mm, 2.7 μm). A gradient program was selected and the mobile phase was composed of 10 mM ammonium formate buffers acidified with 0.1% formic acid in water and methanol, respectively.

Briefly, the extraction of WSVs was based on acid and enzymatic hydrolysis. The acid hydrolysis was conducted by autoclaving the sample with 0.1 N HCl at 100°C for 20 minutes. Then, the pH was adjusted to 4.0 and the enzymatic hydrolysis was followed digesting the sample for 18 hours at 37°C with clara-diastase. FSVs were extracted with a mixture of methanol:dichloromethane (1:1) in ultrasonic bath for 15 minutes. The samples were diluted, filtered, and injected into RRLC-MS/MS. The identification and the qualification were realized for vitamins A, D2, D3, E, K1, K2, B1, B2, B3 (nicotinic acid and nicotinamide), B5, B6, B7, B9, and B12. The method was validated and showed robustness and high sensitivity with a low limit of detection at μg/kg levels (Tayade et al. 2013).

Santos and his group have elaborated a new method for the extraction of fat and water-soluble free forms of vitamins from 12 different types of green leafy vegetables such as lettuce, spinach, pea leaves, and mizuna. Their determination was realized by LC-ESI-MS/MS and LC-DAD. The method was based on the sequential extraction of vitamins B1, B2, B3, B5, B6, B9, C, E, and pro-vitamin A following two experimental procedures. A mixture of ammonium acetate buffer and methanol was used as an extraction solvent. The samples were shaken, sonicated, and centrifuged for 15 minutes, respectively. One milliliter of the extract was injected in LC-MS/MS for the determination of WSVs. For the FSVs, the solid residue was re-extracted twice with ethyl acetate. The organic phase was evaporated, reconstituted with ethyl acetate, and injected into HPLC-DAD. Vitamins A, E, and C were found in the highest concentrations in the vegetables products. Losses in the vitamin content were observed after the 10-days storage at 3°C (Santos et al. 2012).

To measure both FSVs and WSVs in multivitamin tablets and nutritional formulations, Phinney et al. (2011) developed two LC-MS methods with the use of the stable isotope-labeled internal standards when possible. The determined analytes were vitamins B1, B2, B3, B5, B6, A, D2, D3, E, and K1. The analysis of WSVs was performed through LC-ESI-MS extracting the samples with 1% acetic acid in an ultrasonic bath for 30 minutes. A portion of acetonitrile was added and the samples were placed at –20°C in order to coagulate the lipid phase Then, the samples were centrifuged and filtered through RC filters (0.45 μm). The chromatographic separation was achieved on a C18 column (4.6 mm × 250 mm, 3 μm) with ammonium formate (20 mM, pH 4.0, i) and methanol (ii) as mobile phase.

As for FSVs, the samples were incubated at 45°C for 1 hour with 10 mL of 1% EDTA solution and were sonicated for 10 minutes to release the vitamins from the matrices. Then, the vitamins were extracted overnight with 20 mL hexane. The extraction step was repeated four times and the time of extraction was 30 min. For nutritional formulas, the samples were sonicated for 30 minutes, and overnight extraction with ethyl acetated was selected. In a second extraction, samples were sonicated and shaken for 30 minutes, respectively, and the procedure was repeated three more times. Their determination was achieved using an APCI source in positive ionization mode. A C18 column (4.6 mm×250 mm, 5 μm) was used and the mobile phase consisted of (i) methanol and (ii) acetonitrile with 5 mmol/L ammonium acetate (Phinney et al. 2011).

5.5 GAS CHROMATOGRAPHY

GC gives the possibility to identify analytes but firstly laborious experimental and derivatization procedures are necessary. GC is preferred for the determination of analytes in plant origin matrices such as fruits, seeds, and vegetable oils. As carrier gas, helium was used.

The derivatization techniques are used to improve the sensitivity, thermal stability, and volatility of the analytes of interest as well as to reduce the polarity of the analytes. The purpose of this technique is to introduce appropriate chemical groups in order to separate and detect simultaneously various compounds in a GC column. It is important to have as less as the possible water content in the derivatization procedure to prevent the hydrolysis of the derivatization reagents and the complete conversion of the analytes.

The identification of the analytes is conducted by retention time, known as retention index in GC. There are databases for the retention index of compounds in the most common stationary phases but it is not enough for the correct identification of the compound. EI is the technique preferred for the determination of volatile samples. EI technique is based on in-source fragmentation offering important structural information and has the advantage of the repeatability of the fragmentation as the voltage of the electron energy is standard.

In the past, various GC-MS methods have been used for the determination of vitamin D in cow milk (Adachi and Kobayashi 1979), vitamins A and E in biological samples (Smidt, Daniel Jones, and Clifford 1988), vitamin E in tissues (Liebler et al. 1996), vitamin B5 in the blood (Banno 1997), metabolites of vitamin B1 in urine (Amos and Neal 1970), vitamin C in aqueous samples (Deutsch and Kolhouse 1993) and others. Nowadays, they have been replaced by advanced LC methods coupled with MS and MS/MS detectors.

GC is often used as a tool for the determination of vitamin E and its derivatives due to their lipophilic and non-polar characteristics. Several methods have been reported in the literature in various matrices such as olive oil (Aresta and Zambonin 2017), functional foods (Lu et al. 2015), seeds and roasted seeds (Butinar et al. 2011), fruits (Bhatia et al. 2013), and plant (Naz et al. 2014).

For their determination, derivatization techniques such as silylation, trimethylsilylation (one-step), and acylation (two-step) are necessary. The most common columns used for their separation are DB-5, DB-5MS, BPX5, Rtx-5MS, Rtx 5sil-MS, SBP-5, HP-5, and HP-5MS. For the quantitative and quantitative determination of vitamin E, GC-MS in full scan and selected ion monitoring (SIM) mode can be used (Bartosinska Ewa, Buszewska-Forajta, and Siluk 2016; Ardrey 2003).

Hammann et al. (2016) achieved a determination of 182 uncommon polyunsaturated tocopherol isomers in a vitamin E oil capsule combining countercurrent chromatography (CCC) and gas chromatography-tandem mass spectrometry (GC-MS/MS). More specifically, 161 T3 isomers, 18 tetra-unsaturated isomers, 2 tocomonoenol isomers, and some degradation products have been detected. The oil from three capsules was extracted with 3 mL of n-hexane: water (1:1) and was shaken vigorously. The organic phase was collected and stored at 4°C until the analysis.

As a derivatization method, trimethylsilylation was preferred dissolving the oil in pyridine and *N,O*-bis(trimethylsilyl)trifluoroacetamide (BSTFA) /trimethylchlorosilane (TMCS) (99:1). The silylation agent was removed after 2 hours at 60°C and the sample was reconstituted with n-hexane in order to be analyzed by GC-MS. An Rtx-1 column with 100% dimethyl polysiloxane (15 m×0.25 mm×0.1 μm) was used for the free isomers, while for the trimethylsilylated CCC fractions an

HP-5MS column (30 m×0.25 mm×0.25 μm) was preferred. The data were collected for *m/z* 50–800 in full scan mode. The total tocopherol content determined by GC-MS consisted of α-T, γ-T, γ-T3, α-T3, and δ-T and in smaller percentage α-T3 and δ-T3 isomers. This study constitutes an innovative method for the identification of various vitamin E isomers in oil products and can be investigated in other products (Hammann, Kröpfl, and Vetter 2016).

Apart from GC-MS methods, a novel GC-QToF-MS method has been developed for the determination of lipophilic compounds such as derivatives of vitamin E (tocopherols and tocotrienols) from pigmented rice and genetically modified rice. This method constitutes an important analytical tool permitting the identification of 14 compounds of rice grain in 13 minutes. The experimental procedure was based on the saponification with an ethanolic solution of potassium hydroxide (80%, *w/v*), the extraction with hexane, and the silylation with *N*-methyl-*N*-(trimethylsilyl)trifluoroacetamide (MSTFA). The separation was performed with CP-SIL 8 CB column (30 m×0.25 mm). The quantification was conducted by the calibration curve with the use of 5a-cholestane as an internal standard. The GC-QToF-MS method provides a metabolite profile of the different cultivars of rice and can be used for the characterization of other types of rice such as the black rice that is rich in vitamin E (Kim et al. 2012).

Another GC-QToF-MS method was used for the determination of vitamins D2 and D3 in various food matrices such as yogurt, dietary supplements, plant sterol-enriched margarine, and instant milk powder. In brief, the analytes were saponified with 15 mL of a mixture of ethanol–water (1:1) and 1 mL of aqueous solution of potassium hydroxide (50% *w/v*) for 30 minutes at 60°C in a water bath. The samples were extracted with 15 mL of hexane and the organic phase was dried over sodium sulfate. Before the analysis, deuterated internal standards of vitamins D2 and D3 (D2-d3 and D3-d3) were added in order to monitor the system performance and the quantification was done by their pyro isomers. For HPLC, an Allure Si column (2.1 mm×250 mm, 5 μm) was selected, and 0.1% isopropanol in dichloromethane (*v/v*) was used as mobile phase. In GC, the separation of vitamins was conducted on a Rxi-5Sil MS column (30 m× 0.25 mm×0.10 μm) and the oven temperature was programmed from 80°C (6 minutes) to 310°C (6.67 minutes). The total run time was 28 minutes and the vitamins detection with EI at 70 eV was started after 20 minutes. The ToF-MS detector was operated in SIM mode, but full spectra data were collected during the complete chromatogram. The results were compared with HPLC-UV and HPLC-MS/MS and were in agreement between the methods (Nestola and Thellmann 2015).

In addition, an alternative method was developed for the determination of vitamin K1 in human serum after ingestion with deuterium-labeled broccoli. Broccoli is a vegetable rich in phylloquinone and can be considered an appropriate food matrix for this study. It was cultivated using 31 atom % deuterium oxide to label vitamin K1 into the matrix. Then, this deuterium-labeled broccoli was fed to a 23-year-old man and his serum was investigated for phylloquinone's determination. Briefly, the internal standard of broccoli and 0.5 mL of ethanol were added in 0.25 mL of plasma. The sample was mixed well and 0.5 mL of deionized water and 1.5 mL of hexane were added. Then, the sample was vortexed for 2 minutes and the organic layer of the hexane was evaporated. The extract was purified with SPE. Bond Elut silica (500 mg/3 mL) cartridges were conditioned with 2.5 mL of a mixture of ethyl ether:hexane (3.5:96.5), washed with 2.5 mL of hexane, and eluted with 5 mL ethyl ether:hexane. Finally, the sample was reconstituted in 30 μL of methylene chloride and 170 μL of methanol containing 10 mM zinc chloride, 5 mM acetic acid, and 5 mM sodium acetate.

For the determination of vitamin K1 through GC-MS, an HP-5MS silica GC column (30 m×0.25 mm) was chosen. The ionization of vitamin K1 was achieved in negative CI mode with the use of methane (0.5 torr) and there was no need for derivatization of the samples. In the chromatograms were detected not only the endogenous vitamin K1, but also the labeled broccoli vitamin K1. This fact indicates that no loss was observed in the deuterium of labeled broccoli and this can be used successfully for the determination of endogenous vitamin K1 (Dolnikowski et al. 2002).

5.6 CONCLUSIONS

In this chapter, the determination of WSVs and FSVs using MS techniques is reviewed. Vitamins are essential compounds because they perform multiple roles in the body maintaining good health and preventing serious and chronic diseases. Their analysis is important and more and more manufactures want to control the vitamin content of their products. Until now, several methods, both official and in-house, have been developed for their determination in a wide variety of matrices such as cereals, dairy products, vegetables, fruits, beverages, supplements, and biological samples.

Different sample preparation techniques have been widely tested and the selection of the adequate experimental procedure is crucial for the extraction of the vitamins. SFE and PLE are the most promising advanced extraction techniques as they are more environmental friendly and automated, reducing significantly the quantities of the organic solvents. In addition, these extraction techniques require less contribution of the analyst and the extraction is more reproducible. However, LLE methods in combination with SPE are the most commonly used methods until now detecting extremely low concentrations of the analytes of interest as the matrix interferences are removed.

There is no doubt that LC is used more often than GC due to the properties of the vitamins. The use of MS has replaced other classic detectors such as UV and FLD increasing the sensitivity and the selectivity, reducing the analysis times, and detecting other analytes, although their separation is not possible. The latest trend in this field is the use of QToF-MS and other HRMS instruments that permit the detection and the identification of unknown compounds without the use of the analytical standard in contrast to classic LC methods. In the near future, more HRMS methods will be used in routine analysis of the laboratories.

ACKNOWLEDGMENTS

This work was supported by the Hellenic Foundation for Research and Innovation (HFRI) under the HFRI PhD Fellowship grant (Fellowship Number: 744).

REFERENCES

Adachi Atsuko, and Kobayashi Tadashi. 1979. "Identification of Vitamin D3 and 7-Dehydrocholesterol in Cow's Milk by Gas Chromatography-Mass Spectrometry and Their Quantitation by High-Performance Liquid Chromatography." *Journal of Nutritional Science and Vitaminology* 25 (2). (Tokyo): 67–78. doi:10.3177/jnsv.25.67.

Amos William H., and Neal Robert A. 1970. "Gas Chromatography-Mass Spectrometry of the Trimethylsilyl Derivatives of Various Thiamine Metabolites." *Analytical Biochemistry* 36 (2). Academic Press: 332–37. doi:10.1016/0003-2697(70)90368-4.

Aqel Ahmad, Yusuf Kareem, Al-Rifai Asma. 2014. "Vitamin Analysis in Food by UPLC–MS." In *Ultra Performance Liquid Chromatography Mass Spectrometry*, Mu Naushad, Rizwan, Khan Mahammad, Eds., CRC Press: Boca Raton, FL: 243–78.

Ardrey Robert E. 2003. *Liquid Chromatography –Mass Spectrometry: An Introduction* (Vol. 1). John Wiley & Sons, Ltd, Hoboken, NJ.

Aresta Antonella, and Zambonin Carlo. 2017. "Determination of α-Tocopherol in Olive Oil by Solid-Phase Microextraction and Gas Chromatography–Mass Spectrometry." *Analytical Letters* 50 (10). Taylor and Francis Inc.: 1580–92. doi:10.1080/00032719.2016.1238922.

Banno Kiyoshi. 1997. "Measurement of Pantothenic Acid Hopantenic Acid by Chromatography-Mass Spectroscopy." *Methods in Enzymology* 279 (January). Academic Press Inc.: 213–19. doi:10.1016/S0076-6879(97)79025-6.

Bartosinska Ewa, Buszewska-Forajta Magdalena, Siluk Danuta. 2016. "Analysis GC–MS and LC–MS Approaches for Determination of Tocopherols and Tocotrienols in Biological and Food Matrices." *Journal of Pharmaceutical and Biomedical Analysis* 127: 156–69. doi:10.1016/j.jpba.2016.02.051.

Bates Christopher J. 2005. "VITAMINS/Fat-Soluble." In *Encyclopedia of Analytical Science* (Second Edition), Worsfold P., Poole C., Townshend A. and Miró M., Eds., Elsevier, Amsterdam, 159–72.

Bertuzzi Terenzio, Rastelli Silvia, Mulazzi Annalisa, and Rossi Filippo. 2019. “LC-MS/MS Determination of Mono-Glutamate Folates and Folic Acid in Beer.” *Food Analytical Methods* 12 (3): 722–28. doi:10.1007/s12161-018-1396-6.

Bhatia Anil, Bharti Santosh K., Tewari Shri K., Sidhu Om P., and Raja Roy. 2013. “Metabolic Profiling for Studying Chemotype Variations in Withania Somnifera (L.) Dunal Fruits Using GC-MS and NMR Spectroscopy.” *Phytochemistry* 93 (September). Elsevier Ltd: 105–15. doi:10.1016/j.phytochem.2013.03.013.

Blake Christopher J. 2007. “Analytical Procedures for Water-Soluble Vitamins in Foods and Dietary Supplements: A Review.” *Analytical and Bioanalytical Chemistry* 389 (1): 63–76. doi:10.1007/s00216-007-1309-9.

Butinar Bojan, Bučar-Miklavčič Milena, Mariani Carlo, and Raspor Peter. 2011. “New Vitamin E Isomers (Gamma-Tocomonoenol and Alpha-Tocomonoenol) in Seeds, Roasted Seeds and Roasted Seed Oil from the Slovenian Pumpkin Variety ‘Slovenska Golica.’” *Food Chemistry* 128 (2). Elsevier: 505–12. doi:10.1016/j.foodchem.2011.03.072.

Camara Johanna E., Lowenthal Mark S., and Phinney Karen W. 2013. “Determination of Fortified and Endogenous Folates in Food-Based Standard Reference Materials by Liquid Chromatography-Tandem Mass Spectrometry.” *Analytical and Bioanalytical Chemistry* 405 (13): 4561–68. doi:10.1007/s00216-013-6733-4.

Campos-Gimenez Esther, Fontannaz Patric, Triscon Marie Jose I, Kilinc Tamara, Gimenez Catherine, and Andrieux Pierre. 2008. “Determination of Vitamin B12 in Food Products by Liquid Chromatography/UV Detection with Immunoaffinity Extraction: Single-Laboratory Validation.” *Journal of AOAC International* 91 (4): 786–93. doi:10.1093/jaoac/91.4.786.

Careri Maria, Bianchi Federica, and Corradini Claudio. 2002. “Recent Advances in the Application of Mass Spectrometry in Food-Related Analysis.” *Journal of Chromatography A* 970 (1–2): 3–64. doi:10.1016/S0021-9673(02)00903-2.

Chamlagain Bhawani, Sugito Tessa A., Kariluoto Susanna, Varmanen Pekka, and Piironen Vieno. 2018. “In Situ Production of Active Vitamin B12 in Cereal Matrices Using Propionibacterium Freudenreichii.” *Food Science and Nutrition* 6: 67. doi:10.1002/fsn3.528.

Cifuentes Alejandro. 2013. *Foodomics: Advanced Mass Spectrometry in Modern Food Science and Nutrition.* Wiley, Hoboken, NJ.

Demirkaya Fatma, and Kadioglu Yucel. 2007. “Simple GC-FID Method Development and Validation for Determination of α-Tocopherol (Vitamin E) in Human Plasma.” *Journal of Biochemical and Biophysical Methods* 70: 363–68. doi:10.1016/j.jbbm.2006.08.006.

Deutsch John C., and Kolhouse J. Fred. 1993. “Ascorbate and Dehydroascorbate Measurements in Aqueous Solutions and Plasma Determined by Gas Chromatography-Mass Spectrometry.” *Analytical Chemistry* 65. https://pubs.acs.org/sharingguidelines.

Di Stefano Vita, Avellone Giuseppe, Bongiorno David, Cunsolo Vincenzo, Muccilli Vera, Sforza Stefano, Dossena Arnaldo, Drahos László, and Vékey Károly. 2012. “Applications of Liquid Chromatography-Mass Spectrometry for Food Analysis.” *Journal of Chromatography A* 1259. Elsevier B.V.: 74–85. doi:10.1016/j.chroma.2012.04.023.

Dolnikowski Gregory G., Zhiyong Sun, Grusak Michael A., Peterson James W., and Booth Sarah L. 2002. “HPLC and GC/MS Determination of Deuterated Vitamin K (Phylloquinone) in Human Serum after Ingestion of Deuterium-Labeled Broccoli.” *The Journal of Nutritional Biochemistry* 13: 168–74.

Dugo Paola, Herrero Miguel, Giuffrida Daniele, Kumm Tiina, Dugo Giovanni, and Mondello Luigi. 2008. “Application of Comprehensive Two-Dimensional Liquid Chromatography to Elucidate the Native Carotenoid Composition in Red Orange` Essential Oil.” *Journal of Agricultural and Food Chemistry* 56 (10): 3478–85. doi:10.1021/jf800144v.

Dwyer Johanna T., Wiemer Kathryn L., Dary Omar, Keen Carl L., King Janet C., Miller Kevin B., Philbert Martin A. 2015. “Fortification and Health: Challenges and Opportunities.” *Advances in Nutrition* 6: 124–31. doi:10.3945/an.114.007443.micronutrient.

Edelmann Minnamari, Aalto Sanni, Chamlagain Bhawani, Kariluoto Susanna, and Piironen Vieno. 2019. “Riboflavin, Niacin, Folate and Vitamin B12 in Commercial Microalgae Powders.” *Journal of Food Composition and Analysis* 82 (January). Elsevier: 103226. doi:10.1016/j.jfca.2019.05.009.

Eichhorn Peter, Pérez Sandra, and Barceló Damià. 2012. *TOF-MS within Food and Environmental Analysis – Comprehensive Analytical Chemistry* (Vol. 58). Elsevier. doi:10.1016/B978-0-444-53810-9.00009-2.

Engel Rita, Abrankó László and Stefanovits-Bányai Éva. 2010. “Simultaneous Determination of Water Soluble Vitamins in Fortified Food Products.” *Acta Alimentaria* 39 (1): 48–58. doi:10.1556/AAlim.39.2010.1.5.

Fanali Chiara, D'Orazio Giovanni, Fanali Salvatore, and Gentili Alessandra. 2017. "Advanced Analytical Techniques for Fat-Soluble Vitamin Analysis." *TrAC – Trends in Analytical Chemistry* 87. Elsevier Ltd: 82–97. doi:10.1016/j.trac.2016.12.001.

Freisleben Achim, Schieberle Peter, and Rychlik Michael. 2003. "Specific and Sensitive Quantification of Folate Vitamers in Foods by Stable Isotope Dilution Assays Using High-Performance Liquid Chromatography-Tandem Mass Spectrometry." *Analytical and Bioanalytical Chemistry* 376 (2): 149–56. doi:10.1007/s00216-003-1844-y.

GB 5009.82-2016. 2016. "Determination of Vitamin A, D, E in Food. National Food Safety Standard." 1–40.

Hälvin Kristel, Paalme Toomas, and Nisamedtinov Ildar. 2013. "Comparison of Different Extraction Methods for Simultaneous Determination of B Complex Vitamins in Nutritional Yeast Using LC/MS-TOF and Stable Isotope Dilution Assay." *Analytical and Bioanalytical Chemistry* 405 (4): 1213–22. doi:10.1007/s00216-012-6538-x.

Hälvin Kristel, Paalme Toomas, and Nisamedtinov Ildar. 2014. "Comparison of Different Extraction Methods to Determine Free and Bound Forms of B-Group Vitamins in Quinoa." *Analytical and Bioanalytical Chemistry* 406: 7355–66.

Hammann Simon, Kröpfl Alexander, and Vetter Walter. 2016. "More than 170 Polyunsaturated Tocopherol-Related Compounds in a Vitamin E Capsule: Countercurrent Chromatographic Enrichment, Gas Chromatography/Mass Spectrometry Analysis and Preliminary Identification of the Potential Artefacts." *Journal of Chromatography A* 1476 (December). Elsevier B.V.: 77–87. doi:10.1016/j.chroma.2016.11.018.

Herebian Diran, Zühike Sebastian, Lamshöft Marc, and Spiteller Michael. 2009. "Multi-Mycotoxin Analysis in Complex Biological Matrices Using LC-ESI/MS: Experimental Study Using Triple Stage Quadrupole and LTQ-Orbitrap." *Journal of Separation Science* 32 (7): 939–48. doi:10.1002/jssc.200800589.

Herrero Miguel, Cifuentes Alejandro, and Ibáñez Elena. 2012. "Extraction Techniques for the Determination of Carotenoids and Vitamins in Food." *Comprehensive Sampling and Sample Preparation* 4: 181–201. doi:10.1016/B978-0-12-381373-2.00133-2.

Iwase Hiroshi. 2000. "Determination of Vitamin D in Emulsified Nutritional Supplements 2 by Solid-Phase Extraction and Column-Switching High-Performance Liquid Chromatography with UV Detection." *Journal of Chromatography A* 881. www.elsevier.com/locate/chroma.

Jakobsen Jette, Clausen Ina, Leth Torben, and Ovesen Lars. 2004. "A New Method for the Determination of Vitamin D3 and 25-Hydroxyvitamin D3 in Meat." *Journal of Food Composition and Analysis* 17 (6). Academic Press: 777–87. doi:10.1016/j.jfca.2003.10.012.

Jumaah Firas, Larsson Sara, Essén Sofia, Cunico Larissa P., Holm Cecilia, Turner Charlotta, Sandahl Margareta. A rapid method for the separation of vitamin D and its metabolites by ultra-high performance supercritical fluid chromatography–mass spectrometry. *Journal of Chromatography A* 1440 (1 April 2016): 191–200. doi:10.1016/j.chroma.2016.02.043.

Katsa Maria, and Proestos Charalampos. 2019. *Vitamin Analysis in Juices and Nonalcoholic Beverages. Engineering Tools in the Beverage Industry.* Elsevier Inc. doi:10.1016/B978-0-12-815258-4.00005-6.

Katsa Maria, Proestos Charalampos, and Komaitis Efstratios. 2016. "Determination of Fat Soluble Vitamins A and E in Infant Formulas by HPLC-DAD." *Current Research in Nutrition and Food Science* 4 (2). doi:10.12944/CRNFSJ.4.Special-Issue-October.12.

Kienen Vanessa, Costa Willian F., Visentainer Jesuí V., Souza Nilson E., and Oliveira Cláudio C. 2008. "Development of a Green Chromatographic Method for Determination of Fat-Soluble Vitamins in Food and Pharmaceutical Supplement." *Talanta* 75 (1): 141–46. doi:10.1016/j.talanta.2007.10.043.

Kim Jae Kwang, Ha Sun Hwa, Park Soo Yun, Lee Si Myung, Kim Hyo Jin, Lim Sun Hyung, Suh Seok Cheol, Kim Dong Hern, and Cho Hyun Suk. 2012. "Determination of Lipophilic Compounds in Genetically Modified Rice Using Gas Chromatography-Time-of-Flight Mass Spectrometry." *Journal of Food Composition and Analysis* 25 (1). Academic Press: 31–38. doi:10.1016/j.jfca.2011.06.002.

Kumar Sagaya Selva, Chouhan Raghuraj Singh, and Thakur Munna Singh. 2009. "Trends in Analysis of Vitamin B-12 Trends in Analysis of Vitamin B 12." *Analytical Biochemistry* 398 (2). Elsevier Inc.: 139–49. doi:10.1016/j.ab.2009.06.041.

Lebiedzińska Anna, and Szefer Piotr. 2006. "Vitamins B in Grain and Cereal-Grain Food, Soy-Products and Seeds." *Food Chemistry* 95 (1): 116–22. doi:10.1016/j.foodchem.2004.12.024.

Lee Hyeyoung, Lee Joonhee, Choi Kihwan, and Kim Byungjoo. 2017. "Development of Isotope Dilution-Liquid Chromatography/Tandem Mass Spectrometry for the Accurate Determination of Trans- and Cis-Vitamin K1 Isomers in Infant Formula." *Food Chemistry* 221. Elsevier Ltd: 729–36. doi:10.1016/j.foodchem.2016.11.112.

Lee Jung Hoon, Shin Jin-Ho, Park Jung-Min, Kim Ha-Jung, Ahn Jang-Hyuk, Kwak Byung-Man, and Kim Jin-Man. 2015. "Analytical Determination of Vitamin B12 Content in Infant and Toddler Milk Formulas by Liquid Chromatography Tandem Mass Spectrometry (LC-MS/MS)." *Korean Journal for Food Science of Animal Resources* 35 (6): 765–71. doi:10.5851/kosfa.2015.35.6.765.

Li Yanan, Gill Brendon D., Grainger Megan N.C., and Manley-Harris Merilyn. 2019. "The Analysis of Vitamin B12 in Milk and Infant Formula: A Review." *International Dairy Journal* 99. Elsevier Ltd: 104543. doi:10.1016/j.idairyj.2019.104543.

Liebler Daniel C., Burr Jeanne A., Philips Leslie, and. Ham Amy J.L. 1996. "Gas Chromatography-Mass Spectrometry Analysis of Vitamin E and Its Oxidation Products." *Analytical Biochemistry* 236 (1). Academic Press Inc.: 27–34. doi:10.1006/abio.1996.0127.

Lu Dan, Yang Yi, Wu Xin, Zeng Li, Li Yongxin, and Sun Chengjun. 2015. "Simultaneous Determination of Eight Vitamin e Isomers and α-Tocopherol Acetate in Functional Foods and Nutritional Supplements by Gas Chromatography-Mass Spectrometry." *Analytical Methods* 7 (8). Royal Society of Chemistry: 3353–62. doi:10.1039/c4ay02854f.

Melfi Maria Teresa, Nardiello Donatella, Cicco Nunzia, Candido Vincenzo, and Centonze Diego. 2018. "Simultaneous Determination of Water- and Fat-Soluble Vitamins, Lycopene and Beta-Carotene in Tomato Samples and Pharmaceutical Formulations: Double Injection Single Run by Reverse-Phase Liquid Chromatography with UV Detection." *Journal of Food Composition and Analysis* 70. Elsevier Inc.: 9–17. doi:10.1016/j.jfca.2018.04.002.

Nakos Michail, Pepelanova Iliyana, Beutel Sascha, Krings Ulrich, Berger Ralf Günter, and Scheper Tomas H. 2017. "Isolation and Analysis of Vitamin B 12 from Plant Samples." *Food Chemistry* 216. Elsevier Ltd: 301–8. doi:10.1016/j.foodchem.2016.08.037.

Naz Saba, Sherazi Syed T., Talpur Farah N., Kara Huseyin, Uddin Siraj, and Khaskheli Abdul R. 2014. "Chemical Characterization of Canola and Sunflower Oil Deodorizer Distillates." *Polish Journal of Food and Nutrition Sciences* 64 (2): 115–20. doi:10.2478/pjfns–2013–0008.

Ndaw, Sophie, Bergaentzlé Martine, Aoudé-Werner Dalal, and Hasselmann C. 2002. "Enzymatic Extraction Procedure for the Liquid Chromatographic Determination of Niacin in Foodstuffs." *Food Chemistry* 78 (1). Elsevier: 129–34. doi:10.1016/S0308-8146(02)00205-4.

Nestola Marco, and Thellmann Andrea. 2015. "Determination of Vitamins D 2 and D 3 in Selected Food Matrices by Online High-Performance Liquid Chromatography-Gas Chromatography-Mass Spectrometry (HPLC-GC-MS)." *Analytical and Bioanalytical Chemistry* 407: 297–308. doi:10.1007/s00216-014-8123-y.

Nimalaratne Chamila, Sun Chenxing, Wu Jianping, Curtis Jonathan M., and Schieber Andreas. 2014. "Quantification of Selected Fat Soluble Vitamins and Carotenoids in Infant Formula and Dietary Supplements Using Fast Liquid Chromatography Coupled with Tandem Mass Spectrometry." *Food Research International* 66. Elsevier Ltd: 69–77. doi:10.1016/j.foodres.2014.08.034.

Nollet Leo M.L., and Toldrá Fidel. 2013. *Food Analysis*. CRC Press Taylor & Francis Group, LLC. doi:10.5005/jp/books/10642_13.

Nováková Lucie, Solichová Dagmar, Pavlovičová Soňa, and Solich Petr. 2008. "Hydrophilic Interaction Liquid Chromatography Method for the Determination of Ascorbic Acid." *Journal of Separation Science* 31 (9): 1634–44. doi:10.1002/jssc.200700570.

Pakin C., Bergaentzlé Martine, Hubscher Véronique, Aoudé-Werner Dalal, and Hasselmann C. 2004. "Fluorimetric Determination of Pantothenic Acid in Foods by Liquid Chromatography with Post-Column Derivatization." *Journal of Chromatography A* 1035 (1): 87–95. doi:10.1016/j.chroma.2004.02.042.

Phinney Karen W., Rimmer Catherine A., Thomas Jeanice Brown, Sander Lane C., Sharpless Katherine E., and Wise Stephen A. 2011. "Isotope Dilution Liquid Chromatography – Mass Spectrometry Methods for Fat- and Water-Soluble Vitamins in Nutritional Formulations." *Analytical Chemistry* 83 (1): 92–98. doi:10.1021/ac101950r.

Ringling Christiane, and Rychlik Michael. 2017. "Origins of the Difference between Food Folate Analysis Results Obtained by LC–MS/MS and Microbiological Assays." *Analytical and Bioanalytical Chemistry* 409 (7): 1815–25. doi:10.1007/s00216-016-0126-4.

Rodriguez R. San José, Fernández-Ruiz Virginia, Cámara Montana, and Sánchez-Mata Maria de Cortes. 2012. "Simultaneous Determination of Vitamin B 1 and B 2 in Complex Cereal Foods, by Reverse Phase Isocratic HPLC-UV." *Journal of Cereal Science* 55 (3). Elsevier Ltd: 293–99. doi:10.1016/j.jcs.2011.12.011.

Rubert Josep, Zachariasova Milena, and Hajslova Jana. 2015. "Advances in High-Resolution Mass Spectrometry Based on Metabolomics Studies for Food – A Review." *Food Additives and Contaminants - Part A Chemistry, Analysis, Control, Exposure and Risk Assessment* 32 (10): 1685–1708. doi:10.1080/19440049.2015.1084539.

Santos Joana, Mendiola Jose A., Oliveira Maria B.P.P., Ibáñez Elena, and Herrero Miguel. 2012. "Sequential Determination of Fat- and Water-Soluble Vitamins in Green Leafy Vegetables during Storage." *Journal of Chromatography A* 1261: 179–88. doi:10.1016/j.chroma.2012.04.067.

Smidt Carsten R., Daniel Jones A., and Clifford Andrew J. 1988. "Gas Chromatography of Retinol and α-Tocopherol without Derivatization." *Journal of Chromatography B: Biomedical Sciences and Applications* 434 (1 C). Elsevier: 21–29. doi:10.1016/0378-4347(88)80058-6.

Sobhi Hamid Reza, Yamini Yadollah, Esrafili Ali, and Abadi Reza H.H.B. 2008. "Suitable Conditions for Liquid-Phase Microextraction Using Solidification of a Floating Drop for Extraction of Fat-Soluble Vitamins Established Using an Orthogonal Array Experimental Design." *Journal of Chromatography A* 1196–1197 (1–2): 28–32. doi:10.1016/j.chroma.2008.05.005.

Souverain Sandrine, Rudaz Serge, and Veuthey Jean-luc. 2004. "Matrix Effect in LC-ESI-MS and LC-APCI-MS with off-Line and on-Line Extraction Procedures." *Journal of Chromatography A* 1058: 61–66. doi:10.1016/j.chroma.2004.08.118.

Staggs, Cathleen G., Sealey Wendy M., McCabe Beverlry J., Teague April M., and Mock Donald M. 2004. "Determination of the Biotin Content of Select Foods Using Accurate and Sensitive HPLC/Avidin Binding." *Journal of Food Composition and Analysis* 17 (6): 767–76. doi:10.1016/j.jfca.2003.09.015.

Sundarram Ajita, and Murthy Thirupathihalli P.K. 2014. "α-Amylase Production and Applications: A Review." *Journal of Applied & Environmental Microbiology* 2 (4). Science and Education Publishing: 166–75. doi:10.12691/jaem-2-4-10.

Szterk Arkadiusz, Roszko Marek, Malek Krystian, Czerwonkaa Małgorzata, Waszkiewicz-Robak Bożena. 2012. "Application of the SPE Reversed Phase HPLC/MS Technique to Determine Vitamin B12 Bio-Active Forms in Beef." *Meat Science* 91: 408–13. doi:10.1016/j.meatsci.2012.02.023.

Taguchi Kaori, Fukusaki Eiichiro, and Bamba Takeshi. 2014. "Simultaneous Analysis for Water- and Fat-Soluble Vitamins by a Novel Single Chromatography Technique Unifying Supercritical Fluid Chromatography and Liquid Chromatography." *Journal of Chromatography A* 1362. Elsevier B.V.: 270–77. doi:10.1016/j.chroma.2014.08.003.

Tayade Amol B., Dhar Priyanka, Kumar Jatinder, Sharma Manu, Chaurasia Om P., and Srivastava Ravi B. 2013. "Sequential Determination of Fat- and Water-Soluble Vitamins in Rhodiola Imbricata Root from Trans-Himalaya with Rapid Resolution Liquid Chromatography/Tandem Mass Spectrometry." *Analytica Chimica Acta* 789. Elsevier B.V.: 65–73. doi:10.1016/j.aca.2013.05.062.

Turner Charlotta, King Jerry W., and Mathiasson Lennart. 2001. "On-Line Supercritical Fluid Extraction/Enzymatic Hydrolysis of Vitamin A Esters: A New Simplified Approach for the Determination of Vitamins A and E in Food." doi:10.1021/jf000532z.

Turner Charlotta, Persson Mattias, Mathiasson Lennart, Adlercreutz Patrick, and King Jerry W. 2001. "Lipase-Catalyzed Reactions in Organic and Supercritical Solvents: Application to Fat-Soluble Vitamin Determination in Milk Powder and Infant Formula." *Enzyme and Microbial Technology* 29 (2–3): 111–21. doi:10.1016/S0141-0229(01)00359-3.

U.S. Department of Health and Human Services and U.S. Department of Agriculture. 2020. "2015–2020 Dietary Guidelines for Americans. 8th Edition. December 2015." Accessed June 4. http://health.gov/dietaryguidelines/2015/guidelines/.

Vazquez Raphaël, Rotival Romain, Calvez Sophie, Le Hoang My-Dung, and Do Bernard. 2009. "Stability Indicating Assay Method on Vitamins: Application to Their Stability Study in Parenteral Nutrition Admixtures." 7: 629–35. doi:10.1365/s10337-009-0979-1.

Vyas Pathik, and O'Kane Anthony A. 2011. "Determination of Vitamin B12 in Fortified Bovine Milk-Based Infant Formula Powder, Fortified Soya-Based Infant Formula Powder, Vitamin Premix, and Dietary Supplements by Surface Plasmon Resonance: Collaborative Study." *Journal of AOAC International* 94 (4): 1217–26. doi:10.1093/jaoac/94.4.1217.

Wang Xu, Li Kefeng, Yao Liping, Wang Chunling, and Schepdael Ann Van. 2018. "Recent Advances in Vitamins Analysis by Capillary Electrophoresis." *Journal of Pharmaceutical and Biomedical Analysis* 147. Elsevier B.V.: 278–87. doi:10.1016/j.jpba.2017.07.030.

Zhang Haixia, Jha Ambuj B., Warkentin Thomas D., Vandenberg Albert, and Purves Randy W. 2018. "Folate Stability and Method Optimization for Folate Extraction from Seeds of Pulse Crops Using LC-SRM MS." *Journal of Food Composition and Analysis* 71. Elsevier Inc.: 44–55. doi:10.1016/j.jfca.2018.04.008.

Zhang Yuan, Zhou Wei-E., Yan Jia-Qing, Liu Min, Zhou Yu, Shen Xin, Ma Ying-Lin, Feng Xue-Song, Yang Jun, and Li Guo-Hui. 2018. "A Review of the Extraction and Determination Methods of Thirteen Essential Vitamins to the Human Body: An Update from 2010." *Molecules*: 1–25. doi:10.3390/molecules23061484.

Zironi Elisa, Gazzotti Teresa, Barbarossa Andrea, Farabegoli Federica, Serraino Andrea, Pagliuca Giampiero, and Bo Emilia. 2014. "Determination of Vitamin B 12 in Dairy Products by Ultra Performance Liquid Chromatography-Tandem Mass Spectrometry Analytical Conditions Sample Preparation" 3: 254–55. doi:10.4081/ijfs.2014.4513.

Zou Yuchen, Duan Hanying, Li Li, Chen Xuju, and Wang Chao. 2019. "Quantification of Polyglutamyl 5-Methyltetrahydrofolate, Monoglutamyl Folate Vitamers, and Total Folates in Different Berries and Berry Juice by UHPLC–MS/MS." *Food Chemistry* 276 (September 2018): 1–8. doi:10.1016/j.foodchem.2018.09.151.

6 Carbohydrate Analysis by Mass Spectrometry

Alexander O. Chizhov
N.D. Zelinsky Institute of Organic Chemistry,
Russian Academy of Sciences

CONTENTS

6.1 Introduction 85
6.2 Routine Mass Spectrometric Procedures in Application to Carbohydrate Analysis 87
 6.2.1 General Remarks 87
 6.2.2 Determination of Glycosylation Points in Oligo- and Polysaccharides (Methylation Analysis) 88
 6.2.3 Profiling of Small Polysaccharides and Oligosaccharide Mixtures 92
 6.2.4 Qualitative and Quantitative HPLC-MS Analysis of Derivatized Carbohydrates 93
 6.2.5 Structure Elucidation of Oligosaccharides and Glycoconjugates by Tandem Mass Spectrometry 94
6.3 Advanced Techniques in Carbohydrate Mass Spectrometry 97
 6.3.1 *De novo* Determination of the Absolute Configuration of Monosaccharides by Tandem Mass Spectrometry 97
 6.3.2 Separation of Isomeric Carbohydrates by IMS 99
 6.3.3 Determination of Anomeric Configurations by IMS and Gas-Phase IR Spectrometry 101
6.4 Conclusions 102
6.5 List of Abbreviations 102
References 103

6.1 INTRODUCTION

Mass spectrometry (MS) of carbohydrates started in the early 1960s. Like in other organic chemistry and biochemistry branches, MS was applied to the analysis and structure elucidation of carbohydrates due to its high sensitivity; this method provides structural information complementary to infrared (IR) and nuclear magnetic resonance (NMR) spectrometry. The earliest works on carbohydrate MS concerning electron ionization (EI) mass spectra of volatile monosaccharide derivatives were summarized in Ref. [1]. In this chapter, the first classification of fragmentation pathways in the EI mass spectra of glycosides has been reported. For more than 50 years, the number of papers dealing with carbohydrate MS has grown enormously (Figure 6.1). As shown in Figure 6.1, a huge increase in the number of mass spectrometric studies of carbohydrates has begun in 1990, when commercial instruments equipped with electrospray ionization (ESI) and matrix-assisted laser desorption-ionization (MALDI) ion sources became commercially available, and this made the investigation of intact (non-derivatized) complex carbohydrates possible. New mass spectrometric studies of carbohydrates appear every day, and of course, it is almost impossible to review all of them. This chapter gives only a survey of some important books, reviews, and tutorials on MS of carbohydrates and some original papers that seem important, interesting, or curious. To systematize

DOI: 10.1201/9781003091226-8

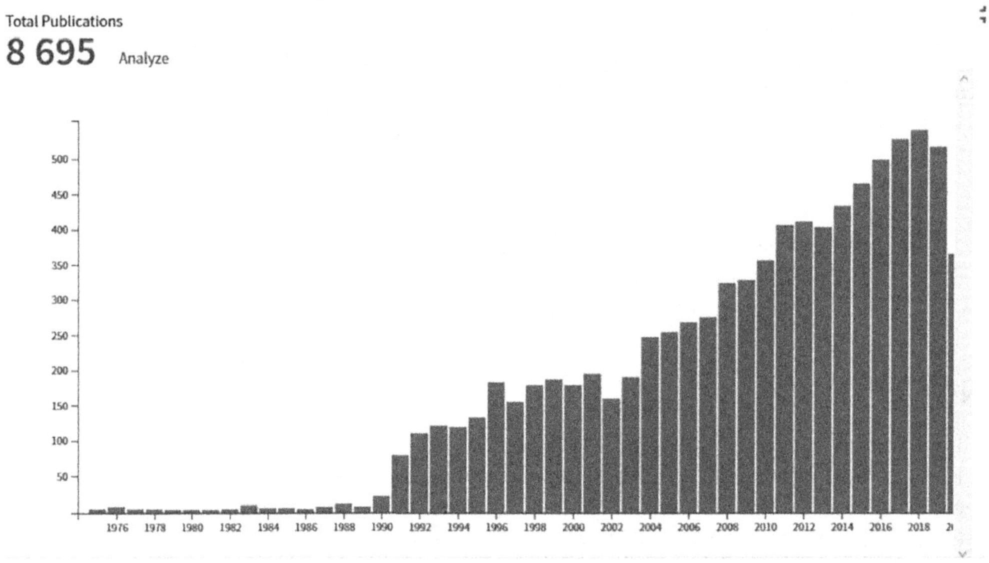

FIGURE 6.1 Diagram of growth of the annual number of publications related to both topics "mass spectr*" and "carbohydr*" (WoS database, Clarivate Analytics, downloaded September 17, 2020). The total no. of publications includes 541 reviews and 120 book chapters).

this tremendous amount of MS information, we have divided MS methods used in the relevant papers into routine (well-developed and widely used) methods and advanced methods (developed in several laboratories or a single research group). We realize that this classification is arbitrarily and temporary: well-developed and reproducible "advanced" procedure may be transformed to "routine" in future, or, in contrast, hardly reproducible and unreliable one may be lost for users, forgotten, and "die." As an example of the latter, one can remember fast-atom bombardment (FAB) ionization widely employed in the 1980s–1990s of the 20th century, which now is almost out of use, being replaced with ESI and MALDI. This is why we do not consider FAB-MS of carbohydrates here, neither as well as high-vacuum chemical ionization (CI), ^{252}Cf plasma desorption, field ionization, and field desorption.

It is needed to discuss what problems of carbohydrate analysis are reasonable to solve using MS approaches. For targeted analysis, for example, monosaccharide composition in materials available in μmole level or higher, chromatography with non-selective detection (GC: flame ionization detector [FID], high-performance liquid chromatography (HPLC): photometric, fluorimetric, or pulsed amperometric detectors) are usually applicable for accurate and reliable measurements. These detectors are substantially cheaper than mass spectrometers. Monosaccharides or their derivatives are fairly identified using their retention times. Quantification procedures are elaborated better for these non-selective detectors than for MS. When mass selection is needed due to chromatographic peaks overlap, or analytes' structures are not known *a priori*, mass spectrometric detection is definitely required.

There are many reviews on applying MS to carbohydrates (see Figure 6.1), and we cannot survey all of them in this chapter. Some useful reviews of general aspects are presented in Refs. [2–13]. Currently updated comprehensive reviews on MALDI-MS analysis of carbohydrates and various glycoconjugates were composed by D.J. Harvey [14–21]. There are many other reviews on particular aspects of the application of MS to the structural analysis of glycans and glycoconjugates, for example, flavonoids [22–24], glycoproteins [25–28], proteoglycans [29–31], plant [32,33], marine [34], and human milk [35] oligosaccharides.

6.2 ROUTINE MASS SPECTROMETRIC PROCEDURES IN APPLICATION TO CARBOHYDRATE ANALYSIS

6.2.1 General Remarks

Carbohydrates are aliphatic, polyhydroxylated carbonyl compounds (glyceraldehyde CHOH homologs and their analogs: amino, deoxy, etc.). They are widespread in nature and present in all living organisms. Classification of carbohydrates includes monosaccharides, oligosaccharides, and polysaccharides. There is no definite board between oligosaccharides and polysaccharides. Monosaccharides have trivial names (glucose, mannose, galactose, xylose, etc.) used in chemical and biochemical literature. Monosaccharides are classified as aldoses (possessing aldehyde group) and ketoses (having internal carbonyl group in a carbon backbone); carbon chain length is reflected in other subdivisions (pentoses, hexoses, etc.). As usual, monosaccharides (monomers) exist in a ring form due to ring-chain tautomerism; the size of the ring may vary. The most frequent are five- and six-membered rings, furanoses, and pyranoses, respectively. Oligosaccharides (oligomers) are formed due to *O*-glycosidic (hemicyclic acetal or ketal) bonds between the carbonyl group of one monosaccharide residue and a hydroxyl group of another one; any free hydroxyl may participate in linkage (giving so-called glycosylation point). Due to multiple glycosylations of a monosaccharide residue, branching of the carbohydrate chain occurs. A new chiral center (anomeric center) brings new uncertainty: it may have two configurations, designated as α and β.

Along with glycosylation, oxygen atoms may be esterified (acetylated, phosphorylated, sulfated, butyrated, etc.) or alkylated (methylated). Compounds in which molecules contain carbohydrate and non-carbohydrate parts are known as glycoconjugates. The chemistry (including nomenclature) and biology of carbohydrates are described in many textbooks and monographs; up to date, the most detailed presentation is given in the four-volume collection [36].

Therefore, the determination of a structure of any complex carbohydrate includes

1. determination of carbohydrate composition (including the absolute configuration of all monosaccharide units);
2. determination of glycosylation points (linkage analysis);
3. sequencing of carbohydrate chains;
4. determination of anomeric configurations of all of the carbohydrate residues; and
5. determination of nature, content, and position(s) of non-carbohydrate components.

If the compound is available in micromole amounts in pure form, one can use one- and two-dimensional NMR methods with no degradation (or minimum, selective degradation). Usually, a complex mixture of oligosaccharides in nano-, pico-, or even femtomole amounts obtained from an object of frontier research is available. Thus, the researcher needs to employ the most sensitive methodology – MS, including hyphenated techniques (for general MS information, see Chapter 1).

Because carbohydrates are polar, non-volatile compounds, two general complementary approaches are evident:

1. to derivatize small carbohydrate components and to use GC-MS-EI/CI and
2. to apply intact (or tagged) components (or their large fragments) to HPLC or CE and to use soft ionization from the condensed phase (ESI and MALDI).

What problems of structure determination of carbohydrates can be solved routinely by MS coupled with chromatography? After total hydrolysis (or other cleavage procedure), the determination of carbohydrate composition can be done usually by chromatography without MS: a carbohydrate mixture may be analyzed intact by HPLC with refractive index detector the form of derivatives (alditol acetates, trimethylsilyl-derivatized alditols) and subjected to GC-FID. Retention times and response factors are tabulated for common monosaccharide derivatives [37,38]. MS is undoubtedly needed if an unknown component is revealed.

6.2.2 Determination of Glycosylation Points in Oligo- and Polysaccharides (Methylation Analysis)

The general approach to the determination of linkage positions in polysaccharides is methylation analysis. Free hydroxyl groups in polysaccharides are transformed to methyl groups using Hakomori (dimethylsulfinyl anion in dimethyl sulfoxide [DMSO], CH_3I) [39] or Kerek (solid NaOH in DMSO, CH_3I) [40] procedures before total hydrolysis. The resulting partially methylated monosaccharides are reduced into alditols (to avoid multiple product formation due to anomeric center) followed by acetylation (Figure 6.2). A mixture of volatile partially methylated alditol acetates (PMAAs) can be analyzed by GC equipped with an FID or mass-selective detector (GC-MS). The latter is preferable; hence, the positions of methyl groups in PMAAs can be determined easily due to known regularities in MS-EI discovered by B. Lindberg's group 5 decades ago [41]. Methylation analysis was extensively developed further [42–45]; see also tutorials [46,47] and reference data on retention times of PMAAs [38,47,48]. Briefly, the pathways of fragmentation of molecular ions of PMAAs are as follows (Figure 6.3):

1. If two methoxyl groups in the carbon backbone are adjacent to each other, the C–C bond between them is ruptured, preferably with the formation of both possible fragment ions.
2. The C–C bond between vicinal methoxyl and acetoxyl groups can also be cleaved; hence, the charge is localized exclusively on a CHOMe side.
3. Rupture of C–C bond between vicinal acetoxyl groups does not occur if at least one methoxyl group in the molecule is present. Such cleavages are realized only for totally acetylated polyols.
4. Secondary fragments are formed by loss of methanol, acetic acid, ketene, and H_2O. Usually, peaks of primary and secondary fragment ions have comparable intensities; moreover, secondary fragments may be more abundant than primary ones.

Some typical mass spectra of PMAAs are presented in Figure 6.4. All of the EI mass spectra of PMAAs are tabulated, and it is easy to determine the structure of any analytically valuable PMAA. The possible linkage analysis problems are listed below, and they are mainly chemical, not mass spectrometric.

1. The most serious problem is undermethylation [49]. Undermethylated hydroxyl groups may be erroneously interpreted as glycosylation points. For avoiding this fault, the polysaccharide sample should be methylated several times until the stable results of GC-FID or GC-MS will be achieved. Oxidative degradation can also occur to a minor extent by methylation in DMSO as a side process [50].
2. "Symmetrization" of methylated carbohydrates under reduction to polyols. Some PMAAs have a symmetric structure. For distinguishing the former anomeric center (e.g., C-1 for aldoses), the reduction is made with $NaBD_4$; single deuterium labeling results in a +1 Da shift for the fragments bearing the former anomeric center [45,51]. Another way is the preparation of partially methylated aldononitrile acetates instead of PMAAs [52]. Evidently, it can be used for aldoses only; the nitrile group is formed from anomeric hemiacetal. Note that these derivatives are much less stable than PMAAs and should be analyzed just after preparation.
3. To avoid uncertainties, uronic acids should be transformed to aldoses; soluble carbodiimide is used to prepare a reducible intermediate [53]. Borodeuteride reduction introduces two deuterium atoms, resulting in a +2 Da shift or the fragments bearing the former carboxyl group.

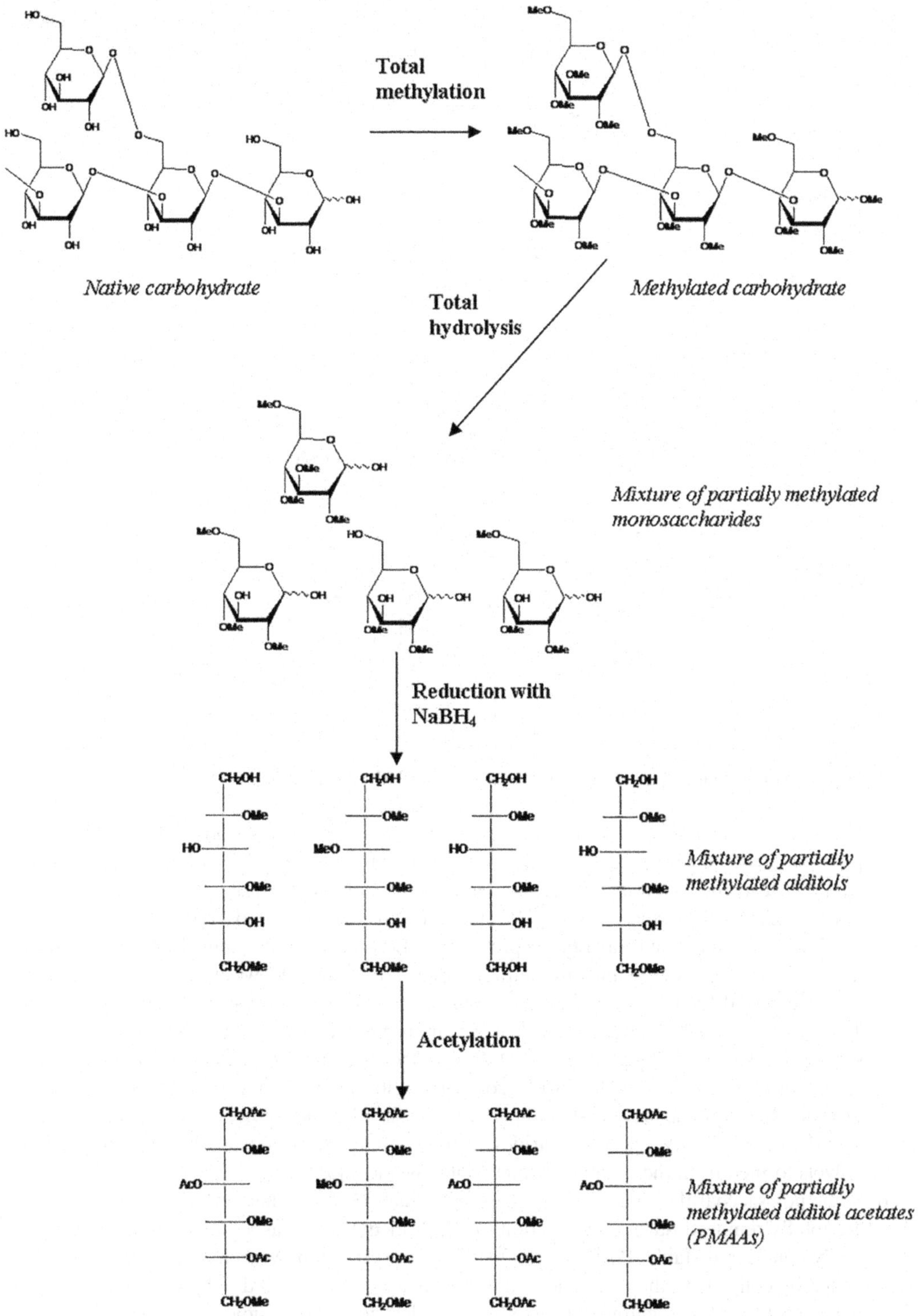

FIGURE 6.2 Scheme of preparation of partially methylated alditol acetates (PMAAs) for linkage analysis.

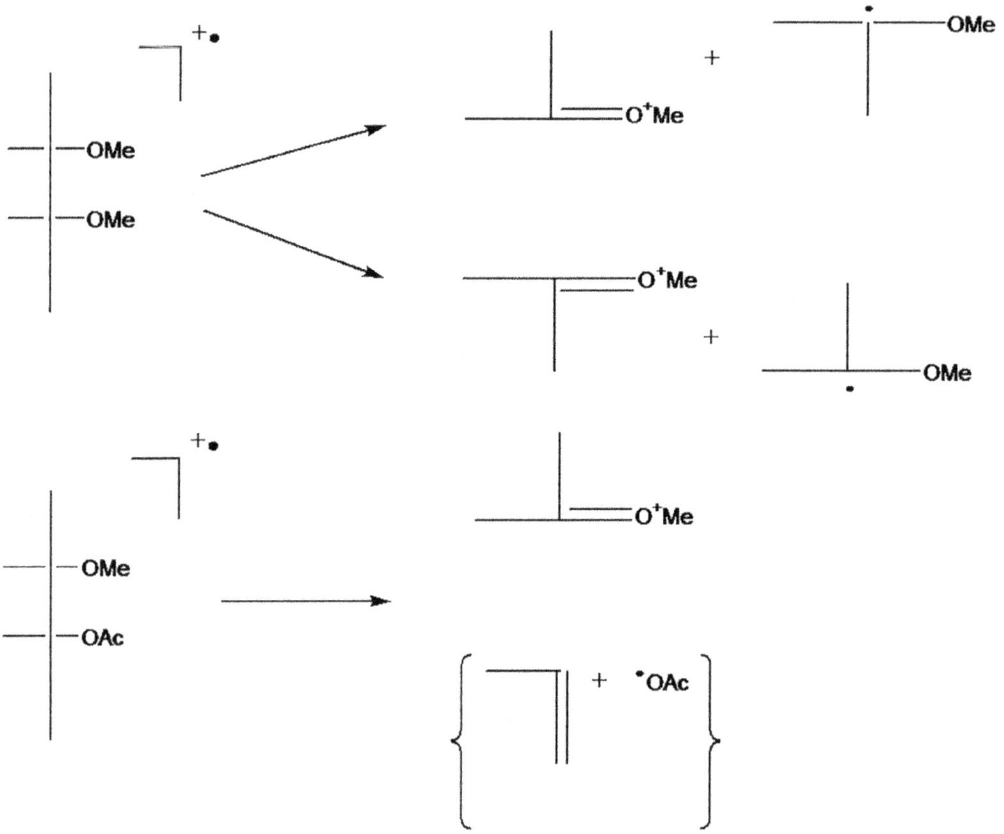

FIGURE 6.3 Fragmentation pathways for molecular ions of PMAAs (scheme).

4. The above procedure cannot answer which acetoxy group corresponds to glycosylation point or cyclic acetal (ketal) bond. So, in the general case, the information on cycle size would be lost. To avoid this problem and retain the size of carbohydrate cycles, G. Gray and coworkers have elaborated the reductive cleavage of methylated oligosaccharides and polysaccharides with triethylsilane in the presence of Lewis acids under anhydrous conditions [54]. Partially methylated anhydroalditol acetates are formed. However, this procedure is laborious and difficult to reproduce since only a few research groups used it episodically.
5. To distinguish methoxy groups that can occur in the native polysaccharide, it is necessary to use deuteromethylation with CD_3I (every trideuteromethyl group gives +3 Da shift) [47,52] or ethylation (+14n Da shift in comparison to methylation, n is a number of ethyl groups). Recently, the GC-MS-based procedure was validated for quantification of 3- and 4-OMe hexoses in algal polysaccharides, borodeuteride reduction was used to distinguish polyols formed from these naturally methylated isomers [55].
6. Some polysaccharides contain *O*-ester groups. Acetoxyl and other carboxyl ester groups are totally removed during methylation in alkali media, whereas sulfate groups are generally considered stable. Positions of *O*-sulfates in sulfated polysaccharides may be estimated by comparing the methylation data for intact and desulfated polysaccharides [56]. Ester substituents are retained during reductive cleavage of glycosidic bonds [54].

Linkage analysis of nitrogen-containing carbohydrates has some peculiarities due to the N(Me)Ac group formation. *N*-Acetyl hexosamines give *N*-methyl-*N*-acetyl PMAAs; their EI-MS spectra are similar to those of PMAAs (tabulated in Ref. [44]). For EI-MS of acetylated/methylated derivatives

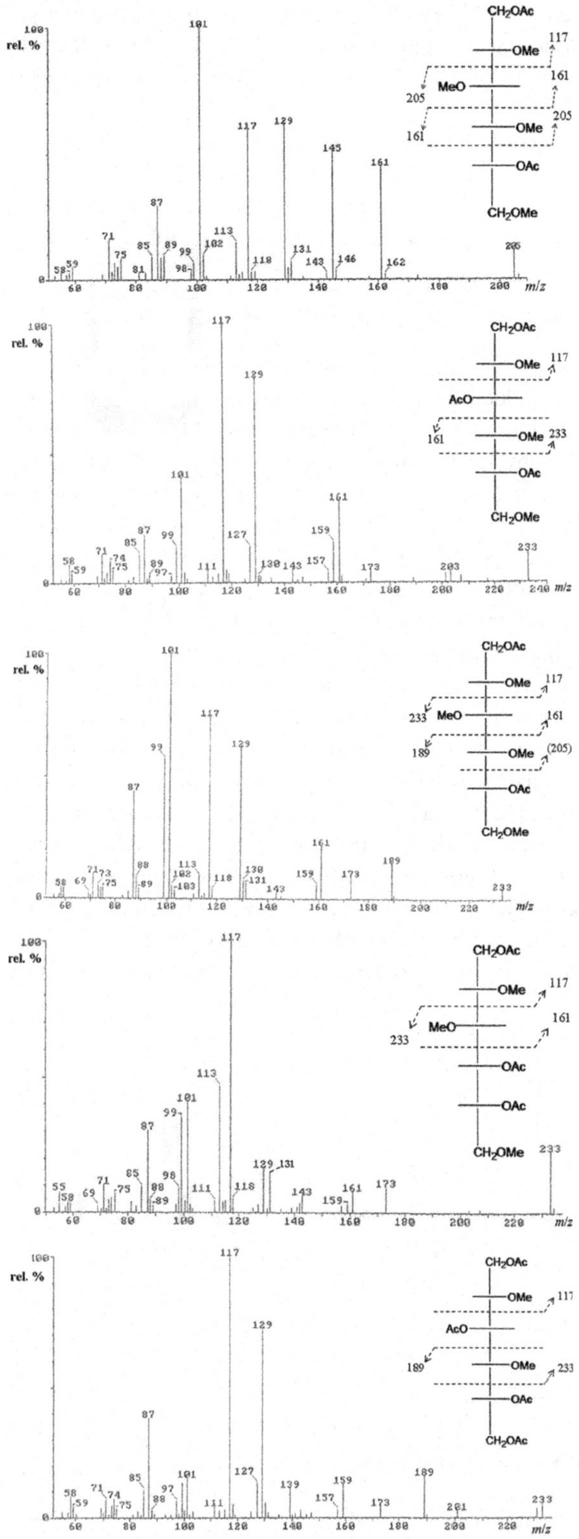

FIGURE 6.4 EI mass spectra of some PMAAs (fragmentations are described in insets). For more examples, see Ref. [51].

of neuraminic acids, see Ref. [57]. MS of aminoalditols obtained by reduction of Schiff bases reversibly formed from monosaccharides and amines are also studied. Conjugation with chiral amine tag makes it possible to distinguish D- and L-isomers of monosaccharides (including methylated) using achiral GC columns [58].

As was mentioned above, the transformation of monosaccharides to alditols is necessary. Although *O*-methylated methyl glycosides were the first carbohydrate derivatives studied systematically [1], they are scarcely used in carbohydrate analysis because of the formation of both pyranoside and furanoside forms; the result strongly depends on methylation conditions [59].

6.2.3 Profiling of Small Polysaccharides and Oligosaccharide Mixtures

Analysis of complex carbohydrates usually includes the estimation of the distribution of components according to their molecular mass. Such profiles can be obtained for intact or derivatized carbohydrates using MS. There are several advantages of MS in comparison to size-exclusion (gel permeation) chromatography (SEC): this is a much more rapid, sensitive, and robust method; it provides real mass-to-charge (*m/z*) values (in contrast to averaged relative volumes of solvated molecules in SEC) of analytes; the order of *m/z* is not reversed (in contrast to exclusion volumes in SEC); and, the last, not the least: by using MS, molecular masses can be calculated accurately. For profiling of carbohydrates, the most frequently used is MALDI–time of flight (ToF) (for the theory of MALDI ionization, see Chapter 1). Although 2,5-dihydroxybenzoic acid (DHB) is usually used for the analysis of neutral carbohydrates, other matrices are also used for different carbohydrate analytes (a choice of a matrix is discussed in Refs. [12,14,15,21,25]). In a positive ion mode, peaks of $[M+H]^+$ or $[M+Met]^+$ (where Met is an ion of metal) ions are observed, and in negative ion mode, $[M-H]^-$ ions are formed. An example of a typical profile of a small polysaccharide, a MALDI-ToF spectrum of laminaran isolated from brown alga *Chorda filum*, is given in Figure 6.5. Each major $[M+Na]^+$ peak is accompanied with a minor $[M+K]^+$ one (16 Th higher) [60].

To distinguish glycoforms, similar profiling is applied for various glycans: lipopolysaccharides and their fragments from Gram-negative bacteria [61], triterpene glycosides [62], *N*-linked glycans [6,12,14–21,25], *O*-linked glycans [12,14–21] of glycoproteins, intact glycoproteins [12], intact proteoglycans and their fragments [29–31], etc. Plant oligosaccharides and glycosides are also widely studied by MALDI-MS; see the corresponding references tabulated in Refs. [14–21].

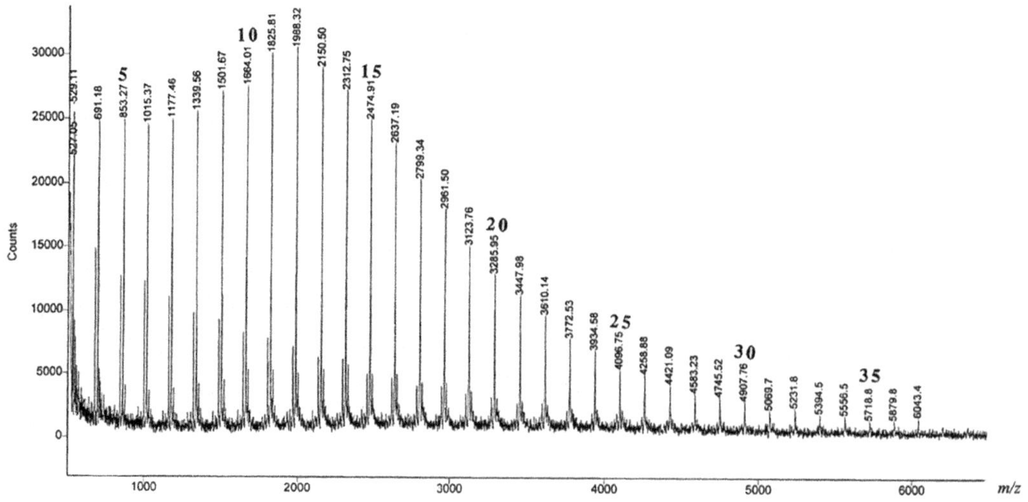

FIGURE 6.5 Profiling of small polysaccharide: MALDI-ToF mass spectrum of the native laminaran from brown seaweed *Chorda filum*, positive ion mode, DHB matrix [60]. (Reproduced with permission of Elsevier.)

ESI (for theory, see Chapter 1) is also used for profiling of glycans [6, 25–35]. However, there are more discrimination factors than those in MALDI-ToF (the most prominent is discrimination of low and/or high masses in an orthogonal accelerator in quadrupole time-of-flight [Q-ToF] instruments) along with the formation of multiply charged ions; hence, a deconvolution procedure is required [63]. Methylated glycans have stronger ESI mass spectra signals than those of corresponding intact carbohydrates [63,64]. Other derivatizations are also often used for ESI-MS analysis of carbohydrates to increase sensitivity and improve quantification (Section 6.2.4).

6.2.4 Qualitative and Quantitative HPLC-MS Analysis of Derivatized Carbohydrates

Chromatographic and electrophoretic methods are widely used for the qualitative and quantitative determination of carbohydrates and glycoconjugates in various complex matrices, including food and beverages [65–69]; mass-selective detection is often used along with spectrophotometric, potentiometric, and other methods of detection (for recent reviews on separations of carbohydrates, see also Ref. [70–73]). To decrease the limits of detection (LOD), increase selectivity, and make quantification more accurate and reliable, various derivatizations of carbohydrate analytes are applied [74,75]. The most often used approach of tagging (adding an easily revealed moiety, i.e., a label) is reductive amination: the reducing carbohydrate is reacted with an excess of amine bearing a functional group, and the resulting Schiff base is reduced with sodium cyanoborohydride or amine–borane complex affording secondary amine (Figure 6.6) [7,75,76 and references therein]. 2-Aminopyridine, aminonaphthalene sulfonic acids, substituted aminobenzenes, and aminoacridone derivatives are the most known reagents for tagging via reductive amination.

There are many chromophore and fluorescent groups (R in Figure 6.6) that are used in carbohydrate analysis; most of them facilitate the formation of ions under MALDI and/or ESI, thus lowering LOD. The best way to make carbohydrates highly visible in ESI and MALDI-MS is to attach a group with a preformed (or permanent) charge center. The fact that ions are readily formed makes one sure that the MS signal of an analyte is independent of a molecule's association with a proton or metal ion (in a positive ion mode) or deprotonation (in a negative ion mode). Such a charge center ("permanent charge") can be introduced by reductive amination with 2-aminoethyltrimethylammonium and 3-amino-2-hydroxypropyltrimethylammonium salts (Figure 6.6, above) or hydrazone formation with Girard's T reagent (Figure 6.7) [76]. For the latter derivatization procedure, good quantification was achieved using ESI and MALDI-ToF instruments for the mixture of maltooligosaccharides [7].

The main problem of reductive amination tagging is incomplete derivatization due to reversible formation of Schiff base; reduction of an open form of aldose affords tag-free alditol irreversibly. Other derivatization approaches yield tagged derivatives almost quantitatively. One is bis-pyrazolone formation from reducing carbohydrate by reaction with 2-pyrazoline-5-ones in alkaline medium. 1-Phenyl-3-methyl-2-pyrazoline-5-one (PMP) group is a strong chromophore, and the corresponding derivatives were analyzed by ESI and MALDI-MS [74]. An example of successful application of PMP derivatization followed by LC-MS detection, measuring of tiny amounts of reducing sugars (10–500 ng/mL) in soil extracts from crop rhizospheres is described in Ref. [77].

Intact oligosaccharides can also be estimated quantitatively by ESI and MALDI. Quantitative aspects of the latter method of ionization were studied [12,78]. Surprisingly, the best results of profiling and quantitation of native glycan chains of glycoproteins were obtained for commonly used DHB as a matrix. Good linearity was achieved from 100 fmol and higher, with no ion signals saturation [78]. Sialic acid residues are often labile during MALDI, so selective protection (esterification) is needed [79]. HPLC-ESI-MS quantitative measurements of non-derivatized glucans are also carried out [80]. However, these measurements in complex matrices such as food products [65] and beverages [68] need caution because of discrimination effects, both positive and negative; for general review and discussion, see Ref. [81]. It is perfectly possible to carry out HPLC-ESI-MS quantitative determinations in MS/MS mode by multiple reaction monitoring; general principles of tandem mass spectra of oligosaccharides and glycoconjugates are discussed in Section 6.2.5.

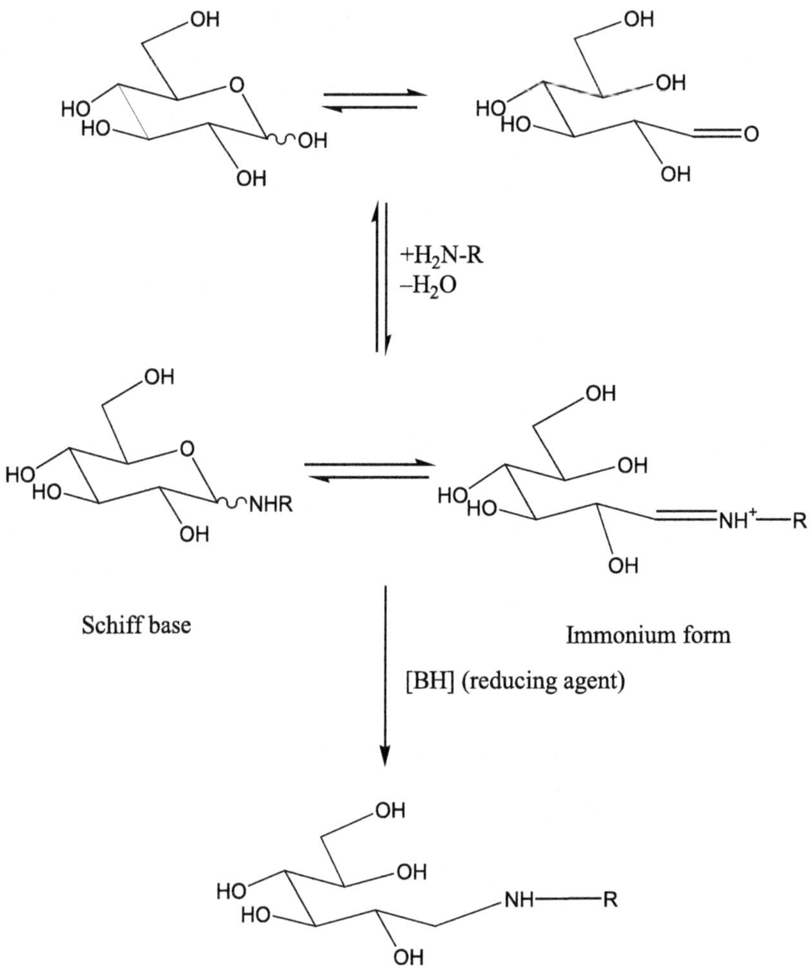

FIGURE 6.6 Tagging of a reducing carbohydrate. R is a chromophore or ionogenic group [76]. (With permission from Elsevier.)

6.2.5 Structure Elucidation of Oligosaccharides and Glycoconjugates by Tandem Mass Spectrometry

General principles of fragmentation of even-electron ions of oligosaccharides were formulated by B. Domon and C. Costello in 1988 [82] (Figure 6.8); despite these regularities were evaluated from FAB-MS, they were found to be applicable for tandem mass spectra of ions generated from oligosaccharides and glycoconjugates by ESI and MALDI and are widely applied in carbohydrate structural analysis [2–35,51,62,64,65,76,78,80,83–87, and many references therein].

In brief, a symbolic representation of fragments in Domon-Costello nomenclature is as follows (Figure 6.8):

1. Fragments that contain "non-reducing" ends are designated by capital letters A, B, and C.
2. Fragments that contain "reducing" ends are designated by capital letters X, Y, and Z.
3. B, C, Y, and Z are glycosidic cleavages, and A and X are cross-ring cleavages.
4. Location of bond cleavages is signed by subscripts and superscripts. Subscript shows the residue cleaved. The enumeration starts from "non-reducing" end for A, B, and C, and from "reducing" end for X, Y, and Z. An ion of non-carbohydrate aglycone is formalized as Y_0.

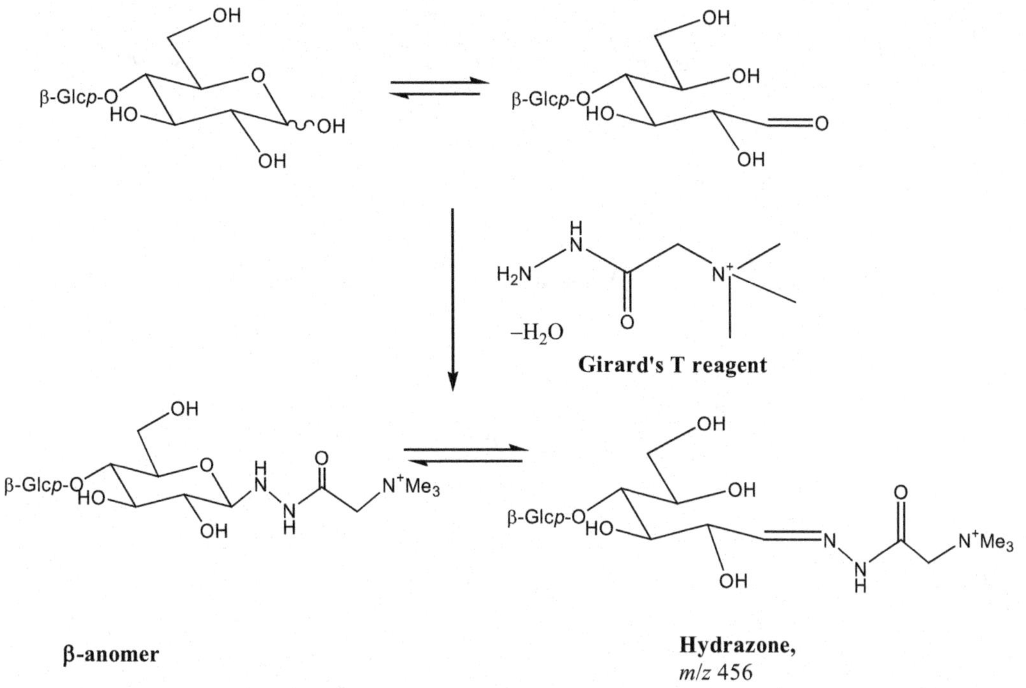

FIGURE 6.7 Tagging of reducing disaccharide (cellobiose) by hydrazone formation of with Girard's T reagent [76]. (Reproduced with permission from Elsevier.)

For cross-residue cleavages, ruptured bonds are signed by superscripts. Bond numbers are assigned as shown in Figure 6.8b and separated by a comma.

5. For branched oligosaccharides and glycosides, "antennae" chains (attached to unbranched "core") are represented by Greek letters (α, β, γ, etc., range by size, α is the largest fragment); further branching is depicted by superscripts in subscripts (prime, ‘, bis, “, tris, “‘, and so on).

It is problematic to classify cleavages in cyclic oligosaccharides (cyclodextrins, cyclofructans, etc.) because there are no "ends" in a cycle, although the chemistry of ion cleavages is similar to that of linear and branched, non-cyclic carbohydrates; an attempt to describe tandem mass spectrometry (MS/MS) of cyclic carbohydrates in terms of Domon-Costello nomenclature leads to confusion and contradictions, see discussion in Ref. [88].

Meanwhile, for most practically valuable glycans, MS/MS can be formalized and explained in Domon-Costello terms. The most important glycans are *N*- and *O*-linked carbohydrate chains of various glycoproteins. Up to date, structure elucidation of carbohydrate chains of glycoproteins is well established, and MS/MS sequencing is a substantial part of the related standard protocol [87]. In brief, isolated glycoprotein (or a mixture of glycopeptides obtained by digestion with endopeptidase) is subjected to PNGase F digestion. The resulting *N*-glycans may be treated with selective exoglycosidases giving glyco fragments by controlled residue loss. *N*-Deglycosylated protein (or peptides) is (are) subjected to reductive cleavage with hydrazine/$NaBH_4$ to release *O*-glycans (in the form of glycosyl polyols). After permethylation, *N*- and *O*-glycans are studied by MALDI-ToF-MS (profiling, Section 6.2.3) and ESI-MS/MS (linkage analysis). Aliquots of glycans are transformed into PMAAs and analyzed by GC-MS (methylation analysis, Section 6.2.2). All three MS analyses give complementary information. An alternative approach for gel permeation chromatography (GPC) separated protein includes successive decoloration, reduction/carboxymethylation of S–S

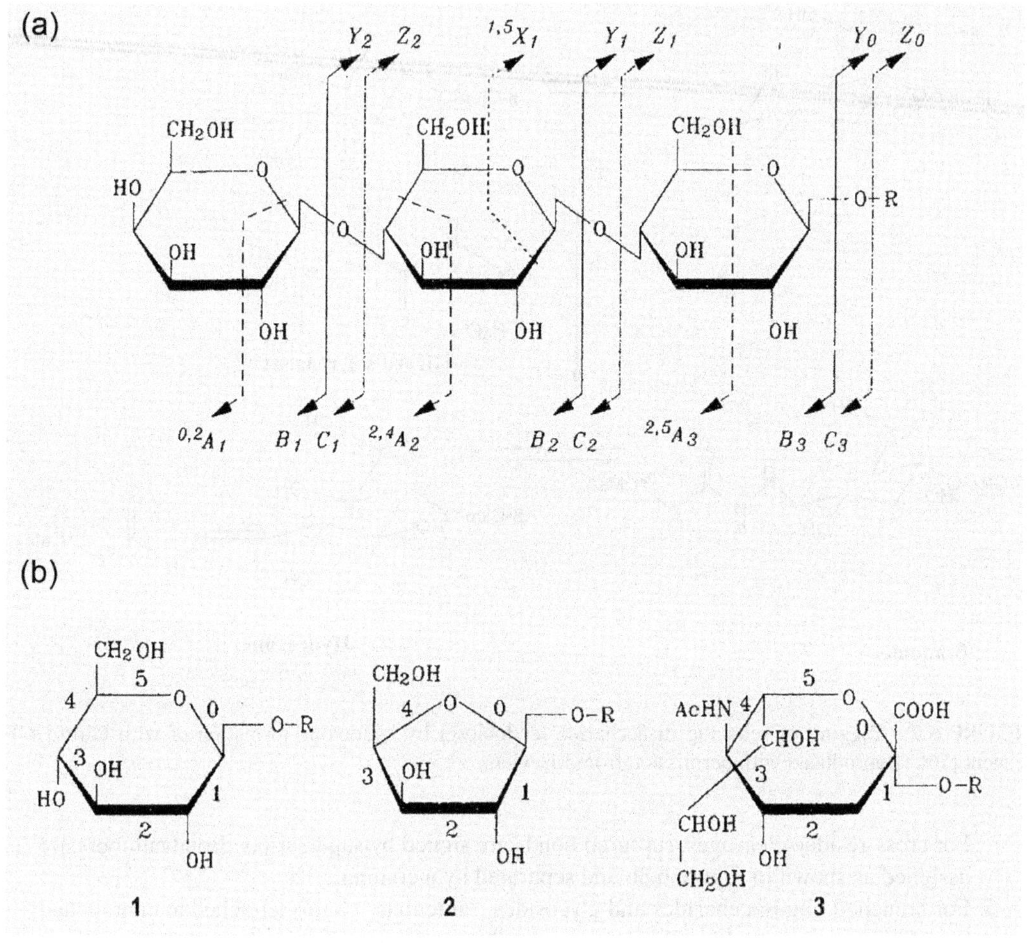

FIGURE 6.8 A scheme of cleavages of oligosaccharide ions in tandem mass spectra by B. Domon and C.E. Costello [82]. (a) Types of cleavages (an arbitrarily taken trisaccharide glycoside is pictured, R is a noncarbohydrate aglycone or hydrogen). (b) Enumeration of bonds in carbohydrate residues; 1: hexopyranoside, 2: hexofuranoside, 3: sialoside. (Reproduced with permission from Springer.)

bonds, PNGase F digestion, chromatographic purification of *N*-glycans, and separation of them on non–porous-graphitized column followed by MALDI-MS(/MS) and nano-ESI-MS/MS studies [87]. Linkage analysis of intact, totally methylated, or specifically tagged glycans now is considered worldwide as a routine procedure that can be done in an automatic manner using specialized computer programs and carbohydrate data banks [3,4,89–91 and refs. therein].

However, it is necessary to keep in mind that the MS patterns of complex carbohydrates strongly depend on the ionization method and polarity mode [2,3,5,7,9–12,51]. For example, oligosaccharides containing acidic groups (carboxylate, phosphate, or sulfate) are readily analyzed using negative ion mode [2,3,5,7]. In MS/MS, MS^n patterns depend on the ion chosen for activation and the method of activation. In contrast to $[M+H]^+$, $[M+Met]^+$ ions produce both glycosidic and cross-ring fragments yielding more structural information (branching, glycosylation points) [2,3,7]. Derivatized oligosaccharides display different fragmentation patterns compared to their underivatized counterparts [2,3,7]. For carbohydrates and glycoconjugates, ion activation is not restricted by low-energy collision-induced dissociation (CID), the most commonly employed fragmentation method implemented in many different MS platforms. Other methods are as follows: higher-energy collisional dissociation (HCD, successfully used to achieve cross-ring cleavages of $[M+Met]^+$), IR multiphoton dissociation (IRMPD), electron-capture dissociation (ECD), electron transfer dissociation (ETD), electron detachment dissociation (EDD), negative electron transfer dissociation (NETD), negative electron-capture dissociation (niECD). EDD and NETD are suited for multiply charged negative ions generated from acidic oligosaccharides and glycopeptides. ECD and ETD are used for fragmentation of multiply charged positive ions. They are useful for glycoprotein structural studies since cleave peptide bonds remaining glycan-peptide glycosidic bonds intact [3]. Ultraviolet photodissociation is used for structural studies of sialylated oligosaccharides and glycopeptides [3]. The combination of the above activation methods provides complementary structural information. For ion activation and fragmentation theory, see Chapter 1.

The general problem in interpreting MS is rearrangements; these are real pitfalls for a researcher seeking easy ways in structure determination. MS of carbohydrates is not an exception: rearrangements were revealed for protonated molecules of permethylated, peracetylated, and intact oligosaccharides. For review, see Ref. [92]. The most often rearrangement type is an internal residue loss (IRL). Other types are trans-antennae carbohydrate residue migration, phosphate migration, and non-carbohydrate internal fragment loss. In an updated Domon-Costello nomenclature, rearranged ions are marked with an asterisk (*) [92]. Until now, no theory can predict the rearrangements of carbohydrate ions. IRLs were not observed for negative ions (deprotonated molecules). Previously, it was considered that they are unlikely for metallated molecules $[M+Met]^+$ [92 and references therein]. However, a single example of IRL for $[M+Na]^+$ was found recently [93]. Internal loss of alkoxyl fragment in alkoxy side chains in cyclic *N*-substituted oligo-β-(1→6)-D-glucosamines was also observed [94].

6.3 ADVANCED TECHNIQUES IN CARBOHYDRATE MASS SPECTROMETRY

6.3.1 *De novo* Determination of the Absolute Configuration of Monosaccharides by Tandem Mass Spectrometry

Most carbohydrates (except symmetric alditols) are chiral compounds, and their absolute configuration is an integral part of the structural description of complex carbohydrates. Routine MS and MS/MS are insusceptible to chirality and cannot simultaneously discriminate diastereomers and enantiomers of a complete set of isomeric monosaccharides. Routinely used chiral derivatizations followed by chromatographic separation of resulting diastereomers [58 and references therein] are time-consuming and laborious. Discrimination of monosaccharides with enantiomers by MS/MS has been performed recently using the fixed ligand kinetic method (FLKM) [95,96]. In brief, the principles of molecular distinguishing in FLKM are as follows. Initial trimeric ion complex formed in the ESI ion source is $[M^{II}(A)(ref)(FL{-}H)]^+$, where M^{II} is a divalent metal cation, A is an analyte (i.e., a monosaccharide), ref is the chiral reference molecule, and FL is the fixed ligand (in this Section, designations are taken from Refs. [95,96]). The fixed ligand molecule should be easily deprotonated and, simultaneously, strongly bound with M^{II} and not be lost under CID. Only two fragmentation pathways are possible (Figure 6.9).

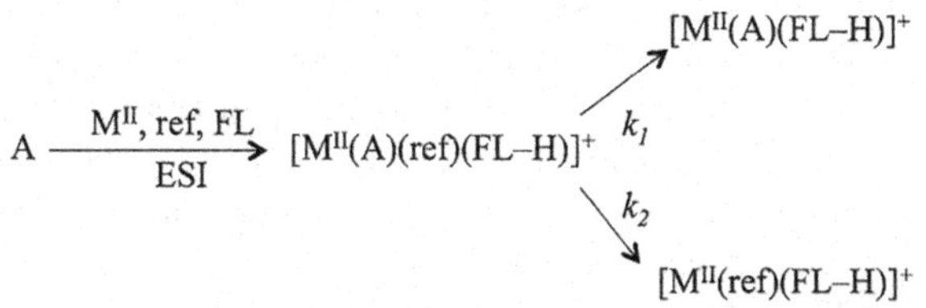

FIGURE 6.9 Formation of divalent metal-bound trimeric ion complex *via* ESI and subsequent fragmentation through CID in FLKM [95]. (Reproduced with permission of ACS Publications.)

(a)

$$R_{fixed} = \frac{[M^{II}(A)(FL\text{-}H)]^{+}}{[M^{II}(ref)(FL\text{-}H)]^{+}}$$

$$R_{D\text{-}fixed} = \frac{[M^{II}(A_D)(FL\text{-}H)]^{+}}{[M^{II}(ref)(FL\text{-}H)]^{+}}$$

$$R_{L\text{-}fixed} = \frac{[M^{II}(A_L)(FL\text{-}H)]^{+}}{[M^{II}(ref)(FL\text{-}H)]^{+}}$$

$$R_{chiral\text{-}fixed} = \frac{R_{D\text{-}fixed}}{R_{L\text{-}fixed}}$$

(b)

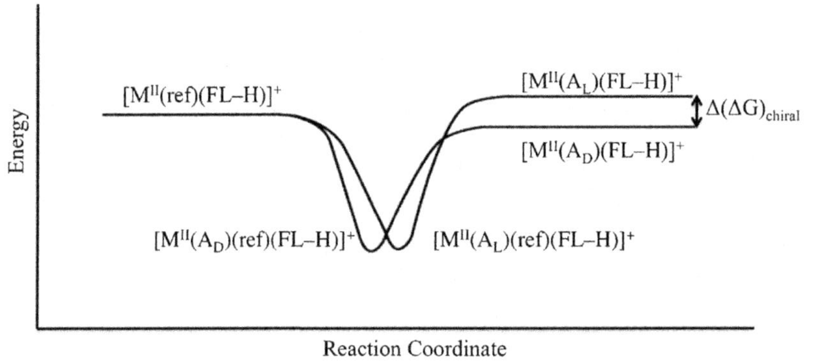

$$\ln\left(R_{chiral\text{-}fixed}\right) = \frac{\Delta(\Delta G)}{RT_{eff}}$$

FIGURE 6.10 (a) Equations describing discrimination of enantiomers in FLKM [95]. (b) Free energy diagram of FLKM that illustrates energetic differences of diastereomeric fragment ion complexes and the corresponding thermodynamic equation which connects $R_{chiral\text{-}fixed}$ value with Δ(ΔG) of CID MS/MS fragments of the trimeric ion complexes. (Reproduced from Ref. [95] with permission of ACS Publications.)

The deprotonation site is confined to FL, which avoids the formation of various isomeric structures. Branching ratio, R_{fixed} is a ratio of relative intensities of the corresponding ion abundances of the fragments (in brackets, Figure 6.10a). Similarly, both ratios are obtained for enantiomers A_D and A_L. Chiral discrimination between an enantiomeric pair can be quantified using $R_{chiral\text{-}fixed}$ term (Figure 6.10a). The latter value is unique for each pair of monosaccharide enantiomers and reflects the difference in Gibbs' free energy Δ(ΔG) of diastereomeric fragments of enantiomers (Figure 6.10b) [95,96].

The application of FLKM is illustrated in Figure 6.11. Analytes are D- and L-lyxose, Ni^{II} is the divalent metal cation, L-aspartic acid is ref, and guanosine monophosphate (5′GMP) is FL. The starting triple complexes are shown in the left lower corner of the scheme, the reference fragment

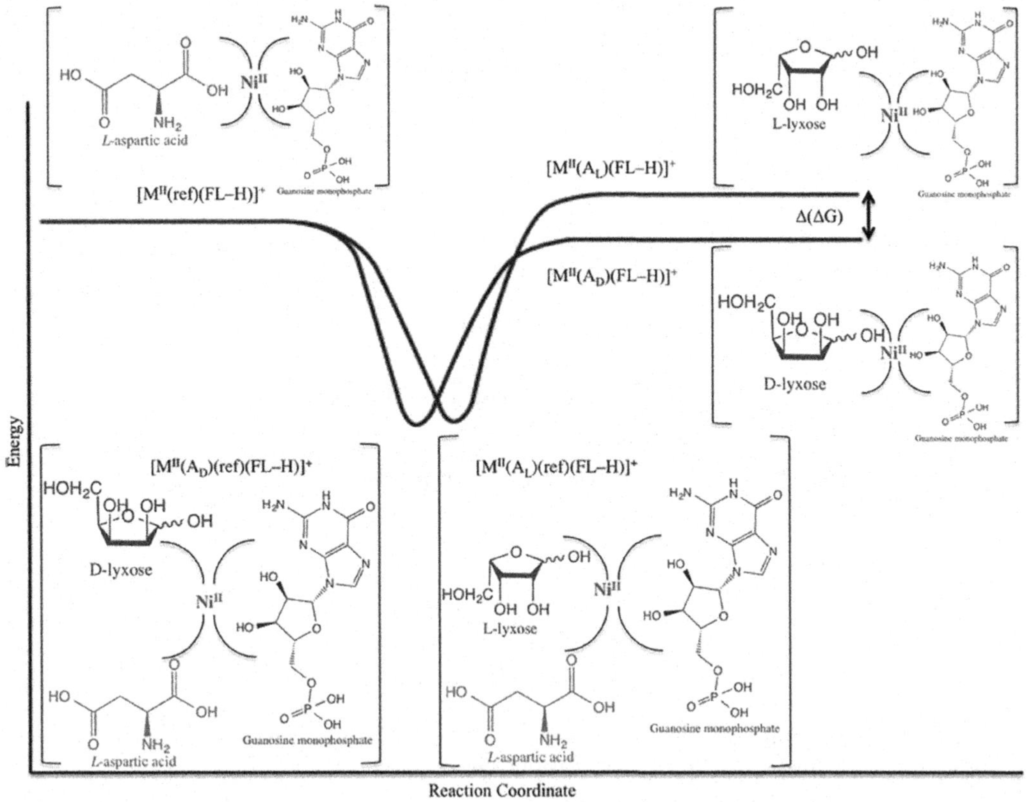

FIGURE 6.11 Free energy diagram depiction in FLKM. For description, see text. (Reproduced from Ref. [96] with permission of ACS Publications.)

is in the left upper corner, and structures of diastereomeric, analyte-containing fragment ions are placed in the right part.

For creating a reliable, non-overlapping set of data, at least two series of trimeric ion complexes must be studied. The results of these studies are presented in two 2D plots. Natural logarithms of R_{fixed} are plotted in abscissa for the first series of triple complexes and in the ordinate for the second one (Figures 6.12 and 6.13). For 24 hexoses (16 aldoses and 8 ketoses), the triple complexes are chosen as follows: (i) Cu^{II} is a divalent metal cation, L-serine is ref, 5′GMP is FL (Figure 6.12, abscissa); (ii) Mn^{II} is a divalent metal cation, L-asparagine is ref, a dipeptide L-phenylalanine-glycine is FL (Figure 6.12, ordinate) [95]. For 12 pentoses (8 aldoses and 4 ketoses), the triple complexes are composed of (i) Cu^{II} as a divalent cation metal, L-serine as ref, 5′GMP as FL (Figure 6.13, abscissa); (2) Ni^{II} as a divalent metal cation, L-asparagine as ref, 5′GMP as FL (Figure 6.13, ordinate) [96]. These two plots contain reference data which could achieve *de novo* identification of all monosaccharide residues in oligo- and polysaccharides. However, there are no more recent data confirming the interlaboratory reproducibility of this methodology. In Ref. [95,96], an orbitrap instrument was applied; no data are available yet on other MS analyzers types.

6.3.2 Separation of Isomeric Carbohydrates by IMS

During chromatographic separation, structurally similar, isomeric compounds are poorly separable due to peak overlap. For improving the separation, ion mobility spectrometry (IMS) was coupled with MS, and a new, hyphenated technique, IM-MS, was created about 3 decades ago. IM-MS is now extensively developed, and it is also used for the separation of isomeric carbohydrates; for

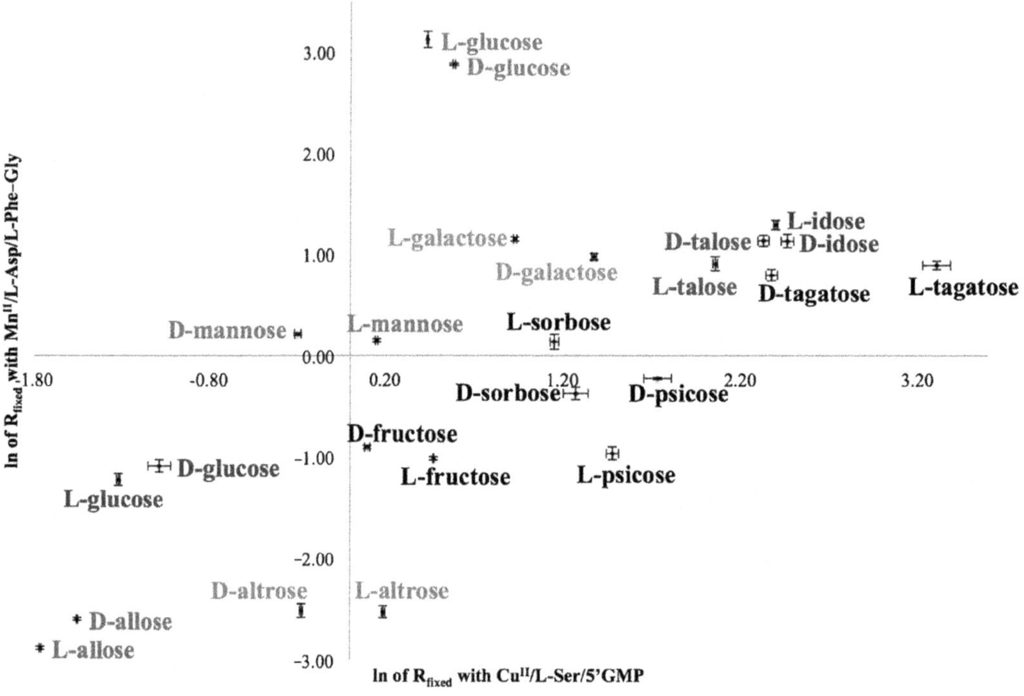

FIGURE 6.12 Two-dimensional plot of ln R_{fixed} for two series of hexose-bearing triple complexes. For details, see text. (Reproduced from Ref. [95] with permission of ACS Publications.)

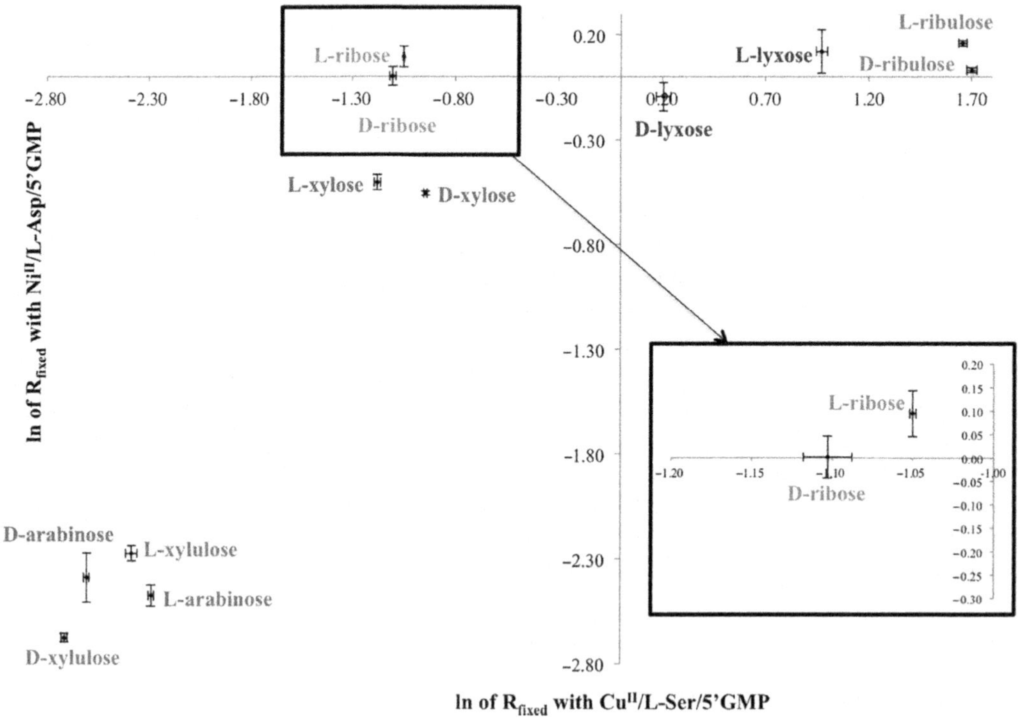

FIGURE 6.13 Two-dimensional plot of ln R_{fixed} for two series of pentose-bearing triple complexes. For details, see text. (Reproduced from Ref. [96] with permission of ACS Publications.)

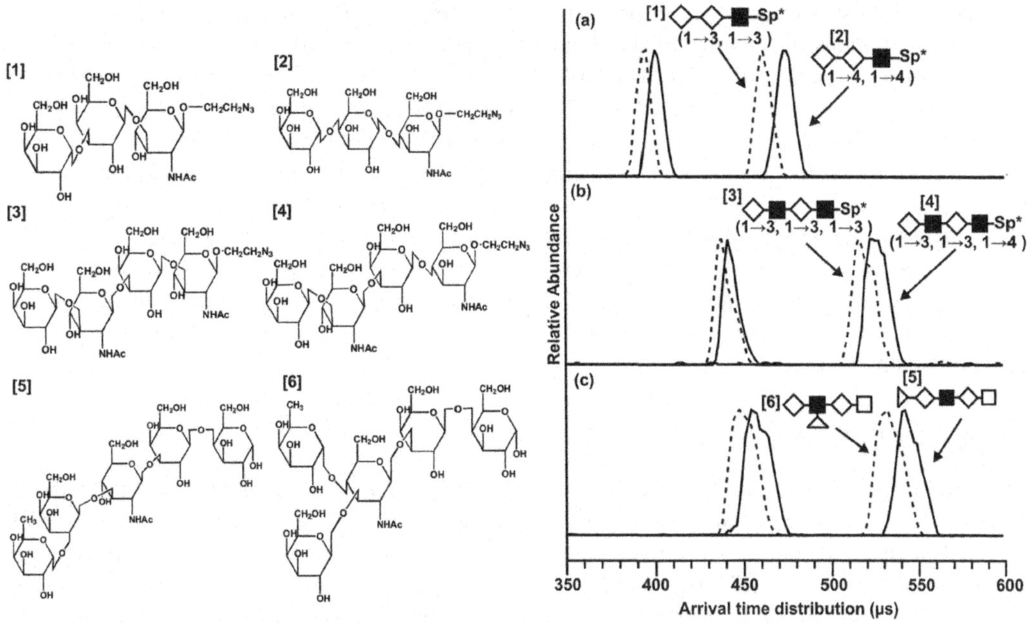

FIGURE 6.14 Separation of isomeric $[M+Na]^+$ ions of oligosaccharides and glycosides by drift tube IMS [103]. (Reproduced with permission of RSC.)

recent reviews, see Ref. [97–102]. In contrast to MS, where their *m/z* values separate ions, the separation of ions in IMS proceed according to the time these ions need to traverse a cell filled with an inert gas under the influence of a weak electric field. This drift time depends on the molecules' shape and size, thus helping to distinguish isomers. Up to date, there are four types of IM-MS commonly used for the analysis of organic compounds, including carbohydrates. These are drift tube, traveling wave, high-field asymmetric waveform ion mobility spectrometry (FAIMS), and trapped ion mobility spectrometry (TIMS). Along with constructive differences, these techniques differ from each other by the nature of the electric field that is used to propel the ions through the ion mobility cell (for a visual representation of these four primary types of IMS systems and IMS theory, see Ref. [98,100,102]). IM-MS is compatible with both ESI and MALDI ion sources. In Figure 6.14, IMS distinguishing three pairs of carbohydrate isomers is presented as an example.

Drift times are related to the ion structure in that larger, more elongated ions experience more collisions with the neutral buffer gas present in the drift cell, causing a longer drift time than that for more compact structures [103]. These spatial characteristics are expressed in averaged values known as the collision cross section. The advantages of combining either pre-IM or post-IM fragmentation before MS analysis are demonstrated for enhanced confidence in carbohydrate identification [103].

IM-MS is a promising technique for glycomics to identify and differentiate the three-dimensional structure of chemically similar carbohydrates and glycoconjugates [100]. Mass spectrometers with IM cells are now commercially available. However, despite numerous works exploiting IM-MS, at the moment, it cannot be regarded as a routine method due to difficulties in operation and indefinite inter-laboratory reproducibility of IM-MS data.

6.3.3 Determination of Anomeric Configurations by IMS and Gas-Phase IR Spectrometry

As mentioned above, routine MS^n is typically blind to stereochemical information, especially for anomeric configurations. Rare exceptions were exclusively mentioned by separate publications; for example, see Ref. [104]. During fragmentation of oligosaccharides, fragments derived from anomers may have only a slight difference in properties. Recently, several methodologies were created to reveal tiny variations in fragmentations of anomers (so-called "anomeric memory") measured by custom-built instruments [105–108]. For CID MS/MS, such differences were revealed in IM and gas-phase IR spectra of fragments [105–107]. For IRMPD activation/gas-phase IR spectroscopy registration, a tunable laser was used to scan excitation [108]. Differences in IR spectra of fragments are too small due to their high temperature. The cooling procedure with superfluid helium drops injection allowed to cool them to temperatures less than 1 K ("cryogenic messenger-tagged IR spectroscopy" coupled with ultrahigh-resolution IMS); very good coincidence with reference IR spectra was achieved, [106 and references therein]. Of course, such sophisticated methodologies can be reproduced only with unique, custom-built instruments. The use of routine instrumentation for regio- and stereoisomeric oligosaccharides [109] looks more promising, and it will be reproduced and validated in the future.

6.4 CONCLUSIONS

MS of carbohydrates is a rapidly expanding area of research. High throughput hybrid MS methodologies of analysis of femto- and even attomole amounts of material in conjunction with powerful data analysis methods put forward tremendous glycomic projects [109,110] and make it possible to score biosimilarity by MS [111]. In conjunction with separation techniques and NMR [112], MS is used adaptively for structural analysis of various polysaccharides [113]. Single-cell membrane glycomic profiling [114], monitoring of lysosomal storage disorders [115], characterizing visible and invisible plant mutant phenotypes [116], vaccines [27], viruses [26], and bacteria [62] – these are some analytical applications of MS, and MS of carbohydrates occupies a substantial part in it. In the nearest future, the advances of MS of carbohydrates will continue to grow exponentially in parallel to the progress of mass spectrometric instrumentation and techniques.

6.5 LIST OF ABBREVIATIONS

A	analyte
Ac	acetyl
AcOH	acetic acid
APCI	atmospheric-pressure chemical ionization
CCS	collision cross section
CE	capillary electrophoresis
CI	chemical ionization
Da	dalton
DHB	2,5-dihydrohybenzoic acid
DMSO	dimethyl sulfoxide
ECD	electron-capture dissociation
EDD	electron detachment dissociation
ETD	electron transfer dissociation
EI	electron ionization
ESI	electrospray ionization
Et	ethyl
FID	flame ionization detector
FL	fixed ligand
FLKM	fixed ligand kinetic method
GC	gas chromatography
5′GMP	guanosine monophosphate
IMMS	ion mobility – mass spectrometry
IMS	ion mobility spectrometry

IR infrared
IRL internal residue loss
IRMPD infrared multiphoton dissociation
LOD limit of detection
MALDI matrix-assisted laser desorption/ionization
Me methyl
MeOH methanol
Met metal, metal ion
MRM multiple reaction monitoring
MS mass spectrometry
MS^n tandem mass spectrometry (n means an order of spectra)
MS/MS second-order tandem mass spectrometry (synonym of MS^2)
NETD negative electron transfer dissociation
niECD negative electron-capture dissociation
PMAAs partially methylated alditol acetates
PMP 1-phenyl-3-methyl-2-pyrazoline-5-one
PNGase F specific glycosidase used for elimination of glycan *N*-chains
ref reference molecule
SEC size-exclusion chromatography
Th Thompson (*m/z* unit)
TMS trimethylsilyl
ToF time-of-flight (analyzer)
UVPD ultraviolet photodissociation

REFERENCES

1. Kochetkov, N.K.; Chizhov, O.S. Mass spectrometry of carbohydrate derivatives. *Adv. Carbohydr. Chem.*, 1966, Vol. 21, pp. 39–93.
2. Zaia, J. Mass spectrometry of oligosaccharides. *Mass Spectrom. Rev.*, 2004, Vol. 23, No. 3, pp. 161–227.
3. Kailemia, M.J.; Ruhaak, L.R.; Lebrilla, C.B.; Amster, I.J. Oligosaccharide analysis by mass spectrometry: A review of recent developments. *Analyt. Chem.*, 2014, Vol. 86, No. 1, pp. 196–212. doi: 10.1021/ac403969n.
4. Mutenda, K.E.; Matthiesen, R. Analysis of carbohydrates by mass spectrometry. *Methods Mol. Biol.*, 2007, Vol. 367, pp. 289–301.
5. Sekiya, S.; Iida, T. Glycan analysis by mass spectrometry. *Trends Glycosci. Glycotechnol.*, 2008, Vol. 20, No. 111, pp. 51–65.
6. Dell, A.; Jang-Lee J.; Pang, P.-C.; Parry, S.; Sutton-Smith, M.; Tissot, B.; Morris, H.R.; Panico, M.; Haslam, S.M. Glycomics and mass spectrometry. In: *Glycoscience*, Fraser-Reid B., Tatsua K., Thiem J. (eds). Springer-Verlag, Berlin-Heidelberg, 2008. pp. 2192–2217.
7. Mischnik, P. Mass spectrometric characterization of oligo- and polysaccharides and their derivatives. *Adv. Polym. Sci.*, 2012, Vol. 248, pp. 105–174. doi: 10.1007/12_2011_134.
8. Compton, B.J.; Siuzdak, G. Mass spectrometry in nucleic acid, carbohydrate and steroid analysis. *Spectroscopy*, 2003, Vol. 17, No. 4, pp. 699–713.
9. Han, L.; Costello, C.E. Mass spectrometry of glycans. *Biochemistry (Moscow)*, 2013, Vol. 78, No. 7, pp. 710–720.
10. Park, Y.; Lebrilla, C.B. Application of Fourier transform ion cyclotron resonance mass spectrometry to oligosaccharides. *Mass Spectrom. Rev.*, 2005, Vol. 24, pp. 232–264.
11. Ruhaak, L.R.; Xu, G.; Li, Q.; Goonatilleke, E.; Lebrilla, C.B. Mass spectrometry approaches to glycomic and glycoproteomic analyses. *Chem. Rev.*, 2018, Vol. 118, pp. 7886–7930.
12. Harvey, D.J. Matrix-assisted laser desorption/ionization mass spectrometry of carbohydrates and glycoconjugates. *Int. J. Mass Spectrom.*, 2003, Vol. 226, pp. 1–35.
13. Reinhold, V.N.; Reinhold, B.B.; Chan, S. Carbohydrate sequence analysis by electrospray ionization–mass spectrometry. *Meth. Enzymol.*, 1996, Vol. 271, pp. 377–402.
14. Harvey, D.J. Matrix-assisted laser desorption/ionization mass spectrometry of carbohydrates. *Mass Spectrom. Rev.*, 1999, Vol. 18, No. 6, pp. 349–450. doi: 10.1002/(SICI)1098-2787(1999)18:6<349::AID-MAS1>3.0.CO;2-H.

15. Harvey, D.J. Analysis of carbohydrates and glycoconjugates by matrix-assisted laser desorption/ionization mass spectrometry: An update covering the period 1999–2000. *Mass Spectrom. Rev.*, 2006, Vol. 25, No. 4, pp. 595–662. doi: 10.1002/mas.20265.
16. Harvey, D.J. Analysis of carbohydrates and glycoconjugates by matrix-assisted laser desorption/ionization mass spectrometry: An update covering the period 2001–2002. *Mass Spectrom. Rev.*, 2008, Vol. 27, No. 2, pp. 125–201. doi: 10.1002/mas.20157.
17. Harvey, D.J. Analysis of carbohydrates and glycoconjugates by matrix-assisted laser desorption/ionization mass spectrometry: An update for the period 2005–2006. *Mass Spectrom. Rev.*, 2011, Vol. 30, No. 1, P. 1–100.
18. Harvey, D.J. Analysis of carbohydrates and glycoconjugates by matrix-assisted laser desorption/ionization mass spectrometry: An update for 2007–2008. *Mass Spectrom. Rev.*, 2012, Vol. 31, No. 2, pp. 183–311.
19. Harvey, D.J. Analysis of carbohydrates and glycoconjugates by matrix-assisted laser desorption/ionization mass spectrometry: An update for 2009–2010. *Mass Spectrom. Rev.*, 2015, Vol. 34, pp. 268–422.
20. Harvey, D.J. Analysis of carbohydrates and glycoconjugates by matrix-assisted laser desorption/ionization mass spectrometry: An update for 2011–2012. *Mass Spectrom. Rev.*, 2016, Vol. 36, pp. 255–422.
21. Harvey, D.J. Analysis of carbohydrates and glycoconjugates by matrix-assisted laser desorption/ionization mass spectrometry. An update for 2013–2014. *Mass Spectrom. Rev.*, 2018, Vol. 37, pp. 353–491.
22. Vukics, V., Guttman, A. Structural characterization of flavonoid glycosides by multi-stage mass spectrometry. *Mass Spectrom. Rev.*, 2010, Vol. 29, No. 1, pp. 1–16.
23. De Villiers, A.; Venter, P; Pasch, H. Recent advances and trends in the liquid chromatography–mass spectrometry analysis of flavonoids. *J. Chromatogr. A*, 2016, Vol. 1430, pp. 16–78. doi: 10.1016/j.chroma.2015.11.077.
24. Kachlicki, P.; Piasecka, A.; Stobiecki, M.; Marczak, L. Structural characterization of flavonoid glycoconjugates and their derivatives with mass spectrometric techniques. *Molecules*, 2016, Vol. 21, No. 1494, pp. 1–21. doi: 10.3390/molecules21111494.
25. Harvey, D.J. Structural determination of *N*-linked glycans by matrix-assisted laser desorption/ionization and electrospray ionization mass spectrometry. *Proteomics*, 2005, Vol. 5, pp. 1774–1786.
26. Harvey, D.J. Mass spectrometric analysis of glycosylated viral proteins. *Exp. Rev. Proteom.*, 2018, Vol. 15, No. 5, pp. 391–412. doi: 10.1080/14789450.2018.1468756.
27. Sharma, V.K.; Sharma, I.; Glick, J. The expanding role of mass spectrometry in the field of vaccine development. *Mass Spectrom. Rev.*, 2020, Vol. 39, No. 1, pp. 83–104. doi: 10.1002/mas21571.
28. Liu, Y.; Cao, Q.; Li, L. Isolation and characterization of glycosylated neuropeptides. *Meth. Enzymol.*, 2019, Vol. 626, pp. 147–202.
29. Zhang, F.; Zhang, Z.; Linhardt, R.L. Glycosaminoglycans. In: *Handbook of Glycomics*, Cummings R.D., Pierce J.M. (eds.) Academic Press – Elsevier, Amsterdam, 2009, pp. 59–80.
30. Zaia, J. Glycosaminoglycan glycomics using mass spectrometry. *Mol. Cell. Proteom.*, 2013, pp. 885–892. DOI: 10.1074/mcp.R112.026294.
31. Wang, Z.; Chi, L. Recent advances in mass spectrometry analysis in low molecular weight heparins. *Chin. Chim. Lett.*, 2018, Vol. 29, No. 1, pp. 11–18. doi: 10.1016/j.cclet.2017.08.050.
32. Matamoros Fernandes, L.E. Introduction to ion trap mass spectrometry: Application to the structural characterization of plant oligosaccharides. *Carbohydr. Polym.*, 2007, Vol. 68, pp. 797–807.
33. Jorge, T.F.; Rodrigues, J.A.; Caldana, C.; Schmidt, R.; van Dongen, J.T.; Thomas-Oates, J.; Antonio, C. Mass spectrometry-based plant metabolomics: Metabolite responses to abiotic stress. *Mass Spectrom. Rev.*, 2016, Vol. 35, No. 5, pp. 620–649.
34. Lang, Y.; Zhao, X.; Liu, L.; Yu, G. Application of mass spectrometry to structural analysis of marine oligosaccharides. *Mar. Drugs*, 2014, Vol. 12, pp. 4005–4030. doi: 10.3390/md12074005.
35. Grabarics, M.; Csernak, O.; Balogh, R.; Beni, S. Analytical characterization of human milk oligosaccharides – potential applications in pharmaceutical analysis. *J. Pharm. Biomed. Anal.*, 2017, Vol. 146, pp. 168–178.
36. Ernst, B., Hart, G.W., Sinay, P. (eds.) *Carbohydrates in Chemistry and Biology*. Part I *Chemistry of Saccharides*, Vols. 1 and 2, Part II: *Biology of Saccharides*, Vols. 3 and 4. Wiley-VCH, Weinheim, 2000.
37. Wrostad, R.E., Acree, T.E., Decker, E.A., Penner, M.H., Reid, D.S., Schwarz, S.J., Shoemaker, C.F., Smith, D.M., Sporns, P. (eds.) *Handbook of Food Analytical Chemistry*. Wiley Interscience, Hoboken, NJ, 2005. Section E: Carbohydrates. pp. 647–756.
38. Churms, S.C., Zweig, G., Sherma, J. (eds.) *CRC Handbook of chromatography. Carbohydrates*. Vol. I. CRC Press, Boca Raton, FL, 1982.

39. Hakomori, S. A rapid permethylation of glycolipid and polysaccharide catalyzed by methylsulfinyl carbanion in dimethyl sulfoxide. *J. Biochem.*, 1964, Vol. 55, pp. 205–208.
40. Ciucanu, I.; Kerek, F. A simple and rapid method for permethylation of carbohydrates. *Carbohydr. Res.*, 1984, Vol. 131, pp. 209–217.
41. Bjőrndal, H.; Hellerqvist, C.G.; Lindberg, B.; Svensson, S. Gas-liquid chromatography and mass spectrometry in methylation analysis of polysaccharides. *Angew. Chem. Internat. Ed.* 1970, Vol. 9, No. 8, pp. 610–619. doi: 10.1002/anie.197006101.
42. Lindberg, B.; Lonngren, J. Methylation analysis of complex carbohydrates: General procedure and application for sequence analysis. *Methods Enzymol.*, 1978, Vol. 50, pp. 3–33.
43. McNeil, M.; Darvill, A.G.; Aman, P.; Franzen, L.E; Albersheim, P. Structural analysis of complex carbohydrates using high-performance liquid chromatography, gas chromatography, and mass spectrometry. *Methods Enzymol.*, 1982, Vol. 83, pp. 3–45.
44. Wong, C.G.; Sung, S.-S.; Sweely, C.C. Analysis and structural characterization of amino sugars by gas-liquid chromatography and mass spectrometry. *Meth. Carbohydr. Chem.*, 1980, Vol. 8, pp. 55–65.
45. Carpita, N.C.; Shea, E.M. Linkage structure of carbohydrates by gas chromatography – mass spectrometry (GC-MS) of partially methylated alditol acetates. In: *Analysis of carbohydrates by GLC and MS.*, Biermann, C.J.; McGinnis, G.D., Ed., CRC, Boca Raton, FL, 1988, pp. 157–216.
46. Pettolino, F.A.; Walsh, C.; Fincher, G.B.; Bacic, A. Determining of polysaccharide composition of plant cell walls. *Nature Protocols*, 2012, Vol. 7, pp. 1590–1607.
47. Sims, I.M.; Carnachan, S.M.; Bell, T.J.; Hinkley, S.F.R. Methylation analysis of polysaccharides: Technical advice. *Carbohyd. Polym.*, 2018, Vol. 188, pp. 1–7.
48. Wang, Z.-F.; He, Y.; Huang, L.-J. An alternative method for the rapid synthesis of partially *O*-methylated alditol acetate standards for GC-MS analysis of carbohydrates. *Carbohydr. Res.*, 2007, Vol. 342, pp. 2149–2151.
49. Needs, P.W.; Selvendran, R.R. An improved methylation procedure for the analysis of complex polysaccharides including resistant starch and the critique of the factors which lead to undermethylation. *Phytochem. Anal.*, 1993, Vol. 4, pp. 210–216.
50. Ciucanu, I.; Costello, C.E. Elimination of oxidative degradation during the per-*O*-methylation of carbohydrates. *J. Am. Chem. Soc.*, Vol. 125, No. 52, pp. 16213–16219.
51. Sassaki, G.L.; Souza, L.M. Mass spectrometry strategies for structural analysis of carbohydrates and glycoconjugates. In: *Tandem Mass Spectrometry – Molecular Characterization*, Coelho, A.V.; de Matos Ferraz Franco, C., Ed., Intech Open Science, Rijeka, Croatia, 2013, pp. 81–115.
52. Seymour, F.R.; Plattner, R.D.; Slodki, M.E. Gas-liquid chromatography – mass spectrometry of methylated and deuteromethylated per-*O*-acetyl-aldononitriles from D-mannose. *Carbohydr. Res.*, 1975, Vol. 44, No. 2, pp. 181–198.
53. Taylor, R.L.; Conrad, H.E. Stoichiometric depolymerization of polyuronides and glycosaminoglycuronans to monosaccharides following reduction of their carbodiimide-activated carboxyl groups. *Biochemistry*, 1972, Vol. 11. No. 8, pp. 1383–1388.
54. Gray, G.R. Linkage analysis using reductive cleavage method. *Meth. Enzymol.* 1990, Vol. 193, P. 573–587. doi: 10.1016/0076-6879(90)93439-R.
55. Pfeifer, L.; Classen, B. Validation of a rapid GC-MS procedure of quantitative distinction between 3-*O*-methyl and 4-*O*-methyl-hexoses and its application to a complex carbohydrate sample. *Separations*, 2020, Vol. 7, 42. doi: 10.3390/separations7030042.
56. Chizhov, A. O.; Dell, A.; Morris, H. R.; Haslam, S. M.; McDowell, R.A.; Shashkov, A.S.; Nifant'ev, N.E.; Khatuntseva, E.A.; Usov, A.I. A study of fucoidan from the brown seaweed *Chorda filum*. *Carbohydr. Res.*, 1999, Vol. 320, pp. 108–119.
57. Kochetkov, N.K.; Chizhov, O.S.; Kadentsev, V.I.; Smirnova, G.P.; Zhukova, I.G. Mass spectra of acetylated derivatives of sialic acids. *Carbohydr. Res.*, 1973, Vol. 27, No. 1, pp. 5–10.
58. Cases, M.R.; Cerezo, A.S.; Storz, C.A. Separation and quantitation of enantiomeric galactoses and their mono-O-methylethers as their diastereomeric acetylated 1-deoxy-1-(2-hydroxypropylamino)alditols. *Carbohydr. Res.*, 1995, Vol. 269, pp. 333–341.
59. Asres, D.D.; Perreault, H. Monosaccharide permethylation products for gas chromatography – mass spectrometry: How reaction conditions can influence isomeric ratios. *Can. J. Chem.*, 1997, Vol. 75, pp. 1385–1392.
60. Chizhov, A.O.; Dell, A.; Morris, H.R.; Reason, A.J.; Haslam, S.M.; McDowell, R.A.; Chizhov, O.S.; Usov, A. I. Structural analysis of laminarans by MALDI and FAB mass spectrometry. *Carbohydr. Res.*, 1998, Vol. 310, No. 3, pp. 203–210.

61. Banoub, J.H.; El Aneed, A.; Cohen, A.M.; Joly, N. Structural investigation of bacterial lipopolysaccharides by mass spectrometry and tandem mass spectrometry. *Mass Spectrom. Rev.*, 2010, Vol. 29, pp. 606–650. doi: 10.1002/mas.20258.
62. Popov, R.S.; Ivanchina, N.V.; Silchenko, A.S.; Avilov, S.A. Kalinin, V.I.; Dolmatov, I.Y.; Stonik, V.A.; Dmitrenok, P.S. Metabolite profiling of triterpene glycosides of the Far Eastern sea cucumber *Eupantacta fraudatrix* and their distribution in various body components using LC-ESI QTOF-MS. *Marine Drugs*, 2017, Vol. 15, 302. doi: 10.3390/md15100302.
63. Read, S.M.; Currie, G.; Bacic, A. Analysis of structural heterogeneity of laminarin by electrospray-ionisation mass spectrometry. *Carbohydr. Res.*, 1996, Vol. 281, pp. 187–201.
64. Dell, A.; Reason, A.J.; Khoo, K.H.; Panico, M.; Mc Dowell, R.A.; Morris, H.R. Mass spectrometry of carbohydrate-containing biopolymers. *Meth. Enzymol.*, 1994, Vol. 230, pp. 108–132.
65. Di Stefano, V.; Avellone, G.; Bongiorno, D.; Cunsolo, V.; Muccili, V.; Sforza, S.; Dossena, A; Drahos, L.; Vekey, K. Application of liquid chromatography – mass spectrometry for food analysis. *J. Chromatogr. A*, 2012, Vol. 1259, pp. 74–85. doi: 10.1016/j.chroma.2012.04.023.
66. Acunha, T.; Ibanes, C.; Garcia-Canas, V.; Simo, C.; Cisuentes, A. Recent advances in the application of capillary electromigration methods for food analysis and foodomics. *Electrophoresis*, 2016, Vol. 37, pp. 111–141.
67. Anderson, H.E.; Santos, I.C.; Hildenbrandt, Z.L.; Shug, K.A. A review of the analytical methods used for beer ingredient and finished product analysis and quality control. *Anal. Chim. Acta*, 2019, Vol. 1085, pp. 1–20.
68. Ahmed, N.; Mirshekar, B.; Kennish, L.; Karachalias, N.; Babaei-Jadidi, R; Thornalley, P.J. Assay of advanced glycation end products in selected beverages and food by liquid chromatography with tandem mass spectrometric detection. *Mol. Nutr. Food Res.*, 2005, Vol. 49, pp. 691–699.
69. Harrubini, G.; Appelblad, P.; Maietta, M.; Papetti, A. Hydrophilic interaction chromatography in food matrices analysis: An updated review. *Food Chemistry*, 2018, Vol. 257, pp. 53–66.
70. Nagy, G.; Peng, T.; Pohl, N.L.B. Recent liquid chromatographic approaches and developments for the separation and purification of carbohydrates. *Anal. Methods*, 2017, Vol. 9, pp. 3579–3593. doi: 10.1039/c7ay01094j.
71. Hajba, L.; Csansky, E.; Guttman, A. Liquid phase separation methods for *N*-glycosylation analysis of glycoproteins of biomedical and biopharmaceutical interest. A critical review. *Analyt. Chim. Acta*, 2016, Vol. 943, pp. 8–16.
72. Losacco, G.L.; Veuthey, J.-L.; Guillarme, D. Supercritical fluid chromatography – mass spectrometry: Recent evolution and current trends. *TrAC*, 2019, Vol. 118, pp. 731–738.
73. Lu, G.; Crichfield, C.L.; Gattu, S.; Veltri, L.M.; Holland, L.A. Capillary electrophoresis separations of glycans. *Chem. Rev.*, 2018, Vol. 118, pp. 7867–7885. doi: 10.1021/acs.chemrev.7b00669.
74. Lamari, F.N.; Kuhn, R.; Karamanos, N.K. Derivatization of carbohydrates for chromatographic, electrophoretic and mass spectrometric structure analysis. *J. Chromatogr. B*, 2003, Vol. 793, pp. 15–36.
75. Harvey, D.J. Derivatization of carbohydrates for analysis by chromatography, electrophoresis, and mass spectrometry. *J. Chromatogr. B*, 2011, Vol. 879, pp. 1196–1225. doi: 10.1016/j.chromb.2010.11.010.
76. Unterieser, I.; Mischnick, P. Labeling of oligosaccharides for quantitative mass spectrometry. *Carbohydr. Res.*, 2011, Vol. 346, No. 1, pp. 68–75.
77. McRae, G., Monreal, C.M., LC-MS/MS quantitative analysis of reducing carbohydrates in soil solutions extracted from crop rhizospheres. *Anal. Bioanal. Chem.*, 2011, Vol. 400, pp. 2205–2215.
78. Harvey D.J. Quantitative aspects of the matrix-assisted laser desorption mass spectrometry of complex oligosaccharides. *Rapid Commun. Mass Spectrom.*, 1993, Vol. 7, No. 7, pp. 614–619.
79. Reiding, K.R.; Blank, D.; Kuijper, D.M.; Deeler, A.M.; Wuhrer, M. High-throughput profiling of protein *N*-glycosylation by MALDI-TOF-MS employing linkage-specific sialic acid esterification. *Anal. Chem.*, 2014, Vol. 86, pp. 5784–5793. doi: 10.1021/ac500335t.
80. Barzen-Hanson, K.A.; Wilkes, R.A.; Aristide, L. Quantitation of carbohydrate monomers and dimers by liquid chromatography coupled with high-resolution mass spectrometry. *Carbohydr. Res.*, 2018, Vol. 468, pp. 30–35. doi: 10.1016/j.carres.2018.08.007.
81. Gosetti, F.; Mazzucco, E.; Zampieri, D.; Gennaro, M.C. Signal suppression/enhancement in high-performance liquid chromatography – tandem mass spectrometry. *J. Chromatogr. A*. 2010, Vol. 1217, No. 25, pp. 3929–3937.
82. Domon, B.; Costello, C. E. A systematic nomenclature for carbohydrate fragmentations in FAB-MS/MS spectra of glycoconjugates. *Glycoconjugate J.*, 1988, Vol. 5, No. 4, pp. 397–409. doi: 10.1007/BF01049915.

83. Veillon, L.; Huang, Y.; Peng, W.; Dong, X.; Cho, B.G.; Mechref, Y. Characterization of isomeric glycan structures by LC-MS/MS. *Electrophoresis*, 2017, Vol. 38, pp. 2100–2114.
84. Gimeno, A.; Valverde, P.; Arda, A.; Jimenes-Barbero, J. Glycan structures and their interactions with proteins. *Curr. Opin. Struct. Biol.* 2020, Vol. 62, pp. 22–30.
85. Harvey, D.J. Fragmentation of negative ions from carbohydrates. Part 3. Fragmentation of hybrid and complex *N*-linked glycans. *J. Am. Soc. Mass Spectrom.*, 2005, Vol. 116, pp. 647–659.
86. Harvey, D.J. Negative ion mass spectrometry for the analysis of *N*-linked glycans. *Mass Spectrom. Rev.*, 2020, Vol. 39, P. 586–679. doi: 10.1002/mas.21622.
87. Morelle, W.; Michalski, J.-C. Analysis of protein glycosylation by mass spectrometry. *Nature Protocols*, 2007, Vol. 2, pp. 1585–1602. doi: 10.1038/nprot.2007.227.
88. Chizhov, A.O.; Tsvetkov, Y.E.; Nifantiev, N.E. Gas-phase fragmentation of cyclic oligosaccharides in tandem mass spectrometry. *Molecules*, 2019, Vol. 24, 2226, doi: 10.3390/molecules24122226.
89. Meitel, S.N.; Apte, A.; Snovida, S.; Rogers, J.C.; Saba, J. Automated mass spectrometry-based quantitative glycomics using aminoxy tandem mass tag reagents with Sim Glycan. *J. Proteomics*, 2015, Vol. 127, pp. 211–222.
90. Ashline, D.L.; Zhang, H.; Reinhold, V.N. Isomeric complexity of glycosylation documented by MS^n. *Anal. Bioanal. Chem.*, 2017, Vol. 409, pp. 439–451. doi: 10.1007/s00216-016-0018-7.
91. Ashwood, C.; Lin, C.-H.; Thajsen-Andersen, M.; Parker, N.H. Discrimination of isomers of released *N*- and *O*-glycans using diagnostic product ions in negative ion PGC-LC-ESI-MS/MS. *J. Am. Soc. Mass Spectrom.*, 2018, Vol. 29, pp. 1194–1209. doi: 10.1007/s13361-018-1932-z.
92. Wuhrer, M.; Deelder, A. M.; van der Burgt, Y. E. M. Mass spectrometric glycan rearrangements. *Mass Spectrom. Rev.*, 2011, Vol. 30, pp. 664–680. doi: 10.1002/mas.20337.
93. Chizhov, A.O.; Filatov, A.V.; Perepelov, A.V.; Knirel, Y.A. A new example of rearrangement observed in the tandem mass spectra of oligosaccharides. *J. Analyt. Chem.* 2020, Vol. 75, No. 17, pp. 1842–1845. doi: 10.1134/S1061934820140075.
94. Chizhov, A.O.; Gening, M.L.; Tsvetkov, Y.E.; Nifantiev, N.E. Tandem electrospray mass spectrometry of cyclic *N*-substituted oligo-β-(1→6)-D-glucosamines. *Int. J. Mol. Sci.*, 2020, Vol. 21, 8284. doi: 10.3390/ijms21218284.
95. Nagy, G.; Pohl, N.L.B. Complete hexose isomer identification with mass spectrometry. *J. Am. Soc. Mass Spectrom.*, 2015, Vol. 26, No. 4, pp. 677–685. doi: 10.1007/s13361-014-1072-z.
96. Nagy, G.; Pohl, N.L.B. Monosaccharide identification as a first step toward *de novo* carbohydrate sequencing: Mass spectrometry strategy for the identification and differentiation of diastereomeric and enantiomeric pentose isomers. *Anal. Chem.*, 2015, Vol. 87, No. 8, pp. 4566–4571. doi: 10.1021/qcs.analchem.5b00760.
97. Hoffmann, J.; Pagel, K. Glycan analysis by ion mobility – mass spectrometry. *Angew. Chem Internat. Ed.*, 2017, Vol. 56, pp. 8342–8349. doi: 10.1002/anie.201701309.
98. Morrison, K.A.; Clowers, B.H. Contemporary glycomic approaches using ion mobility – mass spectrometry. *Curr. Opin. Chem. Biol.* 2018, Vol. 42, pp. 119–129. doi: 10.1016/j.cbpa.2017.11.020.
99. Pagel, K.; Harvey, D.J. Ion mobility – mass spectrometry of complex carbohydrates: Collision cross sections of sodiated *N*-linked glycans. *Anal. Chem.*, 2013, Vol. 85, No. 10, pp. 5138–5145. doi: 10.1021/ac400403d.
100. Gray, C.J.; Thomas, B.; Upton, R.; Migas, L.G.; Eyers, C.E.; Barran, P.E.; Flitsch, S.L. Application of ion mobility mass spectrometry for high throughput, high resolution glycan analysis. *Biochim. Biophys. Acta – Gen. Subj.*, 2016, Vol. 1860, pp. 1688–1709. doi: 10.1016/j/bbagen.2016.02.003.
101. Struwe, W.B; Harvey, D.J. Ion mobility – mass spectrometry of glycoconjugates. *Meth. Mol. Biol.* (book series), 2019, Vol. 2084, pp. 203–219.
102. Mu, Y.; Schulz, B.L.; Ferro, V. Applications of ion mobility - mass spectrometry in carbohydrate chemistry and glycobiology. *Molecules*, 2018, Vol. 23, 2557. doi: 10.3390/molecules23102557.
103. Fenn, L.S.; McLean, J.A. Structural resolution of carbohydrate position and structural isomers based on gas-phase ion mobility – mass spectrometry. *Phys. Chem. Chem. Phys.*, 2011, Vol. 13, No. 6, pp. 2196–2205. doi: 10.1039/C0CP01414A.
104. Ohashi, Y.; Kubota, M.; Hatase, H.; Nakamura, M.; Hirano, T,; Niva, H; Nagai, Y. Distinction of sialyl anomers on ESI- and FAB-MS/MS: Stereospecific fragmentations. *J. Am. Soc. Mass Spectrom.*, 2009, Vol. 20, pp. 394–397. doi: 10.1016/j.jasms.2008.10.020.
105. Gray, C.J.; Migas, L.G.; Barran, P.E.; Pagel, K.; Seeberger, P.H.; Eyers, C.E.; Boons, G.-J.; Pohl, N.L.B.; Compagnon, I.; Widmalm G.; Flitch, S.L. Advancing solutions to the carbohydrate sequencing challenge. *J. Am. Chem. Soc.*, 2019, Vol. 141, No. 37, pp. 14463–14479.

106. Gray, C.J., Compagnon, I.; Flitsch, S. Mass spectrometry hybridized with gas-phase infrared spectroscopy for glucan sequencing. *Curr. Opin. Struct. Biol.* 2020, Vol. 62, pp. 121–131. doi: 10.1016/j.sbi.2019.12.014.
107. Gray, C.J; Schindler, B.; Migas, L.G.; Pičmanova, M.; Allouche, A.R.; Green, A.P.; Mandal, S.; Motawia, M.S.; Sanches-Perez, R.; Bjarnholt, N.; MØller, B.L.; Rijs, A.M.; Barra, P.E.; Compagnon, I.; Eyers, C.E.; Flitsch, S.L. Bottom-up elucidation of glycosidic bond stereochemistry. *Anal. Chem.*, 2017, Vol. 89, pp. 4540–4549. doi: 10.1021/acs.analchem.6b04998.
108. Hernandez, O.; Isenberg, D.; Steinmetz, V.; Glish, G.L.; Maitre, P. Probing mobility-selected saccharide isomers: Selective ion-molecule reaction and wavelength-specific IR activation. *J. Phys. Chem. A*, 2015, Vol. 119, No. 23, pp. 6057–6064. doi: 10.1021/jp511975f.
109. Hsu, H.C.; Liew, C.Y.; Huang S.-P.; Tsai, S.-T. Ni, C.K. Simple method for *de novo* structural determination of underivatized glucose oligosaccharides. *Sci. Reports*, 2018, Vol. 8, 5562. doi: 10.1038/s41598-018-23903-4.
110. Wuhrer, M. Glycomics using mass spectrometry. *Glycoconj. J.*, 2013, Vol. 30, pp. 11–22. doi: 10.1007/s10719-012-9376-3.
111. Yang, Y.; Liu, F.; Franc, V.; Halim, L.A.; Schellekens, H.; Heck, A.J.R. Hybrid mass spectrometry approaches in glycoprotein analysis and their usage in scoring biosimilarity. *Nature Comm.*, 2016, Vol. 7, 13397. doi: 10.1038/ncomms13397.
112. Lammerhardt, N.; Hashemi, P.; Mischnik, P. Comprehensive structural analysis of a set of various branched glucans by standard methylation analysis, ^{1}H NMR spectroscopy, ESI-mass spectrometry, and capillary electrophoresis. *Carbohydr. Res.*, 2020, Vol. 489, 107933. doi: 10.1016/j.carres.2020.107933.
113. Amicucci, M.J.; Galermo, A.G.; Nandita, E.; Vo, T.-T. T.; Liu, Y.; Lee, M.; Xu, G.; Lebrilla, C.B. A rapid-throughput adaptable method for determining the monosaccharide composition of polysaccharides. *Int. J. Mass Spectrom.*, 2019, Vol. 438, pp. 22–28.
114. Xu, G.; Goonatilleke, E.; Wongkham, S.; Lebrilla, C.B. Deep structural analysis and quantitation of *O*-linked glycans on cell membrane reveal high abundances and distinct glycomic profiles associated with cell type and stages of differentiation. *Anal. Chem.*, 2020, Vol. 92, pp. 3758–37768. doi: 10.1021/acs.analchem.9b05103.
115. Sarbu, M.; Cozma, C.; Zamfir, A.D. Structure-to-function relationship to carbohydrates in the lysosomal storage disorders. *Curr. Org. Chem.*, 2017, Vol. 21, pp. 2719–2730.
116. Carpita, N.C.; McCann, M.C. Characterizing visible and invisible cell wall mutant phenotypes. *J. Exp. Bot.*, 2015, Vol. 66, No. 14, pp. 4145–4163.

7 Mass Spectrometry Analysis of Allergens

Behnam Keshavarz
University of Virginia

CONTENTS

7.1 Introduction 109
7.2 Food Allergy and Major Food Allergens 109
7.2.1 Management of Food Allergy 110
7.3 Current Methods of Allergen Detection 110
7.3.1 Enzyme-Linked Immunosorbent Assays (ELISA) 111
7.3.2 Protein Biosensors 111
7.3.3 DNA-Based Methods 112
7.3.4 Mass Spectrometry 112
7.3.4.1 Detection of Food Allergens Using MS 113
7.4 Conclusion 115
References 116

7.1 INTRODUCTION

Food allergy is one of the critical public health issues, which has significantly increased over the past decades (Sampson et al. 2018, Sicherer and Sampson 2018). According to World Allergy Organization, around 220–250 million people may suffer from food allergies globally (Pawankar et al. 2011). Food allergy significantly reduces an individual's quality of life. It affects up to 8% of children, compared to 5% in adults (Moneret-Vautrin and Morisset 2005, Burks et al. 2012, Werfel 2016), and reasons for which some children or adults grow out of their food allergy or achieve tolerance has remained unclear. To date, there is no proven treatment to cure food allergies. In individuals sensitized and who have clinical symptoms, strict avoidance of the offending food is the only effective way to manage and prevent an allergic reaction. As a result, there has been an increased demand in developing different analytical methods to develop highly sensitive and specific techniques for the detection and characterization of food allergens to improve consumer protection and comply with food-labeling regulations (Gendel 2012). Although such methods have been helpful in the detection and quantification of food allergens, there are challenges involved in the development and validation of these techniques, including the nature of food allergens that are usually present in trace amounts, extractability and solubility of allergenic proteins, lack of standardized and reference materials, and presence of food matrix components that can mask food allergens. In this chapter, an overview of available detection methods with the main focus on mass spectrometry (MS) and challenges involved in the detection of food allergens is presented.

7.2 FOOD ALLERGY AND MAJOR FOOD ALLERGENS

Food allergy is a type I immunoglobulin E (IgE)-mediated hypersensitivity reaction mainly caused by allergenic proteins. Almost all food allergens are proteins except for galactose-α-1,3-galactose (α-Gal), which is a carbohydrate and the cause of mammalian meat allergy known as "α-Gal

DOI: 10.1201/9781003091226-9

syndrome" (Platts-Mills et al. 2020). α-Gal is an unusual form of food allergy, and unlike other food allergens, reactions to α-Gal are typically delayed by 2–6 hours (Wilson et al. 2019, Platts-Mills et al. 2020). Proteins that are reported as allergens usually are harmless and have essential roles such as structural or enzymatic functions in plants and animals commonly consumed by humans. However, in some individuals, these proteins are targeted by the immune system following ingestion and exposure due to an abnormal immunologic response (Boyce et al. 2010), leading to the production of IgE antibodies that are specific to these allergens. In addition, these proteins are generally stable in different conditions such as extreme pH, heat, chemical, physical, or enzymatic treatments. Currently, there are about 1,500 known allergenic structures and more than 180 different types of foods that may cause hypersensitivity reactions (Mari et al. 2009, Boyce et al. 2010). Despite all known allergens from different sources, current research is mostly focusing on major food allergens, known as the "big eight" group of food allergens are typically found in cow's milk, egg, soy, wheat, peanut, tree nuts, fish, and shellfish and are the cause of more than 90% of allergic reactions (Sicherer and Muñoz-Furlong 2010). This has helped regulatory authorities issue legislation to protect consumers and provide them with safe food products.

7.2.1 Management of Food Allergy

Data from the National Health Interview Survey from 1997 to 2007 showed that the prevalence of food allergy in the US has increased by 18% among children under age 18 years old (Branum and Lukacs 2009). As noted, there is no treatment for food allergy, and the only possible way to prevent allergic reactions is strict avoidance of offending foods. It is worth noting that there is always the possibility of accidental exposure to "hidden allergens" caused by cross-contamination (e.g. in food manufacturer facilities) or the presence of an allergen in ingredients added during food processing and production (Skypala 2019). Additionally, in some cases, patients with allergies to one particular food may develop allergenic reactions to other foods as well due to cross-reactivity between homologous proteins (usually greater than 70% similarity in amino acid sequence) with conserved IgE-binding epitopes (Gendel 1998).

As such, due to the life-threatening nature of food allergic reactions and the lack of preventative treatments, many countries have enacted food-labeling legislations that mandate food manufacturers to declare and list major allergens or ingredients derived from major allergens on food labels (Thompson et al. 2006). For example, according to the European Union (EU) food-labeling regulations, 14 allergens must be written on the food label. This includes egg, fish, soybeans, peanuts, nuts, milk, mustard, celery, lupine, sesame seeds, molluscs, crustaceans, and gluten-containing cereals (Regulation]EU]) No 1169/2011). In the US, the label of eight foods (sesame has been recently recognized as the ninth major allergen) identified as priority allergens (the "big eight") and responsible for 90% of food allergic reactions, is required by the US Food Labelling and Consumer Protection Act (FDA 2004). To decrease the economic loss of food industry due to food recalls, comply with food regulations, and reduce the risk of food allergy reactions, it is essential to develop robust, sensitive, and reliable *in vitro* detection methods to prevent the incidence of trace allergenic residues in foods.

7.3 CURRENT METHODS OF ALLERGEN DETECTION

To date, several analytical techniques have been developed to detect and quantify food allergens. The recent advancements in technology, molecular biology, immunology, and improvements made in instruments have significantly helped transform information known about food allergens and their detection and quantification. There are several methods and techniques widely used in the detection and identification of different food allergens (Table 7.1), including MS, Enzyme-Linked Immunosorbent Assays (ELISA), protein biosensors, and DNA-based methods, which are widely used in research settings.

TABLE 7.1
General Methods Commonly Used in Detection and Quantification of Food Allergens

Method	Principle	Analytical Target	Characteristics
ELISA	Allergen-specific mono- and polyclonal antibodies	Protein	• Quantitative and qualitative • Rapid • High throughput • Sensitive • No expertise required • Specific mono- and polyclonal antibodies required • Risk of cross-reactivity and non-specific binding • No simultaneous detection • False-negative signal (due to processing condition and/or matrix effect)
PCR	Oligonucleotide primers	DNA	• Quantitative (RT-PCR) and qualitative • High throughput • Specific and sensitive • Simultaneous detection • Some expertise required • Possible DNA contamination • Target is DNA, not protein (possible false-positive signal)
Mass spectrometry		Proteins/Peptides	• Quantitative • High sensitivity and specificity • Simultaneous detection • Low risk of cross-reactivity • High level of expertise required • Laborious

7.3.1 Enzyme-Linked Immunosorbent Assays (ELISA)

Among several immunoassays developed and currently in use, ELISA is one of the most commonly adopted immunoassays, enabling quick and relatively sensitive detection of allergenic proteins. ELISA is a protein-based technique that targets a specific protein (or allergens) in a food product using a specific monoclonal antibody or polyclonal antibody generally raised against the same target protein in animals (Schubert-Ullrich et al. 2009). To date, there are numerous companies and manufacturers that have developed a variety of commercial ELISA test kits (both qualitative and quantitative test kits) for the detection of allergens. These kits usually are provided with straightforward instruction, which makes it possible to run the test without the need for experienced personnel and/or expensive equipment in a short time (Fernandes et al. 2015). The ELISA technique has some limitations compared with other techniques, including reproducibility, the cost associated with the development of high-quality antibodies (particularly monoclonal antibodies), cross-reactivity in the system, and decline in sensitivity and specificity due to the presence of matrix components and in some cases due to food processing and possible changes in the structure of target proteins that may affect the antibody binding properties (van Hengel 2007, Heick et al. 2011, Monaci et al. 2011, Keshavarz et al. 2019).

7.3.2 Protein Biosensors

Another protein-based technique in detecting food allergens is protein biosensors, which are portable real-time devices that make them a good candidate for quick and on-site analyses. This technique works through the binding interaction between a sensing component and a receptor-transducer analytical device. Biosensors can be classified into optical and electrochemical. The optical biosensors work by detecting changes in light absorption, whereas electrochemical biosensors work by using a biochemical receptor in direct contact with an electrochemical transduction element. This method itself has different varieties, such as amperometric, voltammetric, potentiometric, and impedance types. Most of these biosensors are based on surface plasmon resonance (SPR). Different types of SPR biosensors are used to detect food allergens, such as localized SPR, SPR imaging, fiber-optic SPR, and transmission SPR (Zhou et al. 2019). This technique is applied to detect allergenic proteins in different foods, including milk, egg, peanut, fish, and shellfish (Joshi et al. 2014, Pilolli et al. 2015, Vasilescu et al. 2016, Ashley et al. 2018). One of the major drawbacks of using these biosensors is the cost. Other limitations are lack of simultaneous analysis of multiple allergens, need for special instruments, and professionally trained technicians for analysis of the samples.

7.3.3 DNA-Based Methods

This method relies on detecting a part of the DNA stretches, rather than protein in other detection methods, in a food that is a known source of allergens (Poms et al. 2004). In addition, this technique relies on thermal cycling and repeated heating and cooling cycles for DNA replication with Taq polymerase. For quantitative detection of DNA from allergens sources, real-time polymerase chain reaction (PCR) is the preferred method (Heid et al. 1996). However, this technique also has some limitations. Although DNA is more stable during food processing, it may still be affected by thermal treatment, food matrix components, and the presence of matrix impurities can significantly decrease the assay's sensitivity.

Additionally, if low levels of DNA are present in the foods being analyzed, the detection level will be poor, for example, in cases such as milk and egg (Köppel et al. 2009). Additionally, because the allergenicity of food is represented by allergenic protein(s), not the DNA, there can be an increased chance of false-positive findings using PCR (Picariello, Mamone, Addeo, et al. 2011). Despite these limitations, PCR has several advantages are as follows: (i) it is a relatively quick method (compared to ELISA), (ii) it can be adopted in multiplex analysis, (iii) it minimized risk of cross-reactivity, and (iv) it is known to be highly sensitive and specific (Köppel et al. 2009, Słowianek and Majak 2011).

In summary, all these discussed methods have their strengths and limitations. Considering these limitations, it is worth noting that no single method can detect all allergens in different sources and with different physicochemical properties. However, recent developments and advancements in the field of proteomics have provided us with new advancements in using MS in the detection and quantification of food allergens. In Section 7.3.4, an overview of MS methods and an overall explanation of different MS techniques and their applications in detecting food allergens are provided.

7.3.4 Mass Spectrometry

MS method is one of the widely accepted methods for protein identification, characterization, and quantification in food allergen analysis. MS is considered a core technique in proteomics and peptidomics (Picariello et al. 2011) as it has a wide range of applications, including in the detection of allergenic proteins. MS can be used to either accurately determine the molecular mass of a protein/derived-peptides (MS1) or to understand additional structural details (tandem mass spectrometry [MS/MS] or MS^n) as it is also used in research related to glycosylated proteins and posttranslational modifications of proteins (Bunkenborg et al. 2004). Although MS-based methods may have some similar issues to other allergen detection methods, including the effect of food processing, food matrix, and choice of the analyte, it has significantly improved our understanding of proteins and the complex nature of allergens. Throughout the recent years, MS has undergone significant advancements, which have led to the identification and application of different types of

MS methods. In general, the MS method measures the mass-to-charge ratio (*m/z*) of atoms and/or molecules in a given sample and can also differentiate between different isotopes of the same element. The mass spectrometer consists of three inline-coupled components: an ion source, a mass analyzer, and a detector where the ionized analytes enter the gas phase through different ionization methods. Generally, the ion sources for MS include gas-phase methods, spray methods, and desorption methods. The most commonly used ionization technique in proteomic analysis is electrospray ionization and matrix-assisted laser desorption ionization (MALDI) (Faeste et al. 2011). The key component of MS is the mass analyzer. The type of mass analyzers commonly used in proteomics includes quadrupole (Q), time-of-flight (ToF), Orbitrap, quadrupole ion trap (QIT), and Fourier transform ion cyclotron resonance analyzers, from which the ToF, Q, and QIT have been used to identify, characterize, and quantify allergenic proteins. MS can also be combined with other techniques such as gas chromatography, liquid chromatography, hydrogen exchange, and MALDI. Although MS has not been a quantification technique inheritably, recent development in this method has made discovery-based quantitative proteomics possible. There are various quantitative methods, some are relative, and some are absolute quantification methods (Carrera et al. 2020). One of the most commonly used techniques in the precise quantification of a protein in a complex sample is selected reaction monitoring (SRM), also known as multiple reaction monitoring. SRM has been used to quantify proteins from different sources with complex matrices (Croote and Quake 2016).

7.3.4.1 Detection of Food Allergens Using MS

MS has been widely used in the identification of allergenic proteins in different food products and ingredients derived from them such as milk, peanut, egg, wheat, soybean, fish, shellfish, etc. As noted, before, identification of market peptides is a crucial step in the development of robust, reproducible, sensitive, and specific MS-based detection methods. In Table 7.2, marker peptides specific for some of the major protein allergens are listed.

Briefly, the general roadmap in designing an MS experiment for detecting food allergens consists of a selection of a protein target for analysis, sample extraction and preparation, selection of peptide and fragment targets, assessing digestion, peptide recovery, fragmentation, quantitation, specificity, validation, detection, and allergen quantification (Figure 7.1). In protein target selection, ideally the full sequence of the target protein, unique to that particular food, is available and the protein should not be abundant in other ingredients, and it should not be subject to modification. The latter refers to the stability of the protein during food processing and storage. Additionally, the protein should be extractable and reproducibly digested during the preparation step (Faeste et al. 2011).

There are two principal methods used in the identification of proteins: the "top-down" and the "bottom-up" or shotgun method (similar to bottom-up but peptides from a mixture of proteins being analyzed) (Monaci and Visconti 2009, Zhang et al. 2013). In top-down method, the target protein is first isolated and then is analyzed, either intact or treated for enzymatic digestion (using trypsin and/or chymotrypsin), by mass spectrometer (Bondarenko et al. 2002). Bottom-up method has a better sensitivity compared to top-down technique which makes it a preferred approach. The shotgun method is commonly used to identify peptides used as a fingerprint in targeted proteomics investigation. Sample preparation is a critical step for MS analysis which in some cases requires careful defatting steps, sample cleanup, and desalting to improve analyte ionization, decrease matrix effect, and improve sensitivity of MS analysis (Manfredi et al. 2015, Marzano et al. 2020). Additionally, identifying a sensitive and specific marker that can be used to identify and quantify target allergenic proteins is another important step (Li et al. 2008).

Electrophoresis (one-dimensional and/or two-dimensional electrophoresis) coupled with western blot (using serum from an individual with a history of allergy to the food(s) being analyzed) is widely used to identify proteins that may have allergenicity properties. Serum from an allergic individual that contains allergen-specific IgE antibodies is used to screen and identify any IgE-binding

TABLE 7.2
Characteristics of Some of the Marker Peptides in Major Protein Allergens Identified in Mass Spectrometry Analysis

Food	Protein	Marker peptide	Mass-to-charge ratio (*m*/*z*)	Reference
Milk (*Bos taurus*)	α_{s1}-Casein	YLGYLEQLLR	634.2^{2+}	Weber, Raymond et al. (2006)
	β-Casein	DMPIQAFLLYQQPVLGPVR	$2,188.1^{+}$	Miralles, Leaver et al. (2003)
	κ-Casein	VQVTSTAV	805^{+}	Mollé and Léonil (2005)
	α-Lactalbumin	VGINYWLAHK	$1,199.6^{3+}$	Le, Poulsen et al. (2020)
	β-Lactoglobulin	TPEVDDEALEK	623.3^{2+}	Mikołajczak, Fornal et al. (2019)
Egg (*Gallus gallus*)	Ovalbumin	AFKDEDTQAMPFR	519.2^{3+}	Tokarski, Martin et al. (2006)
	Ovotransferrin	TDERPASYFAVAVAR	551.6^{3+}	
	Lysozyme	FESNFNTQATNR	714.8^{2+}	
	Vitellogenin	AGVR	402.3^{+}	
Wheat (*Triticum* spp.)	α-Gliadin	LQLQPFPQPQLPY	784.9^{2+}	Mamone, Addeo et al. (2005)
	γ-Gliadin	VPPECSIMRAPF	$1,396.8^{+}$	
	ω-Gliadin	KELQSPQQSF	$1,190.6^{+}$	
	Glutenin	QQPGQGQQLR	$1,138^{+}$	Qian, Preston et al. (2008)
Peanuts (*Arachis hypogaea*)	Ara h 1	SFNLDEGHALR	629.8^{+}	Shefcheck and Musser (2004)
	Ara h 2.	ANLRPCEQ	494.2^{+}	Chassaigne, Trégoat et al. (2009)
	Ara h 3&4	QIVQNLR	435.8^{+}	
Soybean (*Glycine max*)	Glycinin	KPQQEEDDDEEEQPQCVETD KGCQR	797.8^{4+}	Leitner, Castro-Rubio et al. (2006)
	β-Conglycinin	NFLAGEKDNVVR	681.4^{2+}	Barnes and Kim (2004)

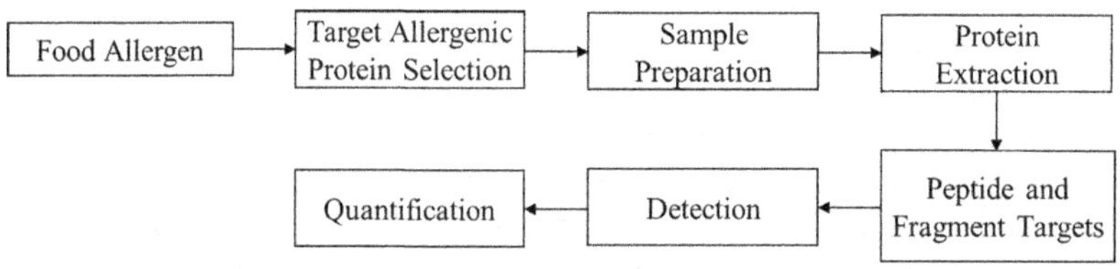

FIGURE 7.1 Experimental design and workflow in food allergen detection using MS.

protein(s) in the extract. Subsequently, the identified protein(s) can be excised from the electrophoresis gel, digested and resulting peptides are separated (e.g., using liquid chromatography), eluted, and fragments are analyzed using MS/MS (Carrera et al. 2020). Data generated by MS are compared with available databases (National Center for Biotechnology Information, UniProt, etc.) if the amino acid sequence of proteins is registered. In cases, the suspect protein is not registered, peptides should be sequenced using *de novo* MS (Shevchenko et al. 1997). It is worth noting that the clinical relevance of identified protein(s) needs to be evaluated through *in vivo* studies (e.g., skin prick test) and available functional assays (e.g., basophil activation test) (Hoffmann-Sommergruber et al. 2015). The overall view and workflow for identification of protein from a food source using MS are illustrated in Figure 7.2.

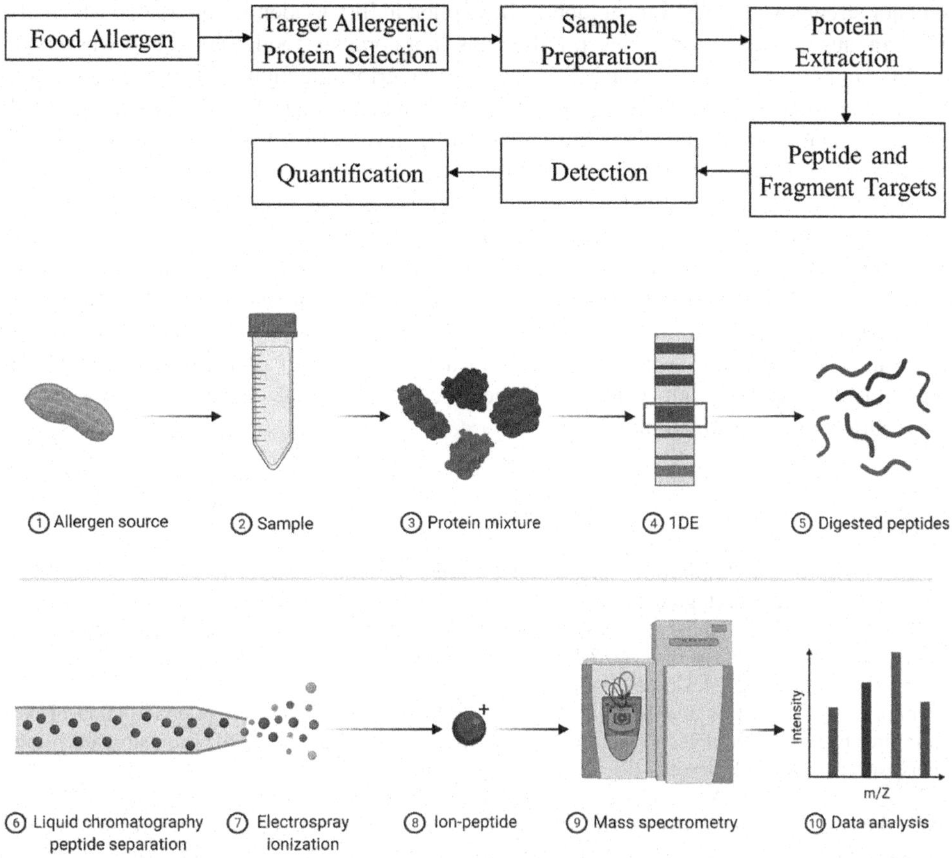

FIGURE 7.2 General steps in characterizing food allergenic proteins using mass spectrometry.

7.4 CONCLUSION

Food allergy prevalence has increased over the past decades, and it has become a significant public health problem. Unfortunately, still no treatment beyond careful avoidance of offending foods exists, and for some allergic patients, ingestion of even a minute amount of allergens can be life-threatening. Although the main issue with the increasing rates of food allergy and demands on health service resources remains, there has been advancement and improvement of the available techniques for detection of allergens that can protect the food allergic individuals. Nonetheless, there are still complications related to analyte conformation changes during and after gastric digestion, low analyte concentration, and complexity of sera proteome, which have made it difficult to study food allergens *in vivo*. Research on proteomics can evolve around the idea of understanding the molecular mechanisms through which one may show allergenic reactions. Although there have been some studies on this idea through the "Immunoproteomics" discipline, there are still lots of questions to be answered. With proteomic approaches and using *in vitro* techniques, a more comprehensive understanding of proteins structure, abundance, modifications, and changes due to protein–protein interactions and protein–matrix components interactions can be achieved. As discussed in this chapter, there are different techniques through which food allergens can be identified and quantified. These techniques are constantly evolving to be more sensitive, specific, and cost-effective. These techniques each have their own cons and pros. Research has shown that MS can be a superior technique compared to other available methods. MS-based proteomics has become an integral part of identification and detection of food allergens, which provides us with rapid technological advances such as modern techniques with unprecedented specificity and sensitivity. One of the major strengths of the MS technique is the explicit ability to identify allergens, which allows for simultaneous quantification of several allergens in a complex food matrix, and when coupled with bioinformatic analysis makes the MS technique a superior detection method compared to other conventional techniques. Lastly, one important question in detection of food allergens is about "how much is too much?" The lowest observed adverse effect level, also called threshold, describes the minimum level of allergens that are required to induce allergic-related symptoms in a sensitized individual. There are challenges in determining a general threshold level which is due to (i) different protocols used in investigational challenges studies, (ii) differences in immunological responses to different allergens among sensitized individuals, and (iii) external factors that may affect threshold levels such as alcohol consumption, sleep deprivation, exercise, medications, etc. (Taylor, Hefle et al. 2002, Remington, Westerhout et al. 2020). Knowing the threshold level will be helpful in the development of assays that are sensitive enough for detection of such allergens and it can also help food manufacturers and regulatory agencies to better control, manage, and avoid the occurrence of allergic reactions in a sensitized individual.

REFERENCES

Ashley, J., R. D'Aurelio, M. Piekarska, J. Temblay, M. Pleasants, L. Trinh, T. L. Rodgers and I. E. Tothill (2018). "Development of a β-Lactoglobulin sensor based on SPR for milk allergens detection." *Biosensors* **8**(2): 32.

Barnes, S. and H. Kim (2004, January 31). Nutriproteomics: Identifying the molecular targets of nutritive and non-nutritive components of the diet. *BMB Reports*. Korean Society for Biochemistry and Molecular Biology. doi: 10.5483/bmbrep.2004.37.1.059.

Bondarenko, P. V., D. Chelius and T. A. Shaler (2002). "Identification and relative quantitation of protein mixtures by enzymatic digestion followed by capillary reversed-phase liquid chromatography–tandem mass spectrometry." *Analytical Chemistry* **74**(18): 4741–4749.

Boyce, J. A., A. Assa'ad, A. W. Burks, S. M. Jones, H. A. Sampson, R. A. Wood, M. Plaut, S. F. Cooper, M. J. Fenton and S. H. Arshad (2010). "Guidelines for the diagnosis and management of food allergy in the United States: report of the NIAID-sponsored expert panel." *Journal of Allergy and Clinical Immunology* **126**(6 Suppl): S1.

Branum, A. M. and S. L. Lukacs (2009). "Food allergy among children in the United States." *Pediatrics* **124**(6): 1549–1555.

Bunkenborg, J., B. J. Pilch, A. V. Podtelejnikov and J. R. Wiśniewski (2004). "Screening for N-glycosylated proteins by liquid chromatography mass spectrometry." *Proteomics* **4**(2): 454–465.

Burks, A. W., M. Tang, S. Sicherer, A. Muraro, P. A. Eigenmann, M. Ebisawa, A. Fiocchi, W. Chiang, K. Beyer and R. Wood (2012). "ICON: food allergy." *Journal of Allergy and Clinical Immunology* **129**(4): 906–920.

Carrera, M., M. Pazos and M. Gasset (2020). "Proteomics-based methodologies for the detection and quantification of seafood allergens." *Foods (Basel, Switzerland)* **9**(8): 1134.

Chassaigne, H., V. Trégoat, J. V. Nørgaard, S. J. Maleki and A. J. van Hengel (2009). "Resolution and identification of major peanut allergens using a combination of fluorescence two-dimensional differential gel electrophoresis, Western blotting and Q-TOF mass spectrometry." *Journal of Proteomics* **72**(3): 511–526. ISSN 1874-3919. doi: 10.1016/j.jprot.2009.02.002.

Croote, D. and S. R. Quake (2016). "Food allergen detection by mass spectrometry: the role of systems biology." *NPJ Systems Biology and Applications* **2**(1): 1–10.

Faeste, C. K., H. T. Ronning, U. Christians and P. E. Granum (2011). "Liquid chromatography and mass spectrometry in food allergen detection." *Journal of Food Protection* **74**(2): 316–345.

FDA (2004). "Food Allergen Labeling and Consumer Protection Act of 2004 (FALCPA) | FDA."

Fernandes, T. J. R., J. Costa, M. B. P. P. Oliveira and I. Mafra (2015). "An overview on fish and shellfish allergens and current methods of detection." *Food and Agricultural Immunology* **26**(6): 848–869.

Gendel, S. M. (1998). "The use of amino acid sequence alignments to assess potential allergenicity of proteins used in genetically modified foods." *Advances in Food and Nutrition Research* **42**: 45–62.

Gendel, S. M. (2012). "Comparison of international food allergen labeling regulations." *Regulatory Toxicology and Pharmacology* **63**(2): 279–285.

Heick, J., M. Fischer, S. Kerbach, U. Tamm and B. Popping (2011). "Application of a liquid chromatography-tandem mass spectrometry method for the simultaneous detection of seven allergenic foods in flour and bread and comparison of the method with commercially available ELISA test kits." *Journal of AOAC International* **94**(4): 1060–1068.

Heid, C. A., J. Stevens, K. J. Livak and P. M. Williams (1996). "Real-time quantitative PCR." *Genome Research* **6**(10): 986–994.

Hoffmann-Sommergruber, K., S. Pfeifer and M. Bublin (2015). "Applications of molecular diagnostic testing in food allergy." *Current Allergy and Asthma Reports* **15**(9): 1–8.

Joshi, A. A., M. W. Peczuh, C. V. Kumar and J. F. Rusling (2014). "Ultrasensitive carbohydrate-peptide SPR imaging microarray for diagnosing IgE mediated peanut allergy." *Analyst* **139**(22): 5728–5733.

Keshavarz, B., X. Jiang, Y. P. Hsieh and Q. Rao (2019). "Matrix effect on food allergen detection - A case study of fish parvalbumin." *Food Chemistry* **274**: 526–534.

Köppel, R., V. Dvorak, F. Zimmerli, A. Breitenmoser, A. Eugster and H.-U. Waiblinger (2009). "Two tetraplex real-time PCR for the detection and quantification of DNA from eight allergens in food." *European Food Research and Technology* **230**(3): 367.

Le, T. T., N. A. Poulsen, G. H. Kristiansen and L. B. Larsen (2020). "Quantitative LC-MS/MS analysis of high-value milk proteins in Danish Holstein cows." *Heliyon* **6**(9): e04620. ISSN 2405-8440. doi: 10.1016/j.heliyon.2020.e04620.

Leitner, A., F. Castro-Rubio, M. Luisa Marina and W. Lindner (2006). "Identification of marker proteins for the adulteration of meat products with soybean proteins by multidimensional liquid chromatography–tandem mass spectrometry." *Journal of Proteome Research* **5**(9): 2424–2430. doi: 10.1021/pr060145q.

Li, Y., L. J. Sokoll, P. E. Barker, H. Zhang and D. W. Chan (2008). "Mass spectrometric identification of proteotypic peptides from clinically used tumor markers." *Clinical Proteomics* **4**(1): 58–66.

Mamone, G., F. Addeo, L. Chianese, A. Di Luccia, A. De Martino, A. Nappo, A. Formisano, P. De Vivo and P. Ferranti (2005). "Characterization of wheat gliadin proteins by combined two-dimensional gel electrophoresis and tandem mass spectrometry." *Proteomics* **5**: 2859–2865. doi: 10.1002/pmic.200401168.

Manfredi, A., M. Mattarozzi, M. Giannetto and M. Careri (2015). "Multiplex liquid chromatography-tandem mass spectrometry for the detection of wheat, oat, barley and rye prolamins towards the assessment of gluten-free product safety." *Analytica Chimica Acta* **895**: 62–70.

Mari, A., C. Rasi, P. Palazzo and E. Scala (2009). "Allergen databases: current status and perspectives." *Current Allergy and Asthma Reports* **9**(5): 376–383.

Marzano, V., B. Tilocca, A. G. Fiocchi, P. Vernocchi, S. Levi Mortera, A. Urbani, P. Roncada and L. Putignani (2020). "Perusal of food allergens analysis by mass spectrometry-based proteomics." *J Proteomics* **215**: 103636.

Miralles, B., J. Leaver, M. Ramos, L. Amigo (2003). "Mass mapping analysis as a tool for the identification of genetic variants of bovine β-casein." *Journal of Chromatography A* **1007**(1–2): 47–53. ISSN 0021-9673. doi: 10.1016/S0021-9673(03)00955-5.

Mollé, D. and J. Léonil (2005). "Quantitative determination of bovine κ-casein macropeptide in dairy products by Liquid chromatography/Electrospray coupled to mass spectrometry (LC-ESI/MS) and Liquid chromatography/Electrospray coupled to tamdem mass spectrometry (LC-ESI/MS/MS)." *International Dairy Journal* **15**(5): 419–428. ISSN 0958-6946. doi: 10.1016/j.idairyj.2004.08.013.

Monaci, L., M. Brohée, V. Tregoat and A. van Hengel (2011). "Influence of baking time and matrix effects on the detection of milk allergens in cookie model food system by ELISA." *Food Chemistry* **127**(2): 669–675.

Monaci, L. and A. Visconti (2009). "Mass spectrometry-based proteomics methods for analysis of food allergens." *TrAC Trends in Analytical Chemistry* **28**(5): 581–591.

Moneret-Vautrin, D. A. and M. Morisset (2005). "Adult food allergy." *Current Allergy and Asthma Reports* **5**(1): 80–85.

Pawankar, R., G. Canonica, S. Holgate, R. Lockey and M. Blaiss (2011). "World Allergy Organisation (WAO) white book on allergy." Wisconsin: World Allergy Organisation.

Picariello, G., G. Mamone, F. Addeo and P. Ferranti (2011). "The frontiers of mass spectrometry-based techniques in food allergenomics." *Journal of Chromatography A* **1218**(42): 7386–7398. doi: 10.1016/j.chroma.2011.06.033.

Pilolli, R., A. Visconti and L. Monaci (2015). "Rapid and label-free detection of egg allergen traces in wines by surface plasmon resonance biosensor." *Analytical and Bioanalytical Chemistry* **407**(13): 3787–3797.

Platts-Mills, T. A. E., S. P. Commins, T. Biedermann, M. van Hage, M. Levin, L. A. Beck, M. Diuk-Wasser, U. Jappe, D. Apostolovic, M. Minnicozzi, M. Plaut and J. M. Wilson (2020). "On the cause and consequences of IgE to galactose-alpha-1,3-galactose: a report from the National Institute of Allergy and Infectious Diseases Workshop on Understanding IgE-Mediated Mammalian Meat Allergy." *Journal of Allergy and Clinical Immunology* **145**(4): 1061–1071.

Platts-Mills, T. A. E., R. C. Li, B. Keshavarz, A. R. Smith and J. M. Wilson (2020). "Diagnosis and management of patients with the alpha-gal syndrome." *Journal of Allergy and Clinical Immunology. In Practice* **8**(1): 15–23 e11.

Poms, R. E., E. Anklam and M. Kuhn (2004). "Polymerase chain reaction techniques for food allergen detection." *Journal of AOAC International* **87**(6): 1391–1397.

Qian, Y., K. Preston, O. Krokhin, et al. (2008). "Characterization of wheat gluten proteins by hplc and MALDI TOF mass spectrometry." *Journal of the American Society for Mass Spectrometry* **19**: 1542–1550. doi: 10.1016/j.jasms.2008.06.008.

Remington, B. C., J. Westerhout, M. Y. Meima, W. Marty Blom, A. G. Kruizinga, M. W. Wheeler, S. L. Taylor, G. F. Houben, J. L. Baumert (2020). "Updated population minimal eliciting dose distributions for use in risk assessment of 14 priority food allergens." *Food and Chemical Toxicology* **139**: 111259. ISSN 0278-6915. doi: 10.1016/j.fct.2020.111259.

Sampson, H. A., L. O'Mahony, A. W. Burks, M. Plaut, G. Lack and C. A. Akdis (2018). "Mechanisms of food allergy." *Journal of Allergy and Clinical Immunology* **141**(1): 11–19.

Schubert-Ullrich, P., J. Rudolf, P. Ansari, B. Galler, M. Fuhrer, A. Molinelli and S. Baumgartner (2009). "Commercialized rapid immunoanalytical tests for determination of allergenic food proteins: an overview." *Analytical and Bioanalytical Chemistry* **395**(1): 69–81.

Shefcheck, K. J. and S. M. Musser (2004). Confirmation of the allergenic peanut protein, Ara h 1, in a model food matrix using liquid chromatography/tandem mass spectrometry (LC/MS/MS). *Journal of Agricultural and Food Chemistry* **52**(10): 2785–2790.

Shevchenko, A., M. Wilm and M. Mann (1997). "Peptide sequencing by mass spectrometry for homology searches and cloning of genes." *Journal of Protein Chemistry* **16**(5): 481–490.

Sicherer, S. H., A. Muñoz-Furlong, J. H. Godbold and H. A. Sampson (2010). "US prevalence of self-reported peanut, tree nut, and sesame allergy: 11-year follow-up." *Journal of Allergy and Clinical Immunology* **125**(6): 1322–1326.

Sicherer, S. H. and H. A. Sampson (2018). "Food allergy: a review and update on epidemiology, pathogenesis, diagnosis, prevention, and management." *Journal of Allergy and Clinical Immunology* **141**(1): 41–58.

Skypala, I. J. (2019). "Food-induced anaphylaxis: role of hidden allergens and cofactors." *Frontiers in Immunology* **10**: 673.

Słowianek, M. and I. Majak (2011). "Methods of allergen detection based on DNA analysis."

Taylor, S. L., S. L. Hefle, C. Bindslev-Jensen, S. Allan Bock, A. Wesley Burks, L. Christie, D. J. Hill, A. Host, J. O'B. Hourihane, G. Lack, D. D. Metcalfe, D. A. Moneret-Vautrin, P. A. Vadas, F. Rance, D. J. Skrypec, T. A. Trautman, I. M. Yman and R. S. Zeiger (2002). "Factors affecting the determination of threshold doses for allergenic foods: How much is too much?." *Journal of Allergy and Clinical Immunology* **109**(1):24–30. ISSN 0091-6749. doi: 10.1067/mai.2002.120564.

Thompson, T., R. R. Kane and M. H. Hager (2006). "Food Allergen Labeling and Consumer Protection Act of 2004 in effect." *Journal of the American Dietetic Association* **106**(11): 1742–1744.

Tokarski, C., E. Martin, C. Rolando and C. Cren-Olivé (2006). "Identification of proteins in renaissance paintings by proteomics." *Analytical Chemistry* **78**(5): 1494–1502. doi: 10.1021/ac051181w.

van Hengel, A. J. (2007). "Food allergen detection methods and the challenge to protect food-allergic consumers." *Analytical and Bioanalytical Chemistry* **389**(1): 111–118.

Vasilescu, A., G. Nunes, A. Hayat, U. Latif and J.-L. Marty (2016). "Electrochemical affinity biosensors based on disposable screen-printed electrodes for detection of food allergens." *Sensors (Basel, Switzerland)* **16**(11): 1863.

Weber, D., P. Raymond, S. Ben-Rejeb and B. Lau (2006). "Development of a liquid chromatography–tandem mass spectrometry method using capillary liquid chromatography and nanoelectrospray ionization–quadrupole time-of-flight hybrid mass spectrometer for the detection of milk allergens." *Journal of Agricultural and Food Chemistry* **54**(5):1604–1610. doi: 10.1021/jf052464s.

Werfel, T. (2016). "Food allergy in adulthood." *Bundesgesundheitsblatt, Gesundheitsforschung, Gesundheitsschutz* **59**(6): 737–744.

Wilson, J. M., A. J. Schuyler, L. Workman, M. Gupta, H. R. James, J. Posthumus, E. C. McGowan, S. P. Commins and T. A. E. Platts-Mills (2019). "Investigation into the alpha-gal syndrome: characteristics of 261 children and adults reporting red meat allergy." *Journal of Allergy and Clinical Immunology. In Practice* **7**(7): 2348–2358 e2344.

Zhang, Y., B. R. Fonslow, B. Shan, M.-C. Baek and J. R. Yates, 3rd (2013). "Protein analysis by shotgun/bottom-up proteomics." *Chemical Reviews* **113**(4): 2343–2394.

Zhou, J., Q. Qi, C. Wang, Y. Qian, G. Liu, Y. Wang and L. Fu (2019). "Surface plasmon resonance (SPR) biosensors for food allergen detection in food matrices." *Biosensors and Bioelectronics* **142**: 111449.

8 Mass Spectrometry of Food Pigments

Laurent Dufossé
Université de La Réunion

Fabio Gosetti
University of Milano-Bicocca

Daria Maria Monti and Antonello Merlino
University of Naples Federico II

Sameer A. S. Mapari
VIVA College [Affiliated to University of Mumbai]

CONTENTS

8.1 Introduction 119
8.2 Mass Spectrometry of C-Phycocyanin 121
8.3 Mass Spectrometry of Carminic Acid 122
8.4 Mass Spectrometry of Azaphilones 125
8.5 Conclusion 132
References 132

8.1 INTRODUCTION

Foods are naturally colored or could be colored by adding a variety of pigments and colorants belonging to many chemical classes such as chlorophylls, anthocyanins, betalains, carotenoids, curcuminoids, azaphilones, anthraquinones, phycocyanins, and caramels (Figure 8.1).

Natural colors are the 1995–2020 marketing trend because consumers are concerned about the safety of artificial food dyes, reinforced by possible health benefits of the natural pigments (Rodriguez-Amaya 2016). However, the replacement of artificial dyes is challenging for food technologists because natural colorants are usually less stable, more costly, are not as easily used as artificial colors, require more material to achieve equivalent color strength, and has a limited range of hues. Huge progress has been made in recent years, and consumers strongly drive the industry.

Mass spectrometry (MS) is a very useful technique to investigate the pigment content of food directly consumed by human beings or sources of colorants extracted to provide colored food ingredients with various purity levels. Quality testing, monitoring of pigments during food processing, and stability during shelf life are among other application fields.

In this chapter, not all chemical classes of food pigments can be covered, and a decision was made to focus on three of them:

i. C-phycocyanin, the trendy blue. Blue is the most difficult color to achieve in food C-phycocyanin is produced at an industrial scale from cyanobacteria. This large compound needs very accurate MS techniques for studying the chromophore and the big proteic part,

DOI: 10.1201/9781003091226-10

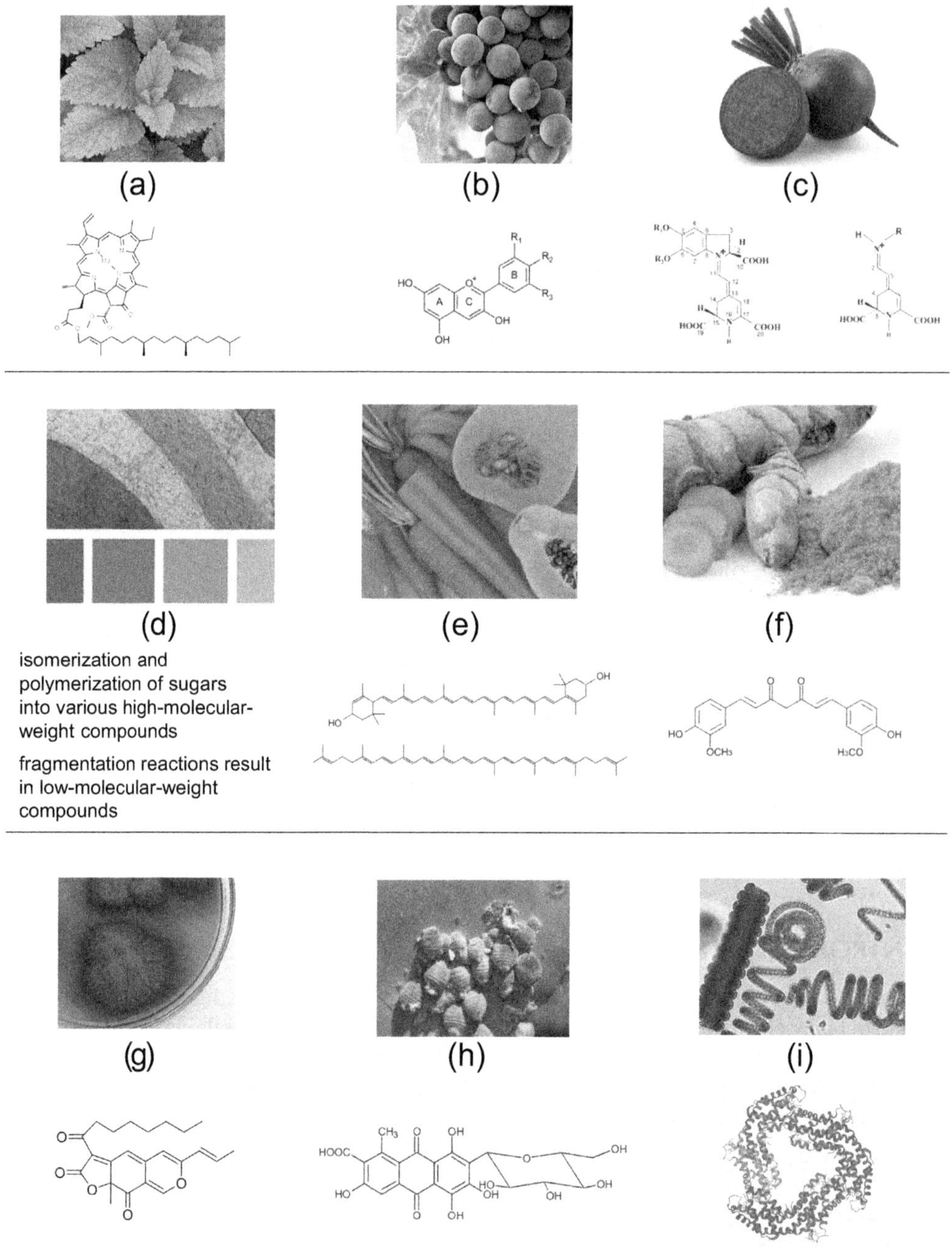

FIGURE 8.1 Food pigments and colorants belonging to many chemical classes: (a) chlorophylls, (b) anthocyanins, (c) betalains, (d) caramels, (e) carotenoids, (f) curcuminoids, (g) azaphilones, (h) anthraquinones, and (i) phycocyanins.

ii. carminic acid (CA), the pure compound extracted from the insect. This anthraquinone is often counterfeited as food red, and MS techniques are used to detect adulteration. In the coming years, biotech CA biosynthesized by yeasts or fungi will appear, triggering new needs for sourcing authentication,
iii. azaphilones, a chemical class incorporated in food in Asia for centuries (red rice or *Monascus* extracts) and not very well known in other parts of the world. Industrial production in

Europe is under progress using strains of non-mycotoxigenic filamentous fungi belonging to *Talaromyces* genus. Start-ups such as Chromologics are in the starting blocks, and here, MS techniques are of special interest as azaphilones are very chemically diverse.

8.2 MASS SPECTROMETRY OF C-PHYCOCYANIN

Phycocyanins (C-PCs) are antenna pigments that collect light energy at wavelengths where chlorophyll *a* poorly absorbs. C-PCs are the major photosynthetic pigments of Cyanobacteria, Rhodophyta, and Cryptophyta. These proteins belong to the family of phycobiliproteins.

Their structure consists of heterodimeric units composed of an α- and a β-chain, which associate to form a ring-shaped $(\alpha\beta)_3$ trimer or a $[(\alpha\beta)_3]_2$ hexamer (Figure 8.2).

Each subunit binds from 1 to 4 chromophore molecules (phycocyanobilin, PCB) with a ring-opening structure (Figure 8.2c). C-PCs are water-soluble, highly stable, and show a maximum of absorbance at 615–630 nm, with a fluorescence emission maximum at approximately 640 nm. Thus, C-PCs have a distinctive deep blue color that makes these proteins a good natural food colorant. Recently, *Arthrospira platensis* has received approval from the Food and Drug Administration and European Food Safety Authority (EFSA) as coloring agent for candies and chewing gums (Code of Federal Regulation 2016). C-PCs from different sources also show antioxidant, anti-inflammatory, and anticancer activities.

In C-PC analysis, MS has been used for *de novo* sequencing, characterization of subunit isoforms and chromophore structure, the oligomeric state's definition, and identification of post-translation modifications. In all these cases, a very high purity grade is mandatory.

In the late 1970s, Fu, Friedman, and Siegelman used proton-transfer chemical-ionization mass spectroscopy to characterize the PCB cleaved from C-PC of *Tolypothrix tenuis* (Fu et al. 1976). Matrix-assisted laser desorption/ionization mass spectrometry (MALDI-MS) (see, *e.g.*, Mancini et al. 2018) has been frequently used for the mass characterization and identification of PCB; peptide–PCB

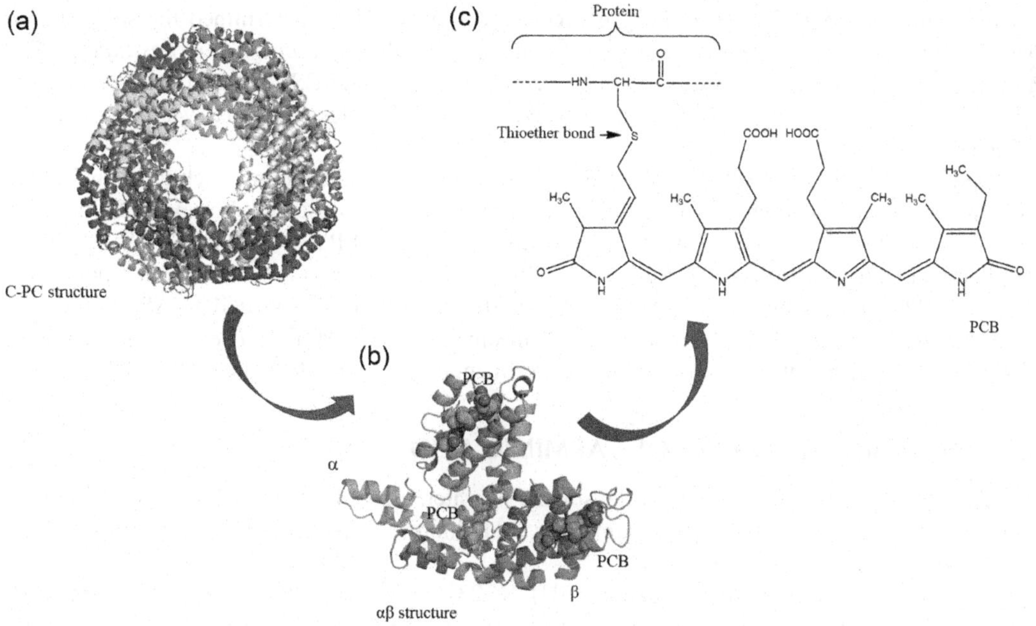

FIGURE 8.2 C-PC structure (a). Crystal structure of the protein from *Galdieria phlegrea* (Ferraro et al. 2020a, PDB code 3Y3D, A) and of its αβ unit (b). One phycocyanobilin (PCB) molecule is bound to the α-subunit and two PCBs are bound to the β-subunit. (c) Chemical structure of PCB, the chromophore responsible for the blue color of the protein.

complex sequences have been often identified by nano high-pressure liquid chromatography-electrospray ionization-OrbiTrap-tandem mass spectrometry (nano-HPLC-ESI-OrbiTrap-MS/MS) (see, *e.g.*, Tong et al. 2020 or Minic et al. 2016, who studied C-PC from commercial *A. platensis*). MALDI-MS studies have also been used to characterize PCB decomposition in the tripyrrole derivative tripyrrole bilin (TPB) (Mancini et al. 2018).

The most frequent use of MS in C-PC studies is for sequence analysis. To ensure good MS data for sequencing, crude algal extracts can be either precipitated using 30%–50% $(NH_4)_2SO_4$ solutions (Nair et al. 2018) or fractionated by using aqueous two-phase extraction (Luo et al. 2016); the pellets obtained should be dissolved in sodium phosphate buffer. Then, protein purification is usually performed by ion-exchange chromatography, size-exclusion chromatography, or ultrafiltration. Before performing MS analyses, sodium dodecyl sulphate-polyacrylamide gel electrophoresis (SDS-PAGE) is run, and then *in situ* digestion is performed using different enzymes, such as papain, dispase, Glu-C, trypsin, or chymotrypsin.

The coverage of the full sequence of the thermophilic C-PC from *Galderia phlegrea* (GpPC) has been recently obtained by hydrolysis with trypsin and chymotrypsin of the two purified subunits of the protein excised from an SDS gel and by subsequent analysis of the obtained peptide mixtures by LC-MS/MS (Ferraro et al. 2020b). Mascot was used for data analysis and identification. The sequences of C-PCs from *Centroceras clavulatum* (macroalga) and *Aphanizomenon gracile* (cyanobacterium) (Nair et al. 2018; Piron et al. 2019) were analyzed using different experimental approaches (different digestions and software). In particular, by digesting the marine red macroalga *C. clavulatum* with trypsin, Nair et al. (2018) found that 15 and 8 peptides were recovered from the two C-PC chains, respectively. The digested peptides were fractionated by high-resolution LC-MS/MS and analyzed by PEAKS and Novor software suites. These analyses showed that the sequence stretches MKTPI/LTEAIA and KCARDI/LGYYLRM were highly conserved in the α-subunit, whereas MLDAFAKVVAQADARGEFLS and SVLDDRCLNGLRETYQAL are highly conserved in the β-subunit.

Combining N-terminal sulfonation of tryptic peptides by 4-sulfophenyl isothiocyanate and MALDI-time-of-flight (ToF)/ToF analyses, Rinalducci et al. (2009) determined the sequence of subunit isoforms of C-PC from the blue-green edible microalga *Aphanizomenon flos-aquae*. The existence of protein isoforms was confirmed by both MALDI- and ESI-MS analyses (Rinalducci et al. 2009). Isoforms of C-PC were also found in CP-C samples from *Acaryochloris marina* (Bar-Zvi et al. 2018).

Oxidation of Met and presence of other post-translational modifications, such as deamidation of Gln residues and methylation of Asn72β, have been frequently observed. For example, the Met residue of the tryptic peptide 63–81 of the α-chain of GpPC, K.FPYTTQMTGPCYASSAIGK.A, has been found to be oxidized (Ferraro et al. 2020b). A common observation has been reported by Nair et al. (2018), who reported oxidation of Met in the peptide FPYTTQMPGPTYASSALGK.

Native MS has also been used to define trimer to hexamer ratio of C-PC from *Thermosynechococcus vulcanus* under different concentrations and experimental conditions (Eisenberg et al. 2017).

8.3 MASS SPECTROMETRY OF CARMINIC ACID

CA is a hydroxyanthraquinone glucoside, soluble in water, alcohol, ester, acid, and alkaline solutions, but insoluble in benzene, petroleum ether, and chloroform (Allevi et al. 1998; Borges et al. 2012; Lloyd 1980).

According to Commission Regulation (EU) N. 231/2012, CA is the main color compound obtained from aqueous, hydroalcoholic, or alcoholic extract from cochineal, that is, extract from dried female insects *Dactylopius coccus* Costa 1835 (synonym *Coccus cacti* Linnaeus 1758) or *Polish cochineal*, that is, *Porphyrophora polonica* Linnaeus 1758 (Sabatino et al. 2012; Wrolstad and Culver 2012; Yamakawa et al. 2009). The term carmine refers to the colorant derived from the boiling of cochineal in ammonium or sodium carbonate, subsequent filtration, and treatment with alum to form calcium–aluminum chelated (lakes) (Dapson 2005; Wrolstad and Culver 2012) with the 10-carbonyl groups of the anthraquinone system of four units of CA (Harris et al. 2009). Commercial preparation of carmine must contain at least 2% of CA in CA extracts and at least 50% in chelates (EFSA 2015; European Commission 2012; JECFA 2006), whereas the Admissible Daily Intake expressed as CA content corresponds to 2.5 mg CA/kg bw/day (EFSA 2015).

Cochineal extract, CA, and carmines are all included under the label E120 in the European Community (EFSA 2015), and Figure 8.3 shows their different chemical structures.

The presence of proteins in cochineal extract derived from insects can cause allergies, and CA can form strong bonds with proteins through both electrostatic and hydrophobic interactions (Nakayama et al. 2015), but further purification step can be useful to overcome this problem

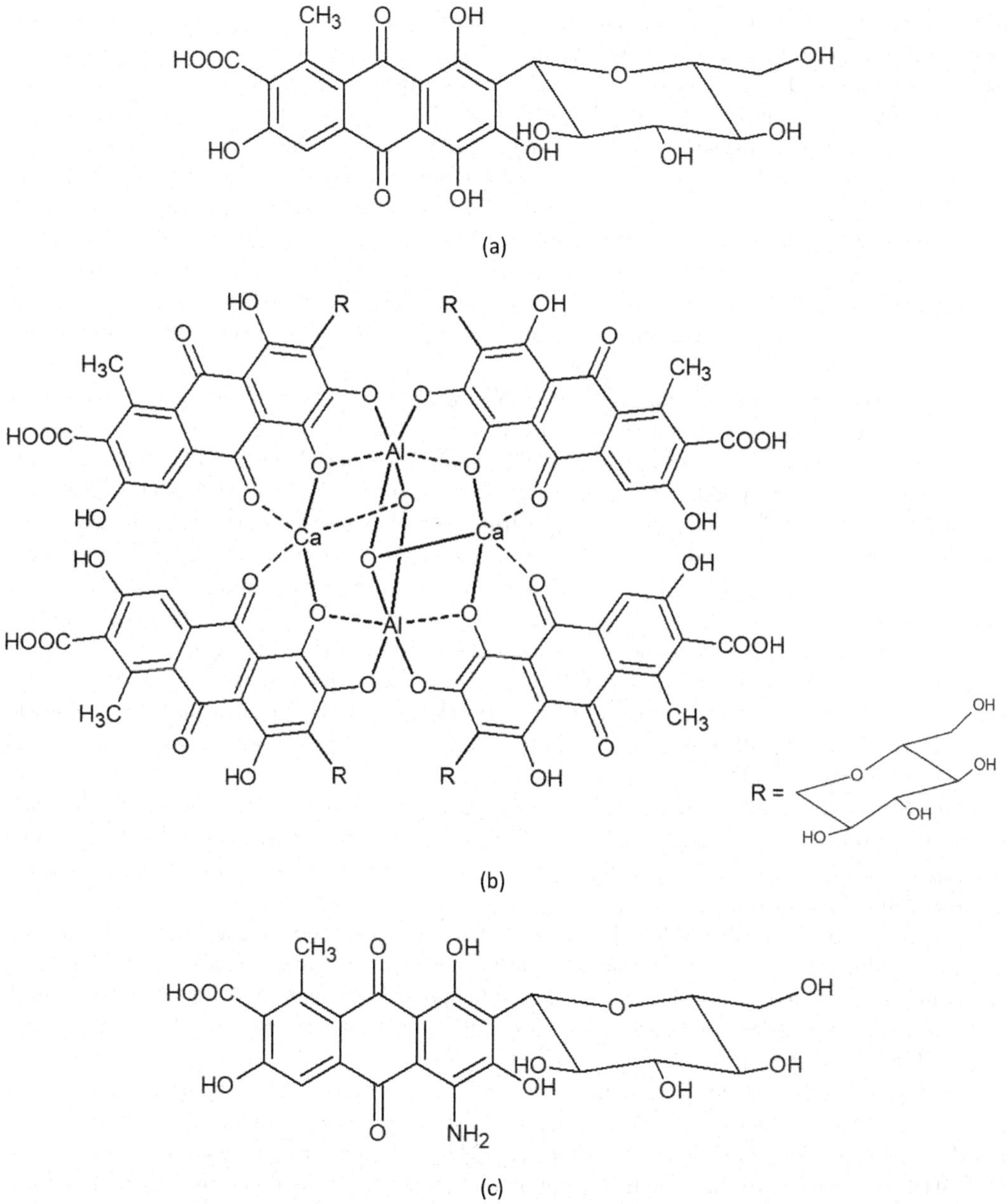

FIGURE 8.3 Chemical structures of carminic acid (a), carmine (b), and 4-aminocarminic acid (c).

(Wrolstad and Culver 2012; Yamakawa et al. 2009). A two-step procedure was used by Ichi et al. (2002). First, enzymatic proteolysis was performed, followed by one or more purification steps, such as adsorption and desorption to a resin, ion-exchange treatment, acid precipitation, extraction by supercritical CO_2, and/or membrane filtration.

During CA production, the use of a cation resin to bind the CA successively released with ethanol was included to remove proteins (EFSA 2015).

Schmidt-Jacobson and Sakstrup-Frandsen (2010) reported a method for the preparation of CA lake in which the presence of protein is necessary for product stabilization. All the original proteins from the insects were removed, whereas other proteins known not to give allergic reactions were added.

Borges et al. (2012) studied two different methods for the extraction of CA from cochineal: pressurized liquid extraction (PLE) and supercritical fluid extraction (SFE). PLE was carried out investigating the effect of extraction temperature, pH, and solvent type (methanol/water, ethanol/water, and ethanol). The best CA extraction from cochineal was obtained using a mixture of methanol/water 60/40 (v/v) as the solvent. When the extraction temperature increased (up to 70°C), the extracted CA concentration increased.

SFE was performed using supercritical CO_2 at pressures between 15 and 30 MPa and at a temperature of 40°C. CA extraction from cochineal resulted significantly higher with PLE (42.4%) and SFE (39.4%) than with conventional solid–liquid extraction (18.5%). Moreover, the authors also showed that adding citric or tartaric acid at levels of more than 0.25 g/L successfully removed all proteins.

It has been reported (Dapson 2005; Gosetti et al. 2015b; Sabatino et al. 2012) the formation of 4-aminocarminic acid (4-ACA) after heating CA with ammonia. This red dye power is stable at a very low pH and is erroneously called acid-stable carmine, but its use is not approved for food. 4-ACA chemical structure is shown in Figure 8.3.

CA is a polar non-volatile small molecule (MW=492.38 g/mol; molecular formula $C_{22}H_{20}O_{13}$) easily ionizable both in positive and negative ion modes with a soft ionization source, in particular ESI, although effective ionizations were obtained by liquid Secondary Ion MS, MALDI (Maier et al. 2004), and ambient-atmospheric pressure chemical-ionization (APCI) source (Kulyk et al. 2019).

The ESI-positive ion of CA gives an MS spectrum with two mass-to-charge (*m/z*) signals corresponding to the quasi-molecular ions at *m/z* 493 $[M+H]^+$ and to a sodium adduct at *m/z* 515 $[M+Na]^+$ (Maier et al. 2004), and a signal at *m/z* 1,007 due to the sodiated dimer cluster. MS/MS spectrum of the precursor ion at *m/z* 493 performed at a collision energy of 20 eV shows the signals at *m/z* 475, 457, 438, 427, and 397 due to the successive losses of water, and a base peak at *m/z* 373 corresponding to $[M\text{-}CO_2H\text{-}2CO\text{-}H_2O]^+$ (Maier et al. 2004). Increasing the collision energy at 40 eV, the product ions at *m/z* 421 $[M+H\text{-}4H_2O]^+$ and 403 $[M+H\text{-}5H_2O]^+$, and a base peak at *m/z* 355 $[M\text{-}CO_2H\text{-}2CO\text{-}2H_2O]^+$ form. MS/MS spectrum of sodiated dimer cluster at *m/z* 1,007 shows the *m/z* at 515 $[M+Na]^+$ (Maier et al. 2004).

The ESI-negative ion of CA shows a quasi-molecular ion at *m/z* 491 $[M\text{-}H]^-$, whose MS/MS spectrum gives rise to the product ions at *m/z* 473 and *m/z* 447 due to neutral loss of H_2O and CO_2, respectively, and *m/z* 327 $[M\text{-}2H\text{-glucose}]^-$ (Gosetti et al. 2015a; Gosetti et al. 2015b; Maier et al. 2004; Sabatino et al. 2012).

CA was also ionized using MALDI in negative ion mode (α-cyano-4-hydroxycinnamic acid as the matrix), although this technique generally is used for cultural heritage samples. The MALDI-MS spectrum shows the *m/z* at 491 and *m/z* 447, due to the quasi-molecular ion $[M\text{-}H]^-$ and the neutral loss of CO_2, as in the ESI NI spectrum, besides of the signals at *m/z* 188 and *m/z* 144 due to the matrix (Maier et al. 2004).

Kulyk et al. (2019) ionized CA by a reactive olfaction ambient MS mechanism with a nano-APCI. The MS spectrum of CA in positive ion mode shows a base peak at *m/z* 495, corresponding to $[M+(3H)]^+$ and another one at *m/z* 509 $[M+OH]^+$. The particular form of the unique ionic species $[M+(3H)]^+$ is attributed to the analyte adsorption on the metal electrode during corona discharge, which facilitates the CA reduction via two sequential reactions with protons and electrons (Kulyk et al. 2019). The reaction is followed by the desorption of products into the gas phase.

Several methods for the determination of CA in foods (raw sausage, beverages, meat, yogurt, ice flavors, candies, jellies, ice creams, ham, milk, and liqueurs) were reported in the literature principally based on spectrophotometric, fluorimetric, or enzyme immunoassays, HPLC-UV–vis, and capillary electrophoresis (Scotter 2011). Only a few methods reported the determination of CA in food using the MS technique. CA was determined in beverages by HPLC coupled with high-resolution Orbitrap MS (Sabatino et al. 2012) and ultra-high-performance liquid chromatography-quadrupole time-of-flight (UHPLC-QToF) (Gosetti et al. 2015a; Gosetti et al. 2015b), in soft drinks by HPLC-triple quadrupole MS (Feng et al. 2011) and wines by UHPLC-Q-Orbitrap (Jia et al. 2014).

Table 8.1 lists the MS methods (conditions and validation parameters) for the analysis of CA in food.

The extraction conditions are generally very simple and involve acid hydrolysis with or without solid-phase extraction. Still, enzymatic digestion can be used for difficult matrices, particularly meat products (EFSA 2015).

On the contrary, when the chemicals to search for are unknown, it is preferable not to perform any sample pre-treatment to avoid the same unknown species' possible losses. This is the case of a study that uses an untargeted approach to determine CA and its photodegradation products formed during the simulated sunlight exposure of 16 beverages of different compositions but all containing the dye (Gosetti et al. 2015a).

A methodology based on Principal Component Analysis coupled with Discriminant Analysis was employed. The method is independent of the beverage composition and allows identifying the degradation of products originating from the CA dye present in each beverage. Retrospective data analysis was useful for identifying 4-ACA and its degradation products in the investigated beverages (Gosetti et al. 2015b).

8.4 MASS SPECTROMETRY OF AZAPHILONES

Azaphilone pigments are hexaketides belonging to the structurally diverse polyketide class of compounds produced as secondary metabolites by filamentous fungi. Structurally, they possess pyrone–quinone moiety with a high-oxygenated bicyclic core and chiral quaternary center (Caro et al. 2015). In addition to their structural diversity, they also exhibit functional diversity, exemplified by their antimicrobial, antiviral, nematicidal, antioxidant, anti-inflammatory, and cytotoxic activities (Gao et al. 2013).

The most well-known azaphilones and/or polyketide pigments are the six characteristic *Monascus* pigments, the traditionally used food colorants in South-East Asia. Among polyketide *Monascus/Talaromyces* pigments, some are azaphilones (monascorubrin, rubropunctatin, and PP-O as examples) characterized by their relatively non-polar aminophilic nature due to which when they react with amino acids (Figure 8.4) or primary amines result in the formation of water-soluble polar counterparts (Jung et al. 2003).

In the past two decades, more and more *Monascus* or *Monascus*-like polyketide and/or azaphilone pigments and/or their derivatives have been discovered to be produced by *Monascus* species or in their commercial products, including food colorants and red rice (Yuliana et al. 2017, Kim and Ku 2018). More interestingly, they are also discovered to be produced by *Penicillium* and/or *Talaromyces* species without the nephrotoxic pigment citrinin (Mapari et al. 2008, Isbrandt et al. 2020, Morales-Oyervides et al. 2020). The discovery of these azaphilones and/or polyketide pigments have been possible due to the advancement, ease, sensitivity, and speed of chromatographic/spectrometric techniques, especially mass spectrometry coupled with liquid chromatography (LC-MS) and rarely gas chromatography (GC-MS) (Mukherjee and Singh 2011) and even (LC-MS/MS) (Srianta et al. 2020) in combination with UV–vis spectroscopy. High-resolution mass spectroscopy (HRMS) has enabled the determination of molecular composition and monoisotopic mass with much higher accuracy (up to four places of decimal) with a room of error of only a few ppm.

TABLE 8.1
LC/MS Methods (Conditions and Validation Parameters) for the Analysis of Carminic Acid (CA) in Food

Analytes	Sample	Instrumentation	Stationary Phase	Mobile Phase	Sample Pre-treatment	LOD/LOQ µg/L	Recovery (%)	ME	Reference
40 food dyes among which CA	Soft drinks	HPLC-MS/MS Source: ESI- Analyzer: QQQ	XB-C18 (100×2.1 mm, 3.0 µm)	Gradient elution: 20 mM ammonium formate+0.1% formic acid and methanol/acetonitrile 7/3 v/v	Solid-phase extraction cartridge; Filtration 0.2 µm	3 / 15	92.1±6.1	no	Feng et al. (2011)
CA and its photodegradation products	Beverages	UHPLC-MS/MS Source: ESI- Analyzer: Q-ToF	Acquit BEH phenyl (100×2.1 mm, 1.7 µm)	Gradient elution: water/acetic acid 99/1 (v/v) and methanol/isopropanol/acetic acid 97/2/1 (v/v/v)	Filtration 0.2 µm	-	-	-	Gosetti et al. (2015a)
4-ACA and its photodegradation products	Beverages	UHPLC-MS/MS Source: ESI- Analyzer: Q-ToF	Acquit BEH phenyl (100×2.1 mm, 1.7 µm)	Gradient elution: water/acetic acid 99/1 (v/v) and methanol/isopropanol/acetic acid 97/2/1 (v/v/v)	Filtration 0.2 µm	1.2 / 3.8	-	-	Gosetti et al. (2015b)
69 food dyes among which CA	Wines	UHPLC-MS/MS Source: ESI- Analyzer: Q-Orbitrap	Accucore C18 aQ (100×2.1 mm, 2.6 µm)	Gradient elution: 4 mM ammonium formate+0.1 formic acid in water and 4 mM ammonium formate+0.1 formic acid in methanol	QuEChERS; Filtration 0.2 µm	Not reported/25	95.8	no	Jia et al. (2014)
CA and 4-ACA	Food additives and beverages	HPLC-PDA-MS/MS Source: ESI- Analyzer: ion trap	Aqualsil C18 (150×2.1 mm, 3 µm)	Gradient elution: water+0.3 formic acid and acetonitrile+0.3% formic acid	Food additives: dilution of the power; Beverage: filtration 0.45 µm	CA: 2,000/4,000 4-ACA: 1,000/3,000	CA: 97±2 4-ACA: 101±3	-	Sabatino et al. (2012)

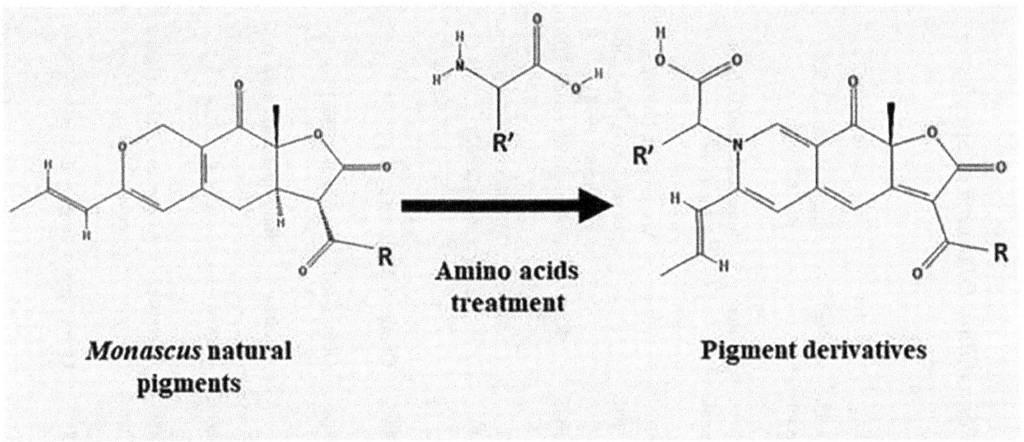

FIGURE 8.4 Representative example of substitution reaction of *Monascus/Talaromyces* azaphilone pigment's oxygen moiety for amino acids' nitrogen moiety. R=C_5H_{11} or C_7H_{15}. R′ is a functional group of amino acids. (Adapted from Kim and Ku 2018.)

Since pigments are colored compounds, mass spectra are often used with UV–vis spectra to properly identify compounds leading to de-replication (Smedsgaard and Frisvad 1996). However, nuclear magnetic resonance (NMR) spectroscopy is a method of choice for structural elucidation of compounds (Elyashberg et al. 2002) including azaphilone/polyketide pigments and their definite identification. A comprehensive list of such compounds has been tabulated (Table 8.2) based on their chromatographic/spectrometric detection and identification methods.

Extraction. Depending on the polarity of *Monascus* pigments, appropriate solvents such as hexane (for yellow pigments: monascin and ankaflavin as examples), ethyl acetate (for orange pigments: monascorubrin and rubropunctatin as examples), and ethanol for red pigments (rubropunctamine and monascorubramine or pigment derivatives of orange pigments as examples) have been used for their extraction from the fermentation broth under agitation conditions (Choe et al. 2012, Jang et al. 2014, Jung et al. 2003, Kim et al. 2006, Daud et al. 2020). Extractive fermentation has been recommended (Shi et al. 2017) to extract the intracellular non-polar *Monascus* pigments to prevent the inhibitory effect of the product inside the cell. In addition, a PLE technique, a greener and faster extraction step of these pigments while preserving their chemical structures and bioactivities, has been reported (Lebeau et al. 2017).

Sample preparation. HRMS, one of the fastest and the most effective methods, can be performed on the filtered crude pigment extracts, thereby minimizing sample preparation steps. As well, analyzing fungal extracts directly with MALDI-ToF-MS has been reported (Shishupala, S. 2008). MALDI-ToF-MS has proven itself an extremely precise, sensitive, and reliable analytical tool for the characterization of fungal biomolecules that can be applied to polyketide and/or azaphilone pigments as well.

Ionization methods. Before MALDI, both positive and negative modes of ESI (Smedsgaard and Frisvad 1996, Nielsen et al. 2004) and rarely chemical-ionization methods (Jung et al. 2003) of MS have been reported for the identification of azaphilone and/or other polyketide pigments in crude pigment extracts. However, positive ESI has been more successful, given its chemical nature.

Data analysis and identification. Data analyses have been performed using various programs, such as MassLynx 4.0. To identify compounds, de-convolution has to be performed, which is to check the ion traces of all ions of interest coming from the same ion. For the positive mode of ESI, it is advisable to look for $[M+H]^+$, $[M+Na]^+$, $[M+K]^+$, $[M+NH_4]^+$, and $[M+Na+CH_3CN]^+$ adducts. In ESI's negative mode, M–H is usually the most common one, but sometimes loss of H_2O is seen, or sometimes adducts with acetic acid are observed. However, de-convoluted mass spectra from

TABLE 8.2
Azaphilone and/or Polyketide Pigments and/or Amino Acid Pigment Derivatives of *Monascus* and/or *Penicillium/Talaromyces* Species Detected and Identified by Instrumental Analysis

S. No.	Pigment Name	Molecular Formula	Molecular Formula weight	Chromatographic/ Spectroscopic methods	Fungal Identity/Media	Color	References
1.	Alanine derivative of Monascorubrin	$C_{26}H_{31}NO_6$	453	APCI-LC-MS/LC-MS	*Monascus* sp. KCCM 10093, liquid medium/commercial *Monascus* pigments	Red	Jung et al. (2003), Sato et al. (1997)
2.	Alanine derivative of Rubropunctatin	$C_{24}H_{27}NO_6$	425	APCI-LC-MS/LC-MS	*Monascus* sp. KCCM 10093, liquid medium/commercial *Monascus* pigments	Red	Jung et al. (2003), Sato et al. (1997)
3.	Ankaflavin	$C_{23}H_{30}O_5$	386	LC-DAD-MS	*Penicillium funiculosum* IBT 3954, Monascus-fermented red rice/ *Monascus* sp.	Yellow	Mapari et al. (2009), Su et al. (2005), Teng and Feldheim (1998)
4.	[a]Atrorosin S	$C_{26}H_{29}NO_9$	499	UHPLC-HRMS-DAD, QToF- ESI, NMR	*Talaromyces atroroseus* IBT 11181/liquid media	Red	Tolborg et al. (2020, 2020a)
5.	[b]Citrinin	$C_{13}H_{14}O_5$	250	LC-DAD-MS	*Monascus purpureus* IBT 9667, 9664 on YES, PDA, MEA agar, *Penicillium citrinum*/solid medium	Yellow	Mapari et al. (2008), Duran et al. (2002)
6.	Cysteine derivative of Monascorubrin	$C_{26}H_{31}NO_6S$	485	APCI-LC-MS	*Monascus* sp. KCCM 10093, liquid medium	Orange-red	Jung et al. (2003)
7.	Cysteine derivative of Rubropunctatin	$C_{24}H_{27}NO_6S$	457	APCI-LC-MS	*Monascus* sp. KCCM 10093, liquid medium	Orange-red	Jung et al. (2003)
8.	FK17-P2B2	$C_{13}H_{16}O_4$	236	ESI-ToF-MS, UV–vis, NMR, IR	Mutant of *Monascus kaoliang* grown on rice	Yellow	Jongrungruangchock et al. (2004)
9	Glycine derivative of Monascorubrin	$C_{25}H_{29}NO_6$	439	APCI-LC-MS	*Monascus* sp. KCCM 10093, liquid medium	Red	Jung et al. (2003)
10.	Glycine derivative of Rubropunctatin or PP-V	$C_{23}H_{25}NO_6$	411	APCI-LC-MS/HR-FAB-MS	*Monascus* sp. KCCM 10093, liquid medium/*Penicillium* sp., liquid medium	Violet-red	Jung et al. (2003), Ogihara et al. (2000)
11.	Isoleucine derivative of Monascorubrin	$C_{29}H_{37}NO_6$	495	APCI-LC-MS	*Monascus* sp. KCCM 10093, liquid medium	Orange-red	Jung et al. (2003)

(Continued)

TABLE 8.2 (*Continued*)
Azaphilone and/or Polyketide Pigments and/or Amino Acid Pigment Derivatives of *Monascus* and/or *Penicillium/Talaromyces* Species Detected and Identified by Instrumental Analysis

S. No.	Pigment Name	Molecular Formula	Molecular Formula weight	Chromatographic/ Spectroscopic methods	Fungal Identity/Media	Color	References
12.	Isoleucine derivative of Rubropunctatin	$C_{27}H_{33}NO_6$	467	APCI-LC-MS	*Monascus* sp. KCCM 10093, liquid medium	Orange-red	Jung et al. (2003)
13.	Leucine derivative of Rubropunctatin	$C_{27}H_{33}NO_6$	467	APCI-LC-MS	*Monascus* sp. KCCM 10093, liquid medium	Orange-red	Jung et al. (2003)
14.	[c]Monankarin A-B	$C_{20}H_{22}O_6$	358	LV-HR-ESI-MS, NMR	*M. anka*/liquid medium	Yellow	Hossain et al. (1996)
15.	[d]Monaphilol A	$C_{23}H_{28}O_5$	384	HR-ESI-MS, UV, IR, Fluorescence spectra	*M. purpureus* NTU 568 in functional food Red Mold Dioscorea	Orange	Hsu et al. (2011)
16.	[e]Monaphilones A	$C_{22}H_{32}O_4$	360	HPLC-ESI-MS, NMR	*M. anka* NTU 568 on rice	Yellow	Hsu et al. (2010)
17.	[f]Monapurones A	$C_{20}H_{26}O_4$	330	UV-HPLC-NMR	*M. purpureus* fermented rice	Yellow	Li et al. (2010)
18.	Monarubrin	$C_{20}H_{26}O_4$	330	LC-ESI-MS, HR-EI-MS, IR, NMR	*Monascus ruber*/liquid medium	Yellow, Blue fluorescence	Loret and Morel (2010)
19.	Monascin	$C_{21}H_{26}O_5$	358	LC-MS, UV–vis	*Monascus* sp./[g]*Penicillium purpurogenum* IBT 3967/ NRRL 1147	Yellow	Fielding et al. (1961), Mapari et al. (2009)
20.	Monascorubramine	$C_{23}H_{27}NO_4$	381	APCI-LC-MS, LC-DAD-ToF-MS	*Monascus* sp. KCCM 10093, liquid medium, 5*Penicillium purpurogenum* 11180 on YES agar	Red	Jung et al, (2003), Mapari et al. (2009)
21.	Monascorubrin	$C_{23}H_{26}O_5$	382	APCI-LC-MS, LC-DAD-ToF-MS	*Monascus* sp. KCCM 10093, liquid medium, *Penicillium aculeatum* IBT 14259/ NRRL 2129 on CYA agar, *P. aculeatum* IBT 14263/ FRR 1802 on YES agar	Orange	Jung et al, (2003), Mapari et al. (2008, 2009)
22.	Monascupiloin	$C_{21}H_{28}O_5$	360	HPLC-UV, ESI-MS, HR-ESI-MS, IR, NMR	*M. pilosus* grown on red mold rice	Yellow	Wu et al. (2015)
23.	[h]Monascuzone A	$C_{13}H_{18}O_5$	254	ESI-ToOF-MS, UV, IR, NMR	*M. kaoliang* grown on rice	Yellow	Jongrungruangchock et al. (2004)

(Continued)

TABLE 8.2 (*Continued*)
Azaphilone and/or Polyketide Pigments and/or Amino Acid Pigment Derivatives of *Monascus* and/or *Penicillium/Talaromyces* Species Detected and Identified by Instrumental Analysis

S. No.	Pigment Name	Molecular Formula	Molecular Formula weight	Chromatographic/ Spectroscopic methods	Fungal Identity/Media	Color	References
24.	[i]Monasfluor A	$C_{22}H_{26}O_4$	354	UV fluorescence spectra, LC-ESI-MS, NMR	Monascus sp. grown on rice	Blue fluorescent	Huang et al. (2008)
25.	Monashexenone	$C_{19}H_{28}O_4$	320	LC-ESI-MS, HR-ESI-MS, IR, NMR	Mutant of *M. purpureus* grown on rice	Yellow	Hsu et al. (2010)
26.	N-glucosyl monascorubramine	$C_{29}H_{37}NO_9$	543	HPLC-FAB-MS, NMR	*Monascus ruber*/liquid medium	Red	Hajjaj et al. (1997)
27.	N-glucosyl rubropunctamine	$C_{27}H_{33}NO_9$	515	HPLC-FAB-MS, NMR	*Monascus ruber*/liquid medium	Red	Hajjaj et al. (1997)
28.	N-glutaryl monascorubramine	$C_{28}H_{33}NO_8$	511	UV, MS, IR, NMR, LC-DAD-MS, ESI-ToF	Monascus sp, TTWWB 6093 liquid medium, *P. purpurogenum* IBT 11181 and IBT 3645/ liquid medium,	Red	Lin et al. (1992), Blanc et al. (1994), Mapari et al. (2009)
29.	N-glutaryl rubropunctamine	$C_{26}H_{29}NO_8$	483	LC-DAD-MS, ESI-ToF	*P. purpurogenum* IBT 11181/liquid medium	Red	Mapari et al. (2009)
30.	PP-O	$C_{23}H_{24}O_7$	412	LC-MS, UHPLC-HRMS-DAD, Q-ToF-ESI, NMR	*Penicillium purpurogenum* IAM 15392/*Talaromyces atroroseus* IBT 11181/ liquid media	Orange	Arai et al. (2015), Isbrandt et al. (2020)
31.	PP-R	$C_{25}H_{31}NO_5$	425	FAB-MS, NMR/ LC-DAD-MS, ESI-ToF	*Penicillium* sp. AZ./*P. purpurogenum* IBT 11180, IBT 23082 on YES agar, IBT 21347 on CYA agar	Red	Ogihara et al. (2001), Mapari et al. (2006, 2009)
32.	Rubropuctin	$C_{22}H_{30}O_4$	358	LC-UV, IR, NMR	*Monascus ruber* in liquid medium	Yellow, Blue Fluorescence	Loret and Morel (2010)
33.	Rubropunctamine	$C_{21}H_{23}NO_4$	353	LC-DAD-TMS, APCI-LC-MS, LC-DAD-ToF-MS	*Monascus anka, Monascus* sp. KCCM 10093/liquid medium, *M. ruber* and *M. purpureus* on solid media	Red	Teng and Feldheim (1998), Jung et al. (2003), Mapari et al. (2008)

(*Continued*)

TABLE 8.2 (*Continued*)
Azaphilone and/or Polyketide Pigments and/or Amino Acid Pigment Derivatives of *Monascus* and/or *Penicillium/Talaromyces* Species Detected and Identified by Instrumental Analysis

S. No.	Pigment Name	Molecular Formula	Molecular Formula weight	Chromatographic/ Spectroscopic methods	Fungal Identity/Media	Color	References
34.	Rubropunctatin	$C_{21}H_{22}O_5$	354	LC-DAD-TMS, APCI-LC-MS	*Monascus anka, Monascus* sp. KCCM 10093, liquid medium	Orange	Teng and Feldheim (1998), Jung et al. (2003)
35.	[j]Red derivative 1	$C_{26}H_{31}NO_6$	453	LC-MS	commercial *Monascus* pigments	Red	Sato et al. (1997)
36.	Red-dione	$C_{21}H_{29}NO_5$	375	UV–vis, HPLC, gas chromatography-MS, IR, NMR	*Monascus purpureus* NFCCI 1756/liquid medium	Red	Mukherjee and Singh (2011)
37.	Threonine derivative of Monascorubrin	$C_{27}H_{33}NO_7$	483	APCI-LC-MS, LC-MS	*Monascus* sp. KCCM 10093/liquid medium, Monascus sp./liquid medium	Red	Jung et al. (2003), Jang et al. (2014)
38.	Threonine derivative of rubropunctatin	$C_{25}H_{29}NO_7$	455	APCI-LC-MS, LC-DAD-ToF-MS	*Monascus* sp. KCCM 10093/ liquid medium *Penicillium aculeatum* IBT 14263 on CYA agar	Red	Jung et al. (2003), Mapari et al. (2008)
39.	Xanthomonascin A	$C_{21}H_{24}O_7$	388	LC-DAD-ToF-MS	*Monascus pilosus*/fermented rice, *Penicillium aculeatum* IBT 14263 on CYA agar	Yellow	Akihisa et al. (2005), Mapari et al. (2008)

[a] Representative example of similar group of compounds called atrorosins; atrorosin-D, atrorosin-E, atrorosin-H, atrorosin-Q, and atrorosin-T. See text.
[b] Nepherotoxic pigment.
[c] Representative example of similar group of compounds namely: Monankarin C-D, Monankarin E, and Monankarin F.
[d] Representative example of similar group of compounds namely: Monaphilols B, C, and D.
[e] Representative example of similar group of compounds namely: Monaphilones B and C.
[f] Representative example of similar group of compounds namely: Monapurones B and C.
[g] Current nomenclature *Talaromyces atroroseus.*
[h] Representative example of similar group of compounds namely, Monascuzone B.
[i] Representative example of similar group of compounds namely, Monasfluor B.
[j] Likely to be valine derivative of rubropunctatin.

the total ion chromatogram are often de-emphasized in the literature related to describing novel or well-known compounds identified by MS. Only data concerning molecular formula and weight and sometimes only $[M+H]^+$ values are reported leading to perplexity in identifying compounds.

Fragmentation patterns. The fragmentation patterns of *Monascus* pigments and their derivatives are identical. The base peaks for *Monascus* pigments and their derivatives have been reported (Teng et al. 1998, Blanc et al. 1994) to be (*m*/*z* 354.7, 312.7, 214.5, 188.7, 146.3, 114.4, and 106.3).

Databases. Antibase (Wiley-VCH, Weinheim, Germany) 2012 is a natural compound identifier that is a database of more than 40,000 natural compounds from microorganisms and higher fungi. The database contains more than 400,000 biological, physical, and physicochemical data records. High-resolution masses for M, M+/–H, and M+Na ions. The data in Antibase have been collected from the primary and secondary literature and then carefully checked and validated. Descriptive data include molecular formula and mass, elemental composition, Chemical Abstracts Service (CAS) registry number, and physicochemical data include melting point and optical rotation. Spectroscopic data UV, ^{13}C-NMR, IR, and mass spectra are also included. A unique feature of Antibase is the use of predicted ^{13}C-NMR spectra for those compounds where no measured spectra are available. Calculated high-resolution molecular masses are included in the new update (Wiley-VCH Verlag GmbH & Co. KGaA. http://www.wiley-vch.de). Another mass spectral database is MassBank, the official database of the Mass Spectrometry Society of Japan (www.massbank.jp).

Practical aspect. Identification of pigments should be based on both the UV–vis spectra and the accurate masses; monoisotopic masses are given in the databases and the available literature references. There exist some discrepancies in the available literature for the identification of azaphilone and/or *Monascus* pigments. For example, the UV–vis spectra of yellow *Monascus* pigments, ankaflavin, and monascin, have been reported (Akihisa et al. 2005) to be in the spectrum of orange pigments. Besides, the $[M+H]^+$ values are often mistaken to be the molecular masses exemplified by Jung et al. (2003). Thus, utmost care should be taken when comparing the UV–vis and/or mass spectral data to the data available in the literature to identify colored compounds.

8.5 CONCLUSION

Is there anything that cannot be done using MS? With new methods published daily, it seems, MS has become one of the most powerful and widely used techniques in analytical science (Doerr 2006), in food science, and the analysis of food pigments.

REFERENCES

Akihisa, T., Tokuda, H., Yasukawa, K., et al. 2005. Azaphilones, furanoisophthalides and amino acids from the extracts of *Monascus pilosus*-fermented rice (red mold rice) and their chemopreventive effects. *J. Agric. Food Chem.* 53:562–565.

Allevi, P., Anastasia, M., Bingham, S., et al. 1998. Synthesis of carminic acid, the colourant principle of cochineal. *J. Chem. Soc. Perkin Trans. I* 1:575–582.

Arai, T., Kojima, R., Motegi, Y., Kato, J., Kasumi, T., Ogihara, J. 2015. PP-O and PP-V, *Monascus* pigment homologues production and phylogenetic analysis in *Penicillium purpurogenum. Fungal Biol.* 119:1226–1236.

Bar-Zvi, S., Lahav, A., Harris, D., Niedzwiedzki, D. M, Blankenship, R.E., Adir, N. 2018. Structural heterogeneity leads to functional homogeneity in A. marina phycocyanin. *Biochim. Biophys. Acta Bioenerg.* 1859:544–553.

Blanc, P. J., Loret, M. O., Santerre, A. L., et al. 1994. Pigments of *Monascus. J. Food Sci.* 59:862–865.

Borges, M.E., Tejera, R.L., Díaz, L., Esparza, P., Ibáñez, E. 2012. Natural dyes extraction from cochineal (*Dactylopius coccus*). New extraction methods. *Food Chem.* 132:1855–1860.

Caro, Y., Venkatachalam, M., Lebeau, J., Fouillaud, M., Dufossé, L. 2015. Pigments and colorants from filamentous fungi. In *Fungal Metabolites.* Merillon, J. -M., Ramawat, K. G. Eds. Springer International Publishing, Switzerland, pp. 1–69.

Chen, F. C., Manchand, P. S., Whalley, W. B. 1971. The chemistry of fungi. *J. Chem. Soc.* 21:3577–3579.

Choe, D., Lee, J., Woo, S., Shin, C. S. 2012. Evaluation of the amine derivatives of *Monascus* pigment with anti-obesity activities. *Food Chem.* 134:315–323.

Code of Federal Regulation. 2016. Code of Federal Regulations (Annual Edition). https://www.govinfo.gov/app/collection/cfr/2016.

Dapson, R.W. 2005. A method for determining identity and relative purity of carmine, carminic acid and aminocarminic acid. *Biotech. Histochem.* 80:201–205.

Daud, N. F. S., Said, F. M., Ramu, M., Yasin, N. M. H. 2020. Evaluation of biored pigment extraction from *Monascus purpureus* FTC 5357. *IOP Conf. Series: Mater. Sci. Eng.* 736:022084.

Doerr, A. 2006. A new avenue for mass spectrometry. *Nat. Methods* 3:72–73.

Duran, N., Tixiera, M. F. S., de Conti, R., Esposito, E. 2002. Ecological friendly pigments from fungi. *Crit. Rev. Food Sci. Nutr.* 42:53–66.

EFSA (European Food Safety Authority). 2015. Scientific Opinion on the re-evaluation of cochineal, carminic acid, carmines (E 120) as a food additive. *EFSA J.* 13:4288–4352.

Eisenberg, I., Harris, D., LeviKalisman, Y., et al. 2017. Concentration-based self-assembly of phycocyanin. *Photosynth. Res.* 134:39–49.

Elyashberg, M. E., Blinov, K. A., Williams, A. J., Martirosian, E. R., Molodtsov, S. G. 2002. Application of a new expert system for the structure elucidation of natural products from the 1D and 2D NMR data. *J. Nat. Prod.* 65:693–703.

European Commission. 2012. Commission regulation 231/2012/EU of 9 March 2012 laying down specifications for food additives listed in Annexes II and III to Regulation No. 1333/2008 of the European Parliament and of the Council. *Off. J. Eur. Union* L83:1–295.

Feng, F., Zhao, Y, Yong, W., Sun, L., Jiang, G., Chu X. 2011. Highly sensitive and accurate screening of 40 dyes in soft drinks by liquid chromatography–electrospray tandem mass spectrometry. *J. Chromatogr. B.* 879:1813–1818.

Ferraro, G., Imbimbo, P., Marseglia, A., Lucignano, R., Monti, D. M., Merlino, A. 2020a. X-ray structure of C-phycocyanin from *Galdieria phlegrea*: determinants of thermostability and comparison with a C-phycocyanin in the entire phycobilisome. *BBA Bioenerg.* 1861:148236. DOI: 10.1016/j.bbabio.2020.148236.

Ferraro, G., Imbimbo, P., Marseglia, A., et al. 2020b. A thermophilic C-phycocyanin with unprecedented biophysical and biochemical properties. *Int. J. Biol. Macromol.* 150:38–51.

Fielding, B. C., Holker, J. S. E., Jones, D. F., et al. 1961. The chemistry of fungi. XXXIX. The structure of monascin. *J. Chem. Soc.* 4579–4589. doi: 10.1039/JR9610004579.

Fu, E., Friedman, L., Siegelman H. W. 1976. Mass-spectral identification and purification of phycoerythrobilin and phycocyanobilin. *Biochem. J.* 179:1–6.

Gao, J-M,. Yang, S-X., Qin, J-C. 2013. Azaphilones chemistry and biology. *Chem. Rev.* 113:4755–4811.

Gosetti, F., Chiuminatto, U., Mazzucco, E., Mastroianni, R., Marengo, E. 2015a. Ultra-high-performance liquid chromatography/tandem high-resolution mass spectrometry analysis of sixteen red beverages containing carminic acid: Identification of degradation products by using principal component analysis/discriminant analysis. *Food Chem.* 167:454–462.

Gosetti, F., Chiuminatto, U., Mastroianni, R., Mazzucco; E., Manfredi, M., Marengo, E. 2015b. Retrospective analysis for the identification of 4-aminocarminic acid photo-degradation products in beverages. *Food Add. Contam.: Part A* 32:285–292.

Hajjaj, H., Klaebe, A., Loret, M. O., Tzedakis, T., Goma, G., Blanc, P. J. 1997. Production and identification of N-glucosylrubropunctamine and N-glucosylmonascorubramine from *Monascus ruber* and occurrence of electron donor-acceptor complexes in thses red pigments. *Appl. Environ. Microbiol.* 63:2671–2678.

Harris, M., Stein, B.K., Tyman, J.H.P., Williams, C.M. 2009. The structure of the colourant/pigment, carmine derived from carminic acid. *J. Chem. Res.* 7:407–409.

Hossain, C. F., Okuyama, E., Yamazaki, M. 1996. A new series of coumarin derivatives having monoamine oxidase inhibitory activity from *Monascus anka. Chem. Pharm. Bull.* 44:1535–1539.

Hsu, Y. W., Hsu, I. C., Liang, Y. H., Ku, O. Y. H., Pan, T. M. 2010. Monaphilones A-C, three new antiproliferative azaphilone derivatives from *Monascus purpureus* NTU 568. *J. Agric. Food Chem.* 58:8211–8216.

Hsu, Y. W., Hsu, I. C., Liang, Y. H., Ku, O. Y. H., Pan, T. M. 2011. New bioactive orange pigments with yellow fluorescence from *Monascus* fermented dioscorea. *J. Agric. Food Chem.* 59:4512–4518.

Huang, Z., Xu, Y., Li, L., Yanping, L. 2008. Two new *Monascus* metabolites with strong blue fluorescence isolated from red yeast rice. *J. Agric. Food Chem.* 56:112–118.

Ichi, T.; Koda, T.; Yokawa, C.; Sakata, M., Sato, H. 2002. Purified cochineal and method for its production. *US 20020058016A1*, United States Patent and Trademark Office.

Isbrandt, T., Tolborg, G., Odum, A., Workman, M., Larsen, T. O. 2020. Atrorosins, a new subgroup of *Monascus* pigments from *Talaromyces atroroseus*. *Appl. Microbiol. Biotechnol.* 104:615–622.

Jang, H., Choe, D., Shin, C. S. 2014. Novel derivatives of *Monascus* pigment having a high CETP inhibitory activity. *Nat. Prod. Res.* 28:1427–1431.

JECFA (Joint FAO/WHO Expert Committee on Food Additives), 2006. Combined Compendium of Food Additive Specifications. Monograph 1. http://www.fao.org/3/a-a0691e.pdf.

Jia, W., Chu, X., Ling, Y., Huang, J., Lin, Y., Chang, J. 2014. Simultaneous determination of dyes in wines by HPLC coupled to quadrupole orbitrap mass spectrometry. *J. Sep. Sci.* 37:782–791.

Jongrungruangchock, S., Kiltakoop, P., Yongsmith, B. 2004. Azaphilone pigments from a yellow mutant of the fungus *Monascus kaoliang*. *Phytochemistry* 65:2569–2575.

Jung, H., Kim, C., Kim, K., Shin, C. S. 2003. Colour characteristics of *Monascus* pigments derived by fermentation with various amino acids. *J. Agric. Food Chem.* 51:1302–1306.

Kim, C., Jung, H., Kim, Y. O., Shin, C. S. 2006. Antimicrobial activities of amino acid derivatives of *Monascus* pigments. *FEMS Microbiol. Lett.* 264:117–124.

Kim, D., Ku, S. 2018. Beneficial effects of *Monascus* sp. KCCM 1093 pigments and derivatives: a mini review. *Molecules* 23:1–15.

Kulyk, D.S., Sahraeian, T., Wan, Q., Badu-Tawiah, K. 2019. Reactive olfaction ambient mass spectrometry. *Anal. Chem.* 91:6790–6799.

Lebeau, J., Venkatachalam, M., Fouillaud, M., Petit, T., Vinale, F., Dufossé, L., Caro, Y. 2017. Production and new extraction method of polyketide red pigments produced by ascomycetous fungi from terrestrial and marine habitats. *J. Fungi* 3:34.

Li, J. J., Shang, X. Y., Li, I. I., Lin, M. T., Zheng, J. Q., Jin, Z. I. 2010. New cytotoxic azaphilone from *Monascus purpureus* fermented rice (red rice). *Molecules* 15:1958–1966.

Lin, T. F., Yakushijin, K., Buchi, G. H., Demain, A. L. 1992. Formation of water soluble *Monascus* pigments by biological and semisynthetic processes. *J. Indus. Microbiol.* 9:173–179.

Lloyd, A. G. 1980. Extraction and chemistry of cochineal. *Food Chem.* 5:91–107.

Loret, M. O., Morel, S. 2010. Isolation and structural characterization of two new metabolites from *Monascus*. *J. Agric. Food Chem.* 58:1800–1803.

Luo, X., Smith, P., Raston, C. L., Zhang, W. 2016. Vortex fluidic device-intensified aqueous two phase extraction of C-Phycocyanin from *Spirulina maxima*. *ACS Sustain. Chem. Eng.* 4:3905–3911.

Maier, M.S., Parera, S.D., Seldes, A.M. 2004. Matrix-assisted laser desorption and electrospray ionization mass spectrometry of carminic acid isolated from cochineal. *Int. J. Mass Spectrom.* 232:225–229.

Mancini, J. A., Sheehan, M., Kodali, G., et al. 2018. *De novo* synthetic biliprotein design, assembly and excitation energy transfer. *J. R. Soc. Interface* 15:20180021. doi: 10.1098/rsif.2018.0021.

Mapari, S. A. S., Hansen, M. E., Meyer, A. S., Thrane, U. 2008. Computerized screening for novel producers of *Monascus*-like food pigments in *Penicillium* species. *J. Agric. Food Chem.* 56:9981–9989.

Mapari, S. A. S., Meyer, A. S., Thrane, U. 2006. Colorimetric characterization for comparative analysis of fungal pigments and natural food colorants. *J. Agric. Food Chem.* 54:7027–7035.

Mapari, S. A. S., Meyer, A. S., Thrane, U., Frisvad, J. C. 2009. Identification of potentially safe promising fungal cell factories for the production of polyketide natural food colorants using chemotaxonomic rationale. *Microb. Cell Factories* 8:24.

Minic, S. L., Stanic-Vucinic, D., Mihailovic, J., Krstic, M., Nikolic, M. R., Cirkovic Velickovic, T. 2016. Digestion by pepsin releases biologically active chromopeptides from C-phycocyanin, a blue-colored biliprotein of microalga Spirulina. *J. Proteom.* 147:132–139.

Morales-Oyervides, L., Ruiz-Sanchez, J. P., Oliveira, J. C., Sousa-Gallagher, M. J., Mendez-zavala, A., Giuffrida, D., Dufossé, L., Montanez, J. 2020. Biotechnological approaches for the production of natural colorants by Talaromyces/Penicillium: A review. *Biotechnol. Adv.* 43:107601.

Mukherjee, G., Singh, S. K. 2011. Purification and characterization of a new red pigment from *Monascus purpureus* in submerged fermentation. *Process Biochem.* 46:188–192.

Nair, D., Krishn, J. G., Panikkar, M.V.N., Nair, B. P., Pai, J. G., Nair, S. S. 2018. Identification, purification, biochemical and mass spectrometric characterization of novel phycobiliproteins from a marine red alga, *Centroceras clavulatum*. *Int. J. Biol. Macromol.* 150:38–51.

Nakayama, N., Ohtsu, Y., Maezawa-Kase, D., Sano, K. 2015. Development of a rapid and simple method for detection of protein contaminants in carmine. *Int. J. Anal. Chem.*, 2015:748056.

Nielsen, K. F., Smedsgaard, J., Larsen, T. O., Lund, F., Thrane, U., Frisvad, J. C. 2004. Chemical identification of fungi: Metabolite profiling and metabolomics. In *Fungal Biotechnology in Agricultural, Food, and Environmental Applications*. Arora, D. K., Eds. Marcel Dekker, Inc., NewYork, pp. 19–35.

Ogihara, J., Kato, J., Oishi, K., Fugimoto, Y. 2001. PP-R, 7-(2-hydroxyethyl)-monascorubramine, a red pigment produced in the mycelia of *Penicillium* sp. *AZ. J. Biosci. Bioeng.* 91:44–47.

Ogihara, J., Kato, J., Oishi, K., Fugimoto, Y., Eguchi. T. 2000. Production and structural analysis of PP-V, a homologue of monascorubramine, produced by a new isolate of *Penicillium* sp. *J. Biosci. Bioeng.* 90:549–554.

Piron, R. Bustamante, T, Barriga, A., Lagos, N. 2019. Phycobilisome isolation and C-phycocyanin purification from the cyanobacterium A*phanizomenon gracile. Photosynthetica* 57:491–499.

Rinalducci, S., Roepstorff, P., Zolla, L. 2009. De novo sequence analysis and intact mass measurements for characterization of phycocyanin subunit isoforms from the blue-green alga *Aphanizomenon flos-aquae. J. Mass Spectrom.* 44:503–515.

Rodriguez-Amaya, D.B. 2016. Natural food pigments and colorants. *Curr. Opin. Food Sci.* 7:20–26. DOI: 10.1016/j.cofs.2015.08.004.

Sabatino, L., Scordino, M., Gargano, M., et al. 2012. Aminocarminic acid in E120-labelled food additives and beverages. *Food Addit. Contam.: Part B* 5:295–300.

Sato, K., Goda, Y., Sakamoto, S. S., Shibata, H., Maitani, T., Yamada, T. 1997. Identification of major pigments containing D-amino acid units in commercial *Monascus* pigments. *Chem. Pharm. Bull.* 45:227–229.

Schmidt-Jacobson, J.F., Sakstrup-Frandsen, R. 2010. Method for the preparation of a carminic acid lake. US 20100061949 A1, United States Patent and Trademark Office.

Scotter, M.J. 2011. Methods for the determination of European Union-permitted added natural colours in foods: a review. *Food Addit. Contam.* 28:527–596.

Shi, K., Tang, R., Huang, T., Wang, L., Zhenqiang, W. 2017. Pigment fingerprint profile during extractive fermentation with *Monascus anka* GIM 3.592. *BMC Biotechnol.* 17:46.

Shishupala, S. 2008. Biochemical analysis of fungi using matrix-assisted laser desorbtion/ionization time-of-flight mass spectrometry (MALDI-TOF-MS). In: Novel techniques and ideas in mycology. Sridhar, K. R, Barlocher, F., Hyde, K.D. Eds. *Fungal Diversity Research Series*, Fungal Diversity Press, NHNBS Bonn, Vol. 20, pp. 327–372.

Smedsgaard, J., Frisvad, J. C. 1996. Using direct electrospray mass spectrometry profiling of crude extracts. *J. Microbiol. Methods* 25;5–17.

Srianta, I., Ristiarins, S., Nugerahani, I. 2020. Pigments extraction from *Monascus-* fermented durian seed. *IOP Conf. Series: Earth Environ. Sci.* 443:012008.

Su, N.-, Lin, Y., Lee, M., Ho, C. 2005. Ankaflavin from Monascus-fermented red rice exhibits selective cytotoxic effect and induces cell death on Hep G2 cells. *J. Agric. Food Chem.* 53:1949–1954.

Teng, S. S., Feldheim, W. 1998. Analysis of anka pigments by liquid chromatography with diode array detection and tandem mass spectrometry. *Chromatographia* 47:529–536.

Tolborg, G., Odum, A. S. R., Isbrandt, T., Larsen, T. O., Workman, M. 2020. Unique processes yielding pure azaphilones in *Talaromyces atroroseus. Appl. Microbiol. Biotechnol.* 104:603–613.

Tolborg, G., Petersen, T. I., Larsen, T. O., Workman, M. 2020a. Process for producing an azaphilone in *Talaromyces atroroseus. US Patent* US2020/0165645 A1, May.

Tong, X., Prasanna, G., Zhang, N., Jing, P. 2020. Spectroscopic and molecular docking studies on the interaction of phycocyanobilin with peptide moieties of C-phycocyanin. *Spectrochim. Acta Part A: Mol. Biomol. Spectrosc.* 236:118316.

Wrolstad, R.E., Culver, C.A. 2012. Alternatives to those artificial FD&C food colorants. *Annu. Rev. Food Sci. Technol.* 3:59–77.

Wu, M. D., Cheng, M. J., Liu, T. W., Chen, Y. L., Chan, H. Y., Chen, H. P., Wu, W. J., Chen K. P., Yuan, G. F. 2015. Chemical constituents of the fungus *Monascus pilosus* BCRC 38093-fermented rice. *Chem. Nat. Comp.* 51:554–556.

Yamakawa, Y., Oosuna, H., Yamakawa, T., Aihara, M., Ikezawa, Z. 2009. Cochineal extract-induced immediate allergy. *J. Dermatol.* 36:72–74.

Yuliana, A., Singgih, M., Julianti, E., Blanc, P. J. 2017. Derivatives of azaphilone *Monascus* pigments. *Agric. Biotechnol.* 9:183–194.

9 Flavour and Mass Spectrometry

Jean-Luc Le Quéré and Géraldine Lucchi
French National Research Institute for Agriculture, Food and Environment (INRAE)

CONTENTS

9.1 Introduction 137
9.2 Aroma and GC-MS 140
    9.2.1 GC-EI-MS 140
    9.2.2 GC-CI-MS 142
    9.2.3 MDGC- and GC×GC-MS 144
    9.2.4 GC-O-MS 145
    9.2.5 Quantification 146
    9.2.6 New Developments: GC-HRMS and GC-IMS 147
    9.2.7 MS Databanks and Software 148
9.3 Taste and HPLC-MS 148
9.4 Flavour Release and Perception: *In Vivo* Analyses 158
    9.4.1 Volatiles, *In Vivo* 158
    9.4.2 Non-Volatiles, *In Vivo* 161
9.5 Global Analyses of Flavour: 'Flavouromics' 162
9.6 Conclusion 167
References 168

9.1 INTRODUCTION

Among the criteria that determine food choice by the consumer, sensory properties of foodstuffs are of prime importance, where flavour plays a prominent role. Flavour is sensed by the integration of sensations in the brain. The perceptions occur in the oral cavity when foods are eaten: taste and odour, feelings of pain, heat and cold (chemesthesis or trigeminal sensitivity) and tactile sensation. However, sensory food characteristics are generally experienced as a unique perception commonly called 'taste'. This familiar 'taste' is a holistic perception of at least odour and taste, generally called 'flavour perception' (Sinding et al., 2021). Therefore, instrumental assessment of flavour should at least include analyses of volatile odourant molecules sensed in the nose at the olfactory receptors either via the orthonasal (odour) or the retronasal (aroma) routes, and of essentially non-volatile sapid compounds sensed on the tongue at the taste buds (taste). These analyses have concentrated on the characterisation of volatile organic compounds (VOCs), primarily because of the major importance of aroma in the overall flavour, as easily demonstrated by the difficulties encountered when attempting to identify a particular flavour if the airflow through the nose is prevented. Second, these volatiles were more amenable to conventional instrumental analyses based on a well-mastered separation technique, namely gas-liquid chromatography (GC). However, some significant efforts have been made recently to develop instrumental procedures based on high-performance liquid chromatography (HPLC) to determine non-volatile components responsible for food taste.

DOI: 10.1201/9781003091226-11

Hyphenated to these separation techniques, mass spectrometry (MS) plays a fundamental role in identifying flavour molecules.

It is not necessary to identify the total volatile and non-volatile content of foods to understand their flavour. In most cases, flavour profiles determined by a panel in sensory evaluation are hardly related to this total content. It appears much more efficient to concentrate the identification efforts on those compounds that significantly contribute to the flavour. This is particularly true for taste components. Therefore, modern methodology used to study flavour compounds includes various steps, the ultimate aim of identifying the compounds that are relevant to the flavour. The aroma compounds are mainly hydrophobic or lipophilic, and consequently, they tend to concentrate on food fat according to their water–fat partition coefficient. This is of considerable importance for their study because they must be separated from the food matrix by a convenient extraction method to be amenable to instrumental analyses suitable for volatile compounds. Taste components, on the contrary, are essentially water-soluble non-volatile compounds, and specific extraction procedures have to be handled. Once a suitable extraction method has been chosen for the volatile fraction, the following steps involve various forms of GC among which gas chromatography–olfactometry (GC-O), that plays a prominent role in determining the key volatile compounds that contribute significantly to food aroma (Leland et al., 2001), and GC-MS, as an essential tool for the identification of those key odourants. The next critical step in analysing food aromas is a precise quantification of all the key volatile components. Two main routes that use MS may be implemented for this task. The first one combines a common standard addition method with increasing quantities of the compounds to be quantified, generally in conjunction with dynamic headspace (DHS) sampling and GC-MS using the selected ion monitoring (SIM) mode. The second route, the stable-isotope dilution assay (SIDA), uses standard compounds labelled with stable isotopes, similar to the unlabelled components to be quantified. These labelled internal standards, the physical and chemical properties of which are identical to those of the substances of interest, are added in known concentrations to the food matrix before extraction. The quantification is effected by measuring the ratio of the unlabelled to the labelled compound by GC-MS. As a final compulsory validation step, the components of interest are incorporated in amounts equivalent to their measured quantities and evaluated together in a model food matrix, preferably not very different from the starting material.

For non-volatiles, the general procedure mimics the one used for volatiles (extraction, separation, identification and possible quantification by HPLC-MS of taste impact compounds finally included in recombined food models for sensory evaluation). The only requirement for the fractionation scheme is to use food-grade solvents, as the fractions have to be sensorially evaluated and, for those that contain key-tastants, used in the final recombination step. The complete approach applied to the active chemosensory key volatile and non-volatile molecules (assigned as 'sensometabolome') has been rationalised by the molecular sensory science concept, so-called 'sensomics' (Schieberle and Hofmann, 2011). The development of efficient analytical procedures and tools to study volatile and non-volatile flavour compounds has allowed research on off-flavours. This research has covered all the steps from raw material to the final product, including manufacture, packaging and storage.

However, relating flavour compounds composition to flavour perception by humans is not straightforward. It is mainly because it is still not completely understood how the various components combine to produce an overall sensory impression. Moreover, cross-modal sensory interactions, such as aroma–taste or flavour–texture interactions, and interactions of trigeminal sensations with taste and aroma, affect perception (Guichard et al., 2017). Perception of flavour is a dynamic process during which the concentration of aroma molecules at the olfactory epithelium, with air, continuously exhaled and inhaled, and of sapid compounds at the taste buds on the tongue varies with time as they are released progressively from the food in the mouth during consumption. The release kinetics depends on the food matrix itself and oral food processing (Salles et al., 2011), such as mastication behaviour and food bolus formation with saliva, for which substantial inter-individual variations exist, due to physiological differences. This renders the acquisition of dynamic flavour release profiles more relevant compared to static methods. Sensory methods, such as time intensity (TI), or the more recent temporal methods: temporal dominance of sensations (TDS) (Pineau et al., 2009) and temporal check-all-that-apply (TCATA) (Castura et al., 2016), are used to account for the dynamic and time-related aspects of flavour perception. MS techniques that measure flavour compounds directly in the mouth or the nose have been developed in the last 20 years to obtain data that are supposed to reflect the pattern of flavour molecules released from food in real-time during consumption, considered to be representative of flavour perception (Taylor, 2002). For sampling aroma from the nose (nose space), techniques based on direct-injection mass spectrometry (DIMS) operated at atmospheric or reduced pressure (atmospheric pressure chemical ionisation [APCI] or proton transfer reaction [PTR] for instance) have been intensively implemented (Beauchamp and Zardin, 2017). Techniques to follow the *in vivo* release of non-volatiles in the mouth have also been developed. Still, they are essentially offline techniques: saliva sampling at different times followed by direct liquid injection mass spectrometry or ionic LC procedures (Taylor and Linforth, 2003).

The methodologies described above implement targeted methods aimed at identifying flavour-active compounds relative to flavour perception. However, specific instrumental MS techniques have been developed for the global analysis of food flavour in a non-targeted manner analogous to the 'metabolomics' approach. Thus, global VOC profiles may be used as fingerprints of food samples, sometimes referred to as 'volatilome', to categorise samples based on specific characteristics such as species, origin, applied pre-harvest treatment, post-harvest and shelf-life, process, and organoleptic properties. Generally, the investigations aim to pinpoint differences between similar products for authentication or adulteration purposes. Still, some studies have also been conducted to correlate the products volatilome with their sensory evaluation (Deuscher et al., 2019). Coupling GC-MS with HS sampling is generally recognised as the gold standard analytical procedure to obtain VOC fingerprinting. HPLC-MS has also been used for fingerprinting with some success (Kuś and van Ruth, 2015; Stark et al., 2020). Despite their performance, the main drawback of chromatography-based methods is the lengthy analysis time detrimental to high-throughput data collection. Evaluation of the volatiles emitted from food using gas sensors arranged in several commercial instruments, including the so-called electronic nose (e-nose), has been demonstrated for more than 20 years (Gebicki, 2016). Although they can produce typical patterns useful for classification purposes, none of these instruments meets the food industry requirements for quality control in terms of precision, specificity, repeatability, reproducibility, sensitivity and stability (Loutfi et al., 2015; Sanaeifar et al., 2017).

Moreover, these instruments can be hardly used to identify single odourants or differentiate samples differing only in subtle, distinctive notes. To conduct classification tasks, fast analytical methods based on MS seem more powerful and reliable. The first method to be cited consists of a global analysis of a total HS sample by a mass spectrometer operated in electron ionisation (EI) mode, without any prior GC separation. This dedicated method is often described as a robust variant of the e-nose, referred to as 'MS-based electronic nose' or 'MS-based e-nose'. As an extension of these MS-based e-noses, other ionisation methods have been made available. They refer to DIMS technologies (Biasioli et al., 2011) or online chemical ionisation mass spectrometry (CIMS) techniques (Beauchamp and Zardin, 2017). They use soft ionisation conducted at low pressure, as in PTR and selected ion flow tube (SIFT) systems, or at atmospheric pressure, as in APCI mode, and give rise to ionised molecular species and limited fragmentations. Although less commonly used, another global method worthwhile to be cited is the pyrolysis-MS technique, where a tiny food sample is pyrolysed at up to 500°C. The resulting volatile fraction, characteristic of the flavour and the matrix, is globally analysed by a mass spectrometer (Aries and Gutteridge, 1987; Pérès et al., 2002). Ambient MS techniques, which allow direct sample analysis without sample preparation, also found applications in VOCs profiling. Thus, direct analysis in real-time (DART) that proved its usefulness in volatile fractions characterisation (Li, 2012; Jastrzembski et al., 2017) was used for species classification (Deklerck et al., 2017). Desorption electrospray ionisation (DESI) (Byliński et al., 2017) and secondary electrospray ionisation (Wang et al., 2016) have also demonstrated some capabilities. With all these analytical profiling methods useful for addressing a 'volatilomics' approach, a mass pattern, or fingerprint, is obtained for each sample, and sophisticated chemometrics should be used to reach the desired samples categorisation (Deuscher et al., 2019).

What're more, biochemical reactions and physical processes determine stable-isotope signatures of natural compounds. As geographical, climatic and agricultural determinants contribute to isotope fractionation, isotope ratios of carbon, nitrogen, oxygen and hydrogen measured by isotope ratio mass spectrometry (IRMS) are helpful indicators for authenticity assessment. Combined for a long time to GC (Matthews and Hayes, 1978) or, more recently, to LC (Godin and McCullagh, 2011), and applied to flavour molecules (Mosandl et al., 1990; Kawashima et al., 2018), IRMS constitutes a robust tool to differentiate between synthetic and natural flavours as well as determine geographical provenances, and as such, to prevent frauds and adulteration.

In all the approaches described above, MS plays a fundamental role illustrated in this chapter dedicated to food flavour.

9.2 AROMA AND GC-MS

GC is the method of choice to identify and quantify VOCs in food. This very mature technology is a reliable, reproducible, robust, selective, sensitive and easy-to-use method to analyse thermostable molecules, with many stationary phases of diverse polarities, for almost all the applications. Minimal sample preparation is needed, and the instruments are now compact and inexpensive. Compared to LC, GC provides a higher chromatographic resolution and peak capacity for complex mixtures, a simpler mobile phase (hydrogen or helium) with fewer problems of solubility and contamination, and an easier separation using a temperature program (McEwen and McKay, 2005). Furthermore, there is less risk of coelution and column bleeding. Headspace-solid phase microextraction (HS-SPME), cumulative SPME sampling, stir bar sorptive extraction (SBSE) analysis and aroma extracts obtained by liquid–liquid extraction are, among other techniques (Roman et al., 2020), methods of choice to detect and identify active compounds that are responsible for odour and aroma in foods by GC-MS techniques (Tranchida, 2020).

9.2.1 GC-EI-MS

When MS is coupled with GC, the resolving power increases since the MS dimension provides additional identification capacity compared to the common flame ionisation detector (FID). In GC-MS, the separated analytes are ionised then analysed by MS. The primary ionisation method is the EI under vacuum conditions (Märk and Dunn, 1985). High kinetic energy (70 eV) is commonly used to obtain ionisation efficiency that provides sensitivity and mass spectral reliability. The electrons pass through the ionisation chamber by cutting the perpendicular flow of analytes. Energy exchanges occur, causing reproducible molecular fragmentations whatever instrument is used. For a wide range of organic compounds, the fragmentation profile, early rationalised in fragmentation rules in dedicated treatises (Budzikiewicz et al., 1964; McLafferty, 1980; McLafferty and Turecek, 1993), combined with GC retention index as an orthogonal method, permits an easy identification. This is facilitated by comparing the results to mass spectral and retention data compiled in numerous existing commercial and in-house libraries. For instance, the determination of volatile compounds in different Italian and Spanish grapes varieties (Vasile-Simone et al., 2017), aroma profiling in white sturgeon caviar at different stages of ripening (Lopez et al., 2020) and characterisation of the key aroma compounds in dog foods (Yin et al., 2020), are just a few examples among very different topics where GC-EI-MS no longer has to prove its effectiveness.

When GC-EI-MS analyses complex samples, coelutions of two or more components can occur, resulting in an overlap of peaks observed in the total ion chromatogram (both species have retention times that differ by less than the resolution of the method). For some organic compounds (i.e., isomers), fragment ions can be less specific, and identification becomes more complicated due to very similar fragmentation patterns. The molecular ion is rarely observed due to intense fragmentations caused by high-energy ionisation; this lack of information cannot help identify. Moreover, if one coeluting molecular species is overexpressed, the low-intensity fragments go entirely unnoticed, even with the best deconvolution methods. The excessive formation of low-mass fragment ions is also a drawback. A large number of these small fragments is common to many chemical classes, and this lack of specificity does not provide any convincing proof to resolve the identification of the coeluting compounds.

To decrease in-source fragmentations and to have more information-rich mass spectra, especially on molecular ions, operating an EI ion source at energies below 70 eV has been reported in a few publications (Tranchida et al., 2019; Pua et al., 2020). The ionisation energy (IE) is the minimum amount of energy required to ionise a neutral molecule, and most of the VOCs have an IE around 10 eV. To avoid a significant reduction of the ionisation efficiency and thus sensitivity, manufacturers have developed low EI (LEI) sources with modified lens geometry (Agilent Technologies) and dedicated ion optics (Markes) (Kranenburg et al., 2020). Furthermore, for electron energy set to 12 eV, lower in-source temperatures are generally correlated to higher molecular ion intensities for early eluting compounds (inverted in late elution times) and different chemical classes. This is what was observed on 22 chemicals at six different concentration levels, including esters, alcohols, terpenes, phenols and ketones, to understand the effects of EI source parameters on molecule ionisation (Lau et al., 2019). Despite the valuable information on molecular ions, the main drawback of using LEI is that the resulting mass spectra can hardly be confronted with the EI-libraries spectral data consensually acquired at 70 eV.

Ionised masses are separated according to their mass-to-charge (*m/z*) ratios, and different mass analysers can be used in GC-MS food analyses (Mellon, 2003). The most simple, relatively small and low-cost system is the single-quadrupole type analyser. An oscillating electrical field created between four parallel rods stabilises or destabilises the ions selectively, depending on the radio frequency (RF) level. For a given RF, only ions with the same *m/z* ratio pass through the quadrupole; the others are deflected and collide against the rods. Therefore, it is easy to scan all the ions from the ionisation source by varying the RF value. A quadrupole mass analyser acts as a mass-selective filter in MS mode analysis only. It is a very commonly used instrument to qualify the content of VOCs inside the samples and compare extraction techniques. Thereby, Moyano et al. (2019) have optimised and validated their dynamic headspace–thermal desorption–GC-MS (DHS-TD-GC-MS) method for their wineomics studies with this strategy. This technique has also shown an excellent olive oil quality classification (Sales et al., 2019).

Triple quadrupole (QQQ) instruments have the advantage to operate in MS/MS mode analysis. Precursor ions are selected thanks to the first quadrupole; the second one leads to fragmentation by collision-induced dissociation (CID). The third quadrupole separates the fragment ions, also called product ions (MS/MS). Then, the third quadrupole may be used to select a specific product ion for multiple reaction monitoring (MRM) detection modes. Fragment ions and their precursor constitute a transition. In most cases, the most intense transition is exploited to quantify the molecular species, while the second most intense transition allows the aroma compound to be identified.

To enable rapid and reliable quantification of selected terpenes, terpenoids, and esters of hop aroma compounds in beer, QQQ has been an excellent technique for quantification across a wide working range (Dennenlohr et al., 2020). Despite getting a good reproducibility, quadrupoles have a limited resolution, peak height vs. mass response is not always linear, and CID efficiency depends strongly on energy, collision gas and pressure.

Ion traps (ITs) operate by using direct current and RF electric fields to store ions in a trap and to eject successive *m/z* ratios to the detector. This compact mass analyser has a high sensitivity but is inadequate for quantification. It is subject to space charge effects and ion-molecule reactions due to the confinement of molecular species for several milliseconds in the trap. ITs are handy for MS/MS analyses. For instance, GC-IT-MS was applied to a sample of Brazilian red wine and allowed the extraction and identification of 60 VOCs, notably the classes of esters, alcohols and fatty acids, among which 55 compounds of importance in merlot red wines from the south of Brazil were quantified (Arcari et al., 2017).

Time-of-flight (ToF) mass analysers measure the time that ions of different masses move from the ion source to the detector. After acceleration, ions of different masses do not have the same kinetic energy. This kinetic energy defines those ions speed. Their speed is inversely proportional to the square root of their *m/z* ratio: smaller ions fly faster than heavier ions. Therefore, according to an adequate calibration, it is possible to determine the mass from the ions 'ToF'. This is the fastest mass analyser with a high ion transmission, and for this reason, it is often paired with two-dimensional (2D) chromatography for comparative studies or profiling analyses. Thus, 288 volatiles from four Chinese traditional fermented fishes have been identified, and comprehensive GC (GC×GC) combined with ToF-MS provides a robust technical method to understand the flavour characteristics of this very complex food matrix (Chen et al., 2021). Profiling of volatile compounds by GC×GC-ToF-MS allowed 61 volatile compounds to be selected for successful discrimination of whiskies (Jelen et al., 2019). ToF can be combined with a quadrupole analyser to benefit from both the simplicity of use of the quadrupole and the performance of the ToF in terms of mass accuracy and resolution (Q-ToF). This kind of hybrid instrument is beneficial to perform identification and quantification simultaneously; thus, recently volatile composition of several blueberry cultivars has been well described, and some esters and terpenoids have been quantified and related to sensory attributes (Cheng et al., 2020).

9.2.2 GC-CI-MS

Some compounds of specific chemical classes (e.g., aliphatic aldehydes and ketones) or possessing unsaturated bonds and cyclic structures have a non-specific fragmentation pattern in EI-MS, or they display exclusively low-mass fragment ions making their identification or their quantification quasi-impossible in complex mixtures. If LEI can help, molecular weight information may be advantageously obtained by chemical ionisation (CI). In CI, a reagent gas is introduced into the source at high pressure and is ionised at high-energy EI to form reactant ions (Harrison, 1992). These ions react with the analytes by proton transfer or charge transfer to afford quasi-molecular ions, sometimes accompanied by characteristic adduct ions. This type of ion-molecule reaction is less energetic than the standard EI at 70 eV. Thus, fragmentations are limited, and molecular weight determination is made easier.

The choice of an ideal CI reagent gas is vital to control fragmentation via the amount of energy transferred to the analytes. Methane, isobutane and ammonia are usually used as protonating reagent gas for CI experiments in positive ionisation (PCI). Methane is the most widely used gas in CI and is the strongest proton donor with a proton affinity (PA) of 544 kJ/mol and IE of 5.7 eV; it allows most molecules to be ionised because its PA is low. However, fragment ions can be observed. The highest PA reagent gas significantly reduces fragmentation. Isobutane (PA 678 kJ/mol and IE 8.5 eV) can be used for many compounds except for some molecules with tiny PA. For softer ionisation, ammonia (PA 854 kJ/mol and IE 9.0 eV) is frequently used and frequently produces adduct ions of the type $[M+NH_4^+]$, of diagnostic significance because it can easily form hydrogen bonds. Fragmentation is not observed, but its greater selectivity does not authorise its use for all chemical classes. CI was used systematically to strengthen the identifications in the recent characterisation of key aroma compounds of dark chocolates differing in organoleptic properties (Deuscher et al., 2020). For example, the mass spectra of 1H-pyrrole-2-carbaldehyde obtained in EI and CI with methane and ammonia as reagent gases are displayed in Figure 9.1. CI spectra revealed unambiguous MH^+ ions together with diagnostic adduct ions $M+C_2H_5^+$ (M+29), and $M+C_3H_5^+$ (M+41) for CH_4-CI and $M+NH_4^+$ (M+18) and $M+N_2H_7^+$ (M+35) for NH_3-CI and CI appeared as a successful method to confirm the identification of impure EI mass spectra (Deuscher et al., 2020).

Many other reagent gas alternatives have been explored for specific applications, such as tetramethylsilane (for analytes containing nucleophilic groups), ethanolamine (for compounds

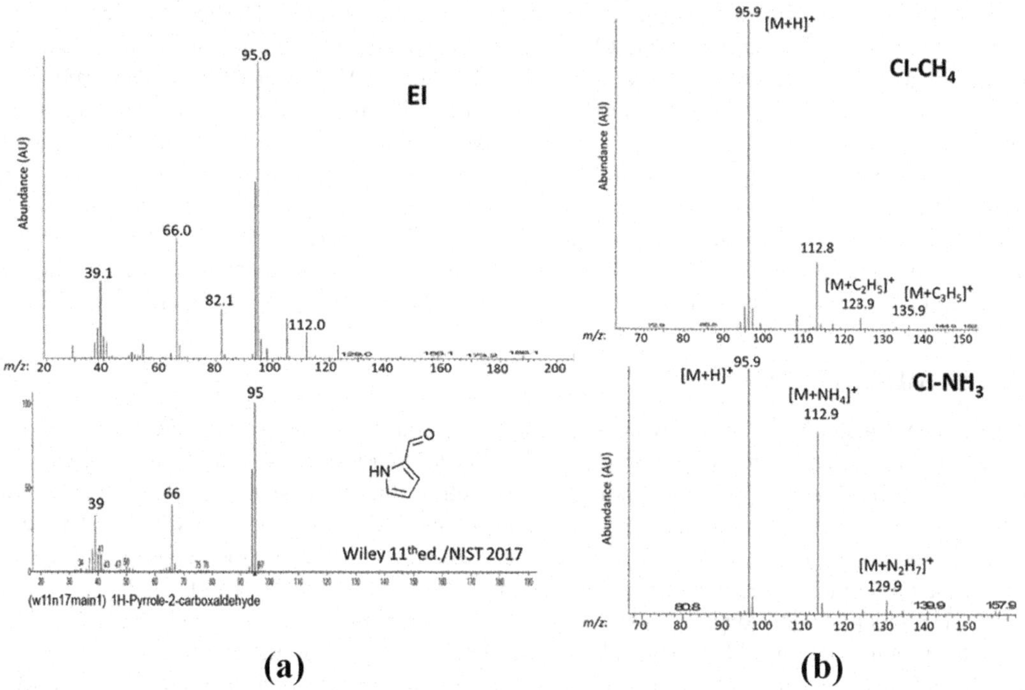

FIGURE 9.1 Mass spectra of 1H-pyrrole-2-carbaldehyde obtained in EI compared to the Wiley 11th Edition/NIST 2017 database reference spectrum (a) and in CI with methane and ammonia as reagent gases (b). Diagnostic ions on both CI spectra are indicated (b). (From Deuscher et al., 2020.)

containing multiple oxygen functions) and ethylene oxide (for alkenes, alkanes, alcohols, aldehydes, dienes and amino dienes) (Vairamani et al., 1990). Reagents in liquid form at standard pressure and temperature (methanol, heptane, dichloromethane, acetone, ethyl acetate, tetrahydrofuran, diethyl ether and acetonitrile) have also been tested for specific applications in numerous publications (Thibon et al., 2015; Allamy et al., 2017). A CI reagent needs to have a high degree of purity to prevent contaminant ions from jamming the spectrum analysis. The CI reagent must generate a limited amount of reactant ions in the proper analytical mass range. Accurate quantification is also more straightforward because PCI generally produces intense $[M+H]^+$ ions (Hjelmeland et al., 2016). The production of unique parent ions enables MS/MS, which makes trace compounds quantification possible.

While PCI is widely used (Moniruzzaman et al., 2014), negative CI (NCI) could be an alternative to improve the signal. First, NCI is very selective and reduces to the minimum the presence of fragment ions in the spectra, thus providing a high signal-to-noise ratio and enhanced sensitivity. Finally, this ionisation mode could add molecule-specific information (Mateo-Vivaracho et al., 2006). For instance, the hydroxyl anion OH^-, having very high PA, reacts efficiently by proton abstraction with any molecules having acidic hydrogen atoms. Thus, the reaction of OH^- with aliphatic alcohols provides very intense and specific molecular alkoxide anions $[M–H]^-$ (Houriet et al., 1980), while the PCI of aliphatic alcohols is not straightforward (Sarris et al., 1985).

For trap instruments, the hybrid chemical ionisation mode should be cited. In this CI mode, ionisation of the reagent gas and the molecule are separated in two steps: reagent ions generated in an external source in EI mode pass through some intermediate lenses to the IT where they react with the analyte eluted from the GC column. The main advantage of this method is the ability to select easily and rapidly a wide range of specific reagent gases. It also minimises the risk of ion-molecule reactions with the neutral reagent and reduces background noise level (Pons et al., 2011).

Finally, a recent complementary innovation is worthwhile to be cited. The ability to interface an atmospheric pressure photo-ionisation (APPI) source between GC and a linear IT-Orbitrap mass spectrometer was found suitable for analysing a wide range of VOCs. In APPI, the absorption of a photon by a molecule leads to the formation of an odd-electron radical cation; hydrogen atom abstraction frequently occurs during in-source collisions and produces a large quantity of protonated molecules MH^+ of diagnostic importance. In this way, dissociative reactions do not occur, with limited secondary ion-molecular interactions, especially when a photo-ionisable dopant is used, working as an intermediate between the analytes and the photons (Robb et al., 2000). However, no application using GC-APPI-HRMS could be found in the field of food sciences, even though a few articles are available in the fragrance domain (Lee et al., 2012; Santerre et al., 2018).

9.2.3 MDGC- and GC×GC-MS

One-dimensional (1D) GC is not always successful in separating a complex mixture. Peak overlapping could occur, and induced coelution phenomena lead to a loss of qualitative and quantitative information in producing complex mixture mass spectra. Multidimensional GC (MDGC) system allows successive separations with a combination of columns of different polarities. The first dimension is generally equipped with FID, even if micro-electron capture, atomic emission, nitrogen-, sulfur-chemiluminescence and nitrogen–phosphorous detection have been reported (Marriott et al., 2012). The second dimension is equipped with a column affording a different selectivity, and an MS detector is commonly used for chemical identification. For improving the interpretation in aroma compounds analyses, dual detection with an olfactometric port is often combined with the MS detector in an MDGC-MS/O arrangement (Mastello et al., 2015).

Bidimensional GC operates in both heart-cutting (MDGC) and comprehensive 2D (GC×GC) systems (Amaral and Marriott, 2019). During a preliminary one-dimensional (1D) separation, MDGC configuration allows one or more specific regions to be chosen on a first column; after a cryogenic refocusing step, the selected fraction is transferred to a second full-length conventional column of different polarity for a better separation of the overlapped region. The advantages of this technology are (i) to focus attention on only one part of the sample and (ii) to possibly accumulate many heart-cuts of the same fraction as an enrichment step in the cryogenic trap to gain sensitivity. GC×GC represents an exciting alternative to resolve all volatiles present in a complex sample in a single analytical run. After a first separation, the eluting compounds are transferred to a modulator to accumulate, refocus and rapidly inject sequentially in fixed time-frames to a second short capillary column. The key points of these experiments are the powerful algorithms and software required to merge the modulated peaks for each component (Nolvachai et al., 2020). At the same time, univariate and multivariate statistics help process and interpret data for improved compound identification. Results are presented as 2D images, providing the best solution for profiling and differential analyses. This technology enables high-speed separations; thus, a very rapid MS detection system is required. Quadrupole can be used as a detector, although its data-acquisition rates are limited. ToF detection combined with GC×GC separation appears as the best choice due to its high data-acquisition rates (Schena et al., 2020), selectivity, deconvolution capacity and ability to scan a broad mass range. Access to these advanced analytical instrumentations is handy for key volatiles investigation, as exemplified in recent studies.

Rice is a high value-added crop since it is the staple food in many countries. Rice aromatic varieties have been largely studied for discrimination purposes. Using GC×GC-ToF-MS and multivariate analysis resulted in the selection of eight key-marker VOCs (i.e., pentanal, hexanal, 2-pentylfuran, 2,4-nonadienal, pyridine, 1-octen-3-ol and (E)-2-octenal) responsible for the specific aroma properties (Setyaningsih et al., 2019). GC×GC-ToF-MS is also used to identify primary and secondary metabolites. Thus, an overview of the aroma potential of hazelnuts was given by using tandem hard (70 eV) and soft (12 eV) EI to display metabolome fingerprints from different cultivars and geographical origins (Rosso et al., 2020). Combining the two ionisation energies improved the confidence in compounds identification and enabled validating the comparative capacity of 2D fingerprints. Thereby, positive correlations between non-volatile precursors and aroma compounds have been highlighted: 3-methylbutanal with isoleucine, 2,3-butanedione/2,3-pentanedione with fructose/glucose derivatives, 2,5-dimethylpyrazine with alanine, and pyrroles with ornithine and alanine derivatives (Rosso et al., 2020).

9.2.4 GC-O-MS

As previously described, GC-MS techniques are powerful methods to obtain structural information on VOCs. However, it is not appropriate to determine the odour-active compounds and their flavour contribution in a food matrix. Moreover, there is no linear relationship between the VOC concentration and its odour intensity. Moreover, some essential odour-active compounds are found in too low concentrations to be detected by GC-MS. GC-O is the best available approach to measure and quantify odours directly through the use of a panel of trained sniffers via a specifically designed odour port placed after GC separation. Several parameters can be evaluated: the duration of the odour activity, the description of the perceived odour and its intensity. Characterisation of the key-odour compounds is possible with reference compounds databases by comparing the retention index and the odour descriptor of the targeted components (Baldovini and Chaintreau, 2020).

Several detection methods have been developed in GC-O. Osme (from the Greek word for odour) is one of the direct intensity measurement methods (McDaniel et al., 1990). The sniffers describe the odours, give their perceived intensities on a predefined scale over the elution time (start, maximum and end). This enables to compile an olfactogram where each maximum odour intensity is represented by the height of the corresponding peak and its duration by its width. This technique was widely used in wine research to discriminate grape varieties and study vine growing conditions and wine-making processes. In recent studies, investigations on syrah (Barbara et al., 2020) and cabernet-sauvignon varieties (Nicolli et al., 2020) can be cited. GC-O (Osme) technique has also been recently used to screen the most odouriferous chemical compounds released in the HS of unifloral honey (Costa et al., 2019). The frequency detection method involves participating in a panel of several participants who sniff the same sample (Pollien et al., 1997). For a given retention time, the percentage of people who have identified the same odour is reported in terms of nasal impact frequency (or peak height of the olfactometric signal) or surface of nasal impact frequency (or peak area). This method is straightforward because no description of the odour is required. For instance, this technique has been used recently to distinguish wild mushroom species (Aisala et al., 2019) and dark chocolates differing in sensory characteristics (Deuscher et al., 2020). Dilution to threshold methods is another answer in detecting potential key odourants in an aroma extract (Acree, 1993; Grosch, 2001). These methods consist of preparing a dilution series of the extract. The threshold below which an odour is no longer perceived enables defining, for each judge, a detection limit. Described as aroma extract dilution analysis (AEDA) or as combined hedonic aroma response measurement, diluted extracts provide flavour dilution (FD) values for each potential key aroma compounds. AEDA is often used as part of the 'sensomics' approach, as exemplified in a recent study on Bavarian gins where terpenes were identified as the dominant key aroma compounds, with additional contribution from aldehydes (Buck et al., 2020).

GC-O is commonly used with complementary GC-MS for compounds identification. A combined GC-MS/O arrangement affords additional advantages as both signals (odour and mass spectral information) are obtained simultaneously for a more straightforward identification step (Song and Liu, 2018). Olfactometry can also be coupled with multidimensional GC and tandem mass spectrometry (MS/MS) techniques in very versatile analytical systems. Thus, new natural flavours generated by biotransformation from fungi material have been identified, combining HS-SPME-GC-MS/MS to an olfactometry port. The combination of olfactometry and the robust chemical identification provided by the MS/MS instrument enabled the detection of 24 new volatile compounds formed by the fungi, from which 15 seemed to be flavour compounds (Ibrahimi et al., 2020). Heart-cutting MDGC-MS/O can resolve coelutions of key odourants. For instance, cooking develops numerous heat-induced odourants in cooked food, and several resulting coelutions obtained in GC make it challenging to identify odour-active compounds. Dynamic HS was used to extract VOCs from cooked meat, and heart-cutting MDGC–MS/O enabled coeluting intense odour zones to be resolved, revealing 15 additional odour-active compounds that were positively identified (Giri et al., 2015). A sophisticated switchable GC/GC×GC-O-MS system has recently been used to characterise key odourants in high-salt liquid-state soy sauce (Wang et al., 2021).

9.2.5 Quantification

Quantification is possible in MS mode with simple quadrupole instruments. Contrary to the full-scan acquisition that results in a typical total ion current, SIM allows the mass spectrometer to focus only on specific ions characteristic of particular compounds. Because single ions are monitored, much more time is spent accumulating signal at these masses of interest. The method enhances sensitivity compared to the full-scan mode and the accuracy of quantitative results due to the acquisition of more points across the chromatographic peak. The principal ion in the mass spectrum is used for quantification, while the second most abundant ion is used as a qualifier. This procedure avoids isotopically labelled standards used in SIDAs (see below) that are expensive, sometimes unstable and not always available commercially or difficult to synthesise. Thus, SIM is a simple technique to quantify specific compounds in a mixture, even in trace amount, combined with GC retention data. Therefore, 3-alkyl-2-methoxypyrazines, key compounds of several fruits and vegetables, imparting herbaceous aroma notes, were monitored using an HS-SPME-GC-SIM-MS efficient approach (Mutarutwa et al., 2018). Standard solutions were spiked in each sample vial for quantification purpose. 3-Isobutyl-2-methoxypyrazine was detected in carrot and cucumber for the first time at a shallow concentration (0.12 and 0.44 ng/g, respectively, while its sensory detection threshold is about 1–2 ng/L in water) (Mutarutwa et al., 2018).

MS/MS coupling with GC has been recently introduced as a major technique to identify and quantify targeted ions in complex matrices. For quantification, QQQ instruments operate in MRM detection mode. Coelutions can quickly be resolved, and consequently, short chromatographic runs can be applied. Thus, fast-GC-MS/MS made it possible to identify and quantitate VOCs of oenological products simultaneously in only 15 minutes (Paolini et al., 2018). Various oenological products were tested to validate the analytical method without any pre-concentration step, and 50 compounds from different chemical classes were selected for quantification in a single separation run, with an excellent retention time stability and repeatability at very low concentration (Paolini et al., 2018). Another example dealt with trace level pyrazines in roasted green tea that play a significant role in the character development of the flavour during consumption. Infrared-assisted extraction coupled to HS-SPME (Yang et al., 2018) brought a low-cost, rapid and efficient extraction method, while QQQ and MRM minimised interferences and provided better efficiency in the characterisation of target analytes (Yang et al., 2020a). Thereby, triple-quadrupole technology combined with MRM detection mode enhances selectivity and sensitivity.

SIDA is an alternative method that does not require MS/MS technology to quantify analytes in liquid samples (e.g., after liquid–liquid extraction). Known amounts of internal standards labelled with stable isotopes (e.g.^{2}H, ^{13}C and ^{15}N) are spiked in the matrix before any treatment. GC-MS can easily determine the isotopic ratio of each labelled/unlabelled pair. A calibration curve for a given concentration range allows quantification. While probably the most accurate quantification method, SIDA suffers from some limitations: labelling stability, availability of labelled molecules, quite an expensive cost and time-consuming experiment setup.

Moreover, since the labelled molecule does not have the same physical properties as the unlabelled one, it is necessary to determine a relative response factor (RF) for each target component obtained by analysing a test sample containing known concentrations of the target analyte and labelled standard. However, RF can be successfully applied to quantify homologues of the target analyte. In the 'sensomics' approach, SIDA is often coupled with odour activity values (OAV), defined as the ratio of concentration to odour threshold. Thus, 42 aroma compounds exhibiting the highest FD factors after AEDA in the GC-O study of a Chinese green tea infusion have been quantified by SIDA (Flaig et al., 2020). To better understand the importance of each aroma compound on the aroma profile, OAVs have been calculated based on odour thresholds in water; 30 odourants have been selected with a concentration equal to or above their odour threshold. By mixing these 30 purified compounds in water at a suitable concentration, the aroma of Chinese tea has been faithfully reconstituted (Flaig et al., 2020).

9.2.6 New Developments: GC-HRMS and GC-IMS

High-resolution mass spectrometry (HRMS) brings a new dimension in molecular analyses thanks to its high resolution, mass accuracy, linear dynamic range and sensitivity. For instance, the high performance found in GC-EI-HRMS provides more reliable identification and is very useful for data classification using fingerprinting analysis of volatile fractions (Belmonte-Sanchez et al., 2018). 2D GC coupled with HRMS (GC×GC-HR-ToF-MS) offers excellent separation and identification power for VOCs from complex samples (Vyviurska and Spanik, 2020). Orbitrap technology is one of the most recent advances in HRMS. The commercial release of a GC-EI-Orbitrap should open new perspectives in the field of aroma research. At the same time, only applications on pesticides or phytosanitary products have been found recently for VOCs analyses (Mol et al., 2016).

GC-ion mobility spectrometry (GC-IMS) is another interesting coupling method for the characterisation of aroma compounds. After a chromatographic run, generated ions are separated in an electric field in the presence of a drift inert gas based on their mobility, depending on their size-to-charge ratio. Therefore, the technique is well suited to differentiate isomers with different conformation and shapes. The technology has been mainly used recently for food fraud cases to perform mapping and classification. For example, HS-GS-IMS was applied on olive oils to classify them as extra virgin, virgin and inferior quality, defective and non-defective, as well as edible and non-edible: IMS allowed finding specific biomarkers (Contreras et al., 2019). In very complex mixtures, IMS allows a good separation of chemical classes and is helpful for fingerprint analyses. Thus, 162 peaks have been detected from edible fungi, from *Eucommia ulmoides* Oliv. leaves that served as growth medium, and from their fermentation products, including alcohols, aldehydes, ketones, acids, and esters (Wang et al., 2020). GC-IMS technique is also applicable as a monitoring system to differentiate processed products, like during the processing of black sesame (Zhang et al., 2020). An important feature should be noticed for the technique: the need for suitable chemometrics to handle the data and mapping and classification purposes. The miniaturisation of GC-IMS instruments, providing handheld and fast devices, makes possible the use of this technology in various applications in an easy-to-use configuration without compromise in sensitivity performances (Ahrens and Zimmermann, 2021). However, two limitations of GC-IMS should be taken into account. First, compound identification is not achievable, and the data have to be backed up by GC-MS and subsequent injection of standards. Second, since IMS detection of aliphatic hydrocarbons is not possible due to their low PA, the joint direct determination of GC retention indices is not possible. Calculation of the retention indices should be based on a homologous series of ketones and extrapolated. Using IMS together with MS in a GC-IMS-MS arrangement should be attractive to provide selective detection. However, technical problems due to the different pressurised ionisation and analysis regions of MS and IMS arise. Recently, a GC-MS-IMS system using an electronic pneumatic control (EPC) of a three-way splitter plate to split the GC effluent between both analytical devices was designed, that used an additional EPC regulator to provide a make-up gas flow of helium to the split point, generating a sufficient gas flow to IMS to reduce peak broadening (Brendel et al., 2021). With prior HS sampling, the obtained combined data were the basis of volatilomics profiling of 47 *Citrus* juices, including grapefruit, blood orange and common sweet orange juices, without requiring any sample pre-treatment. The approach was alleged as a promising and straightforward alternative for future authenticity control of *Citrus* juices (Brendel et al., 2021).

9.2.7 MS Databanks and Software

The MS data obtained in GC-MS, combined with retention indices, make good use of numerous existing spectral libraries to identify VOCs. One of the most useful ones for identifying aroma compounds is the National Institute of Standards and Technology (NIST) Chemistry WebBook that provides convenient access to physical, chemical property and mass spectral data of thousands of chemical species (accessible at https://webbook.nist.gov/chemistry/). In addition to the traditional EI-MS library, MS/MS databases are now available in NIST for high and low-resolution instruments. Recently, the release of NIST Hybrid Search enabled the matching of unknown compounds to similar ones that differ by a modification in a single region of the molecule that does not affect the fragmentation pattern. It combines classical direct peak matching with the equivalent of neutral-loss matching (Chua et al., 2020). The Wiley Registry™ (Hoboken, USA) of Mass Spectral Data, 12th Edition, is the most extensive mass spectral library commercially available, including untargeted GC-MS screening and accurate mass workflows with ToF-MS instruments. Other databases from manufacturers are accessible for common metabolites where aroma compounds could be included. This is the case of the Agilent Fiehn GC/MS Metabolomics RTL Library or the Advanced Mass Spectral Database – *m/z* cloud – from ThermoScientific. Open-source databanks can be easily found on the internet, such as MassBank, METLIN or GNPS, that can be used by various online data search software (i.e., MS-Finder or XCMS). Finally, public and private institutes have developed their homemade banks for their specific needs.

A spectrum derived from a chromatogram at a given retention time is the sum of various signals (matrix, interfering compounds, analytes of different molecular species). Deconvolution allows such interference to be overcome and sensitivity to be improved. A mathematical process is possible to extract the signal of interest from the complex spectra to obtain refined mass spectra to identify better several compounds hiding under a single chromatographic peak. To process the chromatographic runs and the associated mass spectra, from raw data to identification, all MS manufacturers provide their software, often as ‘plug and play’ solutions. Although easy-to-use, such software could be of limited value due to variations in elution time, baseline drifts, unexpected overlapping peaks, or non-Gaussian peak shapes. The most popular software used for spectrum extraction and compound identification is AMDIS – Automatic Mass spectral Deconvolution and Identification System – freely provided by NIST. Some other software as SpectConnect and Gavin, use the AMDIS results files to enhance performances in peak alignment and identification (Morales et al., 2017).

In the same way, Metab R uses the R platform application for data correction, filtering and reshaping datasets initially produced by AMDIS (Sichilongo et al., 2020). Other alternatives to AMDIS can be found in open-source algorithms. PARAllel FACtor analysis 2 (PARAFAC2) requires only a single parameter (number of chemical compounds) to be set by the user for a given retention time interval of several chromatograms simultaneously. An integrated approach called PARAFAC2-based Deconvolution and Identification System (PARADISe) has been developed to integrate this algorithm for easy and robust use (Amigo et al., 2008; Johnsen et al., 2017).

9.3 TASTE AND HPLC-MS

Food odourants elicit a transduction cascade after interacting primarily in a combinatorial code with ca. 400 olfactory receptors in humans. A few hundreds of foodborne volatiles lead to a myriad of odour and aroma perceptual qualities (Dunkel et al., 2014). Contrarily, food tastants interact with specific taste receptors to activate a transduction cascade leading to taste perception. These taste receptors are G-protein-coupled – or ion channels – membrane proteins carried by taste buds on the tongue (Boughter and Munger, 2013). Despite some controversy, it appears that the human perceptual world consists of the same five basic taste qualities: salty, sour, sweet, bitter and umami (Beauchamp, 2019). The recent discovery of long-chain free fatty acid receptors that would be involved in fat detection (Khan and Besnard, 2009; Galindo et al., 2011; Voigt et al., 2014; Besnard et al., 2016) doesn't question this consensual perceptual representation (Beauchamp, 2019).

The salty taste is essentially due to the sodium ion Na^+ notably present in the mineral salt NaCl. The higher molecular weight mineral salts (KCl, $CaCl_2$ and $MgCl_2$) impart bitterness (McSweeney, 1997; Engel et al., 2000a) essentially. The sour taste is essentially due to H_3O^+ resulting from the dissociation of common short-chain fatty acids and other acids such as lactic acid in the aqueous food phase (Le Quéré, 2004). Both stimuli mobilize ion channels taste receptors. Sweetness, primarily due to simple sugars such as sucrose, presents a certain complexity. The human sweet taste receptor is a G-protein-coupled (G-PC) transmembrane heterodimer protein with many different binding sites for sweet ligands. Many agonists with somewhat different chemical structures able to elicit sweetness have been identified (Behrens et al., 2011). Besides carbohydrates, the vast diversity of molecular ligand entities includes foodborne terpenoids, glycosides, saponins, phenolics and proteins (Behrens et al., 2011), not to mention artificial sweeteners such as aspartame, acesulfame K or sucralose. Current research includes structural investigations of new sweet molecules that mobilise MS (Fayad et al., 2020; Pengwei et al., 2020) and numerous *in silico* studies on both entities of the ligand-receptor partners (Chéron et al., 2017; Ben Shoshan-Galeczki and Niv, 2020; Bouysset et al., 2020; Spaggiari et al., 2020; Goel et al., 2021). The umami taste is the typical savour of L-glutamate generally present in food as monosodium glutamate (MSG) with a synergistic effect of 5′-ribonucleotides such as guanosine monophosphate and inosine monophosphate. However, many peptides were described to impart umami taste (Zhang et al., 2017). New umami molecules are revealed regularly and generally identified by HPLC-MS (Kaneko et al., 2006; Rotzoll et al., 2006). The primary human receptor for umami components is a G-PC transmembrane heterodimer protein, sharing a common subunit with the sweet receptor, with possible different ligand-binding sites (Behrens et al., 2011; Zhang et al., 2017) accommodating many types of agonists, objects of current structural investigations (Salger et al., 2019) and in silico approaches (Spaggiari et al., 2020). Sweetness and umami have both positive impacts on the appreciation of food and are linked to appetitive ingestive behaviour (Behrens et al., 2011). The situation is different for bitterness, often perceived negatively in various foods, as the human bitter taste receptor family comprises 25 G-PC transmembrane proteins (Behrens and Ziegler, 2020; Spaggiari et al., 2020). This supports the evolutionary hypothesis that states that harmful dietary substances, generally imparting bitterness, have shaped the diversity of the bitter taste receptor gene repertoire in humans (Li and Zhang, 2013). The bitter taste receptors can be classified according to the number of agonists into (i) broadly tuned ones with numerous agonists, (ii) narrowly tuned ones with very few agonists, receptors interacting with specific chemical classes and (iii) intermediately tuned ones (Behrens and Ziegler, 2020). *In silico* structure-function studies are conducted to enlighten the molecular parameters of tastants (Behrens and Ziegler, 2020; Spaggiari et al., 2020). Moreover, the diversity of bitter foodborne molecules has resulted in numerous investigations to decipher bitterness of many foods, where MS plays a fundamental role for structural characterisation (Dresel et al., 2016; Sebald et al., 2018; Soares et al., 2018; Lang et al., 2020; Sebald et al., 2020).

Coming historically after food aroma characterisation, research on food tastants has benefitted from the reasoning and progress acquired by aroma-oriented study. Thus, the sensory-driven 'sensomics' approach (Schieberle and Hofmann, 2011) was rapidly adopted with the idea to concentrate identification efforts on components of food that are imparting tasty characters. Therefore, sensory activity-guided fractionation schemes were adopted to isolate and characterise the only tasty fractions of food. The underlying idea was to implement procedures analogous to the GC-O approach used for key odorants to identify and quantify key-tastants. Afterwards, before incorporating food recombinants for the final sensory validation step, key-tastants can be characterised by taste activity values (TAV), i.e., ratio of concentration over taste perception threshold (Warmke et al., 1996), a concept analogous to the OAV used for odorants. As most of the tastants are hydrosoluble non-volatile compounds, food water-soluble extracts (WSE) were intensively studied, and filtration methods and gel permeation chromatography (GPC) with food-grade eluents were used to separate the extracts into MW-dependent fractions that could be sensorially evaluated. The sensory assessment of the fractions allowed concentrating identification efforts on the only ones imparting flavour (Aston and Creamer, 1986; Engels and Visser, 1994; Salles et al., 1995). Significant efforts were initially dedicated to food rich in proteins, such as meat, beef broth, milk or cheese, as peptides originating from proteins were engaging scientists due to their specific composition in amino acids known for their sensory properties (Iwaniak et al., 2016). Peptides are also involved in fermentation processes aimed at producing hydrolysates through proteolysis. While peptides may impart sweetness, bitterness or umami, bitter peptides are by far dominant (Iwaniak et al., 2016). Sweet peptides are essentially polypeptides or small proteins such as monellin or brazzein (Iwaniak et al., 2016). Some peptides may also enhance or partially suppress the taste of food and can be viewed as modulators in the perception of taste modalities (Deepankumar et al., 2019).

One of the first examples, which used MS to identify some low-molecular-weight compounds as potential taste constituents, aimed at deciphering the contribution of non-volatile WSE to Comté cheese flavour (Roudot-Algaron et al., 1993). Tasty low-molecular-weight fractions obtained after GPC on Sephadex G15 were separated by HPLC. After collection and purification, the prominent peaks of each HPLC fraction were analysed by MS and MS/MS on a QQQ instrument using desorption chemical ionisation (DCI) with ammonia as reagent gas (Roudot-Algaron et al., 1993). Major constituents were identified as *N*-acyl amino acids, diketopiperazines and some non-peptide compounds. Sensory evaluation of synthetic compounds showed that most were bitter when tasted alone in pure water, particularly the diketopiperazines cyclo(Pro-Phe), cyclo(Pro-Val) and cyclo(Pro-Pro), associated with more complex cheesy and brothy notes. However, none of the identified compounds was found in a concentration above its taste threshold in the WSE that was only slightly bitter. Therefore, it was concluded that the taste of the aqueous fraction of Comté cheese is not due to a single compound but to a complex blend with synergistic or additional effect (Roudot-Algaron et al., 1993).

Further analysis by ammonia DCI-MS/MS of purified components present in other fractions allowed identifying three γ-glutamyl dipeptides (γ-Glu-Phe, γ-Glu-Tyr and γ-Glu-Leu) (Roudot-Algaron et al., 1994). The γ-Glu-Tyr was described as sour and slightly salty, while γ-Glu-Phe had a more intense and complex taste described as umami, somewhat sour, salty and metallic. However, their amounts in the studied cheese were far below their respective taste thresholds (Roudot-Algaron et al., 1994). It was again concluded that those peptides alone could not intervene in cheese flavour, except by additive or synergistic effects with other compounds, such as salts, nucleotides or MSG (Roudot-Algaron et al., 1994). Moreover, the umami taste of Comté cheese was associated with a substantial amount of MSG, which was found in a fraction at a concentration 10 times above its threshold value. At the same time, other amino acids were confirmed to be present well below their detection thresholds (Salles et al., 1995). The authors finally concluded that analysis of interesting flavoured fractions of Comté cheese showed that the taste characteristics are linked more to the presence of salts and amino acids acting in synergy than to the presence of peptides (Salles et al., 1995). However, it was demonstrated that bitterness of Camembert cheese was due to the accumulation of small (MW < 1,000 Da) bitter peptides during ripening, due to the intense proteolytic activity of the strain of *Penicillium camemberti*, specially selected to develop bitterness in this case (Engel et al., 2001). LC-ToF-MS and LC-MS/MS analyses revealed the identification of 16 bitter peptides formed by proteolysis of caseins in Gouda cheese (Toelstede and Hofmann, 2008a). The complete assessment of putative taste-active metabolites in this Gouda cheese was realised. It involved a taste dilution analysis (TDA), giving rise to taste dilution factors, analogous to AEDA and FD factors used for aromas, determination of dose-over-threshold (DoT) factors, i.e., TAV values, followed by the confirmation of the sensory relevance by taste reconstruction and omission experiments, which confirmed the importance of bitter peptides (Toelstede and Hofmann, 2008b).

However, the bitterness of the cheese was found to be induced by $CaCl_2$ and $MgCl_2$, as well as various bitter-tasting free amino acids.

In contrast, bitter peptides were found to influence more the bitterness quality than the cheese's bitter intensity. Strong evidence demonstrated that the sensory contribution of individual bitter peptides is mainly due to the decapeptide YPFPGPIHNS and the nonapeptides YPFPGPIPN and YPFPGPIHN, as well as the tetrapeptide LPQE, originating from β-casein and α_{S1}-casein, respectively (Toelstede and Hofmann, 2008b). All these results questioned the exact role of small peptides in the bitterness of cheese. Thus, following the same procedure, taste-active goat cheese fractions were analysed by combined MS methods including ESI, liquid secondary ionisation and MS/MS to identify and to confirm the sequences of 28 tri- to octapeptides naturally appearing in goat cheese during ripening (Sommerer et al., 1998; Sommerer et al., 2001). Using omission tests, the relative impact of WSE components on the goat cheese taste has been determined. It was found that bitterness resulted entirely from $CaCl_2$ and $MgCl_2$. Amino acids and the identified peptides had no significant impact on the taste properties of the WSE of goats' milk cheese (Engel et al., 2000a; Engel et al., 2000b).

The same controversy also exists on the taste value of some peptides formerly characterised in a beef broth. Thus the octapeptide KGDEESLA isolated in a fraction of a beef extract was named 'delicious' peptide or beefy, meaty peptide (BMP) or savoury taste-enhancing peptide, and was described as possessing meaty, savoury, umami and sour notes. However, after confirming the peptide sequence by ESI-MS, taste evaluation by a trained flavour panel showed that the synthesised octapeptide and some peptide fragments did not have any umami or other taste (Van Wassenaar et al., 1995). Independently, another study aimed at detecting BMP in beef digests and cooked or grilled beef using an online detection by ESI-LC-MS concluded that there is no evidence that BMP exists naturally at detectable levels (Hau et al., 1997). Moreover, tasting the synthetic peptide in beef stock revealed that the compound has little or no discernable flavour or flavour enhancing properties, concluding that BMP cannot be considered a flavour carrier or a potential flavour enhancer (Hau et al., 1997).

No doubt that the emergence of ESI and other atmospheric pressure ionisation technique such as APCI has primarily contributed to establishing LC-MS as a mature, robust technique starting in the late 1990s. Implementation of HPLC-MS and HPLC-MS/MS tools in the 'sensometabolomics' procedure allowed investigating many chemically different taste or taste-modifier/enhancer molecules in many foods. Thus, long-lasting mouthfulness (so-called kokumi flavour) was attributed to γ-glutamyl dipeptides in matured Gouda (Toelstede et al., 2009) and Parmesan (Hillmann and Hofmann, 2016) cheeses, following the action of a γ-glutamyltransferase (Hillmann et al., 2016). However, it was previously concluded that hydrophilic glutamyl di- and tripeptides searched for by hydrophilic interaction liquid chromatography-ESI-MS and MS/MS in savoury fractions were not a precondition for the development of a savoury flavour in mature Cheddar cheese (Andersen et al., 2008). This finding confirmed previous results described above for Comté cheese and those obtained in a survey of alleged umami peptides whose sequences were assessed by MS that concluded that the general occurrence of umami peptides appears to be highly unlikely. (van den Oord and van Wassenaar, 1997). The formation of taste-active amino acids, amino acid derivatives and peptides in food fermentations was recently reviewed (Zhao et al., 2016). The contrasted contribution of amino acids and peptides to food flavour was highlighted, while the necessity of improved knowledge of the interactions between taste-active compounds was underlined (Zhao et al., 2016). Nevertheless, a new approach, so-called 'sensoproteomics', was recently proposed to identify taste-active peptides in fermented foods (Sebald et al., 2018). Aiming to identify key bitter peptides in fermented foods, the approach was developed and applied to fresh cheese samples differing in bitter taste intensity. A targeted proteomics approach was used without the need for intensive activity-guided fractionation aimed at resolving extracts into the individual taste-active peptides before identification (Sebald et al., 2018). The repertoire of ca. 1600 candidate peptides produced from bovine caseins extracted from a literature meta-analysis of dairy products were used. Using TDA, complex fractions of the cheese WSE imparting intense bitter taste were located. These fractions were screened by ultra-performance liquid chromatography (UPLC)-MS/MS using selected reaction monitoring (SRM) methods based on predefined mass transitions on a QQQ linear ITs instrument operated in positive ESI mode. By considering the only peptides located in these bitter-tasting fractions, 17 peptides were identified as bitter candidate ones, whose identification was validated by the synthetic reference peptides. They were quantified in the fractions by the same UPLC-MS/MS instrument using MRM with defined mass transitions, using external calibration conducted with the reference peptides. Fifteen of the 17 target peptides showed significantly low bitter taste thresholds in aqueous solutions adjusted to pH 4.6. Calculation of DoT factors for the 15 peptides revealed two peptides with TAV ≥ 1.0: a β-casein fragment with 18 amino acid residues MAPKHKEMPFPKYPVEPF and a κ-casein fragment with 11 amino acid residues ARHPHPHLSFM. Five additional peptides determined with subthreshold activity value (0.1 ≤ DoT ≤ 1.0) were considered to contribute to the overall bitter taste perception of the fractions as bitter compounds of the same chemical class were reported to coactivate the same bitter taste receptors and mixtures of such agonists found to additively increase the perceived bitterness (Sebald et al., 2018). Although it was argued that the new approach enabled the successful identification of bitter-tasting peptides in fresh cheese (Sebald et al., 2018), the conclusions were only based on DoT considerations and cheese WSE and not on sensory assessment in a cheese model incorporating the alleged bitter peptides.

Among fermented foods and beverages, after cheese, the wine received particular attention. Possessing various properties, tannins have been mainly considered to impart astringency and bitterness to food (Soares et al., 2020). Therefore, they were early studied as essential contributors to wine taste (Arnold et al., 1980; Robichaud and Noble, 1990; Peleg et al., 1999). Lower molecular weight polyphenolic compounds tend to be more bitter than astringent (Peleg et al., 1999; Soares et al., 2020). Isolated fractions of proanthocyanidins were also assessed as possessing mouthfeel properties (Vidal et al., 2003). Polyphenols may also interact with VOCs, inducing significant effects on the release and perception of wine aromas (Pittari et al., 2021). Using the 'sensomics' approach, LC-MS/MS with ESI in positive or negative mode allowed the identification and quantification of phenolic acid ethyl esters and flavan-3-ols in the sensometabolome of red wines as bitterness inducers at subthreshold levels (Hufnagel and Hofmann, 2008a; Hufnagel and Hofmann, 2008b; Frank et al., 2011). Low-molecular weight phenols identified by UPLC-MS were confirmed to impart bitterness to red wines, together with mouthfeel properties (Gonzalo-Diago et al., 2014). Small peptides characterised in white wine were supposed to impart bitterness and umami explained by synergistic effects (Desportes et al., 2001). Low-molecular weight peptides were identified in champagne wine using an online LC-ESI-MS/MS method in positive mode, with SRM monitoring.

The 'sensomics' approach has been used to decipher the taste-active components (key tastants) and taste-modulating compounds in various foods with LC-MS as a tool of primary importance, besides 1D and 2D nuclear magnetic resonance (NMR). Extensive use of the approach by Thomas Hofmann's group in Munich showed the vast diversity of chemicals at the origin of the basic tastes. This sensory-oriented research uses HRMS (LC-ToF-MS) and LC-MS/MS using ESI in the negative (ESI–) or positive (ESI+) mode for identification purposes and LC-MS/MS with MRM for quantification. Thus, besides organic acids, γ-aminobutyric acid and (S)-malic acid 1-O-β-D-glucopyranoside [(S)-morelid] were demonstrated as key tastants of morel mushrooms, the latter imparting sourness and umami while enhancing the taste of MSG and NaCl (Rotzoll et al., 2005, 2006). In black pepper, besides piperine, a series of amides were found significantly pungent and tingling (Dawid et al., 2012). The sweet-bitter tasting hexose acetates 6-*O*-acetyl-α/β-D-glucopyranose and 1-*O*-acetyl-β-D-fructopyranose were identified in Modena's traditional balsamic vinegar, as well as the sweetness modulator 5-acetoxymethyl-2-furaldehyde (Hillmann et al., 2012). In aniseed, trans-anethole revealed sweetness and trans-pseudoisoeugenol 2-methylbutyrate was found bitter (Pickrahn et al., 2014). Besides the amino acids glycine, L-proline and L-alanine, the characteristic seafood-like sweet profile of cooked king prawns was found to be due to the sweet modulatory action of quaternary ammonium compounds, among which betaine, homarine, stachydrin and trimethylamine-*N*-oxide were seen as the key contributors (Meyer et al., 2016). The sensometabolome of the traditional broth 'pot-au-feu' was characterised by minerals, nucleotides/nucleosides, amino acids, organic acids and carbohydrates, together with the dipeptide carnosine, anserine and 1-deoxy-D-fructosyl-N-β-alanyl-L-histidine (Kranz et al., 2018). The taste of orange juice seemed to be created by a rather complex interplay of limonin, limonoid glucosides, polymethoxylated flavones (PMFs), organic acids and sugars when subthreshold concentrations of PMFs were shown to enhance the perceived bitterness of limonoids (Glabasnia et al., 2018).

Moreover, using a matrix-assisted laser desorption/ionisation (MALDI)-ToF/ToF MS/MS instrument, the octapeptide YGGTPPFV imparting umami and sweet taste was identified in a cooked pufferfish fraction targeted by TDA (Zhang et al., 2012). The peptide was an important contributor to the mellowness and tenderness taste of the pufferfish *Takifugu obscurus*. The same MS/MS instrumentation was used to characterise the umami octapeptide SSRNEQSR and the umami-enhancing peptide EGSEAPDGSSR in a peanut hydrolysate (Su et al., 2012) and the umami peptides CCNKSV and AHSVRFY in two different dry-cured hams (Dang et al., 2015).

Considerable efforts have been made to identify particular targets in research that could be considered the quest for bitterness and mouthfeel components. Here again, bitter compounds and kokumi components, the latter defined as molecules that enhance mouthfulness, complexity and long-lastingness of the savoury salty and umami tastes (Dunkel et al., 2007), are represented by a huge diversity of chemicals necessitating deep MS investigations. Only some chosen examples will be further detailed to illustrate this complexity.

Bitterness is a typical taste of coffee, cocoa/chocolate and hop/beer products. The alkaloid theobromine contributes to the characteristic bitter taste of roasted cocoa/chocolate, together with some diketopiperazines (Stark et al., 2005). A sensomics approach conducted on roasted cocoa nibs revealed that the flavan-3-ols epicatechin, catechin and diverse procyanidins were also key components contributing to the bitter taste, as well as to astringency (Stark et al., 2005). A recent study was conducted on cocoa and chocolate polyphenols analysed by UPLC-MS/MS on a QQQ instrument, with MRM conditions targeted at flavan-3-ols and derivatives. It showed that the cocoa beans phenolic composition was a significant factor that explained the classification of dark chocolates differing in sensory properties (Fayeulle et al., 2020). In coffee brew, the alkaloids caffeine and trigonelline partially account for bitterness (Frank et al., 2007). However, the sensomics approach using LC-MS/MS and 2D-NMR led to the unequivocal identification of quinic acid lactone isomers esterified with *p*-coumaric, caffeic, ferulic and quinic acids as strong bitter contributors (Frank et al., 2006). These intensely bitter components, whose representatives are 3,4-*O*-dicaffeoyl-γ-quinide, 3,5-*O*-dicaffeoyl-*epi*-δ-quinide (Figure 9.2) and 4,5-*O*-dicaffeoyl-*muco*-γ-quinide, were formed during coffee roasting (Frank et al., 2006).

Further investigation on coffee brew conducted using LC-MS/MS (ESI-) in MRM mode revealed 1,3-bis(3′,4′-dihydroxy phenyl) butane, trans-1,3-bis(3′,4′-dihydroxy phenyl)-1-butene and eight multiply hydroxylated phenylindanes as important bitterness contributors of coffee (Frank et al., 2007). In a further study on the characterisation of bitter compounds in coffee, the structure of the

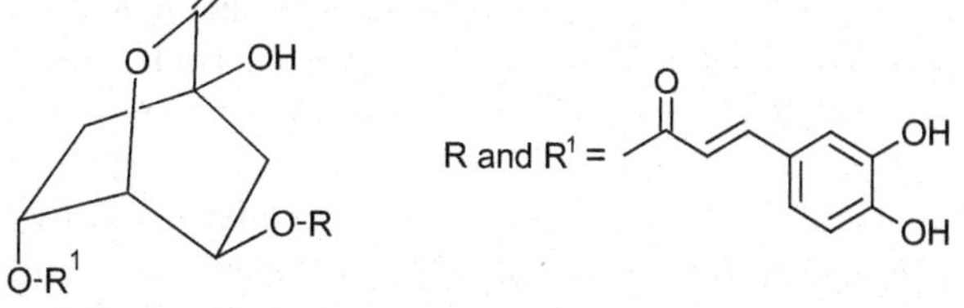

FIGURE 9.2 The key bitter taste compound 3,5-*O*-dicaffeoyl-*epi*-δ-quinide identified in a roast coffee beverage (Frank et al., 2006).

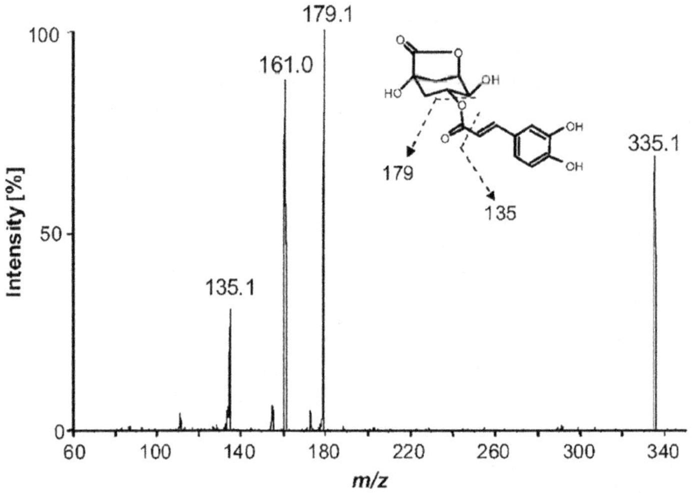

FIGURE 9.3 LC-MS/MS (ESI−; collision energy, −50 V) spectrum of 3-*O*-caffeoyl-*epi*-γ-quinide. (It was reproduced from Frank et al., 2008, with permission.)

bitter lactone 3-*O*-caffeoyl-*epi*-γ-quinide was determined by LC-MS and NMR from thermal treatment of 5-*O*-caffeoylquinic acid (Frank et al., 2008). Its MS/MS spectrum (Figure 9.3) obtained in ESI- mode revealed the daughter ions *m/z* 179, 161 and 135 corresponding to interpreted cleavages of the $[M\text{-}H]^-$ molecular ion at m/z 335.

The occurrence of this bitter lactone, which exhibits a low bitter recognition threshold (58 μmol/L) in aqueous solutions at pH 5.2 in coffee beverages, was confirmed by LC-MS/MS (ESI-) operating in the MRM mode (Figure 9.4). Furthermore, for the mass transition *m/z* 335→161, the target compound 3-*O*-caffeoyl-*epi*-γ-quinide (7) could be unequivocally detected in the coffee brew, as under the chromatographic conditions selected, other interfering lactones (2-6) were not resolved (Frank et al., 2008). Moreover, the identity of this bitter tastant in coffee was confirmed through co-chromatography with the corresponding reference compound (Frank et al., 2008).

Subsequent studies on coffee enlarged the chemical space of bitter tastants in the beverage. Thus, LC-MS/MS allowed verifying the natural occurrence of 4-(furan-2-ylmethyl)benzene-1,2-diol, 4-(furan-2-ylmethyl)benzene-1,2,3-triol, 4-(furan-2-ylmethyl)-5-methylbenzene-1,2-diol, and 3-(furan-2-ylmethyl)-6-methylbenzene-1,2-diol as a novel class of bitter taste compounds in roasted coffee (Kreppenhofer et al., 2011). More recently, a furokaurane glucoside, mozambioside (Figure 9.5), was identified by UPLC-high-definition MS (ESI+, *m/z* 509.2388 for $C_{26}H_{37}O_{10}$, MH^+) and NMR as an Arabica-specific bitter-tasting component in coffee beans (Lang et al., 2015). Although partially degraded upon coffee roasting, mozambioside was quantitatively extracted into the coffee brew, where it was found at a concentration largely above its bitter taste recognition threshold (Lang et al., 2015).

Beer and its hop ingredient have also been the subject of many investigations on bitterness. For centuries, various hop forms have been used as ingredients in beer manufacturing to impart bitterness and attractive aroma to the final beverage. Bitter compounds of hop are prenylated compounds known as α- and β-acids, with humulone and lupulone as respective typical examples (Figure 9.6). A quantitative sensomics HPLC-MS/MS profiling protocol in ESI- mode allowed characterising dozens of α-acid and β-acid derivatives formed during beer manufacturing upon wort boiling or accumulating during beer ageing (Haseleu et al., 2009; Haseleu et al., 2010). Most of them were supposed to contribute to the typical bitter taste of beer (Haseleu et al., 2010), as they were found in a range of commercial beers (Haseleu et al., 2009). Typical examples of these derivatives are shown in Figure 9.6.

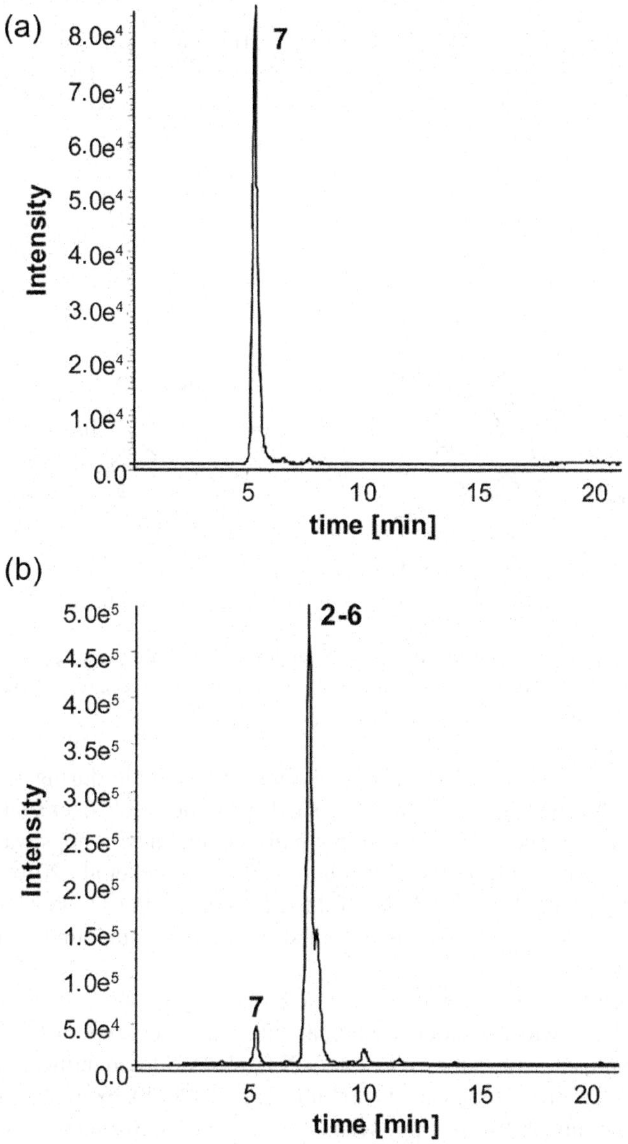

FIGURE 9.4 HPLC-MS/MS analysis (ESI−; MRM, transition *m/z* 335→161) of an aqueous solution of the purified compound (7) [3-*O*-caffeoyl-*epi*-γ-quinide] (a) and of a freshly prepared coffee brew (b). (It was reproduced from Frank et al., 2008, with permission.)

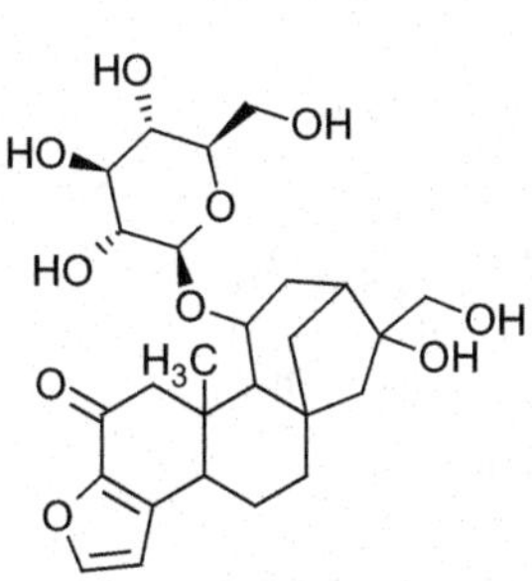

FIGURE 9.5 Structure of mozambioside, a bitter tastant specific to Arabica coffee (Lang et al., 2015).

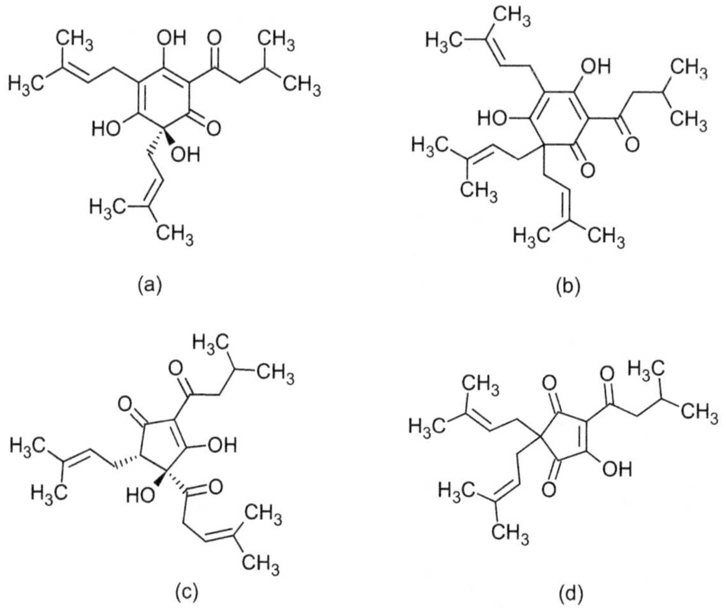

FIGURE 9.6 Humulone (a) and lupulone (b) as examples of hop bitter taste compounds α- and β-acids, respectively; examples of bitter derivatives (c) and (d) formed from (a) and (b), respectively, during beer manufacturing (Haseleu et al., 2009; Haseleu et al., 2010).

A comprehensive HPLC-MS/MS analysis of these compounds during beer storage allowed obtaining a thorough knowledge to deal with the shelf-life of desirable beer's bitter taste (Intelmann et al., 2011). Recently, the chemodiversity of key bitter compounds in hops was investigated using 75 different samples collected from the global hop market (Dresel et al., 2016). A multiparametric HPLC-MS/MS in MRM mode allowed identifying and quantifying 117 key bitter tastants, among which an isoxanthohumol (Figure 9.7) was identified as a marker compound for hop varieties grown in Germany (Dresel et al., 2016).

All the above examples show the huge chemical diversity of food components acting as agonists in good accordance with the extensive human bitter receptors repertoire. Other examples that used a sensomics approach with extensive use of HPLC-MS/MS for identification and quantification include alkynes as bitter off-taste of carrots (Schmiech et al., 2008), oxylipins in processed avocado (Degenhardt and Hofmann, 2010), in poppy seeds (Lainer et al., 2020) and pea protein isolates (Gläser et al., 2020), saponins in asparagus (Dawid and Hofmann, 2014), or a mixture of tryptophan derivatives, 2,3-dihydro-3,5-dihydroxy-6-methyl-4(H)-pyran-4-one and 4-(2-formyl-5-(hydroxymethyl)-1H-pyrrol-1-yl)butanoic acid in the crust of wheat bread (Jiang and Peterson, 2013).

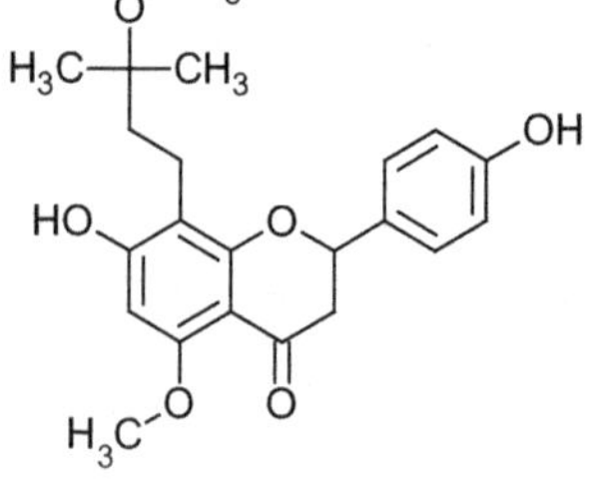

FIGURE 9.7 Isoxanthohumol M identified as a bitter marker compound of hop varieties grown in Germany (Dresel et al., 2016).

Taste modulation or kokumi compounds share with bitter taste components a tremendous chemical diversity. However, they are not true agonists of identified receptors but act as interacting synergistic components with the basic tastes. Thus, initially identified and characterised as a taste enhancer in beef broth by a sensomics approach, the inner salt *N*-(1-carboxyethyl)-6-(hydroxymethyl)pyridinium-3-ol, named alapyridaine, although being tasteless itself, was found as a potent sweetness enhancer (Ottinger and Hofmann, 2003; Ottinger et al., 2003). It was also demonstrated that the (+)-S enantiomer was physiologically active while the (–)-R isomer did not affect sweetness perception (Ottinger et al., 2003). Flavanol-3-glycosides, being themselves astringent, were also described as bitter taste enhancers of tea infusions by amplifying the bitterness of caffeine (Scharbert and Hofmann, 2005). In green tea beverages, the sensomics approach allowed identifying a mixture of components, among which l-theanine, succinic acid, gallic acid and theogallin, as an umami-enhancing fraction (Kaneko et al., 2006). γ-Glutamyl peptides were characterised as key contributors to the kokumi taste of edible beans (*Phaseolus vulgaris* L.) (Dunkel et al., 2007) as well as in a Thai fermented freshwater fish where they were found at subthreshold taste concentrations (Phewpan et al., 2020). Arginyl dipeptides were described as saltiness enhancers in protein digests and fermented fish sauces (Schindler et al., 2011). A strong kokumi effect was attributed to 3-(allylthio)-2-(2-formyl-5-hydroxymethyl-1H-pyrrol-1-yl)propanoic acid, a tasteless component quantified by LC-MS/MS in roasted garlic (Wakamatsu et al., 2016) and to S-((4-amino-2-methyl pyrimidine-5-yl)methyl)-l-cysteine, found to be formed by a Maillard-type reaction of thiamine and cysteine in commercial meat-like process flavours (Brehm et al., 2019). Recent work on overfermented cocoa beans identified numerous short peptides as umami and saltiness enhancers as well as kokumi effectors when applied in a savoury taste matrix (Salger et al., 2019).

Finally, the sensomics approach benefited from the most recent advanced MS methods. Thus, data-independent acquisition using sequential window acquisition of all theoretical fragment ion (SWATH) was used to accelerate the identification of taste-active peptides in cheese (Sebald et al., 2020). SWATH-MS generates independently of any criteria MS/MS spectra of all detectable ions using sequential large isolation windows, typically 20 Da, in the quadrupole of a Q-ToF-MS/MS instrument. These windows step continuously across the selected ToF mass range, generating quantitative high-resolution MRM data. Thus, the SWATH-MS method could be shown for the first time to enable the comprehensive quantitation of all sensorially relevant key bitter peptides with limits of quantification far below each peptide's bitter taste recognition concentration (Sebald et al., 2020).

Furthermore, an ion mobility cell included in a hybrid LC-IMS-MS/MS instrument adds a new selection dimension to conventional devices. After the chromatographic and ionisation steps, ions are first separated based on their respective mobility in the IMS cell, depending on their size-to-charge ratio. IMS has been shown to overcome the challenges encountered in the MS analysis of isobaric compounds, for instance, and IMS-MS/MS instruments appear as a significant breakthrough in proteomics or lipidomics. For example, applied recently in a sensomics approach conducted on chanterelles mushrooms, octadecadien-12-ynoic acids were separated and identified, among which 14,15-dehydrocrepenynic acid and (9*Z*,15*E*)-14-oxooctadeca-9,15-dien-12-ynoic acid were characterised as key kokumi substances (Mittermeier et al., 2020).

As food flavour perception is essentially determined by the concentration of individual aroma and taste molecules, using a single instrumental setup to quantify odourants and tastants of key flavour components should be a breakthrough. High-throughput, accurate, rapid and sensitive quantitative analysis using a high-performance MS/MS instrument was recently demonstrated for key flavour molecules of apple juice (Hofstetter et al., 2019). Thus, using a single UHPLC-MS/MS instrument, it was possible to quantitate odourants and tastants over a large dynamic range of up to 6 orders of magnitude after minimal sample workup. However, aldehydes, ketones and organic acids had to be derivatised with 3-nitrophenylhydrazine before analysis. At the same time, polyphenols and esters were directly analysed by the instrument equipped with a reversed-phase C18 column, sometimes with analyte enrichment using SBSE for volatiles. Although not a one-shot injection analysis, this approach was named 'unified flavour quantitation' and presented as a promising high-throughput analytical method for key flavour molecules (Hofstetter et al., 2019). By applying the sensomics approach, the technique was recently applied to decipher the whole sensometabolome of a dairy milk dessert (Utz et al., 2021). However, volatile sulphur compounds were analysed using HS-SPME-GC, and abundant carbohydrates and organic acids were quantified using ^{1}H NMR (Utz et al., 2021).

9.4 FLAVOUR RELEASE AND PERCEPTION: *IN VIVO* ANALYSES

Even if the 'best' extraction and identification methods are used, and if the 'sensomics' approach, or equivalent procedures, constitutes a breakthrough in identifying flavour-active compounds in food, relating flavour compounds composition to flavour perception by humans is not straightforward in many occurrences. Perception of flavour is a dynamic process (Piggott, 2000). During food consumption, the concentration of retronasal aroma compounds at the olfactory epithelium and sapid compounds at the taste buds on the tongue varies with time as flavour components are released progressively from the food in the mouth. The release kinetics depends on the food matrix itself (Guichard, 2002), but also on in-mouth physiological mechanisms (salivation, mastication and tongue movement) conducting to bolus formation with saliva and subsequent swallowing (Salles et al., 2011), mechanisms for which individual variations are markedly important (Tarrega et al., 2008, 2011). Sensory methods, such as TI (Piggott, 2000), or the more recent TDS (Pineau et al., 2009) and TCATA (Castura et al., 2016), have been used to account for the dynamic- and time-related aspects of flavour perception.

9.4.1 Volatiles, *In Vivo*

Techniques for measuring aroma released under conditions found when humans consume foods have been developed during the last three decades. Significant robust results were obtained for sampling aroma from the nose using a collection of expired air samples on adsorbents then analysed with GC-MS (Linforth and Taylor, 1993; Taylor and Linforth, 1994). A procedure combining in-mouth HS sorptive extraction to TD-GC-MS was also developed for an in-mouth volatile sampling of wine aroma during wine tasting (Pérez-Jiménez and Pozo-Bayón, 2019; Perez-Jimenez et al., 2021). However, real-time *in vivo* flavour release analysis has been obtained using atmospheric or sub-atmospheric pressure ionisation mass spectrometry (Roberts and Taylor, 2000). The two main techniques, i.e., APCI (Linforth et al., 1996; Taylor and Linforth, 1996) and PTR (Lindinger et al., 1993), use H_3O^+ as reagent gas. Volatile compounds that have, in most cases, higher proton affinities than water, ionise by proton transfer from H_3O^+ affording protonated molecular MH^+ ions that are accelerated into a mass spectrometer, initially a quadrupole instrument. APCI sources may also be connected to an MS/MS such as an IT (Sémon et al., 2003; Jublot et al., 2005; Le Quéré et al., 2014), or a QQQ (Hatakeyama and Taylor, 2019) providing sensitivity, selectivity and structural capability benefits of MS/MS. Connecting a PTR source to a ToF instrument, with the benefit of its high-resolution power (Jordan et al., 2009a), has been a significant breakthrough. PTR-MS specificity compared to APCI approaches is that the generation of the reactant ion and the PTR are spatially and temporally separated. Therefore, control of the ionisation process is possible, and individual optimisation and quantification are more accessible (Yeretzian et al., 2000). The same applies to the SIFT-MS. For a long time used in breath analyses (Smith and Španěl, 2015; Smith and Španěl, 2016), SIFT-MS has naturally proven its usefulness in nose-space analysis (Xu and Barringer, 2010; Ozcan and Barringer, 2011; Mirondo and Barringer, 2016; Castada and Barringer, 2019; Langford et al., 2019).

With these techniques, air from the nose (nose space) is sampled directly into the mass spectrometer through a heated interface, making real-time breath-by-breath analysis routinely possible (Taylor et al., 2000; Roberts et al., 2003; Biasioli et al., 2011; Le Quéré et al., 2014). Time-resolved aroma release curves are generally described for each detected ion by the conventional parameters maximum intensity (I_{max}), time to reach the maximum intensity (T_{max}) and area under the curve that reflects the total amount of in-mouth released aroma. From the early beginning, many examples of nose-space analyses and their fundamental advances may be found in dedicated or specialised treatises (Taylor and Linforth, 2010; Le Quéré et al., 2014; Beauchamp and Herbig, 2015; Beauchamp and Zardin, 2017). Some results published recently will be detailed here. Thus, in work conducted with PTR-ToF-MS, coffee brews originating from a single pure Arabica blend (medium roast, dark roast and decaffeinated medium roast) were discriminated against using the nose-space analyses of five panellists (Romano et al., 2014). The nose-space data of more than 40 significant detected ions showed that roast degree effects were better differentiated than decaffeination effects (Romano et al., 2014). In a study conducted on 21 different apple cultivars using PTR-ToF-MS, it was shown that the Tmax parameter, in relation to the first swallow and the number of swallows, was a helpful index to distinguish the cultivars on juiciness (Ting et al., 2016). PTR-ToF-MS was also applied to analyse the nose-space profiles of 10 subjects while eating 10 different chocolates manufactured with beans of different botanical origins (Criollo-Forastero-Trinitario) and geographical origins (Africa-South America-Asia). It appeared that cocoa botanical sources affected in-nose profiles more than geographical origin. It was also evidenced that inter-individual differences were more significant than cocoa beans differences.

Nevertheless, the botanical origin was consistently reflected in the nose-space profile during eating (Acierno et al., 2019). Flavoured model wines differing in tartaric acid and sugar contents were recently investigated by the nose-space analyses of seven panelists using PTR-ToF-MS. If the aroma release time-related parameters allowed distinguishing the samples according to their acid and sugar contents, one main achievement stressed the importance of swallowing in perception (Arvisenet et al., 2019), as also evidenced for aroma release (Figure 9.8). Thus, following the first swallow, the volatile esters isoamyl acetate (MH^+ at *m/z* 131.107) and ethyl hexanoate (MH^+ at *m/z* 145.122) were immediately released to reach their maximum intensities.

A cross-modal interaction between congruent aroma and the taste was also evidenced. However, the post-swallowing decrease of taste resulted in more accurate aroma assessment, i.e., more correlated to *in vivo* aroma release, particularly evidenced in the delayed release of the less volatile ethyl octanoate (MH^+ at m/z 173.154) responsible for a long-lasting fruity aroma perception

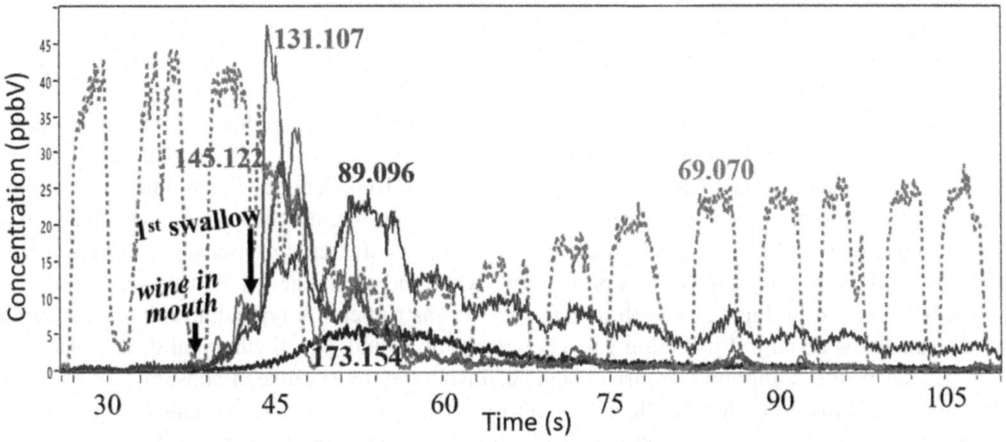

FIGURE 9.8 Typical aroma release curves for one panelist obtained in the nose-space analysis of a flavoured model wine using PTR-ToF-MS (Arvisenet et al., 2019): protonated molecular ions at *m/z* 89.096, $C_5H_{12}OH^+$ (3-methyl butane-1-ol), *m/z* 131.107, $C_7H_{14}O_2H^+$ (isoamyl acetate), *m/z* 145.122, $C_8H_{16}O_2H^+$ (ethyl hexanoate), *m/z* 173.154, $C_{10}H_{20}O_2H^+$ (ethyl octanoate). The ion at *m/z* 69.070 (dashed line) is the protonated isoprene, $C_5H_9^+$ (like acetone, isoprene is a basal metabolite in the breath, allowing the direct measure of breath frequency).

(Arvisenet et al., 2019). Wine aroma persistence was also studied using an *in vivo* PTR-ToF-MS approach with nine participants (Muñoz-González et al., 2019). Results showed that rosé wine aroma persistence was highly compound-dependent. While esters disappeared very fast, other compounds such as linalool remained in the oral cavity for a longer time after the wine was spat out. A low effect of added tannins was observed, except ethyl decanoate that was significantly higher released in the presence of tannins (Muñoz-González et al., 2019). Thus, nose-space analysis appears as a suitable method to distinguish sets of similar samples varying in the production process, origin and composition. Nose-space analysis using SIFT-MS was also used in noticeable studies on deodourisation of garlic breath by various foods (Mirondo and Barringer, 2016; Castada and Barringer, 2019), where the action of polyphenol oxidase and phenolic compounds on garlic-derived volatile sulphur compounds was evidenced (Mirondo and Barringer, 2016).

Combining time-resolved sensory studies with nose-space analysis should be easier to relate temporal parameters of aroma release to perception. Thus, it has been demonstrated for some soft French cheeses (pasteurised Brie, pasteurised and raw milk Camembert) that the perception scored by 15 assessors is consistent with their release of aroma compounds while eating the cheeses (Salles et al., 2003). For instance, perfect correspondence between the sulphury note determined by TI scoring and the APCI-MS release curves of the main sulphur key compounds, previously identified in a GC-O/GC-MS experiment (S-methyl thioacetate, dimethyldisulphide and dimethylsulphide), could be found (Salles et al., 2003; Le Quéré, 2004). In a recent study on espresso coffees, the roasting effect of coffee was evidenced both by TDS and nose-space data obtained simultaneously (Charles et al., 2015). Thus, a change in aroma perception was observed with a switch of dominance from roasted to burnt. At the same time, the nose-space data indicated that more volatiles in higher concentrations were released with increasing roasting degree (Charles et al., 2015).

Moreover, TDS displayed differences in aroma attributes between samples at the middle/end of perception time, while the release of potent odourants showed different behaviours. Thus, volatile markers of the burnt sensory note could be potentially identified as *N*-heterocycles such as substituted methyl-pyrroles and pyridine (Charles et al., 2015). Discontinuous TI sensory evaluation for mint flavour and sweetness perception of chewing gums was recently conducted with simultaneous in-nose release PTR-MS monitoring of VOCs linked to mint aroma: monoterpenes ($C_{10}H_{16}H^+$) at *m/z* 137.133, menthol ($C_{10}H_{19}^+$, dehydrated) at 139.148, menthofuran ($C_{10}H_{14}OH^+$) at *m/z* 151.112 and menthone/1,8-cineole ($C_{10}H_{18}OH^+$) at *m/z* 155.144 (Pedrotti et al., 2019). Significant differences for *in vivo* aroma release were found between Chinese and European panelists, the ethnicity effect being also significant for mint aroma and sweetness perception. In contrast, no gender effect was detected (Pedrotti et al., 2019).

Moreover, it was noticed that physiological parameters (oral cavity volume, salivary flow, breath acetone concentration and fungiform papillae density on the tongue) do not explain the relation between aroma release and perception (Pedrotti et al., 2019). To understand the aroma perception of dark chocolate concerning *in vivo* aroma release, simultaneous dynamic methodologies were implemented to study three dark chocolates differing in sensory characteristics. Sensory evaluation was conducted with both TDS and TCATA procedures, while the nose space of 16 assessors was independently measured in duplicates with a PTR-ToF-MS instrument. A typical aroma release curve for one panelist is displayed in Figure 9.9. The release of 2- and 3-methylbutanal ($C_5H_{10}OH^+$, *m/z* 87.080), a key aroma compound of dark chocolate, during the time course of consumption of 5 g of dark chocolate is displayed for the illustrative purpose (Le Quéré et al., 2021). The three chocolate samples A, B and C were clearly distinguished at the nose-space and sensory levels, whatever the temporal sensory method used (Le Quéré et al., 2021). For instance, sample A was characterised as more fruity and released significantly higher amounts of the protonated molecular ion $C_4H_8O_2H^+$ at *m/z* 89.060, attributed to acetoin and 2-methylpropanoic acid (Figure 9.10). Some predominant ions also characterised this fruity sample at *m/z* 123.092 and 137.107 attributed to substituted pyrazines $C_7H_{11}N_2^+$ and $C_8H_{13}N_2^+$, respectively, and at *m/z* 75.044 attributed to propanoic acid, $C_3H_7O_2^+$ (Le Quéré et al., 2021).

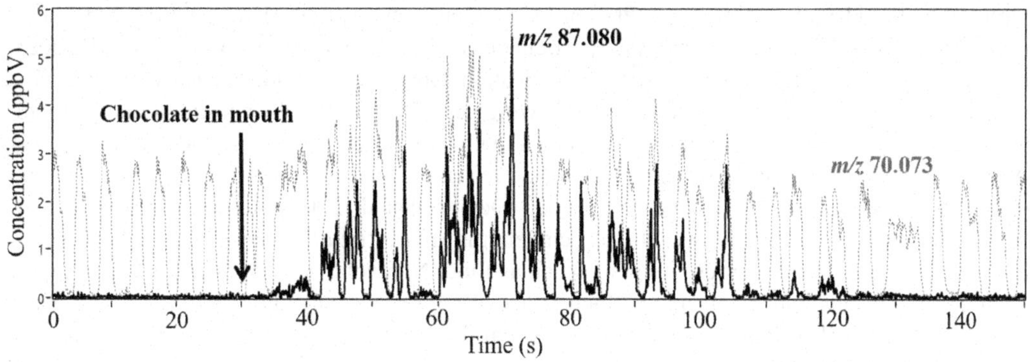

FIGURE 9.9 Typical aroma release curves (nose-space analysis using PTR-TOF-MS) for one panelist consuming 5 g of dark chocolate: protonated molecular ion at *m/z* 87.080, $C_5H_{10}OH^+$ (2- and 3-methylbutanal). The ion at *m/z* 70.073 (dashed line) is the protonated isoprene (^{13}C isotopologue), $[^{13}C]C_5H_9^+$, present in the breath.

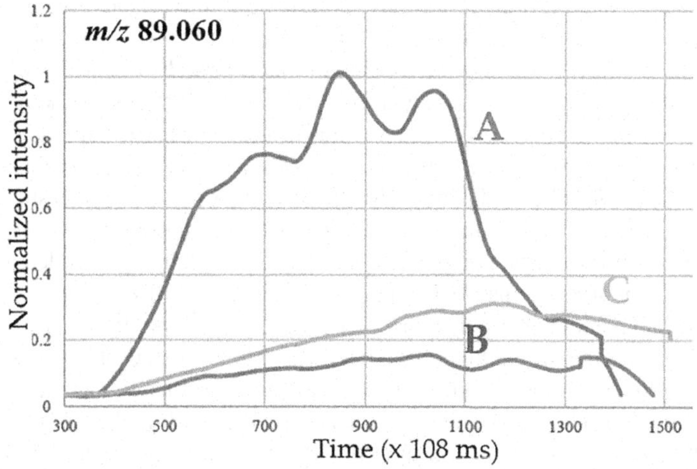

FIGURE 9.10 Smoothed traces (averaged on the panel) of the ion *m/z* 89.060 obtained using PTR-ToF-MS and attributed to the protonated molecular ion $C_4H_8O_2H^+$ of acetoin and 2-methylpropanoic acid, displaying apparent differing release rate in chocolate sample A compared to samples B and C.

Despite significant inter-individual differences largely due to differences in oral parameters (oral volume, respiratory and masticatory values), relationships between *in vivo* aroma release and perceived aroma intensity and quality can be investigated with some success using the paired dynamic methodologies. Early results obtained using APCI and PTR-MS for dairy foods may be found in a specialised treatise (Le Quéré and Cayot, 2013). Interesting examples are discussed in a book dedicated to odours (Beauchamp and Zardin, 2017). A recent review also reports relevant examples obtained with SIFT-MS (Langford et al., 2019).

9.4.2 Non-Volatiles, *In Vivo*

For practical and technical reasons, the development of methods to study flavour release has concentrated mainly on the volatile fraction, while only a limited number of techniques have been devoted to the real-time release of non-volatile compounds in the mouth (Le Quéré, 2004; Le Quéré and Cayot, 2013). One of the most applicable ones consists of saliva sampling using cotton buds coupled with offline liquid analyses, such as HPLC or LC-MS, performed after saliva weighing and extraction in a suitable solvent. For this analysis, participants are asked to take a swab from a specific location on the tongue at different times during mastication using a cotton bud. The feasibility was first demonstrated for the release of sucrose from chewing gums, where sucrose concentration was monitored using liquid-APCI-MS after extraction by a methanol–water solution (Davidson et al., 1999). A continuous sampling technique using a motor-driven ribbon placed across the tongue while a panelist chews a food sample has also been described (Davidson et al., 2000). After a complete swallow of the food sample, the ribbon was cut into 5 cm lengths after estimating the saliva weight adsorbed on the ribbon, each piece representing a particular time. Non-volatile components were extracted from the ribbon cuts with a solvent and their concentration was determined by direct liquid-phase APCI- or ESI-MS (Davidson et al., 2000). Temporal release of sucrose and glucose from biscuits, sodium from crisps, sucrose, glucose and fructose, citric and malic acids from fresh orange, and finally minerals (sodium, calcium and potassium) from Cheddar cheese was monitored successfully (Davidson et al., 2000). The cotton bud technique was applied to a model flavoured cheese, in which aroma and non-volatile compounds consistent with literature data were incorporated, to try to relate taste components release in the mouth during consumption to TI curves of target sensory attributes (Pionnier et al., 2004a; Pionnier et al., 2004b; Pionnier et al., 2005). Time-course release curves for minerals (sodium, calcium, magnesium and potassium), amino acids (leucine, phenylalanine and glutamic acid), organic acids (citric, lactic, propionic and butyric) and phosphoric acid were obtained with offline ESI-MS in negative ionisation mode or/and ion chromatography (Pionnier et al., 2004a). Although large inter-individual differences were noticed in the release behaviour, due to physiological parameters mainly related to mastication and salivation, the results allowed to link the time course of salty and sour attributes to the release of sodium and citric acid, respectively (Pionnier et al., 2004b; Pionnier et al., 2005). Noticeably, DART-MS was also successfully used to monitor the release kinetics of the taste-refreshing compound cyclohexane carboxamide, *N*-ethyl-5-methyl-2-(1-methyl ethyl) from chewing gums into the saliva of subjects sampled at different times during the consumption of the gums (Jeckelmann and Haefliger, 2010).

Flavour release and perception are dynamic processes and must be studied using dynamic methods. Dynamic physicochemical methods have been developed to study the parameters of flavour release when eating foods. Together with an increase in dynamic sensory techniques, they provide a better understanding of food flavour. However, further work, including the study of various interactions arising at different levels in food consumption, is needed to improve our knowledge of flavour perception. Oral physiology and ingestion behaviour appear as key points to link aroma perception to aroma release that is well apprehended by online MS techniques.

9.5 GLOBAL ANALYSES OF FLAVOUR: 'FLAVOUROMICS'

The targeted chemical analyses described above have been used to identify specific chemical traits associated with sensory characteristics. Those resulted in the potential loss of important information. Non-targeted methods, on the other hand, allow getting samples fingerprints that should contain more information, but generally at the cost of more noisy data, and they require extensive data mining and multivariate analysis, with a demanding chemometrics approach. The global study of food flavour using MS methods, in a manner analogous to the 'metabolomics' approach that could be defined as 'flavouromics', has served interesting applications for foodstuffs classification and discrimination. Thus, the volatilome of foodstuffs, i.e., their global VOC profiles, may be used as fingerprints to investigate such food characteristics as species, origin, applied pre-harvest treatment, post-harvest and shelf-life, process, and organoleptic properties (Deuscher et al., 2019). It should be stressed that, as already outlined, traditional analytical methods based on HS sampling coupled with GC-MS or GC×GC-MS are still considered the gold standard for VOC fingerprinting. For instance, the regional origin of Baijiu spirit (He and Jelen, 2021), different types of zaoyu, a traditional Chinese fermented fish (Chen et al., 2021), the regional origin of Chinese dry-cured hams

(Li et al., 2021) or differences in royal jelly produced from different trees (Qi et al., 2021) have been recently characterised. LC-MS giving access, among others, to non-volatile flavour compounds has also been used for fingerprinting in a classical metabolomics approach with some success for food categorisation. Some recent results include wheat species profiling (Stark et al., 2020), Italian red wine terroirs discrimination (Arapitsas et al., 2020), biochemical changes of black garlic during the ageing process (Chang et al., 2020), or bourbon whiskey ageing times (Yang et al., 2020b). However, to overcome the main drawback of chromatography-based methods, i.e., their lengthy analysis time detrimental to high-throughput analyses, DIMS methods that require minimal sample treatment have been implemented.

The first example to be cited consists of a global analysis of an HS sample by a mass spectrometer operated in EI mode, without any GC separation. The method's feasibility, generally referred to as 'MS-based e-nose', was initially demonstrated for rapid classification of four rather different French kinds of cheese (Vernat and Berdagué, 1995). The method originally handled without any GC column may use a GC column in isothermal conditions at 240°C–250°C (Berna et al., 2005; Tran et al., 2015). Alternatively, fingerprints may be obtained by averaging (Berna et al., 2004) or summing (Ratel et al., 2008) the mass spectra of GC-MS measurements. As a consequence of EI on low-resolution mass analysers, the resulting fingerprints are complex spectra of overlapping fragments. Therefore, classification tasks are generally done with a limited number of pertinent mass fragments often selected through chemometrics. This methodology was reviewed (Pérès et al., 2003; Pérez Pavón et al., 2006) and is still in use, for instance, to tackle smoky cocoa off-flavour (Scavarda et al., 2021). Although less commonly used, another method to be cited uses Curie point pyrolysis of tiny food samples as a direct coupling method (Aries and Gutteridge, 1987). The entire volatile fractions resulting from pyrolysed samples at up to 530°C are analysed by MS with low-energy ionisation and characteristic of the flavour and the food matrix breakdown. A pattern or fingerprint is obtained for each sample, and powerful classification/authentication tasks are possible, generally, after several data pre-processing steps to select a reduced number of significant mass fragments from the resulting rather complex mass spectra. Extensive data treatment, either by conventional multivariate statistics or artificial neural networks, allows the construction of maps useful for classification and quality control purposes. The main advantage of the method is that it provides flavour profiles and specific fingerprints of the food matrix, which in the case of cheese, for instance, could be potentially related to textural parameters (Pérès et al., 2002).

In a similar approach, HS samples may be directly connected to DIMS technologies (Biasioli et al., 2011) which are essentially online soft ionisation techniques based on CIMS (Beauchamp and Zardin, 2017). Contrary to the so-called MS-based e-noses, they use APCI-MS or low-pressure (SIFT-MS, PTR-MS) soft ionisation modes giving rise to ionised molecular species and limited fragmentations. Thus, obtained mass spectra bear more explicit chemical information on samples' HS composition. These techniques, as already outlined, use ionisation of the analytes by proton transfer from the reagent ion H_3O^+, as volatile compounds have, in most cases, higher PA than water. Although APCI-MS has been for years the reference technique for *in vivo* and *in vitro* monitoring of aroma compounds, its inherent drawbacks in terms of ionisation control and abilities for monitoring a large number of VOCs in a single run (Deuscher et al., 2019) have weakened its use in profiling experiments. SIFT-MS allows selection of reagent ions giving rise to the high purity of the reagent ion beam before it enters the flow tube where ionisation takes place.

Moreover and noteworthy, the microwave discharge source commonly used to make the reactant ions produces with H_3O^+ other ions such as NO^+, O_2^+ and OH^-. These ions can be easily selected in high purity as reagent ions to be used in charge transfer, hydride or proton abstraction, or in ion-molecule association reactions, all of analytical relevance. Notably, small molecules with PA smaller than the PA of water, such as alkanes or alkenes, can be detected and quantified. Isobaric or isomeric compounds can also be distinguished when their respective pseudo-molecular ions differ (Smith et al., 2011). For example, by using NO^+ as a reactant, aldehydes give $[M-1]^+$ while isomeric ketones give $[M]^+$, allowing their respective identification (Biasioli et al., 2011). With SIFT-MS, absolute quantification of ions is possible thanks to the fine control of ionisation conditions. However, the selection of reagent ions and dilution with the carrier gas entering the flow tube limit the technique's sensitivity, estimated to be similar to APCI (Biasioli et al., 2011). The literature on the operational aspects of SIFT-MS is abundant, and the interested reader may refer to this material (Smith and Španěl, 2005; Španěl and Smith, 2011; Smith and Španěl, 2016). References to some food VOC profiling studies based on SIFT-MS fingerprinting may be found in recent literature (Deuscher et al., 2019) and a dedicated recent review (Langford et al., 2019).

High yield of the reagent hydronium ions resulting in high ionisation efficacy, and admission in the drift-tube of aerial samples without additional dilution resulting in enhanced sensitivity, are the main features that explain the intensive use of PTR-MS in the last years. With high mass resolution, speedy full-scan capabilities and a high dynamic range, the DIMS technique PTR-MS equipped with a ToF mass analyser appears as the ultimate method. Adjunction of a switchable reagent ion source using NO^+ and O_2^+ (Jordan et al., 2009b), as in SIFT-MS, or NH_4^+ (Müller et al., 2020), as reagent ions have broadened their application. A lot of examples can be found in a dedicated textbook (Ellis and Mayhew, 2014) and the recent literature (Beauchamp and Herbig, 2015; Beauchamp and Zardin, 2017; Deuscher et al., 2019), a vast contribution being due to Franco Biasioli's group in Italy (Cappellin et al., 2013; Capozzi et al., 2017). Some results published recently will be detailed here for demonstrative purposes.

Thus, 90 dark chocolate HSs were fingerprinted by PTR-MS (Acierno et al., 2016). Based on 136 masses per sample, it was possible to use chemometrics to reveal significant traits on botanical and geographical origins of the cocoa beans in the chocolates and processing features applied by the different brands (Acierno et al., 2016). A challenging classification of dark chocolates assessed by sensory evaluation in four distinct sensorial categories was recently investigated (Deuscher et al., 2019). The VOC profiles of 206 samples, classified in the four sensory categories, were analysed by PTR-ToF-MS. Based on 143 significant masses per sample, supervised multivariate data analysis of the partial least squares-discriminant analysis (PLS-DA) type allowed the elaboration of a classification model that showed excellent prediction ability with 97% of a test set of 62 samples (of the initial 206) correctly predicted in the sensory categories (Deuscher et al., 2019). The HS volatiles of 100 meat samples obtained from 100 animals of five species/categories (chicken, turkey, pork, veal and beef) were directly investigated using PTR-ToF-MS for the first time (Ni et al., 2020). The results obtained based on 129 mass peaks showed that turkey and chicken had the largest and slightest total concentrations of all VOCs. Of the mammalian meats, veal and beef had greater total VOC concentrations than pork. The proportions of all the individual VOCs differed significantly according to species (Ni et al., 2020).

Moreover, 14 of 17 independent latent explanatory factors (groups of VOCs) identified by multivariate analysis (Bittante et al., 2020) that exhibited significant differences between meat species/categories helped to characterise them (Ni et al., 2020). It was argued that the knowledge of specific VOC profiles paves new avenues for research aimed at describing species through the sensory description, at authenticating species or at identifying abnormalities or fraud (Ni et al., 2020). The volatilome of French wine brandies was recently investigated by PTR-ToF-MS (Malfondet et al., 2021). The obtained HS fingerprints were used to examine the origin of the produced brandies within a limited geographic production area. Thus, based on 82 significant masses, 60 brandies aged 14 years in French oak barrels were successfully classified according to their growth areas using the unsupervised multivariate analysis PCA, principal component analysis (Figure 9.11).

Moreover, models obtained by a supervised PLS-DA multivariate analysis allowed identifying volatile discriminant compounds that were mainly characterised as key aroma compounds of wine brandies (Malfondet et al., 2021). The PTR-MS method could be considered as an objective method to answer the question of French wine brandies terroir, this terroir effect on VOC composition being confirmed at a perceptive level by sensory evaluation (Malfondet et al., 2021).

The tools presented above, which combine analytical instrumentation for global assessment of flavour with multivariate data analyses, have demonstrated their usefulness for classification purposes.

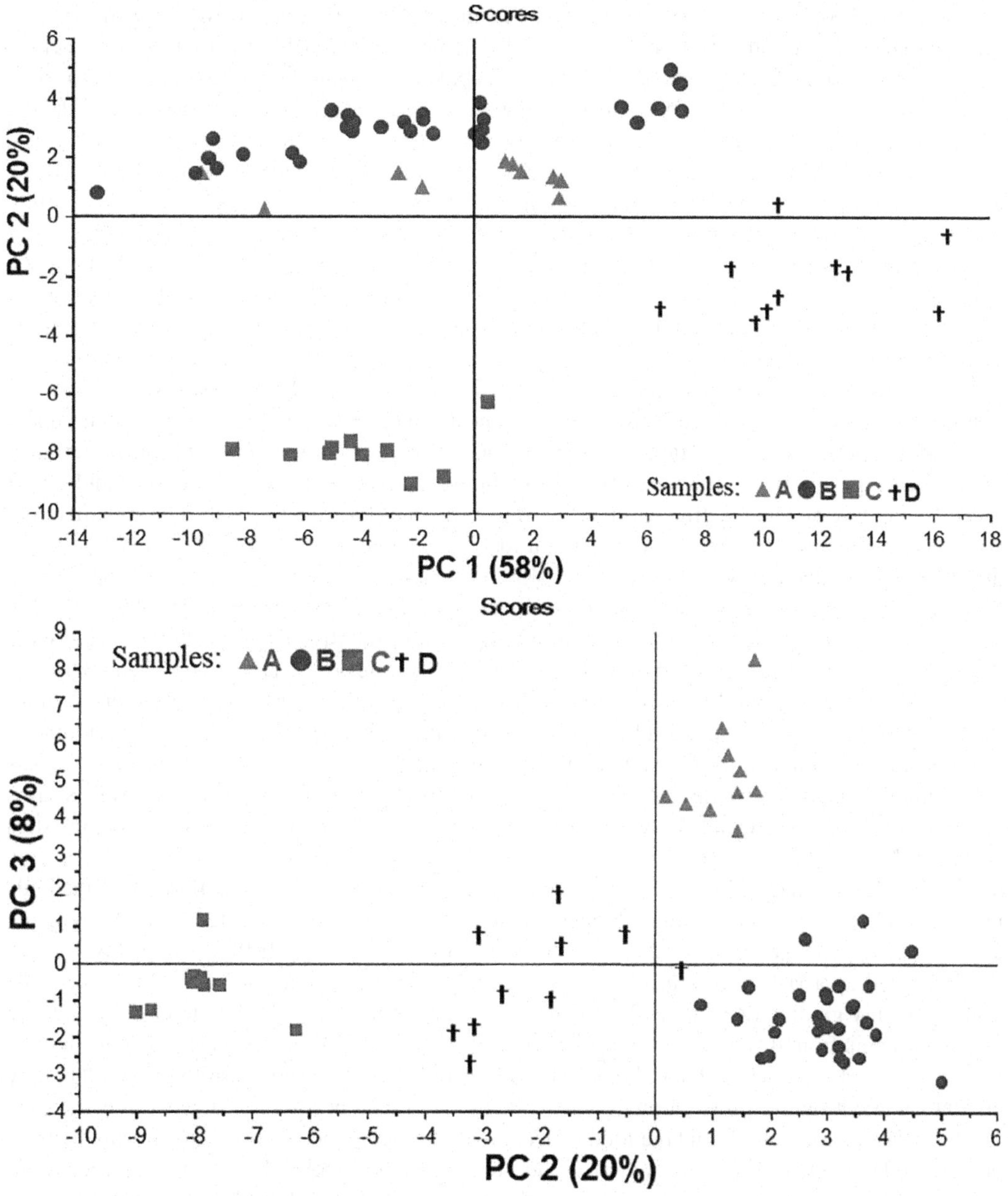

FIGURE 9.11 Principal component analysis (PCA) score plots based on PTR-ToF-MS mass spectra of four types of French wine brandies A, B, C and D, produced within a limited geographic production area, representing the first vs second (top) and second vs third (down) principal components. The HS analyses of two replicates of 60 brandies aged 14 years yielded 82 significant mass peaks; PCA could discriminate their production areas with success.

However, MS-based targeted and untargeted volatilomics is an interesting tool for understanding aroma and flavour chemistry in food process monitoring (Diez-Simon et al., 2019). Some examples of the primary sources of variations (animals, microorganisms, plants, chemical contamination and additives, processing and storage) on the food metabolome may be found in a recent review (Diez-Simon et al., 2019). Some results described in the current literature will be described here for illustrative purpose.

Thus, a protocol based on HS-SPME-GC-MS monitoring of flavour compounds during the different steps of the industrial process of Peruvian chocolate manufacturing was implemented (Michel et al., 2021). Using a retrospective approach, starting from final chocolate bars towards pre-roasted cocoa beans, it was possible to follow the cocoa flavour development involved during each step. The developed methodology allowed distinguishing chocolates from different regions, north and south of Peru, and production lines (Michel et al., 2021). DIMS approaches with a PTR-TOF-MS instrument were used in many instances. Coffee has received a great deal of effort into understanding volatile and aroma compounds formation from the green bean to the coffee brew, including all the critical steps of roasting and grinding, in a time-resolved analytical approach (Yeretzian et al., 2019). Extraction dynamics of aroma compounds could also be investigated during espresso coffee preparation and dynamic evolution of in-mouth coffee aroma followed by a nose-space procedure (Yeretzian et al., 2019). In this context, it was claimed to predict coffee sensory profile from instrumental measurements (Yeretzian et al., 2019). Losses of volatile compounds during the baking process are expected. The aroma intensity of gluten-free bread, known for its weak aroma, is an important issue. The release to the oven of volatile compounds from five gluten-free types of bread (quinoa, teff and rice flours, and corn and wheat starches) and wheat bread during baking and toasting was measured in real-time (Pico et al., 2020). Through different volatile release patterns, it was evidenced that toasting increased VOCs released from the bread matrix and generated new volatiles from the Maillard reaction. Gluten-free bread had higher VOC losses during baking than wheat bread, partially explaining its weaker aroma (Pico et al., 2020). The progress of alcoholic fermentation and investigation of the diversity of volatiles in the wine during fermentation has also been undertaken for the first time with a DIMS approach (Berbegal et al., 2020). VOC variability associated with different combinations of *Saccharomyces*/non-*Saccharomyces* was assessed. Preliminary findings underlined their different behaviours on grape juice and on must, respectively, and confirmed differences among the single yeast strains volatilomes. Notably, it was evidenced that the cooperative use of non-*Saccharomyces* strains can modify the global release of VOCs as a function of the fermented medium characteristics (Berbegal et al., 2020).

As already outlined, IMS is an emerging and cheaper alternative to more traditional methods or DIMS. Thus, a technique using HS-SPME-GC-IMS was recently validated to study dynamic aroma release (Thomas et al., 2021), and static HS coupled with GC-IMS was recently used in an untargeted approach to link aroma chemistry via predictive models to Sauvignon blanc wine sensory quality grading (Zhu et al., 2021). GC-IMS was also used successfully to classify Chinese dry-cured hams from different regions (Li et al., 2021). Ambient MS methods are also involved in these untargeted approaches. They should find strong applicability as their resulting data, generally obtained without sample preparation or with some limited pre-concentration steps, are characteristic of the flavour compounds and the food matrix breakdown components when monitoring thermal processes. Thus, DART and atmospheric solid analysis probe (ASAP)-MS using full-scan mode were recently evaluated for high-throughput authenticity screening of dried Mediterranean oregano and adulteration practices (Damiani et al., 2021). Aided by chemometrics, fingerprinting capabilities were confirmed for classification purpose and adulteration studies. Although both techniques provided satisfactory results, ASAP-MS clearly showed greater potential, leading to reproducible, diagnostic feature-rich mass spectra (Damiani et al., 2021). The use of MALDI- and DESI-MS in this context were reviewed recently (Byliński et al., 2017). Rapid evapourative ionisation MS, where a handheld electrocautery device is used to thermally ablate a food sample and produce an aerosol transported to a mass spectrometer for lipidomic analysis, was used for meat authentication studies (Gatmaitan et al., 2021). The MasSpec Pen technology is another alternative. The MasSpec Pen is a handled device based on a gentle liquid extraction process. A solvent droplet formed at the device tip extracts molecules from the sample upon contacting a sample surface. The solvent droplet is then transported through tubing to the mass spectrometer for immediate analysis (Gatmaitan et al., 2021). Using negative ion mode on a hybrid quadrupole-Orbitrap instrument, meat authentication and classifying venison and grass-fed beef samples adulterated with grain-fed beef were achieved with excellent prediction accuracies (Gatmaitan et al., 2021).

Therefore, a combination of volatilomics with LC-MS or ambient MS techniques that give access to non-volatile profiling appears to step further towards an integrated 'flavouromics' approach.

9.6 CONCLUSION

All the examples presented in this chapter pinpoint the crucial role of MS in flavour research. All the fields of flavour science benefit from the latest developments in MS methods and instrumentation. Thus, key flavour compounds, aroma and taste markers can be identified with various GC-MS and HPLC-MS arrangements that include multidimensional chromatography and MS/MS, with high resolving power for both elements.

Global analyses, which could be rationalised in terms of 'flavouromics', conducted with various DIMS techniques or instruments hyphenated to chromatographic techniques, show powerful classification capabilities, with very often excellent prediction abilities afforded by chemometrics. Thus, MS fingerprints are used to investigate food characteristics such as species, origin, applied pre-harvest treatment, post-harvest and shelf-life, process, and organoleptic properties. In complement, as already evoked in the introduction, it should be noted that IRMS, either hyphenated to GC or HPLC, not included in the present review, can partly play the same role. As natural isotope fractionation occurs as a function of geographical, climatic, agricultural and process determinants, measurement of stable-isotope ratios of carbon, nitrogen, oxygen and hydrogen can be used for authentication purpose. A few recent examples for flavour compounds isotope ratios determination will be cited for illustration. Thus, the country of origin of commercial vegetables was investigated via the determination of isotopic compositions of light elements (^{13}C, ^{15}N, ^{18}O and ^{34}S) using IRMS (Mahne Opatić et al., 2017). The natural origin of vanillin extracted from vanilla pods, commercial flavoured foodstuff and in vanillin samples of different origins was investigated by measuring its carbon isotope ratio (i.e., ^{12}C vs ^{13}C, noted $\delta^{13}C$) using GC-IRMS (Schipilliti et al., 2017) and GC-IRMS was also used for improving the detection of the authenticity of grape must, by measuring the ratios $\delta^{13}C$ and $\delta^{15}N$ of proline as potential geographical markers (Perini et al., 2020). The measurements of the $\delta^{13}C$ values for glucose, fructose, disaccharides, trisaccharides and organic acids using LC-IRMS allowed the classification of honey samples (Kawashima et al., 2018). The $\delta^{13}C$ values of organic acids in sake and wine measured using LC-IRMS helped to identify when organic acids were added to sake and wine and to elucidate the process of their formation (Suto and Kawashima, 2021).

Dynamic methods using various DIMS instruments have investigated the link between *in vivo* aroma release and aroma perception. The simultaneous use of dynamic sensory methods should help in better understand flavour perception. *In vivo* measurements of non-volatile tastants are not straightforward, but some offline MS methods were successfully implemented.

Finally, the content of this chapter dedicated to the use of MS to study food flavour is necessarily not exhaustive, as it results from authors' choice, generally illustrated by the most recent literature. An important part was dedicated to tasting compounds, as, to authors' knowledge, such a comprehensive review was lacking. Interaction of flavour components with the macromolecules of the food matrix that may influence flavour perception has been a recurrent subject in flavour research (Guichard, 2002). As a last significant example, an elegant study on flavour–protein interaction should be cited. While non-covalent interactions between aroma molecules and proteins were studied by static HS GC for thermodynamic binding constants determination (Andriot et al., 2000), and at a molecular level by 2D-NMR methods (Lübke et al., 2002), covalent bond formation leading to covalent adducts was scarcely studied. The formation of covalent adducts between 47 flavour compounds belonging to 13 different chemical classes and the protein β-lactoglobulin was recently evaluated using a UPLC-qTOF/MS instrument in ESI+ mode. Schiff base, Michael addition, and disulphide linkages were evidenced, conducting to covalent protein-odourant adducts characterised by the appearance of significant ions in the mass spectra corresponding to protein+odourant entities (Anantharamkrishnan et al., 2020). Aldehydes, sulphur-containing molecules (especially thiols) and functional group-containing furans were the most reactive flavour compounds. It was argued that the formation of flavour-protein covalent bonding could be potentially responsible for flavour losses, limiting the shelf-life of many foods (Anantharamkrishnan et al., 2020).

REFERENCES

Acierno, V., Liu, N., Alewijn, M., Stieger, M., van Ruth, S. M., 2019. Which cocoa bean traits persist when eating chocolate? Real-time nosespace analysis by PTR-QiToF-MS. *Talanta* 195, 676–682.

Acierno, V., Yener, S., Alewijn, M., Biasioli, F., van Ruth, S., 2016. Factors contributing to the variation in the volatile composition of chocolate: Botanical, and geographical origin of the cocoa beans, and brand-related formulation and processing. *Food Res. Int.* 84, 86–95.

Acree, T. E., 1993. Gas chromatography-olfactometry. In: Ho, C. T., Manley, C. H. (Eds.), *Flavor Measurement.* New York, Dekker, M., pp. 77–94.

Ahrens, A., Zimmermann, S., 2021. Towards a hand-held, fast, and sensitive gas chromatograph-ion mobility spectrometer for detecting volatile compounds. *Anal. Bioanal. Chem.* 413, 1009–1016.

Aisala, H., Sola, J., Hopia, A., Linderborg, K. M., Sandell, M., 2019. Odor-contributing volatile compounds of wild edible Nordic mushrooms analyzed with HS–SPME–GC–MS and HS–SPME–GC–O/FID. *Food Chem.* 283, 566–578.

Allamy, L., Darriet, P., Pons, A., 2017. Identification and Organoleptic Contribution of (Z)-1,5-Octadien-3-one to the Flavor of Vitis vinifera cv. Merlot and Cabernet Sauvignon Musts. *J. Agric. Food Chem.* 65, 1915–1923.

Amaral, M. S. S., Marriott, P. J., 2019. The blossoming of technology for the analysis of complex aroma bouquets—A review on flavour and odorant multidimensional and comprehensive gas chromatography applications. *Molecules* 24, 2080.

Amigo, J. M., Skov, T., Bro, R., Coello, J., Maspoch, S., 2008. Solving GC-MS problems with PARAFAC2. *Trac-Trends Anal. Chem.* 27, 714–725.

Anantharamkrishnan, V., Hoye, T., Reineccius, G. A., 2020. Covalent adduct formation between flavor compounds of various functional group classes and the model protein β-lactoglobulin. *J. Agric. Food Chem.* 68, 6395–6402.

Andersen, L. T., Schlichtherle-Cerny, H., Ardö, Y., 2008. Hydrophilic di- and tripeptides are not a precondition for savoury flavour in mature Cheddar cheese. *Dairy Sci. Technol.* 88, 467–475.

Andriot, I., Harrison, M., Fournier, N., Guichard, E., 2000. Interactions between methyl ketones and β-lactoglobulin: Sensory analysis, headspace analysis, and mathematical modeling. *J. Agric. Food Chem.* 48, 4246–4251.

Arapitsas, P., Ugliano, M., Marangon, M., et al., 2020. Use of untargeted liquid chromatography-mass spectrometry metabolome to discriminate Italian monovarietal red wines, produced in their different terroirs. *J. Agric. Food Chem.* 68, 13353–13366.

Arcari, S. G., Caliari, V., Sganzerla, M., Godoy, H. T., 2017. Volatile composition of Merlot red wine and its contribution to the aroma: Optimization and validation of analytical method. *Talanta* 174, 752–766.

Aries, R. E., Gutteridge, C. S., 1987. Applications of pyrolysis mass spectrometry to food science. In: Gilbert, J. (Ed.), *Applications of Mass Spectrometry in Food Science.* London, Elsevier Applied Science, pp. 377–431.

Arnold, R. A., Noble, A. C., Singleton, V. L., 1980. Bitterness and astringency of phenolic fractions in wine. *J. Agric. Food Chem.* 28, 675–678.

Arvisenet, G., Ballester, J., Ayed, C., et al., 2019. Effect of sugar and acid composition, aroma release, and assessment conditions on aroma enhancement by taste in model wines. *Food Qual. Pref.* 71, 172–180.

Aston, J. W., Creamer, L. K., 1986. Contribution of the components of the water-soluble fraction to the flavour of Cheddar cheese. *N.Z.J. Dairy Sci. Technol.* 21, 229–248.

Baldovini, N., Chaintreau, A., 2020. Identification of key odorants in complex mixtures occurring in nature. *Nat. Prod. Rep.* 37, 1589–1626.

Barbara, J. A., Nicolli, K. P., Souza-Silva, E. A., et al., 2020. Volatile profile and aroma potential of tropical Syrah wines elaborated in different maturation and maceration times using comprehensive two-dimensional gas chromatography and olfactometry. *Food Chem.* 308, 125552.

Beauchamp, G. K., 2019. Basic taste: A perceptual concept. *J. Agric. Food Chem.* 67, 13860–13869.

Beauchamp, J., Herbig, J., 2015. Proton-transfer-reaction time-of-flight mass spectrometry (PTR-TOFMS) for aroma compound detection in real-time: Technology, developments, and applications. In: Guthrie, B., Beauchamp, J., Buettner, A., Lavine, B. K. (Eds.), *The Chemical Sensory Informatics of Food: Measurement, Analysis, Integration*. Washington DC, American Chemical Society, pp. 235–251.

Beauchamp, J., Zardin, E., 2017. Odorant detection by on-line chemical ionization mass spectrometry. In: Buettner, A. (Ed.), *Handbook of Odor*. Cham, Springer International Publishing, pp. 355–408.

Behrens, M., Meyerhof, W., Hellfritsch, C., Hofmann, T., 2011. Sweet and umami taste: Natural products, their chemosensory targets, and beyond. *Angew. Chem. Int. Ed. Engl.* 50, 2220–2242.

Behrens, M., Ziegler, F., 2020. Structure-function analyses of human bitter taste receptors—where do we stand? *Molecules* 25, 4423.

Belmonte-Sanchez, J. R., Gherghel, S., Arrebola-Liebanas, J., et al., 2018. Rum classification using fingerprinting analysis of volatile fraction by headspace solid phase microextraction coupled to gas chromatography-mass spectrometry. *Talanta* 187, 348–356.

Ben Shoshan-Galeczki, Y., Niv, M. Y., 2020. Structure-based screening for discovery of sweet compounds. *Food Chem.* 315, 126286.

Berbegal, C., Khomenko, I., Russo, P., et al., 2020. PTR-ToF-MS for the online monitoring of alcoholic fermentation in wine: Assessment of VOCs variability associated with different combinations of saccharomyces/non-saccharomyces as a case-study. *Fermentation* 6, 55.

Berna, A. Z., Buysens, S., Natale, C. D., et al., 2005. Relating sensory analysis with electronic nose and headspace fingerprint MS for tomato aroma profiling. *Postharvest Biol. Technol.* 36, 143–155.

Berna, A. Z., Lammertyn, J., Saevels, S., Natale, C. D., Nicolai, B. M., 2004. Electronic nose systems to study shelf life and cultivar effect on tomato aroma profile. *Sens. Actuators B* 97, 324–333.

Besnard, P., Passilly-Degrace, P., Khan, N. A., 2016. Taste of fat: A sixth taste modality? *Physiol. Rev.* 96, 151–176.

Biasioli, F., Yeretzian, C., Märk, T. D., Dewulf, J., Van Langenhove, H., 2011. Direct-injection mass spectrometry adds the time dimension to (B)VOC analysis. *Trac-Trends Anal. Chem.* 30, 1003–1017.

Bittante, G., Ni, Q., Khomenko, I., Gallo, L., Biasioli, F., 2020. Rapid profiling of the volatilome of cooked meat by PTR-ToF-MS: Underlying latent explanatory factors. *Foods* 9, 1738.

Boughter, J. D., Munger, S. D., 2013. Taste receptors. In: Lennarz, W. J., Lane, M. D. (Eds.), *Encyclopedia of Biological Chemistry*, 2nd edition. Waltham, Academic Press, pp. 366–368.

Bouysset, C., Belloir, C., Antonczak, S., Briand, L., Fiorucci, S., 2020. Novel scaffold of natural compound eliciting sweet taste revealed by machine learning. *Food Chem.* 324, 126864.

Brehm, L., Jünger, M., Frank, O., Hofmann, T., 2019. Discovery of a thiamine-derived taste enhancer in process flavors. *J. Agric. Food Chem.* 67, 5857–5865.

Brendel, R., Schwolow, S., Rohn, S., Weller, P., 2021. Volatilomic profiling of citrus juices by dual-detection HS-GC-MS-IMS and machine learning—An alternative authentication approach. *J. Agric. Food Chem.* 69, 1727–1738.

Buck, N., Goblirsch, T., Beauchamp, J., Ortner, E., 2020. Key aroma compounds in two bavarian gins. *Appl. Sci.-Basel* 10, 14.

Budzikiewicz, H., Djerassi, C., Williams, A. H., 1964. *Interpretation of Mass Spectra of Organic Compounds*. San Francisco, CA, Holden-Day, Inc.

Byliński, H., Gębicki, J., Dymerski, T., Namieśnik, J., 2017. Direct analysis of samples of various origin and composition using specific types of mass spectrometry. *Crit. Rev. Anal. Chem.* 47, 340–358.

Capozzi, V., Yener, S., Khomenko, I., et al., 2017. PTR-ToF-MS coupled with an automated sampling system and tailored data analysis for food studies: Bioprocess monitoring, screening and nose-space analysis. *J. Vis. Exp.* 123, e54075.

Cappellin, L., Loreto, F., Aprea, E., et al., 2013. PTR-MS in Italy: A multipurpose sensor with applications in environmental, agri-food and health science. *Sensors* 13, 11923.

Castada, H. Z., Barringer, S. A., 2019. Online, real-time, and direct use of SIFT-MS to measure garlic breath deodorization: A review. *Flavour Fragr. J.* 34, 299–306.

Castura, J. C., Antúnez, L., Giménez, A., Ares, G., 2016. Temporal Check-All-That-Apply (TCATA): A novel dynamic method for characterizing products. *Food Qual. Pref.* 47, 79–90.

Chang, W. C., Chen, Y. T., Chen, H. J., Hsieh, C. W., Liao, P. C., 2020. Comparative UHPLC-Q-Orbitrap HRMS-based metabolomics unveils biochemical changes of black garlic during aging process. *J. Agric. Food Chem.* 68, 14049–14058.

Charles, M., Romano, A., Yener, S., et al., 2015. Understanding flavour perception of espresso coffee by the combination of a dynamic sensory method and in-vivo nosespace analysis. *Food Res. Int.* 69, 9–20.

Chen, Z., Tang, H., Ou, C., et al., 2021. A comparative study of volatile flavor components in four types of zaoyu using comprehensive two-dimensional gas chromatography in combination with time-of-flight mass spectrometry. *J. Food Process Preserv.* 45, e15230.

Cheng, K., Peng, B. Z., Yuan, F., 2020. Volatile composition of eight blueberry cultivars and their relationship with sensory attributes. *Flavour Fragr. J.* 35, 443–453.

Chéron, J.-B., Casciuc, I., Golebiowski, J., Antonczak, S., Fiorucci, S., 2017. Sweetness prediction of natural compounds. *Food Chem.* 221, 1421–1425.

Chua, C. K., Lv, Y. B., Zhao, W., Ren, Y., Zhang, H. J., 2020. Improving annotation of known-unknowns with accurately reconstructed mass spectra. *Int. J. Mass Spectrom.* 451, 116321.

Contreras, M. D., Arroyo-Manzanares, N., Arce, C., Arce, L., 2019. HS-GC-IMS and chemometric data treatment for food authenticity assessment: Olive oil mapping and classification through two different devices as an example. *Food Control* 98, 82–93.

Costa, A. C. V., Garruti, D. S., Madruga, M. S., 2019. The power of odour volatiles from unifloral melipona honey evaluated by gas chromatography-olfactometry Osme techniques. *J. Sci. Food Agric.* 99, 4493–4497.

Damiani, T., Dreolin, N., Stead, S., Dall'Asta, C., 2021. Critical evaluation of ambient mass spectrometry coupled with chemometrics for the early detection of adulteration scenarios in Origanum vulgare L. *Talanta* 227, 122116.

Dang, Y., Gao, X., Ma, F., Wu, X., 2015. Comparison of umami taste peptides in water-soluble extractions of Jinhua and Parma hams. *LWT-Food Sci. Technol.* 60, 1179–1186.

Davidson, J. M., Linforth, R. S. T., Hollowood, T. A., Taylor, A. J., 1999. Effect of sucrose on the perceived flavor intensity of chewing gum. *J. Agric. Food Chem.* 47, 4336–4340.

Davidson, J. M., Linforth, R. S. T., Hollowood, T. A., Taylor, A. J., 2000. Release of non-volatile flavor compounds in vivo. In: Roberts, D. D., Taylor, A. J. (Eds.), *Flavor Release*. Washington DC, American Chemical Society, pp. 99–111.

Dawid, C., Henze, A., Frank, O., et al., 2012. Structural and sensory characterization of key pungent and tingling compounds from black pepper (Piper nigrum L.). *J. Agric. Food Chem.* 60, 2884–2895.

Dawid, C., Hofmann, T., 2014. Quantitation and bitter taste contribution of saponins in fresh and cooked white asparagus (Asparagus officinalis L.). *Food Chem.* 145, 427–436.

Deepankumar, S., Karthi, M., Vasanth, K., Selvakumar, S., 2019. Insights on modulators in perception of taste modalities: A review. *Nutr. Res. Rev.* 32, 231–246.

Degenhardt, A. G., Hofmann, T., 2010. Bitter-tasting and kokumi-enhancing molecules in thermally processed avocado (Persea americana Mill.). *J. Agric. Food Chem.* 58, 12906–12915.

Deklerck, V., Finch, K., Gasson, P., et al., 2017. Comparison of species classification models of mass spectrometry data: Kernel Discriminant Analysis vs Random Forest; A case study of Afrormosia (Pericopsis elata (Harms) Meeuwen). *Rapid Commun. Mass Spectrom.* 31, 1582–1588.

Dennenlohr, J., Thorner, S., Manowski, A., Rettberg, N., 2020. Analysis of selected hop aroma compounds in commercial lager and craft beers using HS-SPME-GC-MS/MS. *J. Am. Soc. Brew. Chem.* 78, 16–31.

Desportes, C., Charpentier, M., Duteurtre, B., Maujean, A., Duchiron, F., 2001. Isolation, identification, and organoleptic characterization of low-molecular-weight peptides from white wine. *Am. J. Enol. Vitic.* 52, 376–380.

Deuscher, Z., Andriot, I., Sémon, E., et al., 2019. Volatile compounds profiling by using proton transfer reaction – time of flight – mass spectrometry (PTR-ToF-MS). The case study of dark chocolates organoleptic differences. *J. Mass Spectrom.* 54, 92–119.

Deuscher, Z., Gourrat, K., Repoux, M., et al., 2020. Key aroma compounds of dark chocolates differing in organoleptic properties: A GC-O comparative study. *Molecules* 25, 1809.

Diez-Simon, C., Mumm, R., Hall, R. D., 2019. Mass spectrometry-based metabolomics of volatiles as a new tool for understanding aroma and flavour chemistry in processed food products. *Metabolomics* 15, 41.

Dresel, M., Vogt, C., Dunkel, A., Hofmann, T., 2016. The bitter chemodiversity of hops (Humulus lupulus L.). *J. Agric. Food Chem.* 64, 7789–7799.

Dunkel, A., Köster, J., Hofmann, T., 2007. Molecular and sensory characterization of γ-glutamyl peptides as key contributors to the kokumi taste of edible beans (Phaseolus vulgaris L.). *J. Agric. Food Chem.* 55, 6712–6719.

Dunkel, A., Steinhaus, M., Kotthoff, M., et al., 2014. Nature's chemical signatures in human olfaction: A foodborne perspective for future biotechnology. *Angew. Chem.-Int. Edit.* 53, 7124–7143.

Ellis, A. M., Mayhew, C. A., 2014. *Proton Transfer Reaction Mass Spectrometry. Principles and Applications.* Chichester, Wiley.

Engel, E., Nicklaus, S., Septier, C., Salles, C., Le Quéré, J. L., 2000a. Taste active compounds in a goat cheese water-soluble extract. 2. Determination of the relative impact of water-soluble extract components on its taste using omission tests. *J. Agric. Food Chem.* 48, 4260–4267.

Engel, E., Nicklaus, S., Garem, A., et al., 2000b. Taste active compounds in a goat cheese water-soluble extract. 1. Development and sensory validation of a model water-soluble extract. *J. Agric. Food Chem.* 48, 4252–4259.

Engel, E., Tournier, C., Salles, C., Le Quéré, J. L., 2001. Evolution of the composition of a selected bitter camembert cheese during ripening: Release and migration of taste-active compounds. *J. Agric. Food Chem.* 49, 2940–2947.

Engels, W. J. M., Visser, S., 1994. Isolation and comparative characterization of components that contribute to the flavour of different types of cheese. *Neth. Milk Dairy J.* 48, 127–140.

Fayad, S., Cretin, B. N., Marchal, A., 2020. Development and validation of an LC–FTMS method for quantifying natural sweeteners in wine. *Food Chem.* 311, 125881.

Fayeulle, N., Preys, S., Roger, J. M., et al., 2020. Multiblock analysis to relate polyphenol targeted mass spectrometry and sensory properties of chocolates and cocoa beans. *Metabolites* 10, 311.

Flaig, M., Qi, S. C., Wei, G. D., Yang, X. G., Schieberle, P., 2020. Characterisation of the key aroma compounds in a Longjing green tea infusion (Camellia sinensis) by the sensomics approach and their quantitative changes during processing of the tea leaves. *Eur. Food Res. Technol.* 246, 2411–2425.

Frank, O., Blumberg, S., Krümpel, G., Hofmann, T., 2008. Structure determination of 3-O-Caffeoyl-epi-γ-quinide, an orphan bitter lactone in roasted coffee. *J. Agric. Food Chem.* 56, 9581–9585.

Frank, O., Blumberg, S., Kunert, C., Zehentbauer, G., Hofmann, T., 2007. Structure determination and sensory analysis of bitter-tasting 4-vinylcatechol oligomers and their identification in roastedcoffee by means of LC-MS/MS. *J. Agric. Food Chem.* 55, 1945–1954.

Frank, O., Zehentbauer, G., Hofmann, T., 2006. Bioresponse-guided decomposition of roast coffee beverage and identification of key bitter taste compounds. *Eur. Food Res. Technol.* 222, 492–508.

Frank, S., Wollmann, N., Schieberle, P., Hofmann, T., 2011. Reconstitution of the flavor signature of Dornfelder red wine on the basis of the natural concentrations of its key aroma and taste compounds. *J. Agric. Food Chem.* 59, 8866–8874.

Galindo, M. M., Voigt, N., Stein, J., et al., 2011. G protein–coupled receptors in human fat taste perception. *Chem. Senses* 37, 123–139.

Gatmaitan, A. N., Lin, J. Q., Zhang, J., Eberlin, L. S., 2021. Rapid analysis and authentication of meat using the masspec pen technology. *J. Agric. Food Chem.* 69, 3527–3536.

Gebicki, J., 2016. Application of electrochemical sensors and sensor matrixes for measurement of odorous chemical compounds. *Trac-Trends Anal. Chem.* 77, 1–13.

Giri, A., Khummueng, W., Mercier, F., et al., 2015. Relevance of two-dimensional gas chromatography and high resolution olfactometry for the parallel determination of heat-induced toxicants and odorants in cooked food. *J. Chromatogr. A* 1388, 217–226.

Glabasnia, A., Dunkel, A., Frank, O., Hofmann, T., 2018. Decoding the nonvolatile sensometabolome of orange juice (Citrus sinensis). *J. Agric. Food Chem.* 66, 2354–2369.

Gläser, P., Dawid, C., Meister, S., et al., 2020. Molecularization of bitter off-taste compounds in pea-protein isolates (Pisum sativum L.). *J. Agric. Food Chem.* 68, 10374–10387.

Godin, J.-P., McCullagh, J. S. O., 2011. Review: Current applications and challenges for liquid chromatography coupled to isotope ratio mass spectrometry (LC/IRMS). *Rapid Commun. Mass Spectrom.* 25, 3019–3028.

Goel, A., Gajula, K., Gupta, R., Rai, B., 2021. In-silico screening of database for finding potential sweet molecules: A combined data and structure based modeling approach. *Food Chem.* 343, 128538.

Gonzalo-Diago, A., Dizy, M., Fernandez-Zurbano, P., 2014. Contribution of low molecular weight phenols to bitter taste and mouthfeel properties in red wines. *Food Chem.* 154, 187–198.

Grosch, W., 2001. Evaluation of the key odorants of foods by dilution experiments, aroma models and omission. *Chem. Senses* 26, 533–545.

Guichard, E., 2002. Interactions between flavor compounds and food ingredients and their influence on flavor perception. *Foof Rev. Int.* 18, 49–70.

Guichard, E., Salles, C., Morzel, M., Le Bon, A.-M. (Eds.), 2017. *Flavour: From Food to Perception.* Chichester, West Sussex (UK), John Wiley & Sons, Inc.

Harrison, A. G., 1992. *Chemical Ionization Mass Spectrometry*, 2nd edition. Boca Raton, FL, CRC Press.

Haseleu, G., Intelmann, D., Hofmann, T., 2009. Identification and RP-HPLC-ESI-MS/MS quantitation of bitter-tasting β-acid transformation products in beer. *J. Agric. Food Chem.* 57, 7480–7489.

Haseleu, G., Lagemann, A., Stephan, A., et al., 2010. Quantitative sensomics profiling of hop-derived bitter compounds throughout a full-scale beer manufacturing process. *J. Agric. Food Chem.* 58, 7930–7939.

Hatakeyama, J., Taylor, A. J., 2019. Optimization of atmospheric pressure chemical ionization triple quadropole mass spectrometry (MS Nose 2) for the rapid measurement of aroma release in vivo. *Flavour Fragr. J.* 34, 307–315.

Hau, J., Cazes, D., Fay, L. B., 1997. Comprehensive study of the "beefy meaty peptide". *J. Agric. Food Chem.* 45, 1351–1355.

He, X., Jelen, H. H., 2021. Comprehensive two-dimensional gas chromatography-time of flight mass spectrometry (GCxGC-TOFMS) in conventional and reversed column configuration for the investigation of Baijiu aroma types and regional origin. *J. Chromatogr. A* 1636, 461774.

Hillmann, H., Behr, J., Ehrmann, M. A., Vogel, R. F., Hofmann, T., 2016. Formation of kokumi-enhancing γ-glutamyl dipeptides in parmesan cheese by means of γ-glutamyltransferase activity and stable isotope double-labeling studies. *J. Agric. Food Chem.* 64, 1784–1793.

Hillmann, H., Hofmann, T., 2016. Quantitation of key tastants and re-engineering the taste of parmesan cheese. *J. Agric. Food Chem.* 64, 1794–1805.

Hillmann, H., Mattes, J., Brockhoff, A., et al., 2012. Sensomics analysis of taste compounds in balsamic vinegar and discovery of 5-acetoxymethyl-2-furaldehyde as a novel sweet taste modulator. *J. Agric. Food Chem.* 60, 9974–9990.

Hjelmeland, A. K., Wylie, P. L., Ebeler, S. E., 2016. A comparison of sorptive extraction techniques coupled to a new quantitative, sensitive, high throughput GC–MS/MS method for methoxypyrazine analysis in wine. *Talanta* 148, 336–345.

Hofstetter, C. K., Dunkel, A., Hofmann, T., 2019. Unified flavor quantitation: Toward high-throughput analysis of key food odorants and tastants by means of ultra-high-performance liquid chromatography tandem mass spectrometry. *J. Agric. Food Chem.* 67, 8599–8608.

Houriet, R., Stahl, D., Winkler, F. J., 1980. Negative chemical ionization of alcohols. *Environ. Health Perspect.* 36, 63–68.

Hufnagel, J. C., Hofmann, T., 2008a. Orosensory-directed identification of astringent mouthfeel and bitter-tasting compounds in red wine. *J. Agric. Food Chem.* 56, 1376–1386.

Hufnagel, J. C., Hofmann, T., 2008b. Quantitative reconstruction of the nonvolatilesensometabolome of a red wine. *J. Agric. Food Chem.* 56, 9190–9199.

Ibrahimi, H., Gadzovska-Simic, S., Tusevski, O., Haziri, A., 2020. Generation of flavor compounds by biotransformation of genetically modified hairy roots of Hypericum perforatum(L.) with basidiomycetes. *Food Sci. Nutr.* 8, 2809–2816.

Intelmann, D., Haseleu, G., Dunkel, A., et al., 2011. Comprehensive sensomics analysis of hop-derived bitter compounds during storage of beer. *J. Agric. Food Chem.* 59, 1939–1953.

Iwaniak, A., Minkiewicz, P., Darewicz, M., Hrynkiewicz, M., 2016. Food protein-originating peptides as tastants - Physiological, technological, sensory, and bioinformatic approaches. *Food Res. Int.* 89, 27–38.

Jastrzembski, J. A., Bee, M. Y., Sacks, G. L., 2017. Trace-level volatile quantitation by direct analysis in real time mass spectrometry following headspace extraction: Optimization and validation in grapes. *J. Agric. Food Chem.* 65, 9353–9359.

Jeckelmann, N., Haefliger, O. P., 2010. Release kinetics of actives from chewing gums into saliva monitored by direct analysis in real time mass spectrometry. *Rapid Commun. Mass Spectrom.* 24, 1165–1171.

Jelen, H. H., Majcher, M., Szwengiel, A., 2019. Key odorants in peated malt whisky and its differentiation from other whisky types using profiling of flavor and volatile compounds. *LWT-Food Sci. Technol.* 107, 56–63.

Jiang, D., Peterson, D. G., 2013. Identification of bitter compounds in whole wheat bread. *Food Chem.* 141, 1345–1353.

Johnsen, L. G., Skou, P. B., Khakimov, B., Bro, R., 2017. Gas chromatography - mass spectrometry data processing made easy. *J. Chromatogr. A* 1503, 57–64.

Jordan, A., Haidacher, S., Hanel, G., et al., 2009a. A high resolution and high sensitivity proton-transfer-reaction time-of-flight mass spectrometer (PTR-TOF-MS). *Int. J. Mass Spectrom.* 286, 122–128.

Jordan, A., Haidacher, S., Hanel, G., et al., 2009b. An online ultra-high sensitivity Proton-transfer-reaction mass-spectrometer combined with switchable reagent ion capability (PTR+SRI–MS). *Int. J. Mass Spectrom.* 286, 32–38.

Jublot, L., Linforth, R. S. T., Taylor, A. J., 2005. Direct atmospheric pressure chemical ionisation ion trap mass spectrometry for aroma analysis: Speed, sensitivity and resolution of isobaric compounds. *Int. J. Mass Spectrom.* 243, 269–277.

Kaneko, S., Kumazawa, K., Masuda, H., Henze, A., Hofmann, T., 2006. Molecular and sensory studies on the umami taste of Japanese green tea. *J. Agric. Food Chem.* 54, 2688–2694.

Kawashima, H., Suto, M., Suto, N., 2018. Determination of carbon isotope ratios for honey samples by means of a liquid chromatography/isotope ratio mass spectrometry system coupled with a post-column pump. *Rapid Commun. Mass Spectrom.* 32, 1271–1279.

Khan, N. A., Besnard, P., 2009. Oro-sensory perception of dietary lipids: New insights into the fat taste transduction. *Biochimica et Biophysica Acta (BBA) – Mol. Cell Biol. Lipids* 1791, 149–155.

Kranenburg, R. F., Peroni, D., Affourtit, S., et al., 2020. Revealing hidden information in GC-MS spectra from isomeric drugs: Chemometrics based identification from 15 eV and 70 eV EI mass spectra. *Forensic Chem.* 18, 10025.

Kranz, M., Viton, F., Smarrito-Menozzi, C., Hofmann, T., 2018. Sensomics-based molecularization of the taste of Pot-au-Feu, a traditional meat/vegetable broth. *J. Agric. Food Chem.* 66, 194–202.

Kreppenhofer, S., Frank, O., Hofmann, T., 2011. Identification of (furan-2-yl)methylated benzene diols and triols as a novel class of bitter compounds in roasted coffee. *Food Chem.* 126, 441–449.

Kuś, P. M., van Ruth, S., 2015. Discrimination of polish unifloral honeys using overall PTR-MS and HPLC fingerprints combined with chemometrics. *LWT-Food Sci. Technol.* 62, 69–75.

Lainer, J., Dawid, C., Dunkel, A., et al., 2020. Characterization of bitter-tasting oxylipins in poppy seeds (Papaver somniferum L.). *J. Agric. Food Chem.* 68, 10361–10373.

Lang, R., Klade, S., Beusch, A., Dunkel, A., Hofmann, T., 2015. Mozambioside is an Arabica-specific bitter-tasting furokaurane glucoside in coffee beans. *J. Agric. Food Chem.* 63, 10492–10499.

Lang, T., Lang, R., Di Pizio, A., et al., 2020. Numerous compounds orchestrate coffee's bitterness. *J. Agric. Food Chem.* 68, 6692–6700.

Langford, V. S., Padayachee, D., McEwan, M. J., Barringer, S. A., 2019. Comprehensive odorant analysis for on-line applications using selected ion flow tube mass spectrometry (SIFT-MS). *Flavour Fragr. J.* 34, 393–410.

Lau, H., Liu, S. Q., Tan, L. P., et al., 2019. A systematic study of molecular ion intensity and mass accuracy in low energy electron ionisation using gas chromatography-quadrupole time-of-flight mass spectrometry. *Talanta* 199, 431–441.

Le Quéré, J. L., 2004. Cheese flavour: Instrumental techniques. In: Fox, P. F., McSweeney, P. L. H., Cogan, T. M., Guinee, T. P. (Eds.), Cheese: Chemistry, Physics and Microbiology, 3rd Edition. Volume 1: General Aspects. Amsterdam, Elsevier, pp. 489–510.

Le Quéré, J. L., Cayot, N., 2013. Instrumental assessment of the sensory quality of dairy products. In: Kilcast, D. (Ed.), *Instrumental Assessment of Food Sensory Quality.* Cambridge, UK, Woodhead Publishing Limited, pp. 420–445.

Le Quéré, J. L., Gierczynski, I., Sémon, E., 2014. An atmospheric pressure chemical ionization - ion-trap mass spectrometer for the on-line analysis of volatile compounds in foods: A tool for linking aroma release to aroma perception. *J. Mass Spectrom.* 49, 918–928.

Le Quéré, J. L., Hélard, C., Labouré, H., et al., 2021. Nosespace PTR-MS analysis with simutaneous TDS or TCATA sensory evaluation: Release and perception of the aroma of dark chocolates differing in sensory properties. *American Chemical Society Annual Meeting.* Online, American Chemical Society.

Lee, Y.-J., Smith, E., Jun, J.-H., 2012. Gas chromatography-high resolution tandem mass spectrometry using a GC-APPI-LIT orbitrap for complex volatile compounds analysis. *Mass Spectrom. Lett.* 3, 29–38.

Leland, J. V., Schieberle, P., Buettner, A., Acree, T. E. (Eds.), 2001. *Gas Chromatography-Olfactometry. The State of the Art.* Washington DC, American Chemical Society.

Li, D., Zhang, J., 2013. Diet Shapes the Evolution of the Vertebrate Bitter Taste Receptor Gene Repertoire. *Mol. Biol. Evol.* 31, 303–309.

Li, W., Chen, Y. P., Blank, I., et al., 2021. GC x GC-ToF-MS and GC-IMS based volatile profile characterization of the Chinese dry-cured hams from different regions. *Food Res. Int.* 142, 110222.

Li, Y., 2012. Confined direct analysis in real time ion source and its applications in analysis of volatile organic compounds of Citrus limon (lemon) and Allium cepa (onion). *Rapid Commun. Mass Spectrom.* 26, 1194–1202.

Lindinger, W., Hirber, J., Paretzke, H., 1993. An ion/molecule-reaction mass spectrometer used for on-line trace gas analysis. *Int. J. Mass Spectrom. Ion Proc.* 129, 79–88.

Linforth, R. S. T., Ingham, K. E., Taylor, A. J., 1996. Time course profiling of volatile release from foods during the eating process. In: Taylor, A. J., Mottram, D. S. (Eds.), *Flavour Science: Recent Developments.* Cambridge, MA, The Royal Society of Chemistry, pp. 361–368.

Linforth, R. S. T., Taylor, A. J., 1993. Measurement of volatile release in the mouth. *Food Chem.* 48, 115–120.

Lopez, A., Vasconi, M., Bellagamba, F., et al., 2020. Volatile organic compounds profile in white sturgeon (Acipenser transmontanus) caviar at different stages of ripening by multiple headspace solid phase microextraction. *Molecules* 25, 15.

Loutfi, A., Coradeschi, S., Mani, G. K., Shankar, P., Rayappan, J. B. B., 2015. Electronic noses for food quality: A review. *J. Food Eng.* 144, 103–111.

Lübke, M., Guichard, E., Tromelin, A., Le Quéré, J. L., 2002. NMR spectroscopic study of β-lactoglobulin interactions with two flavor compounds, γ-decalactone and β-ionone. *J. Agric. Food Chem.* 50, 7094–7099.

Mahne Opatić, A., Nečemer, M., Lojen, S., Vidrih, R., 2017. Stable isotope ratio and elemental composition parameters in combination with discriminant analysis classification model to assign country of origin to commercial vegetables – A preliminary study. *Food Control* 80, 252–258.

Malfondet, N., Brunerie, P., Le Quéré, J. L., 2021. Discrimination of French wine brandy origin by PTR-MS headspace analysis using ethanol ionization and sensory assessment. *Anal. Bioanal. Chem.* 413, 3349–3368.

Märk, T., Dunn, G. (Eds.), 1985. *Electron Impact Ionization.* Wien, Springer.

Marriott, P. J., Chin, S.-T., Maikhunthod, B., Schmarr, H.-G., Bieri, S., 2012. Multidimensional gas chromatography. *Trac-Trends Anal. Chem.* 34, 1–21.

Mastello, R. B., Capobiango, M., Chin, S.-T., Monteiro, M., Marriott, P. J., 2015. Identification of odour-active compounds of pasteurised orange juice using multidimensional gas chromatography techniques. *Food Res. Int.* 75, 281–288.

Mateo-Vivaracho, L., Ferreira, V., Cacho, J., 2006. Automated analysis of 2-methyl-3-furanthiol and 3-mercaptohexyl acetate at ngL−1 level by headspace solid-phase microextracion with on-fibre derivatisation and gas chromatography–negative chemical ionization mass spectrometric determination. *J. Chromatogr. A* 1121, 1–9.

Matthews, D. E., Hayes, J. M., 1978. Isotope-ratio-monitoring gas chromatography-mass spectrometry. *Anal. Chem.* 50, 1465–1473.

McDaniel, M. R., Miranda-Lopez, R., Watson, B. T., Michaels, N. J., Libbey, L. M., 1990. Pinot noir aroma: A sensory/gas chromatographic approach. In: Charalambous, G. (Ed.), *Flavors and Off-flavors.* Amsterdam, Elsevier, pp. 23–36.

McEwen, C. N., McKay, R. G., 2005. A combination atmospheric pressure LC/MS: GC/MS ion source: Advantages of dual AP-LC/MS: GC/MS instrumentation. *J. Am. Soc. Mass Spectrom.* 16, 1730–1738.

McLafferty, F. W., 1980. *Interpretation of Mass Spectra.* Mill Valley, CA, University Science Books.

McLafferty, F. W., Turecek, F., 1993. *Interpretation of Mass Spectra*, 4th edition. Mill Valley, CA, University Science Books.

McSweeney, P. L. H., 1997. The flavour of milk and dairy products: III. Cheese: Taste. *Int. J. Dairy Technol.* 50, 123–128.

Mellon, F. A., 2003. Mass spectrometry | Principles and instrumentation. In: Caballero, B. (Ed.), *Encyclopedia of Food Sciences and Nutrition*, 2nd edition. Oxford, Academic Press, pp. 3739–3749.

Meyer, S., Dunkel, A., Hofmann, T., 2016. Sensomics-Assisted Elucidation of the Tastant Code of Cooked Crustaceans and Taste Reconstruction Experiments. *J. Agric. Food Chem.* 64, 1164–1175.

Michel, S., Baraka, L. F., Ibañez, A. J., Mansurova, M., 2021. Mass spectrometry-based flavor monitoring of Peruvian chocolate fabrication process. *Metabolites* 11, 71.

Mirondo, R., Barringer, S., 2016. Deodorization of garlic breath by foods, and the role of polyphenol oxidase and phenolic compounds. *J. Food Sci.* 81, C2425–C2430.

Mittermeier, V. K., Pauly, K., Dunkel, A., Hofmann, T., 2020. Ion-mobility-based liquid chromatography–mass spectrometry quantitation of taste-enhancing octadecadien-12-ynoic acids in mushrooms. *J. Agric. Food Chem.* 68, 5741–5751.

Mol, H. G. J., Tienstra, M., Zomer, P., 2016. Evaluation of gas chromatography - electron ionization - full scan high resolution orbitrap mass spectrometry for pesticide residue analysis. *Anal. Chim. Acta* 935, 161–172.

Moniruzzaman, M., Rodríguez, I., Ramil, M., et al., 2014. Assessment of gas chromatography time-of-flight accurate mass spectrometry for identification of volatile and semi-volatile compounds in honey. *Talanta* 129, 505–515.

Morales, M. L., Callejon, R. M., Ordonez, J. L., Troncoso, A. M., Garcia-Parrilla, M. C., 2017. Comparative assessment of software for non-targeted data analysis in the study of volatile fingerprint changes during storage of a strawberry beverage. *J. Chromatogr. A* 1522, 70–77.

Mosandl, A., Hener, U., Schmarr, H. G., Rautenschlein, M., 1990. Chirospecific flavor analysis by means of enantioselective gas chromatography, coupled on-line with isotope ratio mass spectrometry. *J. High. Resol. Chromatogr.* 13, 528–531.

Moyano, L., Serratosa, M. P., Marquez, A., Zea, L., 2019. Optimization and validation of a DHS-TD-GC-MS method to wineomics studies. *Talanta* 192, 301–307.

Müller, M., Piel, F., Gutmann, R., et al., 2020. A novel method for producing NH4+ reagent ions in the hollow cathode glow discharge ion source of PTR-MS instruments. *Int. J. Mass Spectrom.* 447, 116254.

Muñoz-González, C., Canon, F., Feron, G., Guichard, E., Pozo-Bayón, M. A., 2019. Assessment wine aroma persistence by using an in vivo PTR-ToF-MS approach and its relationship with salivary parameters. *Molecules* 24, 1277.

Mutarutwa, D., Navarini, L., Lonzarich, V., Compagnone, D., Pittia, P., 2018. GC-MS aroma characterization of vegetable matrices: Focus on 3-alkyl-2-methoxypyrazines. *J. Mass Spectrom.* 53, 871–881.

Ni, Q., Khomenko, I., Gallo, L., Biasioli, F., Bittante, G., 2020. Rapid profiling of the volatilome of cooked meat by PTR-ToF-MS: Characterization of chicken, Turkey, pork, veal and beef meat. *Foods* 9, 1776.

Nicolli, K., Biasoto, A., Guerra, C., et al., 2020. Effects of soil and vineyard characteristics on volatile, phenolic composition and sensory profile of cabernet sauvignon wines of Campanha Gaúcha. *J. Braz. Chem. Soc.* 31, 1110–1124.

Nolvachai, Y., McGregor, L., Spadafora, N. D., Bukowski, N. P., Marriott, P. J., 2020. Comprehensive two-dimensional gas chromatography with mass spectrometry: Toward a super-resolved separation technique. *Anal. Chem.* 92, 12572–12578.

Ottinger, H., Hofmann, T., 2003. Identification of the taste enhancer alapyridaine in beef broth and evaluation of its sensory impact by taste reconstitution experiments. *J. Agric. Food Chem.* 51, 6791–6796.

Ottinger, H., Soldo, T., Hofmann, T., 2003. Discovery and structure determination of a novel maillard-derived sweetness enhancer by application of the comparative taste dilution analysis (cTDA). *J. Agric. Food Chem.* 51, 1035–1041.

Ozcan, G., Barringer, S., 2011. Effect of enzymes on strawberry volatiles during storage, at different ripeness level, in different cultivars, and during eating. *J. Food Sci.* 76, C324–C333.

Paolini, M., Tonidandel, L., Moser, S., Larcher, R., 2018. Development of a fast gas chromatography-tandem mass spectrometry method for volatile aromatic compound analysis in oenological products. *J. Mass Spectrom.* 53, 801–810.

Pedrotti, M., Spaccasassi, A., Biasioli, F., Fogliano, V., 2019. Ethnicity, gender and physiological parameters: Their effect on in vivo flavour release and perception during chewing gum consumption. *Food Res. Int.* 116, 57–70.

Peleg, H., Gacon, K., Schlich, P., Noble, A. C., 1999. Bitterness and astringency of flavan-3-ol monomers, dimers and trimers. *J. Sci. Food Agric.* 79, 1123–1128.

Pengwei, G., Song, Q., Li, T., et al., 2020. Confirmative structural annotation for metabolites of (R)-7,3′-Dihydroxy-4′-methoxy-8-methylflavane, a natural sweet taste modulator, by liquid chromatography–three-dimensional mass spectrometry. *J. Agric. Food Chem.* 68, 12454–12466.

Pérès, C., Begnaud, F., Eveleigh, L., Berdagué, J. L., 2003. Fast characterization of foodstuff by headspace mass spectrometry (HS-MS). *Trac-Trends Anal. Chem.* 22, 858–866.

Pérès, C., Viallon, C., Berdagué, J. L., 2002. Curie point pyrolysis-mass spectrometry applied to rapid characterisation of cheeses. *J. Anal. Appl. Pyrolysis* 65, 161–171.

Perez-Jimenez, M., Munoz-Gonzalez, C., Pozo-Bayon, M. A., 2021. Oral release behavior of wine aroma compounds by using in-mouth headspace sorptive extraction (HSSE) method. *Foods* 10, 415.

Pérez-Jiménez, M., Pozo-Bayón, M. Á., 2019. Development of an in-mouth headspace sorptive extraction method (HSSE) for oral aroma monitoring and application to wines of different chemical composition. *Food Res. Int.* 121, 97–107.

Pérez Pavón, J. L., del Nogal Sánchez, M., Pinto, C. G., et al., 2006. Strategies for qualitative and quantitative analyses with mass spectrometry-based electronic noses. *Trac-Trends Anal. Chem.* 25, 257–266.

Perini, M., Strojnik, L., Paolini, M., Camin, F., 2020. Gas chromatography combustion isotope ratio mass spectrometry for improving the detection of authenticity of grape must. *J. Agric. Food Chem.* 68, 3322–3329.

Phewpan, A., Phuwaprisirisan, P., Takahashi, H., et al., 2020. Investigation of kokumi substances and bacteria in Thai fermented freshwater fish (Pla-ra). *J. Agric. Food Chem.* 68, 10345–10351.

Pickrahn, S., Sebald, K., Hofmann, T., 2014. Application of 2D-HPLC/taste dilution analysis on taste compounds in aniseed (Pimpinella anisum L.). *J. Agric. Food Chem.* 62, 9239–9245.

Pico, J., Khomenko, I., Capozzi, V., Navarini, L., Biasioli, F., 2020. Real-time monitoring of volatile compounds losses in the oven during baking and toasting of gluten-free bread doughs: A PTR-MS evidence. *Foods* 9, 1498.

Piggott, J. R., 2000. Dynamism in flavour science and sensory methodology. *Food Res. Internat.* 33, 191–197.

Pineau, N., Schlich, P., Cordelle, S., et al., 2009. Temporal dominance of sensations: Construction of the TDS curves and comparison with time-intensity. *Food Qual. Pref.* 20, 450–455.

Pionnier, E., Chabanet, C., Mioche, L., et al., 2004a. In vivo nonvolatile release during eating of a model cheese: Relationships with oral parameters. *J. Agric. Food Chem.* 52, 565–571.

Pionnier, E., Nicklaus, S., Chabanet, C., et al., 2004b. Flavor perception of a model cheese: Relationships with oral and physico-chemical parameters. *Food Qual. Pref.* 15, 843–852.

Pionnier, E., Le Quéré, J. L., Salles, C., 2005. Real time release of flavor compounds and flavor perception. An application to cheese. In: Spanier, A. M., Shahidi, F., Parliment, T. H., et al. (Eds.), *Food Flavor and Chemistry. Explorations into the 21st Century.* Cambridge, MA, The Royal Society of Chemistry, pp. 13–22.

Pittari, E., Moio, L., Piombino, P., 2021. Interactions between polyphenols and volatile compounds in wine: A literature review on physicochemical and sensory insights. *Appl. Sci.-Basel* 11, 1157.

Pollien, P., Ott, A., Montigon, F., et al., 1997. Hyphenated headspace-gas chromatography-sniffing technique: Screening of impact odorants and quantitative aromagram comparisons. *J. Agric. Food Chem.* 45, 2630–2637.

Pons, A., Lavigne, V., Darriet, P., Dubourdieu, D., 2011. Determination of 3-methyl-2,4-nonanedione in red wines using methanol chemical ionization ion trap mass spectrometry. *J. Chromatogr. A* 1218, 7023–7030.

Pua, A., Lau, H., Liu, S. Q., et al., 2020. Improved detection of key odourants in Arabica coffee using gas chromatography-olfactometry in combination with low energy electron ionisation gas chromatography-quadrupole time-of-flight mass spectrometry. *Food Chem.* 302, 125370.

Qi, D., Ma, C., Wang, W., Zhang, L., Li, J., 2021. Gas chromatography-mass spectrometry analysis as a tool to reveal differences between the volatile compound profile of royal jelly produced from tea and pagoda trees. *Food Anal. Methods* 14, 616–630.

Ratel, J., Berge, P., Berdague, J.-L., Cardinal, M., Engel, E., 2008. Mass spectrometry based sensor strategies for the authentication of oysters according to geographical origin. *J. Agric. Food Chem.* 56, 321–327.

Robb, D. B., Covey, T. R., Bruins, A. P., 2000. Atmospheric pressure photoionisation: An ionization method for liquid chromatography-mass spectrometry. *Anal. Chem.* 72, 3653–3659.

Roberts, D. D., Pollien, P., Antille, N., Lindinger, C., Yeretzian, C., 2003. Comparison of nosespace, headspace, and sensory intensity ratings for the evaluation of flavor absorption by fat. *J. Agric. Food Chem.* 51, 3636–3642.

Roberts, D. D., Taylor, A. J. (Eds.), 2000. *Flavor Release.* Washington DC, American Chemical Society.

Robichaud, J. L., Noble, A. C., 1990. Astringency and bitterness of selected phenolics in wine. *J. Sci. Food Agric.* 53, 343–353.

Roman, S. M. S., Rubio-Breton, P., Perez-Alvarez, E. P., Garde-Cerdan, T., 2020. Advancement in analytical techniques for the extraction of grape and wine volatile compounds. *Food Res. Int.* 137, 109712.

Romano, A., Cappellin, L., Ting, V., et al., 2014. Nosespace analysis by PTR-ToF-MS for the characterization of food and tasters: The case study of coffee. *Int. J. Mass Spectrom.* 365–366, 20–27.

Rosso, M. C., Mazzucotelli, M., Bicchi, C., et al., 2020. Adding extra-dimensions to hazelnuts primary metabolome fingerprinting by comprehensive two-dimensional gas chromatography combined with time-of-flight mass spectrometry featuring tandem ionization: Insights on the aroma potential. *J. Chromatogr. A* 1614, 460739.

Rotzoll, N., Dunkel, A., Hofmann, T., 2005. Activity-guided identification of (S)-malic acid 1-O-d-glucopyranoside (morelid) and γ-aminobutyric acid as contributors to umami taste and mouth-drying oral sensation of morel mushrooms (Morchella deliciosa Fr.). *J. Agric. Food Chem.* 53, 4149–4156.

Rotzoll, N., Dunkel, A., Hofmann, T., 2006. Quantitative studies, taste reconstitution, and omission experiments on the key taste compounds in morel mushrooms (Morchella deliciosa Fr.). *J. Agric. Food Chem.* 54, 2705–2711.

Roudot-Algaron, F., Le Bars, D., Einhorn, J., Adda, J., Gripon, J. C., 1993. Flavor constituents of aqueous fraction extracted from Comte cheese by liquid carbon dioxide. *J. Food Sci.* 58, 1005–1009.

Roudot-Algaron, F., Kerhoas, L., Le Bars, D., Einhorn, J., Gripon, J. C., 1994. Isolation of γ-glutamyl peptides from Comté cheese. *J. Dairy Sci.* 77, 1161–1166.

Sales, C., Portolés, T., Johnsen, L. G., Danielsen, M., Beltran, J., 2019. Olive oil quality classification and measurement of its organoleptic attributes by untargeted GC–MS and multivariate statistical-based approach. *Food Chem.* 271, 488–496.

Salger, M., Stark, T. D., Hofmann, T., 2019. Taste modulating peptides from overfermented cocoa beans. *J. Agric. Food Chem.* 67, 4311–4320.

Salles, C., Septier, C., Roudot-Algaron, F., Guillot, A., Etievant, P. X., 1995. Sensory and chemical analysis of fractions obtained by gel permeation of water-soluble Comte cheese extracts. *J. Agric. Food Chem.* 43, 1659–1668.

Salles, C., Hollowood, T. A., Linforth, R. S. T., Taylor, A. J., 2003. Relating real time flavour release to sensory perception of soft cheeses. In: Le Quéré, J. L., Etiévant, P. X. (Eds.), *Flavour Research at the Dawn of the Twenty-first Century.* Paris, Lavoisier, Tec & Doc, pp. 170–175.

Salles, C., Chagnon, M.-C., Feron, G., et al., 2011. In-mouth mechanisms leading to flavor release and perception. *Crit. Rev. Food Sci. Nutr.* 51, 67–90.

Sanaeifar, A., ZakiDizaji, H., Jafari, A., Guardia, M. D. L., 2017. Early detection of contamination and defect in foodstuffs by electronic nose: A review. *Trac-Trends Anal. Chem.* 97, 257–271.

Santerre, C., Vallet, N., Touboul, D., 2018. Fingerprints of flower absolutes using supercritical fluid chromatography hyphenated with high resolution mass spectrometry. *J. Chromatogr. B* 1092, 1–6.

Sarris, J., Etiévant, P. X., Le Quéré, J. L., Adda, J., 1985. The chemical ionisation mass spectra of alcohols. In: Adda, J. (Ed.), *Progress in Flavour Research 1984.* Amsterdam, Elsevier, pp. 591–601.

Scavarda, C., Cordero, C., Strocchi, G., et al., 2021. Cocoa smoky off-flavour: A MS-based analytical decision maker for routine controls. *Food Chem.* 336, 127691.

Scharbert, S., Hofmann, T., 2005. Molecular definition of black tea taste by means of quantitative studies, taste reconstitution, and omission experiments. *J. Agric. Food Chem.* 53, 5377–5384.

Schena, T., Bjerk, T. R., von Muhlen, C., Caramao, E. B., 2020. Influence of acquisition rate on performance of fast comprehensive two-dimensional gas chromatography coupled with time-of-flight mass spectrometry for coconut fiber bio-oil characterization. *Talanta* 219, 121186.

Schieberle, P., Hofmann, T., 2011. Mapping the combinatorial code of food flavors by means of molecular sensory science approach. In: Jelen, H. (Ed.), *Food Flavors: Chemical, Sensory and Technological Properties.* Boca Raton, FL, CRC Press, pp. 413–438.

Schindler, A., Dunkel, A., Stahler, F., et al., 2011. Discovery of salt taste enhancing arginyl dipeptides in protein digests and fermented fish sauces by means of a sensomics approach. *J. Agric. Food Chem.* 59, 12578–12588.

Schipilliti, L., Bonaccorsi, I. L., Mondello, L., 2017. Characterization of natural vanilla flavour in foodstuff by HS-SPME and GC-C-IRMS. *Flavour Fragr. J.* 32, 85–91.

Schmiech, L., Uemura, D., Hofmann, T., 2008. Reinvestigation of the bitter compounds in carrots (Daucus carota L.) by using a molecular sensory science approach. *J. Agric. Food Chem.* 56, 10252–10260.

Sebald, K., Dunkel, A., Schafer, J., Hinrichs, J., Hofmann, T., 2018. Sensoproteomics: A new approach for the identification of taste-active peptides in fermented foods. *J. Agric. Food Chem.* 66, 11092–11104.

Sebald, K., Dunkel, A., Hofmann, T., 2020. Mapping taste-relevant food peptidomes by means of sequential window acquisition of all theoretical fragment ion-mass spectrometry. *J. Agric. Food Chem.* 68, 10287–10298.

Sémon, E., Gierczynski, I., Langlois, D., Le Quéré, J. L., 2003. Analysis of aroma compounds by atmospheric pressure chemical ionisation - ion trap mass spectrometry. Construction and validation of an interface for in vivo analysis of human breath volatile content. *16th International Mass Spectrometry Conference.* Edinburgh.

Setyaningsih, W., Majchrzak, T., Dymerski, T., Namiesnik, J., Palma, M., 2019. Key-marker volatile compounds in aromatic rice (Oryza sativa) grains: An HS-SPME extraction method combined with GCxGC-TOFMS. *Molecules* 24, 4180.

Sichilongo, K., Padiso, T., Turner, Q., 2020. AMDIS-Metab R data manipulation for the geographical and floral differentiation of selected honeys from Zambia and Botswana based on volatile chemical compositions using SPME-GC-MS. *Eur. Food Res. Technol.* 246, 1679–1690.

Sinding, C., Thibault, H., Hummel, T., Thomas-Danguin, T., 2021. Odor-induced saltiness enhancement: Insights into the brain chronometry of flavor perception. *Neuroscience* 452, 126–137.

Smith, D., Chippendale, T. W. E., Spanel, P., 2011. Selected ion flow tube, SIFT, studies of the reactions of H3O+, NO+ and O2+ with some biologically active isobaric compounds in preparation for SIFT-MS analyses. *Int. J. Mass Spectrom.* 303, 81–89.

Smith, D., Španěl, P., 2005. Selected ion flow tube mass spectrometry (SIFT-MS) for on-line trace gas analysis. *Mass Spectrom. Rev.* 24, 661–700.

Smith, D., Španěl, P., 2015. Pitfalls in the analysis of volatile breath biomarkers: Suggested solutions and SIFT–MS quantification of single metabolites. *J. Breath Res.* 9, 022001.

Smith, D., Španěl, P., 2016. Status of selected ion flow tube MS: Accomplishments and challenges in breath analysis and other areas. *Bioanalysis* 8, 1183–1201.

Soares, S., Silva, M. S., Garcia-Estevez, I., et al., 2018. Human bitter taste receptors are activated by different classes of polyphenols. *J. Agric. Food Chem.* 66, 8814–8823.

Soares, S., Brandao, E., Guerreiro, C., Mateus, N., de Freitas, V., 2020. Tannins in food: Insights into the molecular perception of astringency and bitter taste. *Molecules* 25, 2590.

Sommerer, N., Septier, C., Salles, C., Le Quéré, J. L., 1998. A liquid chromatography purification method to isolate small peptides from goat cheese for their mass spectrometry analysis. *Sci. Aliments* 18, 537–551.

Sommerer, N., Salles, C., Promé, D., Promé, J. C., Le Quéré, J. L., 2001. Isolation of oligopeptides from the water-soluble extract of goat cheese and their identification by mass spectrometry. *J. Agric. Food Chem.* 49, 402–408.

Song, H., Liu, J., 2018. GC-O-MS technique and its applications in food flavor analysis. *Food Res. Internat.* 114, 187–198.

Spaggiari, G., Di Pizio, A., Cozzini, P., 2020. Sweet, umami and bitter taste receptors: State of the art of in silico molecular modeling approaches. *Trends Food Sci. Technol.* 96, 21–29.

Španěl, P., Smith, D., 2011. Progress in SIFT-MS: Breath analysis and other applications. *Mass Spectrom. Rev.* 30, 236–267.

Stark, T., Bareuther, S., Hofmann, T., 2005. Sensory-guided decomposition of roasted cocoa nibs (Theobroma cacao) and structure determination of taste-active polyphenols. *J. Agric. Food Chem.* 53, 5407–5418.

Stark, T. D., Weiss, P., Friedrich, L., Hofmann, T., 2020. The wheat species profiling by non-targeted UPLC–ESI–TOF-MS analysis. *Eur. Food Res. Technol.* 246, 1617–1626.

Su, G., Cui, C., Zheng, L., et al., 2012. Isolation and identification of two novel umami and umami-enhancing peptides from peanut hydrolysate by consecutive chromatography and MALDI-TOF/TOF MS. *Food Chem.* 135, 479–485.

Suto, M., Kawashima, H., 2021. Carbon isotope ratio of organic acids in sake and wine by solid-phase extraction combined with LC/IRMS. *Anal Bioanal Chem* 413, 355–363.

Tarrega, A., Yven, C., Sémon, E., Salles, C., 2008. Aroma release and chewing activity during eating different model cheeses. *Int. Dairy J.* 18, 849–857.

Tarrega, A., Yven, C., Sémon, E., Salles, C., 2011. In-mouth aroma compound release during cheese consumption: Relationship with food bolus formation. *Int. Dairy J.* 21, 358–364.

Taylor, A. J., Linforth, R. S. T., 1994. Methodology for measuring volatile profiles in the mouth and nose during eating. In: Maarse, H., Van der Heij, D. G. (Eds.), *Trends in Flavour Research*. Amsterdam, Elsevier, pp. 3–14.

Taylor, A. J., Linforth, R. S. T., 1996. Flavour release in the mouth. *Trends Food Sci. Technol.* 7, 444–448.

Taylor, A. J., Linforth, R. S. T., Harvey, B. A., Blake, A., 2000. Atmospheric pressure chemical ionisation mass spectrometry for in vivo analysis of volatile flavour release. *Food Chem.* 71, 327–338.

Taylor, A. J., 2002. Release and transport of flavors in vivo: Physicochemical, physiological, and perceptual considerations. *Comp. Rev. Food Sci. Food Safety* 1, 45–57.

Taylor, A. J., Linforth, R. S. T., 2003. Direct mass spectrometry of complex volatile and non-volatile flavour mixtures. *Int. J. Mass Spectrom.* 223–224, 179–191.

Taylor, A. J., Linforth, R. S. T., 2010. On-line monitoring of flavour processes. In: Taylor, A. J., Linforth, R. S. T. (Eds.), *Food Flavour Technology*, 2nd edition. Chichester, Wiley-Blackwell, pp. 266–295.

Thibon, C., Pons, A., Mouakka, N., et al., 2015. Comparison of electron and chemical ionization modes for the quantification of thiols and oxidative compounds in white wines by gas chromatography-tandem mass spectrometry. *J. Chromatogr. A* 1415, 123–133.

Thomas, C. F., Zeh, E., Dorfel, S., Zhang, Y., Hinrichs, J., 2021. Studying dynamic aroma release by headspace-solid phase microextraction-gas chromatography-ion mobility spectrometry (HS-SPME-GC-IMS): Method optimization, validation, and application. *Anal. Bioanal. Chem.* 413, 2577–2586.

Ting, V. J. L., Romano, A., Soukoulis, C., et al., 2016. Investigating the in-vitro and in-vivo flavour release from 21 fresh-cut apples. *Food Chem.* 212, 543–551.

Toelstede, S., Hofmann, T., 2008a. Sensomics mapping and identification of the key bitter metabolites in Gouda cheese. *J. Agric. Food Chem.* 56, 2795–2804.

Toelstede, S., Hofmann, T., 2008b. Quantitative studies and taste re-engineering experiments toward the decoding of the nonvolatile sensometabolome of Gouda cheese. *J. Agric. Food Chem.* 56, 5299–5307.

Toelstede, S., Dunkel, A., Hofmann, T., 2009. A series of kokumi peptides impart the long-lasting mouthfulness of matured Gouda cheese. *J. Agric. Food Chem.* 57, 1440–1448.

Tran, P. D., Van de Walle, D., De Clercq, N., et al., 2015. Assessing cocoa aroma quality by multiple analytical approaches. *Food Res. Internat.* 77, 657–669.

Tranchida, P. Q., Aloisi, I., Giocastro, B., Zoccali, M., Mondello, L., 2019. Comprehensive two-dimensional gas chromatography-mass spectrometry using milder electron ionization conditions: A preliminary evaluation. *J. Chromatogr. A* 1589, 134–140.

Tranchida, P. Q. (Ed.), 2020. *Advanced Gas Chromatography in Food Analysis*. London, The Royal Society of Chemistry.

Utz, F., Kreissl, J., Stark, T. D., et al., 2021. Sensomics-assisted flavor decoding of dairy model systems and flavor reconstitution experiments. *J. Agric. Food Chem.* 69, 6588–6600.

Vairamani, M., Mirza, U. A., Srinivas, R., 1990. Unusual positive ion reagents in chemical ionization mass spectrometry. *Mass Spectrom. Rev.* 9, 235–258.

van den Oord, A. H. A., van Wassenaar, P. D., 1997. Umami peptides: Assessment of their alleged taste properties. *Z. Lebensm. Unters. Forsch.* 205, 125–130.

Van Wassenaar, P. D., Van Den Oord, A. H. A., Schaaper, W. M. M., 1995. Taste of 'Delicious' beefy meaty peptide. Revised. *J. Agric. Food Chem.* 43, 2828–2832.

Vasile-Simone, G., Castro, R., Natera, R., et al., 2017. Application of a stir bar sorptive extraction method for the determination of volatile compounds in different grape varieties. *J. Sci. Food Agric.* 97, 939–948.

Vernat, G., Berdagué, J. L., 1995. Dynamic Headspace-Mass Spectrometry (DHS-MS): A new approach to real-time characterization of food products. In: Etiévant, P., Schreier, P. (Eds.), *Bioflavour 95*. 1995 ed. Dijon, France, INRA Editions, Versailles, pp. 59–62.

Vidal, S., Francis, L., Guyot, S., et al., 2003. The mouth-feel properties of grape and apple proanthocyanidins in a wine-like medium. *J. Sci. Food Agric.* 83, 564–573.

Voigt, N., Stein, J., Galindo, M. M., et al., 2014. The role of lipolysis in human orosensory fat perception. *J. Lipid Res.* 55, 870–882.

Vyviurska, O., Spanik, I., 2020. Assessment of Tokaj varietal wines with comprehensive two-dimensional gas chromatography coupled to high resolution mass spectrometry. *Microchem. J.* 152, 104385.

Wakamatsu, J., Stark, T. D., Hofmann, T., 2016. Taste-active maillard reaction products in roasted garlic (Allium sativum). *J. Agric. Food Chem.* 64, 5845–5854.

Wang, X., Guo, M., Song, H., Meng, Q., Guan, X., 2021. Characterization of key odor-active compounds in commercial high-salt liquid-state soy sauce by switchable GC/GC x GC-olfactometry-MS and sensory evaluation. *Food Chem.* 342, 128224.

Wang, Y., Li, Y., Yang, J., Ruan, J., Sun, C., 2016. Microbial volatile organic compounds and their application in microorganism identification in foodstuff. *Trac-Trends Anal. Chem.* 78, 1–16.

Wang, Z. H., Peng, M. J., She, Z. G., Zhang, M. L., Yang, Q. L., 2020. Development of a flavor fingerprint by gas chromatography ion mobility spectrometry with principal component analysis for volatile compounds from eucommia ulmoides oily. Leaves and its fermentation products. *BioResources* 15, 9180–9196.

Warmke, R., Belitz, H. D., Grosch, W., 1996. Evaluation of taste compounds of Swiss cheese (Emmentaler). *Z. Lebensm. Unters. Forsch.* 203, 230–235.

Xu, Y., Barringer, S., 2010. Comparison of volatile release in tomatillo and different varieties of tomato during chewing. *J. Food Sci.* 75, C352–C358.

Yang, Y.-Q., Yin, H.-X., Yuan, H.-B., et al., 2018. Characterization of the volatile components in green tea by IRAE-HS-SPME/GC-MS combined with multivariate analysis. *PLoS One* 13, e0193393.

Yang, Y. Q., Zhang, M. M., Hua, J. J., et al., 2020a. Quantitation of pyrazines in roasted green tea by infrared-assisted extraction coupled to headspace solid-phase microextraction in combination with GC-QqQ-MS/MS. *Food Res. Int.* 134, 109167.

Yang, K., Somogyi, A., Thomas, C., et al., 2020b. Analysis of barrel-aged kentucky bourbon whiskey by ultra-high resolution mass spectrometry. *Food Anal. Methods* 13, 2301–2311.

Yeretzian, C., Jordan, A., Brevard, H., Lindinger, W., 2000. Time-resolved headspace analysis by proton-transfer-reaction mass-spectrometry. In: Roberts, D. D., Taylor, A. J. (Eds.), *Flavor Release*. Washington DC, American Chemical Society, pp. 58–72.

Yeretzian, C., Opitz, S., Smrke, S., Wellinger, M., 2019. Coffee volatile and aroma compounds - From the green bean to cup. In: Farah, A. (Ed.), *Coffee: Production, Quality and Chemistry*. London, UK, The Royal Society of Chemistry, pp. 726–770.

Yin, M., Shao, S. J., Zhou, Z. L., et al., 2020. Characterization of the key aroma compounds in dog foods by gas chromatography-mass spectrometry, acceptance test, and preference test. *J. Agric. Food Chem.* 68, 9195–9204.

Zhang, M.-X., Wang, X.-C., Liu, Y., Xu, X.-L., Zhou, G.-H., 2012. Isolation and identification of flavour peptides from Puffer fish (Takifugu obscurus) muscle using an electronic tongue and MALDI-TOF/TOF MS/MS. *Food Chem.* 135, 1463–1470.

Zhang, Y., Venkitasamy, C., Pan, Z., Liu, W., Zhao, L., 2017. Novel umami ingredients: Umami peptides and their taste. *J. Food Sci.* 82, 16–23.

Zhang, J. S., Zhang, Z. L., Yan, M. Z., Lin, X. M., Chen, Y. T., 2020. Gas chromatographic-ion mobility spectrometry combined with a multivariate analysis model exploring the characteristic changes of odor components during the processing of black sesame. *Anal. Methods* 12, 4987–4995.

Zhao, C. J., Schieber, A., Gänzle, M. G., 2016. Formation of taste-active amino acids, amino acid derivatives and peptides in food fermentations – A review. *Food Res. Int.* 89, 39–47.

Zhu, W., Benkwitz, F., Kilmartin, P. A., 2021. Volatile-based prediction of sauvignon Blanc quality gradings with static headspace-gas chromatography-ion mobility spectrometry (SHS-GC-IMS) and Interpretable Machine Learning Techniques. *J. Agric. Food Chem.* 69, 3255–3265.

10 Evaluation of Nutraceutical Value

Mariana Martínez-Ávila
Escuela de Ingeniería y Ciencias

Janet A. Gutiérrez-Uribe
Escuela de Ingeniería y Ciencias
Tecnologico de Monterrey

Marilena Antunes-Ricardo
Escuela de Ingeniería y Ciencias

CONTENTS

10.1 Introduction 181
10.2 Sample Preparation for Mass Spectrometry (MS) Analysis 182
10.3 MS Analysis of Phenolic Compounds 183
10.4 MS Analysis of Lipids with Nutraceutical Properties 189
10.4.1 Plant-Based Products 190
10.4.1.1 Characterization of Plant-Based Lipids 190
10.4.1.2 Quality Control 194
10.4.2 Animal Products 194
10.4.2.1 Fermented Products 195
10.4.2.2 Cured Meats 195
10.4.2.3 Marine Lipids 195
10.4.3 Novel Sources of Lipids: Microbial and Insects 196
10.5 Bioactive Peptides 196
10.6 Polysaccharides 198
10.6.1 Plant Polysaccharides 198
10.6.2 Fungi Polysaccharides 198
10.6.3 Marine Polysaccharides 199
10.6.4 Bacterial Exopolysaccharides 199
10.7 Conclusion 199
References 200

10.1 INTRODUCTION

In nature, a wide variety of compounds with biological activities are found. These compounds, known as nutraceuticals, provide medical or health benefits such as modulation of the immune response, antioxidant defenses, cofactors that enhance the stability and absorption of nutrients (Larussa, Imeneo, and Luzza 2017).

Nutraceuticals are dietary compounds that can prevent or treat pathological conditions (Kalra 2003). The interest in nutraceuticals has grown in the last 10 years, and this, in turn, has increased the need to study and characterize them (Daliu, Santini, and Novellino 2018). Nutraceuticals

DOI: 10.1201/9781003091226-12

evaluation is challenging because of the extensive variability in their structures and physicochemical properties, as well as the complexity of their extraction. So, it becomes clear that the analysis of these bioactive components is not a trivial task, given the variety of biological matrices in which they are contained. On the other hand, these compounds are usually present in trace amounts and in the presence of large amounts of macromolecules such as proteins, carbohydrates, and lipids that in most cases bind to and/or trap the bioactive components in food matrices (Yeboah and Konishi 2003).

The continuous evolution of mass spectrometry (MS) techniques has allowed the measurement of a wider variety of analytes, including inorganic or organic compounds, their isotopes, and their fragments. Nutraceuticals include nonvolatile molecules and large organic compounds such as proteins and polysaccharides contained in a complex matrix. Recently, new ambient ionization MS platforms and the development of tandem coupled systems allow simpler, quicker, and more effective manner for the analysis of complex matrices (de Hoffmann 2005; Duncan, Fyrestam, and Lanekoff 2019; Feider et al. 2019). Therefore, it is not surprising that the MS technique is an excellent alternative for the analysis and characterization of many molecules, including a wide variety of nutraceuticals, because of its high resolving power, sensitivity, and accurate mass measurement (Narváez et al. 2020).

10.2 SAMPLE PREPARATION FOR MASS SPECTROMETRY (MS) ANALYSIS

One of the main challenges in the characterization of nutraceutical compounds is related to the extraction from their matrices, concentration, and purification. The sample preparation could have a direct effect on the profile of the extracted nutraceuticals. For example, dissolving analytes from a solid sample or extracting volatile compounds from a liquid or solid sample depends on temperature and pressure. However, aside from the extraction, the sample preparation has other specific objectives, such as the separation of the analytes from other components that affect the subsequent analytical determination. Removing proteins from the matrix for a lipid-focused analysis is critical to avoid encapsulation or emulsification. Sample preparation is also useful to concentrate the nutraceutical compounds when the limits of detection are critical to improving their quantification. Furthermore, the sample preparation also includes the pretreatment step to chemically transform the analytes into smaller species more compatible with analytical techniques (Andrade-Eiroa et al. 2016; Stone 2017).

Extraction methods including dispersive solid-phase, solid-phase, stir bar sorption, and molecularly imprinted polymer might be used for a particular objective (Hernández-Mesa and García-Campaña 2020). Among these techniques, the most widely used for liquid or solid samples is solid-phase extraction (SPE) because it is a technique with high simplicity and flexibility. This technique is based on the separation (i.e., retention) of dissolved or suspended compounds in a liquid mixture via chemical interactions with a solid adsorbent (solid phase). The adsorbent materials used as fillers are very similar to those used in high-performance liquid chromatography (HPLC) columns but with a larger particle size (approximately 40 μm). The selection of the adsorbent material must be based on the nature of the nutraceuticals of interest (Pérez-Fernández et al., 2017).

The general stages of SPE are conditioning of the adsorbent materials, sample loading, washing contaminants, and elution of the analytes of interest. It is important to highlight that for the analysis of some nutraceutical compounds, pretreatment of the sample is needed, including pH adjustment, centrifugation, filtration, dilution, derivatization, and even inclusion of an internal standard.

In the first stage, the *conditioning* of the adsorbent material, a specific solvent or mixture of solvents, is passed through the solid phase (i.e., adsorbent material). The aim is to eliminate impurities and to increase the effective surface area by rinsing the solid phase with a solvent with similar relative polarity to the matrix of the sample that will be loaded (Andrade-Eiroa et al. 2016). Then, in the *sample loading*, the sample is retained in the solid phase through retention mechanisms depending on the nature of the sample. The analytes are retained by the solid phase depending on their polarity through hydrophobic interactions (i.e., reverse phase), hydrophilic interactions (i.e., normal phase), electrostatic attraction of charged functional groups (i.e., ion exchange), or by adsorption (Andrade-Eiroa et al. 2016). In the *washing* stage, a solvent or a mixture of different solvents is used to selectively eliminate unwanted interferences in the sample. The solvent selection is a critical step since the chosen solvent must interact with the interferences to be eliminated; otherwise, there is the risk of eluting the analyte (SUPELCO, 2021). Finally, the *elution of the analytes* is achieved using a solvent that interrupts the interactions between the adsorbent and the analyte. The solvent used should have the maximum interaction with the analyte and minimum interaction with interferences that could have been trapped in the solid phase. At this stage, we must ensure that the elution volume is as small as possible to guarantee the concentration of the analyte (SUPELCO, 2021).

The optimization of the SPE process includes variables such as the solid phase type, the type of eluent, the bed volume, the sampling loading speed, and the porosity of the solid adsorbent phase. Many advances have been made in the SPE process, including the development of different sorbent surfaces, automation, and miniaturization of SPE systems making this process an excellent selection for a suitable MS analysis (Bertolla et al. 2017; Wang et al. 2017; Karki et al. 2021; Rusinga and Weis 2017; Hashemi, Zohrabi, and Shamsipur 2018).

Among the diverse nutraceuticals, phenolic compounds constitute a family that has been widely studied. Likewise, the wide range of fatty acids (FAs) and other nonpolar molecules constitute, today more than ever, a very interesting research niche due to their effects on health, mainly in the prevention of neurologic disorders. Finally, in recent years other complex molecules have been the focus of attention of the scientific community, such as proteins and peptides with biological activity and polysaccharides derived from diverse matrices. The main advances in the analysis of these molecules using various MS techniques will be analyzed below.

10.3 MS ANALYSIS OF PHENOLIC COMPOUNDS

Phenolic compounds are the most abundant secondary metabolites widely distributed in nature which are synthesized by plants and play a significant role in them by regulating growth as an internal physiological regulator or as chemical messengers (Ambriz-Pérez et al., 2016). Synthesis of these compounds is increased as an adaptive response of the plant to stress conditions such as infection, wounding, and UV radiation, among others (Isah 2019). They occur ubiquitously in plants and are a very diversified group of phytochemicals: simple phenols, phenolic acids, coumarins, flavonoids, stilbenes, hydrolyzable and condensed tannins, lignans, and lignins. In plants, phenolics may act as phytoalexins, antifeedants, and attractants for pollinators (Shahidi and Yeo 2016; Hoda, Hemaiswarya, and Doble 2019; Nollet 2018).

Phenolic compounds are considered nutraceutical compounds because of their wide distribution in nature and their proven beneficial effects on health. These characteristics have led to their incorporation into the industry as functional ingredients in foods, beverages, cosmetics, and pharmaceutical products. Since the MS technique is highly reliable and versatile, it has become the perfect analytical tool for the identification and characterization of bioactive components such as phenolics (Figure 10.1).

Many studies have demonstrated that the quadrupole time-of-flight (Q-TOF) has been the most used analyzer in phenolics compounds analysis. Meanwhile, the electrospray ionization (ESI), positive or negative, continues to be the most widely applied ionization method (Table 10.1). On the other hand, other novel MS techniques have been used to analyze phenolic compounds, such as the Fourier transform ion cyclotron resonance mass spectrometry (ESI–FT-ICR-MS) that has been used to identify primary (organic acids and sugars) and secondary metabolites (polyphenolic compounds) in mango (*Mangifera indica* L.) fruits in four distinct maturation stages (Oliveira et al. 2016). These compounds showed chemopreventive effects on Hepa1c1c7 (murine hepatoma) cells by the induction of the potential detoxification phase 2 enzyme.

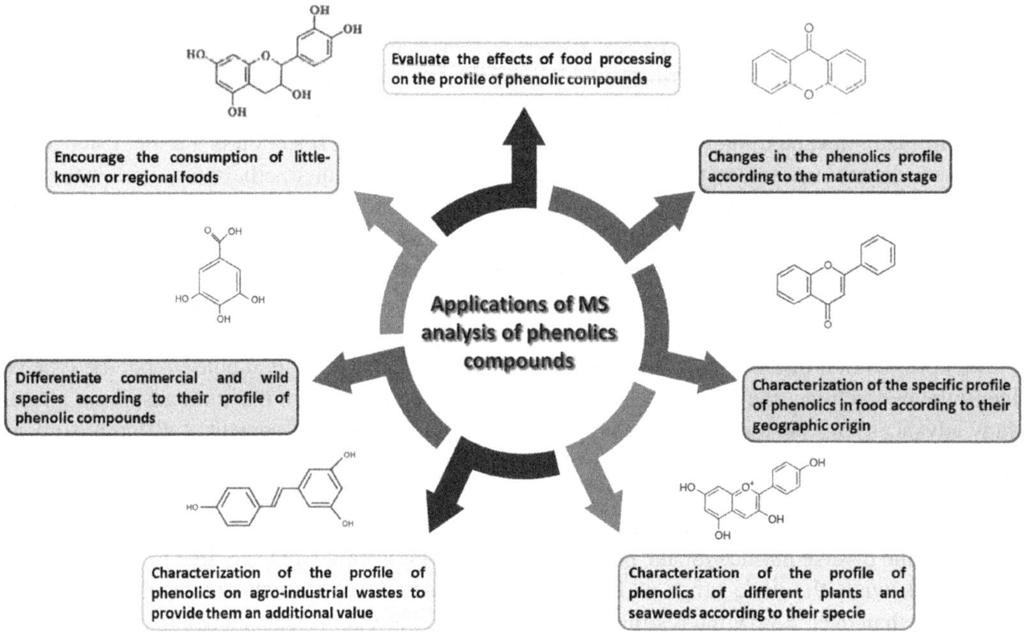

FIGURE 10.1 Summary of the main applications of the phenolics compounds determination by mass spectrometry analysis.

Time-of-flight mass spectrometry (TOF-MS) has been successfully used to identify a wide range of phenolic compounds in different matrices. A wide variety of phenolics and flavonoids have been studied and identified in Algerian honey samples, including 4-hydroxybenzoic acid, caffeic, p-coumaric and vanillic acids; abscisic and syringic acids, benzoic acid, gallic acid, homovanillic acid, apigenin, chrysin, galangin, kaempferol, isorhamnetin, luteolin, pinocembrin, daidzein, and myricetin (Ouchemoukh et al. 2017). Phalsa (*Grewia asiatica* L.), an underutilized crop native to the tropical countries of South Asia, has been identified as a potential functional food due to the wide variety of phenolic compounds (more than 50) identified in it, including flavanols, dihydroflavonols, flavones, flavanols, anthocyanins, isoflavonoids, phenolic acids, flavanones, and other phenolics (Koley et al. 2020). Georgia peach cultivars have also been analyzed for their antioxidant activity, and this activity was strongly related to their content and profile of phenolic compounds such as chlorogenic acid, protocatechuic acid, neochlorogenic acid, (+)-catechin, chlorogenic acid, cyanidin-3-*O*-glucoside, chlorogenic acid derivative, quercetin-3-*O*-galactoside, and quercetin-3-*O*-glucoside (Liao et al. 2020). Q-TOF analysis in combination with nuclear magnetic resonance (NMR) has been used to elucidate for the first time the structure of three phenolic compounds: 2,6-dihydroxyacetophenone (DHAP), 4-hydroxybenzaldehyde, and 4-hydroxyphenethyl alcohol (4-HPEA) in pickled radish. DHAP and 4-HPEA exhibited a strong radical scavenging capacity, and all these phenolic compounds showed antimicrobial activity against *Escherichia coli* and *Candida albicans* (Li et al. 2020b). About 54 phenolic compounds were tentatively identified in eight different seaweed species using liquid chromatography-electrospray ionization quadrupole time-of-flight tandem mass spectrometry (LC–ESI-Q-TOF-MS/MS). Seaweeds *Centroceras* sp., *Ecklonia* sp., and *Caulerpa* sp. showed the largest number of phenolic compounds, while *Ulva* sp. showed the highest content of p-hydroxybenzoic acid (Zhong et al. 2020). Single quadrupole Q and triple quadrupole QqQ MS have been used to identify antioxidant and chemopreventive compounds present in different leaves and fruit, including phenolic acids, flavonoids, anthocyanins, and cathequines (Ding et al. 2020; Tohma et al. 2019; Brahmi-Chendouh et al. 2019; Boeing et al. 2020).

TABLE 10.1
Mass Spectrometry (MS) Techniques in the Analysis of Phenolic Compounds

Mass Analyzer	Ionization Method	Matrix	Identified Compounds	Biological Effects	References
TOF-MS	ESI (–)	Algerian honeys	Phenolic acids, flavonoids	Antioxidant	Ouchemoukh et al. (2017)
TOF-MS	MALDI	Olive waste (leaves and olive mill wastewater)	Secoiridoids flavonoids	N/A	Ventura et al. (2019)
TOF-MS	ESI (–); ESI (+)	Pomegranate peel	Flavonoids, ellagitannins, and proanthocyanidins	N/A	Young et al. (2017)
Q-TOF	ESI (–)	Apple (Malus domestica) peels and pulp	Phenolic acids	N/A	Lee, Chan, and Mitchell (2017)
Q-TOF	ESI (–); ESI (+)	Pickled radish	2,6-dihydroxyacetophenone, 4-hydroxybenzaldehyde, and 4-hydroxyphenethyl alcohol	Antioxidant, antimicrobial	Li et al. (2020b)
Q-TOF	ESI (–); ESI (+)	Brown seaweeds	Phenolics acids, hydroxicinnamc acids	Antioxidant	Zhong et al. (2020)
Q-TOF	ESI (–)	*Phoenix dactylifera* L. (date) seed	Phenolic acids, flavanols, flavonols and flavones.	N/A	Hilary et al. (2020)
Q-TOF	ESI (–); ESI (+)	*Crataegus spp* (Hawthorn)	Phenolic acids, hydroxycinnamic acids, flavonoids, procyanidins	Antioxidant	Li et al. (2020a)
Q-TOF	ESI (–); ESI (+)	Canola (*Brassica napus* L.) seeds waste	Phenolic acids and flavonoids	Antioxidant	Zardo et al. (2020)
Q-TOF	ESI (+)	Peach fruits	Hydroxycinammic acids, flavonoids, procyanidins	Antioxidant and protein glycation inhibition	Liao et al. (2020)
Q-TOF	ESI (–); ESI (+)	*Ficus carica* leaves	Furanocoumarins, flavonoids, phenolic acids	Antihyperglycaemic, antihyperlipidaemic and antioxidant	Belguith-Hadriche et al. (2017)
Q-TOF	ESI (+)	Phalsa (*Grewia asiatica L.*)	Flavonols, dihydroflavonols, flavones, flavanols, anthocyanins, isoflavonoids, phenolic acids, flavanones	N/A	Koley et al. (2020)

(Continued)

TABLE 10.1 (*Continued*)
Mass Spectrometry (MS) Techniques in the Analysis of Phenolic Compounds

Mass Analyzer	Ionization Method	Matrix	Identified Compounds	Biological Effects	References
FT-ICR-MS	ESI (–); ESI (+)	*Vitis vinifera* leaves	Phenolic acids, phytosterols and fatty acids	Antioxidant	Maia et al. (2019)
FT-ICR MS	ESI (–)	Mango (*Mangifera indica* L.)	Gallic acid derivatives (gallotannins), mangiferin	Chemiopreventive	Oliveira et al. (2016)
TQMS	ESI (–)	*Rhus coriaria*	Phenolic acids, flavonoids	Antiproliferative, antidiabetic, anticholinergic activities	Tohma et al. (2019)
TQSM	ESI (–)	*Inula viscosa* leaves:	Phenolic acids, flavonoids	Antioxidant and chemiopreventive	Brahmi-Chendouh et al. (2019)
Triple quadrupole mass spectrometry (TQMS)	Z spray™	*Butia odorata* fruit	Phenolic acids, flavonoid, cathequines	Antioxidant, chemiopreventive	Boeing et al. (2020)
LTQ Orbitrap	ESI (–)	Persimmon (*Diospyros kaki L.*) leaves	Phenolic acids, hydroxybenzoic acids, hydroxycinnamic acids, flavanols, flavonols, flavanones, flavonechalcones, and tyrosols	N/A	Martínez-Las Heras et al. (2017)
LTQ Orbitrap	ESI (–); ESI (+)	Coffee silver skin	Caffeoylquinic acids, lactones and feruloyl-caffeoylquinic acids	Antioxidant	Regazzoni et al. (2016)
LTQ Orbitrap	ESI (–)	Dark Chocolate	Phenolic acid, flavonoids, and procyanidins	N/A	Rusko et al. (2019)
LTQ Orbitrap	ESI (–); ESI (+)	*Averrhoa carambola* L. bark	Phenolic acids, flavonoids, xanthones and terpenoids w	Antioxidant, inhibition of α-glucosidase, elastase, and tyrosinase enzyme activities	Islam et al. (2020)
LTQ Orbitrap	ESI (–); ESI (+)	Brown onion skin	Phenolic acids, anthocyanins, cepaenes, thiosulfinates, and flavonoids	N/A	Campone et al. (2018)
LTQ Orbitrap	ESI (–)	Raisins (dried grapes) *V. vinifera* L.	hydroxybenzoic and hydroxycinnamic acids, flavanoids, flavonoids, flavonols, flavones, and stilbenoids	N/A	Escobar-Avello et al. (2020)

(*Continued*)

TABLE 10.1 ***(Continued)***
Mass Spectrometry (MS) Techniques in the Analysis of Phenolic Compounds

Mass Analyzer	Ionization Method	Matrix	Identified Compounds	Biological Effects	References
QTRAP-MS	ESI (–)	Kale (*Brassica oleracea* L. var. sabellica)	Phenolic acids	Antioxidant	Kasprzak et al. (2018)
QTRAP-MS	ESI (–); ESI (+)	Grape seeds and stems	Phenolic acids, flavonoids, procyanidins	N/A	Nieto et al. (2020)
LC-Ion-Trap MS	ESI (-)	Green pistachio nut	Phenolic acids, cinnamic acids, and flavonoids	Antioxidant	Garavand, Madadlou, and Moini (2017)
LC-Ion-Trap MS	ESI (+)	Peach leaves	Hydroxycinnamic acids and flavonols	Antioxidant	Mokrani et al. (2019)
Q-MS	ESI (–); ESI (+)	Peach fruits	Anthocyanins, flavanols, phenolic acids, and flavonols	Antioxidant	Ding et al. (2020)
LC–ESI–MS–DAD	ESI (+)	Olive leaves	Phenolic acids, flavonoids	Antioxidant	Shirzad et al. (2017)
LC-DAD-MS	APCI; ESI	Apple (*Malus domestica*)	Phenolic acid and triterpenes	Antioxidant and anti-inflammatory	Sut et al. (2019)

In recent years, a clear trend has been observed in the quantification and/or identification of bioactive molecules such as phenolic compounds in agro-industrial waste not only as a strategy to convert them into valuable raw materials but also as part of the development of more sustainable industrial processes. More than 24 phenolic compounds were identified and quantified in canola seed cake using Q-TOF, including sinapic acid derivatives, flavonoid glycosides derivatives, and sinapoyl choline derivatives (Zardo et al. 2020). Likewise, methyl gallate, ethyl gallate, hydroxy phenylacetic acid, three phenylacetic acid isomers, 3-(4-hydroxyphenyl) propionic acid, and homoveratric acid were identified in apple peels and pulp (Lee, Chan, and Mitchell 2017). Meanwhile, the total content and profile of phenolic compounds were analyzed in peels of different pomegranate cultivars (Molla Nepes, Parfianka, Purple Heart, Wonderful, and Vkunsyi). MS analysis revealed that the peel of the pomegranate from the Vkunsyi cultivar was the only one to contain an important compound punicalagin (Young et al. 2017). Leaves of two Tunisian *Ficus carica* cultivars, "Tounsi" and "Temri," were evaluated for their hypoglycaemic, hypolipidaemic, and antioxidative activities in alloxan-induced diabetic rats. Q-TOF analysis was used to identify the presence of different phenolic compounds in both extracts, such as dihydroxybenzoic acid, dipentoside, rutin, psoralen, methoxypsoralen, and oxypeudacin hydrate (Belguith-Hadriche et al. 2017). Hilary et al. (2020) evaluated the bioaccessibility of polyphenol present in *Phoenix dactylifera* L. (date) seeds from Khalas variety in three forms: date seed pita bread, date seed powder, and date seed extract using a human colon (Caco-2) culture model. Q-TOF analysis allowed the identification of up to 29 polyphenols, including phenolic acids, flavonoids, flavonols, and flavones, mainly flavan-3-ols (Hilary et al. 2020). Q-TOF was also used to characterize and discriminate compounds present in the hawthorns from two *Crataegus* species, *Crataegus pinnatifida* Bunge., and *C. pinnatifida* Bge. var. major N. E. The MS analysis allowed the identification of more than 78 compounds in the two species, and 47 of them were saccharides, glycosides, organic acids, phenolics, flavonoids, flavanols, pulp, and triterpenoids (Li et al. 2020a).

Mass spectrometric analysis by liquid chromatography coupled with diode array detection and electrospray ionization (LC–DAD–ESI-MS) and liquid chromatography coupled with atmospheric-pressure chemical ionization (LC–APCI–MS) was used to compare the content of triterpene acids, and phenolic constituents from nine ancient varieties of apple (*Malus domestica*) fruits cultivated in Italy with that phytochemicals profile in four commercial apples ("Golden Delicious," "Red Delicious," "Granny Smith," and "Royal Gala"). The most abundant triterpene in ancient varieties of pulps and peels were pomolic, euscaphic, maslinic, and ursolic acids, while ursolic and oleanolic acids were prevalent in the commercial fruits (Sut et al. 2019). Likewise, Shirzad et al. (2017) used the liquid chromatography coupled with diode array detection and electrospray ionization tandem mass spectrometry (LC–DAD–ESI–MS/MS) for the identification of 27 compounds in olive leaves, including oleuropein as the main compound (Shirzad et al. 2017). Matrix-assisted laser desorption/ionization time-of-flight mass spectrometry (MALDI-TOF-MS) using a binary matrix composed of 1,8-bis(tetramethylguanidino) naphthalene and 9-aminoacridine (1:1 molar ratio) was used to improve ionization properties at low levels of laser energy. This analytical tool has been used to characterize the profile of phenolic compounds present in phenolic-rich extracts obtained from virgin olive oil, olive leaves, and aqueous olive mill wastewater and disclose the chemical nature of these different olive-related materials (Ventura et al. 2019).

Also, ion trap and orbitrap mass spectrometers have been widely used for the identification of phenolic compounds in matrices such as dark chocolate, kale (*Brassica oleracea* L.), and raisins (dried grapes) (Kasprzak et al. 2018; Rusko et al. 2019; Escobar-Avello et al. 2020). However, literature revealed that these technologies had been mainly used in the phytochemical analysis of food waste such as green pistachio nuts, coffee silver skin, grape seeds and steam, *Averrhoa carambola* L. bark, and peach leaves. The phenolic compounds identified in these matrices include phenolic acid, flavonoids, anthocyanins, stilbenoids, xanthones, terpenoids, procyanidins, and tyrosols, many of them with proved antioxidants effects (Garavand, Madadlou, and Moini 2017; Regazzoni et al. 2016; Martínez-Las Heras et al. 2017; Campone et al. 2018; Islam et al. 2020; Mokrani et al. 2019; Nieto et al. 2020).

10.4 MS ANALYSIS OF LIPIDS WITH NUTRACEUTICAL PROPERTIES

Lipids comprise a wide variety of organic molecules present in all living organisms. The Lipid Metabolites and Pathways Strategy (LIPID MAPS, 2021) proposed eight classes of lipids are as follows: fatty acyls, glycerolipids, glycerophospholipids, sphingolipids, sterol lipids, prenol lipids, saccharolipids, and polyketides. Their importance lies in their biological functions including, storing energy in most organisms, cell growth, differentiation, apoptosis, and cell membrane structure, signal recognition and transduction, electron carriers, pigments, and enzyme cofactors (Sun et al., 2020).

According to their structural composition, lipids are classified as simple and complex. Simple lipids are composed of esters of FAs with various alcohols (i.e., fats and waxes) and have predominantly nonpolar properties. In contrast, complex lipids frequently contain three or more chemical compounds (i.e., glycerol, FAs, and sugar, one long-chain base, one nucleoside, one FA, and one phosphate group) and have predominantly polar or amphipathic properties. Some contain only two components but including a sugar moiety. The simple lipids are mostly available in storage tissues, which are easy to extract. In contrast, complex lipids are difficult to extract because of their membrane localization and their association with other biomolecules such as proteins and carbohydrates (Velmurugan and Incharoensakdi 2018).

Triacylglycerols (TAGs) are the main components in oils and fats of plant and animal origin. TAGs are composed of three FAs esterified to a glycerol molecule (Yang and Benning 2018). The FAs from plants, animals, and microbial origin, vary in chain length and presence of functional groups, i.e., *trans*-double bonds, hydroxyl, keto and ether groups, and cyclic rings, which rarely occur in animal tissue. Furthermore, prokaryotic and eukaryotic cells are known for synthesizing straight and branched-chain FAs with odd carbon numbers, polyunsaturated FAs derivatives, and phospholipids (Waktola et al. 2020; Jones et al. 2019). Then, the analysis of TAGs and other nutraceutical lipids is used as authentication method, and these materials are used as biomarkers in metabolomic pathways in biomedical and nutritional assays (Figure 10.2). Furthermore, lipids are recognized as biomarkers associated with abnormal metabolism in several diseases (Hu and Zhang 2017). The diversity of molecular structures and the complexity of their extraction have made lipid analysis a challenge that, thanks to the development of MS technology, has made lipidomic, or the comprehensive analysis of lipids, possible (Wang, Wang, and Han 2019).

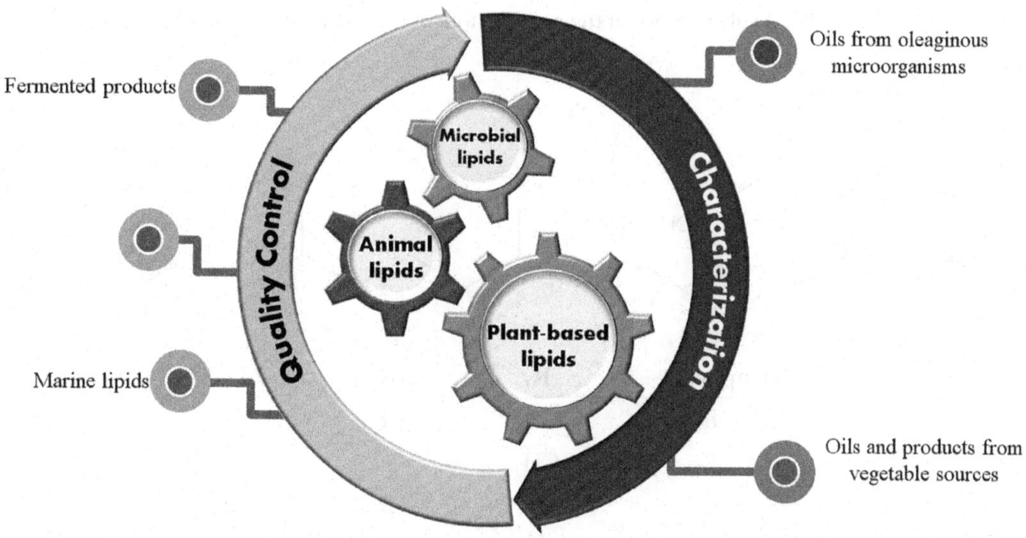

FIGURE 10.2 Main applications of mass spectrometry (MS) in the analysis of lipids with nutraceutical properties.

10.4.1 Plant-Based Products

The main components of the vegetable oils are FAs and di- and triacylglycerides, present around 95%–98% and characteristic minor components including phenolic compounds (Yu et al. 2021), free and conjugated phytosterol (Zhang et al. 2021), phospholipids (Calvano et al. 2020), tocopherols (Li et al. 2020c), carotenoids (Arrizabalaga-Larrañaga et al. 2020), and phytoestrogens (Mornar et al. 2020). The nutritional benefits and FA profile of these oils depend on the plant-based product from which they are extracted. Thus, the FA profile can be used as characterization and authentication of oils (i.e., detection of adulteration) (Indelicato et al. 2017)

10.4.1.1 Characterization of Plant-Based Lipids

Different mass spectrometric and chromatographic separation techniques are suitable for plant-based comprehensive analysis of lipids. Gas chromatography-mass spectrometry (GC–MS) is used in the identification and quantification of volatile and less volatile plant-based lipids. However, for the identification of some nutraceutical compounds such as phytosterols, GC–MS may cause possible sterols degradation or isomerization (Zhang et al. 2021). In contrast, LC–MS/MS has proven high sensitivity for the identification and quantification of trace amounts of phenolic compounds, phytoestrogen, and phospholipids (Calvano et al. 2020; Mornar et al. 2020; Yu et al. 2021). Nevertheless, LC–MS/MS also showed some limitations, such as low ionization efficiency and poor separation characteristics (Yu et al. 2021). These limitations complicate the identification of lipid compounds without the use of an internal standard. Therefore, the selection of the ionization method and mass detector depends on the lipid to be analyzed.

10.4.1.1.1 Phytosterols, Phospholipids, and Tocopherols

Phytosterols form part of the triterpene family, and their structure is similar to cholesterol, the most abundant sterol found in animal cells. Phytosterols are important components of plant cell membranes and play a pivotal role in plant metabolism and development. These compounds have been associated with different health benefits, including anticancer, anti-inflammatory, and especially in lowering blood cholesterol levels; therefore, there is a growing interest in their analysis in foods and food by-products containing intrinsic or added plant sterols/stanols (Nzekoue et al. 2020).

On the other hand, tocopherols are a group of fat-soluble phenolic compounds that consist of a chromanol ring and a 16-carbon phytyl chain and can be classified into α, β, δ, and γ-tocopherols as the number of methylated substituents in the chromanol ring (Figure 10.3). Tocopherols have been

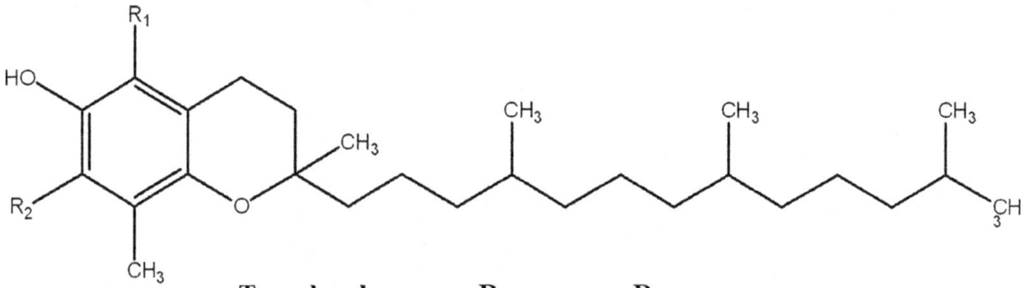

Tocopherol	R_1	R_2
α-	CH_3	CH_3
β-	CH_3	H
γ-	H	CH_3
δ-	H	H

FIGURE 10.3 Tocopherols general chemical structure.

recognized as strong antioxidants exhibiting differences between them based on their chemical structure (Das Gupta and Suh 2016). Due to their strong antioxidant properties, tocopherols have been identified as nutraceutical compounds able to prevent diseases related to oxidative stress such as cancer, hyperglycemia, hyperlipidemia, and inflammation.

Because of its high resolution and reproducibility, GC–MS represents the method of analysis par excellence of nonpolar or few polar compounds such as phytosterols, tocopherols, phospholipids, and other lipids. GC–MS, in combination with SPE, has allowed the identification and quantification of free and esterified phytosterols in food matrices such as almond, pistachio, macadamia nut, walnut, hickory nut, chestnut, cashew, pine nut, hazelnut, peanut, and also their content in food waste such as seeds from the pumpkin, watermelon, and sunflower. It was noticed that β-sitosterol was identified as the major phytosterol present in these samples, followed by α-spinasterol, which was the most abundant phytosterol in pumpkin and watermelon seed kernels representing 57.7% and 67.8% of the total phytosterols detected in these samples, respectively (Wang et al. 2019a). Similarly, Feng et al. (2020) reported the recovery of different phytosterols from the hickory husk, a by-product obtained after the cracking and shelling process of Chinese hickory (*Carya cathayensis* Sarg.). These results are very interesting since these foods are mainly discharged as food waste instead of being considered inexpensive sources for the consumption of nutraceuticals. In the same research line, different phytosterols such as β-sitosterol, campesterol, and brassicasterol, as well as the α-tocopherol have been identified and quantified in the extracts obtained by supercritical CO_2 extraction from the rapeseed oil deodorizer distillate, which is a waste by-product from the vegetable oil industry (Jafarian Asl, Niazmand, and Jahani 2020; Jafarian Asl and Niazmand 2020).

The GC–MS analysis has been applied to analyze the maturity stage of peanut seeds (*Arachis hypogaea*) through the determination of the phytosterols composition in seed components, including the kernel, embryo (heart), and seed coat or skin. It was found that β-sitosterol, campesterol, and stigmasterol represented 82.29% of seed hearts, 86.39% of kernels, and 94.25% of seed coats, respectively. Likewise, the content of stigmasterol was highest in the seed kernel, whereas peanut hearts contained the highest concentration of sterols, possibly as a source of protection and resources for seed development. Concerning the α-tocopherol content, it was found that it increased in peanut hearts during the maturation process, protecting from temperature stress and stability (Zhou et al. 2018).

An important factor in the processing of functional foods is to monitor the changes in concentration and structural conformation of the active molecules in the final product, which will be acquired by the consumers. In this sense, Massimo, Laura, and Ginevra (2020) analyzed, using GC–MS coupled with a flame ionization detector (GC–FID–MS), the content of phytosterols (campesterol, stigmasterol, β-sitosterol, and Δ^5-avenasterol) and the content of these and other phytosterols oxidation products in raw Bronte's pistachios and four industrially processed pistachio products. It was possible to detect beta-sitosterol oxides in raw pistachios, probably for post-harvest processes and storage conditions. Likewise, it was demonstrated that roasting increased the β-sitosterol content with the increase in temperature. 6-Keto-sitostanol was identified as the most abundant β-sitosterol oxide in samples after processing, and it could be used as a molecular marker of thermal stress in pistachio products (Massimo, Laura, and Ginevra 2020). This research provided new insights into the effect of processing on the occurrence of oxidation phenomena in foods.

Another interesting application is the characterization of sterols in the truffle species *Tuber aestivum* and *Tuber borchii* from Italy, Spain, and Romania. This GC–MS characterization allowed the identification of about 27 sterols in both species that were proposed as culinary alternatives to the more expensive black Périgord and white dawn truffles. At the same time, this study established a variety of differences in sterol biosynthesis pathways between the two tuber species that can be used to distinguish these two and possibly other tuber species from each other, providing valuable tools for the identification of new species (Sommer and Vetter 2020). Likewise, in order to classify and discriminate goji berries (*Lycium barbarum* L.) from different production areas, the composition of nutraceutical lipids, such as FAs and phytosterols, of goji berries produced in different countries

(China, Italy, and Mongolia), was investigated. Chinese goji berries showed the highest content of polyunsaturated, monounsaturated, and saturated FAs (61.6%), followed by Mongolian and Italian goji samples with 61.0% and 47.8% of FA, respectively. In addition, Mongolian goji berries also resulted in an interesting source of phytosterols with 130.1 mg/100 g. Interestingly, the phytosterol profile in goji berries changed according to their origin; β-sitosterol was the predominant phytosterol component in Italian goji samples, while Δ5-avenasterol and Δ-5,23-stigmastadienol were more represented in the Chinese and Mongolian goji samples, respectively (Cossignani et al. 2018).

On the other hand, Huang et al. (2019) reported the use of GC–MS for the analysis and identification, for the first time, of 17 compounds in the unsaponifiable fraction extracted from different inbred lines of Djulis (*Chenopodium formosanum* Koidz.) hull, including phytosterols, triterpenes, and other unsaponifiable compounds. The major bioactive compounds identified in almost all inbred lines of Djulis hull studied were γ-sitosterol (phytosterol), cycloartenol (triterpenes), and α-tocopherol (another unsaponifiable compound). Likewise, and to develop breeding programs to produce new germplasm of peppers with enhanced nutritional quality, a complete metabolomic analysis of peppers of different colors (green, red, pale green, and violet) has been performed using the gas chromatography-quadrupole mass spectrometry (GC–QMS) technique through the determination of their content of phenolic acids, carotenoids, phytosterols, policosanols, and capsaicinoids. Interestingly, the tocopherol content showed a positive correlation with the content of carotenoids but a negative correlation with the phytosterol content (Kim et al., 2017).

Free and esterified sterols (cholesterol, campesterol, stigmasterol, and sitosterol) concentrations were detected in human plasma during parenteral nutrition of preterm infants and adults. Plasma concentrations of free cholesterol, free phytosterols, and esterified phytosterols were not different among the groups evaluated. Interestingly, phytosterols were esterified to a lesser extent than cholesterol in the groups evaluated. Likewise, it was found that preterm infants exhibited the highest cholesterol and phytosterol esterification, and this effect was correlated with birth weight and gestational age (Savini et al. 2016). Results strongly supported the notion that the esterification phenomenon depends on the metabolism development/maturation of the subjects, and it can be used to identify hypertriglyceridemic subjects, healthy adults, and phytosterolemic patients on a standard diet. In the same way, GC-electron ionization MS has allowed to detect and quantify relevant interactions between sterols and the enzymes involved in distal cholesterol biosynthesis (DCB). Since DCB has recently been identified as a promising drug target for the treatment of different diseases such as cardiovascular disease, cancer, multiple sclerosis, and Alzheimer's disease, this sterol pattern analysis may improve the inhibitory activity of drug candidates (Müller et al. 2019).

A novel application of LC–MS/MS has made possible the analysis of phytosterols and tocopherols inside complex systems of drug delivery. δ-Tocopherol, β/γ-tocopherol, α-tocopherol, brassicasterol, campesterol, stigmasterol, and β-sitosterol entrapped in a liposomal formulation were detected and simultaneously quantified using a triple quadrupole-linear ion trap mass spectrometer equipped with an APCI in positive mode (Poudel et al. 2020).

10.4.1.1.2 Phytoestrogens

Phytoestrogens are plant-derived molecules whose main characteristic is their structural similarity to 17-β-oestradiol (E2), the primary female sex hormone. Two major phytochemical families have been identified as phytoestrogen molecules, the lignans, and the phenolic compounds. Lignans, present in the cell wall of plants, and the phenolic compounds such as the isoflavones and coumestans, which are found in berries, wine, grains, and nuts, but are most abundant in soybeans and other legumes (Patisaul and Jefferson 2010). These compounds are not considered as nutrients but have been recognized as endocrine disruptors since they have the ability to bind to the estrogen receptors (Rietjens, Louisse, and Beekmann 2017). Physiological effects of phytoestrogens in humans vary according to the life stage and sex. But also, the biological activity of these compounds may be affected by their bioaccessibility and bioavailability (Domínguez-López et al. 2020).

Coumestans and isoflavones such as coumestrol and daidzein, genistein, formononetin, and biochanin A have been quantified by high-performance liquid chromatography-tandem mass spectrometry (HPLC–TQM–ESI-MS/MS) in soy-based meat substitutes used in burgers (Benedetti, Di Carro, and Magi 2018). The authors found that daidzein and genistein were present in samples at high concentration levels, followed by formononetin. The quantification of these compounds in meat substitutes available on the market to make an estimation of their intake for consumers with specific physiological conditions or requirements.

In many studies, the determination of phytoestrogens or their metabolites has been used as a diagnostic tool that allows predicting the prevalence of some diseases from the food intake pattern of the individuals. Phytoestrogens determination has been a useful tool to establish a correlation between the patterns of consumption in women of fruits and vegetables, which represents a source of phytoestrogens, and the reduction in the prevalence of overweight, cancer, and heart disease risk. Urinary samples of about 100 healthy women from 25 to 80 years of age were analyzed by liquid chromatography coupled with MS using the electrospray ionization interface and diode array detector (HPLC-DAD-ESI-MS) to measure isoflavones, lignans, flavonols, coumestrol, resveratrol, naringenin, and luteolin. This study showed a strong correlation between the urinary excretion of total phytoestrogens and their habitual dietary intake and provided valuable information about the potential role of phytoestrogens in reducing the risk of breast cancer and other chronic diseases in the Mexican population (Chávez-Suárez et al. 2017). A similar study was conducted in two different Asian populations, Korea and Vietnam, to evaluate the relationship between phytoestrogen intake and colon cancer risk. Plasma concentrations of genistein, daidzein, and enterolactone were quantified by liquid chromatography-mass spectrometry (LC–MS). As a result of this study, an inverse association between the total isoflavone concentration and the colorectal cancer risk was found; this was especially observed with genistein. Isoflavones consumption showed an important role in the reduction of colorectal cancer risk, mainly in women, despite the different ethnic backgrounds (Ko et al. 2018). Likewise, a low concentration of serum glycitin was associated with an increase of major adverse cardiac events and earlier angina hospitalization in women with suspected ischemic heart disease, while low genistein concentration in serum was associated with an increase in major adverse cardiac events after 6 years (Barsky et al. 2020). Recently, it has also been demonstrated that phytoestrogens could be ideal candidates to combat skin aging and other detrimental effects of hypoestrogenism (Liu et al. 2020).

Phytoestrogens concentrations have also been associated with the quality of semen in a study performed in Chinese men. The parameters analyzed in the semen samples of the volunteers were sperm concentration, sperm count, progressive motility, total motility, volume, and sperm motion, and nine specific phytoestrogens (daidzein, genistein, naringenin, equol, formononetin, secoisolariciresinol, enterodiol, enterolactone, and coumestrol) were analyzed in samples by ultra-performance liquid chromatography and MS/MS. Interestingly, adverse associations were observed between high concentrations of genistein, secoisolariciresinol, and equol in semen samples with the sperm concentration and count. Likewise, sperm motility was negatively associated with a high concentration of secoisolariciresinolin in semen. In contrast, a positive association was found between the concentrations of daidzein and naringenin on sperm motility (Yuan et al. 2019).

Different analytical methods have been developed to quantify the concentrations of phytoestrogens in commercial products available on the market; among these methods, the gas chromatography-tandem mass spectrometry (GC–MS/MS), liquid chromatography coupled with tandem mass spectrometry (LC–MS/MS), and ultra-high-performance liquid chromatography (UHPLC–MS/MS) coupled with triple quadrupole tandem mass spectrometry are the most reported (Socas-Rodríguez et al. 2017; Benedetti, Di Carro, and Magi 2018; Bláhová et al. 2016; Mornar et al. 2020).

The phytoestrogens can also be used as natural biomarkers of the botanical origin of the propolis. The phytoestrogens profile in the red Brazilian propolis was determined by gas chromatography coupled with mass spectrometry (GC–MS) and was constituted mainly for homopterocarpin, medicarpin, 7-*O*-methylvestitol, isovestitol, and formononetin. These compounds allowed the plant

Dalbergia ecastophyllum, a plant belonging to the *Leguminosae* family, as the only source of red propolis (Cassina López et al. 2020).

The content of the biochanin A, formononetin, genistein, ononin, sissotrin, daidzein, and daidzin was analyzed in fresh and wilted legume forage using liquid chromatography-mass spectrometry (LC–MS). The main isoflavones detected in the forage were biochanin A and formononetin, and their content, as well the content of other isoflavones, increased with the wilt. High concentrations of phytoestrogens in forage, or well groundwater discharges, represent a potential risk for the occurrence of reproduction problems and inhibited secretion of animal estrogen (Hloucalová et al. 2016; Thompson et al. 2020). These results provided interesting findings on the care in the dietary intake of animals to guarantee a better quality of products.

Recently, liquid chromatography-tandem mass spectrometry (LC–MS/MS), coupled with different mass detectors and ionization sources, has emerged as a versatile alternative for the analysis of bioactive molecules. In this sense, a frequent approach to the analysis of nutraceutical lipids (FAs, TAGs, phytosterols, tocopherols, phospholipids, and carotenoids) has been the possibility of making simultaneous determinations of different families of these compounds to optimize costs and times, making it a very attractive strategy for the industry (Zhang et al. 2019; Gu et al. 2016; Chang et al. 2020). A specific example of this tendency is the use of the normal-phase liquid chromatography with atmospheric-pressure photoionization-mass spectrometry detection (NPLC–APPI–MS) to simultaneously identify more than 30 sterol oxidation products, which exhibit different polarities, in a single analysis (Grün and Besseau 2016). Likewise, ultra-performance liquid chromatography coupled with quadrupole time-of-flight tandem mass spectrometry (UPLC–Q-TOF-MS/MS) has been useful to determine about 39 different phospholipids and 49 phytosterols present in *Schisandra chinensis* oil (Gao and Wu 2019).

10.4.1.2 Quality Control

Food adulteration has become a more and more frequent phenomenon due to globalization, becoming a real and complex problem which in many cases, jeopardizes the health of the consumer. Because of this issue, FA and TAG composition have been used as a fingerprint for the classification and determination of authenticity, origin, or adulteration evidence for different fats and oils, assessing the quality of the food products and their processing methods (Waktola et al. 2020). Different mass spectrometric and chromatographic techniques are suitable for quality control of the oils, for example, gas chromatography-electrospray ionization mass spectrometry (GC–ESI–MS) and gas chromatography with time-of-flight mass spectrometry (GC–Q-TOF-MS) with EI as ionization method are used in the determination of pesticides in edible vegetable oils (He 2017; Parrilla Vázquez et al. 2016); ultra-high-performance liquid chromatography coupled with triple quadrupole tandem mass spectrometry (UHPLC–QqQ–MS/MS) with ESI as ionization method for detection of mycotoxins in edible vegetable oils (Hidalgo-Ruiz et al. 2019). Within the group of mycotoxins, aflatoxins, including B1, B2, G1, and G2, produced by *Aspergillus* species of fungi, are of main concern for producers. Aflatoxins have been classified as group 1 human carcinogens, and the most dangerous of them is aflatoxin B1, which has been demonstrated to be related to liver cancer (Hidalgo-Ruiz et al. 2019).

10.4.2 Animal Products

The FA profile varies according to its plant or animal origin. Although these natural lipids may have similar functions, the differences are mainly focused on the chain length and the type of functional group present. Bacteria and other organisms can synthesize FAs with odd-numbered and branched chains, such as the presence of FAs with 15 and 17 carbon atoms in ruminants' milk, which is directly associated with their bacterial production in the rumen. However, the most abundant FAs in natural lipids are the straight-chain saturated with even carbon numbers between 14 and 16. Thus, through FA profile analysis, it is possible to characterize and identify fungi, bacteria, and other organisms and their species (Waktola et al. 2020). A specific example is the analysis of the phytosterols profile and content to determine the lard adulteration in cooking oils. In this study, 28 samples of Taiwanese lard were analyzed, and the content of phytosterol was inspected. It was found that adulterated lard showed four specific compounds, belonging to the phytosterol family, and their content ranged from 19.5 to 205.3 μg/g in campesterol and 17.3 to 408.8 μg/g in β-sitosterol and was about 270-fold higher than that of homemade lard or commercial lard (Liao et al. 2017). Another interesting application is in the adulteration of fish oil supplements with plant-based Omega-3 FAs. The increasing development of marine-based supplements has been undermined by adulteration problems. Seafood supplements are of the high value specified by the eicosapentaenoic acid (EPA, 20:5n-3) and docosahexaenoic acid (DHA, 22:6n-3) content. However, the replacement of these almost exclusive Omega-3 FAs from seafood it is becoming more frequent and an important criterion for detecting adulteration (Galuch et al. 2017).

10.4.2.1 Fermented Products

Dairy products are among the most important products for human nutrition. Among the minor lipid components (less than 2%) are found diacylglycerols, cholesterol, phospholipids, and free FAs (Moallem 2018). It is to be expected that fresh and fermented dairy products vary in the amount and type of lipids that they contain. Fermentation has a further impact on the lipid composition of milk, providing volatile FAs (Toral et al., 2018). The beneficial effects of the consumption of fermented dairy products have been linked directly with the pentadecanoic acid (15:0) and the conjugated linoleic acid (CLA) (Jenkins et al. 2018) typically associated with bacterial rather than mammalian metabolism (Furse, Torres, and Koulman 2019). Among the beneficial effects that exhibit the consumption of fermented dairy products are the probiotic abilities such as promoting the secretion of total bile acid and short-chain fatty acid and decreasing serum lipid (Furse, Torres, and Koulman 2019). The unesterified CLA detection is normally preceded by the derivatization of this fatty acid with 4-phenyl-1,2,4-triazoline-3,5-dione prior direct infusion quantification by LC–ESI–MS/MS in the positive ion mode (Mastrogiovanni, Trostchansky, and Rubbo 2020). In general, MS has been a fundamental key in the characterization of lipids from fermented products. This has contributed to enhance the importance and relationship of the microbiome in the expression of some diseases and to elucidate factors that could lead to changes in the composition and function of the human microbiome.

10.4.2.2 Cured Meats

The commercial production of cured products involves the use of high amounts of salt, and lipid oxidation during the manufacturing process is accountable for the characteristic aroma and taste of dry-cured meat products. However, despite the positive impact of lipid oxidation on flavor attributes in cured products, it is also responsible for the development of off-flavors and loss of liposoluble vitamins and other bioactive compounds (Mariutti and Bragagnolo 2017) as a result of poor-quality controls during processing. The presence of higher amounts of polyunsaturated FAs is responsible for increasing the lipid oxidation rate, thus promoting the formation of diverse compounds including pentanal, alcohols, esters, hexanal, and malondialdehyde during the breakdown of the hydroperoxides developed in the first stage of the free-radical reaction. To determine the volatile compositions of the various sample treatments, two volatile analytical instruments, GC–MS and GC-IMS, were deployed (Aheto et al. 2020). MS has become a strong technique in the evaluation of cured meats' quality and safety. Specifically, GC–MS remains the most suitable technique for complex samples such as cured meats.

10.4.2.3 Marine Lipids

Marine products serve as an economical and abundant source of high-quality protein, essential amino acids, ω-3 FAs, vitamins, and minerals. While vegetable oils such as corn oil may contain large amounts of linoleic acid (18:2 ω-6), marine lipids from fish such as cod-liver, tuna, sardine, herring, mackerel, and salmon oils are the major sources of ω-3 polyunsaturated FAs (Loftsson et al. 2016; Toral et al. 2018), mainly EPA (20:5 ω-3) and DHA (22:6 ω-3). EPA and DHA can reduce the risk of coronary heart diseases, and an increase in the dietary ω-3/ω-6 ratio can reduce plasma lipid concentrations and the risk of obesity. EPA and DHA were also reported to decrease the risk of atherosclerosis (Wang et al. 2019b; Uchida et al. 2018). For this reason, controlling the lipid profile of marine products and their content of EPA and DHA are vital to demonstrating a possible adulteration of them (Galuch et al. 2017).

Besides, MS techniques have proven useful in the detection of toxins. In particular, TOF-MS for the identification and quantification of lipophilic marine toxins and contaminants. This study evaluated the presence of natural phenomena contaminants in seawater, suspended matter, and marine sediment in samples collected from Jiaozhou Bay in China. TOF-MS sensitively identified eight marine toxins as okadaic acid, pectenotoxin 2, dinophysistoxin, yessotoxin, azaspiracid, spirolide, and derivatives (Chen et al. 2017). MS has also proven to be reliable in the characterization and elucidation of chemical differences between fatty acid isomers. In particular, through UPLC–Q-TOF-MS reversed-phase analysis, a sensitive separation of EPA and DHA phospholipids and sphingolipids was achieved. In the study, four different fish species were evaluated, and over 700 molecular species from 12 major lipid subclasses were identified using a lipidomic shotgun strategy. To achieve this, a selective lipid extraction with chloroform/methanol and centrifugation methodology was followed (Wang et al. 2019b).

10.4.3 Novel Sources of Lipids: Microbial and Insects

The increasing population worldwide and the use of scarce natural resources more efficiently have led to seeking novel food sources. A wide range of biomass, including plants, animals, and microorganisms, have been found to render food lipids (Béligon et al. 2016). Amongst all these sources, insects hold a special position because some insects reportedly contain up to 75% lipids based on dry weight (Pino Moreno and Ganguly 2016; Paul et al. 2017). Due to their high versatility and rich composition, different applications can be foreseen for oils extracted from the insects, i.e., food and feed applications, energy, surfactants, and materials applications of these lipids.

Oils derived from microorganisms, which are often referred to as microbial oils, microbial lipids, or single-cell oils, can be interesting as a potential source of food supplements since some microorganisms are able to produce polyunsaturated FAs, such as those belonging to the omega-3 and omega-6 series (Béligon et al. 2016; Bellou et al. 2016).

Lipid-accumulating microorganisms, also called oleaginous microorganisms, have the capacity to accumulate lipid content higher than 20% of their dry weight. Several microbial genera, especially molds (*Mucor mucedo*, *Emerisella nidulans*, and *Aspergillus* sp.) and yeasts (*Rhodotorula* sp., *Geotrichum* sp., and *Yarrowia lipolytica*), are known to be high lipid producers under specific culture conditions (Lopes et al. 2018).

10.5 BIOACTIVE PEPTIDES

In the last years, bioactive peptides have attracted the attention of researchers from different areas who have been focused on the identification and production of peptides with biological activities. Recently, many efforts have been made in the obtention and identification of bioactive peptides from industrial waste as a strategy of valorization of by-products. MS has played an indispensable role in the identification of the amino acid sequence of these bioactive molecules for their subsequent identification and potential synthesis.

MS analysis has been used to determine the molecular weight of bioactive peptides, as reported by Lee, Kim, and Nam (2015), who evaluated the anti-inflammatory potential of the bioactive peptide PPY1, composed of five amino acids (K–A–Q–A–D), obtained from the marine algae *Pyropia yezoensis*. MS has played a key role in the identification of amino acid sequences of new peptides. Xue et al. (2015) reported the use of this analytical tool to identify the amino acid sequence (RQSHFANAQP) of an antioxidant peptide (CPe-III) founded in the chickpea albumin hydrolysate. Similarly, Zenezini Chiozzi et al. (2016) reported the use of multidimensional LC and nano-HPLC–high-resolution mass spectrometry (HRMS) for the purification and identification of antioxidant and ACE-inhibitory peptides present in donkey milk. Different peptides with biological activities have been identified by liquid chromatography coupled with a quadrupole-time-of-flight mass spectrometer (LC-Q-TOF-MS) in hydrolysates from Arabica coffee silverskin obtained from three different roasting degrees (light, medium, and dark) and obtained by enzymatic hydrolysis with alcalase, thermolysin, and simulated gastrointestinal digestion (pepsin and pancreatin). About 51 peptides containing between 4 and 12 amino acids were identified in the coffee silverskin hydrolysates, none of them being common to the three different protein hydrolysis employed. Potential biological activities of the peptides obtained were predicted using the BIOPEP database, finding varied activities such as peptides with antibacterial activity, ACE-inhibitory effect, or antioxidant capacity (Pérez-Míguez, Marina, and Castro-Puyana 2019). Likewise, more than 40 peptides with potential biological effects as dipeptidyl peptidase IV inhibitors were identified in cow milk samples by ultra high-performance liquid chromatography–high-resolution mass spectrometry (UHPLC–HRMS)-based untargeted peptidomics approach (Montone et al. 2019).

As mentioned earlier, fermented dairy products have attracted the interest of the scientific community and consumers because of their capacity to apport live probiotic microorganisms with beneficial effects in intestinal microbiota and to produce health-promoting molecules like peptides (Tagliazucchi, Martini, and Solieri 2019). Kefir milk is a fermented product rich in nutraceutical molecules such as amino acids, vitamins, and mineral salts, to which antimicrobial and antitumor activity, immunomodulating effect, and cholesterol-lowering effects have been attributed. Santini et al. (2020) analyzed low-molecular weight polypeptides (8.0–11.0 kDa) derived from the fermentation process of the kefir milk by proteomic analysis, using two-dimensional electrophoresis followed by MS. In this work, MS analysis allowed the identification of different casein-derived polypeptides, mainly from κ-casein, αs1-casein, and αs 2-casein proteolysis. Bioactive peptides have been identified in *Tempe*, a traditional Indonesian fermented soybean, from three different producers, whereas the biological activities of these peptides such as antihypertensive, antidiabetic, antioxidative, and antitumor activities were predicted using BIOPEP databases. Interestingly, more bioactive peptides were identified in those producers having better sanitation levels, and this was associated with the lower microorganism content present in *Tempe*, which may degrade these bioactive molecules (Tamam et al. 2019). Also, proteomics, bioinformatics, and docking analyses have allowed us to identify the antihypertensive peptide AM-1 in the aqueous extract of *Astragalus membranaceus* root, an herb used in traditional medicine. MS analysis allowed identifying of the amino acid sequences of the AM-1 peptide (LVPPHA), whereas the *in vitro* and *in vivo* studies demonstrated that AM-1 peptide regulated the blood pressure through the inhibition of the angiotensin-converting enzyme (ACE) activity (Wu et al. 2020).

The stability of bioactive peptides has been the object of many studies, along with the structural modifications that may improve the biological activity of these molecules. An example is the transepithelial transport of the bioactive peptide VLPVPQK (peptide C), which exhibits antioxidative and ACE inhibition activities (Vij et al. 2016).

Lastly, MS is an essential tool to make more complex analyses such as that reported by Beverly et al. (2019), who performed a peptidomic analysis of milk protein-derived peptides released over three h of gastric digestion in 14 in the preterm infant stomach. Peptides were analyzed with an Orbitrap Fusion Lumos to identify their amino acid sequences. Peptide's bioactivity was predicted by comparing the similarity of their amino acid sequence with known bioactive milk peptides. The profile of milk peptides released during gastric digestion over time results in an essential step to determine which peptides are most likely to be biologically relevant in the infant (Beverly et al. 2019).

10.6 POLYSACCHARIDES

Polysaccharides are naturally occurring biopolymers with a wide structural diversity. In many biological processes, these macromolecules serve as stores of energy (e.g., starch and glycogen) and structural components (e.g., cellulose in plants and chitin in arthropods). In addition, polysaccharides, and their derivatives, participate in signal recognition and cell–cell communication, and also play key roles in the immune system, fertilization, pathogenesis prevention, blood clotting, and regulating cell growth and senescence (Lu et al. 2019; Yao et al. 2020).

Furthermore, polysaccharides and their derivatives have been shown to have anticancer and immunostimulatory activities, among others (Tian et al. 2020). For these reasons, polysaccharides are widely used in medicine, health products, and as materials and functional foods. Furthermore, the relevance of these biopolymers has increased, and MS has played a key role as a selective and powerful analytical technique in the identification and elucidating the structure of polysaccharides.

10.6.1 Plant Polysaccharides

Plant polysaccharides such as pectin, inulin, and gums, must be hydrolyzed or enzymatically degraded into oligosaccharides before analytical analysis. The profiling of oligosaccharide mixtures and the identification of sugar modifications (e.g., methylation, acetylation, and sulfation), as well as wall structure studies of single cells, organelle, or on plant tissue are generally achieved through MALDI-TOF-MS (Lerouxel et al. 2002; Obel et al. 2009) and LC–MS (Ridlova et al. 2008; Royle et al. 2002; Yuan et al. 2005). However, quantification of natural polysaccharides or oligosaccharides by MS is not an efficient method since ionization can differ significantly. In contrast, high-performance size exclusion chromatography coupled with multi-angle laser light scattering and refractive index detector (HPSEC-MALLS-RID) has shown to be an easy and accurate method with no need for individual polysaccharide standards (Yao et al. 2020; Cheong et al. 2016). The monosaccharide composition of plant polysaccharides and their derivatives have been analyzed using mainly GC–MS (Cheong et al. 2016; Liu et al. 2017; Zhang et al. 2017) and GC–FID techniques (Liu et al. 2017; Chen et al. 2013).

10.6.2 Fungi Polysaccharides

Mushrooms have an established history of use by many countries as an edible and medical resource, especially in traditional oriental therapies of some Asian countries. Mushroom-extracted polysaccharides are recognized for anticancer, immunomodulatory, and biological response modifier activities (Zong, Cao, and Wang 2012). Thus, the establishment of full traceability is an important task for quality control. The differentiation of the fingerprint of polysaccharides from different *Ganoderma lucidum* species from different regions in China was possible through hydrophilic-interaction chromatography-evaporative light scattering detection-electrospray time-of-flight mass spectrometry (HILIC-ELSD-ESI-TOF-MS) (Zhao et al. 2020). In recent years, the environmental urgency to replace the use of polymers derived from petroleum with those biopolymers from natural sources has led to the seeking of well-characterized enzymes capable of modifying or degrading natural-derived polysaccharides. In particular, enzymes capable of degrading complex lignocellulosic materials into monomers from which generating biochemicals, high-value products, and second-generation biofuels can be produced (van Munster et al. 2017). Thus, the detection of enzymatic products, like oxidized oligomeric products by MS, was shown to be very useful but challenging, and it is becoming more relevant. The formation and identification of oxidized oligosaccharides products can be determined through MALDI-MS and MALDI-TOF-MS (Filandr et al. 2020; van Munster et al. 2017). Finally, the characterization of structures and bioactivities of secondary metabolites from fungi species can be determined via GC–MS as a sensitive technique for metabolites identification (Salvatore et al. 2018).

10.6.3 Marine Polysaccharides

Among the marine polysaccharides (e.g., fucoidan, chitosan, alginate, agar, and carrageenans), chitin has shown selective cytotoxicity on tumor cells without producing the commonly associated adverse effects, antimicrobial activity in wound infection, and anti-inflammatory activities (Barbosa et al. 2019). The structural elucidation of a low-molecular weight fraction (after depolymerization) of marine polysaccharides can be achieved with MALDI-TOF-MS and LC–ESI Q-TOF (Anastyuk et al. 2009; Dhahri et al. 2020; Kokoulin et al. 2020; Xu, Huang, and Cheong 2017). These analytical methods have the advantage that samples need no further purification before analysis and that provide more information on their structures compared to NMR spectra. Furthermore, the monosaccharide composition analysis of marine polysaccharides can be determined using GC–MS (Kokoulin et al. 2020; Wang et al. 2018).

10.6.4 Bacterial Exopolysaccharides

Bacterial exopolysaccharides (BEPs) are a class of high-molecular-weight extracellular biopolymers produced during the metabolic processes of microorganisms (Chi et al. 2019). BEPs are widely applied in food, cosmetic, and medical industries such as bacterial cellulose, extensively investigated for its potential as a possible replacement for plant-based cellulose, and xantham gum, a well-known food additive used as a thickening agent. Other BEPs, such as sphingans, alginate, glucans, hyaluronan, levan, and succinoglycan, have an important growing industrial and commercial value (Freitas, Alves, and Reis 2011). Thus, to improve their potential applications, their composition and structure must be characterized. The monomers composition of BEPs can be determined using electrospray ionization-collision-induced dissociation mass spectrometry to collision-induced dissociation (ESI-CID-MS) and the electrospray ionization-collision-induced dissociation-tandem mass spectrometry (ESI-CID-MS/MS) to identify the molecular mass and sequence of oligosaccharides (Chi et al., 2019). Also, the glycosyl composition via derivatization of the monosaccharides can be performed by GC–MS (Zhang et al. 2015).

Although MS has been demonstrated to be a very useful analytical method in the identification and elucidation of the structure of macromolecules, there are still limitations for the quantitative analysis of these molecules.

10.7 CONCLUSION

In recent years, people are more aware of what they eat, perhaps inspired by the famous statement for which Hippocrates has been credited with "let thy food be thy medicine and thy medicine be thy food," the truth is that technological advances are the main key. Since ancient times, the knowledge of the beneficial effect of certain products increased their value and promoted their consumption. The validation of such products, the identification of their main compounds as well as the better understanding of their biological role and mechanisms in which they are involved were all possible due to the technological advances focused on analytical techniques. These analytical techniques allowed the comprehension of the broad nutrient content that comprises all food products such as minerals, vitamins, and essential amino acids, but also the presence of a wide variety of compounds with medical or health benefits. These compounds with well-known biological activities, known as nutraceuticals, include phenolic compounds, lipids, bioactive peptides, and polysaccharides. The relevance of these nutraceutical compounds lies in their general acceptance as safer alternatives to conventional drug therapies. Technological advances on higher resolution mass analyzers and novel ionization modes will allow us to overcome the current limitations and will open new possibilities in the discovery of bioactive molecules.

REFERENCES

Aheto, J.H., X. Huang, X. Tian, R. Lv, C. Dai, E. Bonah, and X. Chang. 2020. Evaluation of lipid oxidation and volatile compounds of traditional dry-cured pork belly: the hyperspectral imaging and multi-gas-sensory approaches. *Journal of Food Process Engineering* 43, no. 1. doi:10.1111/jfpe.13092.

Ambriz-Pérez, D.L., N. Leyva-López, E.P. Gutierrez-Grijalva, and J.B. Heredia. 2016. Phenolic compounds: natural alternative in inflammation treatment. A review. *Cogent Food & Agriculture* 2, no. 1. doi:10.1080/23311932.2015.1131412.

Anastyuk, S.D., N.M. Shevchenko, E.L. Nazarenko, P.S. Dmitrenok, and T.N. Zvyagintseva. 2009. Structural analysis of a fucoidan from the brown alga *Fucus evanescens* by MALDI-TOF and tandem ESI mass spectrometry. *Carbohydrate Research* 344, no. 6: 779–787. doi:10.1016/j.carres.2009.01.023.

Andrade-Eiroa, A., M. Canle, V. Leroy-Cancellieri, and V. Cerdà. 2016. Solid-phase extraction of organic compounds: a critical review (Part I). TrAC *Trends in Analytical Chemistry* 80: 641–654. doi:10.1016/j.trac.2015.08.015.

Arrizabalaga-Larrañaga, A., P. Rodríguez, M. Medina, F.J. Santos, and E. Moyano. 2020. Pigment profiles of Spanish extra virgin olive oils by ultra-high-performance liquid chromatography coupled to high-resolution mass spectrometry. *Food Additives & Contaminants: Part A* 37, no. 7: 1075–1086. doi:10.1080/19440049.2020.1753891.

Barbosa, A.I., A.J. Coutinho, S.A. Costa Lima, and S. Reis. 2019. Marine polysaccharides in pharmaceutical applications: fucoidan and chitosan as key players in the drug delivery match field. *Marine Drugs* 17, no. 12: 654. doi:10.3390/md17120654.

Barsky, L., G. Cook-Wiens, M. Doyle, C. Shufelt, W. Rogers, S. Reis, C.J. Pepine, and C. Noel Bairey Merz. 2020. Phytoestrogen blood levels and adverse outcomes in women with suspected ischemic heart disease. *European Journal of Clinical Nutrition*. doi:10.1038/s41430-020-00800-6.

Belguith-Hadriche, O., S. Ammar, M. del M. Contreras, H. Fetoui, A. Segura-Carretero, A. El Feki, and M. Bouaziz. 2017. HPLC-DAD-QTOF-MS profiling of phenolics from leaf extracts of two tunisian fig cultivars: potential as a functional food. *Biomedicine & Pharmacotherapy* 89: 185–193. doi:10.1016/j.biopha.2017.02.004.

Béligon, V., G. Christophe, P. Fontanille, and C. Larroche. 2016. Microbial lipids as potential source to food supplements. *Current Opinion in Food Science* 7: 35–42. doi:10.1016/j.cofs.2015.10.002.

Bellou, S., I.-E. Triantaphyllidou, D. Aggeli, A.M. Elazzazy, M.N. Baeshen, and G. Aggelis. 2016. Microbial oils as food additives: recent approaches for improving microbial oil production and its polyunsaturated fatty acid content. *Current Opinion in Biotechnology* 37: 24–35. doi:10.1016/j.copbio.2015.09.005.

Benedetti, B., M. Di Carro, and E. Magi. 2018. Phytoestrogens in soy-based meat substitutes: comparison of different extraction methods for the subsequent analysis by liquid chromatography-tandem mass spectrometry. *Journal of Mass Spectrometry* 53, no. 9: 862–870. doi:10.1002/jms.4268.

Bertolla, M., L. Cenci, A. Anesi, E. Ambrosi, F. Tagliaro, L. Vanzetti, G. Guella, and A.M. Bossi. 2017. Solvent-responsive molecularly imprinted nanogels for targeted protein analysis in MALDI-TOF mass spectrometry. *ACS Applied Materials & Interfaces* 9, no. 8 (March): 6908–6915. doi:10.1021/acsami.6b16291.

Beverly, R.L., M.A. Underwood, and D.C. Dallas. 2019. Peptidomics analysis of milk protein-derived peptides released over time in the preterm infant stomach. *Journal of Proteome Research* 18, no. 3: 912–922. doi:10.1021/acs.jproteome.8b00604.

Bláhová, L., J. Kohoutek, T. Procházková, M. Prudíková, and L. Bláha. 2016. Phytoestrogens in milk: overestimations caused by contamination of the hydrolytic enzyme used during sample extraction. *Journal of Dairy Science* 99, no. 9: 6973–6982. doi:10.3168/jds.2016-10926.

Boeing, J.S., É.O. Barizão, E.M. Rotta, H. Volpato, C.V. Nakamura, L. Maldaner, and J.V. Visentainer. 2020. Phenolic compounds from *Butia Odorata* (Barb. Rodr.) Noblick fruit and its antioxidant and antitumor activities. *Food Analytical Methods* 13, no. 1: 61–68. doi:10.1007/s12161-019-01515-6.

Brahmi-Chendouh, N., S. Piccolella, G. Crescente, F. Pacifico, L. Boulekbache, S. Hamri-Zeghichi, S. Akkal, K. Madani, and S. Pacifico. 2019. A nutraceutical extract from *Inula viscosa* leaves: UHPLC-HR-MS/MS based polyphenol profile, and antioxidant and cytotoxic activities. *Journal of Food and Drug Analysis* 27, no. 3: 692–702. doi:10.1016/j.jfda.2018.11.006.

Calvano, C.D., M. Bianco, G. Ventura, I. Losito, F. Palmisano, and T.R.I. Cataldi. 2020. Analysis of phospholipids, lysophospholipids, and their linked fatty acyl chains in yellow lupin seeds (*Lupinus luteus* L.) by liquid chromatography and tandem mass spectrometry. *Molecules* 25, no. 4: 805. doi:10.3390/molecules25040805

Campone, L., R. Celano, A. Lisa Piccinelli, I. Pagano, S. Carabetta, R.D. Sanzo, M. Russo, E. Ibañez, A. Cifuentes, and L. Rastrelli. 2018. Response surface methodology to optimize supercritical carbon dioxide/co-solvent extraction of brown onion skin by-product as source of nutraceutical compounds. *Food Chemistry* 269: 495–502. doi:10.1016/j.foodchem.2018.07.042.

Cassina López, B.G., G.A. Bataglion, J.L.P. Jara, E.F. Pacheco Filho, L.S. de Mendonça Melo, J.C. Cardoso, M.N. Eberlin, and A.C.H.F. Sawaya. 2020. Variation of the phytoestrogen composition of red propolis throughout the year. *Journal of Apicultural Research* 59, no. 4: 406–412. doi:10.1080/00218839.2020.1714192.

Chang, M., Z. Wang, T. Zhang, T. Wang, R. Liu, Y. Wang, Q. Jin, and X. Wang. 2020. Characterization of fatty acids, triacylglycerols, phytosterols and tocopherols in peony seed oil from five different major areas in China. *Food Research International* 137: 109416. doi:10.1016/j.foodres.2020.109416.

Chávez-Suárez, K., M. Ortega-Vélez, A. Valenzuela-Quintanar, M. Galván-Portillo, L. López-Carrillo, J. Esparza-Romero, M. Saucedo-Tamayo, et al. 2017. Phytoestrogen concentrations in human urine as biomarkers for dietary phytoestrogen intake in Mexican Women. *Nutrients* 9, no. 10: 1078. doi:10.3390/nu9101078.

Chen, J, D. Yao, H. Yuan, S. Zhang, J. Tian, W. Guo, W. Liang, H. Li, and Y. Zhang. 2013. *Dipsacus asperoides* polysaccharide induces apoptosis in osteosarcoma cells by modulating the PI3K/Akt pathway. *Carbohydrate Polymers* 95, no. 2:780–784. doi:10.1016/j.carbpol.2013.03.009.

Chen, J., X. Li, S. Wang, F. Chen, W. Cao, C. Sun, L. Zheng, and X. Wang. 2017. Screening of lipophilic marine toxins in marine aquaculture environment using liquid chromatography–mass spectrometry. *Chemosphere* 168: 32–40. doi:10.1016/j.chemosphere.2016.10.052.

Cheong, K.-L., D.-T. Wu, Y. Deng, F. Leong, J. Zhao, W.-J. Zhang, and S.-P. Li. 2016. Qualitation and quantification of specific polysaccharides from *Panax* species using GC-MS, saccharide mapping and HPSEC-RID-MALLS. *Carbohydrate Polymers* 153:47–54. doi:10.1016/j.carbpol.2016.07.077.

Chi, Y., H. Ye, H. Li, Y. Li, H. Guan, H. Mou, and P. Wang. 2019. Structure and molecular morphology of a novel moisturizing exopolysaccharide produced by *Phyllobacterium* sp. 921F. *International Journal of Biological Macromolecules* 135: 998–1005. doi:10.1016/j.chroma.2015.04.054.

Cossignani, L., F. Blasi, M.S. Simonetti, and D. Montesano. 2018. Fatty acids and phytosterols to discriminate geographic origin of *Lycium barbarum* berry. *Food Analytical Methods* 11, no. 4: 1180–1188. doi:10.1007/s12161-017-1098-5.

Daliu, P., A. Santini, and E. Novellino. 2018. A decade of nutraceutical patents: where are we now in 2018? *Expert Opinion on Therapeutic Patents* 28, no. 12: 875–882. doi:10.1080/13543776.2018.1552260.

Das Gupta, S., and N. Suh. 2016. Tocopherols in cancer: an update. *Molecular Nutrition & Food Research* 60, no. 6: 1354–1363. doi:10.1002/mnfr.201500847.

Dhahri, M., S. Sioud, R. Dridi, M. Hassine, N.A. Boughattas, F. Almulhim, Z. Al Talla, M. Jaremko, and A.-H.M. Emwas. 2020. Extraction, characterization, and anticoagulant activity of a sulfated polysaccharide from *Bursatella leachii* viscera. *ACS Omega* 5, no. 24:14786–14795. doi:10.1021/acsomega.0c01724.

Ding, T., K. Cao, W. Fang, G. Zhu, C. Chen, X. Wang, and L. Wang. 2020. Evaluation of phenolic components (anthocyanins, flavanols, phenolic acids, and flavonols) and their antioxidant properties of peach fruits. *Scientia Horticulturae* 268: 109365. doi:10.1016/j.scienta.2020.109365.

Domínguez-López, I., M. Yago-Aragón, A. Salas-Huetos, A. Tresserra-Rimbau, and S. Hurtado-Barroso. 2020. Effects of dietary phytoestrogens on hormones throughout a human lifespan: a review. *Nutrients* 12, no. 8: 2456. doi:10.3390/nu12082456.

Duncan, K.D., J. Fyrestam, and I. Lanekoff. 2019. Advances in mass spectrometry based single-cell metabolomics. *The Analyst* 144, no. 3: 782–793. doi:10.1039/C8AN01581C.

Escobar-Avello, D., A. Olmo-Cunillera, J. Lozano-Castellón, M. Marhuenda-Muñoz, and A. Vallverdú-Queralt. 2020. A targeted approach by high resolution mass spectrometry to reveal new compounds in raisins. *Molecules* 25, no. 6: 1281. doi:10.3390/molecules25061281.

Feider, C.L., A. Krieger, R.J. DeHoog, and L.S. Eberlin. 2019. Ambient ionization mass spectrometry: recent developments and applications. *Analytical Chemistry* 91, no. 7: 4266–4290. doi:10.1021/acs.analchem.9b00807.

Feng, S., L. Wang, T. Belwal, L. Li, and Z. Luo. 2020. Phytosterols extraction from hickory (*Carya cathayensis* Sarg.) husk with a green direct citric acid hydrolysis extraction method. *Food Chemistry* 315: 126217. doi:10.1016/j.foodchem.2020.126217.

Filandr, F., P. Man, P. Halada, H. Chang, R. Ludwig, and D. Kracher. 2020. The H_2O_2-dependent activity of a fungal lytic polysaccharide monooxygenase investigated with a turbidimetric assay. *Biotechnology for Biofuels* 13, no. 1: 37. doi:10.1186/s13068-020-01673-4.

Freitas, F., V.D. Alves, and M.A.M. Reis. 2011. Advances in bacterial exopolysaccharides: from production to biotechnological applications. *Trends in Biotechnology* 29, no. 8: 388–398. doi:10.1016/j.tibtech.2011.03.008.

Furse, S., A.G. Torres, and A. Koulman. 2019. Fermentation of milk into yoghurt and cheese leads to contrasting lipid and glyceride profiles. *Nutrients* 11, no. 9: 2178. doi:10.3390/nu11092178.

Galuch, M., F. Carbonera, T. Magon, R. da Silveira, P. dos Santos, J. Pizzo, O. Santos, and J. Visentainer. 2017. Quality assessment of omega-3 supplements available in the Brazilian market. *Journal of the Brazilian Chemical Society*. doi:10.21577/0103-5053.20170177.

Gao, Y., and S. Wu. 2019. Comprehensive analysis of the phospholipids and phytosterols in *Schisandra chinensis* oil by UPLC-Q/TOF- MSE. *Chemistry and Physics of Lipids* 221: 15–23. doi:10.1016/j.chemphyslip.2019.03.003.

Garavand, F., A. Madadlou, and S. Moini. 2017. Determination of phenolic profile and antioxidant activity of pistachio hull using high-performance liquid chromatography–diode array detector–electro-spray ionization–mass spectrometry as affected by ultrasound and microwave. *International Journal of Food Properties* 20, no. 1: 19–29. doi:10.1080/10942912.2015.1099045.

Grün, C.H., and S. Besseau. 2016. Normal-phase liquid chromatography–atmospheric-pressure photoionization–mass spectrometry analysis of cholesterol and phytosterol oxidation products. *Journal of Chromatography A* 1439: 74–81. doi:10.1016/j.chroma.2015.12.043.

Gu, Q., X. Yi, Z. Zhang, H. Yan, J. Shi, H. Zhang, Y. Wang, and J. Shao. 2016. A Facile method for simultaneous analysis of phytosterols, erythrodiol, uvaol, tocopherols and lutein in olive oils by LC-MS. *Analytical Methods* 8, no. 6: 1373–1380. doi:10.1039/C5AY02193F.

Hashemi, B., P. Zohrabi, and M. Shamsipur. 2018. Recent developments and applications of different sorbents for SPE and SPME from biological samples. *Talanta* 187: 337–347. doi:10.1016/j.talanta.2018.05.053.

He, Z. 2017. Determination of 255 pesticides in edible vegetable oils using QuEChERS method and gas chromatography tandem mass spectrometry. *Analytical and Bioanalytical Chemistry* 409, no. 4: 1017–1030. doi:10.1007/s00216-016-0016-9.

Hernández-Mesa, M., and A.M. García-Campaña. 2020. Determination of sulfonylurea pesticide residues in edible seeds used as nutraceuticals by QuEChERS in combination with ultra-high-performance liquid chromatography-tandem mass spectrometry. *Journal of Chromatography A* 1617: 460831. doi:10.1016/j.chroma.2019.460831.

Hidalgo-Ruiz, J.L., R. Romero-González, J.L. Martínez Vidal, and A. Garrido Frenich. 2019. A rapid method for the determination of mycotoxins in edible vegetable oils by ultra-high performance liquid chromatography-tandem mass spectrometry. *Food Chemistry* 288: 22–28. doi:10.1016/j.foodchem.2019.03.003.

Hilary, S., F.A. Tomás-Barberán, J.A. Martinez-Blazquez, J. Kizhakkayil, U. Souka, S. Al-Hammadi, H. Habib, W. Ibrahim, and C. Platat. 2020. Polyphenol characterisation of *Phoenix dactylifera* L. (date) seeds using HPLC-mass spectrometry and its bioaccessibility using simulated *in-vitro* digestion/Caco-2 culture model. *Food Chemistry* 311: 125969. doi:10.1016/j.foodchem.2019.125969.

Hloucalová, P., J. Skládanka, P. Horký, B. Klejdus, J. Pelikán, and D. Knotová. 2016. Determination of phytoestrogen content in fresh-cut legume forage. *Animals* 6, no. 7: 43. doi:10.3390/ani6070043.

Hoda, M., S. Hemaiswarya, and M. Doble. 2019. Phenolic phytochemicals: sources, biosynthesis, extraction, and their isolation. In *Role of phenolic phytochemicals in diabetes management*, eds. Muddasarul Hoda, Shanmugam Hemaiswarya, and Mukesh Doble, pp. 13–44. Singapore: Springer Singapore. doi:10.1007/978-981-13-8997-9_2.

de Hoffmann, E. 2005. Mass spectrometry. In *Kirk-Othmer Encyclopedia of Chemical Technology*. Hoboken, NJ: John Wiley & Sons, Inc.. doi:10.1002/0471238961.1301191913151518.a01.pub2.

Hu, T., and J.-L. Zhang. 2017. Mass-spectrometry-based lipidomics. *Journal of Separation Science*: 41, no. 1: 351–372. doi:10.1002/jssc.201700709.

Huang, C.-Y., Y.-L. Chu, K. Sridhar, and P.-J. Tsai. 2019. Analysis and determination of phytosterols and triterpenes in different inbred lines of Djulis (*Chenopodium formosanum* Koidz.) hull: a potential source of novel bioactive ingredients. *Food Chemistry* 297: 124948. doi:10.1016/j.foodchem.2019.06.015.

Indelicato, S., D. Bongiorno, R. Pitonzo, V. Di Stefano, V. Calabrese, S. Indelicato, and G. Avellone. 2017. Triacylglycerols in edible oils: determination, characterization, quantitation, chemometric approach and evaluation of adulterations. *Journal of Chromatography A* 1515: 1–16. doi:10.1016/j.chroma.2017.08.002.

Isah, T. 2019. Stress and defense responses in plant secondary metabolites production. *Biological Research* 52, no. 1: 39. doi:10.1186/s40659-019-0246-3.

Islam, S., M.B. Alam, A. Ahmed, S. Lee, S.-H. Lee, and S. Kim. 2020. Identification of secondary metabolites in *Averrhoa carambola* L. bark by high-resolution mass spectrometry and evaluation for α-glucosidase, tyrosinase, elastase, and antioxidant potential. *Food Chemistry* 332: 127377. doi:10.1016/j.foodchem.2020.12737.

Jafarian Asl, P., and R. Niazmand. 2020. Modelling and simulation of supercritical CO_2 extraction of bioactive compounds from vegetable oil waste. *Food and Bioproducts Processing* 122: 311–321. doi:10.1016/j.fbp.2020.05.005.

Jafarian Asl, P., R. Niazmand, and M. Jahani. 2020. Theoretical and experimental assessment of supercritical CO_2 in the extraction of phytosterols from rapeseed oil deodorizer distillates. *Journal of Food Engineering* 269: 109748. doi:10.1016/j.jfoodeng.2019.109748.

Jenkins, B., M. Aoun, C. Feillet-Coudray, C. Coudray, M. Ronis, and A. Koulman. 2018. The dietary total-fat content affects the *in vivo* circulating C15:0 and C17:0 fatty acid levels independently. *Nutrients* 10, no. 11: 1646. doi:10.3390/nu10111646.

Jones, A.D., K.L. Boundy-Mills, G.F. Barla, S. Kumar, B. Ubanwa, and V. Balan. 2019. Microbial lipid alternatives to plant lipids in microbial lipid production. In *Methods in Molecular Biology*, ed. V. Balan, pp. 1–32. New York: Springer New York. doi:10.1007/978-1-4939-9484-7_1.

Kalra, E.K. 2003. Nutraceutical-definition and introduction. *AAPS PharmSci* 5, no. 3: 27–28. doi:10.1208/ps050325.

Karki, S., A.K. Meher, E.D. Inutan, M. Pophristic, D.D. Marshall, K. Rackers, S. Trimpin, and C.N. McEwen. 2021. Development of a robotics platform for automated multi-ionization mass spectrometry. *Rapid Communications in Mass Spectrometry* 35, no. S1. doi:10.1002/rcm.8449.

Kasprzak, K., T. Oniszczuk, A. Wójtowicz, M. Waksmundzka-Hajnos, M. Olech, R. Nowak, R. Polak, and A. Oniszczuk. 2018. Phenolic acid content and antioxidant properties of extruded corn snacks enriched with kale. *Journal of Analytical Methods in Chemistry* 2018: 1–7. doi:10.1155/2018/7830546.

Kim, T.J., J. Choi, K.W. Kim, S.K. Ahn, S.-H. Ha, Y. Choi, N.I. Park, and J.K. Kim. 2017. Metabolite profiling of peppers of various colors reveals relationships between tocopherol, carotenoid, and phytosterol content: metabolite profiling of peppers. *Journal of Food Science* 82, no. 12: 2885–2893. doi:10.1111/1750-3841.13968.

Ko, K. P., Y. Yeo, J.H. Yoon, C.S. Kim, S. Tokudome, C. Koriyama, Y.K. Lim, S.H. Chang, H.R. Shin, D. Kang, S.K. Park, C.H. Kang, K.Y. Yoo, 2018. Plasma phytoestrogens concentration and risk of colorectal cancer in two different Asian populations. *Clinical Nutrition* 37, no. 5: 1675–1682. doi:10.1016/j.clnu.2017.07.014.

Kokoulin, Maxim.S., I.N. Lizanov, L.A. Romanenko, and I.V. Chikalovets. 2020. Structure of phosphorylated and sulfated polysaccharides from lipopolysaccharide of marine bacterium *Marinicella litoralis* KMM 3900T. *Carbohydrate Research* 490: 107961. doi:10.1016/j.carres.2020.107961.

Koley, T.K., Z. Khan, D. Oulkar, B. Singh, B.P. Bhatt, and K. Banerjee. 2020. Profiling of polyphenols in phalsa (*Grewia asiatica* L) fruits based on liquid chromatography high resolution mass spectrometry. *Journal of Food Science and Technology* 57, no. 2: 606–616. doi:10.1007/s13197-019-04092-y.

Larussa, T., M. Imeneo, and F. Luzza. 2017. Potential role of nutraceutical compounds in inflammatory bowel disease. *World Journal of Gastroenterology* 23, no. 14: 2483. doi:10.3748/wjg.v23.i14.2483.

Lee, H.-A., I.-H. Kim, and T.-J. Nam. 2015. Bioactive peptide from *Pyropia yezoensis* and its anti-inflammatory activities. *International Journal of Molecular Medicine* 36, no. 6: 1701–1706. doi:10.3892/ijmm.2015.2386.

Lee, J., B.L.S. Chan, and A.E. Mitchell. 2017. Identification/quantification of free and bound phenolic acids in peel and pulp of apples (*Malus domestica*) using high resolution mass spectrometry (HRMS). *Food Chemistry* 215: 301–310. doi:10.1016/j.foodchem.2016.07.166.

Lerouxel, O., T.S. Choo, M. Séveno, B. Usadel, L. Faye, P. Lerouge, and M. Pauly. 2002. Rapid structural phenotyping of plant cell wall mutants by enzymatic oligosaccharide fingerprinting. *Plant Physiology* 130, no. 4: 1754–1763. doi:10.1104/pp.011965.

Li, C.-R., X.-H. Hou, Y.-Y. Xu, W. Gao, P. Li, and H. Yang. 2020a. Manual annotation combined with untargeted metabolomics for chemical characterization and discrimination of two major *Crataegus* species based on liquid chromatography quadrupole time-of-flight mass spectrometry. *Journal of Chromatography A* 1612: 460628. doi:10.1016/j.chroma.2019.460628.

Li, J., S.-Y. Huang, Q. Deng, G. Li, G. Su, J. Liu, and H.-M. David Wang. 2020b. Extraction and characterization of phenolic compounds with antioxidant and antimicrobial activities from pickled radish. *Food and Chemical Toxicology* 136: 111050. doi:10.1016/j.fct.2019.111050.

Li, Y., B. Jiang, Y. Lou, Q. Shi, R. Zhuang, and Z. Zhan. 2020c. Molecular characterization of edible vegetable oils via free fatty acid and triacylglycerol fingerprints by electrospray ionization Fourier transform ion cyclotron resonance mass spectrometry. *International Journal of Food Science & Technology* 55, no. 1: 165–174. doi:10.1111/ijfs.14258.

Liao, C.-D., G.-J. Peng, Y. Ting, M.-H. Chang, S.-H. Tseng, Y.-M. Kao, K.-F. Lin, Y.-M. Chiang, M.-K. Yeh, and H.-F. Cheng. 2017. Using phytosterol as a target compound to identify edible animal fats adulterated with cooked oil. *Food Control* 79: 10–16. doi:10.1016/j.foodcont.2017.03.026.

Liao, X., A.A. Brock, B.T. Jackson, P. Greenspan, and R.B. Pegg. 2020. The cellular antioxidant and anti-glycation capacities of phenolics from Georgia peaches. *Food Chemistry* 316: 126234. doi:10.1016/j.foodchem.2020.126234.

LIPID MAPS. 2021. Lipid Classification System. Retrieved January 12th. 2021, from https://www.lipidmaps.org/data/classification/LM_classification_exp.php.

Liu, J., X. Wang, H. Pu, S. Liu, J. Kan, and C. Jin. 2017. Recent advances in endophytic exopolysaccharides: production, structural characterization, physiological role and biological activity. *Carbohydrate Polymers* 157: 1113–1124. doi:10.1016/j.carbpol.2016.10.084.

Liu, T., N. Li, Y. Yan, Y. Liu, K. Xiong, Y. Liu, Q. Xia, H. Zhang, and Z. Liu. 2020. Recent advances in the anti-aging effects of phytoestrogens on collagen, water content, and oxidative stress. *Phytotherapy Research* 34, no. 3: 435–447. doi:10.1002/ptr.6538.

Loftsson, T., B. Ilievska, G.M. Asgrimsdottir, O.T. Ormarsson, and E. Stefansson. 2016. Fatty acids from marine lipids: biological activity, formulation and stability. *Journal of Drug Delivery Science and Technology* 34: 71–75. doi:10.1016/j.jddst.2016.03.007.

Lopes, M., A.S. Gomes, C.M. Silva, and I. Belo. 2018. Microbial lipids and added value metabolites production by *Yarrowia lipolytica* from pork lard. *Journal of Biotechnology* 265: 76–85. doi:10.1016/j.jbiotec.2017.11.007.

Lu, X., J. Chen, Z. Guo, Y. Zheng, M.C. Rea, H. Su, X. Zheng, B. Zheng, and S. Miao. 2019. Using polysaccharides for the enhancement of functionality of foods: a review. *Trends in Food Science & Technology* 86: 311–327. doi:10.1016/j.tifs.2019.02.024.

Maia, M., A.E.N. Ferreira, G. Laureano, A.P. Marques, V.M. Torres, A.B. Silva, A.R. Matos, C. Cordeiro, A. Figueiredo, and M. Sousa Silva. 2019. *Vitis vinifera* "Pinot Noir" leaves as a source of bioactive nutraceutical compounds. *Food & Function* 10, no. 7: 3822–3827. doi:10.1039/C8FO02328J.

Mariutti, L.R.B., and N. Bragagnolo. 2017. Influence of salt on lipid oxidation in meat and seafood products: a review. *Food Research International* 94: 90–100. doi:10.1016/j.foodres.2017.02.003.

Martínez-Las Heras, R., A. Pinazo, A. Heredia, and A. Andrés. 2017. Evaluation studies of persimmon plant (*Diospyros kaki*) for physiological benefits and bioaccessibility of antioxidants by *in vitro* simulated gastrointestinal digestion. *Food Chemistry* 214: 478–485. doi:10.1016/j.foodchem.2016.07.104.

Massimo, L., D. Laura, and L.-B. Ginevra. 2020. Phytosterols and phytosterol oxides in bronte's pistachio (*Pistacia vera* L.) and in processed pistachio products. *European Food Research and Technology* 246, no. 2: 307–314. doi:10.1007/s00217-019-03343-8.

Mastrogiovanni, M., A. Trostchansky, and H. Rubbo. 2020. Data of detection and characterization of nitrated conjugated-linoleic acid (NO_2-CLA) in LDL. *Data in Brief* 28:105037. doi:10.1016/j.dib.2019.105037.

Moallem, U. 2018. Invited review: roles of dietary n-3 fatty acids in performance, milk fat composition, and reproductive and immune systems in dairy cattle. *Journal of Dairy Science* 101, no. 10: 8641–8661. doi:10.3168/jds.2018-14772.

Mokrani, A., S. Cluzet, K. Madani, E. Pakina, A. Gadzhikurbanov, M. Mesnil, A. Monvoisin, and T. Richard. 2019. HPLC-DAD-MS/MS profiling of phenolics from different varieties of peach leaves and evaluation of their antioxidant activity: a comparative study. *International Journal of Mass Spectrometry* 445: 116192. doi:10.1016/j.ijms.2019.116192.

Montone, C.M., A.L. Capriotti, A. Cerrato, M. Antonelli, G. La Barbera, S. Piovesana, A. Laganà, and C. Cavaliere. 2019. Identification of bioactive short peptides in cow milk by high-performance liquid chromatography on C18 and porous graphitic carbon coupled to high-resolution mass spectrometry. *Analytical and Bioanalytical Chemistry* 411, no. 15: 3395–3404. doi:10.1007/s00216-019-01815-0.

Mornar, A., T. Buhač, D.A. Klarić, I. Klarić, M. Sertić, and B. Nigović. 2020. Multi-Targeted screening of phytoestrogens in food, raw material, and dietary supplements by liquid chromatography with tandem mass spectrometry. *Food Analytical Methods* 13, no. 2: 482–495. doi:10.1007/s12161-019-01653-x.

Müller, C., J. Junker, F. Bracher, and M. Giera. 2019. A Gas Chromatography–Mass Spectrometry-based whole-cell screening assay for target identification in distal cholesterol biosynthesis. *Nature Protocols* 14, no. 8: 2546–2570. doi:10.1038/s41596-019-0193-z.

van Munster, J.M., B. Thomas, M. Riese, A.L. Davis, C.J. Gray, D.B. Archer, and S.L. Flitsch. 2017. Application of carbohydrate arrays coupled with mass spectrometry to detect activity of plant-polysaccharide degradative enzymes from the fungus *Aspergillus niger*. *Scientific Reports* 7, no. 1: 43117. doi:10.1038/srep43117.

Narváez, A., Y. Rodríguez-Carrasco, L. Castaldo, L. Izzo, and A. Ritieni. 2020. Ultra-high-performance liquid chromatography coupled with quadrupole orbitrap high-resolution mass spectrometry for multi-residue analysis of mycotoxins and pesticides in botanical nutraceuticals. *Toxins* 12, no. 2: 114. doi:10.3390/toxins12020114.

Nieto, J.A., S. Santoyo, M. Prodanov, G. Reglero, and L. Jaime. 2020. Valorisation of grape stems as a source of phenolic antioxidants by using a sustainable extraction methodology. *Foods* 9, no. 5: 604. doi:10.3390/foods9050604.

Nollet, Leo M. L., ed. 2018. *Phenolic Compounds in Food: Characterization & Analysis. Food Analysis & Properties.* Boca Raton, FL: CRC Press, Taylor & Francis Group.

Nzekoue, F.K., G. Caprioli, M. Ricciutelli, M. Cortese, A. Alesi, S. Vittori, and G. Sagratini. 2020. Development of an innovative phytosterol derivatization method to improve the HPLC-DAD analysis and the ESI-MS detection of plant sterols/stanols. *Food Research International* 131: 108998. doi:10.1016/j.foodres.2020.108998.

Obel, N., V. Erben, T. Schwarz, S. Kühnel, A. Fodor, and M. Pauly. 2009. Microanalysis of plant cell wall polysaccharides. *Molecular Plant* 2, no. 5: 922–932. doi:10.1093/mp/ssp046.

Oliveira, B.G., H.B. Costa, J.A. Ventura, T.P. Kondratyuk, M.E.S. Barroso, R.M. Correia, E.F. Pimentel, F.E. Pinto, D.C. Endringer, and W. Romão. 2016. Chemical profile of mango (*Mangifera indica* L.) using electrospray ionisation mass spectrometry (ESI-MS). *Food Chemistry* 204: 37–45. doi:10.1016/j.foodchem.2016.02.117.

Ouchemoukh, S., N. Amessis-Ouchemoukh, M. Gómez-Romero, F. Aboud, A. Giuseppe, A. Fernández-Gutiérrez, and A. Segura-Carretero. 2017. Characterisation of phenolic compounds in Algerian honeys by RP-HPLC coupled to electrospray time-of-flight mass spectrometry. *LWT - Food Science and Technology* 85: 460–469. doi:10.1016/j.lwt.2016.11.084.

Parrilla Vázquez, P., E. Hakme, S. Uclés, V. Cutillas, M. Martínez Galera, A.R. Mughari, and A.R. Fernández-Alba. 2016. Large multiresidue analysis of pesticides in edible vegetable oils by using efficient solid-phase extraction sorbents based on quick, easy, cheap, effective, rugged and safe methodology followed by gas chromatography–tandem mass spectrometry. *Journal of Chromatography A* 1463: 20–31. doi:10.1016/j.chroma.2016.08.008.

Patisaul, H.B., and W. Jefferson. 2010. The pros and cons of phytoestrogens. *Frontiers in Neuroendocrinology* 31, no. 4: 400–419. doi:10.1016/j.yfrne.2010.03.003.

Paul, A., M. Frederich, R.C. Megido, T. Alabi, P. Malik, R. Uyttenbroeck, F. Francis, et al. 2017. Insect fatty acids: a comparison of lipids from three orthopterans and *Tenebrio molitor* L. larvae. *Journal of Asia-Pacific Entomology* 20, no. 2: 337–340. doi:10.1016/j.aspen.2017.02.001.

Pérez-Fernández, V., L. Mainero Rocca, P. Tomai, S. Fanali, and A. Gentili. 2017. Recent advancements and future trends in environmental analysis: sample preparation, liquid chromatography and mass spectrometry. *Analytica Chimica Acta* 983: 9–41. doi:10.1016/j.aca.2017.06.029.

Pérez-Míguez, R., M.L. Marina, and M. Castro-Puyana. 2019. High resolution liquid chromatography tandem mass spectrometry for the separation and identification of peptides in coffee silverskin protein hydrolysates. *Microchemical Journal* 149: 103951. doi:10.1016/j.microc.2019.05.051.

Pino Moreno, J.M., and A. Ganguly. 2016. Determination of fatty acid content in some edible insects of Mexico. *Journal of Insects as Food and Feed* 2, no. 1: 37–42. doi:10.3920/JIFF2015.0078.

Poudel, A., G. Gachumi, I. Badea, Z.D. Bashi, and A. El-Aneed. 2020. The simultaneous quantification of phytosterols and tocopherols in liposomal formulations using validated atmospheric pressure chemical ionization- liquid chromatography –tandem mass spectrometry. *Journal of Pharmaceutical and Biomedical Analysis* 183: 113104. doi:10.1016/j.jpba.2020.113104.

Regazzoni, L., F. Saligari, C. Marinello, G. Rossoni, G. Aldini, M. Carini, and M. Orioli. 2016. Coffee silver skin as a source of polyphenols: high resolution mass spectrometric profiling of components and antioxidant activity. *Journal of Functional Foods* 20: 472–485. doi:10.1016/j.jff.2015.11.027.

Ridlova, G., J.C. Mortimer, S.L. Maslen, P. Dupree, and E. Stephens. 2008. Oligosaccharide relative quantitation using isotope tagging and normal-phase liquid chromatography/mass spectrometry. *Rapid Communications in Mass Spectrometry* 22, no. 17: 2723–2730. doi:10.1002/rcm.3665.

Rietjens, I.M.C.M., J. Louisse, and K. Beekmann. 2017. The potential health effects of dietary phytoestrogens: potential health effects of dietary phytoestrogens. *British Journal of Pharmacology* 174, no. 11: 1263–1280. doi:10.1111/bph.13622.

Royle, L., T.S. Mattu, E. Hart, J.I. Langridge, A.H. Merry, N. Murphy, D.J. Harvey, R.A. Dwek, and P.M. Rudd. 2002. An analytical and structural database provides a strategy for sequencing O-glycans from microgram quantities of glycoproteins. *Analytical Biochemistry* 304, no. 1: 70–90. doi:10.1006/abio.2002.5619.

Rusinga, F.I., and D.D. Weis. 2017. Automated strong cation-exchange cleanup to remove macromolecular crowding agents for protein hydrogen exchange mass spectrometry. *Analytical Chemistry* 89, no. 2: 1275–1282. doi:10.1021/acs.analchem.6b04057.

Rusko, J., I. Pugajeva, I. Perkons, I. Reinholds, E. Bartkiene, and V. Bartkevics. 2019. Development of a rapid method for the determination of phenolic antioxidants in dark chocolate using ultra performance liquid chromatography coupled to orbitrap mass spectrometry. *Journal of Chromatographic Science* 57, no. 5: 434–442. doi:10.1093/chromsci/bmz013.

Salvatore, M.M., R. Nicoletti, F. Salvatore, D. Naviglio, and A. Andolfi. 2018. GC–MS approaches for the screening of metabolites produced by marine-derived *Aspergillus*. *Marine Chemistry* 206: 19–33. doi:10.1016/j.marchem.2018.08.003.

Santini, G., F. Bonazza, S. Pucciarelli, P. Polidori, M. Ricciutelli, Y. Klimanova, S. Silvi, V. Polzonetti, and S. Vincenzetti. 2020. Proteomic characterization of kefir milk by two-dimensional electrophoresis followed by mass spectrometry. *Journal of Mass Spectrometry*: e4635. doi:10.1002/jms.4635.

Savini, S., A. Correani, D. Pupillo, R. D'Ascenzo, C. Biagetti, A. Pompilio, M. Simonato, et al. 2016. Phytosterol esterification is markedly decreased in preterm infants receiving routine parenteral nutrition. *Lipids* 51, no. 12: 1353–1361. Doi:10.1007/s11745-016-4197-y.

Shahidi, F., and J. Yeo. 2016. Insoluble-bound phenolics in food. *Molecules* 21, no. 9: 1216. doi:10.3390/molecules21091216.

Shirzad, H., V. Niknam, M. Taheri, and H. Ebrahimzadeh. 2017. Ultrasound-assisted extraction process of phenolic antioxidants from olive leaves: a nutraceutical study using RSM and LC–ESI–DAD–MS. *Journal of Food Science and Technology* 54, no. 8: 2361–2371. doi:10.1007/s13197-017-2676-7.

Socas-Rodríguez, B., D. Lanková, K. Urbancová, V. Krtková, J. Hernández-Borges, M.Á. Rodríguez-Delgado, J. Pulkrabová, and J. Hajšlová. 2017. Multiclass analytical method for the determination of natural/synthetic steroid hormones, phytoestrogens, and *Mycoestrogens* in milk and yogurt. *Analytical and Bioanalytical Chemistry* 409, no. 18: 4467–4477. doi:10.1007/s00216-017-0391-x.

Sommer, K., and W. Vetter. 2020. Gas chromatography with mass spectrometry detection and characterization of 27 sterols in two truffle (tuber) species. *Journal of Food Composition and Analysis* 94: 103650. doi:10.1016/j.jfca.2020.103650.

Stone, J. 2017. Sample preparation techniques for mass spectrometry in the clinical laboratory. In *Mass Spectrometry for the Clinical Laboratory*, pp. 37–62. Elsevier. doi:10.1016/B978-0-12-800871-3.00003-1.

Sun, T., X. Wang, P. Cong, J. Xu, and C. Xue. 2020. mass spectrometry-based lipidomics in food science and nutritional health: a comprehensive review. *Comprehensive Reviews in Food Science and Food Safety* 19, no. 5: 2530–2558. doi:10.1111/1541-4337.12603.

SUPELCO. 2021. Guide to Solid Phase Extraction. Retrieved January 12th. 2021 from https://www.sigmaaldrich.com/Graphics/Supelco/objects/4600/4538.pdf.

Sut, S., G. Zengin, F. Maggi, M. Malagoli, and S. Dall'Acqua. 2019. Triterpene acid and phenolics from ancient apples of Friuli Venezia Giulia as nutraceutical ingredients: LC-MS study and *in vitro* activities. *Molecules* 24, no. 6: 1109. doi:10.3390/molecules24061109.

Tagliazucchi, D., S. Martini, and L. Solieri. 2019. Bioprospecting for bioactive peptide production by lactic acid bacteria isolated from fermented dairy food. *Fermentation* 5, no. 4: 96. doi:10.3390/fermentation5040096.

Tamam, B., D. Syah, M.T. Suhartono, W.A. Kusuma, S. Tachibana, and H.N. Lioe. 2019. Proteomic study of bioactive peptides from tempe. *Journal of Bioscience and Bioengineering* 128, no. 2: 241–248. doi:10.1016/j.jbiosc.2019.01.019.

Thompson, T.J., M.A. Briggs, P.J. Phillips, V.S. Blazer, K.L. Smalling, D.W. Kolpin, and T. Wagner. 2020. Groundwater discharges as a source of phytoestrogens and other agriculturally derived contaminants to streams. *Science of the Total Environment*: 142873. doi:10.1016/j.scitotenv.2020.142873.

Tian, H., H. Liu, W. Song, L. Zhu, T. Zhang, R. Li, and X. Yin. 2020. Structure, antioxidant and immunostimulatory activities of the polysaccharides from *Sargassum carpophyllum*. *Algal Research* 49: 101853. doi:10.1016/j.algal.2020.101853.

Tohma, H., A. Altay, E. Köksal, A.C. Gören, and İ. Gülçin. 2019. Measurement of anticancer, antidiabetic and anticholinergic properties of sumac (*Rhus coriaria*): analysis of its phenolic compounds by LC–MS/MS. *Journal of Food Measurement and Characterization* 13, no. 2: 1607–1619. doi:10.1007/s11694-019-00077-9.

Toral, P.G., F.J. Monahan, G. Hervás, P. Frutos, and A.P. Moloney. 2018. Review: modulating ruminal lipid metabolism to improve the fatty acid composition of meat and milk. Challenges and opportunities. *Animal* 12: s272–s281. doi:10.1017/S1751731118001994.

Uchida, H., Y. Itabashi, R. Watanabe, R. Matsushima, H. Oikawa, T. Suzuki, M. Hosokawa, et al. 2018. Detection and identification of furan fatty acids from fish lipids by high-performance liquid chromatography coupled to electrospray ionization quadrupole time-of-flight mass spectrometry. *Food Chemistry* 252: 84–91. doi:10.1016/j.foodchem.2018.01.044.

Velmurugan, R., and A. Incharoensakdi. 2018. Nanoparticles and organic matter. In *Nanomaterials in Plants, Algae, and Microorganisms*, pp. 407–428. Elsevier. doi:10.1016/B978-0-12-811487-2.00018-9.

Ventura, G., C.D. Calvano, R. Abbattista, M. Bianco, C. De Ceglie, I. Losito, F. Palmisano, and T.R.I. Cataldi. 2019. Characterization of bioactive and nutraceutical compounds occurring in olive oil processing wastes. *Rapid Communications in Mass Spectrometry* 33, no. 21: 1670–1681. doi:10.1002/rcm.851.

Vij, R., S. Reddi, S. Kapila, and R. Kapila. 2016. Transepithelial transport of milk derived bioactive peptide VLPVPQK. *Food Chemistry* 190: 681–688. doi:10.1016/j.foodchem.2015.05.121.

Waktola, H.D., A.X. Zeng, S.-T. Chin, and P.J. Marriott. 2020. Advanced gas chromatography and mass spectrometry technologies for fatty acids and triacylglycerols analysis. *TrAC Trends in Analytical Chemistry* 129: 115957. doi:10.1016/j.trac.2020.115957.

Wang, J., C. Wang, and X. Han. 2019. Tutorial on Lipidomics. *Analytica Chimica Acta* 1061: 28–41. doi:10.1016/j.aca.2019.01.043.

Wang, J., J. Lan, H. Li, X. Liu, and H. Zhang. 2017. Fabrication of diverse pH-sensitive functional mesoporous silica for selective removal or depletion of highly abundant proteins from biological samples. *Talanta* 162: 380–389. doi:10.1016/j.talanta.2016.10.003.

Wang, L., L. Chen, J. Li, L. Di, and H. Wu. 2018. Structural elucidation and immune-enhancing activity of peculiar polysaccharides fractioned from marine clam *Meretrix meretrix* (Linnaeus). *Carbohydrate Polymers* 201: 500–513. doi:10.1016/j.carbpol.2018.08.106.

Wang, M., L. Zhang, X. Wu, Y. Zhao, L. Wu, and B. Lu. 2019a. Quantitative determination of free and esterified phytosterol profile in nuts and seeds commonly consumed in China by SPE/GC–MS. *LWT 100*: 355–361. doi:10.1016/j.lwt.2018.10.077.

Wang, X., H. Zhang, Y. Song, P. Cong, Z. Li, J. Xu, and C. Xue. 2019b. Comparative lipid profile analysis of four fish species by ultraperformance liquid chromatography coupled with quadrupole time-of-flight mass spectrometry. *Journal of Agricultural and Food Chemistry* 67, no. 33: 9423–9431. doi:10.1021/acs.jafc.9b0330.

Wu, J.-S., J.-M. Li, H.-Y. Lo, C.-Y. Hsiang, and T.-Y. Ho. 2020. Anti-hypertensive and angiotensin-converting enzyme inhibitory effects of *Radix astragali* and its bioactive peptide AM-1. *Journal of Ethnopharmacology* 254: 112724. doi:10.1016/j.jep.2020.112724.

Xu, S.-Y., X. Huang, and K.-L. Cheong. 2017. Recent advances in marine algae polysaccharides: isolation, structure, and activities. *Marine Drugs* 15, no. 12: 388. doi:10.3390/md15120388.

Xue, Z., H. Wen, L. Zhai, Y. Yu, Y. Li, W. Yu, A. Cheng, C. Wang, and X. Kou. 2015. Antioxidant activity and anti-proliferative effect of a bioactive peptide from chickpea (*Cicer arietinum* L.). *Food Research International* 77: 75–81. doi:10.1016/j.foodres.2015.09.027.

Yang, Y., and C. Benning. 2018. Functions of triacylglycerols during plant development and stress. *Current Opinion in Biotechnology* 49: 191–198. doi:10.1016/j.copbio.2017.09.003.

Yao, Y.-L., C. Shu, G. Feng, Q. Wang, Y.-Y. Yan, Y. Yi, H.-X. Wang, X.-F. Zhang, and L.-M. Wang. 2020. Polysaccharides from *Pyracantha fortuneana* and its biological activity. *International Journal of Biological Macromolecules* 150: 1162–1174. doi:10.1016/j.ijbiomac.2019.10.125.

Yeboah, F.K., and Y. Konishi. 2003. Mass spectrometry of biomolecules: functional foods, nutraceuticals, and natural health products. *Analytical Letters* 36, no. 15: 3271–3307. doi:10.1081/AL-120026571.

Young, J.E., Z. Pan, H.E. Teh, V. Menon, B. Moderegger, J.J. Pesek, M.T. Matyska, L. Dao, and G. Takeoka. 2017. Phenolic composition of pomegranate peel extracts using a liquid chromatography-mass spectrometry approach with silica hydride columns. *Journal of Separation Science* 40, no. 7: 1449–1456. doi:10.1002/jssc.201601310.

Yu, X., L. Yu, F. Ma, and P. Li. 2021. Quantification of phenolic compounds in vegetable oils by mixed-mode solid-phase extraction isotope chemical labeling coupled with UHPLC-MS/MS. *Food Chemistry* 334: 127572. doi:10.1016/j.foodchem.2020.127572.

Yuan, G., Y. Liu, G. Liu, L. Wei, Y. Wen, S. Huang, Y. Guo, F. Zou, and J. Cheng. 2019. Associations between semen phytoestrogens concentrations and semen quality in Chinese men. *Environment International* 129: 136–144. doi:10.1016/j.envint.2019.04.076.

Yuan, J., N. Hashii, N. Kawasaki, S. Itoh, T. Kawanishi, and T. Hayakawa. 2005. Isotope Tag method for quantitative analysis of carbohydrates by liquid chromatography–mass spectrometry. *Journal of Chromatography A* 1067, no. 1–2: 145–152. doi:10.1016/j.chroma.2004.11.070.

Zardo, I., N.P. Rodrigues, J.R. Sarkis, and L.D. Marczak. 2020. Extraction and identification by mass spectrometry of phenolic compounds from canola seed cake. *Journal of the Science of Food and Agriculture* 100, no. 2:578–586. doi:10.1002/jsfa.10051.

Zenezini Chiozzi, R., A.L. Capriotti, C. Cavaliere, G. La Barbera, S. Piovesana, R. Samperi, and A. Laganà. 2016. Purification and identification of endogenous antioxidant and ace-inhibitory peptides from donkey milk by multidimensional liquid chromatography and nanoHPLC-high resolution mass spectrometry. *Analytical and Bioanalytical Chemistry* 408, no. 20: 5657–5666. doi:10.1007/s00216-016-9672-z.

Zhang, J., T. Zhang, G. Tao, R. Liu, M. Chang, Q. Jin, and X. Wang. 2021. Characterization and determination of free phytosterols and phytosterol conjugates: the potential phytochemicals to classify different rice bran oil and rice bran. *Food Chemistry* 344: 128624. doi:10.1016/j.foodchem.2020.128624.

Zhang, L., S. Wang, R. Yang, J. Mao, J. Jiang, X. Wang, W. Zhang, Q. Zhang, and P. Li. 2019. Simultaneous determination of tocopherols, carotenoids and phytosterols in edible vegetable oil by ultrasound-assisted saponification, LLE and LC-MS/MS. *Food Chemistry* 289: 313–319. doi:10.1016/j.foodchem.2019.03.067.

Zhang, Y., W. Wu, L. Kang, D. Yu, and C. Liu. 2017. Effect of *Aconitum coreanum* polysaccharide and its sulphated derivative on the migration of human breast cancer MDA-MB-435s cell. *International Journal of Biological Macromolecules* 103: 477–483. doi:10.1016/j.ijbiomac.2017.05.084.

Zhang, Z., Y. Chen, R. Wang, R. Cai, Y. Fu, and N. Jiao. 2015. The fate of marine bacterial exopolysaccharide in natural marine microbial communities. Ed. Antonietta Quigg. *PLoS One* 10, no. 11: e0142690. doi:10.1371/journal.pone.0142690.

Zhao, H., C.-J.-S. Lai, Y. Yu, Y. Wang, Y.-J. Zhao, F. Ma, M. Hu, J. Guo, X. Wang, and L. Guo. 2020. Acidic hydrolysate fingerprints based on HILIC-ELSD/MS combined with multivariate analysis for investigating the quality of *Ganoderma lucidum* polysaccharides. *International Journal of Biological Macromolecules* 163: 476–484. doi:10.1016/j.ijbiomac.2020.06.206.

Zhong, B., N.A. Robinson, R.D. Warner, C.J. Barrow, F.R. Dunshea, and H.A.R. Suleria. 2020. LC-ESI-QTOF-MS/MS characterization of seaweed phenolics and their antioxidant potential. *Marine Drugs* 18, no. 6: 331. doi:10.3390/md18060331.

Zhou, W., W. Branch, L. Gilliam, and J. Marshall. 2018. Phytosterol composition of *Arachis hypogaea* seeds from different maturity classes. *Molecules* 24, no. 1: 106. doi:10.3390/molecules24010106.

Zong, A., H. Cao, and F. Wang. 2012. Anticancer polysaccharides from natural resources: a review of recent research. *Carbohydrate Polymers* 90, no. 4: 1395–1410. doi:10.1016/j.carbpol.2012.07.02.

Section 3

MS Analysis of Residues

11 Pesticide Analysis in Food Samples by GC-MS, LC-MS, and Tandem Mass Spectrometry

Jeyabalan Sangeetha
Central University of Kerala

Mahesh Pattabhiramaiah and Shanthala Mallikarjunaiah
Bangalore University

Devarajan Thangadurai, Suraj Shashikant Dabire, Ravichandra Hospet, and Muniswamy David
Karnatak University

Saher Islam
Lahore College for Women University

M. Supriya, Akhil Silla, and R. Preetha
SRM Institute of Science and Technology

Inamul Hasan Madar
Kaohsiung Medical University

Ghazala Sultan
Aligarh Muslim University

Ramachandran Chelliah and Deog-Hwan Oh
Kangwon National University

Anand Torvi
Jain University

Zaira Zaman Chowdhury
University of Malaya

DOI: 10.1201/9781003091226-14

CONTENTS

11.1 Introduction 212
11.2 Significance of Mass Spectrometry in Pesticide Analysis 213
11.3 Analysis of Pesticides by GC-MS 214
11.4 Applications of GC-MS Analysis for Pesticide Determination in Different Food Samples 216
11.5 Estimation of Pesticide Residues by LC–MS 217
11.6 LC–MS/MS for Analysis of Multi-Residue Pesticides 222
11.7 Conclusion 222
References 227

11.1 INTRODUCTION

Pesticides have been extensively used in the past time to produce more enhanced and efficient food for the increasing global population. A pesticide is intentionally released into the environment to avert, deter, control, and destroy populations of insects, weeds, rodents, fungi, or other harmful pests and are classified as (i) insecticides – insects, (ii) herbicides – plants, (iii) rodenticides – rodents (rats and mice), (iv) bactericides – bacteria, (v) fungicides – fungi, and (vi) larvicides – larvae. More than 1,000 compounds and around 860 active substances are formulated in pesticide products (Tomlin 2003) to control undesirable molds, insects, and weeds (Ortelli et al. 2004).

There are more than 500 registered pesticides along with new agrochemicals released continuously worldwide to control pests. These pesticides belong to more than 100 classes, amongst which, most important groups are benzoylureas, carbamates, organophosphorus compounds, pyrethroids, dicarboximides (iprodione and vinclozolin), dinitroaniline (ethalfluralin and trifluralin), dinitrophenol (dinoseb), and dithiocarbamate (triallate), sulfonylureas or triazines. There is public concern about the pesticide residues present in the environment and food, leading to harmful effects on living systems and economic outcomes (Smith et al. 1998; Soler et al. 2007; Lee et al. 2013; Stachniuk and Fornal 2016). Pesticides result in the production of reactive oxygen species, which bring down the antioxidant levels, and in turn, damage the cellular system (Rajveer et al. 2019).

The different pesticide classes are chemically analyzed with gas chromatography (GC) and liquid chromatography (LC) integrated with mass spectrometry (MS). GC is a separation technique for volatile and thermally stable compounds, whereas LC is a separation technique for high molecular weight and non-volatile samples. The necessity for a sample to be analyzed by GC is its volatility and thermal stability, whereas in LC, it is the solubility in the mobile phase. MS is a detector, providing a mass spectrum that will be useful for characterizing the compound added to either GC or LC. GC and LC are crucial tools in today's pesticide residue analysis.

GC can quantify complex chemical mixtures containing hundreds of compounds. GC combined with MS is often attributed as GC–MS suitable for the routine analysis of samples. It has become a preferred traditional analytical technique for less polar pesticides and pesticide residues because of its sensitivity, selectivity, and single-run capability. The general application comprises pesticide analysis in food, environmental samples, and biological samples (Hird et al. 2014). GC–MS offers positive confirmation for various pesticides and pesticide residues, allowing interference-free quantification even with peak coelution. These systems can be activated using either targeted ion monitoring or untargeted full scan acquisition. In GC-MS, the compound is fragmented only once.

In GC-tandem mass spectrometry (MS/MS), a fragmented compound is fragmented further and produces MS/MS spectrum. It enables the user to precisely quantify the target by eliminating any possible matrix interference. Multiresidue pesticide analysis is validated with gas chromatography with tandem mass spectrometry (GC-MS/MS) method. It is a fast, simple, low-cost, and high-throughput technique to limit maximum residue levels (MRLs). EC 396/2005 regulation adopted MRLs for more than 500 different pesticides in over 300 different food commodities in the European Union. GC-MS, GC-MS/MS, LC-MS, LC-MS/MS, and LC-ESI-MS/MS (liquid chromatography electrospray ionization tandem mass spectrometry) are techniques most commonly employed for pesticide analysis owing to their high sensitivity and selectivity, ability to screen many pesticides from various chemical classes in a very composite matrix in a single run (Raina 2011).

In pesticide analysis, LC-MS/MS is regarded as a complementary tool to GC that is predominantly applicable for non-GC-amenable or problematic pesticides (Mol et al. 2015). LC is most preferred over GC because it is capable of detecting more substances than GC. It is estimated that GC can analyze only approximately 20%–30% of compounds, and pesticides that are quite polar, thermally labile, or not easily vaporized, are consequently and are poorly detected by GC (Masia et al. 2014). LC-MS can separate thermolabile compounds, non-volatile, and have not been derivatized, and a much wider range of chemicals, including multi-residue pesticides, can be analyzed (Núñez et al. 2012). In recent times, important chemical pesticides are being analyzed by LC-MS or LC-MS/MS method and often used in formulations with phenyl ureas, phenoxyacid herbicides, sulfonylureas, carbamates, dithiocarbamates, pyrethroids, and azoles. LC-MS/MS is widely used to detect a group of pesticides due to its selectivity and specificity in the analytical process, confirming the quality of the result and providing fast and reliable analysis of about 450 globally important pesticides in various food commodities.

The determination of pesticides by GC-MS/MS and LC-MS/MS carried out by Alder et al. (2006) revealed that LC-MS achieves a broader opportunity and better sensitivity. Limited classes of chemical pesticides, including organophosphates, carbamates, phenoxyacid herbicides, and azoles, have easily been detected by both GC and LC methods with MS. Thus, GC-MS and LC-MS will help develop new analytic tools for the characterization of a pesticide in parts per trillion levels or lower, and the detection of a pesticide residue with a very low threshold will be among the leading option in the future.

11.2 SIGNIFICANCE OF MASS SPECTROMETRY IN PESTICIDE ANALYSIS

The significance of MS-based techniques is an empowering tool for pesticide analysis. Pesticide analysis refers to chemical analysis techniques for the pesticides associated with chemical classes comprehensively analyzed with the MS-mediated methods, namely GC or LC coupled with MS and MS/MS. The utilization of the MS method for pesticide analysis varies based on the chemical constituents of the pesticides. The GC responsive chemical classes of pesticides include organochlorines, organophosphorus pesticides, pyrethroids chloroacetanilides, and triazines, which usually do not require derivatization. The pesticide classes of higher polarity, which show reduced chromatographic performance, poor stability, and ionization mode in GC-MS, the LC-MS and LC-MS/MS methods (Figure 11.1) were developed, minimizing the need for derivatization of these chemicals before proceeding for GC-MS (Raina 2015).

The multi-residues from various sample matrices are extracted by the modified Quick, Easy, Cheap, Effective, Rugged and Safe (QuEChERS) procedure, based on the amalgamation of LC-MS/MS and GC-MS techniques to regulate pesticide residues in a single solvent extract of the sample (Anastassiades et al. 2003; Payá et al. 2007; Rejczak and Tuzimski 2015). Contemporary multi-residue methods use MS/MS employing triple quadrupole (QQQ) instruments coupled with both GC and LC and cover a wide array of GC-MS/MS and LC-MS/MS amenable pesticides. For the determination of compounds at low concentration levels in highly complex matrices, high-end QQQ instruments are frequently used due to high sensitivity, selectivity, and rapid run time (Table 11.1).

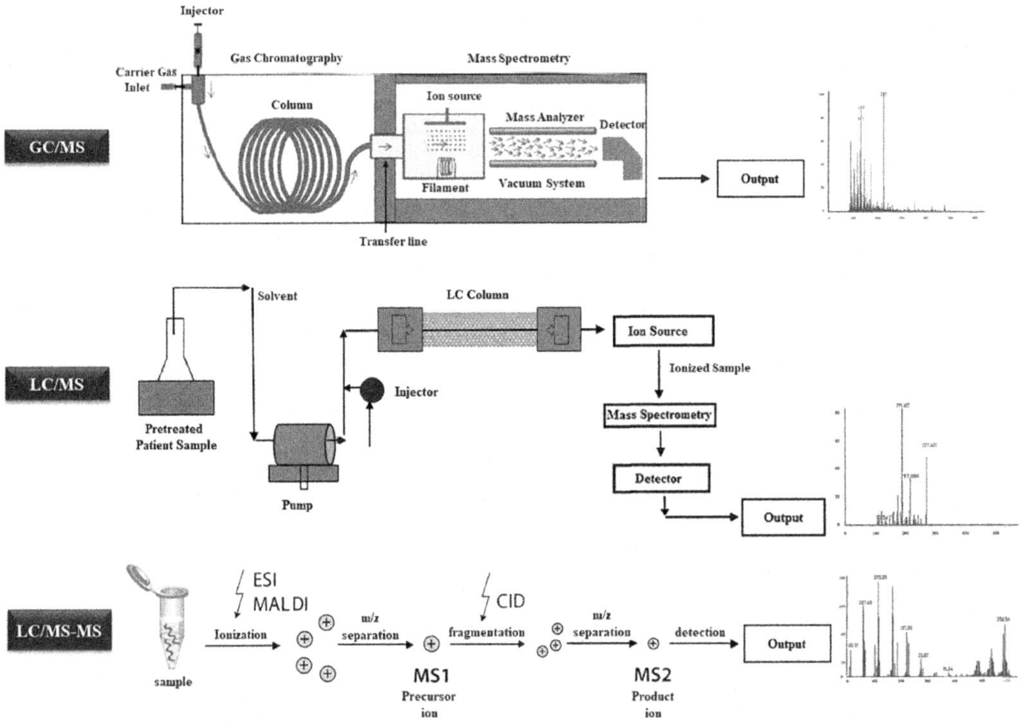

FIGURE 11.1 Mass spectrometry methods: GC-MS, LC-MS, and LC-MS/MS.

11.3 ANALYSIS OF PESTICIDES BY GC-MS

Contamination of food with pesticides has become a global concern, hence involving significant levels of regulation and monitoring. GC-MS is the most preferred approach as it provides positive confirmation of various pesticides in a single run. This technique can analyze more than 350 regulated pesticides. GC columns designed explicitly for analyzing pesticides allow the separation of compounds with high performance, reproducibility, and sensitivity. The separation of GC compounds depends on the volatility and boiling point, and MS identifies the compound with better accuracy. GC-MS technique of determining pesticides requires the selection of an appropriate ionization source. There are chances of coeluting matrix components or pesticides, causing cross-contamination and thus obstructing the analysis. As a result of this, procedures for sample preparation and clean-up are essential in addition to the selection of an appropriate ionization source (Mustapha et al., 2017). The electron impact (EI) ionization and negative chemical ionization (NCI) are the most common ionization modes reported for pesticide determination using MS (Angelika and Marek 2011). If the pesticides are electron-capturing, for example, the ones containing NO_2, halogen, or ester groups, then the pesticides will generally give an increased response with NCI compared with positive chemical ionization or EI. The ionization method selection depends on the type of pesticide residues in the sample. After preparation of the sample, the analytes must be subjected to extraction. The GC-MS technique can be applied for various compounds, including highly acidic compounds and polar pesticides (Kolberg et al. 2010). Sample extraction is carried out by the QuEchERS method with buffered acetonitrile, which is subsequently subjected to phase partitioning and includes a purification technique such as salting out with sodium acetate and magnesium sulfate. Later, cleaning is performed using d-SPE (dispersive solid-phase extraction). Solvent exchange of this acetonitrile extract to acetone/hexane is carried out for split-less injection and detection by selected ion monitoring (SIM) and electron ionization (Noelia et al. 2015; Jyotindrakumar et al. 2019).

TABLE 11.1
Detection of Pesticides in Oil Crop Samples Using GC-MS, GC-MS/MS, and LC-MS/MS-Based Approaches[a]

Technique	Commodity	Clean-Up Process	Determination Method	Analytes and LODs	Recoveries (%)	RSD (%)	Time Required (minutes)	References
QuEChERS	Flaxseeds, peanut	d-SPE with 150 mg PSA and 50 mg C-18	GC-TOF-MS	34 Pesticides: 5–300 μg/kg	22–113	<26	>5 minutes	Koesukwiwat et al. (2010)
	Avocado, almond	d-SPE clean-ups (Z-Sep, Z-Sep+, PSA+C18, and silica)	LC-MS/MS	113 Pesticides: 3–15 μg/kg	70–1200	<20	>15 minutes	Rajski et al. (2013)
	Avocado, almond	d-SPE clean-ups (Z-Sep, Z-Sep+)	GC-MS/MS	166 Pesticides: 3–15 μg/kg	28–159	<20	>15 minutes	Lozano et al. (2014)
Low temperature	Rapeseed, rapeseed oil	12 hours freezing	LC-MS/MS	27 Pesticides: 0.1–6.0 μg/kg	70–118	<27	12 hours	Jiang et al. (2012)
Fat precipitation	Peanut oil	24 hours freezing following by d-SPE with 100 mg MWCNTs and 1 g neutral alumina	GC-MS	9 Pesticides: 0.7–1.6 μg/kg	85.9–114.3	<8.84	>24 hours	Su et al. (2011)
GPC	Olive oil	Gel permeation chromatography (GPC) with ethyl acetate–cyclohexane (1:1) as mobile phase	GC–MS/MS	32 Pesticides: 0.5–20 μg/kg	89–105	<20	>35 minutes	Guardia-Rubio et al. (2006)
SPE	Sesame seeds	MAE+SPE (florisil column)	GC/MS	16 Pesticides: 1.5–3 μg/kg	86–103	<12	>20 minutes	Papadakis et al. (2006)

[a] Reproduced from Lu et al. (2015) with slight modification, © Royal Society of Chemistry.

In GC-MS analysis, the most important parameter is the ionization method, which depends on the type of analysis, i.e., targeted analysis for specific chemicals or multi-residue analysis methods to determine the pesticides (Raina 2011). The selection of ionization mode is followed by sample matrix preparation procedures, including clean-up due to chromatographically similar pesticides; those may interrupt the flow of the analysis if not identified by mass spectra (Húsková et al. 2009; Raina and Hall 2019). The chemical complexes having substances such as β-carbolines, diamines, aromatic amines, and polyamines are carcinogenic and mutagenic. Advancements in analytical approaches can estimate small quantities of toxic compounds and such approaches are useful to assess the threats for human health. The presence of 2-naphthylamine and 4-amino-biphenyl at the concentration level of 23–43 $\mu g/m^3$ analyzed by GC-MS indicates high risk and longtime exposure.

A GC-MS method determined diamines, polyamines, and aromatic amines in port wine and grape juice (Fernandes and Ferreira 2000) simultaneously. A powerful and very effective ion-pair reagent such as bis-2-ethyl phosphate dissolved in chloroform allows enriched purification of the sample extracts. The mass spectrometer in SIM mode and five distinctive internal standards a detection below 10 ng/g was possible and an observation of the derivative amines. For the quantitative assay, target ions and two qualifying ions were selected. This technique's assessment features are in good linearity over three orders of magnitude, noble accuracy with relative standard deviations (RSDs) of 0.7%–16.2%, and high recoveries of 74%–119%.

11.4 APPLICATIONS OF GC-MS ANALYSIS FOR PESTICIDE DETERMINATION IN DIFFERENT FOOD SAMPLES

Organochlorine pesticide residues were found in baby food samples and commercial fruits (Cieslik et al. 2011; Shewta and Jessica 2020). Target compounds were extracted and purified by QuEchERS methodology followed by GC-MS analysis. In this study, the limit of quantitation (LOQ) varied between 0.001 and 0.013 mg/kg (Cieslik et al. 2011). QuEchERS methodology has also been used to investigate the pesticide residues present in grapes. The single quadrupole detector was used for analyzing pesticides in grapes with GC-MS. A scientific approach based on GC was developed and validated by NCI to examine 82 pesticides in samples such as apple and cabbage, which was later performed on grape samples. Two pesticides, pyrifenox-E at 0.27 and pyridaben at 0.32 μg/kg, were detected at levels below present European MRL. The acquisition modes selected for this study were SIM and full scan mass-to-charge ratio (*m/z*) at a range of 50–600. In another study, the QuEChERS approach of extraction was subjected to a modification and applied to quantify pesticide residues in chenpi (dried pericarp of mandarin orange) by performing GC-MS/MS. Primary, secondary amine (PSA) was used in this method. PSA had some reaction on the pesticide ditalimfos that revealed a dissatisfying recovery with a higher amount of PSA because of its weak ion-exchange capacity. As a result, C18 (octadecyl-modified silica) and GCB (graphitized carbon black) were chosen, and aminopropyl was used as an alternative for PSA. This technique indicated that regular monitoring of pesticides is important since five pesticides were analyzed in eight batches of 20 real samples.

Twenty-two pesticides in tea samples were analyzed, and the QuEChERS methodology was used for extraction. Ion-trap GC-MS/MS was used for this purpose. Tea samples were made uniform by homogenizing with water, and the analyte was subjected to extraction with acetonitrile containing 1% of acetic acid. The clean-up was carried out using dSPE. It was further subjected to solvent exchange, and dSPE was used again (David et al. 2010). GC-MS was used to investigate pyrethroid and organophosphate insecticide residues in fruit juices, fruits, and vegetables. The analysis revealed that 5% or 14% of these samples eaten by children were contaminated with pesticide residues (Lu et al. 2010).

Residues of pesticides in processed foods such as frozen corned beef, Chinese dumpling, retort curry, and eel kabayaki were analyzed using an ion trap GC-MS process. The compounds were subjected to extraction using ethyl acetate–cyclohexane along with dry sodium sulfate. After the solvent evaporation, the leftover obtained was re-dissolved using *n*-hexane. Then, lipid was removed

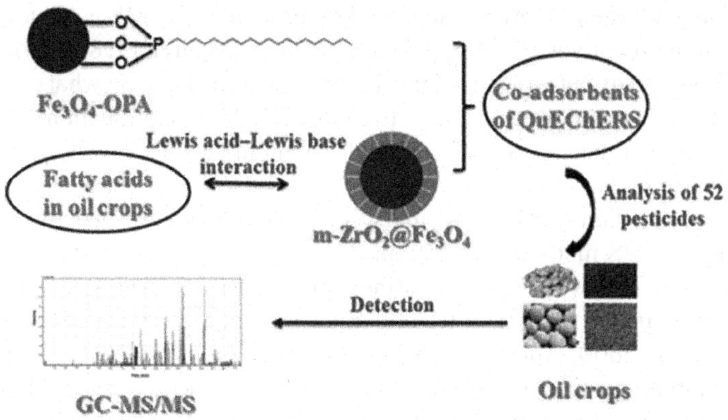

FIGURE 11.2 Determination of pesticide residues using modified QuEChERS and GC–MS/MS in oil crops. (Reproduced from Lu et al. 2015, © Royal Society of Chemistry.)

from the solution by acetonitrile-*n*-hexane partitioning. LOQs were below 0.01μg/g (Makabe et al. 2010). GPC-GC/MS was applied to analyze residues of organochlorine pesticides in milk powder. The concentration of the process was carried out utilizing a multi-position concentrator to make the technique of rotary steam and nitrogen blowing easier. This programmed online approach was utilized to perform sampling analysis along with purification. Results revealed that MDLs were found to be below 0.5 μg/kg. This technique was acceptable for quantitative analysis of pesticides in milk powder (Figure 11.2).

11.5 ESTIMATION OF PESTICIDE RESIDUES BY LC–MS

Most studies focusing on determining pesticide residues in biological samples deal with the organochlorine class of pesticides, including polychlorinated biphenyls. These determinations typically use whole blood, breast milk, serum, etc., and are also quantified in fatty tissues where these pesticides tend to bio-accumulate because of their lipophilic nature. These pesticides are routinely determined in the said matrices in their intact form using GC-based techniques (Aprea et al. 2002; Hernández et al. 2002a, b; Olsson et al. 2003; Pitarch et al. 2003; Hernández et al. 2005). The low polar strength of the majority of these pesticides makes their elimination in urine unlikely.

However, most modern generation pesticides have a polar nature. They do not persist in an intact state for long and do not tend to bio-accumulate. They are rapidly broken down in the body, and the metabolites are excreted usually within 2–3 days either freely or in the form of glucuronide/sulfate-conjugated molecules in urine (Hernández et al. 2005). Any exposure to such pesticides must be confirmed by quantifying their metabolites in urine. These pesticides may be detected in blood only immediately after a high exposure and before being metabolized (Aprea et al. 2002). Few polar organophosphorus insecticides (methamidophos and acephate) and phenoxy herbicides (2,4-dichlorophenoxyacetic acid and 2,4,5-trichlorophenoxyacetic acid) have been detected in urine intact, but most of the analyses involve metabolized products. Few reports are available on the determination of some nonpolar/medium polar pesticides [such as chlorpyrifos (Sancho et al. 2000) and acetamiprid (Marin et al. 2004)] in urine. Important considerations related to the selection of matrix (the biological sample in this case) for analysis of persistent and non-persistent pesticides, their suitability, and associated advantages and disadvantages have been previously discussed in detail (Barr et al. 1999; Olsson et al. 2003).

Reliable quantification techniques are needed to detect parent pesticides and their metabolites, which are often present in very trace amounts in complex biological matrices. Metabolites of contemporary pesticides are generally polar, non-volatile compounds, making LC-MS the most appropriate technique for their determination. At the same time, GC-MS is the technique of choice only for the determination of volatile, non-polar compounds (Maurer 1998). A good understanding of the metabolism pattern of the target pesticide is also essential to decide what compound(s) should be detected during the procedure to achieve the most reliable quantification using LC-MS. The physicochemical nature of the analyte and other interferent molecules already present in the matrix are also to be considered.

Since only polar or medium-polar compounds, capable of being ionized by the LC-MS interface, can be detected by LC-MS methods, reports on the use of this technique to determine non-polar pesticides such as pyrethroids or organochlorines are almost completely absent. GC-MS served as the most routinely applied analytical technique for pesticide bio-monitoring in the recent past (Kumazawa and Suzuki 2000; Aprea et al. 2002; Olsson et al. 2003), mainly due to the problems associated with combining LC and MS. Once the LC-MS interfaces were developed in the 1970s, the potential of this method for pesticide residue analysis has begun to be realized. After the development of two major interfaces: ESI and atmospheric-pressure chemical ionization (APCI), which are types of atmospheric pressure ionization (API) sources, LC-MS became the best alternative to GC-MS. These interfaces have proved to be successful because they can perform the ionization of the metabolites/analytes and remove the mobile-phase solvent (desolvation) simultaneously. The interface characteristics play an important role in determining the efficiency and performance of the LC-MS technique. ESI is the most widely accepted and employed interface in LC-MS-based analyses today due to its potential to efficiently ionize polar and even ionic compounds (Sancho et al. 2000). Second, there are three major mass-analysis techniques for mass separation of the molecules ionized by the interface. These are (i) quadrupole filters, (ii) quadrupole ion traps, and (iii) time-of-flight (TOF). Quadrupole filters can be further operated in two modes: (i) full scan and (ii) SIM, which detect cations or anions generated in the interface, thus being more sensitive. Quadrupole ion traps and TOF can operate in scanning mode only. The use of high-performance TOF and hybrid quadrupole TOF (QTOF) mass analyzers in both single MS and MS/MS is expected to take over contemporary methods soon (Hernández et al. 2005) (Figure 11.3).

Sensitivity or limits of detection (LOD) indicate the applicability of LC-MS for residue analysis, and this is continuously improving in newer mass spectrometers. Correctly optimized LC-MS equipment and detection methods have been employed for the final confirmation of pesticide samples tested positive by other analysis techniques (Driskell et al. 1991; Lyubimov et al. 2000; Vázquez et al. 2002). Coupling two mass analyzers together with LC is termed tandem mass spectrometry (LC-MS/

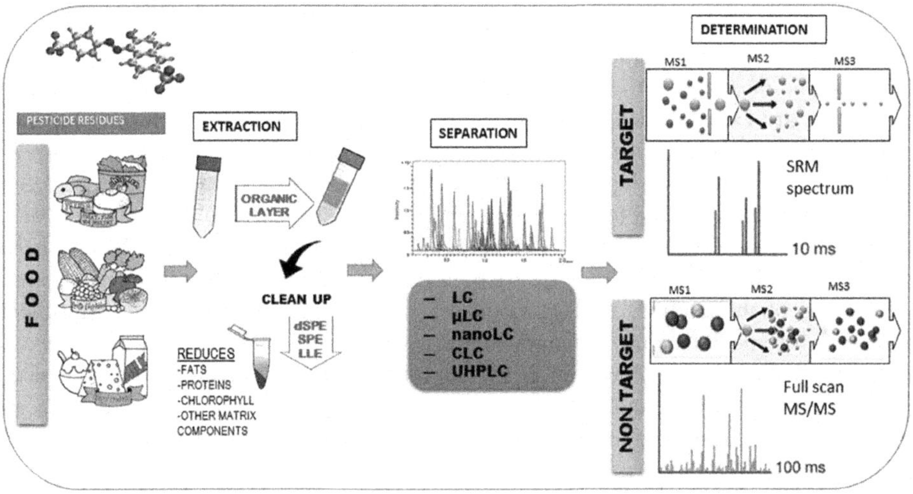

FIGURE 11.3 Determination of pesticide residues in food samples by LC-MS. (Reproduced from Masiá et al. 2016 with slight modification, © Elsevier.)

MS), and this technique provides additional potentials to the overall procedure. It is best suited for determining minute concentrations of polar analyte molecules and offers a LOD of 1 ng/mL or less. This sensitivity can be employed to monitor pesticide exposure in the general population (Aprea et al. 2002). LC-MS methods also generally eliminate the need for extensive sample handling, are compatible with aqueous samples, and facilitate direct sample injection in most cases without preconcentration or solvent exchange steps (Hernández et al. 2002a, b; Hernández et al. 2004).

LC-MS is applied for the analysis of selectable polar metabolites of pesticides in urine samples. However, urine as a matrix for LC has some disadvantages. Urine requires dilution adjustments since its formation is not strictly regulated in the body. It also presents high amounts of urea, salts, glucuronides, and sulfates, all polar, which may interfere with the analysis. This becomes a major issue to address. Although LC-MS possesses desirable qualities required for routine biological monitoring, such as its suitability for non-volatile polar/ionic compounds, direct analysis of aqueous samples, and high throughput, the use of single LC-MS for this purpose has been very limited, MS/MS being preferred for most applications in recent times (Hernández et al. 2005).

Reliable quantification of a pesticide or its metabolites also depends on the elimination of false positives. In the case of complex biological matrices, matrix components coeluting with the analytes from the LC column can interfere with the ionization process in the MS component, either suppressing or enhancing it. This is termed matrix effect (Hernández et al. 2005). The selectivity of the LC-MS technique may not always be sufficient to distinguish between the signal generated by the target analyte and that generated by an interferent in a complex matrix. The differences in urine composition among different subjects also can lead to variable matrix effects (Hernández et al. 2004). Matrix effects may either cause an analyte present in the sample to go undetected altogether or lead to its overestimation/underestimation, which can severely affect the efficiency of LC-MS. Diluting the sample has been demonstrated to help minimize the matrix effects in determining alkyl phosphates in the urine. However, dilution also demands improved sensitivity or limit of detection (Hernández et al. 2002a, b).

Fully polar or ionic pesticide metabolites are directly and freely eliminated in urine [for example, ethylene thiourea (ETU), a metabolite of ethylene bisdithiocarbamate (EBDC) fungicides, such as mancozeb (Sottani et al. 2003)], while medium-polar metabolites need to be conjugated to other moieties (normally glucuronide or sulfate) which increase the polar strength, thus facilitating elimination. Although LC-MS can quantify the intact conjugates, the direct analysis of conjugates has its limitations, mainly the limited availability of commercial reference standards (Hernández et al. 2004). Most workers quantify the target analytes (polar) after an enzymatic or acid hydrolysis procedure that releases the matrix's metabolites. Such free metabolites generally possess physicochemical properties different from interferent substances, thus enabling a coherent clean-up. The determination of free polar metabolites released after hydrolysis has been reported for organophosphorus pesticides including diazinon, malathion, parathion, chlorpyrifos, pirimiphos, for pyrethroid pesticides such as deltamethrin, permethrin cyfluthrin, cypermethrin, and for mercapturic acid conjugates of herbicide such as atrazine, acetochlor, and alachlor (Hernández et al. 2005).

The most commonly used stationary phase in LC-MS is a reversed-phase C18 column. Phenyl columns (Olsson et al. 2004) or polar-embedded columns (Sancho et al. 2002) may be used for specific applications. Care must be taken while using buffering salts or other substances in the mobile phase (generally used for better resolution). These may negatively affect the MS component, mainly while working with highly polar/ionic analytes using an electrospray interface.

Most pesticides and their metabolites present several ions in the ionization step. However, only a single target ion is generally quantified during residue analysis in biological samples by LC-MS for the sake of convenience. The quantification of the most abundant characteristic ion to detect 11 carbamates and 3 benzimidazole pesticides in several patients of different intoxication cases has been reported (Lacassie et al. 2001). Another interesting method is based on the score of identification points (IPs). A score of 3 IPs is considered sufficient to conclude the presence of a legal pesticide in biological samples (EU 2002). This implies that at least three individual ions (of the same species) should be detected with LC-MS (Table 11.2).

TABLE 11.2
Detection of Pesticide Residues in Food Samples Using LC-MS approaches[a]

			Separation		Determination	
Compounds	Matrix	Extraction/Clean-up	Column	Mobile Phase	Detection	Sensitivity ($ng\ g^{-1}/ng\ L^{-1}$)
41 Pesticides	Honey	QuEChERS	LC Narrow bore Phenomenex C_{18} (150×2.0 mm, 3 µm)	Gradient of H_2O and MeOH, both with $HCOONH_4$ 5 mM at 100 µL/min	Qtrap (API 4000, ABSciex) ESI (+), (−), MRM mode	LOQ 10
7 Pesticides	Green and black tea	QuEChERS–dSPE, AcN–dSPE/AcN–HTpSPE/high-throughput planar solid phase extraction (HTpSPE) clean-up	LC-normal-bore chromolith Performance RP-18 end-capped (100×3.0 mm)	Gradient of AcN and 10 mM $HCOONH_4$ with 2% of MeOH	QTRAP (5500, ABSciex) ESI (+) MRM mode	LOQ<2
13 Pesticides	Fish muscle	QuEChERS/dSPE	UHPLC Reversephase Zorbax Eclipse C_{18} (50×2.1 mm, 1.8 µm)	0.0125% (pH 4.04) CH_3COOH and 100% AcN	Qtrap (5500, ABSciex) ESI (+) MRM	LOQ<10
174 Pesticides	Fruit juices	QuEChERS/dSPE	UHPLC Atlantis T3 octadecyl silica (C_{18}) (2.1×100 mm, 3 µm)	(i) 4 mM $HCOONH_4$ and 0.1% HCOOH in H_2O and (ii) 4 mM $HCOONH_4$ and 0.1% HCOOH in AcN at 0.4 mL/min	Qtrap (4000, ABSciex) ESI (+)MRM	LOQ 40–500
400 Pesticides	Sugar beets and their technological product beet molasses	Comparison: MSPD QuEChERS Modified QuEChERS	LC-Normal-bore Kinetex C_{18} (100×2.1, 2.6 µm)	Gradient H_2O and MeOH, both with 0.5% HCOOH and 2mM $HCOONH_4$ at 0.5 mL/min	Qtrap (API 6500, AB Sciex) ESI (+) MRM mode	LOQ 5–10
18 Pesticides	Red wine	Comparison between (i) Membrane-assisted solvent extraction (MASE)/cleaning of the membrane bags using cyclohexane and MeOH (ii) QuEChERS/dSPE	LC-Narrow-bore RP C_{18} Aqua (50×2 mm, 5 µm, 125 Å)	Gradient: (a) 30% MeOH, 70% H_2O and B) 90% MeOH, 10% H_2O; both with 2 mM $HCOONH_4$ at 0.2 mL/min	Qtrap (API 2000, ABSciex) ESI (+) MRM mode	LOQ 3

(Continued)

TABLE 11.2 (*Continued*)
Detection of Pesticide Residues in Food Samples Using LC-MS approaches[a]

			Separation		Determination	
Compounds	**Matrix**	**Extraction/Clean-up**	**Column**	**Mobile Phase**	**Detection**	**Sensitivity ($ng\ g^{-1}$/$ng\ L^{-1}$)**
30 Pesticides	Honey	QuEChERS/dSPE	LC-normal-bore poroshell 120 EC-C_{18} (2.7 µm, 3×100 mm)	Gradient H_2O and MeOH, both with 10mM NH_4OAc at 0.4 mL/min	QQQ (6460, Agilent) ESI (+) MRM mode	LOQ 2.73–75
20 Multiclass pesticides	Cashew	QuEChERS/dSPE	LC-normal-bore ACE C_{18} (150×2.1 mm, 3 µm)	Gradient of (a) MeOH in 1 g/L HCOOH (ii) 5 mM NH4OAc in 1 g/L HCOOH at 0.3 mL/min	QQQ (TSQ Quantum Access, Thermo Scientific) ESI (+) MRM mode	LODs 0.10–0.25 LOQs 0.30–0.75
36 Pesticides	Spinach and cauliflower	QuEChERS/d-SPE with PSA and r-dSPE with MWCNTs	LC-Normal-bore ZORBAX SB-C_{18}, (50×2.1 mm, 3.5 µm)	AcN and H_2O, with 0.1% HCOOH at 0.3 mL/min	QQQ (6460, Agilent) ESI (+) MRM mode	LODs 0.1–5 LOQs 2–30
90 Pesticides	Tomato, pepper, and orange	Citrate buffered QuEChERS/ dSPE	µLC Halo C_{18} (50×0.5 mm, 2.7 µm)	Gradient: AcN and H_2O, with 0.1% HCOOH at 0.03 mL/ min	QTRAP (API 4500, ABSciex) ESI (+) MRM mode	LOQ 5
Pesticide residues	Fruit jams	QuEChERS/dSPE Extracts dilute 30- folds with AcN water	µLC Halo C_{18} (50×0.5 mm, 2.7 µm)	Gradient: AcN and H_2O, with 0.1% HCOOH at 0.03 mL/ min	QTRAP (API 4500, ABSciex) ESI (+) MRM mode	RLs 9 or 45
10 Chiral pesticides	Fruits and vegetables	QuEChERS (citrate buffered)/ dSPE	CHIRAL Lux 3u Cellulous (150×2 mm, 3 µm)	Graadient 2 mM NH_4COO in H_2O and AcN at 0.2 mL/min	QTRAP (API 4500, ABSciex) ESI (+) MRM mode (high sensitivity) EPI MS/ MS *m/z* 50–500	LODs 0.01–0.02

[a] Reproduced from Masiá et al. (2016) with slight modification, © Elsevier.

API, atmospheric pressure ionization; dSPE, dispersive solid-phase extraction; EPI, enhanced product ion; IDA, information-dependent acquisition; MRM, multiple selected reaction monitoring; MWCNTs, multiwalled carbon nanotubes; QQQ, triple quadrupole.

Precisely, LC-MS uses solvent as its mobile phase, while GC-MS uses gases in the same capacity. The LC-MS mediated methods mainly detect phenoxyacid herbicides and other pesticides, having the main chemical classes of pesticides including bromoxynil, phenyl ureas, carbamates, sulfonylureas, pyrethroids, dithiocarbamates, and azoles. LC-MS with ESI method and APCI has studied for β-carboline and toxic amine estimation in food products (Toribio et al. 2000; Diem and Herderich 2001). This method avoids the time-consuming derivatization process necessary preceding GC-MS determination. Analytical methods of LC-MS were developed for the assay of heterocyclic amines (HAs) (Toribio et al., 2000). A distinctive SPE cartridges for HAs were compared from lyophilized food extract. Comprehensive discharge of HAs from the matrix has been detected by accomplishing an alkaline treatment and establishing the required optimal time. Detection limits are 0.4–11.7 ng/g to produce binding effects. LC-MS has been used for analysis of polar pesticides in several kinds of food samples due to its efficient analysis.

11.6 LC–MS/MS FOR ANALYSIS OF MULTI-RESIDUE PESTICIDES

LC-MS/MS is one of the popular methods used for pesticide analysis, with most of the applications target non-GC amenable compounds, polar, thermolabile, and non-volatile pesticides. The chemical classes, i.e., phenoxy acid herbicides, triazines, pyrethroids, and chloroacetanilides, can be analyzed by both GC-MS and LC-MS/MS. But LC-MS/MS is considered more favorable for phenoxy acid herbicides and carbamates, which does not require a derivatization step before analysis. The LC-MS/MS is used over GC-MS for chemical classes to reduce the duration of analysis. Through the LC–MS/MS technique, various pesticides from a wide array of chemical classes can be analyzed. Also, more polar pesticides and their transformation products can be detected effectively. Transformation products that are obtained are less volatile than their base compounds. These have generally no chromatographic performance on non-polar GC-MS. Therefore, to be GC supportive, these products may need derivatization. Also, for LC-MS/MS methods, the polarity difference between base pesticide and transformation products requires various separation conditions or ionization sources for sufficient sensitivity, making simultaneous methods challenging. The LC-MS/MS for pesticide residue analysis is based on systems with atmospheric pressure ionization (API), i.e., APCI or ESI (Table 11.3).

11.7 CONCLUSION

The adaptation of advanced pesticide screening using a chromatographic analysis provides a robust and efficient platform for analyzing various pesticides. The mass spectral data allow positive confirmation and accurate quantification of complex pesticide mixture in a single analytical run. GC-MS, GC-MS/MS, and LC-MS/MS methods target a wide range of chemical classes of pesticides and their transformation products. The combinational approaches of GC and LC were necessary to target maximum pesticides such as carbamates, pyrethroids, phenoxyacid herbicides, acetanilides, triazines, sulfonylureas, and azoles. While developing GC-MS multi-residue methods, it is also necessary to have a GC-EI-MS method that is highly acceptable for acetanilides, atrazine, and triazines transformation products. It is suggested not to analyze azoles with GC-MS due to higher detection limitations when compared to LC-MS/MS methods. Complex multi-residue pesticide analysis methods were discussed using both LC and GC approaches to allow laboratory standards while maintaining required sensitivity and repeatability. A key strategy in the analysis of pesticide residues is incorporating suitable analytical procedures and the results, which can constitute the basis for making correct administrative guidelines regarding food safety. There is a need to introduce significant measures to educate and support farmers to reduce pesticide usage and switch to conventional organic farming practices in which chemical pesticides are replaced gradually by biopesticides. MS, along with GC and LC, is considered an efficient method for pesticide analysis. The screening of pesticides can be on the production status, the existence of regulations

TABLE 11.3
Determination of Pesticide Residues in Food Samples Using LC-MS Based on HRMS[a]

Compounds	Matrix	Extraction/ Clean-Up	Separation: Column	Separation: Mobile Phase	Determination: Detection	Determination: Sensitivity (ng/g, ng/L)	Determination: System
				TOF			
154 Pesticides	Soybean and spinach	AcN extraction	UHPLC ACQUITY BEH C_{18} (100×2.1 mm, 1.7 µm)	Gradient of 10 nM $HCOONH_4$ and MeOH at 0.30 mL/min	Waters LCT Premier ESI (+) (−) R > 10,000 *m/z* 50–1,000	LOQ < 10	Target and suspected analysis: identification based on accurate mass and t_R and confirmation by isotopic profile and CID
288 Pesticides	Fruits, cereals, spices, and oilseeds	Comparison: (i) Aqueous AcN and partition (ii) Aqueous AcN (iii) AcN	UHPLC Acquity UHPLC HSS T3 (100×2.1 mm, 1.8 µm)	Gradient (+): 5 mM $HCOONH_4$ and 0.2% HCOOH in H_2O and MeOH (−): 5 mM NH_4OAc in H_2O and MeOH at 0.3–0.7 mL/min	Waters, LCT Premier XE ESI (+), (−) R > 10,000 m /z 100–1,000	Not reported	Target analysis: different fragments, molecular adduct ions, different ESI polarity – manual identification
60 Pesticides	Vegetables and fruit samples	SPE/clean-up: GCB/PSA	UHPLC[R]BEH C_{18} (2.1×50 mm, 1.7 µm)	Gradient: H_2O and MeOH both 0.1% HCOOH at 0.5 mL/min	Waters TOF/MS ESI + R > 11,000 FWHM *m/z* 50–1,000	MDLs 0.3–3.8 MLOQs 0.8–11.8	Target analysis: EIC of the accurate mass with a ten ppm window
				QTOF			
50 Pesticides	Wine	Direct injection after filtering with 0.22 lm PTEE filter	LC-narrow-bore Halo C_{18} (100×2.5, 2.6 µm, 90Å)	Gradient Milli-Q H_2O with 2 NH_4OAc mM and MeOH at 20 µL/min	AB SCIEX TripleTOF 5600 ESI (+) Full-scan TOF–MS: *m/z* 100–1,000, IDA MS/MS: *m/z* 50–1000	LOQ 50–500	Target and suspected screening: against a database that includes t_R, accurate mass, isotopic pattern, and MS/MS
240 Pesticides	Food	QuEChERS	LC-narrow-bore Restek Ultra Aqueous C_{18} (100×2.1 mm, 3.0 µm)	Gradient H_2O and MeOH both 10 mM $HCOONH_4$/0.1% HCOOH	ABSciex, 5600 ESI (+) R: 30,000 FHWM Full scan *m/z* 50–800 DDA MS/MS, cps. 1,000	LODs < 10	Suspected screening: against a database (includes t_R, accurate mass, isotopic pattern, and EPI MS/MS)

(Continued)

TABLE 11.3 (*Continued*)
Determination of Pesticide Residues in Food Samples Using LC-MS Based on HRMS[a]

			Separation		Determination		
Compounds	Matrix	Extraction/ Clean-Up	Column	Mobile Phase	Detection	Sensitivity (ng/g, ng/L)	System
Isobaric pesticides	Red pepper	Extraction with MeOH/H_2O (80:20)	UHPLC (i) Zorbax C_8 (4.6×150mm, 1.8 and 3.5 µm) (ii) Zorbaz C_{18} (4.6×150mm, 1.8 µm) (iii) Phenyl column (4.6×150mm, 1.8 µm)	Gradient H_2O 0.1% HCOOH and AcN at 0.6 mL/min	Agilent 6540 ESI (+) R>25,000 FHWM Full scan *m/z* 50–1,000 DDA MS/MS (Target Precursor Ion)	Not reported Results in non-spiked samples	Target analysis: EIC with a mass window of 20 mDa, MS/MS Mass Error<2ppm
53 Pesticides	Fruit and vegetable	QuEChERS	LC-Normal-bore Agilent Zorbax Eclipse XDB C_8 (150×4.6mm, 5 µm)	Gradient AcN and H_2O both with 0.1% HCOOH at 0.6 mL/min	Agilent 6530 ESI (+) R>25000 FHWM used as TOF–MS system *m/z* 80–1,200	Study of the matrix effects Results in non-spiked samples	Suspected screening: against a database that includes t_R, accurate mass at 40 mDa window
199 Pesticides	Fruits and vegetable	QuEChERS	UHPLC ACQUITY BEH C_{18} (100×2.1mm, 1.7 µm)	Gradient H_2O and MeOH, both with ten mM NH_4OAc at 0.45 mL/min	Waters, Xevo G2-S ESI (+) R>19,000 FHWM Full scan *m/z* 50–1,200 DIA MS^E mode	SDLs: 10–50	Target screening: database of 504 pesticides (t_R, accurate mass, isotopic pattern, fragments)
>250 Pesticides	Chicken, hen, rabbit, and horse feed	A ‘dilute-and-shoot’ with H_2O and AcN (1% HCOOH) and Florisil clean-up	UHPLC Acquity UPLC BEH C_{18} (2.1×100mm, 1.7 µm)	Gradient of H_2O and MeOH both with 0.1% HCOOH/4mM $HCOONH_4$ at 0.3 mL/min	Waters, Xevo™ ESI (+), (–) R>9000 FWHM Full-scan *m/z* 90–1000 DIA MS$^{E™}$ mode	LOQs 4–200	Suspected screening: in-house database of 364 compounds (t_R, accurate mass, isotopic pattern, fragments)
				Orbitrap			
350 Pesticides	Baby food	Extraction with AcN (0.1% HCOOH)	UHPLC Hypersil GOLD aQ C_{18} (100×2.1mm, 1.7 µm)	Gradient of H_2O and MeOH both with 0.1% HCOOH and 4 Mm HCOONH at 0.3 mL/min	Thermo, Orbitrap ESI (+), (–) R>25,000 FWHM; Full scan *m/z* 70–1,000 DIA AIF mode	LODs 500–50,000 LOQs 1,000–10,000	Suspected screening: database (t_R, accurate mass) and confirmation of isotopic patterns/ fragments

(*Continued*)

TABLE 11.3 (*Continued*)
Determination of Pesticide Residues in Food Samples Using LC-MS Based on HRMS[a]

			Separation		Determination		
Compounds	Matrix	Extraction/ Clean-Up	Column	Mobile Phase	Detection	Sensitivity (ng/g, ng/L)	System
333 Pesticides	Baby food	Extraction with AcN/H_2O with 1% HCOOH. Na_2SO_4, $C_2H_3NaO_2$	LC-Narrow-bore Thermo Accucore C_{18} aQ (100×2.1 mm, 2.6 µm)	Gradient of H_2O and MeOH both with 0.1% HCOOH and 4 Mm $HCOONH_4$ at 0.3 mL/min	Thermo, QExactive ESI (+), (–) R> 70,000 FWHM Full MS *m/z* 70–1,000 DDA MS/MS R: 17,500 FWHM	CC α 0.01–5.3 CCβ 0.01–9.2	Suspected screening: database (t_R, accurate mass) and confirmation of isotopic patterns/ fragments
350 Pesticides and biopesticides	Meat (beef, pork, and chicken)	7.5 mL AcN with 1% HCOOH (v/v)	UHPLC Hypersil GOLD aQ C_{18} (100×2.1 mm, 1.7 µm)	Gradient of H_2O and MeOH both with 0.1% HCOOH and 4 Mm $HCOONH_4$ at 0.3 mL/min	Exactive™, Thermo ESI (+), (–) R: 25 000 FWHM. Full MS *m/z* 70–1,000 DIA AIF mode	LODs 2–16 LOQs 10 (few compounds 50 or 100)	Suspected screening: database (t_R, accurate mass) and confirmation of isotopic patterns/MS/MS
Pesticides	Bakery ingredients and food	Extraction with AcN with 1% CH_3COOH/dSPE with PSA and C_{18}	UHPLC Acquity UPLCTM BEH C_{18} (100×2.1 mm, 1.7 µm	Gradient of 1mM NH_4OAc and MeOH both with 0.05% CH_3COOH at 0.2 mL/min	Orbitrap Exactive™ ESI (+), (–) R: 100.000 FWHM Full scan *m/z* 150–1000	LODs 10	Suspected screening: database (t_R, accurate mass)
350 Pesticides	Honey	7.5 mL AcN with 1% HCOOH (v/v)	UHPLC Hypersil GOLD aQ C_{18} column (100×2.1 mm, 1.7 µm	Gradient of H_2O and MeOH both with 0.1% HCOOH and 4 Mm $HCOONH_4$ at 0.3 mL/min	Orbitrap Exactive™ ESI (+), (–) R: 25,000 FWHM Full scan *m/z* 100–1000 DIA AIF mode R: 10,000 FWHM	LODs 10–50	Suspected screening: database (t_R, accurate mass) and confirmation of isotopic patterns/MS/MS
Pesticides	Fruit and vegetables	QuEChERS	LC-Narrow-bore Accucore aQ C_{18} column (150×2.1 mm, 2.6 µm)	Gradient of H_2O and MeOH both with 0.1% HCOOH and five mM $HCOONH_4$ at 0.3 mL/min	Thermo QExactive™ ESI (+), (–) R: 17,500, 35,000 and 70,000 FWHM Full scan *m/z* 100–800	LODs < 50	Target analysis: comparison to the standard

(Continued)

TABLE 11.3 (*Continued*)
Determination of Pesticide Residues in Food Samples Using LC-MS Based on HRMS[a]

			Separation		Determination		
Compounds	Matrix	Extraction/ Clean-Up	Column	Mobile Phase	Detection	Sensitivity (ng/g, ng/L)	System
166 Pesticides	Fruits	QuEChERS	UHPLC column Acquity UPLC BEH C_{18} (100×2.1 mm, 1.7 μm)	Gradient of (i) 10 mM NH_4OAc in H_2O (ii) AcN at 400 μL/min	Thermo QExactive™ ESI (+) R: 70000 FWHM Full scan *m/z* 460–950 DDA, MS/MS R>1000	LODs <50	Suspected screening: database (t_R, accurate mass); confirmation by isotopic patterns/MS/MS
Pesticides	Vegetables and fruits	Acetate buffered QuEChERS (without dSPE clean up)	LC-normal-bore column Atlantis T3 (100×3 mm, 3 μm)	Gradient of H_2O and MeOH both with two mM $HCOONH_4$ and 0.002% HCOOH at 300 μL/min	Thermo QExactive™ ESI (+) R: 50,000 FWHM Full scan *m/z* 55–1,000 DDA, MS/MS	LOI 10–200	Suspected screening: 56 Pesticide database (t_R, accurate mass)
116 Pesticides	Oranges and hazelnuts	AcN extraction TFC online separation according to MW	HPLC: hypersil GOLD (2.1×100 mm, 1.9 μm	Gradient H_2O and AcN both with 0.1% HCOOH at 0.3 mL/min	Thermo QExactive™ ESI (+) R: 50,000 FWHM Full scan *m/z*: 10,000 DDA, MS/MS R: 10,000 FWHM	LOQ 1–50	Suspected screening: database (t_R, accurate mass); confirmation by isotopic patterns/MS/MS

AIF, all ion fragmentation; DDA, data-dependent acquisition; DIA, data-independent acquisition; MW, molecular weight; TFC, turbo flow chromatography; UHPLC, ultra-high pressure liquid chromatography.

[a] Reproduced from Masiá et al. (2016) with slight modification, © Elsevier.

in food, and residue detection frequency. GC-MS and LC-MS or LC-MS/MS are used for multi-residue methods for pesticides. In comparison of pesticides' applicability and sensitivity obtained with GC-MS, LC-MS, and LC-MS/MS, only for organochlorine pesticides, GC-MS achieved better performance than LC-MS or LC-MS/MS, whereas, for other classes of pesticides, the better results were observed in LC-MS-based analysis.

REFERENCES

Alder, L., Greulich, K., Kempe, G. and Vieth B. 2006. Residue analysis of 500 high priority pesticides: better by GC/MS or LC/MS/MS? *Mass Spectrometry Reviews* 25(6):838–65. doi: 10.1002/mas.20091.

Anastassiades, M., Lehotay, S., Steinbacher, D. and Schenck, F.J. 2003. Fast and easy multiresidue method employing acetonitrile extraction/partitioning and "dispersive solid-phase extraction" for the determination of pesticide residues in produce. *Journal of AOAC International* 86(2):412–31.

Angelika, W. and Marek, B. 2011. Determination of pesticide residues in food matrices using the QuEChERS methodology. *Food Chemistry* 125(3):803–12.

Aprea, C., Colosio, C. and Mammone, T., et al. 2002. Biological monitoring of pesticide exposure: a review of analytical methods. *Journal of Chromatography B* 769:191–219.

Barr, J.R., Driskell, W.J., Hill, R.H., et al. 1999. Strategies for biological monitoring of exposure for contemporary-use pesticides. *Toxicology and Industrial Health* 15(1–2):169–80. doi:10.1177/074823379901500114.

Cieslik, E., Sadowska-Rociek, A., Ruiz, J. M., et al. 2011. Evaluation of QuEChERS method for the determination of organochlorine pesticide residues in selected group of fruits. *Food Chemistry* 125(2):733–8.

David, S., Gulping, L., Jessie, B., et al. 2010. Determination of multiresidue pesticides in green tea by using a modified QuEChERS extraction and Ion-Trap Gas Chromatography/Mass Spectrometry. *Journal of AOAC International* 93(4):1169–79.

Diem, S. and Herderich, M. 2001. Reaction of tryptophan with carbohydrates: identification and quantitative determination of novel beta-carboline alkaloids in food. *Journal of Agricultural and Food Chemistry* 49(5):2486–92.

Driskell, W.J., Groce, D.F. and Hill, R.H. 1991. Methomyl in the blood of a pilot who crashed during aerial spraying. *Journal of Analytical Toxicology* 15(6):339–40. doi:10.1093/jat/15.6.339.

EU. 2002. European Commission Decision 2002/657/EC of 12 August 2002 implementing Council Directive 96/23/EC concerning the performance of analytical methods and the interpretation of results (Text with EEA relevance) (notified under document number C(2002)3044. Publications Office of the European Union, Luxembourg.

Fernandes, J.O. and Ferreira, M.A. 2000. Combined ion-pair extraction and gas chromatography-mass spectrometry for the simultaneous determination of diamines, polyamines, and aromatic amines in Port wine and grape juice. *Journal of Chromatography A* 886(1–2):183–95.

Guardia-Rubio, M., Cordova, M.L.F., Ayora-Canada, M.J. and Ruiz-Medina, A. 2006. Simplified pesticide multiresidue analysis in virgin olive oil by gas chromatography with thermionic specific, electron-capture and mass spectrometric detection. *Journal of Chromatography A* 1108:231–9.

Hernández, F., Pitarch, E., Serrano, R., et al. 2002a. Multiresidue determination of endosulfan and metabolic derivatives in human adipose tissue using automated liquid chromatographic clean-up and gas chromatographic analysis. *Journal of Analytical Toxicology* 26(2):94–103. doi:10.1093/jat/26.2.94.

Hernández, F., Sancho, J.V. and Pozo, O.J. 2002b. Direct determination of alkyl phosphates in human urine by liquid chromatography/electrospray tandem mass spectrometry. *Rapid Communications in Mass Spectrometry* 16(18):1766–73. doi:10.1002/rcm.790.

Hernández, F., Sancho, J.V. and Pozo, O.J. 2004. An estimation of the exposure to organophosphorus pesticides through the simultaneous determination of their main metabolites in urine by liquid chromatography-tandem mass spectrometry. *Journal of Chromatography B.* 808(2):229–39.

Hernández, F., Sancho, J.V. and Pozo, O.J. 2005. Critical review of the application of liquid chromatography/mass spectrometry to the determination of pesticide residues in biological samples. *Analytical and Bioanalytical Chemistry* 382:934–46. doi:10.1007/s00216-005-3185-5.

Hird, S.J., Lau, B.P.Y., Schuhmacher, R. and Krska, R. 2014. Liquid chromatography-mass spectrometry for the determination of chemical contaminants in food. *Trends in Analytical Chemistry* 59:59–72.

Húsková, R., Matisová, E., Hrouzková, S. and Svorc, L. 2009. Analysis of pesticide residues by fast gas chromatography in combination with negative chemical ionization mass spectrometry. *Journal of Chromatography A.* 1216(35):6326–34. doi:10.1016/j.chroma.2009.07.013.

Jiang, Y., Li, Y., Jiang, Y., Li, J. and Pan, C. 2012. Determination of multiresidues in rapeseed, rapeseed oil, and rapeseed meal by acetonitrile extraction, low-temperature clean-up, and detection by liquid chromatography with tandem mass spectrometry. *Journal of Agricultural and Food Chemistry* 60:5089–98.

Jyotindrakumar, J.B., Gajera, H.P., Daya, B.D., et al. 2019. Residual pesticides analysis of various vegetables by GC-MS. *Research Journal of Life Sciences, Bioinformatics, Pharmaceutical and Chemical Sciences* 5(1):53.

Koesukwiwat, U., Lehotay, S.J., Mastovska, K., Dorweiler, K.J. and Leepipatpiboon, N. 2010. Extension of the QuEChERS method for pesticide residues in cereals to flaxseeds, peanuts, and doughs. *Journal of Agricultural and Food Chemistry* 58:5950–8.

Kolberg, D.I., Prestes, O.D., Adaime, M.B., et al. 2010. A new gas chromatography/mass spectrometry/(GC-MS) method for the multiresidue analysis of pesticides in bread. *Journal of the Brazilian Chemical Society* 21(6):1065–70.

Kumazawa, T. and Suzuki, O. 2000. Separation methods for amino group-possessing pesticides in biological samples. *Journal of Chromatography B.* 747(1–2):241–54. doi:10.1016/s0378-4347(00)00117-1.

Lacassie, E., Marquet, P., Gaulier, J.M., et al. 2001. Sensitive and specific multiresidue methods for the determination of pesticides of various classes in clinical and forensic toxicology. *Forensic Science International* 121(1–2):116–25. doi:10.1016/s0379-0738(01)00461-3.

Lee, H., Park, S.S., Lim, M.S., Lee, H.S., Park, H.J., Hwang, H.S., Park, S.Y. and Cho, D.H. 2013. Multiresidue analysis of pesticides in agricultural products by a liquid chromatography/tandem mass spectrometry based method. *Food Science and Biotechnology* 22:1205–16.

Lozano, A., Rajski, L., Ucles, S., Belmonte-Valles, N., Mezcua, M. and Fernandez-Alba, A.R. 2014. Evaluation of zirconium dioxide-based sorbents to decrease the matrix effect in avocado and almond multiresidue pesticide analysis followed by gas chromatography-tandem mass spectrometry. *Talanta* 118:68–83.

Lu, C., Schenck, F.J., Pearson, M.A., et al. 2010. Assessing children's dietary pesticide exposure: direct measurement of pesticide residues in 24-hr duplicate food samples. *Environmental Health Perspectives* 118(11):1625–30.

Lu, W., Peng, X., Shen, J., Hu, X., Peng, L. and Feng, Y. 2015. Development and validation of a modified QuEChERS method based on magnetic zirconium dioxide microspheres for the determination of 52 pesticides in oil crops by gas chromatography-tandem mass spectrometry. *Analytical Methods* 7:8663–72.

Lyubimov, A.V., Garry, V.F., Carlson, R.E., et al. 2000. Simplified urinary immunoassay for 2,4-D: validation and exposure assessment. *Journal of Laboratory and Clinical Medicine* 136:116–124.

Makabe, Y., Miyamoto, F., Hashimoto, H., et al. 2010. Determination of residual pesticides in processed foods manufactured from livestock foods and foods using ion trap GC/MS. *Journal of the Food Hygiene of Japan* 51(4):182–95.

Marin, A., Martinez Vidal, J.L., Egea Gonzalez, F.J., et al. 2004. Assessment of potential (inhalation and dermal) and actual exposure to acetamiprid by greenhouse applicators using liquid chromatography-tandem mass spectrometry. *Journal of Chromatography B.* 804:269–75.

Masia, A., Blasco, C. and Pico, Y. 2014. Last trends in pesticide residue determination by liquid chromatography-mass spectrometry. *Trends in Analytical Chemistry* 2:11–24.

Masiá, A., Suarez-Varela, M.M., Llopis-Gonzalez, A. and Picó, Y. 2016. Pesticides and veterinary drug residues determination in food by liquid chromatography-mass spectrometry: A review. *Analytica Chimica Acta* 936:40–61.

Maurer, H.H. 1998. Liquid chromatography-mass spectrometry in forensic and clinical toxicology. *Journal of Chromatography B.* 713(1):3–25. doi:10.1016/s0378-4347(97)00514-8.

Mol, H.G.J., Zomer, P., García, L.M., Fussell, R.J., Scholten, J. and de Kok, A. 2015. Identification in residue analysis based on liquid chromatography with tandem mass spectrometry: experimental evidence to update performance criteria. *Analytica Chimica Acta* 873:1–13.

Mustapha, F.A.J., Dawood, G.A., Mohammed, S.A., et al. 2017. Monitoring of pesticide residues in commonly used fruits and vegetables in Kuwait. *International Journal of Environmental Research and Public Health* 14(8):833.

Noelia, B.V., Samanta, V., Natalia, B., et al. 2015. Analysis of pesticide residues in fruits and vegetables using gas-chromatography-high resolution time-of-flight mass spectrometry. *Analytical Methods* 7(5):2162–71.

Núñez, O., Gallart-Ayala, H., Ferrer, I., Moyano, E. and Galceran, M.T. 2012. Strategies for the multi-residue analysis of 100 pesticides by liquid chromatography-triple quadrupole mass spectrometry. *Journal of Chromatography A* 1249:164–80.

Olsson, A.O., Baker, S.E., Nguyen, J.V., et al. 2004. A liquid chromatography–tandem mass spectrometry multiresidue method for quantification of specific metabolites of organophosphorus pesticides, synthetic pyrethroids, selected herbicides, and DEET in human urine. *Analytical Chemistry* 76(9):2453–61. doi:10.1021/ac0355404.

Olsson, A.O., Nguyen, J.V., Sadowski, M.A., et al. 2003. A liquid chromatography/electrospray ionization–tandem mass spectrometry method for quantification of specific organophosphorus pesticide biomarkers in human urine. *Analytical and Bioanalytical Chemistry* 376:808–15. doi:10.1007/s00216-003-1978-y.

Ortelli, D., Edder, P. and Corvi, C. 2004. Multiresidue analysis of 74 pesticides in fruits and vegetables by liquid chromatography-electrospray-tandem mass spectrometry. *Analytica Chimica Acta* 520(1–2): 33–45.

Papadakis, E.N., Vryzas, Z. and Papadopoulou-Mourkidou, E. 2006. Rapid method for the determination of 16 organochlorine pesticides in sesame seeds by microwave-assisted extraction and analysis of extracts by gas chromatography-mass spectrometry. *Journal of Chromatography A* 1127:6–11.

Payá, P., Anastassiades, M., Mack, D., Sigalova, I., Tasdelen, B., Oliva, J. and Barba, A. 2007. Analysis of pesticide residues using the Quick Easy Cheap Effective Rugged and Safe (QuEChERS) pesticide multiresidue method in combination with gas and liquid chromatography and tandem mass spectrometric detection. *Analytical and Bioanalytical Chemistry* 389(6):1697–714. doi:10.1007/s00216-007-1610-7.

Pitarch, E., Serrano, R., López, F.J., et al. 2003. Rapid multiresidue determination of organochlorine and organophosphorus compounds in human serum by solid-phase extraction and gas chromatography coupled to tandem mass spectrometry. *Analytical and Bioanalytical Chemistry* 376:189–97.

Raina, R. 2011. Chemical analysis of pesticides using GC/MS, GC/MS-MS and LC/MS/MS. In *Pesticides – Strategies for Pesticides Analysis*, InTech Open: Rijeka. doi:10.5772/13242.

Raina, R. 2015. Determination of neonicotinoid insecticides and strobilurin fungicides in particle phase atmospheric samples by liquid chromatography-tandem mass spectrometry. *Journal of Agricultural and Food Chemistry* 63(21):5152–62. doi:10.1021/acs.jafc.5b01347.

Raina, R. and Hall, P. 2019. Comparison of gas chromatography-mass spectrometry with electron ionization and negative-ion chemical Ionization for analyses of pesticides at trace levels in atmospheric samples. *Analytical Chemistry Insights* 3:111–25.

Rajski, L., Lozano, A., Ucles, A., Ferrer, C. and Fernandez-Alba, A.R. 2013. Determination of pesticide residues in high oil vegetal commodities by using various multi-residue methods and clean-ups followed by liquid chromatography-tandem mass spectrometry. *Journal of Chromatography A* 1304:109–20.

Rajveer, K., Gurjot, K., Mavi. and Shweta, R. 2019. Pesticides classification and its impact on the environment. *International Journal of Current Microbiology and Applied Sciences* 8(3): 1889–97.

Rejczak, T. and Tuzimski, T. 2015. A review of recent developments and trends in the QuEChERS sample preparation approach. *Open Chemistry* 13:980–1010.

Sancho, J.V., Pozo, O.J., Hernández, F. 2000. Direct determination of chlorpyrifos and its main metabolite 3,5,6-trichloro-2-pyridinol in human serum and urine by coupled-column liquid chromatography/electrospray-tandem mass spectrometry. *Rapid Communications in Mass Spectrometry* 14:1485–90.

Sancho, J.V., Pozo, O.J., Lopez, F.J. and Hernández, F. 2002. Different quantitation approaches for xenobiotics in human urine samples by liquid chromatography/electrospray tandem mass spectrometry. *Rapid Communications in Mass Spectrometry* 16(7):639–45.

Shewta, B. and Jessica, G. 2020. Investigating the presence of pesticides in fruits and vegetables by the chromatographic techniques: a review. *Journal of Applied Pharmaceutical Research* 8(4): 1–9.

Smith, M.E., van, E.O., Ravenswaay, and Thompson, R.S. 1998. Sales loss determination in food contamination incidents: an application to milk bans in Hawaii. *American Journal of Agricultural Economics* 70(3): 513–20.

Soler, C., James, K.J. and Pico, Y. 2007. Capabilities of different liquid chromatography-tandem mass spectrometry systems in determining pesticide residues in food. Application to estimate their daily intake. *Journal of Chromatography A* 1157:73–84.

Sottani, C., Bettinelli, M., Fiorentino, M.L. and Minoia, C. 2003. Analytical method for the quantitative determination of urinary ethylene thiourea by liquid chromatography/electrospray ionization tandem mass spectrometry. *Rapid Communications in Mass Spectrometry* 17:2253–9.

Stachniuk, A. and Fornal, E. 2016. Liquid chromatography-mass spectrometry in the analysis of pesticide residues in food. *Food Analytical Methods* 9:1654–65.

Su, R., Xu, X., Wang, X.H., Li, D., Li, X.Y., Zhang, H.Q. and Yu, A.M. 2011. Determination of organophosphorus pesticides in peanut oil by dispersive solid-phase extraction gas chromatography-mass spectrometry. *Journal of Chromatography B* 879:3423–8.

Tomlin, C.D.S. 2003. *The Pesticide Manual – A World Compendium*, 13th edn. British Crop Protection Council (BCPC): Hampshire.

Toribio, F., Moyano, E., Puignou, L. and Galceran, M. T. 2000. Comparison of different commercial solid-phase extraction cartridges used to extract heterocyclic amines from a lyophilised meat extract. *Journal of Chromatography A* 880(1–2):101–12.

Vázquez, P.P., Fernández, J.M. and Gil García, M.D. 2002. Determination of azole pesticides in human serum by coupled column reversed-phase liquid chromatography using ultraviolet absorbance and mass spectrometric detection. *Journal of Liquid Chromatography and Related Technologies* 25(19):3045–58. doi:10.1081/JLC-120015890.

12 Mass Spectrometry of Food Contact Materials

Deepthi Eswar
Government of Tamil Nadu

Selvi Chellamuthu
Tamil Nadu Agricultural University

CONTENTS

12.1 Introduction 231
12.2 Food Contact Materials 233
12.2.1 Plastics 233
12.2.2 Paper and Cardboard 233
12.2.3 Metals and Alloys 233
12.2.4 Ceramics 235
12.2.5 Glass and Cellulose 235
12.2.6 Rubber 235
12.2.7 Inks and Photoinitiators 236
12.2.8 Biomaterials – Wood and Fibre 236
12.3 Mass Spectrometry Methods 236
12.3.1 Gas Chromatography-Mass Spectrometry 236
12.3.2 Liquid Chromatography-Mass Spectrometry 240
12.3.2.1 Ultra-High-Performance Liquid Chromatography-Mass Spectrometry 241
12.3.3 Inductively Coupled Plasma Mass Spectrometry (ICP-MS) 241
12.3.4 Other Mass Spectrometry Techniques 241
12.4 Conclusion 242
References 242

12.1 INTRODUCTION

Food contact materials (FCMs) are materials in contact with food during production, processing, storage and food preparation before being consumed. Various materials such as packaging materials, including plastics, paper and cardboard, metals, glass, tableware, kitchenware, processing machines and storage containers, are classified as FCMs (EC No. 1935/2004). The submicroscopic process in which these materials release minuscule amounts of their constituents when they come in contact with food is termed 'migration'. Migration of chemicals into food is of concern in terms of both food safety and food quality. The migrated chemical constituents can be harmful to human health when ingested regularly or in large quantities (Hahladakis et al., 2018) and reduce consumer appeal by impairing the colour or odour of food substances.

Chemical migration of FCMs into food depends on the nature of food substance, size and shape of packaging materials, and the extent of contact and temperature. EU Commission Regulation (2011) defines two types of migration limits for FCMs based on their behaviour and toxicity – specific migration limit (SML) and overall migration limit. The former specifies the maximum

DOI: 10.1201/9781003091226-15

amount of permitted migration of a given substance into food, and the latter specifies the maximum amount of non-volatile substances migration permitted. They are expressed as mg of substance per kg of food. The common FCMs such as plastics, paper and cardboard, metals, glass, ceramics, cellulose, rubber, wax and wood materials (Figure 12.1) are discussed further in this chapter.

The effects of FCMs on human health are yet to be sufficiently researched. It is generally considered to pose negligible risks assuming the exposure to such chemicals as limited. FCMs can be classified as (i) intentionally added substances (IASs) and (ii) non-intentionally added substances (NIASs). The effects of the latter are difficult to determine as identification is strenuous and they are not available as pure chemicals making the testing process expensive (Muncke et al., 2020). It is estimated that nearly 12,000 IASs and up to 100,000 NIASs from FCMs migrate into food (McCombie 2018; Groh et al., 2020). Generally, impurities of raw substances, reaction byproducts during the supply chain, or break-down products are classified as NIASs (Koster et al., 2014). Grob et al. (2006) reported that the concentration of chemical constituents in FCMs exceeds the concentration of other residues like pesticides in food. Also, food packaging is the major source of exposure to plasticisers in humans (Biryol et al., 2017). The migration of chemicals from FCMs is usually evaluated individually in residue studies, but migration in mixtures needs to be considered.

Mass spectrometry (MS) offers high sensitivity, selectivity and less response time, and hence, it is one of the widely used technologies for foods and FCMs analysis (Klampfl, 2013). Mass spectrometers possess three main components: an ion source, mass analyser and detector system. The commonly available mass spectrometer analysers are quadrupole (Q), time-of-flight (TOF), ion-trap (IT) and magnetic mass spectrometer. They are also used in combinations such as Q-TOF, IT-TOF, or triple quadrupole (QQQ), but using tandem mass spectrometry (MS/MS) is expensive when compared to using analysers individually. The selection of methods of separation and final determination mainly depends on the target

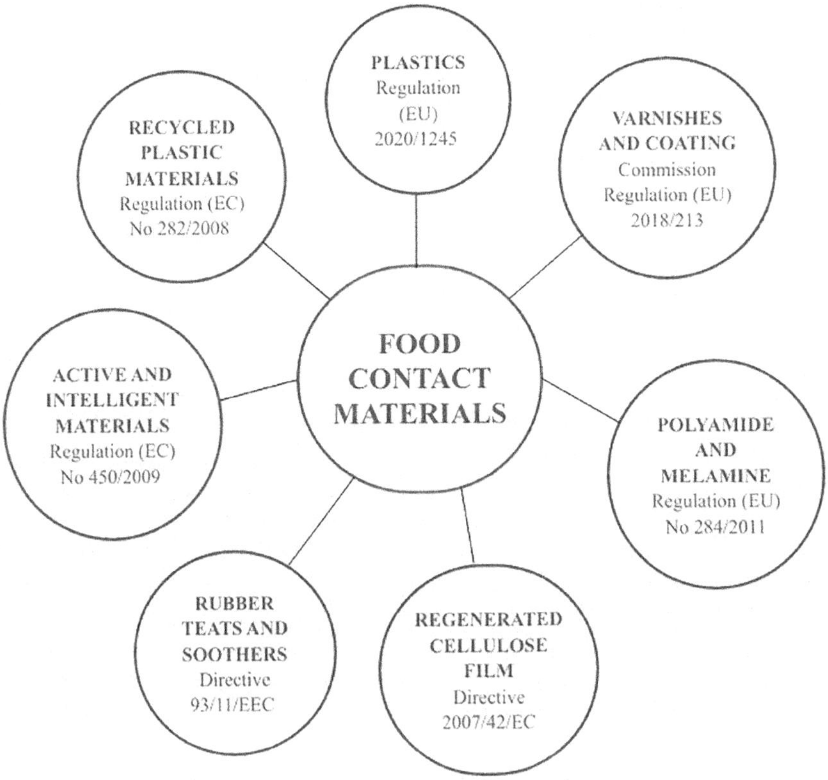

FIGURE 12.1 Food contact materials – types and legislations. (EURL Food Contact Materials – Legislations, https://ec.europa.eu/jrc/en/eurl/food-contact-materials/legislation [Accessed 29 December 2020].)

compounds to be analysed. The analytes can be separated using gas chromatography (GC), suitable for less polar and more volatile compounds, and liquid chromatography (LC) is suitable for less volatile compounds with a higher polarity such as NIASs, perfluorinated compounds and primary aromatic amines. However, bisphenol A (BPA) compounds separation requires both gas and LC (Sanchis et al., 2017). The final determination can be achieved using MS-MS/MS. This chapter outlines the various MS techniques available for the determination of FCMs.

12.2 FOOD CONTACT MATERIALS

12.2.1 Plastics

Plastics are popularly used FCMs that offer advantages such as protecting food from damage and preserving freshness. Global plastic production is estimated to be around 359 million tonnes in 2018, and nearly 40% of the plastics produced are used for packaging purposes (Plastics Europe, 2019). There are many types of commonly used plastics (Table 12.1), and numerous compounds are used in the manufacturing of these polymers either intentionally or unintentionally. The commonly used additives in plastics are plasticisers, flame retardants, stabilisers and antioxidants, curing agents and colourants. The migration of chemicals that are used in plastic production and as additives or plasticisers have been widely researched. In recent years, nanomaterials have been developed for use as additives in polymers for manufacturing food-packaging materials. The migration of these nanomaterials and migration of these compounds into food is of huge concern. Food products such as apple, cheese, butter, milk powder, and juice packed in nanosilver polyethene containers and nanosilver-coated polyethene cling film were investigated for migration by Metak et al. (2015). They reported 29 ppb of silver migration from the cling film into food after being stored at 400°C for 10 days.

12.2.2 Paper and Cardboard

Paper and cardboard are widely used to packaging dry food or as secondary packing due to their availability, renewability, cost-effectiveness and degradation capacity. Paper packaging materials contain chemical substances originating from printing inks, adhesives, plasticisers, photoinitiators or mineral oils used to manufacture paper and additives (Geueke et al., 2018). Additionally, several NIASs can be present in recycled paper and may migrate into food when used for packaging. Several studies (Van Den Houwe et al., 2016; Jung and Simat, 2014, Cai et al., 2017; Xue et al., 2019) have reported the migration of photoinitiators from recycled paperboard into dry foods including cereals, biscuits, rice, noodles and pasta and fatty foodstuffs and migration of phthalates and BPA in vegetables, ice-cream, chocolate, eggs and fruits. Naphthalene, diisobutyl phthalate, methyl hexadecanoate, 2,6-diisopropyl naphthalene (Vápenka et al., 2016) and pentadecanoic acid (Blanco-Zubiaguirre et al., 2020) were reported to migrate from paper-based packaging into food. Vavrouš et al. (2016) reported 0.43 and 31 mg/kg DiBP migration from paper packages of wheat flour and bakery products, respectively.

12.2.3 Metals and Alloys

Metals are used in alloys to make processing equipment and household utensils, which are FCMs. The commonly used metal alloys are stainless steel (iron and chromium), brass (copper and zinc), bronze (copper and tin) and German silver (copper, nickel and zinc). Metals are used as packaging material in the form of food and beverage cans, ends, closures (either steel or aluminium), foils and layers, cutlery and trays. Metal food-packaging materials offer the advantages of long-term storage stability, resistance to abuse, extended shelf-life and safety from external contamination. The metal tins and cans that are used to pack beverages or food are mostly heat-processed after filling to ensure product stability and prevent microbial contamination providing an opportunity for migration of metals into the packed food (Abdel-Rahman, 2015). Trace metals such as tin, lead, chromium, cadmium and nickel

TABLE 12.1
Types of Plastic Food-Packaging Materials

				Migration into Food			References
S. No.	Type of Plastic	Characteristics	Usage	Food Contact Material	Compound	Reported Findings	
1	Polyethene terephthalate (PETE or PET)	Lightweight; semi-rigid or rigid; impact resistant	Packaging for water, soft drinks, ketchup and vegetable oil	Bottled water – stored at −18 up to 40°C, for 24 hours, 10, 30 and 45 days	Dibutyl phthalate (DBP) n-Butyl benzyl phthalate (BBP) Di(2-Ethylhexyl) phthalate (DEHP)	79–303 ng/L Bdl −13 ng/L 217–917 µg/L	Jeddi et al. (2015)
2	High-density polyethene (HDPE)	Hard; opaque; lightweight; strong	Packaging for juice and milk, vinegar, chocolate syrup and grocery bags	Water exposed to polyolefin bottles at ambient temperature for 72 hours	Isophorone 2,4-di-tert-butylphenol	4 µg/L 25 µg/L	Skjevrak et al. (2005)
3	Polyvinyl chloride (PVC)	Biologically and chemically resistant	Packaging for over-the-counter medications, breath-mints or gums	Hot fills in baby bottles at 70°C and 2 hours contact time	Diisobutyl phthalate (DiBP) and Dibutyl phthalate (DBP)	50–150 µg/kg	Simoneau et al. (2012)
				Three types of local cheeses (Feta, Gouda and Cheddar) stored at 5°C for 14 days	Di-(2-Ethylhexyl) adipate (DEHA)	Feta −97.3 mg/kg; Gouda −134.3 mg/kg and cheddar −194.3 mg/kg after 14 days	Aly (2016)
4	Low-density polyethene (LDPE)	Thin; high heat resilience; tough; flexible	Coffee can lids, bread bags, fruit and vegetable bags	Posho wrapped in black polyethylene bag heated at 65°C, 80°C and 95°C for 5 hours	Lead (Pb) Cadmium(Cd) Chromium (Cr) Cobalt (Co)	120.60 ppm 12.25 ppm 9.45 ppm 15.42 ppm at 95°C after 5 hours of heating	Musoke et al. (2015)
5	Polypropylene (PP)	Stiff; less brittle; high melting point	Containers for yoghurt, maple syrup, cream cheese and sour cream and coffee capsules	Coffee capsules kept at 60°C in the oven for 10 days	N,N-Bis(2-hydroxyethyl) trydecylamine	40–1,050 µg/kg	Otoukesh et al. (2020)
6	Polystyrene (PS)	Colourless; hard	Used to make plastic cups, bakery trays, fast food containers and lids, and egg cartons	yoghurt, cookies, candies	Styrene monomer	2.6–163 ng/g	Genualdi et al. (2014)

from metal containers often migrate into food. Arvanitoyannis and Kotsanopoulos (2014) reported migration of trace levels of tin in beverages packed in tinplate cans. In addition to trace metals, epoxy resins containing bisphenol compounds that are used to coat food cans, migrate into food of aqueous nature. Bisphenol A diglycidyl ether (BADGE) is the common bisphenol compound formed through hydrolysis when the coating comes in contact with food (Wagner et al., 2018). Several factors such as the nature of food, type of metal, coating and lacquer properties influence the migration process (Buculei et al., 2014; Parkar and Rakesh, 2014; Wagner et al., 2018).

12.2.4 Ceramics

Ceramic plates, casseroles, stewing pottery and mugs, generally used for preparation and serving of food, release metals especially lead, when they come in contact with food (Jones et al., 2013; Lynch et al., 2008). The main source of metals for migration into food is from the corrosion of ceramic glaze and ion exchange in the presence of water molecules. The migration of metals from ceramics depends on the glaze composition, contact temperature and duration of contact. Lehman (2002) reported that adding alkali oxides (sodium oxide and potassium oxide) and copper as pigment increase the solubility of lead in food. Ceramicware, when exposed to temperature over 600°C, exhibits dissolution of the ceramic network, causing migration of metals (Beldì et al., 2016). The migration limits of lead and cadmium from hollow ceramic ware are 1.5 and 0.1 mg/L, respectively, and for flatware 0.8 mg/dm^2 (lead) and 0.07 mg/dm^2 (cadmium) (Council of the European Communities, 1984). Other metals such as aluminium, iron, calcium and magnesium also migrate from ceramicware into food. Cakste et al. (2017) reported migration of iron and aluminium when ceramic ware containing acid food stimulant was exposed to 20 ± 2 °C for more than 60 minutes and migration of calcium and magnesium at 180°C±5°C after 90 minutes of exposure.

12.2.5 Glass and Cellulose

The migration of alkali metals from glass jars and plasticisers from gasket of glass jar lids into food has been reported (McCombie et al., 2012). The glass surface releases alkali when it comes in contact with water molecules at elevated pressure and temperature. The main alkali metals that migrate are sodium, potassium and lithium. Similarly, plasticisers from gaskets of glass jar lids migrate into food. Hammarling et al. (1998) reported the migration of epoxidised soya bean oil (2 mg/kg), a common plasticiser into baby food stored in glass jars. Elss et al. (2004) studied glass jars used for storing baby food and fruit juices and reported the migration of 2-ethyl hexanoic acid at the range of 0.25–3.2 mg/kg in baby foods and 0.01–0.59 mg/L in fruit juices. Regenerated cellulose films have been widely used to packaging dry foods such as flour, biscuits, tea, coffee and cheeses. De Potter (1987) reported the migration of plasticisers such as dibutyl phthalate and dicyclohexyl phthalate (DCHP) from regenerated cellulose film packed cakes, biscuits, chocolates and meat.

12.2.6 Rubber

Usage of rubber as food contact material can be observed in the dairy industry, brewery, food canning industries and meat packaging. The types of rubber commonly used are natural rubber, nitrile rubber, fluorocarbon rubber and silicone rubber. Silicone rubber is widely used to make baking cans, pressure cooker seals, cake moulds, baby bibs, spoons and bowls, as it is flexible, has nonstick property and withstands high temperature (Helling et al., 2009; Mojsiewicz-Pieńkowska et al., 2016). Liu et al. (2020) reported migration of octamethylcyclotetrasiloxane, decamethylcyclopentasiloxane, dodecamethylcyclohexasiloxane from silicone rubber baking mould into food stimulants (olive oil and Tenax®) and cakes. Also established that these compounds migrate easily to oily foods due to their lipid affinity. Bouma and Schothorst (2003) identified potential migrants such as alkanes, alkenes and nitrosamines in natural rubber meat nettings using GC-MS.

12.2.7 Inks and Photoinitiators

Inks are used for printing on the outer side of FCMs, and among them, ultraviolet (UV) curable inks are the most commonly used. They are a mixture of resins or oligomers, additives, pigments and photoinitiators, which absorb UV energy and form chain reactions (Lemos et al., 2017). Cai et al. (2017) and Rosenmai et al. (2017) noticed that inks migrate easily into paper and cardboard packaging owing to their porous nature. Asensio et al. (2019) reported the migration of volatile compounds from printing inks, low-density polyethene (LDPE) coating and cardboard used for making gourmet cardboard cups. In addition, photoinitiators used in recycled paperboard have been reported to migrate into dry food (cereals, noodles, pasta or oat flakes) by Van Den Houwe et al. (2016), and fatty foodstuffs (milk powder and beverages) by Cai et al. (2017).

12.2.8 Biomaterials – Wood and Fibre

There is a growing interest in developing sustainable food-packaging options, and biodegradable materials such as wood, bamboo and vegetable fibres are being explored to make food-packaging materials. Although they have the advantage of environmental sustainability, their barrier properties, and ability to conserve the quality and safety need to be researched further (Calva-Estrada et al., 2019). Asensio et al. (2020) studied the migration of compounds from dishes made from wheat pulp and wood when they come in contact with liquid food simulants such as ethanol (10%), acetic acid (3%) and ethanol (95%) and established that the migration was below the SMLs.

12.3 MASS SPECTROMETRY METHODS

Identifying migrants from FCMs is complex as most migrant compounds are present in very low concentrations, not included in chemical databases. Information on ingredients used for manufacturing is not readily available. The potential chemical migrants from FCMs include antioxidants, oligomers, nanoparticles, IASs and NIASs. For identifying and quantifying these migrant compounds, GC and LC methods coupled with mass spectrometry have been widely used. MS is an analytical technique that can be used to quantify known compounds and identify unknown compounds (Llopis-Gonzalez and Pico, 2016) based on the principle of conversion of the sample into gaseous ions. The ions are then identified and characterised depending on their mass-to-charge ratio. Mass spectrometers possess three main components: an ion source, mass analyser and detector system (Figure 12.2). The identification and quantification of known compounds using data libraries can be achieved using low-resolution mass analysers such as single quadrupole and QQQ or IT (Kühne et al., 2018). In contrast, identification of NIASs or unknown migrants requires high-resolution mass analysers such as TOF or Orbitrap (Gómez-Ramos et al., 2019). The commonly used food simulants for food contact material analysis are 95% ethanol, 3% acetic acid (liquid) and Tenax® (solid). Solid-phase microextraction (SPME) followed by GC-MS method and ultra-high performance liquid chromatography (UHPLC) with Q-TOF-MS is the extensively used methods of analysis for food stimulants such as ethanol (10%, 20%, 50% and 95%), acetic acid (3%) and Tenax® (Aznar et al., 2019; Ubeda et al., 2019; Song et al., 2019; Da Silva Oliveira et al., 2019). Various MS techniques used to determine migrants from FCMs are listed in Table 12.2.

12.3.1 Gas Chromatography-Mass Spectrometry

GC-MS is a commonly used analytical method for identifying and quantifying volatile organic compounds (VOCs) and odorous compounds in FCMs. The widely used mass spectrometer combinations are GC-MS, atmospheric pressure gas chromatography (APGC)-Q-TOF with alternating low-energy collision-induced dissociation and high-energy collision-induced dissociation (MSE) and GC-Q-Orbitrap-MS. These instruments are highly sensitive and allow detecting analytes at trace levels (Lucera et al., 2019; Cruz et al., 2019). Typically, GC-MS uses electron impact ionisation

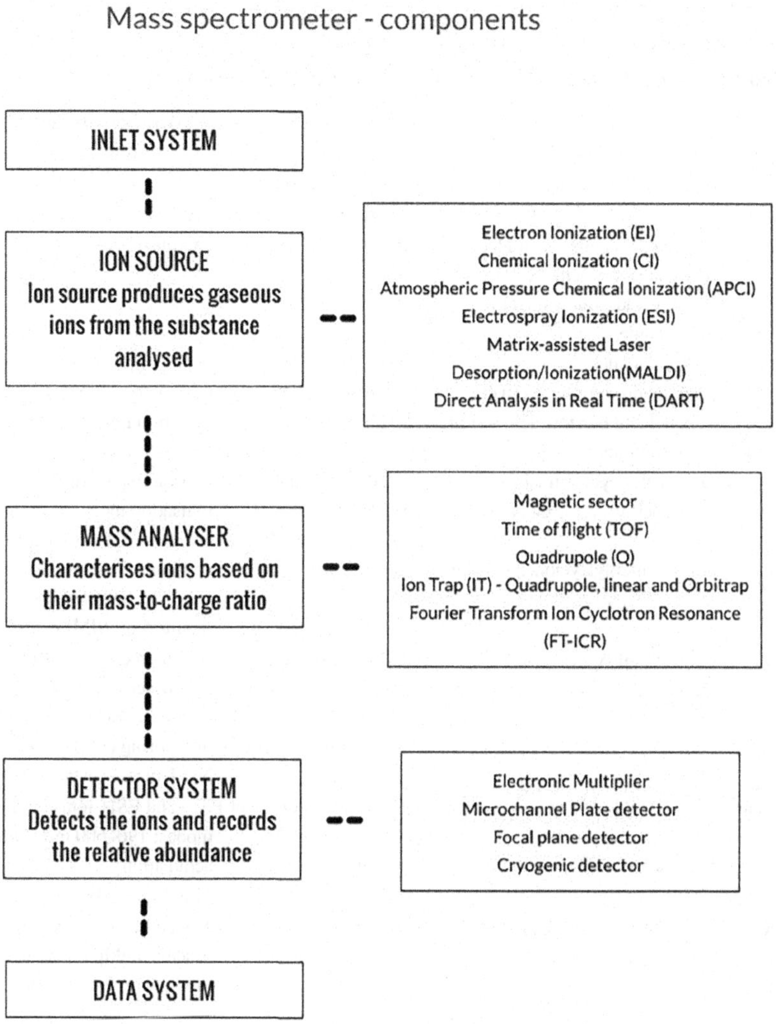

FIGURE 12.2 Mass spectrometer components.

TABLE 12.2
Mass Spectrometry Analysis for Various Food Contact Materials

S. No.	Test Material	Analytes	Method of Analysis	Mass Spectrometer Configuration	References
1	Food contact materials	Methyl, ethyl, propyl and butylparaben, tert-butylhydroquinone (TBHQ), N-butylbenzenesulfonamide (NBBS), tris(2-butoxyethyl) phosphate (TBEP), 2,6-di-tert-butyl-4-methyl-phenol (BH), 3-tert-butyl-4-hydroxyanisole (BHA)	GC-MS/MS	Electron impact (EI); ionisation energy – 70 eV; m/z range 35–700; mass spectral library NIST (version 2.0, updated 19.05.2011)	Žnideršič et al. (2020)
			HRMS	LTQ Orbitrap XL (Thermo Fisher) equipped with heated electrospray ionisation (HESI-II); m/z range 100–800 following positive ESI	

(Continued)

TABLE 12.2 (*Continued*)
Mass Spectrometry Analysis for Various Food Contact Materials

S. No.	Test Material	Analytes	Method of Analysis	Mass Spectrometer Configuration	References
2	Wheat pulp and wood dishes	Nonanal, 4-allylanisole, decanal, phthalic anhydride, 1-dodecanol, ethyl dodecanoate, tetradecanal, tetradecanoic acid, hexadecanoic acid, octadecanoic acid and erucamide	SPME – GC-MS (MS 5975B mass spectrometer detector)	Electron impact (EI) ionisation mode; ionisation energy – 70 eV; 25–250 m/z	Asensio et al. (2020)
3	Cardboard cups for hot beverages	Benzaldehyde, butylhydroxytoluene (BHT), phenol and benzoic acid – from printing inks	SPME – GC-MS (mass spectrometer MS 5975B detector)	50–600 m/z; NIST Chemistry WebBook spectrum library	Asensio et al. (2019)
4	silicone rubber baking mould	Octamethylcyclotetrasiloxane, decamethylcyclopentasiloxane and dodecamethylcyclohexasiloxane	HS-GC/MS and GC/MS (5975C quadrupole mass selective detector)	Electron impact (EI) ionisation mode; ionisation energy – 70 eV; transfer line and ion source temperature – 250°C; Selected ion monitoring (SIM) mode	Liu et al. (2020)
5	Rubber teats	Alkanes, siloxanes, esters, aldehydes	GC-MS (ISQ mass selective detector)	Ionisation energy – 70 eV; ion source temperature – 250°C; selected ion monitoring (SIM) mode; NIST mass spectra library	Feng et al. (2016)
6	Coated tinplate cans	Metals and bisphenol compound	UHPLC (time-of-flight mass spectrometry detector)	ESI– and ESI+ ionisation modes; 190–600 m/z; Data acquisition – MassLynxTMversion 4	El Moussawi et al. (2019)
7	Polystyrene and polyvinyl chloride (PVC) food packaging	Benzoxazolyl-based substances	UPLC-MS/MS	ESI– and ESI+ ionisation modes; multiple-reaction monitoring (MRM) mode	Wu et al. (2016)
8	Plastic baby bottles	Alkanes, phthalates, silicones, antioxidants	GC-MS (triple quadrupole (QQQ) mass spectrometer)	Electron impact (EI) ionisation; selected ion monitoring (SIM) mode; quadrupole temperature – 150°C and an ion source temperature – 230°C; multiplier voltage – 2,200 V	Onghena et al. (2014)
9	Plastic packaging for baby food	Bisphenols	GC-MS/MS (triple quadrupole mass spectrometer)	Electron impact (EI) ionisation; selected reaction monitoring method (SRM); ionisation energy – 70 eV; ion source temperature – 220°C	García-Córcoles et al. (2018)
10	Plastic food contact materials	84 Substances (hexadecyltrimethylammonium bromide; dibutyl phthalate; methylparaben; styrene and so on)	GC-MS (single quadrupole mass detector)	Electron impact (EI) ionisation; ionisation energy – 70 eV	Tsochatzis et al. (2020)

(*Continued*)

TABLE 12.2 (*Continued*)
Mass Spectrometry Analysis for Various Food Contact Materials

S. No.	Test Material	Analytes	Method of Analysis	Mass Spectrometer Configuration	References
11	Starch-based biopolymer	Glycerol, tetradecanoic acid, ethyl ester, 2-palmito glycerol and 9,12-octadecadienoic acid, methyl ester	GC-EI-MS (quadrupole mass spectrometry detector – 5977D, Agilent)	Electron impact (EI) ionisation; ionisation energy – 70 eV; SCAN mode; 50–450 m/z; MSD ChemStation software	Osorio et al. (2019)
		Tris(2,4-ditert-butylphenyl) phosphite	APGC-Q/ToF (hybrid quadrupole/ time-of-flight mass spectrometer (Xevo G2-XS Q-ToF, Waters))	Atmospheric pressure chemical ionisation (APCI+) mode; ion source temperature -150°C; 50–650 m/z; UNIFI Scientific Information System	
12	General-purpose polystyrene (GPPS) cups and bowls for hot beverages	Styrene monomer	TSQ Quantum XLS GC-MS	Electron impact (EI) mode; ionisation energy – 70 eV; selected ion monitoring (SIM) mode; 104.06 m/z	Naziruddin et al. (2020)
13	Paper and cardboard food packaging	68 Organic contaminants (phthalates, polycyclic aromatic hydrocarbons, photoinitiators, bisphenols and polyfluorinated compounds)	HPLC-MS/MS (triple quadrupole MS/MS Agilent 6490)	Jetstream electrospray ionisation (ESI) source – negative polarity; multiple-reaction monitoring (MRM) mode	Vavrouš et al. (2016)
			GC-MS/MS (TSQ Quantum XLS Ultra triple quadrupole MS (Thermo Scientific))	Electron ionisation (EI); ionisation energy – 70 eV; source temperature -150°C; full scan mode; 50–650 m/z	
14	Polyamide kitchenware	64 Substances (plasticisers, antioxidants, phthalates)	UPLC-Q-TOF-MS (quadrupole time-of-flight-mass spectrometry(Waters Xevos G2-S))	ESI+ mode at 3.0 kV in and ESI^- mode at 2.5 kV; 50 to 1200 m/z; ion source temperature –120°C; UNIFI (V 1.9.3)	Hu et al. (2020)
15	Metals cans for food and beverages	Bisphenol A diglycidyl ether (BADGE), novolac-glycidyl ethers (NOGEs)	LC-MS/MS (TRIPLE QUAD TQ4500 tandem mass spectrometer (ABSCIEX, MA, USA))	Electrospray interface (ESI) – positive mode; ion source temperature – 500°C, spray voltage – 5,500 V	Hwang et al. (2020)
16	PolyEthylene Terephthalate (PET), Low-Density Polyethylene (LDPE) and isotactic polypropylene films	Polyethene oxide (PEO), oligomers	LESA-nanoESI-MS (Orbitrap Fusion LUMOS mass spectrometer (Thermo Scientific, Waltham, MA))	Electrospray ionisation (ESI); IonSense DART source (Danvers, MA)	Issart et al. (2019)

(*Continued*)

TABLE 12.2 (*Continued*)
Mass Spectrometry Analysis for Various Food Contact Materials

S. No.	Test Material	Analytes	Method of Analysis	Mass Spectrometer Configuration	References
17	Plastic food packaging	Inks, adhesives, plasticisers, emulsifiers	UPLC-Q-TOF-MS (Xevo G2 Q-TOF mass spectrometer (Waters))	Positive ionisation; sensitivity mode; capillary at 2.5 kV; sampling cone at 30 V; extraction cone at 4 V; source temperature at 120°C; MassLynx v4.1 software	Aznar et al. (2016)
			IM-Q-TOF-MS (SYNAPT G2-Si HDMS (Waters))	Positive ionisation; sensitivity mode; capillary at 3 kV; sampling cone at 30 V; extraction cone at 4 V; source temperature at 120°C; desolvation temperature at 450°C; MassLynx v4.1 and UNIFI v1.8 software	
			DESI-MS (Xevo G2 XS Q-ToF (Waters))	electrospray voltage – 3.5 kV; gas pressure – 6 bar; solvent flow – 2 μL; Waters HDI 1.3.5 software	

mode operated at 70 electron volt (eV). It ensures a high degree of molecule fragmentation and allows the creation of databases for individual molecules with their mass spectra characteristics. High-resolution mass spectrometers (HRMS) are highly efficient in determining unknown compounds and are often coupled with GC-MS. APGC improves GC where the degree of fragmentation of analytes can be controlled by controlling ionising electrons. APGC is commonly combined with HRMS such as Q-TOF to identify migrants from FCMs (Wrona and Nerín, 2020; Su et al., 2019).

Žnideršič et al. (2020) developed a GC-MS/MS method using tandem mass spectrometer 7000B and a multi-purpose autosampler to quantify compounds such as methylparaben, ethylparaben, butylparaben, *N*-butylbenzene sulfonamide and 3-tert-butyl-4-hydroxyanisole found in FCMs (EU Commission Regulation, 2011). A linear ion trap LTQ Orbitrap XL (Thermo Fisher Scientific Company, Villebon, France) with heated electrospray ionisation-II was used to record high-resolution mass spectra. Osorio et al. (2019) evaluated odorous compounds migrating from starch-based polymers used for packaging fish, mushroom, spices and honey using headspace (HS) SPME-GC with olfactometry (O) coupled with a mass detector.

12.3.2 Liquid Chromatography-Mass Spectrometry

LC-MS/MS is a widely used technique for analysing perfluorinated compounds, primary aromatic amines and photoinitiators used in plastic packaging, utensils, and paper and board packaging. LC-MS can be effectively used to analyse non-VOCs in FCMs compared to conventional MS techniques that require the degradation of the high-molecular-weight binding systems. Atmospheric pressure ionisation interfaces are commonly used in LC-MS and LC-MS/MS analyses. LC technique reduces the need for complex sample clean-up procedures, minimises error and provides high recoveries. LC is highly suitable for food contact compounds with high molecular weight (oligomers) or polar functional groups (polyamides, polyurethanes and polycarbonates) and that are thermally labile.

Similarly, LC is well suited for analysing the migration of FCMs using food stimulants such as distilled water, aqueous ethanol (10% v/v) and aqueous acetic acid (3% w/v) when compared with GC. Single ion recording and multiple-reaction monitoring modes are widely used to quantify non-volatile compounds. Hwang et al. (2020) developed and validated an LC-MS/MS method to determine and quantify the levels of compounds such as BADGE and novolac-glycidyl ethers migrating from food and beverage cans into food simulants.

12.3.2.1 Ultra-High-Performance Liquid Chromatography-Mass Spectrometry

Ultra-high-performance liquid chromatography (UHPLC) is an advancement of LC that provides higher efficiency, reduces analysis time, offers high resolution and is compatible with MS detection. UHPLC-Q-TOF-MSE is highly efficient for analysing non-volatile compounds from FCMs as it produces ions with high sensitivity and high mass accuracy (Vera et al., 2018).

Ubeda et al. (2019) studied the migration of 39 different cyclic oligomers from polylactic acid (PLA) biopolymer using UHPLC-Q-TOF-MSE. Similarly, Aznar et al. (2019) reported UHPLC -Q-TOF-MS to be a suitable analytical method of analysis of non-volatile components in biodegradable PLA-polyester blend films and pellets and 2.2′-(tridecylimino)*bis*-ethanol and its derivatives in polypropylene baby bottles (Da Silva Oliveira et al., 2019). Ultra-high-performance liquid chromatography coupled with triple quadrupole mass spectrometry (UHPLC-QQQ-MS) provided very high sensitivity. It was used to analyse the migration of adhesive acrylates from recycled polyethene terephthalate (PET; Otoukesh et al., 2019). The method was also used to analyse migrants from PLA films and pellets into food simulates such as ethanol 95% (v/v), ethanol 10% (v/v) and acetic acid 3% (w/v) and 14 compounds that were used as plasticisers and additives were detected. Recently, meat preservation with green tea active packaging is gaining popularity, and it is necessary to analyse the migration of compounds from the active packaging into the meat. Wrona et al. (2017) established the migration of catechins and caffeine from tea into meat on preservation with active packaging using UPLC-Q-TOF-MSE. Qualitative analysis of non-volatile migrants from FCMs requires multiple levels of fragmentation, high mass resolution and accuracy and hybrid linear IT-high-resolution MS (LTQ Orbitrap), a combination of linear IT mass spectrometer and Orbitrap mass analyser, is highly beneficial (Martínez-Bueno et al., 2017).

12.3.3 Inductively Coupled Plasma Mass Spectrometry (ICP-MS)

In recent times, nanoparticles (NPs) are widely used to improve polymers' mechanical properties, making them sustainable and as a substitute for petroleum-based materials in biopolymers (Souza and Fernando, 2016). Inductively coupled plasma mass spectrometry (ICP-MS) is the most commonly used technique for analysing inorganic NPs in food simulants. Mackevica et al. (2016) reported migration of silver (up to 0.31 mg/dm^2) into the acidic medium from commercial polymeric nanocomposites using ICP-MS and single-particle ICP-MS). Similarly, Addo Ntim et al. (2016) and Bott et al. (2014) measured silver NP migration in 3% acetic acid after 10 days, and Artiaga et al. (2015) established migration of silver chloride and silver sulphide in food simulants such as water and 3% acetic acid.

12.3.4 Other Mass Spectrometry Techniques

Recently, MS techniques such as atmospheric solids analysis probe MS, direct analysis in real-time MS, desorption electrospray ionisation (DESI)-MS and liquid extraction surface analysis nano-electrospray mass spectrometry (LESA-nano-ESI-MS) has been gaining attention for migration analysis in FCMs and food simulants (Wrona and Nerín, 2020). These techniques are beneficial for quick screening and analysis of known analytes. Assert et al. (2019) developed a LESA-nanoESI-MS method for quantifying the migration of compounds from polymers such as PET LDPE and isotactic polypropylene films. Ambient ionisation–accurate MS was used by Bentayeb et al. (2012)

to analyse the migration of compounds from set-off inks on food packaging into food. DESI-MS and DESI-HRMS have been employed to quantify migrants from set-off inks on polymeric FCMs (Aznar et al., 2016) and melamine from melamine tableware (Mattarozzi et al., 2012).

12.4 CONCLUSION

Migration of compounds from FCMs is gaining attention in recent times. Several substances are added intentionally or non-intentionally into the manufacture of FCMs, and many of these compounds migrate into food. There is an urgent need to develop and validate analytical techniques to determine and quantify food contact substances, especially when there are some unidentified NIASs. Specialised analytical methods are mandatory for each type of food or food stimulant. Food and food stimulants are hydrophilic, lipophilic or amphiphilic based on their chemical properties. MS techniques are efficient in quantifying migrants, and many studies have been carried out in the recent past. The food-packaging industry needs to keep in mind the possibility of migration of chemical compounds into food and minimising migration. The curbing of migration of chemical substances into food is important to determine food quality and deter illnesses caused in humans.

REFERENCES

Abdel-Rahman, N.A.G. 2015. Tin-plate corrosion in canned foods. *Journal of Global Biosciences*, 4(7):2966–2971.

Addo Ntim, S., Thomas, T.A., Noonan, G.O. 2016. Influence of aqueous food simulants on potential nanoparticle detection in migration studies involving nanoenabled food-contact substances. *Food Additives & Contaminants: Part A*, 33(5):905–912.

Aly, S.S. 2016. Studying the migration of plasticizers from plastic packaging to local processed cheese. *Sciences*, 6(4):957–963.

Artiaga, G., Ramos, K., Ramos, L., Cámara, C., Gómez-Gómez, M. 2015. Migration and characterisation of nanosilver from food containers by AF4-ICP-MS. *Food Chemistry*, 166:76–85.

Arvanitoyannis, I.S., Kotsanopoulos, K.V. 2014. Migration phenomenon in food packaging. Food–package interactions, mechanisms, types of migrants, testing and relative legislation – A review. *Food and Bioprocess Technology*, 7(1):21–36.

Asensio, E., Montañés, L., Nerín, C. 2020. Migration of volatile compounds from natural biomaterials and their safety evaluation as food contact materials. *Food and Chemical Toxicology*, 142: 111457.

Asensio, E., Peiro, T., Nerín, C. 2019. Determination the set-off migration of ink in cardboard-cups used in coffee vending machines. *Food and Chemical Toxicology*, 130:61–67.

Aznar, M., Alfaro, P., Nerín, C., Jones, E. and Riches, E. 2016. Progress in mass spectrometry for the analysis of set-off phenomena in plastic food packaging materials. *Journal of Chromatography A*, 1453:124–133.

Aznar, M., Ubeda, S., Dreolin, N., Nerín, C. 2019. Determination of non-volatile components of a biodegradable food packaging material based on polyester and polylactic acid (PLA) and its migration to food simulants. *Journal of Chromatography A*, 1583:1–8.

Beldì, G., Jakubowska, N., Peltzer, M. A., Simoneau, C. 2016. Testing approaches for the release of metals from ceramic articles – In support of the revision of the Ceramic Directive 84/500/EEC, EUR 28363 EN. https://doi.org/10.2788/402683.

Bentayeb, K., Ackerman, L.K., Begley, T.H. 2012. Ambient ionization–accurate mass spectrometry (AMI-AMS) for the identification of nonvisible set-off in food-contact materials. *Journal of Agricultural and Food Chemistry*, 60(8):1914–1920.

Biryol, D., Nicolas, C.I., Wambaugh, J., Phillips, K., Isaacs, K. 2017. High-throughput dietary exposure predictions for chemical migrants from food contact substances for use in chemical prioritization. *Environment International*, 108:185–194.

Blanco-Zubiaguirre, L., Zabaleta, I., Prieto, A., Olivares, M., Zuloaga, O., Elizalde, M.P. 2020. Migration of photoinitiators, phthalates and plasticizers from paper and cardboard materials into different simulants and foodstuffs. *Food Chemistry*, 344:128597.

Bott, J., Störmer, A., Franz, R. 2014. A model study into the migration potential of nanoparticles from plastics nanocomposites for food contact. *Food Packaging and Shelf Life*, 2(2):73–80.

Bouma, K., Schothorst, R.C. 2003. Identification of extractable substances from rubber nettings used to package meat products. *Food Additives & Contaminants*, 20(3):300–307.
Buculei, A., Amariei, S., Oroian, M., Gutt, G., Gaceu, L., Birca, A. 2014. Metals migration between product and metallic package in canned meat. *LWT-Food Science and Technology*, 58(2):364–374.
Cai, H., Ji, S., Zhang, J., Tao, G., Peng, C., Hou, R., Zhang, L., Sun, Y., Wan, X.2017. Migration kinetics of four photo-initiators from paper food packaging to solid food simulants. *Food Additives & Contaminants: Part A*, 34(9):1632–1642.
Cakste, I., Kuka, M., Kuka, P. 2017. Migration of iron, aluminium, calcium, magnesium and silicon from ceramic materials into food simulant. *11th Baltic Conference on Food Science and Technology "Food Science and Technology in a Changing World" FOODBALT*, Jelgava, 160–163.
Calva-Estrada, S.J., Jiménez-Fernández, M., Lugo-Cervantes, E. 2019. Protein-based films: Advances in the development of biomaterials applicable to food packaging. *Food Engineering Reviews*, 11(2):78–92.
Castle, L., Mercer, A.J., Startin, J.R., Gilbert, J. 1988. Migration from plasticized films into foods 3. Migration of phthalate, sebacate, citrate and phosphate esters from films used for retail food packaging. *Food Additives & Contaminants*, 5(1):9–20.
Commission Regulation (EU) No. 10/2011 of 14 January 2011 on plastic materials and articles intended to come into contact with food, n.d. 89. https://eur-lex.europa.eu/legal-content/EN/TXT/?uri=celex:32011R0010.
Council Directive 84/500/EEC of 15 October 1984 on the approximation of the laws of the Member States relating to ceramic articles intended to come into contact with foodstuffs (OJ L277, 20.10.1984, p. 12).
Cruz, R.M.S., Rico, B.P.M., Vieira, M.C. 2019. 9-Food packaging and migration. In *Food Quality and Shelf Life*; Galanakis, C.M., Ed.; Academic Press: San Diego, CA.
Da Silva Oliveira, W., Ubeda, S., Nerín, C., Padula, M., Godoy, H.T. 2019. Identification of non-volatile migrants from baby bottles by UPLC-Q-TOF-MS. *Food Research International*, 123:529–537.
El Moussawi, S.N., Karam, R., Cladière, M., Chébib, H., Ouaini, R., Camel, V. 2018. Effect of sterilisation and storage conditions on the migration of bisphenol A from tinplate cans of the Lebanese market, Part A Chemistry, analysis, control, exposure & risk assessment. *Food Additives & Contaminants: Part A*, 35(2):377–386. https://doi-org.eres.qnl.qa/10.1080/19440049.2017.1395521.
El Moussawi, S.N., Ouaini, R., Matta, J., Chébib, H., Cladière, M., Camel, V. 2019. Simultaneous migration of bisphenol compounds and trace metals in canned vegetable food. *Food Chemistry*, 288:228–238.
Elss, S., Grünewald, L., Richling, E., Schreier, P. 2004. Occurrence of 2-ethylhexanoic acid in foods packed in glass jars. *Food Additives and Contaminants*, 21(8):811–814.
EC No. 1935/2004. 2004. European Union Regulation (EC) No. 1935/2004 on materials and articles intended to come into contact with food and repealing Directives 80/590/EEC and 89/109/EEC.
EU. 2011. Commission Regulation (EU) No. 10/2011 of 14 January 2011 on plastic materials and articles intended to come in contact with food. *Official Journal of European Union*, L12:1–89.
EURL Food Contact Materials – Legislations. https://ec.europa.eu/jrc/en/eurl/food-contact-materials/legislation. Accessed 29 December 2020.
Feng, D., Yang, H., Qi, D., Li, Z. 2016. Extraction, confirmation, and screening of non-target compounds in silicone rubber teats by purge-and-trap and SPME combined with GC-MS. *Polymer Testing*, 56:91–98.
García-Córcoles, M.T., Cipa, M., Rodríguez-Gómez, R., Rivas, A., Olea-Serrano, F., Vílchez, J.L., Zafra-Gómez, A. 2018. Determination of bisphenols with estrogenic activity in plastic packaged baby food samples using solid-liquid extraction and clean-up with dispersive sorbents followed by gas chromatography tandem mass spectrometry analysis. *Talanta*, 178:441–448.
Genualdi, S., Nyman, P., Begley, T. 2014. Updated evaluation of the migration of styrene monomer and oligomers from polystyrene food contact materials to foods and food simulants. *Food Additives & Contaminants: Part A*, 31(4):723–733.
Geueke, B., Groh, K., Muncke, J. 2018. Food packaging in the circular economy: Overview of chemical safety aspects for commonly used materials. *Journal of Cleaner Production*, 193:491–505.
Gómez-Ramos, M.M., Ucles, S., Ferrer, C., Fernández-Alba, A.R., Hernando, M.D. 2019. Exploration of environmental contaminants in honeybees using GC-TOF-MS and GC-Orbitrap-MS. *Science of The Total Environment*, 647:232–244.
Grob, K., Biedermann, M., Scherbaum, E., Roth, M., Rieger, K. 2006. Food contamination with organic materials in perspective: Packaging materials as the largest and least controlled source? A view focusing on the European situation. *Critical Reviews in Food Science and Nutrition*, 46(7):529–535.
Groh, K., Geueke, B., Muncke, J. 2020. FCCdb: Food Contact Chemicals database. Zenodo. https://doi.org/10.5281/zenodo.3240108. Accessed 7 December 2020.

Guerreiro, T.M., de Oliveira, D.N., Melo, C.F.O.R., de Oliveira Lima, E., Catharino, R.R. 2018. Migration from plastic packaging into meat. *Food Research International*, 109:320–324.

Hahladakis, J.N., Velis, C.A., Weber, R., Iacovidou, E., Purnell, P. 2018. An overview of chemical additives present in plastics: Migration, release, fate and environmental impact during their use, disposal and recycling. *Journal of Hazardous Materials*, 344:179–199.

Hammarling, L., Gustavsson, H., Svensson, K., Karlsson, S., Oskarsson, A. 1998. Migration of epoxidized soya bean oil from plasticized PVC gaskets into baby food. *Food Additives & Contaminants*, 15(2):203–208.

Helling, R., Mieth, A., Altmann, S., Simat, T.J. 2009. Determination of the overall migration from silicone baking moulds into simulants and food using 1H-NMR techniques. *Food Additives & Contaminants: Part A*, 26(3):395–407.

Hu, Y., Du, Z., Sun, X., Ma, X., Song, J., Sui, H., Debrah, A.A. 2020. Non-targeted analysis and risk assessment of non-volatile compounds in polyamide food contact materials. *Food Chemistry,* 128625.

Hwang, J.B., Lee, S., Lee, J.E., Choi, J.C., Park, S.J., Kang, Y. 2020. LC-MS/MS analysis of BADGE, NOGEs, and their derivatives migrated from food and beverage metal cans. *Food Additives & Contaminants: Part A*, 37(11): 1974–1984.

Issart, A., Godin, S., Preud'Homme, H., Bierla, K., Allal, A., Szpunar, J. 2019. Direct screening of food packaging materials for post-polymerization residues, degradation products and additives by liquid extraction surface analysis nanoelectrospray mass spectrometry (LESA-nESI-MS). *Analytica Chimica Acta*, 1058:117–126.

Jeddi, M.Z., Rastkari, N., Ahmadkhaniha, R., Yunesian, M. 2015. Concentrations of phthalates in bottled water under common storage conditions: Do they pose a health risk to children?. *Food Research International*, 69:256–265.

Jung, T. and Simat, T.J. 2014. Multi-analyte methods for the detection of photoinitiators and amine synergists in food contact materials and foodstuffs – Part II: UHPLC-MS/MS analysis of materials and dry foods. *Food Additives & Contaminants: Part A*, 31(4):743–766.

Klampfl, C.W. 2013. Mass spectrometry as a useful tool for the analysis of stabilizers in polymer materials. *TrAC Trends in Analytical Chemistry*, 50:53–64.

Koster, S., Rennen, M., Leeman, W., Houben, G., Muilwijk, B., van Acker, F., Krul, L. 2014. A novel safety assessment strategy for non-intentionally added substances (NIAS) in carton food contact materials. *Food Additives & Contaminants: Part A*, 31(3):422–443.

Kühne, F., Kappenstein, O., Straßgütl, S., Weese, F., Weyer, J., Pfaff, K., Luch, A. 2018. N-nitrosamines migrating from food contact materials into food simulants: Analysis and quantification by means of HPLC-APCI-MS/MS. *Food Additives & Contaminants: Part A*, 35(4):793–806.

Lehman, R.L. 2002. *Lead Glazes for Ceramic Foodware*; The International Lead Management Center Research: Triangle Park, NC, 8–108.

Lemos, A.B., Perez, M.Â.F., Bordin, M.R., Silva, L.B., Godoy, H.T., Padula, M. 2017. Set-off: Development of a simulation press and analytical approach to study the phenomenon. *Packaging Technology and Science*, 30(8): 495–504.

Liu, Y.Q., Yu, W.W., Jiang, H., Shang, G.Q., Zeng, S.F., Wang, Z.W., Hu, C.Y. 2020. Variation of baking oils and baking methods on altering the contents of cyclosiloxane in food simulants and cakes migrated from silicone rubber baking moulds. *Food Packaging and Shelf Life*, 24:100505.

Lynch, R., Elledge, B., Peters, C. 2008. An assessment of lead leachability from lead-glazed ceramic cooking vessels. *Journal of Environmental Health*, 70(9):36–41.

Mackevica, A., Olsson, M.E., Hansen, S.F. 2016. Silver nanoparticle release from commercially available plastic food containers into food simulants. *Journal of Nanoparticle Research*, 18(1):5.

Martínez-Bueno, M.J., Hernando, M.D., Uclés, S., Rajski, L., Cimmino, S., Fernández-Alba, A.R.2017. Identification of non-intentionally added substances in food packaging nano films by gas and liquid chromatography coupled to orbitrap mass spectrometry. *Talanta*, 172:68–77.

Masiá, A., Suarez-Varela, M.M., Llopis-Gonzalez, A., Picó, Y. 2016. Determination of pesticides and veterinary drug residues in food by liquid chromatography-mass spectrometry: A review. *Analytica Chimica Acta*, 936:40–61.

Mattarozzi, M., Milioli, M., Cavalieri, C., Bianchi, F., Careri, M. 2012. Rapid desorption electrospray ionization-high resolution mass spectrometry method for the analysis of melamine migration from melamine tableware. *Talanta*, 101:453–459.

McCombie G. 2018. Enforcement's perspective. In: European Commission, Directorate General Health and Food Safety. Available: https://ec.europa.eu/food/sites/food/files/safety/docs/cs_fcm_evalworkshop_20180924_pres07.pdf. Accessed 7 December 2020.

McCombie, G., Harling-Vollmer, A., Morandini, M., Schmäschke, G., Pechstein, S., Altkofer, W., Biedermann, M., Biedermann-Brem, S., Zurfluh, M., Sutter, G., Landis, M. 2012. Migration of plasticizers from the gaskets of lids into oily food in glass jars: A European enforcement campaign. *European Food Research and Technology*, 235(1):129–137.

Metak, A.M., Nabhani, F., Connolly, S.N. 2015. Migration of engineered nanoparticles from packaging into food products. *LWT-Food Science and Technology*, 64(2):781–787.

Mojsiewicz-Pieńkowska, K., Jamrógiewicz, M., Szymkowska, K., Krenczkowska, D. 2016. Direct human contact with siloxanes (silicones)–safety or risk part 1. Characteristics of siloxanes (silicones). *Frontiers in Pharmacology*, 7:132.

Muncke, J., Andersson, A.M., Backhaus, T., Boucher, J.M., Almroth, B.C., Castillo, A.C., Chevrier, J., Demeneix, B.A., Emmanuel, J.A., Fini, J.B., Gee, D. 2020. Impacts of food contact chemicals on human health: A consensus statement. *Environmental Health*, 19(1): 1–12.

Musoke, L., Banadda, N., Sempala, C., Kigozi, J. 2015. The migration of chemical contaminants from polyethylene bags into food during cooking. *The Open Food Science Journal*, 9(1):416.

Naziruddin, M.A., Sulaiman, R., Lim, S.A.H., Jinap, S., Nurulhuda, K., Sanny, M. 2020. The effect of fat contents and conditions of contact in actual use on styrene monomer migrated from general-purpose polystyrene into selected fatty dishes and beverage. *Food Packaging and Shelf Life*, 23:100461.

Onghena, M., van Hoeck, E., Vervliet, P., Scippo, M.L., Simon, C., van Loco, J., Covaci, A. 2014. Development and application of a non-targeted extraction method for the analysis of migrating compounds from plastic baby bottles by GC-MS. *Food Additives & Contaminants: Part A*, 31(12):2090–2102.

Osorio, J., Aznar, M., Nerín, C. 2019. Identification of key odorant compounds in starch-based polymers intended for food contact materials. *Food chemistry*, 285:39–45.

Otoukesh, M., Nerín, C., Aznar, M., Kabir, A., Furton, K.G., Es'haghi, Z. 2019. Determination of adhesive acrylates in recycled polyethylene terephthalate by fabric phase sorptive extraction coupled to ultra performance liquid chromatography-mass spectrometry. *Journal of Chromatography A*, 1602:56–63.

Otoukesh, M., Vera, P., Wrona, M., Nerin, C., Es'haghi, Z. 2020. Migration of dihydroxyalkylamines from polypropylene coffee capsules to Tenax® and coffee by salt-assisted liquid–liquid extraction and liquid chromatography–mass spectrometry. *Food Chemistry*, 321:126720.

Parkar, J., Rakesh, M. 2014. Leaching of elements from packaging material into canned foods marketed in India. *Food Control*, 40:177–184.

Plastics Europe. 2019. An analysis of European plastics production, demand and waste data. Retrieved from https://plasticseurope.org/application/files/9715/7129/9584/FINAL_web_version_Plastics_the_facts2019_14102019.pdf on 9 December 2020.

Rosenmai, A.K., Bengtström, L., Taxvig, C., Trier, X., Petersen, J.H., Svingen, T., Binderup, M.L., Dybdahl, M., Granby, K., Vinggaard, A.M. 2017. An effect-directed strategy for characterizing emerging chemicals in food contact materials made from paper and board. *Food and Chemical Toxicology*, 106:250–259.

Sanchis, Y., Yusà, V., Coscollà, C.2017. Analytical strategies for organic food packaging contaminants. *Journal of Chromatography A*, 1490:22–46.

Skjevrak, I., Brede, C., Steffensen, I., Mikalsen, A., Alexander, J., Fjeldal, P., Herikstad, H. 2005. Non-targeted multi-component analytical surveillance of plastic food contact materials: Identification of substances not included in EU positive lists and their risk assessment. *Food Additives and Contaminants*, 22(10):1012–1022. https://doi.org/10.1080/02652030500090877.

Song, X.C., Wrona, M., Nerin, C., Lin, Q.B., Zhong, H.N. 2019. Volatile non-intentionally added substances (NIAS) identified in recycled expanded polystyrene containers and their migration into food simulants. *Food Packaging and Shelf Life*, 20:100318.

Souza, V.G.L. and Fernando, A.L.2016. Nanoparticles in food packaging: Biodegradability and potential migration to food – A review. *Food Packaging and Shelf Life*, 8:63–70.

Su, Q.Z., Vera, P., Van de Wiele, C., Nerín, C., Lin, Q.B., Zhong, H.N. 2019. Non-target screening of (semi-) volatiles in food-grade polymers by comparison of atmospheric pressure gas chromatography quadrupole time-of-flight and electron ionization mass spectrometry. *Talanta*, 202:285–296.

Tsochatzis, E.D., Lopes, J.A., Hoekstra, E., Emons, H. 2020. Development and validation of a multi-analyte GC-MS method for the determination of 84 substances from plastic food contact materials. *Analytical and Bioanalytical Chemistry*, 412(22):5419–5434.

Ubeda, S., Aznar, M., Alfaro, P., Nerín, C. 2019. Migration of oligomers from a food contact biopolymer based on polylactic acid (PLA) and polyester. *Analytical and bioanalytical chemistry*, 411(16):3521–3532.

Van Den Houwe, K., Van Heyst, A., Evrard, C., Van Loco, J., Bolle, F., Lynen, F., Van Hoeck, E., 2016. Migration of 17 photoinitiators from printing inks and cardboard into packaged food–results of a belgian market survey. *Packaging Technology and Science*, 29(2):121–131.

Vápenka, L., Vavrouš, A., Votavova, L., Kejlova, K., Dobias, J., Sosnovcova, J. 2016. Contaminants in the paper-based food packaging materials used in the Czech Republic. *Journal of Food and Nutrition Research*, 55 (4):361–373.

Vavrouš, A., Vápenka, L., Sosnovcová, J., Kejlová, K., Vrbík, K., Jírová, D. 2016. Method for analysis of 68 organic contaminants in food contact paper using gas and liquid chromatography coupled with tandem mass spectrometry. *Food Control*, 60:221–229.

Wagner, J., Castle, L., Oldring, P.K., Moschakis, T., Wedzicha, B.L. 2018. Factors affecting migration kinetics from a generic epoxy-phenolic food can coating system. *Food Research International*, 106:183–192.

Wrona, M., Nerín, C. 2020. Analytical approaches for analysis of safety of modern food packaging: A review. *Molecules*, 25(3):752.

Wrona, M., Nerín, C., Alfonso, M.J., Caballero, M.Á. 2017. Antioxidant packaging with encapsulated green tea for fresh minced meat. *Innovative Food Science & Emerging Technologies*, 41:307–313.

Wu, Z., Xu, Y., Li, M., Guo, X., Xian, Y., Dong, H. 2016. Simultaneous determination of fluorescent whitening agents (FWAs) and photoinitiators (PIs) in food packaging coated paper products by the UPLC-MS/MS method using ESI positive and negative switching modes. *Analytical Methods*, 8(5):1052–1059.

Xue, M., Chai, X.S., Li, X., Chen, R. 2019. Migration of organic contaminants into dry powdered food in paper packaging materials and the influencing factors. *Journal of Food Engineering*, 262:75–82.

Žnideršič, L., Mlakar, A., Prosen, H. 2020. Data on the optimisation of GC-MS/MS method for the simultaneous determination of compounds from food contact material. *Data in Brief*, 28:105060.

13 Mass Analysis in Veterinary Drugs

Semih Ötleş
Ege University

Vasfiye Hazal Özyurt
Near East University

CONTENTS

13.1 Introduction 247
13.2 Mass Spectrometry Techniques 248
13.2.1 Single Mass Spectrometry Approach 248
13.2.2 Multidimensional Mass Spectrometry Approach 248
13.2.3 High-Resolution Mass Spectrometry Approach 248
13.3 Applications 249
13.4 Conclusion 252
References 252

13.1 INTRODUCTION

Veterinary drugs are classified as antibiotics or dyes, and drugs exhibiting growth-promoting properties such as steroids, β-agonist compounds, thyrostats, or growth hormones (Lillenberg, Herodes, Kipper, & Nei, 2010). The antibiotics administered to food-producing animals are tetracyclines, sulfonamides, fluoroquinolones, macrolides, lincosamides, aminoglycosides, beta-lactams, cephalosporins, and others (Table 13.1) (Andreu, Blasco, & Picó, 2007). These drugs are used inappropriately and abusively in food-producing animals. Analytical methods are used to monitor the remaining amount of these drugs in food-producing animals for the protection of the consumers from undesirable health problems. The remaining of these substances in animal-derived food has been regulated with strict regulation by the European Union for more than 20 years since the presence of veterinary drug residues in food is a potential health risk. Some of the risks are allergies and antimicrobial-resistant microorganisms through excessive use of veterinary antibiotics. The maximum residue limits are defined as the maximum concentration of a residue, resulting from the registered use of an agricultural or veterinary chemical that is recommended to be legally permitted or recognized as acceptable in or on a food, agricultural commodity, or animal feed (Cycoń, Mrozik, & Piotrowska-Seget, 2019). EU has prepared two main groups for veterinary drugs presented in Table 13.1.

The present analysis methods rely on targeted analytical methodology and have focused on the detection of residues of the administered compounds or their metabolites in different kinds of feed, food, or biological matrices. This chapter is focusing on the main mass analysis, single mass, multidimensional, and high-resolution spectrometry to identify and quantify veterinary drugs in food.

DOI: 10.1201/9781003091226-16

TABLE 13.1
Classification of Veterinary Drug According to EU

Banned Veterinary drug			Authorized Veterinary Drug	
Hormones	**β-Agonists**	**Others**	**Antibacterials**	**Other veterinary drugs**
Resorcylic acid lactones including zeranol		Chloramphenicol	β-Lactams	Anthelmintics
Steroids		Chlorpromazine	Aminoglycosides	Anticoccidiostats, including nitroimidazoles
Stilbenes and derivatives		Dapsone	Macrolides	Carbamates and pyrethroids
Thyreostats		Dimetridazole	Quinolones	Carbodox and olaquindox
		Metronidazole	Sulfonamides and trimethoprim	Sedatives
		Nitrofurans (including furazolidone)	Tetracyclines	Non-steroidal anti-inflammatory drugs (NSAIDs)
		Ronidazole		

13.2 MASS SPECTROMETRY TECHNIQUES

Mass spectrometry (MS) combined either with gas chromatography or liquid chromatography (LC) is frequently used as a confirmatory technique. This chapter won't be a detailed discussion on MS techniques but will focus on the possibilities and limitations of the MS approach applied to veterinary drugs.

13.2.1 Single Mass Spectrometry Approach

In the 1980s, MS was considered as a revolution regarding the fight against the misuse of forbidden compounds due to its outstanding specificity and sensitivity. However, today single MS approaches in the low-resolution mode are not good enough to analyze the remaining compounds. It needed numerous possible ionization interfaces, ion transmissions, mass analyzers, and ion detection systems. Moreover, MS systems only cannot overcome problems such as the matrix effect in the analysis of veterinary drugs at trace levels. Because of this reason, adequate sample preparation steps are developed.

13.2.2 Multidimensional Mass Spectrometry Approach

Tandem or multidimensional MS (mass analyzers typically triple quadrupole, QQQ) have currently an extremely wide application range in veterinary drug residue analysis at trace levels in complex matrices in terms of its sensitivity, selectivity, and linear range. Tandem mass spectrometry (MS/MS) combines various types of mass analyzers, such as quadrupole-ion trap or quadrupole-linear ion trap to have adequate sensitivity and reliability to perform quantitative trace analysis.

13.2.3 High-Resolution Mass Spectrometry Approach

High-resolution mass spectrometry (HRMS) is increasingly acquiring fame at trace levels and offers several advantages such as high mass accuracy (between 2 and 5 ppm), high resolving power (between 10,000 and 100,000), the ability to analyze a theoretically unlimited number of compounds in a sample, the detection and identification of other untargeted or unknown compounds in complex matrices without reinjecting the samples. Time-of-flight (TOF) is the simplest mass analyzer but it gives positive and false negative results because of unresolved matrix interferences and coelution of isobaric compounds. The latest developed mass analyzer is Orbitrap, but it is expensive.

13.3 APPLICATIONS

Several hundreds of papers have been published on the analysis of veterinary drugs in food using MS (Table 13.2).

Bogialli, Coradazzi, Di Corcoa, Lagana, and Sergi (2007) researched the residues of four widely used tetracycline antibiotics and determined three of the four epimers in cheese. The developed method consists of matrix solid-phase dispersion technique followed by liquid chromatography–tandem mass spectrometry (LC-MS/MS). After samples were prepared, MS data acquisition was performed in multiple reaction monitoring (MRM), selecting two precursor ion-to-product ion transitions for each target compound. Limits of quantitation (LOQs) were calculated as 1–2 ng/g (Bogialli, Coradazzi, Di Corcoa, Lagana, & Sergi, 2007).

Martins, Kussumi, Wang, and Lebre (2007) developed a method to identify and quantify 14 antibiotics, including five β-lactams, four sulfonamides, three tetracyclines, one macrolide, and one

TABLE 13.2
Selected Studies for the Determination of Veterinary Drug Residues in Food Samples

Sample	Matrix	Instrumentation	References
10 Drugs from β-lactams (penicillins & cephalosporins)	Milk	LC-MS/MS	Kantiani et al. (2009)
4 Drugs from tetracyclines	Cheese	LC-MS/MS	Bogialli et al. (2007)
120 Drugs All veterinary drug families	Fresh egg	LC-MS/MS	Piatkowska, Jedziniak, and Zmudzki (2016)
16 Drug from sulfonamides	Beeswax	LC-MS/MS	Mitrowska and Antczak (2015)
46 Drug from tetracyclines, sulfonamides, fluoroquinolones, β-lactams, cephalosporins, macrolides	Bovine milk and bovine, swine, poultry, equine, fish, and shrimp meat samples	LC-MS/MS	Jank et al. (2017)
Minocycline	Porcine muscle	LC-MS/MS	Park et al. (2016)
14 Drugs from β-lactams, sulphonamides, tetracycline, macrolides, cephalosporin	Milk	LC-MS/MS	Martins et al. (2007)
Enrofloxacin and tetracycline	Chicken Pork	LC-MS	Kim et al. (2013)
38 drugs from flumequine, sulfapyridine, sulfamethoxazole, lincomycin	Milk	UPLC-MS/MS	Han et al. (2015)
Ciprofloxacin, ofloxacin, norfloxacin, sulfadimethoxine, and sulfamethoxazole	Lettuce root	LC-MS	Lillenberg et al. (2010)
>100 Veterinary drugs, including benzimidazoles, macrolides, penicillins, quinolones, sulphonamides, pyrimidines, tetracylines, nitroimidazoles, tranquilizers, ionophores, amphenicols, and non-steroidal anti-inflammatory agents	Milk	UPLC–ToF-MS)	Stolker et al. (2008)

(Continued)

TABLE 13.2 (*Continued*)
Selected Studies for the Determination of Veterinary Drug Residues in Food Samples

Sample	Matrix	Instrumentation	References
<100 Veterinary drug, benzimidazoles, macrolides, penicillines, quinolones, sulphonamides, pyrimidine, tetracyclines, nitroimidazolen, tranquilizers, coccidiostat, ionophores, amphenicols, and non-steroidal anti-inflammatory agents	Meat, egg, and fish	HRLC-TOF-MS	Peters et al. (2009)
150 Veterinary drugs residues and metabolites, including avermectines, benzimidazoles, beta-agonists, beta-lactams, corticoides, macrolides, nitroimidazoles, quinolones, sulfonamides, tetracyclines, and some others	Raw milk	UPLC-TOF-MS	Ortelli et al. (2009)
90 Veterinary drugs, including lincomycins, macrolides, sulfonamides, quinolones, tetracyclines, β-agonists, β-lactams, sedatives, β-receptor antagonists, sex hormones, glucocorticoids, nitroimidazoles, benzimidazoles, nitrofurans, and some others	Milk	UPLC-QTOF-MS	Zhang et al. (2015)
Various nitrofuran and chloramphenicol	Muscle, kidney, liver, fish, honey, eggs, and milk	UHPLC-HRMS	Kaufmann et al. (2015)
32 β-lactam antibiotic residues from 12 penicillins, 14 cephalosporins, five carbapenems, and faropenem	Milk	UHPLC-MS/MS	Di Rocco et al. (2020)
Tigecycline, minocycline, chlortetracycline, and tetracycline	Broiler chicken	HPLC-MS/MS	Ziółkowski et al. (2020)
Diverse veterinary drug residues including sulfonamides and potentiators, macrolides, lincosamides, nitrofurans, nitroimidazoles, amphenicols, quinolones and fluoroquinolones, one triphenylmethane dye and its leuco-metabolite, and tetracyclines	Freshwater and marine seafood	UHPLC-MS/MS	Dinh et al. (2020)
78 Veterinary drugs from seven different classes	Eggs	UHPLC-MS/MS	Wang et al. (2021)
180 Veterinary drugs	Milk	UHPLC-Q-Orbitrap	Sun et al. (2021)

cephalosporin, in milk using reversed-phase LC with electrospray ionization (ESI) and QQQ-MS (Martins, Kussumi, Wang, & Lebre, 2007).

Stolker et al. (2008) developed a method for screening and determination of more than 100 veterinary drugs, including benzimidazoles, macrolides, penicillins, quinolones, sulphonamides, pyrimidines, tetracyclines, nitroimidazoles, tranquilizers, ionophores, amphenicols, and non-steroidal anti-inflammatory agents, in milk using ultra-performance liquid chromatography combined with time-of-flight mass spectrometry (UPLC-TOF-MS) (Stolker et al., 2008).

Peters, Bolck, Rutgers, Stolker, and Nielen (2009) used high-resolution liquid chromatography combined with time-of-flight mass spectrometry (HRLC-TOF-MS) for the identification of about 100 veterinary drugs in different food samples (meat, fish, and egg) (Peters, Bolck, Rutgers, Stolker, & Nielen, 2009).

Ortelli, Cognard, Jan, and Edder (2009) searched the usage of UPLC coupled with orthogonal acceleration TOF-MS for the identification of 150 veterinary drug residues and metabolites, avermectins, benzimidazoles, beta-agonists, beta-lactams, corticoids, macrolides, nitroimidazoles, quinolones, sulfonamides, tetracyclines, and some others, in raw milk (Ortelli, Cognard, Jan, & Edder, 2009).

Kantiani et al. (2009) developed a fully automated method for the detection of β-lactam antibiotics, including six penicillins (amoxicillin, ampicillin, cloxacillin, dicloxacillin, oxacillin, and penicillin G) and four cephalosporins (cefazolin, ceftiofur, cefoperazone, and cefalexin) in bovine milk samples. As a method, online solid-phase extraction–liquid chromatography/electrospray ionization–tandem mass spectrometry (SPE-LC-ESI-MS/MS) was used. After sample preparation steps, two MRM transitions were acquired for each analyte in the positive electrospray ionization mode (ESI(+)). LOQs in milk were between 0.09 and 1.44 ng/mL (Kantiani et al., 2009).

Lillenberg et al. (2010) used LC-MS to demonstrate if drugs can be a source of the contamination of food plants. The presence of ciprofloxacin, ofloxacin, norfloxacin, sulfadimethoxine, and sulfamethoxazole into lettuce root was searched. The result of this study showed that although no limits have been set for pharmaceutical residues in plant products, some residue values have been found (Lillenberg et al., 2010).

Kim et al. (2013) used LC coupled with MS to confirm the presence of two groups of antibiotics in animal production (Kim et al., 2013).

Kaufmann, Butcher, Maden, Walker, and Widmer (2015) developed a method to detect various nitrofuran and chloramphenicol residues in food using ultra-high-performance liquid chromatography coupled with high-resolution mass spectrometry (UHPLC-HRMS) and the results were compared with tandem quadrupole mass spectrometry (Kaufmann, Butcher, Maden, Walker, & Widmer, 2015).

Mitrowska and Antczak (2015) determined 16 sulfonamides in beeswax. After extraction, the determination was achieved by LC-MS/MS. The limits of detection (LODs) and limits of quantification were found to be from 1 to 2 μg/kg and from 2 to 5 μg/kg, respectively.

Han et al. (2015) determined 38 veterinary antibiotic residues in raw milk by UPLC-MS/MS, simultaneously. The limit of quantification for all antibiotics was 0.03–10 μg/kg (Han et al., 2015).

Zhang et al. (2015) developed a method for the determination of 90 veterinary drugs, including lincomycins, macrolides, sulfonamides, quinolones, tetracyclines, β-agonists, β-lactams, sedatives, β-receptor antagonists, sex hormones, glucocorticoids, nitroimidazoles, benzimidazoles, nitrofurans, and some others, in milk by ultra-performance liquid chromatography coupled with quadrupole time-of-flight mass spectrometry (UPLC-QTOF-MS). The limit of quantification for the drugs in the milk was found between 0.10 and 17.30 μg/kg (Zhang et al., 2015).

Piatkowska, Jedziniak, and Zmudzki (2016) developed a method to determine 120 analytes (6 of tetracyclines, 11 of fluoroquinolones, 17 of sulphonamides, 9 of nitroimidazoles, 2 of amphenicols, 7 of cephalosporins, 8 of penicillins, 8 of macrolides, 20 of benzimidazoles, 14 of coccidiostats, 3 of insecticides, 12 of dyes, and 3 of others) in fresh eggs, simultaneously.

Park et al. (2016) developed an analytical method for the detection of minocycline residues in porcine muscle and milk using liquid chromatography–triple quadrupole tandem mass spectrometry (LC-QQQ-MS/MS) with ESI. Limit of quantification was found as 10 μg/kg (Park et al., 2016).

Jank et al. (2017) developed a screening method for analysis of 46 antibiotics residues, belonging to different classes, such as tetracyclines, sulfonamides, fluoroquinolones, β-lactams, cephalosporins, macrolides, and other minority groups in bovine milk and bovine, swine, poultry, equine, fish, and shrimp meat samples. After sample preparation, instrumental analysis was performed using LC-MS/MS system. A total of 3,833 samples were analyzed using this method (Jank et al., 2017).

Ziółkowski et al. (2020) researched tigecycline, minocycline, chlortetracycline, and tetracycline in broiler chicken using HPLC-MS/MS (Ziółkowski et al., 2020).

Dinh et al. (2020) screened for diverse veterinary drug residues including sulfonamides and potentiators, macrolides, lincosamides, nitrofurans, nitroimidazoles, amphenicols, quinolones and fluoroquinolones, one triphenylmethane dye and its leuco-metabolite, and tetracyclines in freshwater and marine seafood products using ultra-high-performance liquid chromatography–tandem mass spectrometry (UHPLC-MS/MS). The LOD was found between 0.002 and 3 μg/kg (Dinh et al., 2020).

Di Rocco et al. (2020) developed an analytical method for 32 β-lactam antibiotic residues (12 penicillins, 14 cephalosporins, 5 carbapenems, and 1 faropenem) in milk samples using UHPLC-MS/MS. LODs and LOQs were found as between 0.0090 and 1.5 μg/kg and from 0.030 to 5.0 μg/kg, respectively (Di Rocco et al., 2020).

Wang et al. (2021) developed a method to identify and quantify 78 veterinary drugs, including homosulfanilamide, sulfacetamide, sulfaguanidine, sulfapyridine, sulfadiazine, sulfamethoxazole, sulfathiazole, diaveridine, sulfamerazine, sulfisoxazole, sulfamoxol, sulfamethizole, sulfabenzamide, sulfisomidine, sulfamethazine, sulfamonomethoxine, sulfamethoxypyridazine, sulfameter, sulfalozine, sulfachloropyridazine, trimethoprim, sulfaquinoxaline, sulfadoxine, sulfadimethoxine, sulfaphenazole, azaperone, xylazine, chlorpromazine, ronidazole, metronidazole, dimetridazole, ipronidazole, danofloxacin, ofloxacin, enoxacin, 1-(2-hydroxyethyl)-2-hydroxymethyl-5-nitroimidazole, 2-hydroxymethyl-1-methyl-5-nitroimidazole, 1-methyl-2-(2′-hydroxyisopropyl)-5-nitroimidazole, tiamulin, valnemulin, cefalexin, promazine, promethazine, carazolol, acepromazine, azaperol, propionpromazine, triflupromazine, albendazole, fenbendazole, mebendazole, thiabendazole, oxfendazole, 5-hydroxythiabendazole, albendazole sulfoxide, albendazole sulfone, fenbendazole sulfone, flubendazole, oxibendazole, 2-amino-albendazole sulfone, droperidol, haloperidol, nitrazepam, estazolam, oxazepam, perphenazine, diazepam, norfloxacin, pefloxacin, lomefloxacin, enrofloxacin, sarafloxacin, difloxacin, marbofloxacin, cefapirin, cefotaxime, cefquinome, and cefpirome, in eggs using UHPLC-MS/MS. The limit of quantification was between 0.1 and 1 μg/kg, which was sufficient to support surveillance monitoring (Wang et al., 2021).

Sun et al. (2021) developed a new approach, integrated data-dependent and data-independent acquisition method, using ultra-high-performance liquid chromatography coupled with quadrupole Orbitrap high-resolution mass spectrometry (UHPLC-Q-Orbitrap-HRMS) to analyze 180 veterinary drugs in milk (Sun et al., 2021).

13.4 CONCLUSION

Mass analysis is an important method for the identification and quantification of veterinary drug residues in food in terms of food security. The advances in mass analysis field are the monitoring of several compounds simultaneously with high sensitivity and accuracy. LC, either HPLC or UHPLC in combination with mass analyzer, are common equipment for the analysis of veterinary drug residues in animal-based food. Several researchers developed various methodologies to determine veterinary drug residues in milk, honey, eggs, meat, and chicken. These methodologies involve the simultaneous determination of various veterinary drug residues. However, the high cost of these instrumentations makes them the most commonly used analyzers in only routine laboratories.

REFERENCES

Andreu, V., Blasco, C., & Picó, Y. (2007). Analytical strategies to determine quinolone residues in food and the environment. *TrAC – Trends in Analytical Chemistry*, *26*(6), 534–556. https://doi.org/10.1016/j.trac.2007.01.010

Bogialli, S., Coradazzi, C., Di Corcoa, A., Lagana, A., & Sergi, M. (2007). A rapid method based on hot water extraction and liquid chromatography-tandem mass spectrometry for analyzing tetracycline antibiotic residues in cheese. *Journal of AOAC International*, *90*(3), 864–871. https://doi.org/10.1093/jaoac/90.3.864

Cycoń, M., Mrozik, A., & Piotrowska-Seget, Z. (2019). Antibiotics in the soil environment—degradation and their impact on microbial activity and diversity. *Frontiers in Microbiology*, *10*(Mar). https://doi.org/10.3389/fmicb.2019.00338

Di Rocco, M., Moloney, M., Haren, D., Gutierrez, M., Earley, S., Berendsen, B., … Danaher, M. (2020). Improving the chromatographic selectivity of β-lactam residue analysis in milk using phenyl-column chemistry prior to detection by tandem mass spectrometry. *Analytical and Bioanalytical Chemistry*, *412*(18), 4461–4475. https://doi.org/10.1007/s00216-020-02688-4

Dinh, Q. T., Munoz, G., Vo Duy, S., Tien Do, D., Bayen, S., & Sauvé, S. (2020). Analysis of sulfonamides, fluoroquinolones, tetracyclines, triphenylmethane dyes and other veterinary drug residues in cultured and wild seafood sold in Montreal, Canada. *Journal of Food Composition and Analysis*, *94*(August). https://doi.org/10.1016/j.jfca.2020.103630

Han, R. W., Zheng, N., Yu, Z. N., Wang, J., Xu, X. M., Qu, X. Y., … Wang, J. Q. (2015). Simultaneous determination of 38 veterinary antibiotic residues in raw milk by UPLC-MS/MS. *Food Chemistry*, *181*, 119–126. https://doi.org/10.1016/j.foodchem.2015.02.041

Jank, L., Martins, M. T., Arsand, J. B., Motta, T. M. C., Feijó, T. C., dos Santos Castilhos, T., … Pizzolato, T. M. (2017). Liquid chromatography–tandem mass spectrometry multiclass method for 46 antibiotics residues in milk and meat: development and validation. *Food Analytical Methods*, *10*(7), 2152–2164. https://doi.org/10.1007/s12161-016-0755-4

Kantiani, L., Farré, M., Sibum, M., Postigo, C., De Alda, M. L., & Barceló, D. (2009). Fully automated analysis of β-lactams in bovine milk by online solid phase extraction-liquid chromatography-electrospray-tandem mass spectrometry. *Analytical Chemistry*, *81*(11), 4285–4295. https://doi.org/10.1021/ac9001386

Kaufmann, A., Butcher, P., Maden, K., Walker, S., & Widmer, M. (2015). Determination of nitrofuran and chloramphenicol residues by high resolution mass spectrometry versus tandem quadrupole mass spectrometry. *Analytica Chimica Acta*, *862*, 41–52. https://doi.org/10.1016/j.aca.2014.12.036

Kim, D. P., Degand, G., Douny, C., Pierret, G., Delahaut, P., Ton, V. D., … Scippo, M.-L. L. (2013). Preliminary evaluation of antimicrobial residue levels in marketed pork and chicken meat in the Red River Delta region of Vietnam. *Food and Public Health*, *3*(6), 267–276. https://doi.org/10.5923/j.fph.20130306.02

Lillenberg, M., Herodes, K., Kipper, K., & Nei, L. (2010). Plant uptake of some pharmaceuticals from fertilized soils. *Proceedings of 2009 International Conference on Environmental Science and Technology*, 23–25 April, Bangkok, Thailand.

Martins, H. A., Kussumi, T. A., Wang, A. Y., & Lebre, D. T. (2007). A rapid method to determine antibiotic residues in milk using liquid chromatography coupled to electrospray tandem mass spectrometry. *Journal of the Brazilian Chemical Society*, *18*(2), 397–405. https://doi.org/10.1590/s0103-50532007000200023

Mitrowska, K., & Antczak, M. (2015). Determination of sulfonamides in beeswax by liquid chromatography coupled to tandem mass spectrometry. *Journal of Chromatography B: Analytical Technologies in the Biomedical and Life Sciences*, *1006*, 179–186. https://doi.org/10.1016/j.jchromb.2015.10.040

Ortelli, D., Cognard, E., Jan, P., & Edder, P. (2009). Comprehensive fast multiresidue screening of 150 veterinary drugs in milk by ultra-performance liquid chromatography coupled to time of flight mass spectrometry. *Journal of Chromatography B: Analytical Technologies in the Biomedical and Life Sciences*, *877*(23), 2363–2374. https://doi.org/10.1016/j.jchromb.2009.03.006

Park, J. A., Jeong, D., Zhang, D., Kim, S. K., Cho, S. H., Cho, S. M., … Shin, H. C. (2016). Simple extraction method requiring no cleanup procedure for the detection of minocycline residues in porcine muscle and milk using triple quadrupole liquid chromatography-tandem mass spectrometry. *Applied Biological Chemistry*, *59*(2), 297–303. https://doi.org/10.1007/s13765-016-0158-7

Peters, R. J. B., Bolck, Y. J. C., Rutgers, P., Stolker, A. A. M., & Nielen, M. W. F. (2009). Multi-residue screening of veterinary drugs in egg, fish and meat using high-resolution liquid chromatography accurate mass time-of-flight mass spectrometry. *Journal of Chromatography A*, *1216*(46), 8206–8216. https://doi.org/10.1016/j.chroma.2009.04.027

Piatkowska, M., Jedziniak, P., & Zmudzki, J. (2016). Multiresidue method for the simultaneous determination of veterinary medicinal products, feed additives and illegal dyes in eggs using liquid chromatography-tandem mass spectrometry. *Food Chemistry*, *197*, 571–580. https://doi.org/10.1016/j.foodchem.2015.10.076

Stolker, A. A. M., Rutgers, P., Oosterink, E., Lasaroms, J. J. P., Peters, R. J. B., Van Rhijn, J. A., & Nielen, M. W. F. (2008). Comprehensive screening and quantification of veterinary drugs in milk using UPLC-ToF-MS. *Analytical and Bioanalytical Chemistry*, *391*(6), 2309–2322. https://doi.org/10.1007/s00216-008-2168-8

Sun, F., Tan, H., Li, Y., De Boevre, M., Zhang, H., Zhou, J., … Yang, S. (2021). An integrated data-dependent and data-independent acquisition method for hazardous compounds screening in foods using a single UHPLC-Q-Orbitrap run. *Journal of Hazardous Materials*, *401*(June). https://doi.org/10.1016/j.jhazmat.2020.123266

Wang, C., Li, X., Yu, F., Wang, Y., Ye, D., Hu, X., … Xia, X. (2021). Multi-class analysis of veterinary drugs in eggs using dispersive-solid phase extraction and ultra-high performance liquid chromatography-tandem mass spectrometry. *Food Chemistry*, *334*(July), 127598. https://doi.org/10.1016/j.foodchem.2020.127598

Zhang, Y., Li, X., Liu, X., Zhang, J., Cao, Y., Shi, Z., & Sun, H. (2015). Multi-class, multi-residue analysis of trace veterinary drugs in milk by rapid screening and quantification using ultra-performance liquid chromatography-quadrupole time-of-flight mass spectrometry. *Journal of Dairy Science*, *98*(12), 8433–8444. https://doi.org/10.3168/jds.2015-9826

Ziółkowski, H., Jasiecka-Mikołajczyk, A., Madej-Śmiechowska, H., Janiuk, J., Zygmuntowicz, A., & Dąbrowski, M. (2020). Comparative pharmacokinetics of chlortetracycline, tetracycline, minocycline, and tigecycline in broiler chickens. *Poultry Science*, *99*(10), 4750–4757. https://doi.org/10.1016/j.psj.2020.06.038

14 Multi-Target Analysis and Suspect Screening of Xenobiotics in Milk by UHPLC-HRMS/MS

Mikel Musatadi
University of the Basque Country (UPV/EHU)

Belén González-Gaya, Mireia Irazola, Ailette Prieto, Nestor Etxebarria, Maitane Olivares, and Olatz Zuloaga
University of the Basque Country (UPV/EHU)
University of the Basque Country (PiE-UPV/EHU)

CONTENTS

14.1 Introduction 256
14.2 Materials and Methods 257
14.2.1 Reagents 257
14.2.2 Milk Samples 258
14.2.3 Extraction of Xenobiotics 258
14.2.4 Clean-Up 259
14.2.5 UHPLC-qOrbitrap Analysis 259
14.2.6 Target Analysis and Suspect Screening 260
14.2.7 Method Validation 260
14.3 Results and Discussion 261
14.3.1 Protein Precipitation Optimization 261
14.3.2 Fluoroquinolones, EDTA, and Oasis HLB Optimization 261
14.3.3 Addition of Salts and EDTA to the Extraction Solvent 262
14.3.4 Validation 264
14.3.4.1 Instrumental and Procedural Limits of Quantification (LOQs) 264
14.3.4.2 Linearity Ranges and Determination Coefficients (*r*2) 264
14.3.4.3 Absolute and Apparent Recoveries 265
14.3.4.4 Instrumental and Procedural Repeatability 266
14.3.4.5 Instrumental and Procedural Limits of Identification (LOIs) 267
14.3.5 Target Analysis of Commercial and Breast Milk Samples 267
14.3.6 Suspect Analysis of Commercial and Breast Milk Samples 270
14.4 Conclusions 270
Supplementary Materials 273
Author Contributions 274

Open Access
Separations **2021**, *8*(2), 14.

DOI: 10.3390/separations8020014-17

Funding 274
Institutional Review Board Statement 274
Informed Consent Statement 274
Conflicts of Interest 274
References 274

14.1 INTRODUCTION

Unexpected increase of health-related issues in humans at different life stages have raised concern about exposure to chemical compounds, since around 300 million tons of synthetic compounds are used in industrial and consumer products annually [1]. Moreover, the list of compounds with emerging interest is rapidly growing due to the metabolites and/or transformation products of the chemicals, alongside the newly synthesized ones [1]. Therefore, all living organisms are exposed to an overwhelming number of chemical compounds that can potentially trigger adverse health effects [2]. This way, the concept of "exposome", which engages all non-genetic factors that can be linked to adverse health outcomes [3,4], has gained considerable recognition [5].

At present, it is well established that around three-quarters of human diseases are related to exposure to chemical compounds [6]. Monitorization of the exposome could turn out as a useful tool to evaluate potential health risks and open new frontiers in the comprehension of external, internal, and non-specific exposures and their consequences on the health of living organisms, especially of humans [7]. Monitorization of biological fluids and tissues to find relevant biomarkers from the epidemiological point of view could identify potential subpopulations to suffer adverse health effects [8,9]. In this context, biomonitorization is gaining importance in epidemiological studies [10] and the European Union is boosting the biomonitorization of chemical compounds in humans to inform the understanding of exposure–response relationships [11].

Concerning biological fluids, urine and blood are typically selected [12]. However, bioaccumulation of organic molecules in breast milk has also drawn special attention in the last decade, since it is the major exposure source of contaminants to breastfed newborns [12] and a well-established risk factor for breast cancer in women [13]. Moreover, the increasing demand for human breast milk has caused a rapid growth of milk banks that do not follow regulations regarding organic micropollutants due to the absence of appropriate analytical tools, focusing only on the elimination of microorganisms [14,15]. Since breast is mainly built up with adipose tissue and the lipid content of the milk can go up to 5%, very persistent lipid-soluble compounds are likely to accumulate in breast milk [16]. Anyhow, due to the considerable water content of milk, more polar and water-ionizable compounds can also be present there [16]. Besides human breast milk, animal-origin milk has also been the center of attention in the last decade, especially bovine milk. Milk is a major constituent of the human diet worldwide and the widespread use of drugs and pesticides in dairy farming and agricultural practices, usually contaminate it with their residues [17]. Moreover, the increasing usage of illegal or off-license drugs and pesticides in dairy production further increases health risks to consumers [18].

Considering the undefinable amount of organic xenobiotics that can be present in milk and could have potential adverse health effects, the development of non-target methods is compulsory to properly identify all of them [19,20]. However, there is a lack of literature covering a wide range of chemical compounds in milk as far as polarity and hydrophilicity are concerned. In fact, most of the analyses have been limited to the classical non-polar priority compounds [21–25]. The determination of more polar compounds in a complicated matrix like milk, with variable quantities of lipids, proteins, sugars, vitamins, or minerals, remains a challenging task [25,26]. In recent literature, most works focus specifically on selected target compounds [27], such as selected pharmaceuticals [17,28], antibiotics [29–31], or phthalates [26], instead of analyzing the multiple xenobiotics present [23].

In the analytical procedures to determine xenobiotics in milk, liquid–liquid extraction (LLE) using non-polar solvents (e.g., diethyl ether, hexane, or dichloromethane) is one the most used extraction technique [32,33]. By adding ethanol to the non-polar solvents or using more polar solvents such as acetonitrile (AcN) or methanol (MeOH), extraction of the more polar xenobiotics is favored [24]. Moreover, the addition of salts such as anhydrous magnesium sulfate ($MgSO_4$) or sodium chloride (NaCl) enhances phase separation and ensures higher recoveries by salting-out effect. According to the literature, the effectiveness of LLE is improved by vigorously handshaking or using vortex [17,25]. Following the mentioned approaches, some works have employed high-speed solvent extraction procedures, which also use solvent mixtures to get medium polarities (e.g., acetone/hexane mixtures) and salts (e.g., anhydrous sodium sulfate [Na_2SO_4]) to extract non-polar and slightly polar compounds [34,35].

Under those non-selective extraction conditions, a significant amount of proteins and lipids is co-extracted alongside the organic xenobiotics [32,33]. Although a protein precipitation step is often performed just after the extraction step [25,33,36], a further clean-up of the extracts is mandatory to minimize the matrix effect at the detection step. In this sense, solid-phase extraction (SPE) has been mainly investigated for the removal of interferences. Considering the wide polarity of analytes, polymeric-based sorbents such as Oasis hydrophilic–lipophilic balance (HLB) have been widely used [29,30,37]. Clean-up mechanisms based on size exclusion such as miniaturized gel permeation chromatography [38] or Captiva Non-Drip (ND)-Lipids filters [25] are adequate for removing big biomolecules as well.

In the determination of polar organic compounds, liquid chromatography (LC) is mostly used since it allows the separation of a broad spectrum of compounds as far as polarity is concerned, which can be interesting in the analysis of emerging compounds and their metabolites and/or transformation products [39,40]. As for the detection step, mass spectrometry (MS) is the selected option in most recent works since it solves coelution problems that other detectors have [32,41]. Electrospray ionization (ESI) is preferably used to couple the MS with LC since it is capable of analyzing ionizable compounds within a large molecular weight range [42]. High-resolution mass spectrometry has shown to be a very powerful tool to identify unknown compounds present in milk and to get a more holistic understating of the exposome [23,25]. Hybrid detectors such as quadrupole time-of-flight (qTOF) or quadrupole-Orbitrap (qOrbitrap) allow performing tandem MS at high resolution obtaining both the MS1 (pseudomolecular ion and isotopic profile) and MS2 (fragmentation spectra), which allows the elucidation of unknown compounds [43].

According to the literature, the full scan data-dependent (dd)-MS2 acquisition mode has been used in a wide variety of samples without previous selection of suspects (discovery mode), in particular, in the analysis of river water [44], fish muscle [45], packaging materials [46], or sediments [47]. The main advantage of this acquisition mode is that the fragments in MS2 can be directly linked to their respective precursor in MS1 being the identification easier. On the downside, not all precursors are fragmented but only the selected (confirmation) or the most intense ones (discovery), as mentioned above [10,27].

Taking all into consideration, the analytical challenge nowadays is to develop methods to simultaneously detect the major number of polar xenobiotics present in human milk. The objective of the present work was to optimize the extraction and clean-up of a multi-target method (245 diverse analytes) for the determination of emerging xenobiotics in milk samples, and then, to extend the method to suspect screening with the aim of increasing the number of compounds (approximately 17,800 compounds) monitored in the samples, thereby gaining a better understanding of the exposome.

14.2 MATERIALS AND METHODS

14.2.1 Reagents

The 245 target analytes were selected to mimic as realistically as possible real exposure to xenobiotics that living organisms suffer throughout their entire lifespan, including diverse analytes in terms of polarity, acidity/basicity, functional groups, structures, molecular weight, and usage. The selected compounds are listed in Supplementary Table S1 together with the commercial vendors, purity of the standards, and the solvents used for preparing the stock solution of each individual compound.

Standard stock solutions were prepared in the 100–10,000 μg/g range using methanol MeOH (99.9%, ultra-high-performance liquid chromatography [UHPLC]-MS quality, Scharlab, Barcelona, Spain), AcN (ChromAR HPLC, Macron Fine Chemicals, Avantor, Radnor Township, PA, USA), acetone (ChromAR HPLC), ethanol (EtOH) (ChromAR HPLC), dimethyl sulfoxide (DMSO, Applichem, Panreac, Barcelona, Spain), and/or Milli-Q water (H_2O <0.05 μS/cm, Millipore 185, Millipore, Burlington, MA, USA), depending on the target compound. Solutions up to 2 μg/g containing all the target compounds (214 in optimization and 245 at validation) were prepared in MeOH and kept at 20°C in the darkness. A surrogate mixture solution of 1 μg/g containing [$2H_5$]-atrazine, [$13C_3$]-caffeine, [$2H_8$]-ciprofloxacin, [$2H_6$]-diuron, and [$2H_5$]-enrofloxacin was separately prepared in MeOH and stored under the same conditions as the target analytes. All solutions were freshly prepared according to the specific experimentation requirements.

AcN, MeOH, Milli-Q water, acetic acid (HOAc, 100%, Merck, Darmstadt, Germany), trifluoroacetic acid (TFA, >99.5%, Sigma-Aldrich, Darmstadt, Germany), trichloroacetic acid (TCA, >99.5%, Sigma-Aldrich), formic acid (HCOOH, 98.0, Honeywell, Fluka, Muskegon, MI, USA), anhydrous $MgSO_4$ (99.5%, Alfa Aesar, Haverhill, MA, USA), anhydrous Na_2SO_4 (100%, Panreac, Barcelona, Spain), and NaCl (100%, Merck, Darmstadt, Germany) were used during extraction and/or clean-up procedures. Ethylenediaminetetraacetic acid (EDTA, 99%, Panreac, Barcelona, Spain), NaOH pellets (99%, Merck, Darmstadt, Germany), and hydrochloric acid (HCl, 36%, Merck, Darmstadt, Germany) were used for preparing 30 mM EDTA (pH 4.0), 1 M NaOH (pH 13.0), and 0.1 mM HCl (pH 4.0) solutions, respectively. Ammonium chloride (NH_4Cl, 25%, Panreac, Barcelona, Spain) and ammonia (NH_3, 25%, AppliChem, Panreac, Barcelona, Spain) were also used for preparing 0.5 M ammonia buffer (NH_4+/NH_3, pH 9.0). For the clean-up procedures, Captiva Non-Drip (ND)-Lipid (100 mg, 3 mL, Agilent Technologies, Santa Clara, CA, USA) filters and Oasis HLB SPE cartridges (200 mg, 6 mL, Waters, Milford, MA, USA) were tested. Nitrogen (N_2, 99.999%, Air Liquide, Paris, France) was used for evaporating the extracts. Finally, HCOOH, Milli-Q water, and AcN (UHPLC-MS grade) used as mobile phase in the UHPLC-qOrbitrap were provided by Fischer Scientific (Merelbeke, Belgium). N_2 gas (99.999%), provided by Air Liquide (Madrid, Spain), was used as both nebulizer and drying gas.

14.2.2 Milk Samples

Several commercial and human breast milk samples were used in order to optimize the method and detect possible xenobiotics. For method optimization, treated (pasteurized) whole bovine (*Bos taurus*) milk was used. For method application, freeze-dried milk powder and untreated raw bovine milk were also used. All commercial milk samples were purchased from a local market. As for the breast milk samples, they were provided by four healthy and primiparous mothers from Biscay and anonymized for ethical reasons according to the Bioethics Committee rules of the University of the Basque Country (CEISH-UPV/EHU, BOPV 32, 17/2/2014, M10_2020_230). All milk samples were stored at −20°C until the analysis.

14.2.3 Extraction of Xenobiotics

Before extraction, all milk samples were thawed at room temperature. The optimization of the extraction step was carried out by spiking bovine milk samples with 214 target analytes to get concentrations around 300 ng/g in the final extract. Extractions were performed using vortex at maximum speed for 1 min, and the extraction solvents tested were: (i) AcN, (ii) AcN with different combinations of $MgSO_4$, Na_2SO_4, and NaCl, (iii) AcN:Milli-Q water (95:5, *v/v*) with 0.1% EDTA, (iv) MeOH:HOAc (95:5, *v/v*), and (v) MeOH with TFA or TCA (80:20, *v/v*).

Under optimal conditions, 1 mL of whole liquid milk and 3 mL of AcN were placed in polypropylene falcon tubes (40 mL, Deltalab, Barcelona, Spain), and 0.5 g of Na_2SO_4 and 0.1 g of NaCl were added to the mixture while the extraction was accelerated using vortex at maximum speed for 1 minute.

14.2.4 Clean-Up

Once the extraction step was over, samples were centrifuged at low temperature (4°C) for 15 minutes at 10,000 rpm (Centrifuge Allegra X-30R, F2402H, Beckman Coulter, Wycombe, UK) and the supernatant was quantitatively recovered. To enhance protein precipitation, the collected fractions were kept in the freezer at 20°C overnight. After protein precipitation, the supernatant was quantitatively recovered into a glass test tube for a further clean-up step.

For the optimization of the additional clean-up step, Captiva ND-Lipids filters and Oasis HLB cartridges at different conditions were tested individually and combined. For the optimal usage of the Captiva ND-Lipid filters, the recommendations of the supplier were followed [48]. The supernatant recovered after protein precipitation was evaporated to ~500 μL under a gentle stream of N_2 in a Turbovap LV Evaporator (Zymark, Biotage, Uppsala, Sweden) at 35°C. Then, 1,500 μL of AcN acidified with 0.1% of HCOOH was added to the cartridge as crash solvent followed by the ~500 μL of extract. The mixture was homogenized with the help of a pipette and filtered through the cartridge for biomolecules removal. Finally, the Captiva filters were dried under vacuum.

When Oasis HLB cartridges were used either for clean-up purposes or for solvent exchange after the addition of EDTA, the procedure explained hereinafter was carried out. The extracts obtained either from Captiva ND-Lipid filters or protein precipitation were evaporated to dryness at 35°C under an N_2 stream in the Turbovap and were reconstituted in 5 mL EDTA (30 mM, pH 4.0) before their loading onto the Oasis HLB cartridge. The pH of the solution was adjusted in each experiment using 0.5 and 1.0 mL of 0.5 M NH_4+/NH_3 buffer to obtain pH 6.0 and 9.0, respectively. The extracts were loaded onto Oasis HLB cartridges that were previously conditioned with 5 mL of MeOH and equilibrated with 5 mL of Milli-Q water adjusted at the corresponding pH (4.0, 6.0, or 9.0). After loading the samples, 5 mL of Milli-Q water were used as washing solution and the cartridges were completely dried under vacuum before the elution. The analytes were recovered using 5 mL of MeOH.

Under optimal conditions, the supernatant recovered from protein precipitation at low temperature was evaporated to ~500 μL under a gentle stream of N_2 at 35°C and the Captiva ND-Lipids protocol was followed.

14.2.5 UHPLC-qOrbitrap Analysis

The extracts obtained after clean-up were evaporated to dryness and re-dissolved in 200 μL MeOH. All samples and solutions were filtered before the analysis using 0.22 μm polypropylene filters (Membrane Solutions) in chromatography vials and kept in the freezer at 20°C until analysis. A Dionex Ultimate 3000 UHPLC (Thermo Fisher Scientific, Waltham, MA, USA) coupled with a high-performance Q Exactive Focus Orbitrap (qOrbitrap, Thermo Fisher Scientific, Waltham, MA, USA) mass analyzer with a heated electrospray ionization source (HESI, Thermo Fisher Scientific, Waltham, MA, USA) was used for the analysis of the xenobiotics. For instrumental control, Xcalibur 3.1 (Thermo Fischer Scientific, Waltham, MA, USA) was used.

Analyte separation was performed in an ACE UltraCore XB-C18 (2.1 mm, 150 mm, 1.7 μm) column with a pre-filter (2.1 mm ID, 0.2 μm) from Phenomenex. Milli-Q water (A line) and AcN (B line) were used as mobile phase, both containing 0.1% HCOOH and 5 mM ammonium acetate for positive and negative ionization modes, respectively. Column flow was set at 0.3 mL/min and the temperature was maintained at 50°C. Gradient elution started with 13% B that changed to 50% B in 10 minutes. Then, the composition of the B line was increased to 95% in 3 minutes and kept for 3 minutes. Finally, the mobile phase composition was changed to the initial conditions in 3 minutes.

Regarding the HESI parameters, spray voltage was set at 3.2 kV for positive and 3.5 kV for negative ionization modes. For positive ionization, the capillary temperature was set at 320°C, the sheath gas at 40 arbitrary units (au), the auxiliary gas at 15 au and 310°C, and the sweep gas at 1 au. For negative ionization, the capillary temperature was set at 300°C, the sheath gas at 40 arbitrary units (au), the auxiliary gas at 15 au and 280°C, and the sweep gas at 1 au.

External calibration of the qOrbitrap mass analyzer was performed every 3 days using Pierce linear ion trap LTQ ESI (Thermo Fisher Scientific, Waltham, MA, USA) calibration solutions. Measurements were performed in negative and positive ionization modes in the full scan – data-dependent MS2 (full MS-ddMS2) discovery acquisition mode in the mass-to-charge ratio (*m/z*) 70–1050 Da range. After a complete scan at 70,000 FWHM resolution at *m/z* 200, three scans were performed in the *m/z* 100–600 Da range at 17,500 FWHM at *m/z* 200 with an isolation window of 3.0 *m/z* with a stepped collision energy of 10, 45, and 90. The ddMS2 scans were run with an automatic intensity threshold and dynamic exclusion. Automated gain control (ACG) target was set at 5e4 and its minimum was set at 8.00e3.

14.2.6 Target Analysis and Suspect Screening

Target analysis and quantification were performed using the TraceFinder 5.0 software (Thermo Fischer Scientific, Waltham, MA, USA), which contained a homemade database including the retention time, exact mass (included in Supplementary Table S1), isotopic pattern, and characteristic MS2 fragments of each target compound. Regarding the criteria for target identification and subsequent quantification, a 60-second window was permitted for the retention times, while a 5 ppm error was allowed for monoisotopic masses and fragment ions. Moreover, 70% fitting was accepted for experimental and theoretical isotopic patterns. Suspect screening was performed using the Compound Discoverer 3.1 program (Thermo Fisher Scientific, Waltham, MA, USA), and as suspect list, compounds included in the mzCloud library were used (approximately 17,800 compounds). From the detected xenobiotics, endogenous compounds were discarded using The Human Metabolome Database (https://hmdb.ca/). To identify suspects, first, only features with a Lorentzian chromatographic peak shape and a minimum peak area of 107 were considered. Moreover, the feature should be present in the three replicates performed for each sample and the group variance should be lower than 30%. The ratio with respect to the procedural blanks should be equal to or higher than 10 as well. The Compound Discoverer 3.1 program provided all the features that, according to their exact mass and isotopic profile, matched with one or several of the compounds in the suspect list. Then, fitting higher than 70% in the case of the fragmentation spectrum was considered using the *mz* Cloud library. Finally, retention time was considered before confirmation: (i) when the pure standard was available, a deviation of 0.1 minute was admitted, and (ii) when not available, an estimation of the theoretical retention time was performed using the retention time index platform (http://rti.chem.uoa.gr/).

14.2.7 Method Validation

The analytical method was validated at two concentration levels after spiking bovine whole milk samples with the target analytes at 10 and 40 ng/g. Apart from the xenobiotics used in the optimization (214 compounds), 31 new compounds were added in order to have an even wider variety of analytes and mimic more realistically real cases (Supplementary Table S1). Extractions were performed in triplicate to calculate procedural repeatability. Procedural blanks were also analyzed in triplicate to check for possible cross-contamination or contamination through the process. All samples were spiked at 25 ng/g with a surrogate mixture containing [$2H_5$]-atrazine, [$13C_3$]-caffeine, [$2H_8$]-ciprofloxacin, [$2H_6$]-diuron, and [$2H_5$]-enrofloxacin for recovery correction.

For absolute recovery calculation, an external calibration consisting of eight points was built between the instrumental limit of quantification (LOQ_{inst}) and 300 ng/g. Calibration points corresponding to 2, 5, 10, 25, and 50 ng/g were injected in triplicate for LOQ_{inst} calculation. Instrumental repeatability was determined from the 50 ng/g calibration point injected in triplicate as well. Instrumental limits of identification (LOI_{inst}) were determined from the external calibration.

The apparent recoveries were determined by two different strategies: (i) surrogate correction and (ii) matrix-matched calibration. As for the surrogate correction, the absolute recoveries were corrected with the recoveries of the corresponding surrogates. Regarding the matrix-matched calibration, analyte-free commercial milk samples were spiked at seven concentration levels with 245 target compounds between the procedural limit of quantification (LOQ_{proc}) and 60 ng/g in milk. Calibration points corresponding to 1, 2, 5, and 10 ng/g in milk were injected in triplicate for LOQ_{proc} calculation. Procedural limits of identification (LOI_{proc}) were calculated from the matrix-matched calibration as well. Both external and matrix-matched calibrations were injected twice, at the beginning and the end of the sequence, in order to examine possible signal drift. MeOH was injected every six injections along the sequence to check for possible carryover.

14.3 RESULTS AND DISCUSSION

14.3.1 Protein Precipitation Optimization

Centrifugation at low temperature [36] and sample freezing in the presence of an organic solvent [25] are reported to be effective for protein precipitation after performing the extraction. These strategies can be combined with Captiva ND-Lipids filters for further clean-up to minimize matrix effect [25,45]. Other alternative methods, such as the addition of strong acids such as TFA and TCA also promote protein precipitation, and they have been widely used combined with MeOH as extraction solvent [29,30]. By quantitatively precipitating proteins, Captiva ND-Lipid filters could be avoided. However, considering the risk that those acids suppose, extremely precautious handling is needed. Therefore, a weaker acid like HOAc was also tested as a safer alternative.

In this sense, four experiments were performed employing different extractants: (i) AcN, (ii) MeOH:HOAc (95:5, *v/v*), (iii) MeOH:TFA (80:20, *v/v*), and (iv) MeOH:TCA (80:20, *v/v*). In all the cases, centrifugation and cold protein precipitation were performed, whereas in the assays done with pure AcN a further clean-up using the Captiva ND-Lipids filters was performed. To estimate the recovery of the procedures, external calibration was used to calculate the concentration of spiked milk samples ($n=3$) at 50 ng/g concentration in milk. As for precision, it was estimated in terms of relative standard deviation (RSD) of the three replicates. To establish a criterion for discarding analytes, only analytes with recoveries in the 10%–180% range and RSDs <30% were considered as "detected analytes" in all optimization experiments.

TFA and TCA in MeOH proved to be effective to promote severe protein precipitation and the use of Captiva ND-Filters could be avoided. However, HOAc was not strong enough to quantitatively precipitate proteins, being the procedure inviable (data not shown). The introduction of TFA and TCA, however, led to the loss of several analytes diverse in terms of polarity and acidity/basicity due to the extraordinarily strong acidic media set. In fact, recoveries around 40%–60% and higher RSD values (mean 12%) were obtained for the detected analytes (less than 80% in both approaches). Using AcN as extractant and Captiva ND-Lipids filters for clean-up instead, recoveries between 40% and 108% (mean 76%) and RSDs in the 2%–16% (mean 9%) range were obtained for the detected analytes (82%), respectively.

Bearing in mind that the proposed methodology should be useful for suspect screening analysis, the use of those strong acids was discarded since they allowed the extraction of a lower number of targets with lower recovery values. Taking those reasons into a consideration, AcN was chosen as extraction solvent and Captiva ND-Lipids filters for protein removal. However, in all the experiments fluoroquinolones (FQs) were not properly detected, probably due to their chelation to free calcium ions (Ca2+) in the liquid milk [29,31], so different strategies were tested to favor their detection.

14.3.2 Fluoroquinolones, EDTA, and Oasis HLB Optimization

According to literature, EDTA is useful for breaking down the chelation between Ca2+ and FQs since it creates very stable chelates with polyvalent cations [29,30]. Therefore, the extracts obtained from Captiva ND-Lipid filters were dried and reconstituted in EDTA. In that context, reverse-phase SPE was needed to perform solvent change from water to MeOH and preconcentrate analytes.

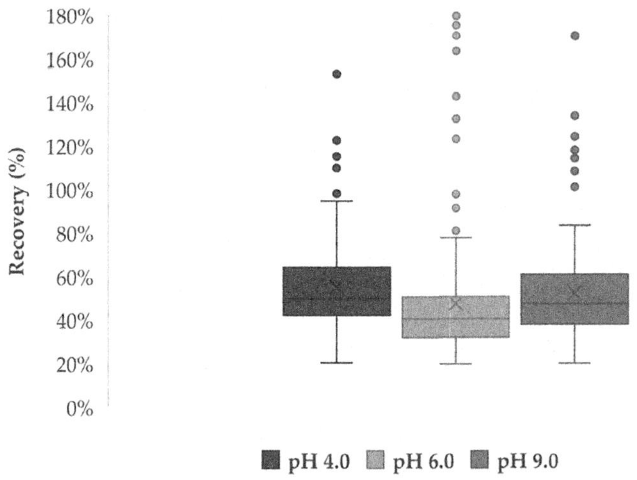

FIGURE 14.1 Boxplots of the recoveries (%) for Oasis HLB clean-up at pH 4.0, 6.0, and 9.0, showing minimum and maximum values (whiskers), percentiles (Q_1 and Q_3), average (x), median (line), and outliers (points).

According to the literature, Oasis HLB was chosen to fulfill that aim. Moreover, the introduction of Oasis HLB combined with Captiva ND-Lipids filters would provide even cleaner extracts, since Oasis HLB has been widely used to clean-milk extracts [29,30,37].

In a first approach, Oasis HLB was used at different pH values (4.0, 6.0, and 9.0) in order to check which pH allowed the preconcentration and subsequent detection of the maximum of compounds with the highest recovery. The recoveries for each condition are shown in Figure 14.1 as box plots.

As can be observed, the recoveries for the procedures using Oasis HLB at the tested pH values were acceptable, being most of them between 40% and 65%. The highest percentage of detected analytes was at pH 4.0 (76%) and pH 6.0 (75%) compared to pH 9.0 (64%). However, overall higher recoveries were obtained at pH 4.0 than at pH 6.0. Moreover, FQs were correctly detected at all of the experiments, showing that they were detached from Ca2+ ions and well retained at the Oasis HLB independently of the pH. Taking all the observations into account, pH 4.0 was chosen as the optimal pH for Oasis HLB use.

14.3.3 Addition of Salts and EDTA to the Extraction Solvent

Going further into the optimization, the influence of additives such as salts and EDTA in the extraction solvent was studied. According to the applications found in the literature, $MgSO_4$, Na_2SO_4, and NaCl are the most used salts in the extraction of xenobiotics in milk [17,25,34]. Therefore, different combinations of extractants with salts were tested are as follows: (i) AcN with NaCl, (ii) AcN with Na_2SO_4 and NaCl, and (iii) AcN with $MgSO_4$ and NaCl. However, since magnesium could also interfere with FQs in the same way as the calcium in the milk, separate extractions were set up by (iv) extracting first with AcN and subsequently with AcN, $MgSO_4$, and NaCl. This way, the extraction of the maximum number of compounds was ensured. In all the cases, centrifugation, cold protein precipitation, and Captiva ND-Lipids filtering steps were included. Moreover, all the approaches were also tested with Oasis HLB at pH 4.0 after reconstructing the dried extract from Captiva ND-Lipids filters in EDTA (30 mM, pH 4.0).

Furthermore, an additional experiment was carried out by adding EDTA to AcN in the extraction solvent to check whether it would be useful for breaking down the chelation between the Ca2+ ions and the FQs before extraction. The extractant used was (v) AcN:Milli-Q water 0.1% EDTA (95:5, *v/v*), and therefore the use of Oasis HLB was not necessary in that case. Table 14.1 summarizes the recoveries of the whole procedure for each experiment with the % of detected compounds.

TABLE 14.1
Results of the Experiments to Optimize the Addition of Salts

	Without Oasis HLB		With Oasis HLB	
Experiment	**Detected (%)**	**Recoveries (%)[a]**	**Detected (%)**	**Recoveries (%)[a]**
(i) AcN+NaCl	88	63–92 (80)	88	9–93 (46)
(ii) AcN+Na_2SO_4+NaCl	92	68–110 (90)	82	9–122 (44)
(iii) AcN+$MgSO_4$+NaCl	84	50–110 (78)	78	30–111 (69)
(iv) 1. AcN, 2. AcN+$MgSO_4$+NaCl	89	58–101 (81)	79	26–118 (50)
(v) AcN:Milli-Q water 0.1% EDTA (95:5, *v/v*)	81	7–85 (47)	Not performed	Not performed

[a] Recoveries are represented as the range and the mean recovery (in brackets) of all detected compounds.

The addition of EDTA in the extraction solvent allowed the detection of FQs by breaking down the chelation between the compounds and Ca2+ in the solution. Nevertheless, the water content in the extraction solution increased the evaporation time of the extracts. Not only that, but the recoveries were also, in general, worse than in the rest of the experiments for the rest of the compounds. Based on those observations and with the main aim of detecting as many compounds as possible, the addition of EDTA to the extraction solvent was discarded.

However, the addition of salts improved the recoveries due to the salting-out effect. Regarding method throughput, the addition of anhydrous Na_2SO_4 or $MgSO_4$ turned out as compulsory to remove moisture and to make the procedures much less time-consuming. Comparing the results obtained with both salts, Na_2SO_4 provided slightly better results since it allowed the detection of more compounds than $MgSO_4$, including FQs. Moreover, in the experiments where consecutive extractions were performed, better results were obtained comparing to a single extraction with $MgSO_4$.

As it can be concluded from the experiments using Oasis HLB, lower recoveries were obtained due to the extended procedure, leading to the loss of more analytes. To accurately assess the effect of Oasis HLB in the additional clean-up, matri effect at detection was studied. Matrix effect at the detection was calculated by spiking at 300 ng/g just after the whole treatment of non-spiked milk and calculating the concentration using external calibration. The results are gathered in Table 14.2, and values close to 100% represent a lack of matrix effect. However, the protocols using EDTA in the extraction solvent and a single extraction with $MgSO_4$ were discarded since the other approaches provided more promising results.

The outcomes of the experiments showed that a similar matrix effect was observed at the detection with and without Oasis HLB. However, this additional clean-up lowered absolute recoveries of the whole treatment and also led to a reduced number of detected compounds, whereas the addition of Na_2SO_4, promoted a higher number of compounds detected with reasonably good recoveries. Overall, it was the best approach considering laboratory viability, number of detected analytes, and

TABLE 14.2
Matrix Effect (%) at Detection

	Without Oasis HLB	With Oasis HLB
Experiment	**Recoveries (%)[a]**	**Recoveries (%)[a]**
(i) AcN+NaCl	85–115 (109)	72–108 (97)
(ii) AcN+Na_2SO_4+NaCl	92–119 (109)	81–112 (101)
(iv) 1. AcN 2. AcN+$MgSO_4$+NaCl	95–128 (114)	83–113 (102)

[a] Recoveries are represented as the range and the mean recovery (in brackets) of all detected compounds.

their respective recoveries with almost no matrix effect. Based on all those observations, method validation was performed under those optimal conditions that consisted of performing the extraction with AcN, NaCl, and Na_2SO_4 by vortex, and protein removal by centrifugation, precipitation at low temperature, and filtration through Captiva ND-Lipids filters.

14.3.4 Validation

The optimized analytical procedure to determine polar organic xenobiotics in milk was validated at two concentration levels, by spiking whole bovine milk at 10 and 40 ng/g for 245 compounds (214 used in the optimization plus 31 new compounds introduced for validation). All the results were corrected using the signals obtained from the procedural blanks, which included several phthalates leached from the plastic material used throughout the analytical process. All the results are individually collected as shown in Supplementary Table S2. The figures of merit of the optimized method are described hereinafter.

14.3.4.1 Instrumental and Procedural Limits of Quantification (LOQs)

Low-concentration points from the external (2, 5, 10, 25, and 50 ng/g in the extract) and matrix-matched (1, 2, 5, and 10 ng/g in the sample) calibration curves were injected in triplicate for calculating instrumental (LOQ_{inst}) and procedural (LOQ_{proc}) LOQs, respectively (see Section 14.2.7). In order to set the limits, precision and systematic error were considered. As for the precision, it was determined as the RSD (%) of the three replicates. With regard to the systematic error, it was calculated as the difference of the calculated and the real concentration (%) of the calibration points. LOQs were, therefore, set as the lowest concentration value fitting into a linear calibration curve with RSD and systematic error values lower than 30% (see Supplementary Table S2 for LOQs).

For most analytes (213 of 245, 87%), excellent LOQ_{inst} (below 20 ng/g) were obtained. The worst values, ranging between 100 and 200 ng/g, were obtained for bisphenol A, flutamide, meclocycline, and metribuzin. Considering the LOQ_{proc}, almost all analytes (90%, 221/245) provided results lower than 10 ng/g. The analytes with the highest LOQ_{proc} values (below 55 ng/g) were captopril, methotrexate, ciprofloxacin, di-octyl phthalate (DOP), gabapentin, bis(2-ethylhexyl)phthalate (DEHP), meclocycline, and hydroxychloroquine. According to literature, LOQ_{proc} lower than 30 ng/g has been obtained in the analysis of xenobiotics in milk samples [17,24,49]. However, it is worth mentioning that, in most cases, softer criteria were chosen to set LOQ values, with the only requirement being the signal-to-noise ratio (S/N) to be higher than 10. Moreover, in those works less analytes were studied, most of them focusing only on specific xenobiotics families, such as drugs or endocrine disruptors [17,24,49].

14.3.4.2 Linearity Ranges and Determination Coefficients ($r2$)

Linearity ranges for external and matrix-matched calibrates were studied by the determination coefficients ($r2$) that are collected at Supplementary Table S2 for each compound. The external calibrations, built between LOQ_{inst} and 300 ng/g, provided high $r2$ values since only seven compounds (3%) had lower values than 0.9500. These compounds were DEHP, ciprofloxacin, DOP, enoxacin, meclocycline, norfloxacin, and nonylphenol, for which semiquantitative analysis was considered.

In the case of the $r2$ values of the matrix-matched calibrations built between LOQ_{proc} and 60 ng/g in the sample, slightly worse values were obtained as expected, since each concentration point of the calibrates underwent separately the analytical procedure. In this case, 94% of the analytes provided $r2$ values higher than 0.9500 and for the rest (6%), semiquantitative analysis was considered. These compounds were nonylphenol, DOP, dibutyl phthalate, ofloxacin, enoxacin, metribuzin, gabapentin, parathion, enrofloxacin, ranitidine, caprolactam, metformin, imatinib, bis(2-ethylhexyl)adipate, and DEHP.

14.3.4.3 Absolute and Apparent Recoveries

Absolute recoveries (%) were determined as the ratio of the concentrations calculated from the external calibration and the spiked concentration. As for the apparent recoveries, those were calculated using two different approaches: (i) correction using isotopically labeled surrogates and (ii) matrix-matched calibration. The correction using surrogates was applied to each absolute recovery value with every surrogate, and the corrected apparent recovery closest to 100% with the lowest RSD value was chosen for each target analytes when possible. Which surrogate was chosen for each analyte whenever possible is also collected in Supplementary Table S2. In the case of the matrix-matched calibration, the apparent recoveries were determined by the ratio of the concentrations calculated from the curve and the spiked concentration.

Absolute and apparent recoveries obtained at low (10 ng/g) and high (40 ng/g) concentrations are shown as box plots in Figure 14.2a,b, respectively, while the results for each analyte are gathered

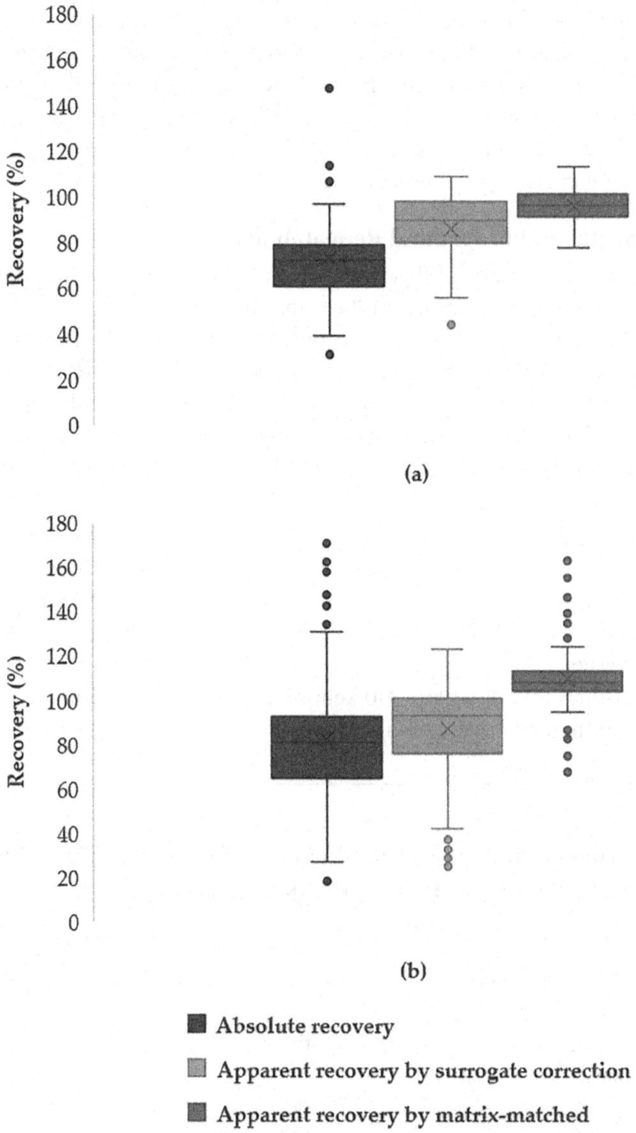

FIGURE 14.2 Absolute and apparent (surrogate correction and matrix-matched) recoveries (%) spiking milk samples at (a) 10 ng/g and (b) 40 ng/g.

in Supplementary Table S2. Recoveries between 70% and 130% were considered satisfactory while semiquantitative analysis was considered for compounds out of this recovery range.

At low concentration (Figure 14.2a), absolute recoveries were properly determined for 88 analytes (36%), since most of them showed values below 70% (mean value 73%). As for the apparent recoveries, considerably better results were obtained. With surrogate correction, absolute recoveries for 127 analytes (52%) were corrected with the mean value of the apparent recoveries being 86% while with matrix-matched calibration, apparent recoveries for 210 (86%) were properly determined (mean value 96%).

At high concentration (Figure 14.2b), higher absolute recoveries were obtained since 119 analytes were within the satisfactory range (mean 83%). With surrogate correction, the apparent recoveries for 140 analytes were acceptable (mean value 87%), and using matrix-matched good apparent recoveries between 70% and 130% (mean 110%), were obtained for 220 analytes.

Comparing the apparent recoveries obtained from both strategies, matrix-matched calibration rendered better results than surrogate correction. In order to improve the results obtained using surrogate correction, a larger number of labeled standards is necessary. In this sense, even though matrix-matched calibration requires more laboratory work, it is more profitable since expensive surrogates can be avoided. On the downside, the sample in which matrix-matched calibration curves are built can differ from the samples analyzed and this drawback can be avoided when proper surrogates are available for each target compound.

14.3.4.4 Instrumental and Procedural Repeatability

Instrumental and procedural repeatability was calculated in terms of RSD of the three injections of the 50 ng/g point from the external calibration and the three replicates analyzed at each spiked concentration level, respectively. For the procedural repeatability, RSD values at low and high concentrations were determined in the case of both absolute and apparent recoveries (four repeatability values for each compound). Since apparent recoveries were calculated by surrogate correction or matrix-matched, the highest value (worst scenario) was set for each analyte in each case. The results are gathered in Table 14.3 (see the specific values for each analyte in Supplementary Table S2).

As can be seen in Table 14.3, most analytes provided instrumental RSD values lower than 10%, showing the repetitiveness of the UHPLC-qOrbitrap measurements. Considering procedural values, higher RSDs were obtained as expected. Nevertheless, the repeatability was still acceptable with most analytes rendering values below 20%. Moreover, slightly better results were obtained at high concentrations as foreseen.

In a similar multi-target work where 200 xenobiotics were analyzed with UHPLC-qOrbitrap, also RSD values lower than 20% were obtained for instrumental repeatability [45]. For procedural

TABLE 14.3
Instrumental and Procedural Repeatability Values in Terms of RSD (%) for Absolute (abs.) and Apparent (app.) Recoveries at Low and High Concentrations

			Number of Analytes			
Repeatability	**Concentration**		**<10%**	**10–20%**	**20–50%**	**>50%**
Instrumental	50 ng/g (final extract)		234	8	2	1
	10 ng/g	abs. rec.	183	37	16	9
Procedural	(milk)	app. rec.	191	34	13	7
	40 ng/g	abs. rec.	197	26	15	7
	(milk)	app. rec.	195	28	15	7

repeatability, also RSD values below 20% have been obtained while determining xenobiotics in milk samples, although no other work has analyzed more than 200 compounds [17,23,25,36].

14.3.4.5 Instrumental and Procedural Limits of Identification (LOIs)

For the calculation of LOIs, the external and matrix-matched calibration points (instrumental, LOI_{inst}, and procedural, LOI_{proc}, values, respectively) were considered under a suspect screening approach using Compound Discoverer 3.1 (see Section 14.3.5). The lowest concentration values where the analytes could be identified were set as the LOIs (see Supplementary Table S2).

Overall, the LOIs for most compounds were below 10 ng/g. Regarding instrumental LOIs, analytes with higher values than 50 ng/g were naproxen, indomethacin, meclocycline, ethion, 2,6-di-tert-butyl-4-(dimethylaminomethyl)phenol, benzothiazole, fenthion, ciprofloxacin, enoxacin, pendimethalin, oryzalin, danofloxacin, enrofloxacin, glycitin, genistin, celecoxib, and hydrochlorothiazide. As for the procedural LOIs, the following analytes provided values higher than 20 ng/g: *bis*(methylglycol)phthalate, acyclovir, sotalol, captopril, naproxen, pyrantel, DOP, ethion, indomethacin, ranitidine, meclocycline, montelukast, enoxacin, erythromycin, oryzalin, 2,6-di-tert-butyl-4-(dimethylaminomethyl)phenol, hydroxychloroquine, and glycitin.

Since very few works performing suspect screening or non-target analysis of water-soluble organic xenobiotics in milk are available, identification limits are poorly investigated. Some works use limits of decision to set minimum concentration values at which xenobiotics need to be present to identify them [31]. However, this parameter is calculated following the EU 2002/657/EC regulation [50] that does not require suspect screening approaches to set the values. Some works have determined LOIs for these kinds of xenobiotics but in other complex matrices such as swine manure and fish muscle, with the instrumental values being lower than 30 ng/g [45,51].

14.3.5 Target Analysis of Commercial and Breast Milk Samples

Target analysis was performed in three commercial (whole bovine milk, raw bovine milk, and freeze-dried milk powder) and four breast milk (labeled A, B, C, and D) samples, following the details mentioned in Section 14.2.6. The concentration values were corrected by surrogates or matrix-matched whenever possible, taking into account the results from the validation, and expressed with 95% confidence level (2s, s being the standard deviation of the three procedural replicates). Only results that provided lower RSD values than 35% are shown in Table 14.4 (commercial milk) and Table 14.5 (breast milk).

Although several compounds were detected in treated whole and untreated bovine milk samples, most of them were below LOQs, except for acetaminophen (anti-inflammatory drug) and DOP (plasticizer) in the untreated milk, which have been previously determined in bovine milk [52,53]. In the freeze-dried milk powder samples, instead, several diverse compounds were quantified such as benzothiazole (food additive), caffeine (stimulant), genistein, genistin and glycitin (phytoestrogens), and 2-ethylhexyl-4-dimethylaminobenzoate (EHDAB, UV filter).

As far as breast milk samples are concerned, caffeine was quantified at high concentration (33–1,011 ng/g) in all of the samples, which was not surprising since it is common in several foods and drinks that are regularly taken by humans through the diet, and it has been detected in breast milk in other works as well [25,36]. Moreover, compounds such as DEHP and DOP (plasticizers) were quantified in half of the samples analyzed, while benzophenone and EHDAB (UV filters), orlistat (pharmaceutical), caprolactam and triethyl phosphate (industrial chemicals), and benzothiazole (food additive) were punctually detected. Several other xenobiotics were detected below LOQs.

TABLE 14.4
Target Analysis Results: Concentration of Xenobiotics (95%, 2 seconds) in Commercial Milk Samples

Compound	Whole Bovine Milk (ng/g)	Raw Bovine Milk (ng/g)	Milk Powder (ng/g)
Acetaminophen[a]	n.d	3.2±0.4	n.d
Benzothiazole[b]	n.d	n.d	11±6
Caffeinec	n.d	n.d	2.0±0.1
EHDABa	n.d	n.d	0.70±0.04
Genistein	n.d	n.d	20.1±0.3
Genistin	n.d	n.d	29±1
Glycitin	n.d	n.d	4.7±0.6
2-Hydroxybenzothiazole	<LOQ	n.d	<LOQ
Azoxystrobin	n.d	<LOQ	n.d
Benzethonium	n.d	<LOQ	n.d
Bicalutamide	<LOQ	<LOQ	<LOQ
BEHA	n.d	<LOQ	n.d
BEHP	n.d	<LOQ	n.d
Butylparaben	<LOQ	<LOQ	<LOQ
Carbendazim	<LOQ	n.d	n.d
Cortisone	<LOQ	<LOQ	<LOQ
Cotinine	<LOQ	n.d	<LOQ
Crotamiton	n.d	<LOQ	n.d
Diethyl Toluamide	<LOQ	<LOQ	<LOQ
DOPa	n.d	<LOQ	n.d
Enrofloxacin	<LOQ	n.d	n.d
Fenpropimorph	<LOQ	<LOQ	<LOQ
Finasteride	n.d	n.d	<LOQ
Ifosfamide	n.d	<LOQ	<LOQ
Methylparaben	<LOQ	n.d	n.d
Pirimicarb	<LOQ	<LOQ	<LOQ
Pirimiphos-methyl	n.d	n.d	<LOQ
Primidone	<LOQ	<LOQ	<LOQ
Progesterone	<LOQ	n.d	n.d
Propiconazole	n.d	<LOQ	n.d
Pyrazophos	<LOQ	n.d	n.d
Sulfamethazine	<LOQ	n.d	n.d
Sulfamethoxazole	<LOQ	n.d	n.d
Terbutryn	<LOQ	<LOQ	<LOQ

Data abbreviations: (<LOQ), Below limit of quantification; (n.d), non-detected. *Compounds abbreviations*: (EHDAB), 2-ethylhexyl-4-dimethylaminobenzoate; (BEHA), bis(2-ethylhexyl)adipate; (BEHP), bis(2-ethylhexyl)phthalate; (DOP), di-n-octyl phthalate.

[a] matrix-matched.

[b] $[^2H_5]$-atrazine.

[c] $[^{13}C_3]$-caffeine.

TABLE 14.5
Target Analysis Results: Average Concentration of Xenobiotics (95%, 2 seconds) in Breast Milk Samples

Compound	A (ng/g)	B (ng/g)	C (ng/g)	D (ng/g)
Benzophenone	n.d	n.d	4±1	n.d
Benzothiazole	45±4	n.d	n.d	n.d
BEHP[a]	45±30	n.d	48±29	n.d
Caffeine[b]	988±127	35±2	33±3	1,011±210
Caprolactam	27±2	n.d	n.d	n.d
DOP[a]	39±27	n.d	41±26	n.d
EHDAB[a]	n.d	n.d	n.d	0.58±0.01
MBHB[b]	<LOQ	n.d	4±2	n.d
Orlistat	n.d	n.d	2.3±0.4	n.d
Triethyl phosphate	n.d	n.d	20±8	n.d
2,4-Dinitrophenol	<LOQ	n.d	n.d	n.d
Azoxystrobin	n.d	<LOQ	n.d	n.d
Benzethonium	n.d	<LOQ	n.d	n.d
Bicalutamide	n.d	<LOQ	<LOQ	<LOQ
BEHA	n.d	n.d	<LOQ	n.d
BPA	<LOQ	n.d	n.d	n.d
Butylparaben	<LOQ	<LOQ	<LOQ	<LOQ
Cotinine	<LOQ	<LOQ	<LOQ	n.d
Crotamiton	<LOQ	n.d	<LOQ	<LOQ
DBP	n.d	n.d	<LOQ	n.d
Diethyl toluamide	<LOQ	n.d	<LOQ	<LOQ
Exemestane	n.d	n.d	<LOQ	n.d
Fluvoxamine	n.d	n.d	<LOQ	n.d
MBP	<LOQ	n.d	n.d	n.d
Medroxyprogesterone	<LOQ	<LOQ	<LOQ	n.d
Methylparaben	<LOQ	n.d	n.d	n.d
Norfloxacin	n.d	n.d	<LOQ	n.d
Pindolol	n.d	<LOQ	n.d	<LOQ
Pirimicarb	<LOQ	<LOQ	<LOQ	<LOQ
Propiconazole	<LOQ	<LOQ	<LOQ	<LOQ
Sotalol	<LOQ	n.d	n.d	<LOQ
Sulfamethoxazole	<LOQ	n.d	n.d	n.d
Tamoxifen	<LOQ	n.d	<LOQ	n.d
Trimethoprim	n.d	n.d	<LOQ	n.d
Triphenylphosphate	<LOQ	n.d	n.d	n.d

Data abbreviations: Below limit of quantification (<LOQ), non-detected (n.d). *Compounds abbreviations*: bis(2-ethylhexyl) phthalate (BEHP), di-n-octyl phthalate (DOP), 2-ethylhexyl-4-dimethylaminobenzoate (EHDAB), methyl 3,5-di-tert-butyl-4-hydroxybenzoate (MBHB), bis(2-ethylhexyl)adipate (BEHA), bisphenol A (BPA), dibutyl phthalate (DBP), methylbenzophenone (MBP).

[a] Matrix-matched.

[b] $[^{13}C_3]$-caffeine.

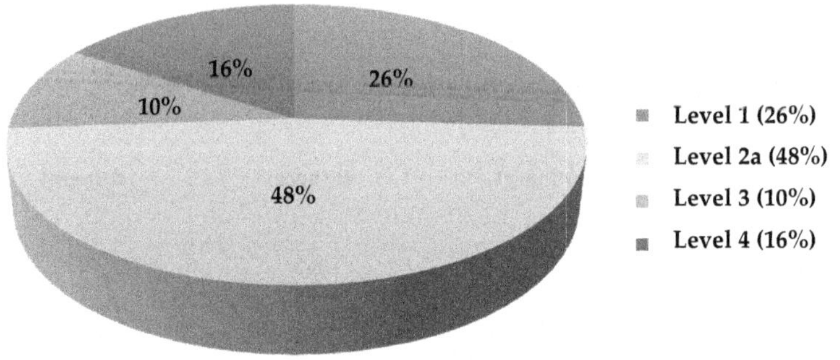

FIGURE 14.3 Distribution of the identification confidence levels of the xenobiotics by suspect screening.

14.3.6 Suspect Analysis of Commercial and Breast Milk Samples

Suspect screening was performed under the conditions described in Section 14.2.6, which allowed the identification of 50 xenobiotics of diverse origin (natural or artificial) at different identification levels [54] in the milk samples. The distribution of the confidence levels for the xenobiotics' identification is represented in Figure 14.3.

Among the identified compounds, 13 (26%) were identified at level 1 by confirmation via pure standard, precisely, the compounds quantified via target analysis. A wide variety of xenobiotics (24%) were confirmed at level 2a since library data from *mz*Cloud were used for MS2 confirmation. Some ubiquitous xenobiotics in commercial milk samples were 4-indolecarbaldehyde (plant metabolite), *bis*(2-ethylhexyl) amine (surfactant), isoquinoline (alkaloid). or saccharin (artificial sweetener), whereas in breast milk samples was frequent to find xenobiotics such as 2-(8-hydroxy-4a,8-dimethyldecahydro-2-naphthalenyl)acrylic acid (sesquiterpene), 2,5-di-tert-butylhydroquinone (industrial chemical), 4-indolecarbaldehyde (algal metabolite), avobenzone (UV filter), piperine (alkaloid), saccharin (artificial sweetener), or shogaol (plant component).

Several other xenobiotics like indoles or quinolones were identified in commercial and breast milk but it was not possible to differentiate between isomers (level 3, 10%). Indoles are industrially used as synthetic favoring compounds for perfume fixative roles, or they can be directly present in tobacco smoke [55]. As for the quinolones, they are broad-spectrum antibiotics, which have been previously detected in breast milk samples [56]. For other detected xenobiotics, it was not possible to annotate their MS2 (level 4, 16%).

As it has been observed, some xenobiotics were derived from plants, and are likely less harmful than artificial ones. All the xenobiotics identified at each level are collected in Table 14.6 indicating the ones with natural origin, while the extracted ion chromatograms of xenobiotics are identified at levels 1 and 2a in each commercial, and breast milk samples are gathered in Supplementary Figures S1–S9.

14.4 CONCLUSIONS

In this work, a multi-target method able to determine more than 200 xenobiotics in commercial and breast milk samples was developed, which was later extended to perform suspect screening using a database that contains almost 18,000 xenobiotic compounds. Although in most works dealing with contaminants in milk samples, mainly non-polar compounds are assessed, in this work, the monitoring of compounds with a wide range of polarities was carried out. Bearing in mind that aim, the optimal conditions for the analytical procedure were set to detect as many compounds as possible in the milk samples with an efficient lipid and protein removal.

In the optimized procedure, the importance of salts, such as anhydrous Na_2SO_4 and NaCl, was highlighted, which not only made the procedure straightforward due to moisture removal, but also enhanced the recoveries because of the salting-out effect. Moreover, Captiva ND-Lipids filters

TABLE 14.6
Suspect Screening of Commercial and Breast Milk: Xenobiotics Identified at Levels 1, 2a, 3 and 4

				Commercial Milk			Breast Milk			
Annotated Compound $[M+H]^+$	**Molecular Formula**	**Exact Mass**	t_R **(min)**	**1**	**2**	**3**	**A**	**B**	**C**	**D**
Level 1										
Acetaminophen	$C_8H_9NO_2$	151.06333	1.77	X	✓	X	X	✓	X	X
BEHP	$C_{24}H_{38}O_4$	390.27701	15.90	X	X	X	✓	X	✓	X
Benzophenone	$C_{13}H_{10}O$	182.07316	11.75	X	X	X	X	X	✓	X
Benzothiazole	C_7H_5NS	135.01427	6.50	X	X	✓	✓	X	X	X
Caffeine	$C_8H_{10}N_4O_2$	194.08038	2.50	X	X	✓	✓	✓	✓	✓
Caprolactam	$C_6H_{11}NO$	113.08406	2.35	X	X	X	✓	X	X	X
DOP	$C_{24}H_{38}O_4$	390.27701	15.90	X	<LOI	X	<LOI	X	<LOI	X
EHDAB	$C_{17}H_{27}NO_2$	277.20418	14.50	X	X	<LOI	X	X	X	<LOI
Genistein[a]	$C_{15}H_{10}O_5$	270.05282	7.55	X	X	✓	X	X	X	X
Genistin[a]	$C_{21}H_{20}O_{10}$	432.10565	4.30	X	X	✓	X	X	X	X
Glycitin[a]	$C_{22}H_{22}O_{10}$	446.12	130	3.43	X	X	<LOI	X	X	X
MBHB	$C_{16}H_{24}O_3$	264.17256	13.60	X	X	X	<LOI	X	<LOI	X
Triethyl phosphate	$C_6H_{15}O_4P$	182.0708	5.70	X	X	X	X	X	✓	X
Level 2a										
1,2,3,9-Tetrahydro-4 H-carbozol-4-one oxime	$C_{12}H_{12}N_2O$	200.09496	4.99	✓	X	X	X	X	X	X
3-Acetyl-2,5-dimethylfuran	$C_8H_{10}O_2$	138.06808	3.12	X	X	X	X	X	X	✓
3-Amino-2-phenyl-2H- pyrazolo [4,3-c]pyridine-4,6-diol	$C_{12}H_{10}N_4O_2$	242.08038	5.07	✓	✓	✓	✓	X	X	X
3-Methyl-5-(5,5,8a-trimethyl-2-methylene-7-oxodecahydro-1-naphthalenyl)pentyl acetate	$C_{22}H_{36}O_3$	348.26645	13.70	X	✓	X	✓	✓	✓	X
4-Indolecarbaldehyde[a]	C_9H_7NO	145.05276	3.05	✓	✓	✓	X	✓	✓	X
4-Methoxycinnamic acid[a]	$C_{10}H_{10}O_3$	178.06299	7.04	X	X	X	✓	X	X	X
5-Hydroxymethyl-2-furaldehyde	$C_6H_6O_3$	126.03169	1.63	X	X	✓	X	X	X	X
7-Methyl-2-phenylquinoline-4-carboxylic acid	$C_{17}H_{13}NO_2$	263.09463	12.09	✓	✓	X	X	X	X	X
Amfepramone	$C_{13}H_{19}NO$	205.14666	5.66	X	X	X	✓	X	X	X
Avobenzone	$C_{20}H_{22}O_3$	310.15689	14.40	X	X	X	X	✓	✓	X
Bis(2-ethylhexyl) amine	$C_{16}H_{35}N$	241.27695	11.34	✓	X	✓	X	X	X	X
Carvone	$C_{10}H_{14}O$	150.10447	11.72	X	X	X	X	X	✓	X
Citroflex A-4	$C_{20}H_{34}O_8$	402.22537	14.00	X	✓	X	X	X	X	X
Didecyldimethylammonium	$C_{22}H_{47}N$	325.37085	12.60	X	X	X	✓	X	X	X
Isoquinoline[a]	C_9H_7N	129.05785	3.80	✓	✓	✓	X	X	X	X
Nootkatone	$C_{15}H_{22}O$	218.16707	12.95	X	X	X	X	X	✓	X
OPEO	$C_{16}H_{26}O_2$	250.19328	13.07	X	X	X	X	✓	✓	✓
Piperine[a]	$C_{17}H_{19}NO_3$	285.13649	11.81	X	X	✓	X	X	✓	✓
Shogaol[a]	$C_{17}H_{24}O_3$	276.17254	10.70	X	✓	X	✓	✓	✓	X
Tetramethylene sulfoxide	C_4H_8OS	104.02959	1.46	X	X	X	✓	X	X	X

(Continued)

TABLE 14.6 (*Continued*)
Suspect Screening of Commercial and Breast Milk: Xenobiotics Identified at Levels 1, 2a, 3 and 4

Level 3										
2-Oxindole/4-hydroxyindole/5-hydroxyindole	C_8H_7NO	133.05276	1.54	✓	✓	✓	X	X	X	X
2-Hydroxyquinoline/8-hydroxyquinoline[a]	C_9H_7NO	145.05276	3.95	✓	X	✓	✓	X	X	X
1,5-Isoquinolinediol/2,4-quinolinediol	$C_9H_7NO_2$	161.04768	4.17	✓	✓	✓	X	X	X	X
1,5-Isoquinolinediol/2,4-quinolinediol	$C_9H_7NO_2$	161.04768	4.48	✓	X	X	X	X	X	X
Paraxanthine/theophylline/theobromine[a]	$C_7H_8N_4O_2$	180.06473	1.69	X	X	X	✓	✓	✓	✓
Level 4										
2-Oxindole/4-hydroxyindole/5-hydroxyindole/6-methylbenzoxazole	C_8H_7NO	133.05276	4.68	✓	X	✓	✓	✓	✓	✓
Pulegone/D,L-camphor/citral[a]	$C_{10}H_{16}O$	152.12012	7.81	X	X	X	✓	✓	✓	X
DL-2-(acetylamino)-3-phenylpropanoic acid/4-morpholinobenzoic acid	$C_{11}H_{13}NO_3$	207.08954	3.86	X	✓	X	X	X	X	X
2-(8-Hydroxy-4a,8-dimethyldecahydro-2-naphthalenyl) acrylic acid/2-[(2R,4aR,8R,8aR)-8-hydroxy-4a,8-dimethyl-decahydronaphthalen-2-yl] prop-2-enoic acid/polygodial	$C_{15}H_{24}O_3$	252.17254	11.83	X	X	X	X	X	✓	X
5-O-Methylgenistein/5,7-dihydroxy-3-(4-methoxyphenyl)-4H-chromen-4-one/biochanin A/glycitein/wogonin	$C_{16}H_{12}O_5$	284.06847	6.46	X	X	✓	X	X	X	X
1,4a-dimethyl-9-oxo-7-(propan-2-yl)-1,2,3,4,4a,9,10,10a-octahydrophenanthrene-1-carboxylic acid/Kahweol	$C_{20}H_{26}O_3$	314.18819	12.45	X	X	✓	X	X	✓	X
N-Benzylformamide/phenacylamine	C_8H_9NO	135.06841	3.13	✓	✓	✓	X	X	✓	✓
[M−H]⁻										
Level 2a										
2-(8-Hydroxy-4a,8-dimethyldecahydro-2-naphthalenyl)acrylic acid	$C_{15}H_{24}O_3$	252.17254	10.82	X	X	X	✓	X	✓	X
2,5-di-tert-butylhydroquinone	$C_{14}H_{22}O_2$	222.16198	12.49	X	X	X	✓	✓	✓	✓
Myristyl sulfate	$C_{14}H_{30}O_4S$	294.18648	12.22	X	X	X	✓	X	X	✓
Saccharin	$C_7H_5NO_3S$	182.99901	1.38	✓	✓	✓	✓	✓	✓	✓
Level 4										
Chrysin/daidzein/abietic acid	$C_{15}H_{10}O_4$	254.05791	6.09	X	X	✓	X	X	X	X

[a] Natural origin.

successfully removed co-extracted lipids and proteins ensuring high recoveries for the rest of the analytes, without the need of a further clean-up using SPE HLB cartridges.

The optimal procedure to determine xenobiotics in milk was validated at two concentration levels by spiking the milk samples with the target 245 analytes at 10 and 40 ng/g. Most analytes (220) were validated at high concentrations showing apparent recoveries between 70% and 130% that were determined by surrogate correction and/or matrix-matched calibration. As for the rest of figures of merit, satisfactory results were obtained considering the complexity of the matrix, the wide variety of analytes, and the small sample amount (1 mL) used. However, the use of isotopically labeled compounds should be further studied in the future since only five surrogates were used in this work.

The optimized multi-target procedure was successfully applied to quantify xenobiotics in three different commercial milk samples: treated and untreated raw bovine milk and freeze-dried milk powder. The liquid milk samples turned out to be almost free of the targeted xenobiotics since only acetaminophen (anti-inflammatory) and DOP (plasticizer) were quantified in the raw milk samples. However, several compounds were quantified in the milk powder, such as benzothiazole (food additive), caffeine (stimulant), genistein, genistin and glycitin (phytoestrogens), and EHDAB (UV filter).

Regarding breast milk samples, even though caffeine (stimulant) was quantified at the highest concentration in all of the samples, other xenobiotics such as DEHP and DOP (plasticizers), benzophenone, and EHDAB (UV filters) were also found. Apart from those xenobiotics, a wide variety of other xenobiotics that were not included in the target list were identified by suspect screening at different confidence levels including personal and pharmaceutical products, food additives, and disinfectants.

Although the presence of non-polar compounds is often expected in milk samples, the screening of compounds with a broad range of polarities is advisable in biomonitoring programs in order to have more information about the exposome. The suitability of the multi-target and suspect screening method for the analysis of milk samples opens the opportunity to extend the optimized method to other biological fluids such as blood or urine, to get a more holistic understanding of the exposome and find out the relationship between exposure to xenobiotics and adverse health effects.

SUPPLEMENTARY MATERIALS

The following are available online at https://www.mdpi.com/2297-8 739/8/2/14/s1, Figure S1: Extracted ion chromatograms (EICs) of the identified compounds at confidence levels 1 and 2a in commercial treated milk by suspect screening in positive ionization modes, Figure S2: Extracted ion chromatograms (EICs) of the identified compounds at confidence levels 1 and 2a in commercial raw milk by suspect screening in positive ionization modes, Figure S3: Extracted ion chromatograms (EICs) of the identified compounds at confidence levels 1 and 2a in commercial powder milk by suspect screening in positive ionization modes, Figure S4: Extracted ion chromatograms (EICs) of the identified compounds at confidence level 1 in breast milk A by suspect screening in positive ionization modes, Figure S5: Extracted ion chromatograms (EICs) of the identified compounds at confidence level 2a in breast milk A by suspect screening in positive ionization modes, Figure S6: Extracted ion chromatograms (EICs) of the identified compounds at confidence levels 1 and 2a in breast milk B by suspect screening in positive ionization modes, Figure S7: Extracted ion chromatograms (EICs) of the identified compounds at confidence level 1 in breast milk C by suspect screening in positive ionization modes, Figure S8: Extracted ion chromatograms (EICs) of the identified compounds at confidence level 2a in breast milk C by suspect screening in positive ionization modes, Figure S9: Extracted ion chromatograms (EICs) of the identified compounds at confidence levels 1 and 2a in breast milk D by suspect screening in positive ionization modes, Table S1: Information and physicochemical properties of the 245 target analytes, Table S2: Validation results for the target 245 analytes: instrumental (UHPLC-qOrbitrap) and procedural (Vortex–Captiva ND-Lipids–UHPLC-qOrbitrap) figures of merit.

AUTHOR CONTRIBUTIONS

Conceptualization: M.O. and O.Z.; methodology: M.M., M.O., and A.P.; software: M.I. and O.Z.; validation: M.M. and M.O.; formal analysis: M.M. and B.G.-G.; investigation: M.I., M.O., and O.Z.; resources: N.E.; data curation: M.O. and O.Z.; writing–original draft preparation: M.M.; writing–review and editing: M.M., M.O., and O.Z.; visualization: B.G.-G.; supervision: M.O. and O.Z.; project administration: N.E.; funding acquisition: N.E. All authors have read and agreed to the published version of the manuscript.

FUNDING

This research was funded by the Agencia Estatal de Investigación (AEI) of Spain, the European Regional Development Fund through projects CTM2017-84763-C3-1-R and CTM2017-90890-REDT (AEI/FEDER, EU), and the Basque Government through the financial support as consolidated group of the Basque Research System (IT1213-19). B.G. acknowledges a *Juan de la Cierva-Formación* fellowship by the Spanish Ministry for the Economy, Industry and Competitiveness (MINECO).

INSTITUTIONAL REVIEW BOARD STATEMENT

The study was conducted according to the guidelines of the Declaration of Helsinki, and approved by the Ethics Committee of the University of the Basque Country (CEISH-UPV/EHU, BOPV 32, 17/2/2014, M10_2020_230).

INFORMED CONSENT STATEMENT

Informed consent was obtained from all subjects involved in the study.

CONFLICTS OF INTEREST

The authors declare no conflicts of interest. The funders had no role in the design of the study; in the collection, analyses, or interpretation of data; in the writing of the manuscript, or in the decision to publish the results.

REFERENCES

1. Aceña, J.; Heuett, N.; Gardinali, P.; Pérez, S. Chapter 12: Suspect screening of pharmaceuticals and related bioactive compounds, their metabolites and their transformation products in the aquatic environment, biota and humans using LC-HR-MS techniques. In *Applications of Time-of-Flight and Orbitrap Mass Spectrometry in Environmental, Food, Doping, and Forensic Analysis*; Pérez, S., Eichhorn, P., Barceló, D., Eds.; Elsevier: Amsterdam, 2016; pp. 357–378.
2. Chiaia-Hernandez, A.C.; Krauss, M.; Hollender, J. Screening of lake sediments for emerging contaminants by liquid chromatography atmospheric pressure photoionization and electrospray ionization coupled to high resolution mass spectrometry. *Environ. Sci. Technol.* **2013**, *47*, 976–986.
3. Rappaport, S.M. Implications of the exposome for exposure science. *J. Expo. Sci. Environ. Epidemiol.* **2011**, *21*, 5–9.
4. Wild, C.P. Complementing the genome with an "exposome": The outstanding challenge of environmental exposure measurement in molecular epidemiology. *Cancer Epidemiol. Biomark. Prev.* **2005**, *14*, 1847–1850.
5. Dennis, K.K.; Jones, D.P. The exposome: A new frontier for education. *Am. Biol. Teach.* **2016**, *78*, 542–548.
6. Cui, Y.; Balshaw, D.M.; Kwok, R.K.; Thompson, C.L.; Collman, G.W.; Birnbaum, L.S. The exposome: Embracing the complexity for discovery in environmental health. *Environ. Health Perspect.* **2016**, *124*, A137–A140.

7. Escher, B.I.; Hackermüller, J.; Polte, T.; Scholz, S.; Aigner, A.; Altenburger, R.; Böhme, A.; Bopp, S.K.; Brack, W.; Busch, W.; et al. From the exposome to mechanistic understanding of chemical-induced adverse effects. *Environ. Int.* **2017**, *99*, 97–106. [CrossRef]
8. Patel, C.J.; Bhattacharya, J.; Butte, A.J. An Environment-Wide Association Study (EWAS) on type 2 diabetes mellitus. *PLoS ONE* **2010**, *5*, e10746.
9. Vineis, P.; Chadeau-Hyam, M.; Gmuender, H.; Gulliver, J.; Herceg, Z.; Kleinjans, J.; Kogevinas, M.; Kyrtopoulos, S.; Nieuwenhui-jsen, M.; Phillips, D.H.; et al. The exposome in practice: Design of the EXPOsOMICS project. *Int. J. Hyg. Environ. Health* **2017**, *220*, 142–151.
10. Andra, S.S.; Austin, C.; Patel, D.; Dolios, G.; Awawda, M.; Arora, M. Trends in the application of high-resolution mass spectrometry for human biomonitoring: An analytical primer to studying the environmental chemical space of the human exposome. *Environ. Int.* **2017**, *100*, 32–61.
11. Exposure and Health. HBM4EU—Science and Policy for a Healthy Future. Available online: https://www.hbm4eu.eu/the-project/exposure-and-health/ (accessed on 2 July 2020).
12. Yusa, V.; Millet, M.; Coscolla, C.; Roca, M. Analytical methods for human biomonitoring of pesticides. A review. *Anal. Chim. Acta* **2015**, *891*, 15–31.
13. Siddique, S.; Kubwabo, C.; Harris, S.A. A review of the role of emerging environmental contaminants in the development of breast cancer in women. *Emerg. Contam.* **2016**, *2*, 204–219.
14. García Lara, N.R.; Peña Caballero, M. Riesgos asociados al uso no controlado de la leche materna donada. *An. Pediatría* **2017**, *86*, 237–239.
15. Irazusta, A. High-pressure homogenization and high hydrostatic pressure processing of human milk: Preservation of immuno-logical components for human milk banks. *J. Dairy Sci.* **2020**, *103*, 14.
16. Stefanidou, M.; Maravelias, C.; Spiliopoulou, C. Human exposure to endocrine disruptors and breast milk. *Endocr. Metab. Immune Disord. Drug Targets* **2009**, *9*, 269–276.
17. Jadhav, M.R.; Pudale, A.; Raut, P.; Utture, S.; Shabeer, T.A.; Banerjee, K. A unified approach for high-throughput quantitative analysis of the residues of multi-class veterinary drugs and pesticides in bovine milk using LC-MS/MS and GC–MS/MS. *Food Chem.* **2019**, *272*, 292–305.
18. Beyene, T. Veterinary drug residues in food-animal products: Its risk factors and potential effects on public health. *J. Vet. Sci. Technol.* **2015**, *7*, 1–7.
19. Díaz, R.; Ibáñez, M.; Sancho, J.V.; Hernández, F. Target and non-target screening strategies for organic contaminants, residues and illicit substances in food, environmental and human biological samples by UHPLC-QTOF-MS. *Anal. Methods* **2012**, *4*, 196–209.
20. Plassmann, M.M.; Brack, W.; Krauss, M. Extending analysis of environmental pollutants in human urine towards screening for suspected compounds. *J. Chromatogr. A* **2015**, *1394*, 18–25.
21. Nickerson, K. Environmental contaminants in breast milk. *J. Midwifery Women Health* **2006**, *51*, 26–34.
22. Cano-Sancho, G.; Alexandre-Gouabau, M.-C.; Moyon, T.; Royer, A.-L.; Guitton, Y.; Billard, H.; Darmaun, D.; Rozé, J.-C.; Boquien, C.-Y.; Le Bizec, B.; et al. Simultaneous exploration of nutrients and pollutants in human milk and their impact on preterm infant growth: An integrative cross-platform approach. *Environ. Res.* **2020**, *182*, 109018.
23. Tran, C.D.; Dodder, N.G.; Quintana, P.J.E.; Watanabe, K.; Kim, J.H.; Hovell, M.F.; Chambers, C.D.; Hoh, E. Organic contaminants in human breast milk identified by non-targeted analysis. *Chemosphere* **2020**, *238*, 124677.
24. Rodríguez-Gómez, R.; Jiménez-Díaz, I.; Zafra-Gómez, A.; Ballesteros, O.; Navalón, A. A multiresidue method for the determination of selected endocrine disrupting chemicals in human breast milk based on a simple extraction procedure. *Talanta* **2014**, *130*, 561–570.
25. Baduel, C.; Mueller, J.F.; Tsai, H.; Gomez Ramos, M.J. Development of sample extraction and clean-up strategies for target and non-target analysis of environmental contaminants in biological matrices. *J. Chromatogr. A* **2015**, *1426*, 33–47.
26. Adenuga, A.A.; Ayinuola, O.; Adejuyigbe, E.A.; Ogunfowokan, A.O. Biomonitoring of phthalate esters in breast-milk and urine samples as biomarkers for neonates' exposure, using modified quechers method with agricultural biochar as dispersive solid-phase extraction absorbent. *Microchem. J.* **2020**, *152*, 104277.
27. Pourchet, M.; Debrauwer, L.; Klanova, J.; Price, E.J.; Covaci, A.; Caballero-Casero, N.; Oberacher, H.; Lamoree, M.; Damont, A.; Fenaille, F.; et al. Suspect and non-targeted screening of chemicals of emerging concern for human biomonitoring, environmental health studies and support to risk assessment: From promises to challenges and harmonisation issues. *Environ. Int.* **2020**, *139*, 105545.
28. Shishov, A.; Nechaeva, D.; Bulatov, A. HPLC-MS/MS determination of non-steroidal anti-inflammatory drugs in bovine milk based on simultaneous deep eutectic solvents formation and its solidification. *Microchem. J.* **2019**, *150*, 104080.

29. Herrera-Herrera, A.V.; Hernández-Borges, J.; Rodríguez-Delgado, M.Á. Fluoroquinolone antibiotic determination in bovine, ovine and caprine milk using solid-phase extraction and high-performance liquid chromatography-fluorescence detection with ionic liquids as mobile phase additives. *J. Chromatogr. A* **2009**, *1216*, 7281–7287.
30. Meng, Z.; Shi, Z.; Liang, S.; Dong, X.; Li, H.; Sun, H. Residues investigation of fluoroquinolones and sulphonamides and their metabolites in bovine milk by quantification and confirmation using ultra-performance liquid chromatography–tandem mass spectrometry. *Food Chem.* **2015**, *174*, 597–605.
31. Kantiani, L.; Farré, M.; Barceló, D. Rapid residue analysis of fluoroquinolones in raw bovine milk by online solid phase extraction followed by liquid chromatography coupled to tandem mass spectrometry. *J. Chromatogr. A* **2011**, *1218*, 9019–9027.
32. Lopes, B.R.; Barreiro, J.C.; Cass, Q.B. Bioanalytical challenge: A review of environmental and pharmaceuticals contaminants in human milk. *J. Pharm. Biomed. Anal.* **2016**, *130*, 318–325.
33. Jiménez-Díaz, I.; Vela-Soria, F.; Rodríguez-Gómez, R.; Zafra-Gómez, A.; Ballesteros, O.; Navalón, A. Analytical methods for the assessment of endocrine disrupting chemical exposure during human fetal and lactation stages: A review. *Anal. Chim. Acta* **2015**, *892*, 27–48.
34. Devanathan, G.; Subramanian, A.; Sudaryanto, A.; Takahashi, S.; Isobe, T.; Tanabe, S. Brominated flame retardants and polychlorinated biphenyls in human breast milk from several locations in India: Potential contaminant sources in a municipal dumping site. *Environ. Int.* **2012**, *39*, 87–95.
35. Asante, K.A.; Adu-Kumi, S.; Nakahiro, K.; Takahashi, S.; Isobe, T.; Sudaryanto, A.; Devanathan, G.; Clarke, E.; Ansa-Asare, O.D.; Dapaah-Siakwan, S.; et al. Human exposure to PCBs, PBDEs and HBCDs in Ghana: Temporal variation, sources of exposure and estimation of daily intakes by infants. *Environ. Int.* **2011**, *37*, 921–928.
36. López-García, E.; Mastroianni, N.; Postigo, C.; Valcárcel, Y.; González-Alonso, S.; Barceló, D.; López de Alda, M. Simultaneous LC–MS/MS determination of 40 legal and illegal psychoactive drugs in breast and bovine milk. *Food Chem.* **2018**, *245*, 159–167.
37. Calafat, A.M.; Slakman, A.R.; Silva, M.J.; Herbert, A.R.; Needham, L.L. Automated solid phase extraction and quantitative analysis of human milk for 13 phthalate metabolites. *J. Chromatogr. B* **2004**, *805*, 49–56.
38. Malarvannan, G.; Kunisue, T.; Isobe, T.; Sudaryanto, A.; Takahashi, S.; Prudente, M.; Subramanian, A.; Tanabe, S. Organohalogen compounds in human breast milk from mothers living in Payatas and Malate, the Philippines: Levels, accumulation kinetics and infant health risk. *Environ. Pollut.* **2009**, *157*, 1924–1932.
39. Jamin, E.L.; Bonvallot, N.; Tremblay-Franco, M.; Cravedi, J.-P.; Chevrier, C.; Cordier, S.; Debrauwer, L. Untargeted profiling of pesticide metabolites by LC-HRMS: An exposomics tool for human exposure evaluation. *Anal. Bioanal. Chem.* **2014**, *406*, 1149–1161.
40. Gago-Ferrero, P.; Schymanski, E.L.; Hollender, J.; Thomaidis, N.S. Chapter 13—Nontarget analysis of environmental samples based on liquid chromatography coupled to high resolution mass spectrometry (LC-HRMS). In *Comprehensive Analytical Chemistry*; Pérez, S., Eichhorn, P., Barceló, D., Eds.; Elsevier: Amsterdam, 2016; pp. 381–403.
41. Hermo, M.P.; Nemutlu, E.; Kır, S.; Barrón, D.; Barbosa, J. Improved determination of quinolones in milk at their MRL levels using LC–UV, LC–FD, LC–MS and LC–MS/MS and validation in line with regulation 2002/657/EC. *Anal. Chim. Acta* **2008**, *613*, 98–107.
42. Gross, J.H. Mass spectrometry. A textbook. *Anal. Bioanal. Chem.* **2005**, *381*, 1319–1320.
43. Cortéjade, A.; Kiss, A.; Cren, C.; Vulliet, E.; Buleté, A. Development of an analytical method for the targeted screening and multi-residue quantification of environmental contaminants in urine by liquid chromatography coupled to high resolution mass spectrometry for evaluation of human exposures. *Talanta* **2016**, *146*, 694–706.
44. Menger, F.; Ahrens, L.; Wiberg, K.; Gago-Ferrero, P. Suspect screening based on market data of polar halogenated micropollutants in river water affected by wastewater. *J. Hazard. Mater.* **2021**, *401*, 123377.
45. Musatadi, M.; González-Gaya, B.; Irazola, M.; Prieto, A.; Etxebarria, N.; Olivares, M.; Zuloaga, O. Focused ultrasound-based extraction for target analysis and suspect screening of organic xenobiotics in fish muscle. *Sci. Total Environ.* **2020**, *740*, 139894.
46. Blanco-Zubiaguirre, L.; Zabaleta, I.; Usobiaga, A.; Prieto, A.; Olivares, M.; Zuloaga, O.; Elizalde, M.P. Target and suspect screening of substances liable to migrate from food contact paper and cardboard materials using liquid chromatography-high resolution tandem mass spectrometry. *Talanta* **2020**, *208*, 120394.
47. Weiss, J.M.; Simon, E.; Stroomberg, G.J.; de Boer, J.; de Boer, R.; van der Linden, S.C.; Leonards, P.E.G.; Lamoree, M.H. Identification strategy for unknown pollutants using high-resolution mass spectrometry: Androgen-disrupting compounds identified through effect-directed analysis. *Anal. Bioanal. Chem.* **2011**, *400*, 3141–3149.

48. Agilent Captiva ND and Captiva ND Lipids Method Guide. Available online: https://www.crawford-scientific.com/media/wysiwyg//Literature/Sample_Prep/Agilent/Captiva-ND-and-ND-Lipids-Method-Guide.pdf (accessed on 25 April 2019).
49. Dualde, P.; Pardo, O.; Fernández, S.F.; Pastor, A.; Yusà, V. Determination of four parabens and bisphenols A, F and S in human breast milk using QuEChERS and liquid chromatography coupled to mass spectrometry. *J. Chromatogr. B* **2019**, *1114–1115*, 154–166.
50. Available online: http://europa.eu.int/eur-lex/pri/en/oj/dat/2002/l_221/l_22120020817en00080036.pdf (accessed on 19 January 2021).
51. Solliec, M.; Roy-Lachapelle, A.; Sauvé, S. Development of a suspect and non-target screening approach to detect veterinary antibiotic residues in a complex biological matrix using liquid chromatography/high-resolution mass spectrometry. *Rapid Commun. Mass Spectrom.* **2015**, *29*, 2361–2373.
52. Peng, T.; Zhu, A.-L.; Zhou, Y.-N.; Hu, T.; Yue, Z.-F.; Chen, D.-D.; Wang, G.-M.; Kang, J.; Fan, C.; Chen, Y.; et al. Development of a simple method for simultaneous determination of nine subclasses of non-steroidal anti-inflammatory drugs in milk and dairy products by ultra-performance liquid chromatography with tandem mass spectrometry. *J. Chromatogr. B* **2013**, *933*, 15–23.
53. Lin, J.; Chen, W.; Zhu, H.; Wang, C. Determination of free and total phthalates in commercial whole milk products in different packaging materials by gas chromatography-mass spectrometry. *J. Dairy Sci.* **2015**, *98*, 8278–8284.
54. Schymanski, E.L.; Jeon, J.; Gulde, R.; Fenner, K.; Ruff, M.; Singer, H.P.; Hollender, J. Identifying small molecules via high resolution mass spectrometry: Communicating confidence. *Environ. Sci. Technol.* **2014**, *48*, 2097–2098.
55. PubChem. Hazardous Substances Data Bank (HSDB): 599. Available online: https://pubchem.ncbi.nlm.nih.gov/source/hsdb/599#section=Interactions-(Complete) (accessed on 6 December 2020).
56. Yalçin, S.S.; Günes¸, B.; Yalçin, S. Incredible pharmaceutical residues in human milk in a cohort study from Şanlıurfa in Turkey. *Environ. Toxicol. Pharmacol.* **2020**, *80*, 103502.

15 Analytical Determination of Persistent Organic Pollutants from Food Sources Using High-Resolution Mass Spectrometry

Jeyabalan Sangeetha
Central University of Kerala

Mahesh Pattabhiramaiah and Shanthala Mallikarjunaiah
Bangalore University

Devarajan Thangadurai, Lokeshkumar Prakash, Ravichandra Hospet, and Muniswamy David
Karnatak University

Inamul Hasan Madar
Kaohsiung Medical University

Saher Islam
Lahore College for Women University

Akhil Silla, M. Supriya, and R. Preetha
SRM Institute of Science and Technology

Ghazala Sultan
Aligarh Muslim University

Ramachandran Chelliah and Deog-Hwan Oh
Kangwon National University

Anand Torvi
Jain University

Zaira Zaman Chowdhury
University of Malaya

DOI: 10.1201/9781003091226-18

CONTENTS

15.1 Introduction 280
15.2 GC-MS for POPs Analysis 282
15.3 Application of GC-MS for POPs Determination in Different Food Samples 282
15.4 Source Identification of Persistent Organic Pollutants Using GC-MS/MS 283
15.5 LC-MS for Analysis of Emerging POPs 286
15.6 Conclusion 287
Reference 287

15.1 INTRODUCTION

In recent decades, the focus has been on a subset of hazardous organic chemicals, primarily of anthropogenic origin, commonly known as persistent organic pollutants (POPs) (Jones and de Voogt 1999; Loganathan and Masunaga 2009; El-Shahawi et al. 2010; Ashraf 2017; Gaur et al. 2018). POPs are a class of carbon-based organic hazardous chemicals that are persistent, moderate to low volatility, prone to deterioration, bioaccumulative, deliberately developed for different purposes, or produced as byproducts of combustion or manufacturing processes, and have the capacity for long-range transport. Currently, roughly 1,209 pesticides and their metabolites and degradation products belonging to more than 100 chemical groups are found in the environment. Three types of POPs are present in the environment: (i) pesticides, in general, organochlorine pesticides (OCPs) including some dichlorodiphenyltrichloroethane (DDT) and its metabolites; (ii) industrial and technical chemicals, which also include polychlorinated biphenyls (PCBs), polybrominated diphenyl ethers (PBDEs), and perfluorooctane sulfonate; and (iii) byproducts of industrial processes, such as polychlorinated dibenzo-*p*-dioxins (PCDDs), polychlorinated dibenzofurans (PCDFs), and polyaromatic hydrocarbons (PAHs) (Gaur et al. 2018). PAHs do not directly contain POPs and are recognized only as POPs in the Aarhus Protocol as they can be metabolized quickly and thus prevent further bioaccumulation (UNECE 1998). PAHs are, however, often categorized as POPs in several studies due to their lipophilicity and continuous release (Ashraf 2017; Alharbi et al. 2018). Reputable POPs among the contaminants intended for removal under the Stockholm Convention include OCPs, PCBs, PCDDs, and PCDFs. POPs enter our foods as contaminants and are linked to a wide range of health conditions, including hormonal distortion, cancer, cardiovascular diseases, and obesity (El-Shahawi et al. 2010; Ashraf 2017; Alharbi et al. 2018; Gaur et al. 2018).

Mass spectrometry (MS) is the most appropriate and widely used technique for analyzing and detecting POPs in sample arrays due to its advantage of high sensitivity, selectivity, and throughput. MS has been a crucial scientific technique for great discoveries in history since its discovery by Joseph John Thompson in the early 1990s, revealing, for example, the presence of isotopes. This technique has been widely explored in various fields, such as proteomics and metabolomics, pesticide identification, and the assessment of the elementary biological structure. The recent advancements in MS and the availability of sophisticated analytical instruments combined with revolutionary methods of sample preparation allow analytical chemists to identify a wide variety of POPs in food and environmental samples to analyze and record substances at considerably lower concentrations than in previous samples of high matrix complexity. Many technologies are linked to high-resolution chromatography instruments combined with high-resolution or tandem mass spectrometry (MS/MS) instrumentation, improved ionization sources, greater mass analyzer sensitivity, and better acquisition software. To determine the trace amounts (very low concentrations) of POPs and their metabolites in the environment, it is essential to use mass spectrometric methods to follow a sequence of accurate quantification operations. The series of operations include (i) isolation, extraction, and segregation of POPs from the sample matrix (food, air, water, sediment, living things, etc.); (ii) sample clean-up, including the separation and purification of POP residues from co-extracted, non-target chemicals; (iii) the sample's concentration; and (iv) detection by MS of POP residues (Kailasa et al. 2013).

POPs are a group of toxic chemicals found in the environment from various sources and considered highly poisonous for the human health when it reaches the body. POPs have carcinogenic properties and have bad effects on the nervous system, reproductive system, and immune systems. Therefore, the identification and separation of POPs from food sources or surroundings is an immediate need. Various methods have been used to remove contamination, such as liquid-liquid extraction, solid-phase extraction (SPE), gas chromatography (GC), and liquid-phase chromatography. MS has proved to be an effective and strong tool for POPs tracking food products. MS is extensively used when coupled with gas chromatography (GC-MS), liquid chromatography (LC-MS), and tandem mass spectrometry (GC×GC/MS) with higher speed and powerful automation for the identification of POPs in contaminated food. Orbitrap coupled with LC is also considered one of the most powerful instruments due to high sensitivity, selectivity, specificity, and throughput.

MS coupled with chromatography is the most widely used tool for POP quantification in food and environmental matrices (Hagberg 2009; Portoles et al. 2016; Geng et al. 2017; Rivera-Austrui et al. 2017). MS coupling with efficient separation techniques such as GC-MS, LC-MS, and two-dimensional gas chromatography (GC-GC-MS) has been commonly used to detect POPs in the environment due to its quick automation and high speed (Campo and Picó 2015). MS-based methods for wide-ranging POP landscape analysis have traditionally used GC and electron ionization (EI) in the magnetic field for limited and sensitive monitoring of legacy POPs. Gas chromatography coupled to mass spectrometry (GC-MS) has been accepted as the confirmatory tool for determining POPs and measuring volatile and thermally stable POPs among the methods of analysis.

MS caters to the need to identify ultra-trace-level POPs in complex matrices, especially compared to other MS analyzers such as time-of-flight (TOF)-MS, single quadrupole MS, and quadrupole ion storage TOF-MS (Focant et al. 2005b). Confirmatory research approaches to check conformity with maximum levels of PCDDs, PCDFs, dioxin-like, and non-dioxin-like PCBs in foodstuff include either the traditionally established high-resolution mass spectrometry (HRMS) of the GC-magnetic sector (GC-HRMS) or, more recently, GC-triple quadrupole (QQQ)-MS (Franchina et al. 2019). GC-MS is highly selective, so it is preferred to be used for POP detection. The mass spectrometric detector must be operated in selected ion mode to obtain sensitivities ranging from 1 to 10 pg in most of the compounds. Ion-trap tools are as sensitive as quadrupole detectors and can be used in full-scan mode. Moreover, recent technical developments in QQQ-MS have made these analyzers a feasible alternative to GC-HRMS, particularly when used in tandem mode for selective trace-level analysis (Ingelido et al. 2012; García-Bermejo et al. 2015). For both untargeted and targeted analyses, the use of QQQ-MS was also documented, using a simultaneous scan and multiple reaction monitoring (Tranchida et al. 2013a, 2013b; Franchina et al. 2015). Aerts et al. (2019) analyzed human breast milk samples using GC–electron capture negative ion ionization–MS, and GC-EI-MS/MS revealed POPs viz., organochlorine compounds (*p,p′*-DDT, *p,p′*-DDE, hexachlorobenzene [HCB], and β-HCH).

However, some recent and emerging POPs need LC due to properties that prevent effective separation with GC in their native state. Sources of ionization of atmospheric pressure promote this coupling to MS and significantly increase the identification of new POPs. Single quadrupole MS is widely used in addition to ionization sources, but QQQ-MS displays greater instrumental sensitivity by eliminating matrix interference (Losada et al. 2010; Fulara and Czaplicka 2012). MS/MS is documented to provide low background noise due to background removal in the second step of MS/MS analysis compared to MS with possible interference from chlorinated compounds (Losada et al. 2010). GC-MS and LC-MS are appropriate techniques for detecting organic environmental pollutants due to GC and LC's separation capacity combined with MS's selectivity and sensitivity. This combination of chromatography and mass selective detection enables certain contaminants to be systematically calculated in parts per trillion.

In recent years, the applicability of LC-MS in POP research has further increased with advances in chromatography, such as the advent of ultra-high-performance liquid chromatography (UHPLC), using the same separation technique as standard HPLC, but using smaller columns filled with smaller particulate matter (usually about two µm of particle size or less). In the field of MS-based POPs analysis, numerous challenges and opportunities still exist and can be summarized in part as the need for improved sensitivity, selectivity, and confidence recognition challenges for compounds identified during non-targeted studies.

15.2 GC-MS FOR POPs ANALYSIS

In 2001 at the Stockholm Convention on POPs, it was suggested to eliminate or reduce the release of POPs into the surrounding and end the POPs to be used for commercial purposes. The Stockholm Convention also suggested that World Health Organization and United Nations Environment Programme initiate a monitoring program for POP analysis in breast milk to monitor POPs exposure to humans. Later these monitoring programs were applied to food items to know the source of exposure of POPs to humans. This monitoring program became an integral part of the CODEX Alimentarius Commission as POPs presence in a food item became a potential risk factor (Hubert 2013). POPs analysis is a multistep process that includes clean-up, sample preparation, separation, and detection. Clean-up is done to avoid cross-contamination. The sample preparation method is selected based on the characteristics of the food (Stephen and Benedict 2011). For solid food sample analysis, commonly reported sample preparation methods are SPE (Aurea et al. 2015) and liquid-liquid extraction (Andrew et al. 2011). If the food sample is in a liquid state, then the easiest way for sample preparation is liquid-liquid microextraction (Weiguang et al. 2013).

SPE is the most commonly used method among these techniques, as it needs less solvent for extraction. Moreover, it is less expensive, and operating procedures for this technique are elementary compared with other procedures. Supercritical fluid extractions (Walter et al. 1999), ultrasound-assisted extractions, microwave-assisted extractions (Angelika and Marek 2010), and pressurized liquid extraction (Camino et al. 2011) were reported commonly for solid-state sample preparation of food samples. GC separates the compounds based on their boiling point and depends on their affinity toward their stationary phase. Almost all the POPs are non-polar or moderately polar and semi-volatile. Hence it is effectively separated and analyzed using GC-MS (Guillaume et al. 2016). There are two different ionization modes in MS, EI (Portoles et al. 2016), and negative chemical ionization (NCI) (Dawei et al. 2016). Almost all the POPs are semi-volatile. If POPs are ionized by hard ionization techniques such as the EI method, then the compounds can lose their selectivity. MS can analyze only PCB and PBDE in EI mode as they both are non-volatile compounds (Wenjn et al. 2019). In the NCI mode of MS, most POPs are detected, including PCB and PDBE. For detection of POPs, after separation by GC, the MS works in two detection modes, namely QQQ and TOF (Wenjn et al. 2019).

15.3 APPLICATION OF GC-MS FOR POPs DETERMINATION IN DIFFERENT FOOD SAMPLES

The majority of POPs are present in the environment as PCB and OCPs. If agrochemical industrial waste and agriculture waste with PCB and OCs finally end up in water bodies, aquatic animals, including fishes, are exposed to different POPs as the water bodies are contaminated with PCB and OCs. There are many reports on POPs analysis in aquaculture products (Sara et al. 2019). PCB/ Dioxin accidentally contaminated the whole animal feed supply in Belgian incidence in January 1999 (Alfred et al. 2002). After this incident, the scientific communities speculated that there is a chance that PCB can be there in the milk supply system through the domesticated animal feed (Janice and David 2005). Milk is generally contaminated by PCB and PCDD/F. Sample preparation is done by liquid-liquid extraction, where the fat is removed from the milk because the POPs are fat-soluble compounds. The extracted milk fat is cleaned up by gel permeation chromatography to remove PCB and PCDD/F from fat. Cleaned up sample is then passed through GC, coupled with

MS at NCI mode and TOF detection mode. This method can analyze PCB with the precision of 40% and accuracy of 60%. PCDD/F was analyzed with a precision of 25% and an accuracy of 83% (Jeffery and Roy 2017).

To collect the nectar, honey bees cover a large area; hence they are exposed to materials including POPs contaminated water bodies and pollens; hence honey will also get contaminated (Chiesa et al. 2016). The POPs present in honey are PBDE, OP (Organophosphate), PCB, and OC (Organochlorinated). The GC-MS technique commonly analyses the POPs in the honey. The honey samples were extracted and cleaned up by the accelerated solvent extraction method. The sample extract was then passed through a GC column coupled with MS at EI mode and QQQ detectors. This method was evaluated for its limits of quantification, detection limits, and repeatability for PCB analysis, PDBE, OCs, and OPs (Chiesa et al. 2016) (Table 15.1).

15.4 SOURCE IDENTIFICATION OF PERSISTENT ORGANIC POLLUTANTS USING GC-MS/MS

POPs are a group of semi-volatile, toxic, chemically, and thermally stable organic compounds found widely in the environment and are hazardous to human health and environments (Figure 15.1). They undergo slow metabolic degradation and are highly lipophilic and hydrophobic (Mussalo-Rauhamaa 1991). POPs find their entry into the human body by sources such as absorption through the skin, breathing, and predominantly by the ingestion of contaminated food. The primary sources of POPs contamination in food are seafood such as crab, fish, and shrimp (Brown et al. 1987). A study conducted on the US population found that over 13% of the US population may contain a minimum of ten types of POPs (Pumarega et al. 2016).

POPs are extremely degenerative to human health. They are highly toxic to the nervous system, have carcinogenic properties, and exert adverse effects on the reproductive and immune systems.

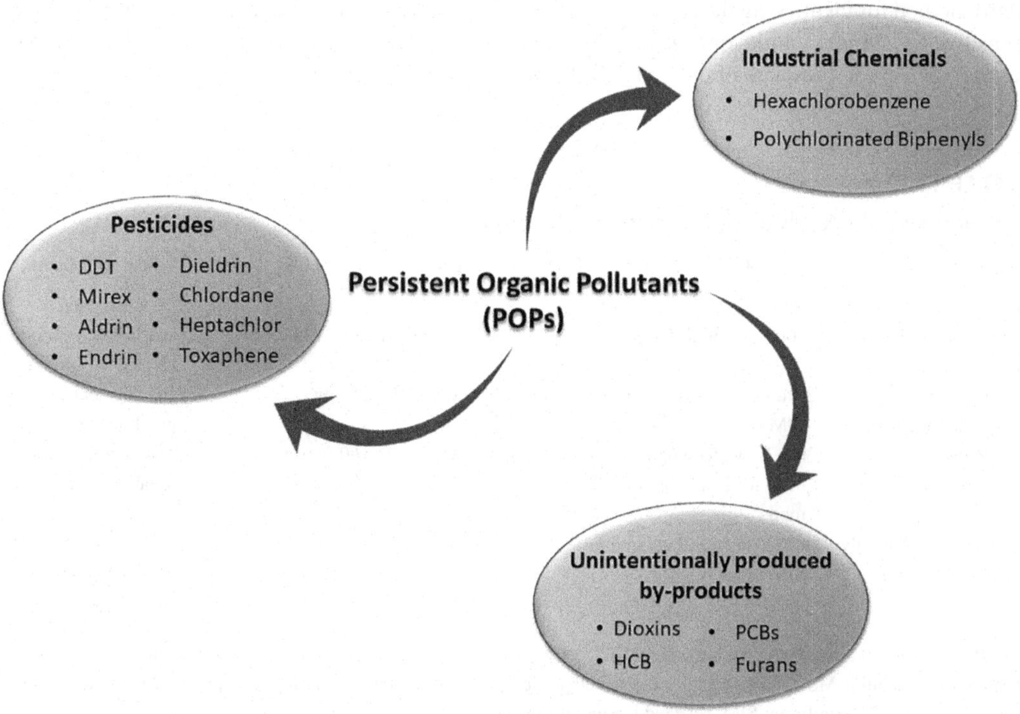

FIGURE 15.1 Sources of persistent organic pollutants (POPs).

They are also associated with hormonal disruptions, cardiovascular diseases (CVDs), and cancer (Ashraf 2017; Omar et al. 2018). Liquid-liquid extraction, SPE, chromatographic techniques such as GC, and liquid-phase chromatography are some of the classical analytical tools used for the separation of pollutants from food (Bayat et al. 2019). The various analytical methods for the detection of POPs in food are given in Table 15.2.

In recent years, MS has proved to be a very efficient and powerful instrument for POPs monitoring in food products. Mass spectrometry imaging is a 2D ionization technology that facilitates the detection of particular compounds with good efficiency (Guo et al. 2019; Yukihiro and Zaima 2020). Orbitrap, a mass analyzer for carrying out the mass spectrometric analysis, was developed by Makarov (2000). Over the years, it was widely used for LC-MS/MS analysis of complex compounds. With further advancements in the following years, Orbitrap is currently used as a part of the hybrid mass spectrometer, operating at a high speed of up to 20 Hz with a resolving power of 500,000 (at $m/z = 200$). By coupling with LC, it is widely used to analyze highly complex compounds (Andjelković et al., 2017). It is considered one of the most powerful instruments for POPs detection due to its selectivity, specificity, high sensitivity, and throughput. MS is also coupled with other classical separation techniques such as GC-MS, LC-MS, and two-dimensional (2D) gas chromatography (GC×GC-MS). It is extensively used to identify POP contamination in food products with a much higher speed and automation (Mázor 1959).

Another widely used technique to detect POPs in the food samples is GC-MS in the selected ion mode. This technique focuses on a selected number of relevant masses that correspond to the analysts and thus improves the selectivity of the process (Portoles et al. 2016). TOF-MS coupled with comprehensive 2D gas chromatography (GC×GC-TOF-MS) was effectively used for the detection of POPs such as dioxins and PCBs in the food samples (Focant et al. 2005a). In the studies performed recently, the POPs in food samples such as fish and vegetable oils were detected using high-performance QQQ-MS/MS coupled with GC with higher efficiency and selectivity, and sensitivity (L'Homme et al. 2014). ^{13}C-labeled isotope dilution HRMS coupled with GC is counted as a standard method for identifying distinctive POPs such as dioxins and furans (Focant et al. 2005b). High-sensitive GC-MS/MS multi-residue method along with liquid-liquid extraction is also reported to be used to detect the presence of chlorinated POPs in trout with great specificity.

TABLE 15.1
GC and GC-MS Analysis of POPs in Food Samples[a]

Analytes	GC Parameters	Detector	LOD	Pros and Cons
Ortho-PCBs and OCPs, except toxaphene	DB-1, DB-5 or equivalent columns (30 or 60 m)	ECD	0.1–1 pg for DDT/DDE; 0.5 pg TEQ for PCDD/PCDFs	Low cost; good sensitivity; cannot identify co-elution
PCBs and OCPs, not good for toxaphene	30 m Low bleeding column; SIM mode	EI-LRMS	10–150 pg	Less misidentification; good sensitivity
All POPs except toxaphene	30 m Low bleeding column; 60 m for PCDD/PCDFs; SIM mode	EI-HRMS	10–160 fg of various POPs	Most sensitive and reliable; expensive
Toxaphene and highly chlorinated POPs	30 m Low bleeding column; SIM mode	ECNI-MS	0.1–1 pg for OCPs; 10 pg for toxaphene	Less misidentification; good sensitivity and specificity

ECNI, Electron capture negative ion ionization; ECD, electron capture detector; HRMS, high-resolution mass spectrometry; LRMS, low resolution MS; OCPs, Organic chlorinated pesticides; PCB, polychlorinated biphenyl; PCDD/PCDFs, polychlorinated dibenzo-p-dioxin/furan; SIM, selected ion monitoring.

[a] Reproduced with permission from Xu et al. (2013). © Elsevier.

TABLE 15.2
Analytical Approaches in the Detection of POPs in Food Samples[a]

Type	Method	Description
Extraction	Soxhlet extraction (SOX)	Suitable for solid samples; efficient but time-consuming and possible low analyte recovery
	Solid-liquid extraction (SLE)	Suitable for solid samples; expensive and uses large volumes of organic solvents
	Pressurized liquid extraction (PLE)	Suitable for solid samples; highly automated but need expensive specialized equipment
	Supercritical fluid extraction (SFE)	Suitable for solid matrices; high efficiency, selectivity, and low solvent volume, but need clean-up step
	Microwave-assisted extraction (MAE)	Suitable for solid samples; high efficiency but need clean-up step
	Ultrasonic-assisted extraction (UAE)	Suitable for solid samples; require low solvent volumes but need to optimize different operating factors
	Matrix solid-phase dispersion (MSPD)	Suitable for solid, semi-solid, and viscous sample matrices; combines extraction and clean-up within a single step but needs trials and errors to pick the right sorbent
	Liquid-liquid extraction (LLE)	Suitable for liquid/aqueous sample; high efficiency and selectivity but tedious and requires large amounts of organic solvents
	Solid-phase extraction (SPE)	Suitable for aqueous/liquid samples; requires large sample volumes
	Stir sorptive bar extraction (SBSE)	Suitable for liquid/aqueous samples; simple and solvent-less, but not suitable for polar compounds
	Solid-phase microextraction (SPME)	Suitable for liquid/aqueous samples; simple, solvent-less, less sample loss and contamination, but may need a clean-up process
Separation	Gas chromatography (GC)	Good separation potential but restricted to use on more volatile compounds, e.g., high-resolution gas chromatography (HRGC), Atmospheric Pressure Gas Chromatography (APGC).
	Liquid chromatography (LC)	Good for a polar water-soluble class of chemicals; poor separation potential, e.g., High-Pressure Liquid Chromatography (HPLC)
	GC×GC	Good separation potential but restricted on more volatile compounds
Detection	Electron capture detector	The most commonly used detection method with low detection limits
	Mass spectrometry (MS) in the negative chemical ionization mode	Better sensitivity but restricted on non-polar POPs
	MS in the electron ionization mode	Better sensitivity and selectivity due to abundant fragmentation but restricted on non-polar POPs
	MS in the selected ion monitoring mode	Better sensitivity, but the selected ion window may need to be monitored
	High-resolution mass spectrometry (HRMS)	High sensitivity but expensive
	MS/MS	Improves sensitivity and selectivity compared to single quadrupole MS, e.g., ion-trap MS/MS; triple quadrupole MS/MS

[a] Adapted from Guo et al. (2019), Creative Commons Attribution 4.0 International.

15.5 LC-MS FOR ANALYSIS OF EMERGING POPs

Several analytical techniques have been proposed to detect POPs scheduled under Stockholm Convention. The majority of them depends upon GC-MS (Zhang and Rhind 2011), HRMS (Salihovic et al. 2013), electron capture detector (Grimalt et al. 2010), TOF-MS (Focant et al. 2004), and HPLC coupled with turbo ion-spray source–mass spectrometry (TIS-MS), and even UHPLC coupled with TIS-MS for instrumental analysis (Keller et al. 2010). LC-MS analysis is found to be more sensitive in analyzing polar molecules than non-polar molecules. POPs, which are low polarity chemicals need to be analyzed in an appropriate instrumental technique (Kucklick et al. 2002). Recent studies pointed out LC-MS as a convenient way of POP detection in the food sample.

The optimized acetonitrile extraction method with weak anion exchange with SPE clean-up assisted in the extraction of 28 Perfluorinated compounds (PFCs) in the sample (Yeung et al. 2009). The development of the SPE-LC-MS method aided in the extraction of 12 PFCs in human blood; this method offered reliable results as it skips the evaporation step, which is crucial in quantification (Kärrman et al. 2005). The conduct of LC-MS analysis with a small amount of human serum produced a limit of detection ranging from 0.027 ng/ml and limit of quantification starting from 0.071 ng/ml (Koponen et al. 2013). LC–atmospheric pressure photoionization-MS was found to be more effective than atmospheric chemical ionization, as it can achieve sufficient retention and separation of HCB and pentachlorophenol (Osaka et al. 2009). Since there are various analytical procedures specific for each class of compounds, there was a necessity to develop an appropriate methodology for the detection of a large number of compounds.

A study was reported of the simultaneous detection of 46 analytes belonging to different chemical classes in which most of which are POPs (Buiarelli et al. 2017). Application of LC-MS in detecting perfluorinated acids (PFCAs) and perfluorinated sulfonates (PFASs), brominated flame retardants in fish and shellfish are well documented (Surma et al. 2015; Tavoloni et al. 2020). Emerging PFAS detection via LC-MS is of principal interest. To analyze them, a few studies followed conventional LC-ESI-MS methods. LC-MS's detection of hexafluoropropylene oxide-dimer acid was attempted

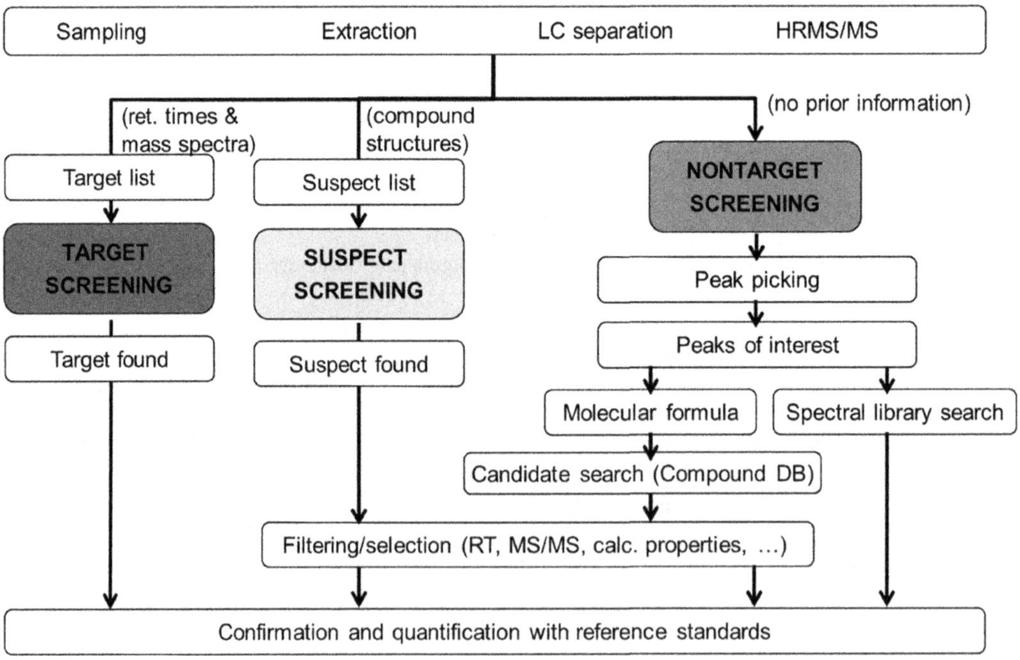

FIGURE 15.2 Detection of POPs from food sources using high-resolution mass spectrometry. (Adapted from Brack et al. 2019, Creative Commons Attribution 4.0 International.)

by LC-MS and appreciable results were found. This study encountered several analytical challenges, including dimer formation and fragmentation (Mullin et al. 2019). This technique is less useful in non-polar emerging POPs, and the sample contamination at the time of sample processing cannot be neglected as it might interfere with the final results. However, LC-MS method could be implemented successfully to detect PFAs in food samples as they are polar substances (Lorenzo et al. 2018) (Figure 15.2).

15.6 CONCLUSION

POPs are hazardous chemicals with moderate to low volatility that resist degradation and accumulate in living tissues. The key benefits of GC-MS and LC-MS/MS analysis include a significantly extended range of toxic compounds amenable for the analysis, uniform response for improved quantification and measurement of chemical reaction yields, improved selectivity, and reduced matrix interference on molecular ions. The combination of high-throughput qualitatively superior techniques improves the analysis of complex POPs in the samples. To ensure food safety and human health, it is necessary to understand the POPs pathways into food and develop effective strategies to minimize human exposure. POPs are observed as a global concern. National and international regulations and legislations play a vital role in reducing the use of POPs. Extensive research activities are necessary to develop new analytical methods to investigate these emerging pollutants in food and environmental samples. Regarding the health effects of POPs, more studies are needed to understand the mode of toxicological mechanisms. Although GC-MS and LC-MS/MS chromatographic techniques are effective, the research concerning POPs removal is far from enough, and further investigations are needed to develop novel eco-friendly and human-friendly analytical techniques.

REFERENCES

Aerts, R., Van Overmeire, I., Colles, A., et al. 2019. Determinants of persistent organic pollutant (POP) concentrations in human breast milk of a cross-sectional sample of primiparous mothers in Belgium. *Environment International* 131:104979. doi:10.1016/j.envint.2019.104979.

Alfred, B., Fabrice, B., Geert, D.P., et al. 2002. The Belgian PCB/Dioxin incident: analysis of the food chain contamination and health risk evaluation. *Environmental Research* 88(1):1–18.

Alharbi, O.M.L., Basheer, A.A., Khattab, R.A. and Ali, I. 2018. Health and environmental effects of persistent organic pollutants. *Journal of Molecular Liquids* 263:442–53. doi:10.1016/j.molliq.2018.05.029.

Andjelković, U., Šrajer Gajdošik, M., Gašo-Sokač, D., Martinović, T. and Josić, D. 2017. Foodomics and food safety: where we are. *Food Technology and Biotechnology* 55(3):290–307. doi:10.17113/ftb.55.03.17.5044.

Andrew, Q., Wayne, C. and Damian, C. 2011. Dispersive liquid-liquid microextraction in the analysis of milk and dairy products: a review. *Journal of Chemistry* 2016(1):167–76.

Angelika, M.W. and Marek, B. 2010. Rapid method for the determination of organochlorine pesticides and PCBs in fish muscle samples by microwave-assisted extraction and analysis of extract by GC-ECD. *Journal of AOAC International* 93(6):1987–94.

Ashraf, M.A. 2017. Persistent organic pollutants (POPs): a global issue, a global challenge. *Environmental Science and Pollution Research International* 24:4223–7.

Aurea, A.E., Moisés, C.L., Valérie, L.C., et al. 2015. Solid phase extraction of organic compounds: a critical review – Part I. *Trends in Analytical Chemistry* 80(1):641–54.

Bayat, M., Tehrani, M.S., Kobarfard, F., Husain, S.W. and Yazdanpanah, H. 2019. Validation of an analytical method for simultaneous determination of 18 persistent organic pollutants in trout using LLE extraction and GC-MS/MS. *Iran J Pharm Res.* 18(3):1224–38. doi:10.22037/ijpr.2019.1100748.

Brack, W., Hollender, J., de Alda, M.L. et al. 2019. High-resolution mass spectrometry to complement monitoring and track emerging chemicals and pollution trends in European water resources. *Environmental Sciences Europe* 31:62. doi:10.1186/s12302-019-0230-0.

Brown, J.F. Jr., Bedard, D.L., Brennan, M.J., Carnahan, J.C., Feng, H. and Wagner, R.E. 1987. Polychlorinated biphenyl dechlorination in aquatic sediments. *Science* 236(4802):709–12. doi:10.1126/science.236.4802.709.

Buiarelli, F., Di Filippo, P., Pomata, D., Riccardi, C. and Bartocci, M. 2017. A liquid chromatography-tandem mass spectrometry method for simultaneous analysis of 46 atmospheric particulate-phase persistent organic pollutants and comparison with gas chromatography/mass spectrometry. *International Journal of Environmental Analytical Chemistry* 97:797–818. doi:10.1080/03067319.2017.1369535.

Camino, F.J., Alberto, Z.G., Marques, J.C., et al. 2011. Validation of a GC–MS/MS method for simultaneous determination of 86 persistent organic pollutants in marine sediments by pressurized liquid extraction followed by stir bar sorptive extraction. *Chemosphere* 84(7):869–81.

Campo, J. and Picó, Y. 2015. Emerging contaminants. In *Comprehensive Analytical Chemistry*, ed. Y. Picó. Volume 68, 515–578. Elsevier: Barcelona.

Chiesa, L.H., Labella G.F., Giorgi, A., et al. 2016. The occurrence of pesticides and persistent organic pollutants in Italian organic honeys from different productive areas in relation to potential environmental pollution. *Chemosphere* 154(11):482–90.

Dawei, G., Ingrid, E.J., Jody, D., et al. 2016. Gas chromatography/atmospheric pressure chemical ionization/mass spectrometry for the analysis of organochlorine pesticides and polychlorinated biphenyls in human serum. *Journal of Chromatography A* 1453(23):88–98.

El-Shahawi, M.S, Hamza, A., Bashammakh, A.S. and Al-Saggaf, W.T. 2010. An overview on the accumulation, distribution, transformations, toxicity, and analytical methods for the monitoring of persistent organic pollutants. *Talanta* 80:1587–97 doi:10.1016/j.talanta.2009.09.055.

Focant, J.F., Eppe, G., Scippo, M.L., Massart, A.C., Pirard, C., Maghuin-Rogister, G. and De Pauw E. 2005a. Comprehensive two-dimensional gas chromatography with isotope dilution time-of-flight mass spectrometry for the measurement of dioxins and polychlorinated biphenyls in foodstuffs. Comparison with other methods. *Journal of Chromatography A*. 1086(1–2):45–60. doi:10.1016/j.chroma.2005.05.090.

Focant, J.F., Pirard, C., Eppe, G. and De Pauw, E. 2005b. Recent advances in mass spectrometric measurement of dioxins. *Journal of Chromatography A*. 1067:265–75.

Focant, J.-F., Sjödin, A., Turner, W.E. and Patterson, D.G. 2004. Measurement of selected polybrominated diphenyl ethers, polybrominated and polychlorinated biphenyls, and organochlorine pesticides in human serum and milk using comprehensive two-dimensional gas chromatography isotope dilution time-of-flight mass spectrometry. *Analytical Chemistry*. 76:6313–20. doi:10.1021/ac048959i.

Franchina, F.A., Lazzari, E., Scholl, G. and Focant, J.F. 2019. Assessment of a new GC-MS/MS system for the confirmatory measurement of PCDD/Fs and (N)DL-PCBs in food under EU regulation. *Foods* 18(8):302. doi:10.3390/foods8080302.

Franchina, F.A., Machado, M.E., Tranchida, P.Q., Zini, C.A., Caramão, E.B. and Mondello, L. 2015. Determination of aromatic sulphur compounds in heavy gas oil by using (low-) flow modulated comprehensive two-dimensional gas chromatography-triple quadrupole mass spectrometry. *Journal of Chromatography A* 1387:86–94.

Fulara, I. and Czaplicka, M. 2012. Methods for determination of polybrominated diphenyl ethers in environmental samples – review. *Journal of Separation Science* 35(16):2075–87. doi:10.1002/jssc.201200100.

García-Bermejo, Á., Ábalos, M., Sauló, J., Abad, E., González, M.J. and Gómara, B. 2015. Triple quadrupole tandem mass spectrometry: a real alternative to high resolution magnetic sector instrument for the analysis of polychlorinated dibenzo-p-dioxins, furans, and dioxin-like polychlorinated biphenyls. *Analytica Chimica Acta* 189:156–65.

Gaur, N., Narasimhulu, K. and PydiSetty, P. 2018. Recent advances in the bioremediation of persistent organic pollutants and their effect on the environment. *Journal of Cleaner Production* 198:1602–31. doi:10.1016/j.jclepro.2018.07.076.

Geng, D., Kukucka, P. and Jogsten, I. E. 2017. Analysis of brominated flame retardants and their derivatives by atmospheric pressure chemical ionization using gas chromatography coupled to tandem quadrupole mass spectrometry. *Talanta* 162:618–24. doi:10.1016/j.talanta.2016.10.060.

Grimalt, J.O., Howsam, M., Carrizo, D., Otero, R., de Marchi, M.R.R. and Vizcaino, E. 2010. Integrated analysis of halogenated organic pollutants in sub-millilitre volumes of venous and umbilical cord blood sera. *Analytical Bioanalytical Chemistry* 396:2265–72. doi:10.1007/s00216-010-3460-y.

Guillaume, T. D., Igor, C. P., Georges, S., et al. 2016. The performance of atmospheric pressure gas chromatography-tandem mass spectrometry compared to gas chromatography – high resolution mass spectrometry for the analysis of polychlorinated dioxins and polychlorinated biphenyls in food and feed sample. *Journal of Chromatography A* 1477(16):76–90.

Guo, W., Pan, B., Sakkiah, S., et al. 2019. Persistent organic pollutants in food: contamination sources, health effects and detection methods. *International Journal of Environmental Research and Public Health* 16(22):4361. doi:10.3390/ijerph16224361.

Hagberg, J. 2009. Analysis of brominated dioxins and furans by high resolution gas chromatography/high resolution mass spectrometry. *Journal of Chromatography A*. 1216:376–84. doi:10.1016/j.chroma.2008.10.022.

Hubert, P.T. 2013. Recent development in analysis of persistent organic pollutants under Stockholm convention. *TrAC Trends in Analytical Chemistry* 45(16): 48–66.

Ingelido, A.M., Brambilla, G., Abballe, A., et al. 2012. PCDD, PCDF, and DL-PCB analysis in food: Performance evaluation of the high-resolution gas chromatography/low-resolution tandem mass spectrometry technique using consensus-based samples. *Rapid Communications in Mass Spectrometry* 26:236–42.

Janice, K.H. and David, J.J. 2005. Laboratory and on-farm studies on the bioaccumulation and elimination of dioxins from a contaminated mineral supplement fed to dairy cows. *Journal of Agricultural and Food Chemistry* 53(6):2362–70.

Jeffery, C.A. and Roy, G.J. 2017. Automated milk fat extraction for the analyses of persistent organic pollutants. *Journal of Chromatography B* 1041–1042(8):70–6.

Jones, K.C. and de Voogt, P. 1999. Persistent organic pollutants (POPs): State of the science. *Environmental Pollution* 100:209–21.

Kailasa, S.K., Wu, H.F. and Huang, S.D. 2013. Recent developments on mass spectrometry for the analysis of pesticides in wastewater. In *Wastewater – Treatment Technologies and Recent Analytical Developments*, ed. F.S.G. Einschlag. IntechOpen: London. doi:10.5772/51929.

Kärrman, A., van Bavel, B., Järnberg, U., Hardell, L. and Lindström, G. 2005. Development of a solid-phase extraction-HPLC/single quadrupole MS method for quantification of perfluorochemicals in whole blood. *Analytical Chemistry* 77:864–70. doi:10.1021/ac049023c.

Keller, J.M., Calafat, A.M., Kato, K., et al. 2010. Determination of perfluorinated alkyl acid concentrations in human serum and milk standard reference materials. *Analytical Bioanalytical Chemistry* 397:439–51. doi:10.1007/s00216-009-3222-x.

Koponen, J., Rantakokko, P., Airaksinen, R. and Kiviranta, H. 2013. Determination of selected perfluorinated alkyl acids and persistent organic pollutants from a small volume human serum sample relevant for epidemiological studies. *Journal of Chromatography A* 1309:48–55. doi:10.1016/j.chroma.2013.07.064.

Kucklick, J.R., Struntz, W.D.J., Becker, P.R., York, G.W., O'Hara, T.M. and Bohonowych, J.E. 2002. Persistent organochlorine pollutants in ringed seals and polar bears collected from northern Alaska. *Science of the Total Environment* 287:45–59. doi:10.1016/S0048-9697(01)00997-4.

L'Homme, B., Scholl, G., Eppe, G. and Focant, J.F. 2014. Validation of a gas chromatography-triple quadrupole mass spectrometry method for confirmatory analysis of dioxins and dioxin-like polychlorobiphenyls in feed following new EU Regulation 709/2014. *Journal of Chromatography A* 1376:149–58 doi:10.1016/j.chroma.2014.12.013.

Loganathan, B.G. and Masunaga, S. 2009. PCBs, dioxins, and furans: human exposure and health effects. In *Handbook of Toxicology of Chemical Warfare Agents*, ed. R.C. Gupta. 245–253. Academic Press: San Diego, CA.

Lorenzo, M., Campo, J. and Picó, Y. 2018. Analytical challenges to determine emerging persistent organic pollutants in aquatic ecosystems. *TrAC Trends in Analytical Chemistry* 103:137–55. doi:10.1016/j.trac.2018.04.003.

Losada, S., Parera, J., Abalos, M., Abad, E., Santos, F.J. and Galceran, M.T. 2010. Suitability of selective pressurized liquid extraction combined with gas chromatography–ion-trap tandem mass spectrometry for the analysis of polybrominated diphenyl ethers. *Analytica Chimica Acta* 678(1):73–81.

Makarov, A. 2000. Electrostatic axially harmonic orbital trapping: a high-performance technique of mass analysis. *Analytical Chemistry* 72(6):1156–62. doi:10.1021/ac991131p.

Mázor, L. 1959. *Comprehensive Analytical Chemistry.* Vol. 15. Elsevier Scientific Publishing Company: New York.

Mullin, L., Katz, D.R., Riddell, N., Plumb, R., Burgess, J.A., Yeung, L.W.Y. and Jogsten, I.E. 2019. Analysis of hexafluoropropylene oxide-dimer acid (HFPO-DA) by liquid chromatography-mass spectrometry (LC-MS): review of current approaches and environmental levels. *TrAC Trends in Analytical Chemistry* 118:828–39. doi:10.1016/j.trac.2019.05.015.

Mussalo-Rauhamaa, H. 1991. Partitioning and levels of neutral organochlorine compounds in human serum, blood cells, and adipose and liver tissue. *Sci of the Total Environment* 103(2–3):159–75. doi:10.1016/0048-9697(91)90142-2.

Omar, M.L.A., Al Arsh, B., Rafat, A.K. and Imran, A. 2018. Health and environmental effects of persistent organic pollutants. *Journal of Molecular Liquids* 263:442–53.

Osaka, I., Yoshimoto, A., Nozaki, K., Moriwaki, H., Kawasaki, H. and Arakawa, R. 2009. Simultaneous LC/MS analysis of hexachlorobenzene and pentachlorophenol by atmospheric pressure chemical ionization (APCI) and photoionization (APPI) methods. *Analytical Science* 25:1373–6. doi:10.2116/analsci.25.1373.

Portolés, T., Sales, C., Abalos, M., Saulo, J. and Abad, E. 2016. Evaluation of the capabilities of atmospheric pressure chemical ionization source coupled to tandem mass spectrometry for the determination of dioxin-like polychlorobiphenyls in complex-matrix food samples. *Analytica Chimica Acta* 937:96–105. doi:10.1016/j.aca.2016.06.038.

Pumarega, J., Gasull, M., Lee, D.H., López, T. and Porta, M. 2016. Number of persistent organic pollutants detected at high concentrations in blood samples of the United States population. *PLoS One* 11(8):e0160432. doi:10.1371/journal.pone.0160432.

Rivera-Austrui, J., Martinez, K., Abalos, M., et al. 2017. Analysis of polychlorinated dibenzo-p-dioxins and dibenzofurans in stack gas emissions by gas chromatography-atmospheric pressure chemical ionization-triple-quadrupole mass spectrometry. *Journal of Chromatography A* 1513:245–9 doi:10.1016/j.chroma.2017.07.039.

Salihovic, S., Nilsson, H., Hagberg, J. and Lindström, G. 2013. Trends in the analysis of persistent organic pollutants (POPs) in human blood. *TrAC Trends in Analytical Chemistry* 46:129–38. doi:10.1016/j.trac.2012.06.009.

Sara, P., Luca, C., Gabriele, G., et al. 2019. Persistent organic pollutant in fish: biomonitoring and cocktail effect with implication for food safety. *Food Additives and Contaminants* 36(4):601–11.

Stephen, W.C.C. and Benedict, L.S.C. 2011. Determination of organochlorine pesticide residues in fatty foods: a critical review on the analytical methods and their testing capabilities. *Journal of Chromatography A* 1218(33):5555–67.

Surma, M., Wiczkowski, W., Zieliński, H. and Cieślik, E. 2015. Determination of selected perfluorinated acids (PFCAs) and perfluorinated sulfonates (PFASs) in food contact materials using LC-MS/MS: PFCs in food contact materials. *Packaging Technology and Science* 28:789–99. doi:10.1002/pts.2140.

Tavoloni, T., Stramenga, A., Stecconi, T., Siracusa, M., Bacchiocchi, S. and Piersanti, A. 2020. Single sample preparation for brominated flame retardants in fish and shellfish with dual detection: GC-MS/MS (PBDEs) and LC-MS/MS (HBCDs). *Analytical and Bioanalytical Chemistry* 412:397–411. doi:10.1007/s00216-019-02250-x.

Tranchida, P.Q., Franchina, F.A., Zoccali, M., Bonaccorsi, I., Cacciola, F. and Mondello, L. 2013a. A direct sensitivity comparison between flow-modulated comprehensive 2D and 1D GC in untargeted and targeted MS-based experiments. *Journal of Separation Science* 36:2746–52.

Tranchida, P.Q., Franchina, F.A., Zoccali, M., Pantò, S., Sciarrone, D., Dugo, P. and Mondello, L. 2013b. Untargeted and targeted comprehensive two-dimensional GC analysis using a novel unified high-speed triple quadrupole mass spectrometer. *Journal of Chromatography A* 1278:153–9.

UNECE. 1998. *Protocol to the 1979 Convention on Long-Range Transboundary Air Pollution on Persistent Organic Pollutants.* United Nations Economic Commission for Europe: Aarhus. http://www.unece.org/env/lrtap/pops_h1.html.

Walter, F., John, W.P., Robert, A.G., et al. 1999. Supercritical fluid extraction of organochlorine pesticides in eggs. *Journal of Agriculture and Food Chemistry* 47(1):206–11.

Weiguang, X., Xian, W. and Zongwei, C. 2013. Analytical chemistry of the persistent organic pollutions identified in the Stockholm convention – a review. *Analytica Chimica Acta* 790(6):1–13.

Wenjn, G., Bohu, P., Sugunadevi, S., et al. 2019. Persistent organic pollutant in food: contamination sources, health effects and detection methods. *International Journal of Environmental Research and Public Health* 16(22):43–61.

Xu, W., Wang, X. and Cai, Z. 2013. Analytical chemistry of the persistent organic pollutants identified in the Stockholm convention: a review. *Analytica Chimica Acta* 790:1–13 doi:10.1016/j.aca.2013.04.026.

Yeung, L.W.Y., Taniyasu, S., Kannan, K., Xu, D.Z.Y., Guruge, K.S., Lam, P.K.S. and Yamashita, N. 2009. An analytical method for the determination of perfluorinated compounds in whole blood using acetonitrile and solid-phase extraction methods. *Journal of Chromatography A* 1216:4950–6. doi:10.1016/j.chroma.2009.04.070.

Yukihiro, Y. and Zaima, N. 2020. Application of mass spectrometry imaging for visualizing food components. *Foods* 9(5):575. doi:10.3390/foods9050575.

Zhang, Z. and Rhind, S.M. 2011. Optimized determination of polybrominated diphenyl ethers and polychlorinated biphenyls in sheep serum by solid-phase extraction–gas chromatography-mass spectrometry. *Talanta* 84:487–93 doi:10.1016/j.talanta.2011.01.042.

16 MS and Food Forensics

Leo M.L. Nollet
University College Ghent

CONTENTS

16.1 Introduction 292
16.2 Origin Testing 292
16.2.1 Introduction 292
16.2.2 Olive Oil 292
16.2.3 Honey 293
16.2.4 Wine 295
16.2.5 Beer 296
16.2.6 Distillers Dried Grains and Solubles 296
16.2.7 Gelatine 297
16.2.8 Baijiu 297
16.2.9 Truffles 297
16.2.10 Adzuki Bean 298
16.2.11 Durum Wheat 298
16.3 Organically versus Conventionally Grown Foodstuffs 299
16.3.1 Introduction 299
16.3.2 Analysis of Organically versus Conventionally Grown Foodstuffs 299
16.3.3 Analysis of Organic Products 302
16.4 Speciation 304
16.4.1 Chemical Speciation 304
16.4.2 Arsenic 304
16.4.3 Selenium 307
16.4.4 Lead 308
16.4.5 Chromium 308
16.4.6 Organotin 308
16.4.7 Silver 309
16.4.8 Multi-Elemental Speciation 309
16.5 Adulteration 316
16.5.1 What Is Adulteration? 316
16.5.2 Adulteration in Protein-Containing Foods 316
16.5.3 Adulterants in Functional Foods, Health Foods, and Herbal Medicines 317
16.5.4 Adulteration of Dairy Products 317
16.5.5 Adulteration in Oils 319
16.5.6 Adulteration in Honey 320
16.5.7 Adulteration in Meat and Meat Products 320
16.5.8 Non-Targeted Screening 321
References 322

DOI: 10.1201/9781003091226-19

16.1 INTRODUCTION

It is necessary to verify the composition, processing, or origin of food. This is a difficult task but is important to protect consumers and enforce food laws.

This chapter has four parts. The first one is dealing with origin testing of food. Correspond the contents of the tested foodstuff to the geographic origin printed on the label?

In the second section, mass spectrometry (MS) methods are discussed comparing organically and conventionally grown food items.

Chemical speciation is the determination of the individual concentrations of the various chemical forms of an element together making up the total concentration of that element in a sample. These are the contents of Section 16.3.

In the last section, MS methods of adulteration of food are enumerated.

The reader is also directed to Chapter 21 of this book.

16.2 ORIGIN TESTING

16.2.1 Introduction

From the location in which plants grow or animals are reared, they acquire an 'environmental fingerprint'. This profile is specific to the country, or even the region of production.

Food is considered authentic if the product or its contents correspond to the original condition. Authentic goods are free from adulteration with regard to geographical origin, among other parameters.

Certification of the geographic origin of food products is a useful tool for the protection of the quality of products.

Modern instruments used for origin testing of food products include plasma laser-induced breakdown spectroscopy, Fourier-transform infrared spectroscopy, near-infrared (NIR) and mid-infrared spectroscopy, chromatographic techniques as gas chromatography-mass spectrometry (GC-MS) or high-performance liquid chromatography (HPLC), nuclear magnetic resonance, electron spin resonance, polymerase chain reaction (PCR), enzyme-linked immunosorbent assay (ELISA), techniques DNA-based and isotope ratio mass spectrometry (IRMS).

Each method has advantages and disadvantages.

In the subsequent paragraphs, a number of mass spectrometric techniques for origin testing of olive oil, honey, wines, beers, distillers dried grains and solubles (DDGS), gelatin, Baijiu, truffles, Adzuki beans and durum wheat are discussed.

In a great number of studies, the results of MS analyses are processed by chemometrics.

The reader will find further MS methods on origin testing for sesame seeds [1], rice [25], Chinese cabbage [26], Chinese tea [27], cocoa beans [28], and hazelnuts [29].

16.2.2 Olive Oil

The combined use of data obtained from different analytical instruments is a complex problem.

In a study by Casale et al. [2], the potential of coupling three analytical techniques for building a class model for extra virgin (e.v.) olive oil from Liguria, was investigated.

A sampling design for the Ligurian e.v. olive oil was developed by the selection of a representative subset from all possible e.v. olive oil samples of the Liguria region.

In order to choose this subset with uniform distribution on the production area and representative of the production density, two algorithms for sampling have been used: Kennard–Stone and Potential Function.

The samples were analysed by headspace MS (electronic nose), ultraviolet (UV)–visible and NIR spectroscopy.

The electronic nose used in this study is not a traditional electronic nose based on sensors but it is a headspace mass spectrometer assembled in the laboratory of the authors.

The headspace autosampler (HT200H, HTA: hi-tech applications, Brescia, Italy) contains a tray with 40 slots and a six-position oven to generate the headspace (150°C max) [3]. The syringe (2.5 or 5 mL), whose temperature is controlled by the autosampler, samples the headspace and injects it directly into the ionisation chamber of the mass spectrometer, using nitrogen as cleaning inert gas and helium as carrier gas. The injector (OPTIC 3 from ATAS GL, Veldhoven, the Netherlands) incorporates temperature and gas flow control for sample introduction. The direct interface (Abreg, Alessandria, Italy) between the injector and the detector is ensured by a transfer line, which consists of a 15 cm long empty retention gap placed in an oven with a temperature controller. The transfer line directly enters into the mass spectrometer (HP 5973 MSD, Agilent Technologies).

In particular, the exceptional possibility provided by chemometrics to effectively extract and combine (fusion) the information from these multivariate and non-specific data and to build a class model for Liguria e.v. olive oil was studied.

Selected ion flow tube mass spectrometry (SIFT-MS) in combination with chemometrics was used to authenticate the geographical origin of Mediterranean virgin olive oils produced under geographical origin labels [4]. SIFT-MS is a quantitative mass spectrometry technique for trace gas analysis, which involves the chemical ionisation of trace volatile compounds by selected positive precursor ions during a well-defined time period along a flow tube.

In particular, 130 oil samples from six different Mediterranean regions (Kalamata [Greece]; Toscana [Italy]; Meknès and Tyout [Morocco] and Priego de Córdoba and Baena [Spain]) were considered. The headspace volatile fingerprints were measured by SIFT-MS in full scan with H_3O^+, NO^+ and O_2^+ as precursor ions and the results were subjected to chemometric treatments. Principal component analysis (PCA) was used for preliminary multivariate data analysis, and partial least squares-discriminant analysis (PLS-DA) was applied to build different models (considering the three reagent ions) to classify samples according to the country of origin and regions (within the same country). The multiclass PLS-DA models showed very good performance in terms of fitting accuracy (98.90%–100%) and prediction accuracy (96.70%–100% accuracy for cross-validation and 97.30%–100% accuracy for external validation [test set]). Considering the two-class PLS-DA models, the one for the Spanish samples showed 100% sensitivity, specificity and accuracy in calibration, cross-validation and external validation; the model for Moroccan oils also showed very satisfactory results (with perfect scores for almost every parameter in all the cases).

16.2.3 Honey

The geographical origin of three Slovenian unifloral honey types (black locust, lime and chestnut) was investigated by analysis of some physicochemical parameters, the elemental content using total reflection X-ray fluorescence spectrometry, and the stable carbon and nitrogen isotope ratios using IRMS [5]. All analyses were performed on a Europa Scientific 20-20 continuous flow mass spectrometer with an Anca-SL solid–liquid preparation module. The results were interpreted by chemometric methods. A total of 122 samples of Slovenian black locust, lime and chestnut honeys were collected from domestic beekeepers all over Slovenia for 3 years. Slovenia is a small country by area, but paedologically and climatically diverse, therefore offering interesting possibilities for studying geographical influences. The combination of the investigated parameters offers the possibility of distinguishing among samples of specific honey types from the four different Slovenian natural geographical macroregions, namely the Alpine, Dinaric, Pannonian and Mediterranean regions. Lime honey samples were 100% correctly classified, whereas the success rates for black locust and chestnut honeys were slightly lower at 98.2% and 94.6%, respectively.

The aim of the study by Karabagias et al. [6] was to characterise and classify Greek pine honey according to geographical origin, based on the determination of volatile compounds and physicochemical parameters using multivariate analysis of variance (MANOVA) and linear discriminant analysis (LDA). Thirty-nine pine honey samples were collected during the harvesting period 2011 from four different regions in Greece known to produce good quality pine honey. The analysis of volatile compounds was performed by headspace solid-phase microextraction (HS-SPME)-GC-MS. An Agilent 7890 A GC unit coupled with an Agilent 5975 MS detector was used to analyse the honey solutions. A DB-5MS (cross-linked 5% PH ME siloxane) capillary column (60 m 320 μm i.d., 1 μm film thickness) was used, with helium as the carrier gas (purity 99.999%), at 1.5 mL/min flow rate. The injector and MS-transfer line were maintained at 250°C and 270°C, respectively. For the SPME analysis, oven temperature was held at 40°C for 3 minutes, and increased to 260°C at 8°C/min and held for 6 minutes. Electron-impact mass spectra were recorded at the 50–550 mass range. An electron ionisation system was used with ionisation energy of 70 eV.

Fifty-five volatile compounds were tentatively identified and semiquantified. Physicochemical parameter analysis included the determination of pH, free, lactonic and total acidity, electrical conductivity, moisture, ash, lactonic/free acidity ratio and colour parameters L^*, a^*, and b^*. Using 8 selected volatile compounds and 11 physicochemical parameters, the honey samples were satisfactorily classified according to geographical origin using volatile compounds (84.6% correct prediction), physicochemical parameters (79.5% correct prediction) and the combination of both (74.4% correct prediction).

The effect of the geographical origin on the volatile composition and sensory properties of chestnut honeys produced in different areas was investigated [7]. An Agilent 6890N gas chromatograph coupled with a 5,973 inert mass selective detector was used. Two microliters of extracts were injected in splitless mode (0.6 minutes) on a polyethylene glycol capillary column BP-21 (50 m × 0.32 mm × 0.25 μm of film thickness). Oven temperature was programmed to remain at 60°C for 3 minutes and then increased at 2°C/min to 200°C and held for 30 minutes. Helium was used as carrier gas at a flow rate of 0.8 mL/min. Injector and transfer line temperatures were 250°C and 280°C, respectively. Ionisation was performed by electron-impact mode at 70 eV. Mass spectrum acquisition was performed in scan mode (40–450 *m/z* range). Samples from the Spanish northeast presented significantly higher concentrations of aldehydes, alcohols, lactones and volatile phenols which are associated with herbaceous, woody and spicy notes detected by the assessors. Chestnut honeys from the northwest area showed superior levels of terpenes, esters and some benzene derivatives, closely related with honey-like, floral and fruity notes. Finally, chestnut honeys produced in the southeast area were characterised by the norisoprenoids content and showed sensory characteristics between the two previous locations.

Multivariate statistical analysis revealed variations among production zones allowing the successful differentiation of chestnut honeys from different geographical areas according to their volatile composition and sensory description. Additionally, the presence of 2-hydroxyacetophenone and eucarvone is proposed as useful chemical markers that contribute to typify this botanical source.

The possibility of verifying the geographical origin of honeys based on the profiles of volatile compounds was examined [8]. A headspace solid-phase microextraction (SPME) combined with comprehensive two-dimensional gas chromatography–time-of-flight mass spectrometry (GC × GC-ToF-MS) was used to analyse the volatiles in honeys with various geographical and floral origins.

A Pegasus 4D system consisting of an Agilent 6890N gas chromatograph equipped with a split/splitless injector (Agilent Technologies, Palo Alto, CA, USA), an MPS2 autosampler for automated SPME (Gerstel, Mülheim an der Ruhr, Germany) and a Pegasus III high-speed ToF mass spectrometer (Leco Corp., St. Joseph, MI, USA), was used. Once the analytical data were collected, supervised pattern recognition techniques were applied to construct classification/discrimination rules to predict the origin of samples based on their profiles of volatile compounds. Specifically, LDA, soft independent modelling of class analogies (SIMCAs), discriminant partial least squares (DPLS) and support vector machines (SVM) with the recently proposed Pearson VII universal kernel (PUK) were used in this study to discriminate between Corsican and non-Corsican honeys. Although DPLS and LDA provided models with high sensitivities and specificities, the best performance was achieved by the SVM using PUK. The results of this study demonstrated that GC×GC-ToF-MS combined with methods such as LDA, DPLS and SVM can be successfully applied to detect mislabelling of Corsican honeys.

Multi-element analysis of honey samples was carried out with the aim of developing a reliable method of tracing the origin of honey [6]. Forty-two chemical elements were determined (Al, Cu, Pb, Zn, Mn, Cd, Tl, Co, Ni, Rb, Ba, Be, Bi, U, V, Fe, Pt, Pd, Te, Hf, Mo, Sn, Sb, P, La, Mg, I, Sm, Tb, Dy, Sd, Th, Pr, Nd, Tm, Yb, Lu, Gd, Ho, Er, Ce and Cr) by inductively coupled plasma mass spectrometry (ICP-MS). The determination of the elements was performed by ICP-MS (Elan DRC II, Perkin-Elmer, Norwalk, CT, USA). Then, three machine-learning tools for classification and two for attribute selection were applied in order to prove that it is possible to use data mining tools to find the region where honey originates. These results clearly demonstrate the potential of SVM, multilayer perceptron (MLP) and random forest (RF) chemometric tools for honey origin identification. Moreover, the selection tools allowed a reduction from 42 trace element concentrations to only 5.

The mineral content of 55 honey samples, which represented three different types of honey, namely honeydew, buckwheat and rape honey from different areas in Poland, was evaluated [9]. Determination of 13 elements (Al, B, Ba, Ca, Cd, Cu, K, Mg, Mn, Na, Ni, Pb and Zn) was performed using ICP-MS (Agilent 7500 ce with Octople Reaction System ICP mass spectrometer).

The authors tried to prove that the analysis of quality and quantity of honey elements could be used to define honey origin by using ICP-MS as a technique for simultaneous determination of elements. Chemometric methods, such as correspondence analysis (CA) and PCA, were applied to classify honey according to mineral content. CA showed three clusters corresponding to the three botanical origins of honey. PCA permitted the reduction of 13 variables to four principal components explaining 77.19% of the total variance. The first most important principal component was strongly associated with the value of K, Al, Ni and Cd. This study revealed that CA and PCA appear to be useful tools for differentiation of honey samples authenticity using the profile of mineral content and they highlighted the relationship between the elements distribution and honey type.

The article of Kaškonienė et al. [10] is aimed at summarising the studies carried out in the past two decades. An attempt is made to find useful chemical markers for unifloral honey, based on the analysis of the compositional data of honey volatile compounds, phenolic acids, flavonoids, carbohydrates, amino acids and some other constituents. This review demonstrates that currently, it is rather difficult to find reliable chemical markers for the discrimination of honey collected from different floral sources because the chemical composition of honey also depends on several other factors, such as geographic origin, collection season, mode of storage, bee species and even interactions between chemical compounds and enzymes in the honey. Therefore, some publications from the reviewed period have reported different floral markers for honey of the same floral origin. In addition, the results of chemical analyses of honey constituents may also depend on sample preparation and analysis techniques. Consequently, a more reliable characterisation of honey requires the determination of more than a single class of compounds, preferably in combination with modern data management of the results, for example, PCA or cluster analysis.

Other strategies for origin testing of honeys may be found in Refs. [11–14,30].

16.2.4 Wine

ICP-MS and optical emission spectroscopy were used to determine the multi-element composition of 272 bottled Slovenian wines [14]. Samples were analysed without preliminary treatment, such as microwave digestion or UV decomposition. No pretreatment is simpler, quicker and minimises possibilities of sample contamination compared to digestion, moreover, analytical limits of detection are reported to be better [15].

To minimise possible matrix effects, suitable dilutions were made using 2% (*v/v*) nitric acid: 10-fold for ICP-MS and 3-fold for ICP- optical emission spectrometry (OES) measurements. To correct for instrumental drift, internal standards were added, Y for ICP-OES and Rh and Tl for ICP-MS.

Agilent Technologies 7500ce ICP-MS instrument, equipped with a micromist glass concentric nebuliser and a Peltier-cooled, Scott-type spray chamber was used. To suppress polyatomic interferences originating from the sample matrix, an octopole reaction system with 5 mL/min He as collision gas and kinetic energy discrimination mode was used (collision mode). For ICP-OES measurements a Varian 715-ES radial ICP-OES, equipped with a V-groove pneumatic nebuliser and a Sturman-Masters spray chamber was used.

To achieve the geographical classification of the wines by their elemental composition, PCA and counter-propagation artificial neural networks (CPANN) have been used. From the 49 elements measured, 19 were used to build the final classification models. CPANN was used for the final predictions because of its superior results. The best model gave 82% correct predictions for the external set of the white wine samples. For the red wines, which were mostly represented from one region, even subregion classification was possible with great precision. From the level maps of the CPANN model, some of the most important elements for classification were identified.

The combined effects of vineyard origin and winery processing have been studied in 65 red wines samples [16]. Grapes originating from five different vineyards within 40 miles of each other were processed in at least two different wineries. Sixty-three different elements were determined with ICP-MS, and wines were classified according to vineyard origin, processing winery, and the combination of both factors. An ICP-MS instrument from Agilent (7700x, Santa Clara, CA, USA) was used for the elemental profiling of 63 trace elements in a mass range between 7 and 238 *m/z*. Vineyard origin and winery processing have an impact on the elemental composition of wine, but each winery and each vineyard change the composition to a different degree. For some vineyards, wines showed a characteristic elemental pattern, independent of the processing winery, but the same was found for some wineries, with a similar elemental pattern for all grapes processed in these wineries, independent of the vineyard origin.

Differentiation of wines according to grape variety and geographical origin based on volatiles profiling using SPME-MS and SPME-GC/MS methods is discussed also in Ref. [17].

16.2.5 Beer

A metabolomic fingerprinting/profiling generated by ambient MS employing direct analysis in real-time (DART) ion source coupled with high-resolution ToF-MS was employed as a tool for beer origin recognition [18]. For DART-ToF-MS analyses, the system consisted of an ambient DART ion source (IonSense, Danvers, MA, USA), an AccuToF LP high-resolution ToF mass spectrometer (JEOL (Europe) SAS, Croissy sur Seine, France) and an AutoDART HTC PAL autosampler (Leap Technologies, Carrboro, NC, USA). In a first step, the DART-ToF-MS instrumental conditions were optimised to obtain the broadest possible representation of ionisable compounds occurring in beer samples (direct measurement, no sample preparation). In the next step, metabolomic profiles (mass spectra) of a large set of different beer brands (Trappist and non-Trappist specialty beers) were acquired. In the final phase, the experimental data were analysed using PLS-DA, LDA and artificial neural networks with MLP with the aim of distinguishing (i) the beers labelled as Rochefort 8; (ii) a group consisting of Rochefort 6, 8, 10 beers; and (iii) Trappist beers. The best prediction ability was obtained for the model that distinguished the group of Rochefort 6, 8, 10 beers from the rest of the beers. In this case, all chemometric tools employed provided ≥95% correct classification.

16.2.6 Distillers Dried Grains and Solubles

In recent years DDGS, co-products of the bioethanol and beverages industries have become globally traded commodities for the animal feed sector. As such, it is becoming increasingly important to be able to trace the geographical origin of commodities in case of a contamination incident or authenticity issue arises. A total of 137 DDGS samples from a range of different geographical origins (China, USA, Canada and European Union) were collected and analysed [19]. IRMS was used to analyse the DDGS for $^{2}H/^{1}H$, $^{13}C/^{12}C$, $^{15}N/^{14}N$, $^{18}O/^{16}O$ and $^{34}S/^{32}S$ isotope ratios that can vary depending on geographical origin and processing. Univariate and multivariate statistical techniques were employed to investigate the feasibility of using the IRMS data to determine the botanical and geographical origin of the DDGS. The results indicated that this commodity could be differentiated according to their place of origin by the analysis of stable isotopes of hydrogen, carbon, nitrogen and oxygen but not with sulphur. By adding data to the models produced in this study, potentially an isotope databank could be set up for traceability procedures for DDGS, similar to the one established already for wine which will help in feed and food security issues arising worldwide.

16.2.7 Gelatine

Gelatine is a component of a wide range of foods. It is manufactured as a byproduct of the meat industry from bone and hides, mainly from bovine and porcine sources. Accurate food labelling enables consumers to make informed decisions about the food they buy. Since labelling currently relies heavily on due diligence involving a paper trail, there could be benefits in developing a reliable test method for the consumer industries in terms of the species origin of gelatine. A method to determine the species origin of gelatines by peptide MS methods was presented [20]. An evaluative comparison is also made with ELISA and PCR technologies.

Samples were loaded onto a nanoAcquity UPLC system (Waters Corp., Manchester, UK) equipped with a nanoAcquity Symmetry C_{18}, 5 μm trap (180 μm×20mm, Waters Corp.) and a nanaAcquity BEH 130 1.7 μm C_{18} capillary column (75 μm×250mm, Waters). The nanoLC system was interfaced with a maXis LC-MS/MS System (Bruker Daltonics, Coventry, UK). Positive ESI-MS and MS/MS spectra were acquired using AutoMSMS mode.

Commercial gelatines were found to contain undeclared species. Furthermore, undeclared bovine peptides were observed in commercial injection matrices. This analytical method could therefore support the food industry in terms of determining the species authenticity of gelatine in foods.

16.2.8 Baijiu

A metabolomics strategy was developed to differentiate strong aroma-type baijiu (SAB) (distilled liquor) from the Sichuan basin (SCB) and Yangtze-Huaihe River Basin (YHRB) through liquid–liquid extraction coupled with GC×GC-ToF-MS [21].

GC×GC-ToF-MS analysis was performed on a Pegasus 4D instrument (LECO; St. Joseph, MI) equipped with a 7890B gas chromatograph connected to a ToF mass spectrometer. A DB-WAX column (30 m×0.25mm i.d., 0.25 μm thickness, Agilent, USA) was used as the first-dimensional column and a DB-5 column (2 m×0.1mm i.d., 0.1 μm thickness, Agilent, USA) was used as the second column. Mass spectra (35–450 *m/z*) were acquired in an electron ionisation mode.

PCA separated the samples effectively from these two regions. The PLS-DA training model was excellent, with explained variation and predictive capability values of 0.988 and 0.982, respectively. As a result, the model demonstrated its ability to perfectly differentiate all the unknown SAB samples. Twenty-nine potential markers were located by variable importance in projection values, and 24 of them were identified and quantitated. Discrimination ability is closely correlated to the characteristic flavour compounds, such as acid, esters, furans, alcohols, sulfides and pyrazine. Most of the marker compounds were less abundant in the SCB samples than in the YHRB samples. The quantitated markers were further processed using hierarchical cluster analysis for targeted analysis, indicating that the markers had great discrimination power to differentiate the SAB samples.

16.2.9 Truffles

The aim of the study of Segelke et al. [22] was to develop a protocol for the authentication of truffles using ICP-MS. Multi-elemental analyses were performed on an HR-ICP-MS Element2

(Thermo Fisher, Inc., Waltham, MA, USA) and an ICP-Q-MS Agilent 7700x (Agilent Technologies, Inc., Santa Clara, CA, USA).

The price of the different truffle species varies significantly, and because the visual differentiation is difficult within the white truffles and within the black truffles, food fraud is likely to occur. Thus, in the context of this work, the elemental profiles of 59 truffle samples of five commercially relevant species were analysed and the resulting element profiles were evaluated with chemometrics. Classification models targeting the species and the origins were validated using nested cross-validation and were able to differentiate the most expensive *Tuber magnatum* from any other examined truffle. For the black truffles, an overall classification accuracy of 90.4% was achieved, and most importantly, a falsification of the expensive *Tuber melanosporum* by *Tuber indicum* could be ruled out. With regard to the geographical origin, for Italy and Spain, one-versus-all classification models were calculated each to differentiate truffle samples from any other origins by 75.0% and 86.7%, respectively. The prediction was still possible according to an internal mathematical normalisation scheme using only the element ratios instead of the absolute element concentrations.

16.2.10 Adzuki Bean

Adzuki beans (*Vigna angularis*) are cultivated worldwide, but there are no tools for accurately discriminating their geographical origin. The study by Kim et al. [23] aimed to develop a method for discriminating the geographical origin of adzuki beans through targeted and non-targeted metabolite profiling with GC-ToF-MS combined with multivariate analysis. Orthogonal PLS-DA showed clear discrimination between adzuki beans cultivated in Korea and China. Non-targeted metabolite profiling showed better separation than targeted profiling. Furthermore, citric acid and malic acid were the most notable metabolites for discriminating adzuki beans cultivated in Korea and China. The geographical discrimination method combining non-targeted metabolite profiling and Pareto-scaling showed excellent predictability ($Q^2=0.812$). Therefore, it is a suitable prediction tool for the discrimination of geographical origin and is expected to be applicable to the geographical authentication of adzuki beans.

GC-ToF-MS analysis was performed using an Agilent 7890A gas chromatograph (Agilent, Atlanta, USA) coupled to a Pegasus HT ToF mass spectrometer (LECO) with a CP-SIL 8 CB column (30 m length, 0.25 mm diameter and 0.25 μm thickness, Agilent). The split ratio, injector temperature and helium gas flow were 1:25, 230°C and 1.0 mL/min, respectively. The column temperature was held for 2 minutes at 80°C, increased at 15°C/min to 320°C and then maintained for 10 minutes. The temperatures of the transfer line and ion sources were 250°C and 200°C, respectively, and scan mode was used with a mass range of 85–600 *m/z*.

16.2.11 Durum Wheat

The assessment of durum wheat's geographical origin is an important and emerging challenge, due to the added value that a claim of origin could provide to the raw material itself, and subsequently to the final products (i.e., pasta).

Up to now, the typical approach presented in literature is the evaluation of different isotopic ratios of the elements, but other techniques could represent an interesting and even more powerful alternative.

In the study of Cavanna et al. [24], using a non-targeted high-resolution MS approach, a selection of chemical markers related to the geographical origin of durum wheat was provided. Samples of the 2016 wheat campaign were used to set up the model and to select the markers, while samples from the 2018 campaign were used for model and markers validation.

The chromatographic separation was performed with a Dionex Ultimate 3000 UHPLC (Thermo Fisher Scientific, Inc, Waltham, MA) equipped with a BEH C18 150×2.1 mm, 1.7 μm particle size analytical column (Waters, Milford, MA, USA) maintained at 40°C. Formic acid (FA) and ammonium formate (AF) were used as mobile-phase modifiers during the gradient elution (flow rate: 0.3 mL/min).

Gradient elution was: after 1 minute with 95% of mobile phase A (0.1% FA and 5 mM AF in water) and 5% of mobile phase B (0.1% FA and 5 mM AF in methanol), the percentage of solvent B increased to 100% in 8 minutes and then maintained at this percentage for 5 minutes before column re-equilibration (7 minutes).

MS detection was performed with a benchtop Q Exactive™ Hybrid Quadrupole-Orbitrap™ mass spectrometer (Thermo Fisher Scientific, Waltham, MA) equipped with a heated electrospray ionisation interface (ESI) (Thermo Fisher Scientific, Waltham, MA). Two analytical sequences (one with positive and one with negative ionisation mode) were executed.

Including in the samples set different geographies across different continents, discrimination through Italian, European and not-European samples is now possible.

16.3 ORGANICALLY VERSUS CONVENTIONALLY GROWN FOODSTUFFS

16.3.1 Introduction

Organic foods can include fruits, vegetables, grains, dairy foods, eggs and to some extent, meats and poultry. Organic foods are defined as foods that are grown without the use of synthetic fertilisers, sewage sludge, irradiation, genetic engineering, pesticides or drugs.

In the first part of this chapter are discussed methods comparing organically and conventionally grown food products. There are different approaches in developing methods. The obtained data are processed by chemometrics.

In the second part, some analysis methods of compounds in organically grown food products are detailed.

16.3.2 Analysis of Organically versus Conventionally Grown Foodstuffs

Four Rabbiteye blueberry cultivars (i.e., Powderblue, Climax, Tifblue and Woodward) grown organically and conventionally were compared regarding their chemical profiles and antioxidant capacity in terms of total phenolic content (TPC), total anthocyanin content (TAC) and oxygen radical antioxidant capacity (ORAC) activity [31]. Regardless of the high TPC, TAC and ORAC in both organically and conventionally grown blueberries, not all the organic berries showed significantly higher TPC, TAC and ORAC than the conventional berries. The chemical profiles (i.e., the free phenolic compounds and anthocyanins) were determined with aid of HPLC-MS, by which 13 individual anthocyanins and 7 free phenolics were identified.

HPLC electrospray ionisation mass spectrometry (HPLC-ESI-MS) method with positive detective mode was used to identify individual anthocyanins in both organically and conventionally grown samples of the Powderblue and Climax blueberry. The peak identification was primarily based on the comparison of the elution order with published data, stock standards, the detected molecular weight and characteristic MS spectra. The anthocyanins belonged to five main groups of anthocyanidins, which include cyanidin (Cy), delphinidin (Dp), peonidin (Pn), malvidin (Mv) and petunidin (Pt). After the identification of anthocyanins using HPLC-ESI-MS, the levels of individual anthocyanins per 100 g fresh sample were quantified.

For chemical analysis of the phenolic-conjugates before the acid hydrolysis, HPLC-ESI-MS method with negative detective mode was used to analyse for both organically and conventionally grown samples of the Powderblue and Climax blueberry. Using the same hydrolytic conditions for anthocyanins, the phenolics do not undergo hydrolysis. Therefore, the HPLC chromatograms showed many peaks that were hard to be identified. After the comparison of the HPLC peak retention time and MS spectrum of each peak with the authentic standards, some phenolics such as quercetin3-galactoside (*m/z*, 463) and quercetin-3-glucoside (*m/z*, 463) in samples were identified and quantified. Based on the results, the organic and conventional environments showed different effects on the content of quercetin-3-galactoside and quercetin-3-glucoside in the plant. Quercetin-3-glucoside was more varied for the organically grown compared with the conventionally grown in both Powderblue and Climax blueberries. Furthermore, the contents of both quercetin-3-galactoside and quercetin-3-glucoside were significantly less than individual anthocyanins in blueberries.

Subtle differences of bioactive phytochemicals between the organically and conventionally grown berries were demonstrated.

A rapid and sensitive analytical method for the quantification of polyacetylenes in carrot roots was developed by Søltoft et al. [32]. The traditional extraction method (stirring) was compared to a new ultrasonic liquid processor (ULP)-based methodology using high-performance liquid chromatography–ultraviolet (HPLC-UV) and MS for identification and quantification of three polyacetylenes.

The mobile phases used were as A and B eluents Milli-Q water and MeCN, respectively, The separation of analytes was performed at 40°C on an Acquity HSS C18 column, 2.1 × 100 mm, 1.8 μm (Waters, Milford, MA). The HPLC gradient was scaled according to the column dimensions, and gradient flow rates were adjusted to the UPLC mode. The gradient programme was as follows (0.25 mL/min): 70% B for 1.7 minutes, a linear gradient to 86% B for 4.3 minutes, a linear gradient to 95% B for 0.7 minute, isocratic elution for 2.9 minutes, followed by a 1.2 minutes ramp back to 70% B and re-equilibration for 3.6 minutes, giving a total run time of 14.4 minutes.

The ToF-MS, LCT Premier/XE mass spectrometer (Waters, Milford, MA) was operated in electrospray ionisation positive-ion mode (ESI^+).

ULP was superior because a significant reduction in extraction time and improved extraction efficiencies were obtained. After optimisation, the ULP method showed good selectivity, precision [relative standard deviations (RSDs) of 2.3%–3.6%], and recovery (93% of falcarindiol) of the polyacetylenes. The applicability of the method was documented by comparative analyses of carrots grown organically or conventionally in a 2-year field trial study. The average concentrations of falcarindiol, falcarindiol-3-acetate and falcarinol in year 1 were 222, 30 and 94 μg of falcarindiol equiv/g of dry weight, respectively, and 3%–15% lower in year 2. The concentrations were not significantly influenced by the growth system, but a significant year–year variation was observed for falcarindiol-3-acetate.

A HS-SPME method coupled with gas chromatography–ion-trap mass spectrometry has been developed and applied for profiling of volatile compounds released from five *Ocimum basilicum* L. cultivars grown under both organic and conventional conditions [33]. Comprehensive two-dimensional gas chromatography coupled with time-of-flight mass spectrometry (GCGC-ToF-MS) was employed for confirmation of the identity of volatiles extracted from the basil headspace by SPME. Linalool, methyl chavicol, eugenol, bergamotene and methyl cinnamate were the dominant volatile components, the relative content of which was found to enable differentiating between the cultivars examined. The relative content of some sesquiterpenes, hydrocarbons benzenoid compounds and monoterpene hydrocarbons was lower in dried and frozen leaves as compared to fresh basil leaves. Sensory analysis of all examined samples proved the differences between evaluated cultivars.

A Pegasus 4D instrument consisting of an Agilent 6890N gas chromatograph with a split/splitless injector (Agilent, USA), an MPS2 autosampler for automated SPME (Gerstel, Germany) and a Pegasus III high-speed ToF-MS (Leco Corp., USA) was used for analyses of some samples. Inside the GC oven, a cryogenic modulator and the secondary oven were mounted (Leco Corp., USA). Resistively heated air was used as a medium for (two) hot jets, while (two) cold jets were supplied by gaseous nitrogen-cooled by liquid nitrogen. An orthogonal gas chromatographic system consisting of a DB-5 ms capillary column (30 m length 0.25 mm I.D. 0.25 μm film thickness [Agilent, USA]) coupled with a SUPELCOWAX 10 capillary column (1.25 m 0.10 mm I.D. 0.10 μm film thickness [Supelco, USA]) was operated under following conditions: primary oven temperature programme: 45°C (0.2 minute), 10°C/min to 200°C, 30°C/min to 245°C (1.80 minutes); secondary oven temperature: +20°C above the primary oven temperature; modulator offset: +35°C above the primary oven temperature; modulation period: 3 seconds (hot pulse 0.6 seconds); carrier gas: helium;

column flow: 1.3 mL/min. The conditions of ToF-MS were as follows: electron ionisation mode (70 eV); ion source temperature: 220°C; mass range: 25–300 *m/z*; acquisition rate: 300 spectra/s; detector voltage: 1,750 V. ChromaToF (LECO Corp.) software (v. 2.31) was used for instrument control, data acquisition and data processing.

Basil plants cultivated by organic and conventional farming practices were accurately classified by pattern recognition of GC-MS data [34].

All the data were collected on a Shimadzu GC-MS-QP2010SE gas chromatograph/mass spectrometer equipped with an AOC-20i autoinjector and an AOC-20S autosampler (Shimadzu Scientific Instruments, Columbia, MD). The GC/MS system was controlled by the GCMSsolution software version 2.70 (Shimadzu Scientific Instruments Inc., Columbia, MD). The GC separation was accomplished on a 30 m×0.25 mm×0.25 μm 5% diphenyl/95% dimethyl polysiloxane cross-linked capillary column (SHRXI-5MS, Shimadzu Scientific Instruments Inc., Columbia, MD). The injector temperature was 260°C, and the injection volume was 1 μL with a split ratio of 1:10. The temperature programming was as follows: 75°C, hold for 4 minutes; ramp 10°C/min and 280°C, hold for 10 minutes. The interface temperature was 260°C, and the ion source temperature was 200°C. The carrier gas helium (99.99% purity) was maintained at a flow rate of 1.5 mL/min throughout the experiment. A novel extraction procedure was devised to extract characteristic compounds from ground basil powders. Two in-house fuzzy classifiers, i.e., the fuzzy rule-building expert system (FuRES) and the fuzzy optimal associative memory (FOAM) for the first time, were used to build classification models. Two crisp classifiers, i.e., SIMCA and PLS-DA, were used as control methods. Before data processing, baseline correction and retention time alignment were performed. Classifiers were built with the two-way datasets, the total ion chromatogram representation of datasets, and the total mass spectrum representation of datasets, separately. Bootstrapped Latin partition was used as an unbiased evaluation of the classifiers. By using two-way datasets, average classification rates with FuRES, FOAM, SIMCA and PLS-DA were 100%±0%, 94.4%±0.4%, 93.3%±0.4% and 100%±0%, respectively, for 100 independent evaluations. The established classifiers were used to classify a new validation set collected 2.5 months later with no parametric changes except that the training set and validation set were individually mean-centred. For the new two-way validation set, classification rates with FuRES, FOAM, SIMCA and PLS-DA were 100%, 93%, 97%, and 100%, respectively. Thereby, the GC-MS analysis was demonstrated as a viable approach for organic basil authentication. It is the first time that a FOAM has been applied to classification. A novel baseline correction method was used also for the first time. The FuRES and the FOAM are demonstrated as powerful tools for modelling and classifying GC-MS data of complex samples, and the data pretreatments are demonstrated to be useful to improve the performance of classifiers.

Ultra-performance liquid chromatography-mass spectrometry (UPLC-MS), flow injection mass spectrometry (FIMS) and headspace-GC combined with multivariate data analysis techniques were examined and compared in differentiating organically grown oregano from that grown conventionally [35].

Waters Acuity UPLC-Xevo G2 QToF-MS system was selected for both UPLC-MS and FIMS analyses. Mobile phase A consisted of 0.1% FA in the water, and mobile phase B consisted of 0.1% FA in acetonitrile (ACN) for both UPLC-MS and FIMS analyses. For UPLC-MS, a Waters BEH C18 column (2.1 mm i.d.×100 mm, 1.7 μm) was used at 40°C; the elution gradient started with 5% phase B, changed linearly to 50% in 10 min, increased linearly to 90% B for 15 minutes, was maintained for 2 minutes and then returned to its initial conditions for 2 minutes to re-equilibrate the column for the next injection. The flow rate was 0.4 mL/min with an injection volume of 2 μL.

An MSE method was used with a mass range from 100 to 1,000 *m/z* in ESI negative mode. For FIMS, the mobile phase of methanol/water in a ratio of 1:1 (*v/v*) was used, and the flow rate was 0.2 mL/min. Oregano extracts were diluted ten times before injection, and the injection volume was 2 μL. MS spectra were collected from 0.05 to 0.55 minute, and MS conditions were the same as those used in UPLC-MS analysis. Masslynx 4.1 software (Waters) was used for the alignment of peaks and accurate mass weight calculation and identification.

The results indicated that UPLC-MS, FIMS and headspace-GC-FID fingerprints with OPLS-DA were able to effectively distinguish oreganos under different growing conditions, whereas with PCA, only FIMS fingerprint could differentiate the organically and conventionally grown oregano samples. UPLC fingerprinting provided detailed information about the chemical composition of oregano with a longer analysis time, whereas FIMS finished a sample analysis within 1 min. On the other hand, headspace GC-FID fingerprinting required no sample pretreatment, suggesting its potential as a high-throughput method in distinguishing organically and conventionally grown oregano samples. In addition, chemical components in oregano were identified by their molecular weight using QToF-MS and headspace-GC-MS.

HPLC and flow injection ESI with ion-trap MS (FIMS) fingerprints combined with PCA were examined for their potential in differentiating commercial organic and conventional sage samples [36]. The individual components in the sage samples were also characterised with an ultra-performance liquid chromatograph with a quadrupole-time-of-flight mass spectrometer (UPLC Q-ToF-MS). The results suggested that both HPLC and FIMS fingerprints combined with PCA could differentiate organic and conventional sage samples effectively. FIMS may serve as a quick test capable of distinguishing organic and conventional sages in 1 minute and could potentially be developed for high-throughput applications, whereas HPLC fingerprints could provide more chemical composition information with a longer analytical time.

An Agilent 1100 HPLC system was used for HPLC fingerprinting (Agilent Technologies, Palo Alto, CA, USA) and connected with an LCQ Deca ion-trap mass spectrometer for FIMS (Agilent Technologies, Palo Alto, CA, USA). ESI in negative ion mode was used.

A Waters symmetry C18 column (2.1 mm i.d. × 150 mm, 3.5 μm) (Waters, Milford, MA, USA) was used for HPLC-UV fingerprinting, mobile phase A consisted of 0.1% FA in H_2O and mobile phase B consisted of 0.1% FA in ACN. The elution began with 20% phase B; changed linearly to 50% B in 20 minutes, increased linearly to 95% B in 35 minutes and washed at this ratio for 5 minutes. After washing, the mobile phase was returned to its initial conditions for 5 minutes to re-equilibrate the column for the next injection. The flow rate was 0.2 mL/min, with an injection volume of 10 μL and a detection wavelength of 280 nm.

For FIMS analysis, no analytical column but a two-way valve was used to minimise potential contamination in the MS system. The mobile phase A had 0.1% FA in H_2O (*v/v*) and mobile phase B contented 0.1% FA in ACN (*v/v*) with isocratic elution at a ratio of phase A to phase B at 60:40 (*v/v*) and a flow rate of 0.5 mL/min. Sage extractions were diluted ten times with water and the injection volume was 2 μL. MS spectra were collected from 0.05 to 0.55 minute and the mass range was from 150 to 1,000 *m/z*.

A UPLC-Q-ToF-MS system (Waters, Milford, MA, USA) was used for the identification of major peaks in the sage extraction. A BEH C18 column (2.1 mm i.d. × 100 mm, 1.7μm) (Waters, Milford, MA, USA) was used at 40°C. UPLC conditions were the same as those for HPLC fingerprint. The MS conditions were: capillary voltage 3.00 kV; sampling cone voltage 30 V; extraction cone voltage 4.0 V; source temperature 120°C; and desolvation temperature 450°C. The cone gas flow rate was 50 L/h and the desolvation gas flow was 800 L/h. An MS^E method (a scan model to get full information for both parent ion and daughter ion in one injection using tandem mass spectrometry) was used with a mass range from 100 to 1,000 *m/z*. MassLynx 4.1 software (Waters, Milford, MA, USA) was used for alignment of peaks and accurate mass weight identification.

16.3.3 Analysis of Organic Products

A method for the simultaneous determination of pesticides, biopesticides and mycotoxins from organic products was developed [37]. Extraction of more than 90 compounds was evaluated and performed by using a modified QuEChERS-based sample preparation procedure. The method was based on a single extraction with acidified ACN, followed by partitioning with salts, avoiding any clean-up step before the determination by UHPLC-MS/MS.

Chromatographic analyses were performed in an ACQUITY UPLCTM system (Waters, Milford, MA, USA) equipped with a binary solvent delivery system, degasser, autosampler and column heater. Chromatographic separation was performed using an Acquity UPLC BEH C18 column (100 mm × 2.1 mm), with 1.7 μm particle size, from Waters. MS/MS detection was performed using an Acquity TQD tandem quadrupole mass spectrometer (Waters, Manchester, UK), equipped with an ESI operating in positive-ion mode.

A gradient programme was used consisting of methanol (eluent A) and an aqueous solution of AF (5 mM) (eluent B). The gradient profile started at 20% of eluent A and increased linearly up to 95% in 11 minutes, keeping constant for 0.5 minute before being returned to the initial conditions in 0.5 minute. The column was re-equilibrated for 1.0 minute at the initial mobile-phase composition, obtaining a total run time of 13.0 minutes. The flow rate was set at 0.45 mL/min, and the column temperature was kept at 30°C. The injection volume was 5 μL and, the autosampler was flushed with a methanol/water solution (1:9 *v/v*) before sample injection to avoid carry-over.

For MS/MS detection, source parameters were as follows: capillary voltage, 3.0 kV; extractor voltage, 3 V; source temperature, 120°C and desolvation gas temperature, 350°C. Desolvation gas and cone gas (both nitrogen) flow rates were set at 50 and 600 L/h, respectively. Collision-induced dissociation was performed using argon as collision gas at a pressure of 4.0×10^{-3} mbar in the collision cell. For instrument control, data acquisition and processing, MassLynx and QuanLynx software version 4.1 (Waters) were used.

Validation studies were carried out in wheat, cucumber and red wine as representative matrices. Recoveries of the spiked samples were in the range between 70% and 120% (with intra-day precision, expressed as RSD, lower than 20%) for most of the analysed compounds, except picloram and quinmerac. Inter-day precision, expressed as RSD, was lower than 24%. Limits of quantification were lower than 10 μg/kg.

Practical and easy control of the authenticity of organic sugarcane samples based on the use of machine-learning algorithms and trace elements determination by ICP-MS is proposed [38].

All analyses were carried out by inductively coupled plasma quadrupole mass spectrometry (ICPQ-MS). A Perkin-Elmer ELAN DRC (e) instrument was used with a Meinhart nebuliser and silica cyclonic spray chamber and continuous nebulisation [39]. The following instrumental parameters of Elan DRC-e spectrometer were used: 0.92 L/min Nebulizer Gas Flow (NEB), 1.2 L/min Auxiliary Gas Flow, 15 L/min Plasma Gas Flow, 1100 W. ICP RF Power, 0.0 Quadrupole Rod Offset, 70.00 Discriminator Threshold. Reference ranges for 32 chemical elements in 22 samples of sugarcane (13 organic and 9 non-organic) were established and then two algorithms, Naive Bayes (NB) and RF, were evaluated to classify the samples. Accurate results (>90%) were obtained when using all variables (i.e., 32 elements). However, accuracy was improved (95.4% for NB) when only eight minerals (Rb, U, Al, Sr, Dy, Nb, Ta, Mo), chosen by a feature selection algorithm, were employed.

Euterpe oleracea (açaí) has been reported to be rich in health-beneficial chemical constituents. Mulabagal and Calderón [40] have developed LC/MS-based fingerprinting and mass profiling methods to identify fatty acids, anthocyanins and non-anthocyanin polyphenols in three processed raw materials; non-organic açaí powder (ADSR-1), raw-organic açaí powder (ADSR-2) and freeze-dried açaí powder (ADSR-3) that are used in the preparation of botanical dietary supplements. For LC/MS analysis of fatty acids and non-anthocyanin polyphenols, the açaí samples were extracted sequentially with dichloromethane followed by methanol. To study fingerprinting analysis of anthocyanins, açaí samples were extracted with acidic methanol–water. An Agilent 6520 Q-ToF mass spectrometer equipped with a 1220 RRLC system (Agilent Technologies, Little Falls, DE) and electrospray ion (ESI) source was used for fingerprinting and mass profiling of açaí samples.

Chromatographic analysis of fatty acids and non-anthocyanin polyphenols in dichloromethane and methanol extracts of açaí samples was achieved on a 2.1 × 100 mm, 1.8 μm ZORBAX Eclipse plus C18 column (Agilent Technologies, New Castle, DE). Liquid chromatographic analysis was carried out using a gradient mobile phase consisting of (A) 0.1% FA in water and (B) 0.1% FA in methanol at 40°C. The flow rate was 0.2 mL/min. A linear gradient was run as follows: 0 minute, 30% B; 2 minutes, 30% B; 25 minutes, 99% B; 27 minutes, 99% B; 30 minutes, 30% B; and 35 minutes, 30% B.

LC separation for anthocyanins was carried out using a gradient mobile phase containing (A) 0.1% FA in water and (B) 0.1% FA in methanol and ACN (50:50, *v/v*) with an initial condition of 1% B for 2 minutes. The mobile phase was then linearly increased to 99% B in 25 minutes and maintained the same until 27 minutes. The conditions were back to 1% B at 30 minutes and equilibrated further for 5 minutes. The flow rate was 0.2 mL/min.

Full-scan MS analysis was performed over a mass range of 100–1,000 *m/z*. MS analysis for fatty acids and non-anthocyanin polyphenols was carried out using an electrospray ion source in negative mode. MS analysis for anthocyanins was performed in positive mode.

LC/MS analysis of dichloromethane extracts of (ADSR-1), (ADSR-2) and (ADSR-3) açaí powders have shown to contain fatty acids, γ-linolenic acid, linoleic acid, palmitic acid, and oleic acid. Whereas, the fingerprinting analysis of methanol extracts of ADSR-1, ADSR-2 and ADSR-3 led to the identification of phenolic acids, anthocyanin and non-anthocyanin polyphenols.

Rice can be a major source of inorganic arsenic (As_i) for many sub-populations. Rice products are also used as ingredients in prepared foods, some of which may not be obviously rice-based. Organic brown rice syrup (OBRS) is used as a sweetener in organic food products as an alternative to high-fructose corn syrup.

The concentration and speciation of As in commercially available brown rice syrups and in products containing OBRS, including toddler formula, cereal/energy bars, and high-energy foods used by endurance athletes, were determined [41].

As_{total} was determined by ICP-MS (model 7700x; Agilent, Santa Clara, CA) using helium as collision gas at a flow rate of 4.5 mL/min. Samples were analysed by either external calibration or the method of standard additions. As speciation of the 1% HNO_3 extracts was determined by ion chromatography coupled to ICP-MS using a Hamilton PRP X100 anion exchange column (Hamilton Company, Reno, NV) and a 20 mM ammonium phosphate eluant at pH 8.

It was found that OBRS can contain high concentrations of As_i and dimethyl-arsenate (DMA). An organic toddler milk formula containing OBRS as the primary ingredient had As_{total} concentrations up to six times the U.S. Environmental Protection Agency safe drinking water limit. Cereal bars and high-energy foods containing OBRS also had higher As concentrations than equivalent products that did not contain OBRS. As_i was the main As species in most food products tested in this study.

16.4 SPECIATION

16.4.1 Chemical Speciation

A definition of chemical speciation may be the determination of the individual concentrations of the various chemical forms of an element together making up the total concentration of that element in a sample.

The preferred analysis method is ICP-MS. ICP-MS is a type of mass spectrometry that uses an ICP to ionise the sample. It atomises the sample and creates atomic and small polyatomic ions to be detected.

16.4.2 Arsenic

Dufailly et al. [42] present a fully validated and rapid quantitative method for the determination of inorganic arsenic [arsenite, As(III) and arsenate, As(V)] and organic arsenic species (methylarsonic acid, dimethylarsinic acid, and arsenobetaine) by ion chromatography paired with ICP-MS after ultrasonic-assisted enzymatic extraction (UAEE) in rice- and seafood-based raw materials and finished products.

Series 200 (Perkin-Elmer) HPLC system equipped with a Series 200 quaternary pump, autosampler, vacuum degasser, injection valve with a 100 mL polyether ether ketone (PEEK) sample loop, and data software (Chromera, Perkin-Elmer). An IonPac AG7 guard column (50′ 4 mm, 10 μm particles) and an IonPac AS7 anion exchange column (250×4 mm, 10 μm particles; both from Dionex, Sunnyvale, CA) were used for the chromatographic separation of compounds. The analytical column was connected to the nebuliser of the ICP/MS instrument using 50 cm long PEEK tubing (0.17 mm i.d.). The spectrometer was an Elan DRC II (Perkin-Elmer) ICP/MS instrument equipped with a mini cyclonic spray chamber (Cinabar) fitted with a micromist nebuliser, 0.2 mL/min (Glass Expansion, Melbourne, Australia). For total As analysis, the sample solutions were pumped by a peristaltic pump from tubes arranged on an ASX 500 Model 510 autosampler (CETAC, Omaha, NE) and aspirated into the argon plasma.

This method gives toxicological meaning to arsenic analysis, since the sum of the toxic chemical forms As(III) and As(V) can be determined. In contrast to classical water–methanol extraction, UAEE enables drastic acceleration of sample extraction (5 minutes instead of several hours), while total arsenic extraction efficiency is improved without species conversion. Validation was performed to evaluate the method for selectivity, linearity, LOD/LOQ (0.007–0.020 mg/kg), trueness, precision (HorRat values, 0.2–0.6), recovery (93–122%), and uncertainty. The method was also satisfactorily tested using two proficiency tests. Performance characteristics are reported for four certified reference materials, standard reference material (SRM) 1568a (rice flour), Institute for Reference Materials and Measurements 804 (rice flour), SRM 2976 (mussel tissue), certified reference material-627 (tuna fish), and several commercial food samples populating five AOAC triangle food sectors. The results indicated that this speciation method is cost-efficient, time-saving, and accurate, as well as fit-for-purpose, according to International Organization for Standardization/International Electrotechnical Commission 17025:2005 standard, and could be used for routine analysis.

Arsenic species have been investigated in *Anemonia sulcata*, which is a frequently consumed food staple in Spain battered in wheat flour and fried with olive oil [43]. Speciation in tissue extracts was carried out by anion/cation exchange chromatography with inductively coupled plasma mass spectrometry (HPLC-(AEC/CEC)–ICP-MS). Three methods for the extraction of arsenic species were investigated (ultrasonic bath, ultrasonic probe and focused microwave) and the optimal one was applied. Arsenic speciation was carried out in raw and cooked anemone and the dominant species are dimethylarsinic acid (DMA^V) followed by arsenobetaine (AB), As^V, monomethylarsonic acid (MA^V), tetramethylarsonium ion (TETRA) and trimethylarsine oxide (TMAO). In addition, arsenocholine (AsC), glyceryl phosphorylarsenocholine (GPAsC) and dimethylarsinothioic acid (DMAS) were identified by liquid chromatography coupled to triple quadrupole mass spectrometry (HPLC-MS). These results are interesting since GPAsC has been previously reported in marine organisms after experimental exposure to AsC, but not in natural samples.

Elemental detection was performed with an ICP mass spectrometer model Agilent 7500ce (Agilent Technologies, Tokyo, Japan) equipped with an octopole collision cell and a micromist nebuliser.

The separation of the arsenic species was performed in a 25 cm×4.1 mm Hamilton PRP X-100 column (Hamilton, Reno, NV, USA). The mobile phase contained 50.0 mM of $(NH_4)_2CO_3$ at pH 8.5 (adjusted with 5 N ammonium hydroxide) and was flushed at 0.8 mL/min through the column. Since AB, TETRA (tetramethylarsonium ion) and TMAO coelute in this column, the separation was carried out using a cation exchange column (Supelcosil LC-SCX, 25 cm×4.6 mm, 5 μm). The mobile phase used was 20 mM pyridine at pH 2.5 as the mobile phase at 1 mL/min.

For the standardless identification of arsenic species, HPLC was coupled to a triple quadrupole instrument model API 2000 (AB/MDS Sciex, Concord, Canada) via an ESI interface using a PEEK tube. The flow rate was reduced to 300 μL/min using a post-column split. To enhance sensitivity, 10% (*v*/*v*) of FA was added to the anion exchange chromatographic mobile phase and the source temperature was maintained at 400°C. Data were collected in positive-ion mode (ES^+) and the mass analyser scanned from 50 to 400 *m*/*z*.

Data were treated with the manufacturer's software (Analyst 4.1).

Total arsenic in rice was determined by ICP-MS; arsenite, arsenate, monomethylarsonic acid, and dimethyarsinic acid were quantified by HPLC-ICP-MS [44]. Methods using nitric acid and malonic acid were validated at various extraction conditions and mobile-phase systems.

The ICP-MS used in this study was an Elan 6100 Dynamic Reaction Cell II (Perkin-Elmer Sciex). It had a high-efficiency sample introduction desolvating system equipped with a quartz cyclonic spray chamber and an additional mixing peristaltic pump (Apex-IR). The operating conditions were a forward power of 1.35 kW with argon gas flow rates of 16.00 L/min (plasma), 1–1.3 L/min (auxiliary), and 1–1.07 L/min (nebuliser). The argon gas was of spectral purity (99.9998%). Before measurements, the instrument was tuned for daily performance using an aqueous multi-element standard solution of Li, Y, Co, Ce and Tl for consistent sensitivity (^{7} Li, 89Y, and^{205}Tl) and minimum doubly charged and oxide species (^{140}Ce).

A Shimadzu HPLC (LC-VP series) was coupled with ICP-MS for arsenic speciation. The column was a Hamilton PRP X-100 (4.1×250 mm, 10 μm) with an injection volume of 50 μL at room temperature. The quadrupole mass analyser was operated in the single ion monitoring mode (75 *m/z*) for detecting arsenic. Data evaluation was performed using the instrument software (version 3.3). The quantification was based on peak height by external calibration using standard solutions of arsenic species.

The linear dynamic range, limit of detection, precision, fortification and analysis of a white rice flour certified reference material (CRM-7503-a) were evaluated for quality assurance. The use of 5 mM malonic acid for extraction with an isocratic mobile phase was optimised for extraction time and temperature and employed for arsenic speciation in rice. The concentrations of total arsenic, arsenite, arsenate, monomethylarsonic acid and dimethyarsinic acid were low compared to the provisional tolerable weekly intakes specified by the Food and Agriculture Organization/World Health Organization Joint Expert Committee on food additives and European food safety authority and thus do not pose a threat to consumers.

Ultrasound- and microwave-assisted extraction of arsenic in wheat and wheat-based food using different solvents or enzymes was investigated in terms of extraction yield and species stability [45]. Four extraction procedures were selected for the study of arsenic speciation in wheat and wheat products by anion exchange HPLC-ICP-MS using a PRP-X100 column with 10 mM $NH_4H_2PO_4$, 10 mM NH_4NO_3, and 2% CH_3OH at pH 5.5 as the mobile phase. Total arsenic in the samples ranged from 8.6 to 29.8 ng/g dry weight. About 95% of the arsenic was found to be present in inorganic form with As^{III} as the most abundant species, whereas the remainder was mainly DMA. Microwave-assisted extraction with HNO_3 was the most effective in liberating the arsenic species, which were then satisfactorily recovered from the chromatographic column. The LODs achieved, i.e., 0.35–0.46 ng/g dry weight, were suitable for the determination of arsenic species at the low levels found in sample extracts.

Qu et al. [46] report an analytical methodology for the quantification of common arsenic species in rice and rice cereal using capillary electrophoresis coupled with inductively coupled plasma mass spectrometry (CE–ICP-MS) (Agilent 7700x, Agilent Technology, Santa Clara, CA).

CE was performed on a 7100 CE system (Agilent Technology, Santa Clara, CA). A 60 cm long-coated fused silica capillary (Molex, Phoenix, AZ) with a 100 μm inner diameter and a 360 μm outer diameter was used for sample introduction. The separation of five arsenic species was carried out under basic conditions through the use of Na_2CO_3 (8 mM, pH 11) buffer.

An enzyme (i.e., α-amylase)-assisted water-phase microwave extraction procedure was used to extract four common arsenic species, including dimethylarsinic acid (DMA), monomethylarsonic acid (MMA), arsenite [As(III)], and arsenate [As(V)] from the rice matrices. The addition of the enzyme α-amylase during the extraction process was necessary to reduce the sample viscosity, which subsequently increased the injection volume and enhanced the signal response. o-Arsanilic acid was added to the sample solution as a mobility marker and internal standard. The obtained repeatability (i.e., RSD%) of the four arsenic analytes of interest was less than

1.23% for elution time and 2.91% for peak area. The detection limits were determined to be 0.15–0.27 ng/g. Rice standard reference materials SRM 1568b and CRM 7503-a were used to validate this method. The quantitative concentrations of each organic arsenic and summed inorganic arsenic were found within 5% difference of the certified values of the two reference materials.

An ICP-MS was used as an ion chromatographic (IC) detector for the speciation analysis of arsenic in edible oil [47]. The arsenic species studied include arsenite, arsenate, monomethylarsonic acid, dimethylarsinic acid, arsenobetaine and AsC. Gradient elution using $(NH_4)_2CO_3$ and methanol at pH 8.5 allowed the chromatographic separation of all species in less than 8 minutes. Effluents from the IC column were delivered to the nebuliser of ICP-MS for the determination of arsenic.

An ELAN 6100 DRC II ICP-MS (Perkin-Elmer SCIEX, Concord, ON, Canada) was used for these experiments.

The HPLC system had as stationary phase a Hamilton PRP-X100 anion exchange column (10 µm–4.1 mm i.d.–250 mm length) and as mobile phase A: 0.5 mmol/L $(NH_4)_2CO_3$, 1% *v/v* methanol (pH 8.5) and B: 50 mmol/L $(NH_4)_2CO_3$, 1% *v/v* methanol (pH 8.5). Gradient programme 0–0.1 minute: 100% A; 0.1–7 minutes: 100% B. Mobile-phase flow rate 1.0 mL/min.

The concentrations of arsenic species have been determined in several used and fresh vegetable oil samples. In this study, a microwave-assisted extraction method was used for the extraction of arsenic species from oil samples. The extraction efficiency was better than 92% and the recoveries from spiked samples were in the range of 90%–105%. The precision between sample replicates was better than 8% for all determinations. The limits of detection were in the range of 0.008–0.024 ng/mL for various arsenic species based on peak height, which corresponded to 0.08–0.24 ng/g in the original oil sample.

16.4.3 Selenium

Bañuelos et al. [48] investigated Se in *Opuntia ficus-indica* using ICP-MS, microfocused X-ray fluorescence elemental and chemical mapping (mXRF), Se K-edge X-ray absorption near-edge structure (XANES) spectroscopy, and liquid chromatography-mass spectrometry (LC-MS).

Total Se was analysed by an ICP mass spectrometer (Agilent 7500cx). LC-MS Dual Shimadzu LC-10AD pumps, a Shimadzu SC-10A photodiode array detector (PDA), a Phenomenex Hyperclone 5-mm ODS (C18) 120A 250–3 2.00-mm column at 20°C, and an Applied Biosystems QSTAR XL high-resolution quadrupole time-of-flight mass spectrometer were used. A column flow of 0.5 mL/min was composed of two eluents: A, 10 mM AF in deionised water; B, 10 mM AF in a 10%:90% deionised water: ACN mixture. The mobile-phase composition (in terms of percentage ratios) was isocratic at 100 A:0 B for 5 minutes, ramped to 60 A:40 B over 2 minutes, ramped to 25 A:75 B over 3 minutes, held for 5 minutes, ramped down to 100 A:0 B over 3 minutes, and held for 2 minutes. Column effluent was routed through the PDA and to the mass spectrometer. After the PDA, however, a splitter directed approximately 80% of the flow to the fraction collector and approximately 20% (100 µL/min) to the mass spectrometer. Spectra over (+) 70–1,300 *m/z* were obtained for all organic Se compounds, while spectra over (–) 40–700 *m/z* were obtained for two inorganic Se compounds using an Ion Spray source with a 4.5-mm i.d. capillary sample tube and a spray voltage of 64,750 V. The declustering potential was 670 V, the focusing potential was 6,230 V, and the ion release delay and width were 6 and 5 (arbitrary units), respectively. Parameters for nitrogen source and curtain gas were 15 p.s.i. and 1.31 L/min, respectively. An initial detection or screening programme method was then created using the five major isotopes and a 0.1-atomic mass unit window of appropriate masses for either positive or negative mode. Retention times in minutes for Se standards were as follows: SeO_3^{-2} = 1.83, SeO_4^{-2} = 1.85, SeCystine = 1.54, SeCys = 1.93, methyl-SeCys = 2.14, γ-glutamyl methyl-SeCys = 2.42, and SeMet = 3.12 (2–3 seconds).

mXRF showed Se concentrated inside small conic, vestigial leaves (cladode tips), the cladode vasculature, and the seed embryos. Se K-edge XANES demonstrated that approximately 96% of total Se in cladode, fruit juice, fruit pulp, and seed is carbon-Se-carbon (C-Se-C). Micro and bulk XANES analysis showed that cladode tips contained both selenate and C-Se-C forms. ICP-MS quantification of Se in HPLC fractions followed by LC-MS structural identification showed selenocystathionine-to-selenomethionine (SeMet) ratios of 75:25, 71:29, and 32:68, respectively, in cladode, fruit, and seed. Enzymatic digestions and subsequent analysis confirmed that Se was mainly present in a "free" nonproteinaceous form inside cladode and fruit, while in the seed, Se was incorporated into proteins associated with lipids. mXRF chemical mapping illuminated the specific location of Se reduction and assimilation from selenate accumulated in the cladode tips into the two LC-MS-identified C-Se-C forms before they were transported into the cladode mesophyll.

16.4.4 Lead

Chen et al. [49] reported an environment-friendly microwave-assisted extraction used to extract trace lead compounds from marine animals and an ultrasensitive method for the analysis of Pb^{2+}, trimethyl lead chloride (TML), and triethyl lead chloride (TEL) by using CE-ICP-MS. The extraction method is simple and has a high extracting efficiency. It can be used to completely extract both inorganic lead and organolead in marine animal samples without altering its species.

The CE-ICP-MS system consists of a CEI-SP20 CE-interface system (Reeko Instrument Co., Xiamen, China) and an Agilent 7500ce ICP-MS (Agilent Technologies, USA).

The analytical method has a detection limit as low as 0.012–0.084 ng Pb/mL for Pb^{2+}, TML, and TEL, and can be used to determine ultratrace Pb^{2+}, TML, and TEL in marine animals directly without any preconcentration. With the help of the above methods, Pb^{2+}, TML, and TEL were successfully determined in a clam and oyster tissue within 20 min with an RSD ($n=6$)<5% and a recovery of 91%–104%. The results showed that Pb^{2+} was the main species of lead in clam and oyster, and organolead (TML) was only found in oyster.

16.4.5 Chromium

Titanium dioxide nanotubes (TDNTs) were used as a solid-phase extraction adsorbent for chromium species by a packed microcolumn coupled with ICP-MS, including total, suspended and soluble chromium as well as Cr(III) and Cr(VI) in tea leaves and tea infusion [50]. The experimental results indicated that Cr(III) was quantitatively retained on TDNTs in the pH range of 5.0–8.0, while Cr(VI) remained in the solution. The total chromium was determined after reducing Cr(VI) to Cr(III). The concentration of Cr(VI) is calculated by the difference between total chromium and Cr(III). Under optimal conditions, the detection limits of this method were 0.0075 ng/mL for Cr(III). The RSD was 3.8% ($n=9$, $c=1.0$ ng/mL).

16.4.6 Organotin

An environmentally friendly method for the determination of seven organotin compounds in honey and wine samples, using headspace solid-phase microextraction (HS-SPME) and GC, was developed [51]. The analytes were derivatised *in situ* with sodium tetraethylborate, and the derivatisation and preconcentration steps were optimised. A 100 µm polydimethylsiloxane fibre was the most suitable for preconcentrating the derivatised analytes from the headspace of an aqueous solution containing the sample. When microwave-induced plasma atomic emission (MIP-AED) and MS detection were compared, higher sensitivity was attained for all compounds with MS, although MIP-AED showed more specific chromatograms. Using MS, detection limits ranged roughly from

0.3 to 4.3 pg (Sn)/g, depending on the compound for honey samples, and from 0.1 to 2 pg (Sn)/mL for wine samples.

GC-MS instruments and conditions were: capillary column: HP-5MS (30 m×0.25 mm i.d. 0.25 μm); electron-impact mode ionisation; Agilent 6890N gas chromatograph (Agilent, Waldbronn, Germany); G5973 microwave-induced quadrupole mass selective spectrometer (Agilent).

16.4.7 Silver

Silver migration from a commercial baby feeding bottle and a food box containing AgNPs (nanoparticles), as confirmed by SEM-EDX analysis, was evaluated using food simulant solutions [i.e., water, 3% (v/v) acetic acid, and 10% and 90% (*v/v*) ethanol] [52].

Silver release was investigated at temperatures in the 20°C–70°C range using contact times of up to 10 days. Migration of silver from the food box was in all cases 2–3 orders of magnitude higher than that observed for the baby bottle, although the total silver content in the original box material was half of that found in the baby bottle. As expected, for both food containers, silver migration depended on both the nature of the tested solution and the applied conditions. The highest release was observed for 3% acetic acid at 70°C for 2 hours, corresponding to 62 ng dm^2 and 1887 ng/dm of silver for the baby bottle and the food box, respectively.

Single particle (SP)-ICP-MS (7700x Series Agilent Technologies, USA) was used to characterise and quantify AgNPs in the food simulants extracts. SP-ICP-MS has been revealed as a powerful technique for the characterisation of metallic NPs size distribution at ultratrace level. Sample preparation was optimised to preserve AgNPs integrity. The experimental parameters affecting AgNPs detection, sizing and quantification by SP-ICP-MS were also optimised. Analyses of water and acidic extracts revealed the presence of both dissolved silver and AgNPs. Small AgNPs (in the 18–30 nm range) and particle number concentrations within the 4–1,510 × 10^6/L range were detected, corresponding to only 0.1%–8.6% of the total silver released from these materials. The only exception was AgNPs migrated into water at 40°C and 70°C from the food box, which accounted for as much as 34% and 69% of the total silver content, respectively.

16.4.8 Multi-Elemental Speciation

It is interesting to develop reliable multi-elemental speciation methodologies, to reduce costs, waste and time needed for the analysis. Separation and detection of species of several elements in a single analytical run can be accomplished by HPLC-ICP-MS. The review of Marcinkowska et al. [53] assembles articles concerning multi-elemental speciation determination of As, Se, Cr, Sb, I, Br, Pb, Hg, V, Mo, Te, Tl, Cd and W in environmental, biological, food and clinical samples analysed with HPLC/ICP-MS. It addresses the procedures in terms of the following issues: sample collection and pretreatment, selection of optimal conditions for elements species separation by HPLC and determination using ICP-MS as well as metrological approach.

Zhang et al. [54] establish a hyphenated methodology coupling HPLC with ICP-MS for simultaneous speciation analysis of arsenic, mercury and lead for the first time. Four arsenicals (As(III), DMA, MMA and As(V)), four mercurials (Hg(II), MeHg, EtHg and PhHg) and three lead compounds (Pb(II), TML and TEL) were simultaneously analysed within only 8 min with acceptable resolution (2.0–8.2 for As, 1.6–6.1 for Hg and 2.7–4.0 for Pb). The detection limits were 0.036–0.20 for As species, 0.023–0.041 for Hg species, and 0.0076–0.14 μg/L for Pb species. The developed method was applied for the measurement of five lotus seed samples, indicating the presence of DMA (19.6–28.2 μg/kg), TML (1.4–2.9 μg/kg), MeHg (1.2–4.8 μg/kg) and EtHg (0.8–2.2 μg/kg).

More references on multi-element speciation may be found in Table 16.1 [53].

TABLE 16.1
Analytical and Metrological Aspects Concerning Multi-elemental Speciation Analysis by HPLC/ICP-MS [54]

	Sample	Chromatographic Separation	HPLC Operating Systems	ICP-MS Detection System	Interferences	Figures of Merits	References
	1. Analytes – inorganic/organic compounds 2. Matrix	1. Column 2. Elution type 3. Mobile phase	1. Mobile-phase flow rate (mL/min) 2. Column temperature (°C) 3. Injection volume (μL) 4. Retention time (min) 5. Total analysis time (min)	1. Type of detection system 2. Monitored isotopes	1. Type of interferences 2. Method of elimination	1. Ld values (ùg/L) 2. Precision as a RSD of peak area (%) 3. Recoveries (%)	
Pb, Hg	1. Pb^{II}-lead(II)nitrate, trimethyl Pb – trimethyl lead(II)chloride, Hg^{II}-mercury(II)nitrate, methyl-Hg-methylmercury(II)chloride, ethyl-Hg-ethylmercury(II)chloride 2. Fish samples	1. Alltech microbore column C18, (150 mm × 1 mm, 5μm) 2. Isocratic 3. 0.2% (*v/v*) 2-ercaptoethanol, 174.2 mg/L 1-pentanesulfonate, 12% (*v/v*) MeOH and 1 mg/L EDTA, pH 2.8	1. 0.12 2. Room temperature 3. 5 4. 0.67 for Pb^{II}, 3.17 for trimetyl-Pb, 14.83 for triethyl-Pb, 8.08 for Hg^{II}, 5.75 for methyl-Hg and 15.08 for ethyl-Hg 5. 16.7	1. Perkin-Elmer SCIEX ELAN 6100 DRCII with high-efficiency nebuliser (DIHEN, 170-AA) 2. ^{200}Hg, ^{202}Hg, ^{206}Pb, ^{207}Pb, ^{208}Pb	1. No data 2. No data	1. 0.1 for Pb^{II}, 0.1 for trimethylPb, 0.3 for triethylPb, 0.2 for Hg^{II}, 0.2 for methyl-Hg, 0.3 for etyl-Hg^{b} 2. <7 for Pb species and <5 for Hg species 3. 98–99 for pb^{II}, 95–99 for trimethyl Pb, 96–99 for triethyl-IICysPb, 96–97 for Hg^{II}, 94–99 for methyl-Hg and 94–95 for methyl-Hg	[68]
Se, Te	1. SeI^{V} – disodium selenite, Se^{VI} – sodium selenite, $MeSe^{II}Cys$ – Se-methylselenocysteine, $Se^{II}Met$ – seleno-DL-methionine TeI^{V} – sodium tellurite, Te^{VI} – sodium tellurate 2. Extracts from milk powder and rice flour samples	1. Hamilton PRP-X100 (4.1 mm × 250 mm, 10 μm) 2. Gradient 3. A: 0.5 mM/L ammonium citrate in 2% methanol (pH 3.7), B: 20 mM/L ammonium citrate in 2% methanol (pH 8.0)	1. 1.0 2. NO data 3. 200 4. 2.5 for Te^{VI}, 3.16 for $MeSe^{II}Cys$, 4.83 for $Se^{II}Met$, 6.92 for Se^{IV}, 9.00 for Se^{VI}, 10.50 for Te^{IV}	1. Perkin-Elmer Sciex ELAN 6100DRCII 2. ^{78}Se, ^{80}Se, ^{82}Se,^{1} ^{30}Te	1. $^{38}Ar^{40}Ar^{+}$ ions interfering with $^{78}Se^{+}$ and $^{80}Se^{+}$ respectively 2. DRC with methane as a reaction gas	1. 0.01 for $MeSe^{II}Cys$, 0.03 for $Se^{II}Met$, 0.01 for Se^{IV}, 0.02 for Se^{VI}, 0.08 for Te^{IV}, 0.01 for Te^{VI} 2. <5.4 for all species 3. 99–105 for Se^{VI}, 96–106 for Te^{IV}, 97–106 for TE^{VI} 1. < 5.4 for all species 3. 95–102 for, 96–104 for, 98–103 for, 95–105 for, 96–106 for, 97–106 for	[69]

(*Continued*)

TABLE 16.1 (*Continued*)
Analytical and Metrological Aspects Concerning Multi-elemental Speciation Analysis by HPLC/ICP-MS [54]

	Sample	Chromatographic Separation	HPLC Operating Systems	ICP-MS Detection System	Interferences	Figures of Merits	References
As, Se	1. Se^{IV} – selenious acid, Se^{VI} – selenic acid· As^{III} – arsenite, $Se^{II}Met$ – seleno-DL-methionine, $As^{III}B$ – arsenobetaine, $(Se^{II}Cys)_2$ – seleno-DL-cystine, $AS^{III}C$ – arsenocholine, $TMeAs^{III}$ – tetramethylarsonium iodide 2. Brine samples and extract of canned tuna fish	1. A reversed phase C18 column Phenomenex, Torrance, CA (250 mm×4.6 mm) 2. Isocratic 3. 10 mM sodium hexanesulfonate and 0.1% methanol (pH 3.5)	1. 1.2 2. 70 3. 20 4. 2.00 for SeI^{V}, 0.012 for ^{I}2.17 for Se^{VI}, 3.00 for As^{III}, 3.17 for $Se^{II}Met$, 4.17 for $As^{III}B$, 5.17 for $(Se^{II}Cys)_2$, 10.00 for $As^{III}C$, 15.00 for $TMeAs^{III}$	1. A VG PlasmaQuad 2 Turbo Plus ICP-MS (VG Elemental) 2. ^{75}As, ^{77}Se	1. $^{40}Ar^{35}Cl^+$ and $^{40}Ar^{37}Cl$ ions interfering with $^{75}As^+$ and $^{77}Se^+$ 2. Chromatographic separation	1. 7 for Se^{IV}, 10 for Se^{VI}, 12 for $Se^{II}Met$, 11 for $(Se^{II}Cys)_2$	[70]
Br, I	1. BrO_3^- – potassium bromate, Br^- – potassium bromide, IO_3^- – potassium iodate, I^- – potassium iodide 2. Drinking water	1. ICS-A23 ion-exchange column 2. Isocratic 3. 0.03 M ammonium carbonate at pH 8	1. 0.8 2. No data 3. 1,000 4. 2.61 for BrO_3^-, 3.43 for Br^-, 2.18 for IO_3^-, 9.03 for I^-	1. 7500a ICP-MS (Agilent Technologies, Santa Clara, CA, USA) 2. ^{127}I, ^{79}Br	1. $^{40}Ar^{39}K^+$ 2. Chromatographic separation	1. 0.032 for BrO_3^-, 0.063 for Br^-, 0.008 for IO_3^-, 0.012 for I^- [a] 2. for BrO_3^-, 3.3 for Br^-, 2.2 for IO_3^-, 3.2 for I^-	72
As, Cr	1. As^{III} –sodium (meta)arsenite, As^{V} – arsenate – ammonium dichromate 2. Drinking water	1. PRP-X100 (4.6 mm×150 mm, 5 μm) 2. Isocratic 3. Procedure A: 22 mM $(NH_4)_2HPO_4$, 25 mM NH_4NO_3 at pH 9.2 (with ammonia) Procedure B: 22 mM $(NH_4)_2HPO_4$, 66 mM NH_4NO_3 at pH 9.2 (with ammonia)	1. 1.4 2. Ambient 3. 100 4. Procedure A: 1.3 for As^{III}, 2. For As^{V}, 5.0 for Cr^{VI}, Procedure B: 1.2 for $As^{III,}$ 1.6 for As^{V}, 2.5 for Cr^{VI} 5. Procedure A: 6; Procedure B: 3	1. PE Sciex ELAN 6100 DRC II 2. ^{91}AsO, ^{52}Cr	1. Polyatomic spectral interference for $^{75}As^+$ 2. DRC with oxygen as a reaction gas	1. Procedure A: 0.16 for As^{III}, o,o9 for As^{V}, 0.073 for Cr^{VI} Procedure B: 0.04 for As^{III}, 0.062 for As^{V}, 0.15 for Cr^{VI} [d] 2. Recoveries values were in the range in procedure A: 96–100 for As^{III}, 96–102 for As^{V}, 96–102 for Cr^{VI} and in procedure B: 98–101 for As^{III}, 94–99 for As^{V}, 96–100 for Cr^{VI} 3. Procedure A: 4.2 for As^{III}, 7.4 for As^{V}, 7.6 for Cr^{VI} Procedure B: 2.4 for As^{III}, 2.3 for As^{V}, 2.0 for Cr^{VI}	[72]

(Continued)

TABLE 16.1 (*Continued*)
Analytical and Metrological Aspects Concerning Multi-elemental Speciation Analysis by HPLC/ICP-MS [54]

	Sample	Chromatographic Separation	HPLC Operating Systems	ICP-MS Detection System	Interferences	Figures of Merits	References
As, Sb	1. As^{III}-sodium (meta)arsenite, As^{V} – disodium hydrogen arsenate heptahydrate, MMA^{V} – monomethylarsenic acid, DMA^{V} – dimethylarsinic acid, $As^{III}B$ – arsenbetaine, $TMeAs^{V}O$ – trimethylarsine oxide, $TMeAs^{III}$ – tetramethylarsonium odide, $As^{III}C$ – arsenocholine bromide, Sb^{III} – diantimonate dipotassium trihydrate, Sb^{V} – potassium hexahydroxoantimonate 2. Hot spring water, extract from fish samples	1. Develosil C30-UG-5 (4.6 mm×250 mm) 2. Isocratic 3. 10 mM sodium butanesulfonate, 4 mM malonic acid, 4 mM tetramethylammonium hydroxide, 0.01% (*v/v*) methanol, 20 mM ammonium tartrate (pH 2.0)	1. 0.75 2. Room temperature 3. 10 4. for As^{V}, 5.67 for As^{III} 6.17 for MMA^{V}, 7.33 for DMA^{V}, 8.50 for $As^{III}B$, 9.83 for $TMeAs^{III}O$, 10.28 for $TMeAs^{III}$, 10.67 for $As^{II}C$, 5.00 for Sb^{III}, 7.95 for Sb^{V} 5. 11.7	1. Agilent 7500c from Yokogawa Analytical Systems 2. ^{75}As, ^{121}Sb	1. No data 2. Reaction mode with He	1. ~0.2 for As species and ~0.5 for Sb species 2. <2 for As species and <3 for Sb species 3. The determined value (32.4±0.5 mg/kg) of arsenobetaine for CRM7402-a was in good agreement with the certified value (33.1±1.5 mg/kg)	[73]
As, Cr	1. As^{III} – arsenic trioxide, As^{V} – sodiumarsenate, DMA^{V} – cacodylic acid sodium salt, MMA^{V} – dimethylarsinic acid, $As^{III}B$ – arsenobetaine, Cr^{VI} – sodium chromate 2. Drinking and waste water	1. Waters IC-Pak A HC (150 mm×4.6 mm, 10 μm) 2. Isocratic 3. 20 mM KNO_3 at pH 9.8	1. 2 2. No data 3. 250 4. 0.93 for $As^{III}B$, 1.47 for DMA^{V}, 2.00 for As^{III}, 3.13 for MMA^{V}, 3.73 for As^{V}, 6.83 for Cr^{VI} 5. 8.0	1. Fisons Plasma Quad PQII+ (VG Elemental, Winsford, UK 2. ^{52}Cr, ^{75}As, $^{53}Cr^{c}$	1. $^{40}Ar^{12}C^{+}$, $^{40}Ar^{35}Cl^{+}$ 2. Chromatographic separation, ^{40}Ar $^{35}Cl^{+}$ was negligible when Cl concentration was below 1,000 mg/L	1. 0.5 for As^{III}, As^{V}, 0.4 for Cr^{VI} as a^{52}Cr, 0.5 for Cr^{VI} as a ^{52}Cr 2. <7% for each of the samples 3. 93–106 for Cr^{VI}, 93–103 for As^{III}, 96–107 for As^{V}	[74]
As, Se	1. As^{III} – sodium(meta)arsenite, As^{V} – disodium hydrogen arsenate heptahydrate, SeI^{VI} – sodium selenate 2. Extracts of coal fly ash and sediment, tap, well, river water	1. Capcell C18 RP column (250 mm×4.6 mm) 2. Isocratic 3. 5 mM of butane sulfonic acid, 2 mM malonic acid, 0.3 mM hexane sulfonic acid and 0.5% methanol of pH 2.5	1. 1 2. No data 3. 25 4. 2.4 for Se^{VI}, 2.6 for As^{V}, 3.0 for As^{III}, 3.1 for Se^{IV} 5. 4.0	1. A Perkin-Elmer Elan 6000 ICP-MS instrument 2. ^{75}As, ^{77}Se	1. ^{40}Ar $^{35}Cl^{+}$, ^{40}Ar $^{35}Cl^{+}$ interfering with $^{75}As^{+}$ and $^{77}Se^{+}$ 2. Monitoring the concentration of Cl^{-} and keeping them below 1,000 mg/L	1. 0.07 for As^{V}, 0.08 for As^{III}, 0.63 for Se^{VI}, 0.77 for Se^{IV} 3. 95–105 for each of analytes	[75]

(*Continued*)

TABLE 16.1 (*Continued*)
Analytical and Metrological Aspects Concerning Multi-elemental Speciation Analysis by HPLC/ICP-MS [54]

	Sample	Chromatographic Separation	HPLC Operating Systems	ICP-MS Detection System	Interferences	Figures of Merits	References
As, Se	1. As^{III} – sodium arsenite(III), As^{V} – arsenate, MMA^{V} – disodium methyl arsenate, MMA^{V} – disodium methyl arsenate, DMA^{V} – dimethylarsinic acid, SeI^{V} – disodium selenite, Se^{VI} – sodium seleselenite^{III}B – arsenobetaine 2. Urine samples and extracts from fish samples	1. Hamilton PRP-X100 (4.1 mm×250 mm, 10 µm) 2. Gradient 3. A: 10 mM $(NH_4)_2CO_3$ in 2% *v/v* methanol (pH 9.0), B: 50 mM $(NH_4)_2CO_3$ in 2% *v/v* methanol (pH 9.0)	1. 1 2. No data 3. 200 4. 2.83 for $As^{III}B$, 3.42 for As^{III}, 4.58 for DMA^{V}, 6.75 for MMA^{V}, 8.00 for SeI^{V}, 8.08 for As^{V}, 10.57 for Se^{VI} 5. 12.0	1. Perkin-Elmer SCIEX ELAN 6100DRC 2. ^{75}As ^{12}CHH, ^{75}Se, ^{80}Se	1. $^{38}Ar^{40}Ar^+$, $^{40}Ar^{40}Ar^+$, $^{40}Ar^{35}Cl^+$ interfering with the $^{78}Se^+$ and $^{80}Se^+$ 2. DRC with 0.6 mL/min CH_4 as a reaction gas	1. 0.01 for As^{III}, 0.002 for As^{V}, 0.003 for MMA^{V}, 0.004 for DMA^{V}, 0.003 for $As^{III}B$, 0.01 for Se^{IV}, 0.02 for Se [b] 2. 2–4 for each of the analytes 3. 90–105 for As^{III}, 93–104 for As^{V}, 102–104 for MMA^{V}, 93–105 for DMA^{V}, 95–104 for $As^{III}B$, 94–101 for Se^{IV}, 95 for Se^{VI}	[76]
As, Se	1. As^{III} – sodium arsenite, As^{V} – arsenate, MMA^{V} – disodium methyl arsenate, DMA^{V} – dimethylarsinic acid, Se^{V} – disodium selenite, Se^{VI} – sodium selenate, $MeSe^{II}Cys$ – methylselenocysteine, $(Se^{II}Cys)_2$ – selenocystine, $Se^{II}Met$ – selenomethionine 2. Extracts from cereal samples	1. Hamilton PRP-X100 (4.1 mm×150 mm, 10 µm) 2. Gradient 3. A: 0.5 mM ammonium citrate in 1% methanol (pH 4.5), B: 15 mM ammonium citrate in 1% methanol (pH 8.0)	1. 1 2. No data 3. 200 4. 2.42 for As^{III}, 3.33 for DMA^{V}, 7.08 for MMA^{V}, 7.58 for As^{V}, 2.58 for $(Se^{II}Cys)_2$, 3.42 for $MeSe^{II}Cys$, 5.83 for $Se^{II}Met$, 7.50 for Se^{IV}, 9.33 for Se^{VI} 5. 11.0	1. Perkin-Elmer SCIEX ELAN 6100DRC 2. ^{75}As ^{12}CHH, ^{78}Se, ^{80}Se	1. ^{38}Ar $^{40}Ar^+$ and ^{40}Ar $^{40}Ar^+$ interfering ions with $^{78}Se^+$ and $^{80}Se^+$ 2. DRC with 10 mL/min CH_4 as a reaction gas	1. 0.009 for As^{III}, 0.009 for DMA^{V}, 0.006 for MMA^{V}, 0.006 for As^{V}, 0.01 for $(Se^{II}Cys)_2$, 0.01 for Me $Se^{II}Cys$, 0.03 for $Se^{II}Met$, 0.009 for Se^{IV}, 0.01 for Se^{VI} [b] 2. <5 for each of the analytes 3. 98–105 for As^{III}, 94–100 for DMA^{V}, 96–97 for MMA^{V}, 99–102 for As^{V}, 95–98 for, $(Se^{II}Cys)_2$, 96–99 for Me $Se^{II}Cys$, 94–100 for $Se^{II}Met$, 95–98 for $Se^{IV,}$ 96–103 for Se^{VI}	[77]
Cr, As	1. As^{III} – no data, As^{V} – no data, Cr^{VI} – no data, Cr^{III} – no data 2. Lake and river water, bottled water	1. Pecosphere C8, 3 µm particles, 3 cm 2. Isocratic 3. 1 mM TBAH, 0.5 mM EDTA (potassium salt), 5% methanol at pH 7.2	1. 1.5 2. No data 3. 50 4. 0.3 for As^{III}, 1.2 for As^{V}, 1.9 for Cr^{VI}, 1.1 for Cr^{III} 5. 2.5	1. Elan DRC II (Perkin-Elmer, SCIEX) 2. ^{52}Cr, ^{91}AsO	1. $^{40}Ar^{12}C^+$, $^{40}Ar^{35}Cl^+$, $^{40}Ca^{35}Cl^+$ interfering with $^{52}Cr^+$, $^{75}As^+$ 2. DRC with O_2 as a reaction gas	1. No data 2. No data 3. No data	[78]

(*Continued*)

TABLE 16.1 (*Continued*)
Analytical and Metrological Aspects Concerning Multi-elemental Speciation Analysis by HPLC/ICP-MS [54]

	Sample	Chromatographic Separation	HPLC Operating Systems	ICP-MS Detection System	Interferences	Figures of Merits	References
As, Se	1. As^{III} – arsenite, As^{V} –arsenate, $MMMA^{V}$ – sodium methylarsenate hexahydrate, DMA^{V} – dimetylarsinic acid sodium salt trihydrate, $SeI^{V}Met$ – selenomethionine, $As^{III}B$ – arsenobetaine 2. Fish tissues	1. Metrosept Anion Dual 3 column (100 mm×4 mm, 6 μm) 2. Gradient 3. A: 5 mM NH_4NO_3, B: 50 mM NH_4NO_3, 2% *v/v* methanol, pH 8.7	1. 1 2. Ambient 3. 100 4. 1.03 for $As^{III}B$, 1.62 for As^{III}, 1.93 for $Se^{II}Met$, 2.48 for DMA^{V}, 4.50 for MMA^{V}, As^{V}, 5.41 for SeI^{V}, 5.91 for Se^{VI} 5. 7.0	1. Agilent HP-4500 2. ^{75}As, ^{77}Se, ^{82}Se	1. No data 2. Matrix matching	1. 0.1 for As species, 0.7 for Se species 2. <5 for each of the analytes (as a reproducibility) 3. No data	[79]
As, Se	1. As^{III} – no data, As^{V} – no data, MMA^{V} – no data, DMA^{V} – no data, SeI^{V} – no data, Se^{VI} – no data, $Se^{II}Me$ – no dta, $(Se^{II}Cys)_2$ – no data, $As^{III}B$ – no data, $As^{III}C$ – no data 2. Animal/plant derived foodstuff	1. C18 column (Waters Symmetry, 4.6 mm×250 mm, 5 μm) 2. Gradient 3. A: 0.2% (*v/v*) 1-hexyl-3-mrthylimidaolium tetrafluoroborate $[HMIM]BF_4$ in 5% (v/v) methanol, pH 6.0; B: 0.4% (v/v)) $[HMIM]BF_4$ in 5% (v/v) methanol, pH 6.0	1. 1 2. 35 3. 20 4. 2.8 for As^{III}, 9.3 for As^{V}, 6.7 for MMA^{V}, 5.1 for DMA^{V}, 3.8 for SeI^{V}, 12.4 for Se^{VI}, 5.9 for $Se^{II}Me$, 2.6 for $(Se^{II}Cys)_2$, 3.3 for $As^{III}B$, 1.9 for $As^{III}C$ 5. 20.0	1. Agilent 7500x 2. ^{75}As, ^{82}Se, 7V8Se	1. $^{40}Ar^{35Cl}$, $^{40}Ca^{35}Cl^{+}$ interfering with $^{75}As^{+}$ 2. Collision Cell Technique (CCT) with helium as a collision gas and interference equation EPA200.8	1. 0.034 for As^{III}, 0.41 for As^{V}, 0.86 for MMA^{V}, 0.83 for DMA^{V}, 1.23 for Se^{IV}, 1.52 for Se^{VI}, 1.1 for $Se^{II}Me$, 0.94 for $(Se^{II}Cys)_2$, 0.65 for $As^{III}B$, 0.74 for $As^{III}C^{b}$	[80]
As, Cr, Cd	1. As^{III} – no data, As^{V} – no data, MMA^{V} – methyl arsenic acid mono sodium salt hydrate, DMA^{V} – cacodylic acid sodium salt, Cr^{III} – chromium(III)chloride hexahysrate, Cr^{VI} – potassium dichromate, Cd^{II} – cadmium nitrate tetrahydrate 2. Surface and drinking water	1. Hamilton PRP-X100 (4.1 mm×250 mm, 10 μm) 2. Isocratic 3. 40 mM NH_4NO_3, pH 8.6	1. 1 2. 22–27 3. 50 4. 2.52 for As^{III}, 6.08 for As^{V}, 3.9 for MMA^{V}, 3.17 for DMA^{V}, 12.17 for Cr^{III}, 11.00 for Cr^{VI}, 11.50 for Cd^{II} 5. 12.0	1. Agilent 7700x 2. ^{52}Cr, ^{53}Cr, ^{75}As, ^{111}Cd	1. $^{40}Ar^{12}C$, $^{35}Cl^{16}O^{1}H^{+}$, $^{40}Ar^{35}Cl^{+}$ interfering with $^{75}As^{+}$, $^{52}Cr^{+}$ 2. Collision<reaction cell with hell as a collision gas	1. 0.08 for As^{III}, 0.07 for As^{V}, 0.08 for MMA^{V}, 0.11 for DMA^{V}, 0.10 for Cr^{III}, 0.07 for Cr^{VI}, 0.12 for $Cd^{II\,b}$ 2. 6.2 for As^{III}2.9 for As^{V}, 4.1 for MMA^{V}, 7.4 for DMA^{V}, 4.4 for Cr^{III}, 3.6 for Cr^{VI}, 6.8 for Cd^{II} 3. 93–108 for As^{III}, 96–100 for As^{V}, 89–94 for MMA^{V}, 88–105 for DMA^{V}, 91–98 for Cr^{III}, 93–102 for Cr^{VI}, 102–116 for As^{V}	[81]

(*Continued*)

TABLE 16.1 (*Continued*)
Analytical and Metrological Aspects Concerning Multi-elemental Speciation Analysis by HPLC/ICP-MS [54]

	Sample	Chromatographic Separation	HPLC Operating Systems	ICP-MS Detection System	Interferences	Figures of Merits	References
As, Sb, Cr	1. As^{III} – diarsenic trioxide, As^{V} – arsenate, Cr^{VI} – ammonium dichromate, Sb^{III} – potassium antimony(III)tartrate hydrate, Sb^{V}-potassium hexahydroxyantimonate 2. Drinking, bottled water samples	1. Hamilton PRP-X100 (4.6 mm×150 mm, 10 μm) 2. Gradient 3. A: 36 mM NH_4NO_3, 3 mM $EDTANa_2$, pH 4.6, B: 36 mM NH_4NO_3 3 mM $EDTANa_2$, pH 9.0	1. 1.5 2. 20 3. 100 4. 1.5 for As^{III}, 2.1 for As^{V}, 4.6 for Sb^{III}, 7.1 for Cr^{VI}, 1.7 for Sb^{V} 5. 15.0 with column re-equilibration	1. PE Sciex ELAN 6100 DRC II 2. ^{91}AsO, ^{52}Cr, ^{121}Sb	1. $^{40}Ar^{12}C^{+}$, $^{35}Cl^{16}O^{1}H^{+}$, $^{40}Ar^{35}Cl^{+}$ interfering with $^{75}As^{+}$, $^{52}Cr^{+}$ 2. DRC, O_2 as a reaction gas	1. 0.067 for As^{III}, 0.068 for As^{V}, 0.098 for Cr^{VI}, 0.083 for Sb^{III}, 0.038 for Sb^{V} [d] 2. for As^{III}, 2.0 for As^{V}, 1.3 for Cr^{VI}, 2.4 for Sb^{III}, 2.1 for Sb^{V} 3. 110 for As^{III}, 93 for As^{V}, 98 for Cr^{VI}, 99 for Sb^{III}, 94 for Sb^{V} at concentration 0.5 ppb level μg/L, 98 for As^{III}, 98 for Cr^{VI}, 97 for Sb^{III}, 95 for Sb^{V} at 7 ppb concentration level μg L^{-1}	[82]
As, Se, Sb, Te	1. As^{III} – diarsenic trioxide, As^{V} – arsenate, MMA^{V} – monomethylarsonic acid disodium salt trihydrate, $As^{III}B$ – arsenobetaine, PAA – phenylarsonic acid, Se^{IV} – selenite, Se^{VI} – sodium selenite decahydrate, $Se^{II}Met$ – seleno-L-methionine, Sb^{III} – potassium antimonyl tartrate hemihydrate and antimony(III)chloride, Sb^{V} – potassium hexahydroxyantimonate and potassium antimonite, Te^{VI}-tellric(V)acid 2. Water, urine, fish and soil extracts	1. Anion exchange column IonPak AS14 (250 mm×4 mm) 2. Gradient 3. 2 mM ammonium hydrogen carbonate, 2.2–45 mM tartaric acid, pH 8.2	1. No data 2. No data 3. 50 4. 2.58 for As^{III}, 9.83 for As^{V}, 2.25 for $As^{III}B$, 3.33 for DMA^{V}, 4.5 For MMA^{V}, 10.50 for PAA, 7.00 for SeI^{V}, 10.40 for Se^{VI}, 5.16 for $Se^{II}Met$, 11.58 for Sb^{III}, 3.17 for Sb^{V}, 3.00 for Te^{VI}5. 13.0	ICP-MS Elan 5000, Perkin-Elmer, SCIEX 2. No data	1. No data 2. No data	1. 0.36 for As^{III}, 0.34 for As^{V}, 0.47 for $As^{III}B$, 0.48 for DMA^{V}, 0.26 for MMA^{V}, 0.36 for PAA, 3.7 for Se^{IV}, 2.7 for Se^{VI}, 4.3 for $Se^{II}Met$, 1.7 for Sb^{III}, 0.14 for Sb^{V}, 0.41 for Te ^{VI} [e] 2. 3.3 for As^{III}, 6.4 for As^{V}, 5.02 for $As^{III}B$, 3.6 for DMA^{V}, 5.3 for MMA^{V}, 6.3 for PAA, 7.0 for Se^{IV}, 6.9 for Se^{VI}, 7.5 for $Se^{II}Met$, 5.5 for Sb^{III}, 3.3 for Sb^{V}, 5.0 for Te^{VI} 3. 85–115 for each of the analytes	[25–83]

[a] As the concentration giving a signal-to-noise ratio of 3.
[b] Based on the concentration (as element) necessary to yield a net signal equal to three times the standard deviation of the background.
[c] Ass×20 blank/slope of calibration curve (IUPAC, k=3).
[d] As 3× standard deviation of independent measurements for samples/matrix with addition of quantifiable amount of analytes.
[e] As 3× standard deviation of blank

16.5 ADULTERATION

16.5.1 What Is Adulteration?

An adulterant is a substance found in food that compromises the safety or effectiveness of this foodstuff. It isn't normally present in any specification or declared contents of the substance, and may not be legally allowed. The addition of adulterants is called adulteration. The adulterants may be harmful or reduce the potency of the product, or they may be harmless.

The reason for adulteration is usually the cost of the ingredients.

A wide range of analysis methods have been developed, each with its advantages and disadvantages.

16.5.2 Adulteration in Protein-Containing Foods

A method was developed to determine the presence of six compounds (cyromazine, triuret, biuret, dicyandiamide, melamine, amidinourea, urea) with the potential to be used in economic adulteration to enhance the nitrogen content in milk products and bulk proteins [60]. Residues were extracted from the matrices (skim milk, skim milk powder, soy protein, wheat flour, wheat gluten, and corn gluten meal) with 2% FA, after which ACN was added to induce precipitation of the proteins. Extracts were analysed by liquid chromatography using a ZIC-HILIC column with tandem mass spectrometry (LC-MS/MS) using ESI.

A Shimadzu Prominence UFLC XR liquid chromatography system (Shimadzu, Columbia, MD) with a SeQuant ZIC-HILIC (150 mm×2.1 mm, 5 μm) PEEK HPLC column (EMD Chemicals/Merck, Gibbstown, NJ) was used for the LC separation. An initial flow rate of 400 L/min of 100% mobile phase A (95:5 ACN:0.1% FA/10 mM AF in H_2O) for the first 5 minutes followed by a linear ramp to 25% mobile phase A/75% mobile phase B (50:50 ACN:0.1% FA/10 mM AF in H_2O) at 12.8 minutes, holding at 25% mobile phase A until 15.8 minutes, and returning to 100% mobile phase A with an increased flow rate to 600 L/min in 16 minutes, remaining at that composition and flow rate until returning to initial conditions of 100% mobile phase A and 400 μL/min in 24.90 minutes, and remaining at initial conditions until 25 minutes. The injection volume was 20 μL. An AB Sciex 4000 QTRAP with an ESI source in positive-ion mode with Analyst 1.5 software (AB SciEx, Foster City, CA). Single-laboratory method validation data were collected in six matrices fortified at concentrations down to 1.0 μg/g (ppm). Average recoveries and average RSDs using spiked matrix calibration standard curves were the following: cyromazine 95.9% (7.5% RSD), dicyandiamide 98.1% (5.6% RSD), urea 102.5% (8.6% RSD), biuret 97.2% (6.6% RSD), triuret 97.7% (5.7% RSD), and amidinourea 93.4% (7.4% RSD).

A method for the rapid quantification of nine potential nitrogen-rich economic adulterants (dicyandiamide, urea, biuret, cyromazine, amidinourea, ammeline, amidinourea, melamine, and cyanuric acid) in five milk and soy-derived nutritional ingredients, i.e. whole milk powder, nonfat dry milk, milk protein concentrate, sodium caseinate, and soy protein isolate has been developed and validated for routine use [55]. The samples were diluted tenfold with water followed by treatment with 2% FA and ACN to precipitate proteins. Sample extracts were analysed using hydrophilic interaction chromatography and tandem mass spectrometry (HILIC-MS/MS) under both positive and negative modes.

Aw Waters Acquity UPLC system with an Acquity BEH HILIC (2.1 mm×1,500 mm×1.7 μm) UPLC column (Waters Corp.) was used for separation. A Waters TQD LC/MS system with an ESI source in both positive and negative mode controlled by Masslync™ software was used.

Stable isotope-labelled internal standards were used to ensure accurate quantification. In multi-day validation experiments, the average accuracies, RSDs, and method detection limits (MDL) for all analytes in whole milk powder were 82%–101%, 6%–13%, and 0.1–7 mg/kg, respectively. The retention times of the analytes in matrix spiked controls were within ±0.06 minute of the average retention times of the corresponding analytes in calibration standards. The validated method was proven to be rugged for routine use to quantify the presence of nine nitrogen-rich compounds in milk and soy-derived ingredients and to provide a defence against economically motivated adulteration.

16.5.3 Adulterants in Functional Foods, Health Foods, and Herbal Medicines

Eight adulterants including two appetite suppressants, two energy expenditure-enhancing drugs, one diuretic and three cathartics (ephedrine, norpseudoephedrine, fenfluramine, sibutramine, clopamide, emodin, rhein, and chrysophanol) in slimming functional foods were simultaneous determined by high-performance liquid chromatography with electrospray ionisation–tandem mass spectrometry (HPLC-ESI-MS/MS) [56]. After samples were ultrasonically extracted with 70% (*v/v*) methanol aqueous solution and centrifuged, the components of the eight adulterants in the sample solution were separated by a Hypersil Gold column (2.1 mm×150 mm, 5 μm) using programmed gradient elution. A mobile phase consisting of 0.02% (*v/v*) FA– AF buffer solution (pH 3.50) and methanol was used for elution with a flow rate set at 250 μL/min and column temperature of 25°C. Qualitative determination was based on characteristic ion pairs and retention time of the targeted compounds using SRM (selective reaction monitoring) mode. Clenbuterol and ibuprofen were internal standards in positive and negative ionisation mode, respectively. The internal standard curves were used for quantification measurement. The average recoveries of three different concentrations were from 80.2% to 94.5%. The limits of detection (LODs) were from 0.03 to 0.66 mg/kg (except chrysophanol 1.6 mg/kg). The linear dynamic range covered from 1 to 500 μg/L (except chrysophanol 50–5,000 μg/L) for the 12 samples analysed. Adulterants in four different kinds of slimming functional foods were determined by this developed method, and satisfactory results were obtained. These experimental results showed that adulteration of sibutramine or/and fenfluramine were the major adulterating components with contents varying from 6.1 to 1.3×10^3 and 1.9 to 9.7×10^3 mg/kg, respectively. In addition, three cathartic compounds were detected in six of those tested samples, and ephedrine, norpseudoephedrine and clopamide were not detected in all samples.

An ultra-fast LC-ESI-MS/MS method was developed and validated to simultaneously screen, confirm, and determine 18 illegal adulterants in herbal medicines and health foods for male sexual potency [57]. The separation was achieved on a Shim-Pack XR-ODS II column (2.0×100 mm, 2.2 μm) with ACN and aqueous solution (12 mmol/L AF, 0.01% acetic acid) as mobile phase at a flow rate of 0.4 mL/min with a gradient elution. The column temperature was maintained at 40°C and the run time was within 18 minutes. The 18 illegal adulterants were detected in ESI positive mode by multiple-reaction monitoring. All the calibration curves showed good linearity with a correlation coefficient higher than 0.996 within the tested concentration ranges. The extraction recoveries and relative recoveries were in the range of 79.5%–114% and 82.0%–120%, respectively. The RSD of repeatability and intermediate precision was all less than 18% and the accuracy was in the range of 81.7%–118%. The intra-day and inter-day stability was in the range of 86.8%–110%. The validated method was successfully applied to screen, confirm, and determine 16 samples. Nine products were confirmed to contain illegal adulterants and the contents of adulterants were related to the therapeutic dosages.

16.5.4 Adulteration of Dairy Products

The multiple-reaction monitoring (MRM) approach, traditionally applied in biomedical research, is particularly suitable for the detection of food adulteration and for the verification of authenticity to assure food safety and quality, both recognised as top priorities by the European Union Commission.

A UPLC-ESI-MS/MS methodology based on MRM has been developed for the sensitive and selective detection of buffalo mozzarella adulteration [59]. The targeted quantitative analysis was performed by monitoring specific transitions of the phosphorylated β-casein f33–48 peptide, identified as a novel species-specific proteotypic marker. The high sensitivity of MRM-based MS and the wide dynamic range of triple quadrupole spectrometers have proved to be a valuable tool for the analysis of food matrices such as dairy products, thus offering new opportunities for monitoring food quality and adulterations.

For methodology development, mass spectrometry analysis was performed using a quadrupole ToF (ESI-Q-ToF) mass spectrometer equipped with an ESI source. Tryptic peptides were separated using a modular CapLC system (Waters, Manchester, UK). Samples were loaded onto a C18 precolumn (5 mm length 300 mm i.d.) at a flow rate of 20 mL/min and desalted for 5 minutes with a solution of 0.1% FA. Peptides were then directed onto a Symmetry-C18 analytical column (10 cm 300 mm i.d.) equilibrated with 2% mobile phase A (2% CH_3CN/98% H_2O/0.1% FA). The chromatographic separation was carried out using a linear gradient from 2% mobile phase A to 55% mobile phase B (95% CH_3CN/5% H_2O/0.1% FA) over 60 minutes at a flow rate of 5 mL/min. Shorter chromatographic separations were also performed by using the same gradient over 10 min at a flow rate of 5 mL/min. The precursor ion masses and associated fragment ion spectra of the tryptic peptides were mass measured with the mass spectrometer directly coupled to the chromatographic system. The ToF analyser of the mass spectrometer was externally calibrated with a multi-point calibration using selected fragment ions that resulted from the collision-induced decomposition of human [Glu1]-fibrinopeptide B [500 fmol/mL in $CH_3CN:H_2O$ (50:50), 0.1% FA] at an infusion rate of 5 mL/min in the ToF-MS/MS mode. The instrument resolution in MS/MS mode for the [Glu1]-fibrinopeptide B fragment ion at 684.3469 *m/z* was found to be above 5,000 full width at half maximum (FWHM). Electrospray mass spectra and MS/MS data were acquired on the Q-ToF mass spectrometer operating in the positive-ion mode with a source temperature of 80°C and with a potential of 3,500 V applied to the capillary probe. MS/MS data on tryptic peptides were acquired in the data-directed analysis (DDA) MS/MS mode. Charge state recognition was used to select doubly and triply charged precursor ions for the MS/MS experiments, which also includes the automated selection of the collision energy based on both charge and mass. A maximum of three precursor masses was defined for concurrent MS/MS acquisition from a single MS survey scan. MS/MS fragmentation spectra were collected from 50 to 1,600 *m/z*. The MS/MS data were processed automatically, and de novo sequencing was obtained using the Biolynx application of MassLynx 4.0 software (Waters, Manchester, UK). All MS/MS spectra leading to protein identification were manually double checked to verify sequence assignments.

The quantitative methodology based on the monitoring of markers identified by the untargeted analysis was developed by using an Acquity H-Class UPLC coupled with a Xevo TQ-S triple quadrupole mass spectrometer equipped with an ESI source (Waters, Manchester, UK). Chromatographic separations of tryptic digests (10 mL, ~50 mg of total proteins) were carried out on an Acquity HSS T3 column packed with a trifunctional C18 alkyl phase (100 mm length 2.1 mm ID, 1.8 mm particle size), maintained at 40°C (Waters, Manchester, UK). The optimised mobile phase consisted of 0.1% FA in water (solvent A) and 0.1% FA in CH_3CN (solvent B). The elution was carried out by using a linear gradient from 2% to 55% solvent B over 5 minutes and from 55% to 95% solvent B over 0.5 minutes at a flow rate of 0.5 mL/min. A washing step at 95% B for 1 minute and a re-equilibration step at 2% B for 2.5 minutes was always performed before the next injections (total run time 9 minutes). The ESI source operating parameters were set as follows: positive ionisation mode; capillary voltage 1.0 kV; source temperature 150°C; desolvation temperature 450°C; desolvation nitrogen gas flow rate 1,000 L/h; cone gas 250 L/h; sampling cone voltage 25 V. For quantitative determinations, acquisitions were made in MRM mode with a dwell time of 50 ms by following the transitions at 688.3 > 977.4 *m/z* and 1032.1 > 982.9 *m/z* by using collision energies 20 and 30 eV, respectively, diagnostic for the presence of the bovine marker. In addition, transitions at 698.3 > 1007.2 *m/z* and 1047.1 > 997.7 *m/z* by using collision energies of 20 and 30 eV, respectively, were also monitored as diagnostic for the presence of the buffalo marker. In addition, product ion spectra for the doubly charged ions at 1047.1 *m/z* (*Bubalus bubalis*) and 1032.1 *m/z* (*Bos taurus*) were also acquired in Product Ion Confirmation (PIC) mode. Each sample was analysed in duplicate, and blank runs were introduced after each injection to prevent carry-over between separate runs. Data processing was performed by means of MassLynx 4.1 software (Waters Corporation, Manchester, UK.

Method validation was performed by studying the linear dynamic range and testing the instrumental sensitivity of standard mozzarella samples. The linear dynamic range was evaluated by injecting serial dilutions (i.e. 1:10, 1:100, 1:1,000, 1:10,000) of the standard mozzarella sample prepared with a known percentage (i.e. 10%) of bovine milk. The reproducibility of the analysis was evaluated by performing duplicate injections of the standard mozzarella samples and evaluating the RSD of the peak areas and retention times of bovine and buffalo markers. The developed methodology was applied to the analysis of seven buffalo mozzarella cheeses declared as Mozzarella di Bufala Campana DOP (samples DOP A-G) and of two mozzarella samples produced with cow milk (samples BOV A-B) used as negative controls. Bovine and buffalo milk samples were also analysed to assess the applicability of the methodology to the raw material.

Czerwenka et al. [58] used the whey protein β-lactoglobulin as a marker for adulteration. LC-MS analysis of the whey protein fraction was performed using an Agilent 1100 series HPLC instrument (Agilent Technologies, Waldbronn, Germany) coupled to an Agilent 1100 series LC/MSD Trap SL quadrupole ion-trap mass spectrometer equipped with a pneumatically assisted ESI source. The LC separation was carried out on a Supelco Discovery Bio Wide Pore C8 column (150, 2.1 mm, 3 μm) from Sigma–Aldrich (Vienna, Austria). Elution was performed at a temperature of 40°C and a flow rate of 0.25 mL/min using water containing 0.5% (*v/v*) acetic acid as eluent A and ACN containing 0.5% (*v/v*) acetic acid as eluent B. For the separation of the whey proteins a linear gradient from 35%B to 50%B in 16 minutes was employed, followed by a washing step at 80%B. For the milk samples, 5 μL of 1:5 dilutions of the whey protein fractions were injected, whereas 3–15 μL of the undiluted whey protein fractions were injected for the mozzarella samples.

MS detection was carried out employing positive ionisation with a capillary voltage of 3,700 V, a nebuliser pressure of 206.8 kPa (30 psi), a drying gas flow rate of 8 L/min and a drying gas temperature of 325 C. The ion-trap was operated in the standard mass range/enhanced resolution mode (50–2,200 *m/z*, It offers a rapid determination combined with unequivocal identification of the marker protein in every run. An in-depth discussion of the subsequent data analysis highlights the potential problems of obtaining quantitative information on the level of adulteration. In an examination of 18 commercial buffalo mozzarella samples, three products were found to be adulterated with high levels of cow's milk.

A fast and simple matrix-assisted laser desorption/ionisation time-of-flight mass spectrometry (MALDI-ToF-MS)-based methodology (MALDI-ToF micro MX, Waters, Manchester, UK) has been developed for detecting the adulteration or contamination of buffalo ricotta with bovine milk [61].

A fast procedure for digesting milk proteins and identified a novel specific proteotypic marker, corresponding to region 149-162 of β-lactoglobulin, as highly diagnostic for the presence of cow milk within ricotta matrices was optimised.

16.5.5 Adulteration in Oils

Flaxseed oil is popular edible oil and an important additive in functional foods and feeds. Recently, economically motivated adulteration is a type of oil fraud become an emerging risk [62]. The fatty acid profiles of flaxseed oil were analysed by GC-MS operating in selected ion monitoring mode and then used to detect adulterated flaxseed oil with the help of multivariate statistical methods including PCA, and recursive SVM (R-SVM).

CC-MS analysis was conducted by an Agilent GC 7890 gas chromatograph interfaced with an Agilent 5973 mass spectrometer. The GC system was equipped with a fused silica capillary column DB-23 (30 m×0.25 mm x 0.25 μm). Helium gas (99.999% purity) was used as the carrier gas at a flow rate of 1.2 mL/min. Mass spectrometry was in electron ionisation mode at 70 eV. The selected ion monitoring mode was at 55, 67, 74, and 79 *m/z*.

The detection results indicate that the discriminant model built with 28 fatty acids can identify adulterated flaxseed oil samples (10%) with high accuracy of 95.6%. Therefore, fatty acid profiles based adulteration detection for flaxseed oil is an important strategy for preventing customers far from adulterated flaxseed oil.

121 samples of olive oil, non-adulterated and adulterated with different portions of sunflower and olive-pomace oils were analysed [63].

The apparatus used to measure the patterns of volatiles of the oil samples was a Chemical Sensor HP4440 A system from Agilent Technology (Waldbronn, Germany). This comprises a headspace sampler (HP 7694) with a tray for 44 consecutive samples, an oven, where the headspace is generated, and a sampling system comprising a stainless steel needle, a 316-SS six-port valve with a nickel loop and two solenoid valves (for pressurisation and to recognise complex mixtures of volatiles and venting). The headspace sampler is coupled to a quadrupole mass spectrometer (HP 440, based on the HP 5973 MSD) by a transfer line. Data collection was performed with Pirouette 2.6 software from Infometrix.

Application of the LDA technique to the data from the signals was sufficient to differentiate the adulterated from the non-adulterated oils and to discriminate the type of adulteration. The results obtained revealed 100% success in classification and close to 100% in prediction. The main advantages of the proposed methodology are the speed of analysis (since no prior sample preparation steps are required), low cost, and the simplicity of the measuring process.

16.5.6 Adulteration in Honey

Honey adulteration with sugar syrups is a widespread problem. Several types of syrups have been used in honey adulteration, and there is no available method that can simultaneously detect all of these adulterants. Du et al. [64] generated a small-scale database containing the specific chromatographic and mass spectrometry information on sugar syrup markers and developed a simple, rapid, and effective ultra high-performance liquid chromatography/quadrupole time-of-flight mass spectrometry (UHPLC/Q-ToF-MS) method for the detection of adulterated honey.

UHPLC was performed in an Agilent 1260 series UHPLC (Agilent, Palo Alto, CA, USA) equipped with an autoinjector and a quaternary UHPLC pump. An Acquity UPLC BEH Amide column (2.1 mm × 100 mm, 1.7 µm) was used. The mobile-phase gradient consisted of 30% water (solvent A) and 70% ACN (solvent B) for the first 5 minutes. Solvent B was reduced to 45% ACN over 5 minutes and held at this concentration for 5 minutes. Subsequently, solvent B was adjusted to 70% over 0.1 minute and held at this concentration for 2.9 minutes to clean the column. The total run time was 18 minutes, with a flow rate of 0.3 mL/min, and the injection volume was 5 µL. The column temperature was maintained at 50°C.

MS was performed in an Agilent 6530 ESI-Q-TOF.

Corn syrup, high-fructose corn syrup, inverted syrup, and rice syrup were used as honey adulterants; polysaccharides, difructose anhydrides, and 2-acetylfuran-3-glucopyranoside were used as detection markers. The presence of 10% sugar syrup in honey could be easily detected in <30 minutes using the developed method.

16.5.7 Adulteration in Meat and Meat Products

The review of Stachniuk et al. [65] offers an overview of the current status and the most recent advances in liquid chromatography-mass spectrometry (LC-MS) techniques with both high-resolution and low-resolution tandem mass analysers applied to the identification and detection of heat-stable species-specific peptide markers of meat in highly processed food products. The authors present sets of myofibrillar and sarcoplasmic proteins, which turned out to be the source of 105 heat-stable peptides, detectable in processed meat using LC-MS/MS. A list of heat-stable species-specific peptides was compiled for eleven types of white and red meat including chicken, duck, goose, turkey, pork, beef, lamb, rabbit, buffalo, deer, and horse meat, which can be used as markers for meat authentication. Among the 105 peptides, 57 were verified by MRM, enabling the identification of each species with high specificity and selectivity. The most described and monitored species by LC-MS/MS so far are chicken and pork with 26 confirmed heat-stable peptide markers for each meat. In thermally processed samples, myosin, myoglobin, haemoglobin, L-lactase dehydrogenase A and β-enolase are the main protein sources of heat-stable markers.

Meat adulteration was examined using a well-defined proteogenomic annotation, carefully selected surrogate tryptic peptides and high-resolution mass spectrometry [66]. Selected mammalian meat samples were homogenised, and the proteins were extracted and digested with trypsin.

The HPLC system was a Thermo Scientific UltiMate 3000 Rapid Separation UHPLC system (San Jose, CA, USA). Chromatography was achieved using a gradient mobile phase along with a microbore column, Thermo Biobasic C8 100×1 mm, with a particle size of 5 µm. The initial mobile-phase condition consisted of ACN and water (both fortified with 0.1% of FA) at a ratio of 5:95. From 0 to 1 min, the ratio was maintained at 5:95. From 1 to 31 min, a linear gradient was applied up to a ratio of 50:50 and maintained for 2 minutes. The mobile-phase composition ratio was reverted to the initial conditions and the column was allowed to re-equilibrate for 14 minutes for a total run time of 47 minutes. The flow rate was fixed at 75 µL/min and 2 µL of samples were injected.

A Thermo Scientific Q Exactive Orbitrap Mass Spectrometer (San Jose, CA, USA) was interfaced with a Thermo Scientific UltiMate 3000 Rapid Separation UHPLC system using a pneumatic-assisted heated electrospray ion source. MS detection was performed in positive-ion mode and operating in scan mode at high-resolution and accurate mass (HRAM). The mass spectrometer was operated in full-scan HRAM using resolving powers of 140,000 and 17,500 (FWHM) in full-scan MS and MS/MS, respectively. Data independent acquisition (DIA) mode was used including 12 DIA MS/MS scans to cover the mass range of 600–1,200 *m/z*. Methodically *in silico* analyses of myoglobin, myosin-1, myosin-2 and β-haemoglobin sequences allow for the identification of species-specific tryptic peptide mass lists and theoretical MS/MS spectra.

The analyses of meat samples were performed using a hybrid Quadrupole-Orbitrap mass spectrometer operating in MS at a resolution of 140,000 (FWHM) and in MS2 at a resolution of 17,500 (FWHM).

Following comprehensive MS, MS/MS or DIA analyses, the method was capable of the detection and identification of very specific tryptic peptides for all four targeted proteins for each animal species tested with observed *m/z* below 3 ppm compared with the theoretical *m/z*. The analyses were successfully performed with raw and cooked meat. Specifically, the method was capable of detecting 1% (*w/w*) of pork or horse meat in a mixture before and after cooking (71°C internal temperature).

16.5.8 Non-Targeted Screening

The majority of analytical methods for food safety monitor the presence of a specific compound or defined set of compounds. Non-targeted screening methods are complementary to these approaches by detecting and identifying unexpected compounds present in food matrices that may be harmful to public health. However, the development and implementation of generalised non-targeted screening workflows are particularly challenging, especially for food matrices due to inherent sample complexity and diversity and a large analyte concentration range. One approach that can be implemented is LC coupled to high-resolution mass spectrometry, which serves to reduce this complexity and is capable of generating molecular formulae for compounds of interest [67]. Current capabilities, strategies and challenges are reviewed for sample preparation, mass spectrometry, chromatography and data processing workflows. Considerations to increase the accuracy and speed of identifying unknown molecular species are also addressed, including suggestions for achieving sufficient data quality for non-targeted screening applications.

The data analysis workflow for non-targeted screening with LC/HR-MS is shown in Figure 16.1.

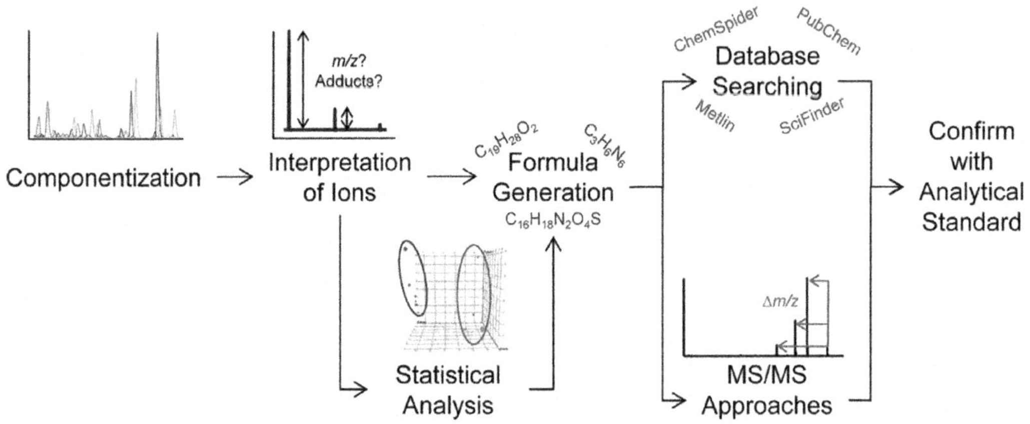

FIGURE 16.1 Data analysis workflow for non-targeted screening with LC/HR-MS [67].

REFERENCES

1. Choi Y.H., Hong C.K., Kim M., Jung S.O., Park. J., Oh J.H., Kwon J.-H. Multivariate analysis to discriminate the origin of sesame seeds by multi-element analysis inductively coupled plasma-mass spectrometry. *Food Science and Biotechnology*, 2017, 26, 375–379. https://doi.org/10.1007/s10068-017-0051-0
2. Casale M., Casolino C., Oliveri P., Forina M. The potential of coupling information using three analytical techniques for identifying the geographical origin of Liguria extra virgin olive oil. *Food Chemistry*, 2010, 118, 163–170. https://doi.org/10.1016/j.foodchem.2009.04.091
3. Oliveros C.C., Boggia R., Casale M., Armanino C., Forina M. Optimisation of a new headspace mass spectrometry instrument Discrimination of different geographical origin olive oils. *Journal of Chromatography A*, 2005, 1076, 7–15. https://doi.org/10.1016/j.chroma.2005.04.020
4. Bajoub A., Medina-Rodriguez S., Ajal E, Cuadros-Rodriguez L., Monasterio R.P., Vercammen J., Fernández-Gutiérrez A., Carrasco-Pancorbo A., A metabolic fingerprinting approach based on selected ion flow tube mass spectrometry (SIFT-MS) and chemometrics: A reliable tool for Mediterranean origin-labeled olive oils authentication. *Food Research International*, 2018, 106, 232–242. https://doi.org/10.1016/j.foodres.2017.12.027
5. Kropf U., Korošec M., Bertoncelj J., Ogrinc N., Nečemer M., Kump P., Golob T. Determination of the geographical origin of Slovenian black locust, lime and chestnut honey. *Food Chemistry*, 2010, 121 (3), 839–846. https://doi.org/10.1016/j.foodchem.2009.12.094
6. Karabagias I.K., Badeka A., Kontakos S., Karabournioti S., Kontominas M.G. Characterisation and classification of greek pine honeys according to their geographical origin based on volatiles, physicochemical parameters and chemometrics. *Food Chemistry*, 2014, 146, 548–557. https://doi.org/10.1016/j.foodchem.2013.09.105
7. Castro-Vázquez L., Díaz-Maroto M.C., de Torres C., Pérez-Coello M.S. Effect of geographical origin on the chemical and sensory characteristics of chestnut honeys. *Food Research International*. 2010, 43, 2335–2340. https://doi.org/10.1016/j.foodres.2010.07.007
8. Stanimirova I., Üstün B., T. Cajka, Riddelova K., Hajslov J., Buydens L.M.C., Walczak B. Tracing the geographical origin of honeys based on volatile compounds profiles assessment using pattern recognition techniques. *Food Chemistry*, 2010, 118 (1), 171–176. https://doi.org/10.1016/j.foodchem.2009.04.079
9. Chudzinska M., Baralkiewicz D. Estimation of honey authenticity by multielements characteristics using inductively coupled plasma-mass spectrometry (ICP-MS) combined with chemometrics. *Food and Chemical Toxicology*, 2010, 48 (1), 284–290. https://doi.org/10.1016/j.fct.2009.10.011
10. Kaškonienė V., Venskutonis P.R. Floral markers in honey of various botanical and geographic origins: A review. *Comprehensive Reviews in Food Science and Food Safety*. 2010, 9, 620–634. https://doi.org/10.1111/j.1541–4337.2010.00130.x
11. Lippolis V., De Angelis E., Fiorino G.M., Di Gioia A., Arlorio M., Logrieco A.F., Monaci L. Geographical origin discrimination of monofloral honeys by direct analysis in real time ionization-high resolution mass spectrometry (DART-HRMS). *Foods*, 2020, 9, 1205. https://doi.org/10.3390/foods9091205

12. Sipteri M., Dubin E., Cotton J., Poirel M., Corman B. Jamin E., Lees M., Rutletge D. Data fusion between high resolution ^{1}H-NMR and mass spectrometry: A synergetic approach to honey botanical origin characterization. *Analytical and Bioanalytical Chemistry*, 2016, 408, 4389–4401. https://doi.org/10.1007/s00216-016-9538-4
13. Ballabio D., Robotti E., Grisoni F., Quasso F., Bobba M. Vercelli S., Gosetti F., Calabrese G., Sangiorgi E., Orlandi M., Marengo E. Chemical profiling and multivariate data fusion methods for the identification of the botanical origin of honey. *Food Chemistry*, 2018, 266, 79–89. https://doi.org/10.1016/j.foodchem.2018.05.084
14. Šelih V.S., Šala M. Drgan V. Multi-element analysis of wines by ICP-MS and ICP-OES and their classification according to geographical origin in Slovenia. *Food Chemistry*, 2014, 153, 414–423. https://doi.org/10.1016/j.foodchem.2013.12.081
15. Coetzee P.P., Steffens F.E., Eiselen R.J., Augustyn O.P., Balcaen L., Vanhaecke F. Multi-element analysis of south african wines by ICP–MS and their classification according to geographical origin. *Journal of Agricultural and Food Chemistry*, 2005, 53 (13), 5060–5066. https://doi.org/10.1021/jf048268n
16. Hopfer H., Nelson J., Thomas S., Collins T.S., Heymann H., Ebeler S.E. The combined impact of vineyard origin and processing winery on the elemental profile of red wines. *Food Chemistry*, 2015, 175, 486–496. https://doi.org/10.1016/j.foodchem.2014.09.113
17. Ziółkowska A., Wąsowicz E., Jeleń H.H. Differentiation of wines according to grape variety and geographical origin based on volatiles profiling using SPME-MS and SPME-GC/MS methods. *Food Chemistry*, 2016, 213, 714–720. https://doi.org/10.1016/j.foodchem.2016.06.120
18. Cajka T., Riddellova K., Tomaniova M., Haislova J. Ambient mass spectrometry employing a DART ion source for metabolomic fingerprinting/profiling: A powerful tool for beer origin recognition. *Metabolomics*, 2011, 7, 500–508. https://doi.org/10.1007/s11306-010-0266-z
19. Nietner T., Haughey S.A., Ogle N., Fauhl-Hassek C., Elliott T. Determination of geographical origin of distillers dried grains and solubles using isotope ratio mass spectrometry. *Food Research International*, 2014, 60, 146–153. https://doi.org/10.1016/j.foodres.2013.11.002
20. Grundy H.H., Reece P., Buckley M., Solazzo C.M., Dowle A.A., Ashford D., Charlton A.J., Wadsley M.K., Collins M.J. A mass spectrometry method for the determination of the species of origin of gelatine in foods and pharmaceutical products. *Food Chemistry*, 2016, 190, 276–284. https://doi.org/10.1016/j.foodchem.2015.05.054
21. Song S., Jing S., Zhu L., Ma C., Song T., Wu J., Zhao Q., Zheng F., Zhao M., Chen F. Untargeted and targeted metabolomics strategy for the classification of strong aroma-type baijiu (liquor) according to geographical origin using comprehensive two-dimensional gas chromatography-time-of-flight mass spectrometry. *Food Chemistry*, 2020, 314, 126098. https://doi.org/10.1016/j.foodchem.2019.126098
22. Segelke T., von Wuthenau K., Neitzke G., Müller M.-S., Fischer M. Food Authentication: Species and origin determination of truffles (*Tuber* spp.) by inductively coupled plasma mass spectrometry and chemometrics. *Journal of Agricultural and Food Chemistry*, 2020, 68, 14374–14385. https://doi.org/10.1021/acs.jafc.0c02334
23. Kim T.J., Park J.G., Ahn S.K., Kim K.W., Chi J., Kim H.Y., Ha S.-H., Seo W.D., Kim J.K. Discrimination of Adzuki Bean (*Vigna angularis*) geographical origin by targeted and non-targeted metabolite profiling with gas chromatography time-of-flight mass spectrometry. *Metabolites*, 2020, 10 (3), 112–128. https://doi.org/10.3390/metabo10030112
24. Cavanna D., Loffi C., Dall'Asta C., Suman M. A non-targeted high-resolution mass spectrometry approach for the assessment of the geographical origin of durum wheat. *Food Chemistry*, 2020, 317, 123366. https://doi.org/10.1016/j.foodchem.2020.126366
25. Majone C., Batista B.L., Campiglia A.D. Barbosa F., Barbosa R.M. Classification of geographic origin of rice by data mining and inductively coupled plasma mass spectrometry. *Computers and Electronics in Agriculture*, 2016, 121, 101–107. https://doi.org/10.1016/j.compag.2015.11.009
26. Lee W.H., Choi S., Oh I.N., Shim J.Y., Lee K.S., An G., Park J.-T. Multivariate classification of the geographic origin of Chinese cabbage using an electronic nose-mass spectrometry. *Food Science and Biotechnology*, 207, 26, 603–609. https://doi.org/10.1007/s10068-017-0102-6
27. Zhang D., Wu W., Qiu X., Li X., Zhao F., Ye N. Rapid and direct identification of the origin of white tea with proton transfer reaction time-of-flight mass spectrometry. *Rapid Communications in Mass Spectrometry*, 2020, 34 (20), e8830. https://doi.org/10.1002/rcm.8830
28. Kumar S., D'Souza R.N., Behrends B., Corno M., Ullrich M.S., Kuhnert N., Hütt M.-T. Supervised and unsupervised classification of cocoa bean origin and processing using liquid chromatography-mass spectrometry. *BioRxiv*, 2020. https://doi.org/10.1101/2020.02.09.940577

29. Klockmann S., Reiner E., Bachmann R., Hackl T., Fischer M. Food Fingerprinting: Metabolomic approaches for geographical origin discrimination of hazelnuts (*Corylus vellana*) by UPLC-QTOF-MS. *Journal of Agricultural and Food Chemistry*, 2016, 64 (48), 9253–9262. https://doi.org/10.1021/acs.jafc.6b04433
30. Wang H., Cao X., Han T., Pei H., Ren H., Stead S. A novel methodology for real-time identification of the botanical origins and adulteration of honey by rapid evaporative ionization mass spectrometry. *Food Control*, 2019, 106, 106753. https://doi.org/10.1016/j.foodcont.2019.106753
31. You Q., Wang B., Chen F., Huang Z., Wang X., Luo P.G. Comparison of anthocyanins and phenolics in organically and conventionally grown blueberries in selected cultivars. *Food Chemistry*, 2011, 125 (1), 201–208. https://doi.org/10.1016/j.foodchem.2010.08.063
32. Søltoft M., Eriksen M.R., Brændholt Träger A.W., Nielsen J., Laursen K.H., Husted S., Halekoh U., Knuthsen P. Comparison of polyacetylene content in organically and conventionally grown carrots using a fast ultrasonic liquid extraction method. *Journal of Agricultural and Food Chemistry*, 2010, 58 (13), 7673–7679. https://doi.org/10.1021/jf101921v
33. Klimánková E., Holadová K., Hajšlová J., Čajka T., Poustka J., Koudela M. Aroma profiles of five basil (*Ocimum basilicum* L.) cultivars grown under conventional and organic conditions. *Food Chemistry*, 2008, 107 (1), 464–472. https://doi.org/10.1016/j.foodchem.2007.07.062
34. Wang Z., Chen P., Yu L., Harrington P.B. Authentication of organically and conventionally grown basils by gas chromatography/mass spectrometry chemical profiles. *Analytical Chemistry*. 2013, 85(5), 2945–2953. https://doi.org/10.1021/ac303445v
35. Gao B., Lu W., Ding T., Chen Y., Lu W., Yu L. Differentiating organically and conventionally grown oregano using ultraperformance liquid chromatography mass spectrometry (UPLC-MS), headspace gas chromatography with flame ionization detection (Headspace-GC-FID), and flow injection mass spectrum (FIMS) fingerprints combined with multivariate data analysis. *Journal of Agricultural and Food Chemistry*, 2014, 62 (32), 8075–8084. https://doi.org/10.1021/jf502419y
36. Gao B., Lu Y., Sheng Y., Chen P., Lu L. Differentiating organic and conventional sage by chromatographic and mass spectrometry flow injection fingerprints combined with principal component analysis. *Journal of Agricultural and Food Chemistry*, 2013, 61 (12), 2957–2963. https://doi.org/10.1021/jf304994z
37. Romero-González R., Garrido Frenich A., Martínez Vidal J.L., Prestes O.D., Grio S.L. Simultaneous determination of pesticides, biopesticides and mycotoxins in organic products applying a quick, easy, cheap, effective, rugged and safe extraction procedure and ultra-high performance liquid chromatography–tandem mass spectrometry. *Journal of Chromatography A*, 2011, 1218 (11), 1477–1485. https://doi.org/10.1016/j.chroma.2011.01.034
38. Barbosa R.M., Batista B.L., Barião C.V., Varrique R.M., Coelho V.A, Campiglia A.D., Barbosa Jr F. A simple and practical control of the authenticity of organic sugarcane samples based on the use of machine-learning algorithms and trace elements determination by inductively coupled plasma mass spectrometry. *Food Chemistry*, 2015, 184, 154–159. https://doi.org/10.1016/j.foodchem.2015.02.146
39. Voica S., Dehelean A., Kovacs M.H. The use of inductively coupled plasma mass spectrometry (ICP-MS) for the determination of toxic and essential elements in different types of food sample. *Food Chemistry*, 2009, 112 (3), 727–732. https://doi.org/10.1016/j.foodchem.2008.06.010
40. Mulabagal V., Calderón A.I. Liquid chromatography/mass spectrometry based fingerprinting analysis and mass profiling of *Euterpe oleracea* (açaí) dietary supplement raw materials. *Food Chemistry*, 2012, 134 (2), 1156–1164. https://doi.org/10.1016/j.foodchem.2012.02.123
41. Jackson B.P., Taylor V.F., Karagas M.R., Punshon T., Cottingham K.L. Arsenic, organic foods, and brown rice syrup. *Environmental Health Perspectives*, 2012, 120 (5), 623–626. https://doi.org/10.1289/ehp.1104619
42. Dufailly V., Nicolas M;, Richoz-Payot J., Poitevin E. Validation of a method for arsenic speciation in food by ion chromatography-inductively coupled plasma/mass spectrometry after ultrasonic-assisted enzymatic extraction. *Journal of AOAC International*, 2011, 94 (3), 947–958.
43. Contreras-Acuña M., García-Barrera T., García-Sevillano M.A., Gómez-Ariza J.L. Speciation of arsenic in marine food (*Anemonia sulcata*) by liquid chromatography coupled to inductively coupled plasma mass spectrometry and organic mass spectrometry. *Journal of Chromatography A*, 2013, 1282, 133–141. https://doi.org/10.1016/j.chroma.2013.01.068
44. Choi J.Y., Khan N., Nho E.Y., Choi H., Park K.S., Cho M.J. Speciation of arsenic in rice by high-performance liquid chromatography–inductively coupled plasma mass spectrometry. *Analytical Letters*, 2016, 49 (12), 1926–1937. https://doi.org/10.1080/00032719.2015.1125912

45. D'Amato M., Aureli F., Ciardullo S., Raggi A., Cubadda F. Arsenic speciation in wheat and wheat products using ultrasound- and microwave-assisted extraction and anion exchange chromatography-inductively coupled plasma mass spectrometry. *Journal of Analytical Atomic Spectrometry*, 2011, 26, 207–213. https://doi.org/10.1039/C0JA00125B
46. Qu H., Mudalige T.K., Linder S.W. Arsenic speciation in rice by capillary electrophoresis/inductively coupled plasma mass spectrometry: Enzyme-assisted water-phase microwave digestion. *Journal of Agricultural and Food Chemistry*, 2015, 63 (12), 3153–3160. https://doi.org/10.1021/acs.jafc.5b00446
47. Chu Y.-L., Jiang S.-J. Speciation analysis of arsenic compounds in edible oil by ion chromatography–inductively coupled plasma mass spectrometry. *Journal of Chromatography A*, 2011, 1218 (31), 5175–5179. https://doi.org/10.1016/j.chroma.2011.05.089
48. Bañuelos G.S., Fakra S.C., Walse S.S., Marcus M.A., Yang S.I., Pickering I.J., Pilon-Smits E.A.H., Freeman J.L. Selenium accumulation, distribution, and speciation in spineless prickly pear cactus: A drought- and salt-tolerant, selenium-enriched nutraceutical fruit crop for biofortified foods. *Plant Physiology*, 2011, 155, 315–327. https://doi.org/10.1104/pp.110.162867
49. Chen Y., Huang L., Wu W., Ruan Y., Wu Z., Xue Z., Fu F. Speciation analysis of lead in marine animals by using capillary electrophoresis couple online with inductively coupled plasma mass spectrometry. *Electrophoresis*, 2014, 35 (9), 1346–1352. https://doi.org/10.1002/elps.201300410
50. Chen S., Zu S., He Y., Lu D. Speciation of chromium and its distribution in tea leaves and tea infusion using titanium dioxide nanotubes packed microcolumn coupled with inductively coupled plasma mass spectrometry. *Food Chemistry*, 2014, 150, 254–259. https://doi.org/10.1016/j.foodchem.2013.10.150
51. Campillo N., Viñas P., Peñalver R., Cacho J.I., Hernandez-Córdoba M. Solid-phase microextraction followed by gas chromatography for the speciation of organotin compounds in honey and wine samples: A comparison of atomic emission and mass spectrometry detectors. *Journal of Food Composition and Analysis*, 2012, 25 (1), 66–73. https://doi.org/10.1016/j.jfca.2011.08.001
52. Ramos K., Gómez-Gómez M.M., Cámara C., Ramos L. Silver speciation and characterization of nanoparticles released from plastic food containers by single particle ICPMS. *Talanta*, 2016, 151, 83–90. https://doi.org/10.1016/j.talanta.2015.12.071
53. Marcinkowska M., Baralkiewicz D. Multielemental speciation analysis by advanced hyphenated technique – HPLC/ICP-MS: A review. *Talanta*, 2016, 161, 177–204. https://doi.org/10.1016/j.talanta.2016.08.034
54. Zhang D., Yang S., Ma Q., Sun J., Cheng H., Wang Y., Liu J. Simultaneous multi-elemental speciation of As, Hg and Pb by inductively coupled plasma mass spectrometry interfaced with high-performance liquid chromatography. *Food Chemistry*, 2020, 313, 126119. https://doi.org/10.1016/j.foodchem.2019.126119
55. Draher D;, Pound V., Reddy T.M. Validation of a rapid method of analysis using ultrahigh-performance liquid chromatography–tandem mass spectrometry for nitrogen-rich adulterants in nutritional food ingredients. *Journal of Chromatography A*, 1373, 2014, 106–113. https://doi.org/10.1016/j.chroma.2014.11.019
56. Shi Y., Sun C., Gao B., Sun A. Development of a liquid chromatography tandem mass spectrometry method for simultaneous determination of eight adulterants in slimming functional foods. *Journal of Chromatography A*, 1218, 2011, 7644–7662. https://doi.org/10.1016/j.chroma.2011.08.038
57. Ren Y., Wu C., Zhang J. Simultaneous screening and determination of 18 illegal adulterants in herbal medicines and health foods for male sexual potency by ultra-fast liquid chromatography-electrospray ionization tandem mass spectrometry. *Journal of Separation Science*, 35 (21), 2012, 2847–2857. https://doi.org/10.1002/jssc.201200280
58. Czerwenka C., Műller L., Lindner W. Detection of the adulteration of water buffalo milk and mozzarella with cow's milk by liquid chromatography–mass spectrometry analysis of β-lactoglobulin variants. *Food Chemistry*, 122 (3), 2010, 901–908. https://doi.org/10.1016/j.foodchem.2010.03.034
59. Russo R., Severino V., Mendez A., Lliberia J., Parente A., Chambery A. Detection of buffalo mozzarella adulteration by an ultra-high performance liquid chromatography tandem mass spectrometry methodology. *Journal of Mass Spectrometry*, 2012, 47, 1407–1414. https://doi.org/10.1002/jms.3064
60. MacMahon S., Begley T.H., Diachenko G.W., Stromgren S.A. A liquid chromatography–tandem mass spectrometry method for the detection of economically motivated adulteration in protein-containing foods. *Journal of Chromatography A*, 1220, 2012, 101–107. https://doi.org/10.1016/j.chroma.2011.11.066
61. Russo R., Rega C., Chambery A. Rapid detection of water buffalo ricotta adulteration or contamination by matrix-assisted laser desorption/ionisation time-of-flight mass spectrometry. *Rapid Communications in Mass Spectrometry*, 2016, 30 (4), 497–503. https://doi.org/10.1002/rcm.7463

62. Sun X., Zhang L., Li P., Xu B., Ma F., Zhang Q., Zhang W. Fatty acid profiles based adulteration detection for flaxseed oil by gas chromatography mass spectrometry. *LWT – Food Science and Technology*, 63 (1), 2015, 430–436. https://doi.org/10.1016/j.lwt.2015.02.023
63. Lorenzo I.M., Pérez Pavón L.J., Fernández Laespada M.E., Pinto C.G., Cordero B.M. Detection of adulterants in olive oil by, headspace–mass spectrometry. *Journal of Chromatography A*, 2002, 9145 (1–2), 221–230. https://doi.org/10.1016/S0021–9673(01)01502–3
64. Du B., Wu L., Xue X., Chen L., Li Y., Zhao J., Cao W. Rapid screening of multiclass syrup adulterants in honey by ultrahigh-performance liquid chromatography/quadrupole time of flight mass spectrometry. *Journal of Agricultural and Food Chemistry*, 2015, 63 (29), 6614–6623. https://doi.org/10.1021/acs.jafc.5b01410
65. Stachniuk A., Sumara A., Montowska M. Fornal E. Liquid chromatography–mass spectrometry bottom-up proteomic methods in animal species analysis of processed meat for food authentication and the detection of adulterations. *Mass Spectrometry Reviews. Early View.* https://doi.org/10.1002/mas.21605
66. Ruiz Orduna A., Husby E., Yang C.T., Ghosh D., Beaudry F. Detection of meat species adulteration using high-resolution mass spectrometry and a proteogenomics strategy. *Food Additives & Contaminants: Part A*, 2017, 34 (7), 1110–11120. https://doi.org/10.1080/19440049.2017.1329951
67. Knolhoff A.M., Croley T.R. Non-targeted screening approaches for contaminants and adulterants in food using liquid chromatography hyphenated to high resolution mass spectrometry. *Journal of Chromatography A*, 1428, 2016, 86–96. https://doi.org/10.1016/j.chroma.2015.08.059
68. Acon B.W., McLean J.A. A direct injection high efficiency nebulizer interface for microbore high-performance liquid chromatography-inductively coupled plasma mass spectrometry. *Journal of Analytical Atomic Spectroscopy, 2001, 8, 852–857.* https://doi.org/10.1039/B103085J
69. Kuo, C.Y., Jiang S.J. Determination of selenium and tellurium compounds in biological samples by ion chromatography dynamic reaction cell inductively coupled plasma mass spectrometry. *Journal of Chromatography A,* 2008, 1181 (1–2), 60–66. https://doi.org/10.1016/j.chroma.2007.12.065
70. Le X.C., Li X.F., Ma M., Yalcin S., Feldmann J. Simultaneous speciation of selenium and arsenic using elevated temperature liquid chromatography separation with inductively coupled plasma mass spectrometry detection. *Spectrochimica Acta Part B: Atomic Spectroscopy*, 1998, 53 (6–8), 899–909. https://doi.org/10.1016/S0584-8547(98)00105-0
71. Liu W., Yang H., Li B., Xu S. Determination of bromine and iodine speciation in drinking water using high performance liquid chromatography-inductively coupled plasma-mass spectrometry. *Geostandards and Geoanalytical Research*, 2011, 35 (1), 69–74. https://doi.org/10.1111/j.1751-908X.2010.00033.x
72. Marcinkowska M., Komorowicz I., Barałkiewicz D. Study on multielemental speciation analysis of Cr(VI), As(III) and As(V) in water by advanced hyphenated technique HPLC/ICP-DRC-MS. Fast and reliable procedures. *Talanta*, 2015, 144, 233–240. https://doi.org/10.1016/j.talanta.2015.04.087
73. Morita Y., Kobayashi T., Kuroiwa T., Narukawa T. Study on simultaneous speciation of arsenic and antimony by HPLC–ICP-MS. *Talanta*, 2007, 73 (1), 81–86. https://doi.org/10.1016/j.talanta.2007.03.005
74. Szopa S., Michalski R. Simultaneous determination of inorganic forms of arsenic, antimony, and thallium by HPLC-ICP-MS. *Spectroscopy*, 2015, 30(2), 54–63.
75. Sathrugnan K., ShizukoHirata S. Determination of inorganic oxyanions of As and Se by HPLC–ICPMS. *Talanta*, 2004, 64 (1), 237–243. https://doi.org/10.1016/j.talanta.2004.02.016
76. Wang R.-Y., Hsu Y.-L., Chang L.-F., Jiang S.-J. Speciation analysis of arsenic and selenium compounds in environmental and biological samples by ion chromatography–inductively coupled plasma dynamic reaction cell mass spectrometer. *Analytica Chimica Acta*, 2007, 590 (2), 239–244. https://doi.org/10.1016/j.aca.2007.03.045
77. Tsai C.-Y., Jiang S.-J. Microwave-assisted extraction and ion chromatography dynamic reaction cell inductively coupled plasma mass spectrometry for the speciation analysis of arsenic and selenium in cereals. *Analytical Science*, 2011, 27, 271–276. https://doi.org/10.2116/analsci.27.271
78. Neubauer K.R., Reuter W.M., Perrone P.A., Grosser Z.A. Simultaneous arsenic and chromium speciation by HPLC/ICP-MS in environmental waters. PerkinElmer Appl. Note, 2003. https://www.perkinelemr.com.
79. Reyes L.H., Mar J.L.G., Rahman G.M.M., Seybert B., Fahrenholz T., Kingston H.M.S. Simultaneous determination of arsenic and selenium species in fish tissues using microwave-assisted enzymatic extraction and ion chromatography–inductively coupled plasma mass spectrometry. *Talanta*, 2009, 78 (3), 983–990. https://doi.org/10.1016/j.talanta.2009.01.003

80. Fang G., Lv Q., Liu C., Huo M., Wang S. An ionic liquid improved HPLC-ICP-MS method for simultaneous determination of arsenic and selenium species in animal/plant-derived foodstuffs. *Analytical Methods*, 2015, 20 (7), 8617–8625. https://doi.org/10.1039/C5AY02021B
81. Sun J., Yang Z., Lee H., Wang L. Simultaneous speciation and determination of arsenic, chromium and cadmium in water samples by high performance liquid chromatography with inductively coupled plasma mass spectrometry. *Analytical Methods*, 2015, 7, 2653–2658. https://doi.org/10.1039/C4AY02813A
82. Marcinkowska M., Komorowicz I., DanutaBarałkiewicz D. New procedure for multielemental speciation analysis of five toxic species: As(III), As(V), Cr(VI), Sb(III) and Sb(V) in drinking water samples by advanced hyphenated technique HPLC/ICP-DRC-MS. *Analytica Chimica Acta*, 2016, 920, 102–111. https://doi.org/10.1016/j.aca.2016.03.039
83. Lindemann T., Prange A., Dannecker W., Neidhart B. Stability studies of arsenic, selenium, antimony and tellurium species in water, urine, fish and soil extracts using HPLC/ICP-MS. *Fresenius' Journal of Analytical Chemistry*, 2000, 368, 214–220. https://doi.org/10.1007/s002160000475

17 Detecting Food Pathogens through Mass Spectrometry Approaches

Saher Islam
Lahore College for Women University

Devarajan Thangadurai, Ravichandra Hospet, Zabin Khoje, Lokeshkumar Prakash, and Muniswamy David
Karnatak University

Namita Bedi and Neeta Bhagat
Amity University

Umar Farooq and Khizar Hayat
MNS University of Agriculture

Jeyabalan Sangeetha
Central University of Kerala

Adil Hussain
Abdul Wali Khan University

Zaira Zaman Chowdhury
University of Malaya

CONTENTS

17.1 Introduction 330
17.2 Liquid Chromatography (LC)-MS for Identification of Food Pathogens 332
17.3 Detection of Foodborne Pathogens Using Matrix-Assisted Laser Desorption/Ionization (MALDI)-MS 337
17.4 Electrospray Ionization-Mass Spectrometry (ESI-MS) for Identification of Food-Spoilage Microorganisms 338
17.5 PCR-Coupled MS for Food Authentication 340
17.6 Gas Chromatography (GC)-MS in Food Microbe Analysis 341
17.7 Capillary Electrophoresis-MS for Foodborne Pathogen Detection 341
17.8 Conclusion 342
References 342

DOI: 10.1201/9781003091226-20

17.1 INTRODUCTION

Microbes are ubiquitously present in the environment and responsible for spoilage of food and foodborne pathogenesis (Li and Zhu 2017). Almost 48 million cases of foodborne illness are reported each year (CDC 2016). Bacterial species such as *Escherichia coli*, *Salmonella* sp., *Listeria monocytogenes*, *Cyclospora* sp., *Clostridium* sp., *Campylobacter* sp., *Vibrio* sp., *Shigella* sp., *Pseudomonas* sp., *Acinetobacter* sp., *Aeromonas caviae*, *Aeromonas hydrophila*, *Aeromonas sobria*, vancomycin-resistant enterococcus, *Mycobacterium* spp., *Helicobacter pylori*, *Enterobacter sakazakii*, and Shiga toxin-producing *E. coli* cause spoilage of food rendering them unsuitable for consumption by changing the biochemical properties of the food (Lee and Raghunath 2018; Feng et al. 2020). Foodborne diseases may result in dehydration, vomiting, diarrhea or bloody diarrhea, abdominal pain, fever and chills, hemolytic uremic syndrome, enterohemorrhagic disease, paralysis, bacillary dysentery, and other complicated acute gastroenteritis (Asuming-Bediako et al. 2019).

These foodborne diseases are widespread worldwide amongst all age groups and gender (Pinu 2016). The emergence and prevalence of multidrug-resistant (MDR) foodborne pathogens in contaminated food pose more serious threats to public health and food safety globally (WHO 2014; FAO 2016; Lee et al. 2018; Islam et al. 2019). Many countries have made it mandatory to report pathogenic microorganisms in food samples (Gwida and El-Gohary 2015). So, newer and rapid, robust, reliable, and accurate methods for rapid detection of pathogens are required for preventing and controlling the outbreak of foodborne diseases. Early pathogen detection not only reduces the risk of food outbreaks, but also assures food quality.

Even today, food laboratories are dependent on conventional culture-based microscopic and biochemical methods (Davis 2014; Hameed et al. 2018). Immunological techniques (Nagaraj et al. 2016) and molecular techniques such as nucleic acid sequence-based detection (Moriconi et al. 2017) also offer good results. Various molecular techniques such as real-time polymerase chain reaction (RT-PCR), DNA microarrays, Fourier transform infrared (FTIR), and biosensors have replaced the traditional methods and have offered a rapid and sensitive way of microbial identification (Bahadir and Sezginturk 2017; Feng et al. 2020). At the same time, omics-based approaches such as proteomics and metabolomics widened and improved the scope of pathogen detection techniques (Singhal et al., 2015; Pinu, 2016).

Mass spectrometry (MS) has emerged as a powerful, sophisticated analytical tool that measures the mass to charge (*m/z*) ratio of one or more molecules in a sample with improved specificity. In a mass analyzer, an electric or magnetic field accelerates the ions and separates them based on the *m/z* ratio. The signal will be translated into spectra, compared with commercial or in-house databases, and provide a rapid and reliable identification at a low cost and high precision. MS works on two techniques: ESI and matrix-assisted laser desorption/ionization (MALDI) (Pavlovic et al. 2013).

MS can be applied to analyze pathogens by both a culture and a non-culture approaches. In culture approaches, microbial biomarkers are analyzed directly with MALDI-MS or extracted/digested, separated by chromatography, and identified with MS (Huber et al. 2018). In non-culture approaches, cell enrichment is done using various physical, chemical, or biochemical interactions with target cells, followed by MS analysis (Ho and Reddy 2011). Matrix-assisted laser desorption/ionization mass spectrometry (MALDI-TOF-MS) has been recognized as an effective technique with high specific spectral profiles in the low-mass range of 1.5–20 kDa of intact bacterial cells (De Koster and Brul 2016). Various proteins have been searched using proteome database to use them as potential biomarkers such as ribosomal proteins for MALDI-TOF-MS for bacterial species identification (Karlsson et al. 2015). This method has obtained universality as a microbial biotyping tool due to its rapidity, cost-effectiveness, simplicity, and relevance for a vast range of microbes (Fall et al. 2015). Quéro et al. (2019) used the MALDI-TOF-MS technique to quickly identify the food-spoilage molds and created a database based on the mass spectra of 618 fungal strains belonging to 136 species obtained by MS. Moothoo-Padayachie et al. (2013) proved that the MALDI-TOF-MS technique is a much better biotyping tool for the identification of industrial yeast strains than molecular methods such as PCR amplification using delta elements and contour-clamped homogenous electric field gel electrophoresis. MALDI-TOF technique was used for the routine identification of foodborne pathogens such as *S. aureus*, *B. cereus*, *L. monocytogenes*, and *Clostridium botulinum* (Böhme et al. 2012; Huber et al. 2018). This technique has shown an infinite potential in identifying antimicrobial resistance in food pathogens (Welker et al., 2019). MALDI-TOF-MS databases contain more than 4,000 strains of pathogenic bacteria, fungi, mycobacteria identified with an accuracy of 90% at the species level and strain typing (Jadhav 2019).

Although there are limitations of the MALDI-TOF-MS technique in taxonomic resolution, at the infraspecies level and when dealing with differentiation of genetically closely related species such as variation between *Shigella* and *E. coli* or between *B. anthracis* and *B. cereus* (Lasch et al. 2020). Further limiting factors of the MALDI-TOF-MS method are the relatively low resolution that usually results in reduced selectivity and less dynamic sensitivity, specifically lowered detectable protein signals over a broad concentration range (Lucia et al. 2019). The differences of bacterial species of an unknown strain to a reference database of spectral profiles helped develop a bacterial strain fingerprint (Elbehiry et al. 2017). Among MS methods, ESI, and rapid evaporative ionization MS have shown a great future in identifying varied microbes (Kailasa et al. 2019). It is a gentle ionization process where gas-phase ions are generated by applying a high electric field, yielding ions with little or no fragmentation. ESI-MS technique was efficiently used to identify methicillin-resistant *Staphylococcus aureus* (MRSA) in cow milk samples (Perreten et al. 2013) and *Megasphaera*, *Pectinatus* bacteria in beer samples (Řezanka et al. 2015). ESI-MS can rapidly detect various microbial metabolites, phospholipids, and proteins with minimal sample preparation. These proteins act as biomarkers for the discriminative identification of pathogenic and non-pathogenic strains of various bacteria from food samples (Dworzanski et al. 2010).

To further improve the implementation of ESI-MS and MALDI-TOF MS, it is coupled with various chromatography-based methods. Separation techniques including LC, GC, and CE are combined with various detection units such as FTIR, NMR, PDA, and MS, enabling the detection of analytes present in raw samples. Liquid chromatography-tandem MS (LC-MS^2) has an application in microbial characterization utilizing numerous distinguishing peptides that enable recognition at species/strain level. LC-MS^2 normally detects considerable signals at high resolution along with great mass accurateness in a single run (Lucia et al. 2019). Moreover, coupling MS with chromatographic separation (LC) has increased the dynamic sensitivity for and detection of low abundant peptides. Jabbour et al. (2010) have reported the effective characterization of *Bacillus cereus* proteins from poisoned food using LC-ESI-MS/MS.

PCR is coupled with ESI-MS to accurately identify pathogens, bacterial DNA, and protein sequences in food samples (Sampath et al. 2007a,b). Further application of PCR/ESI-MS has been studied novel foodborne pathogen (FBP) plate for the rapid differentiation and identification of enteric pathogens from the mixtures of microorganisms that contained more than 800 bacterial isolates (Pierce et al. 2012). PCR combined with ESI-MS approaches have successfully validated the classification, identification, and discrimination of a broad range of pathogens with a high degree of success rates over the traditional molecular biology methods, making the potential technique in diagnostics. Amplified DNA by PCR is purified by fully automated, high-throughput ion-pair reversed-phase high-performance LC with monolithic capillary columns (Jeng et al. 2012). Another efficient method, capillary electrophoresis coupled with MS (CE-MS), has been developed for the fast separation and identification of microorganisms (Petr et al. 2009). In this technique, microorganisms transfer the charged elements to their exterior surface, and then an electric paired layer gets created when charged microorganisms come across with the aqueous solution (background electrolyte). So, under electric field influence, bacterium shows electrophoretic mobility due to charge on surface and paired electric layer. CE allows simultaneous and rapid analysis of various microorganisms in a single sample that includes identification and quantification. Inductively coupled plasma mass spectrometry (ICP-MS) has been used in trace element detection in microbial samples (Beauchemin 2006). FTIR MALDI-TOF MS supplemented with FTIR offers better accuracy for identifying and typing microorganisms (Feng et al. 2020).

This chapter discusses the advantage of MS and MS-associated techniques for rapid, accurate, cost-effective, sensitive, time-saving approaches for identifying genus or species-specific biomarkers for food pathogens.

17.2 LIQUID CHROMATOGRAPHY (LC)-MS FOR IDENTIFICATION OF FOOD PATHOGENS

Identification of food pathogens by quick and robust techniques is of utmost importance in the food industry. Timely detection and identification of pathogens are handy in reducing the spread of dangerous microorganisms, thus avoiding health hazards. The analysis of pathogens in food samples is very challenging because of diversity and complexity. Besides, the use of classical techniques to detect food pathogens is limited due to time consumption and complications of the method. Therefore, several techniques like biochemical, spectroscopic, and genomic analyses have been developed and are currently available in the market to identify food pathogens in various food samples. Such development in instrumentation (separation and mass analysis), ionization techniques, and biological methodologies helped to accurately analyze the pathogens in food samples (Figures 17.1 and 17.2). Most recently, mass spectroscopic techniques (desorption electrospray ionization [DESI], MALDI, ESI, and bioaerosol MS) have appeared to be the most rigorous and cost-effective tools for the detection of microbes. However, sample pretreatment is required due to the complexity of microbial biomarkers, considered as the most important steps in the isolation and characterization of pathogens. Thorough sample pretreatment is indeed a milestone to attain satisfactory detection ability and selectivity. The complexity in microbial biomarkers can be controlled by various chromatographic techniques, such as gas chromatography (GC), LC, and affinity chromatography coupled with mass spectroscopy.

The MALDI-TOF MS method can identify the unknown pathogen at the genus and species level by matching the microbial spectra with the known microbial spectral libraries in the database. Nevertheless, MALDI-TOF MS has limitations at the infraspecies level. MALDI-TOF MS has difficulty in the differentiation of genetically closely related species such as *E. coli* and *Shigella*. On the other hand, LC-MS can differentiate between closely related species due to its capacity to process numerous signals with high precision and resolution. LC-MS is one of the most widely used methods for separating and identifying proteins, peptides, nucleotides, and metabolites present in biological samples.

Furthermore, MS coupled with chromatographic separation (LC) has revolutionized the detection and identification of pathogens in food samples based on microbial proteomics. The method is very sensitive, and its dynamic sensitivity allows the detection of even less abundant peptides of food pathogens. The selective proteolytic-peptide analysis by the LC method can identify multiple pathogens in a single MS run. Presently, the improved LC-MS approach comprises efficient extraction of proteins from cultivated microbial cells, digestion by proteolytic enzyme(s), and LC-MS measurements. The generated MS data can then be extracted and systematically compared with the existing peptide libraries (for example, UniProt Knowledgebase Swiss-Prot and TrEMBL databases) to confirm the foodborne pathogen's identity (Figures 17.3 and 17.4).

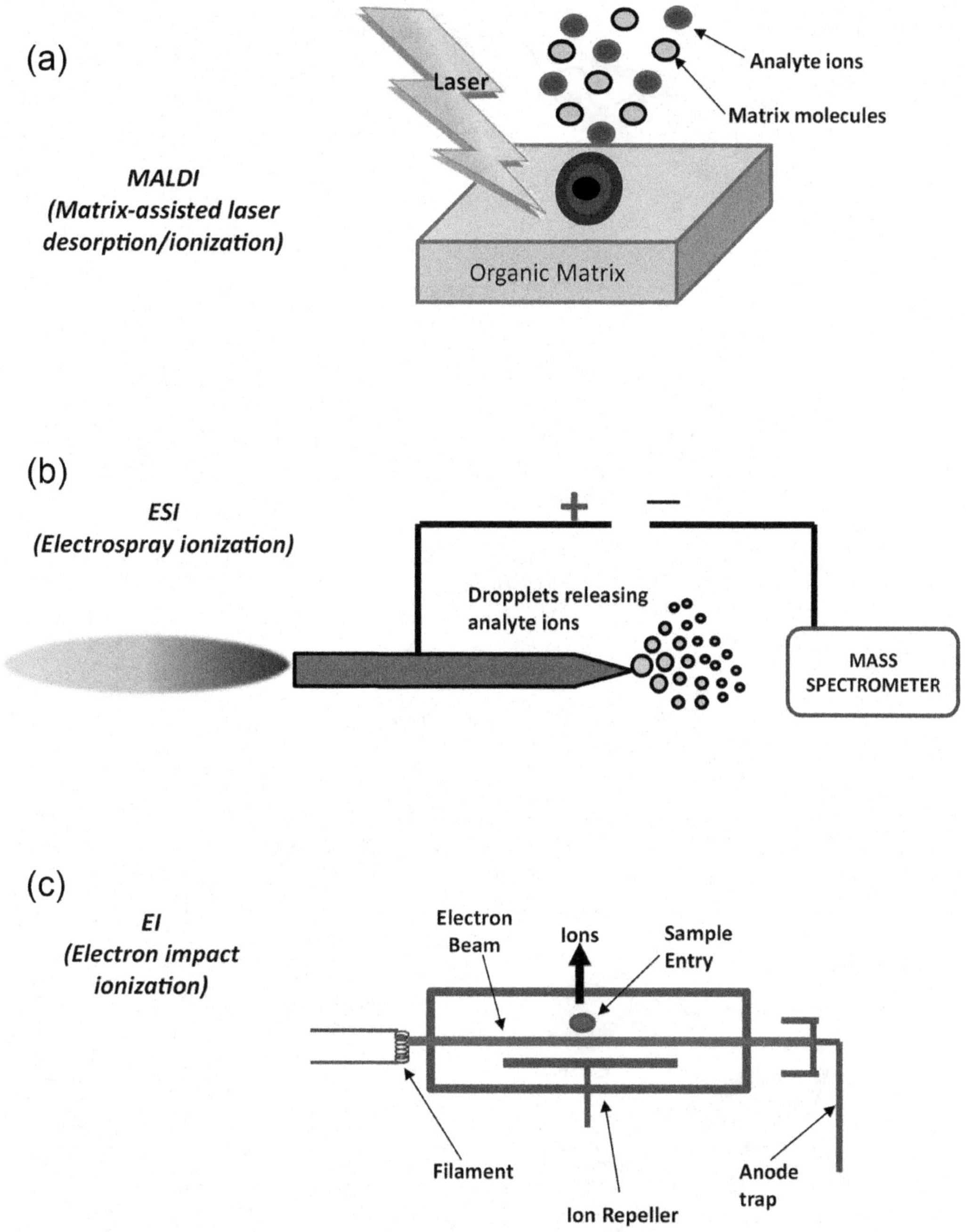

FIGURE 17.1 Most commonly used ionization sources in mass spectrometry: (a) MALDI, (b) ESI, and (c) EI ionization. (Reproduced with permission from Picó 2015, © Elsevier.)

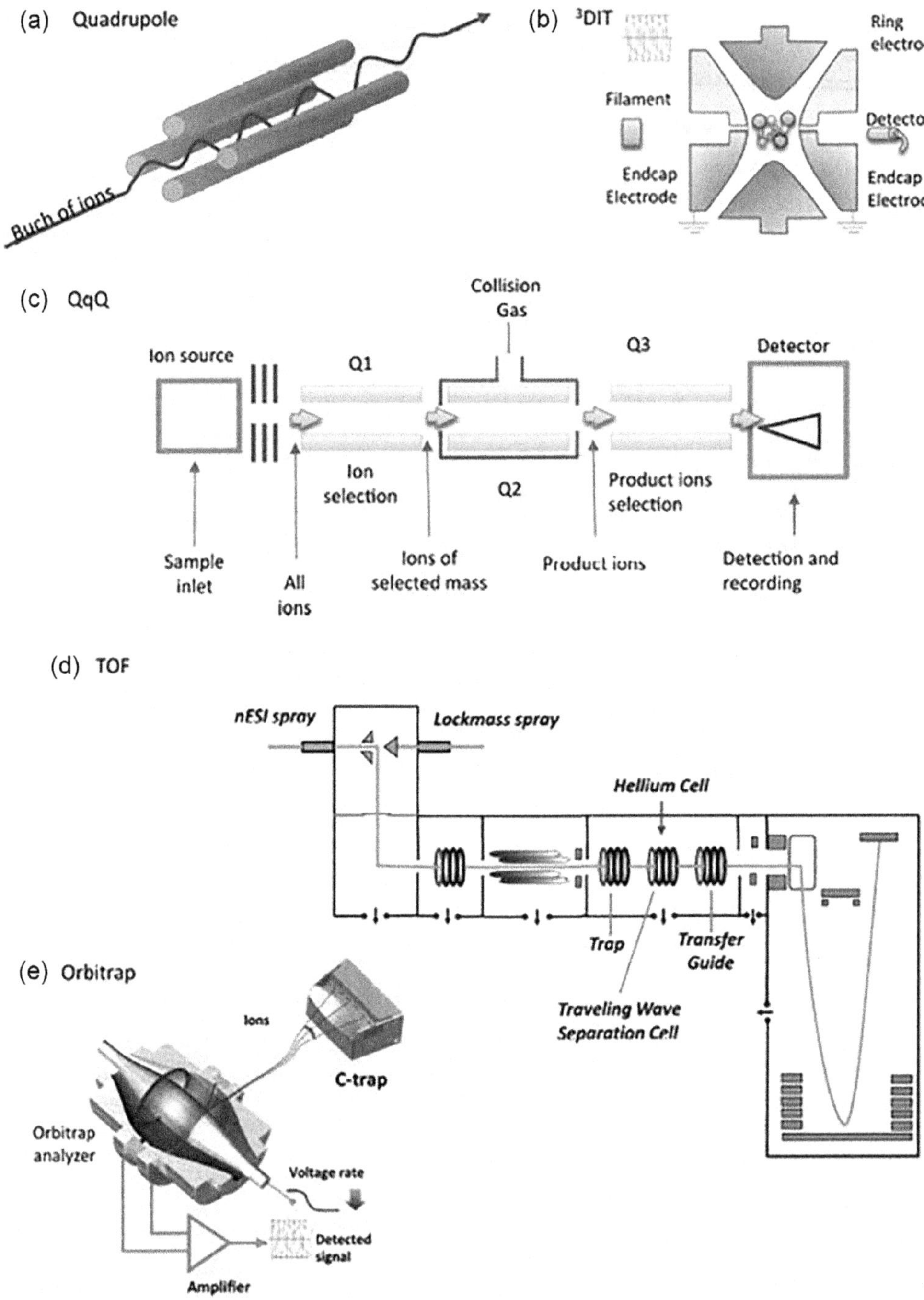

FIGURE 17.2 Most common mass analyzers used in mass spectrometry: (a) Quadrupole, (b) 3D ion trap, (c) triple quadrupole, (d) quadrupole time-of-flight, and (e) Orbitrap. (Reproduced with permission from Picó 2015, © Elsevier.)

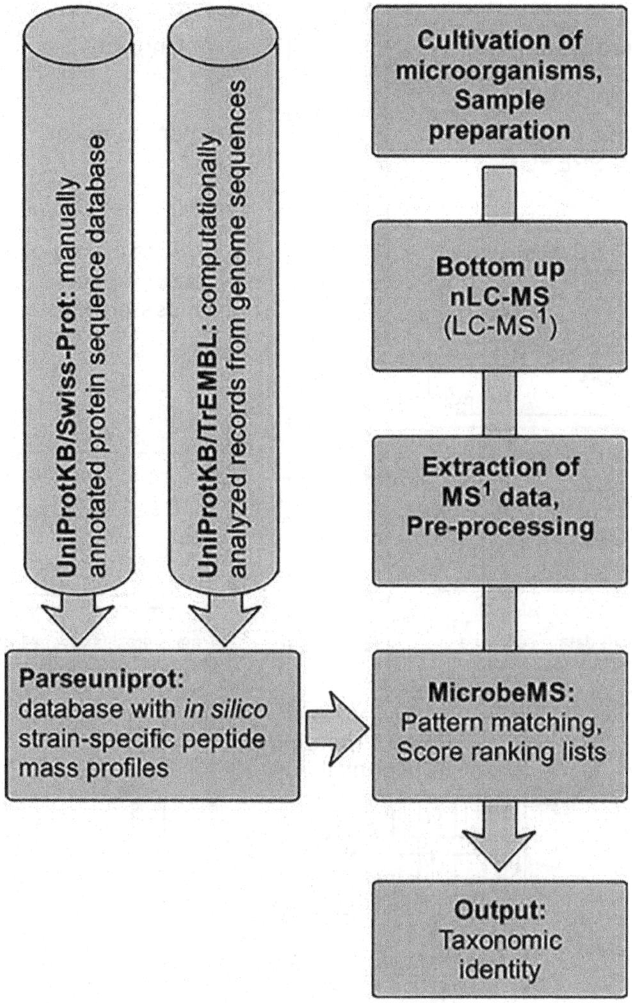

FIGURE 17.3 Schematic representation of foodborne pathogenic microbial identification using LC-MS^1. (Adapted from Lasch et al. 2020, Creative Commons Attribution 4.0 International.)

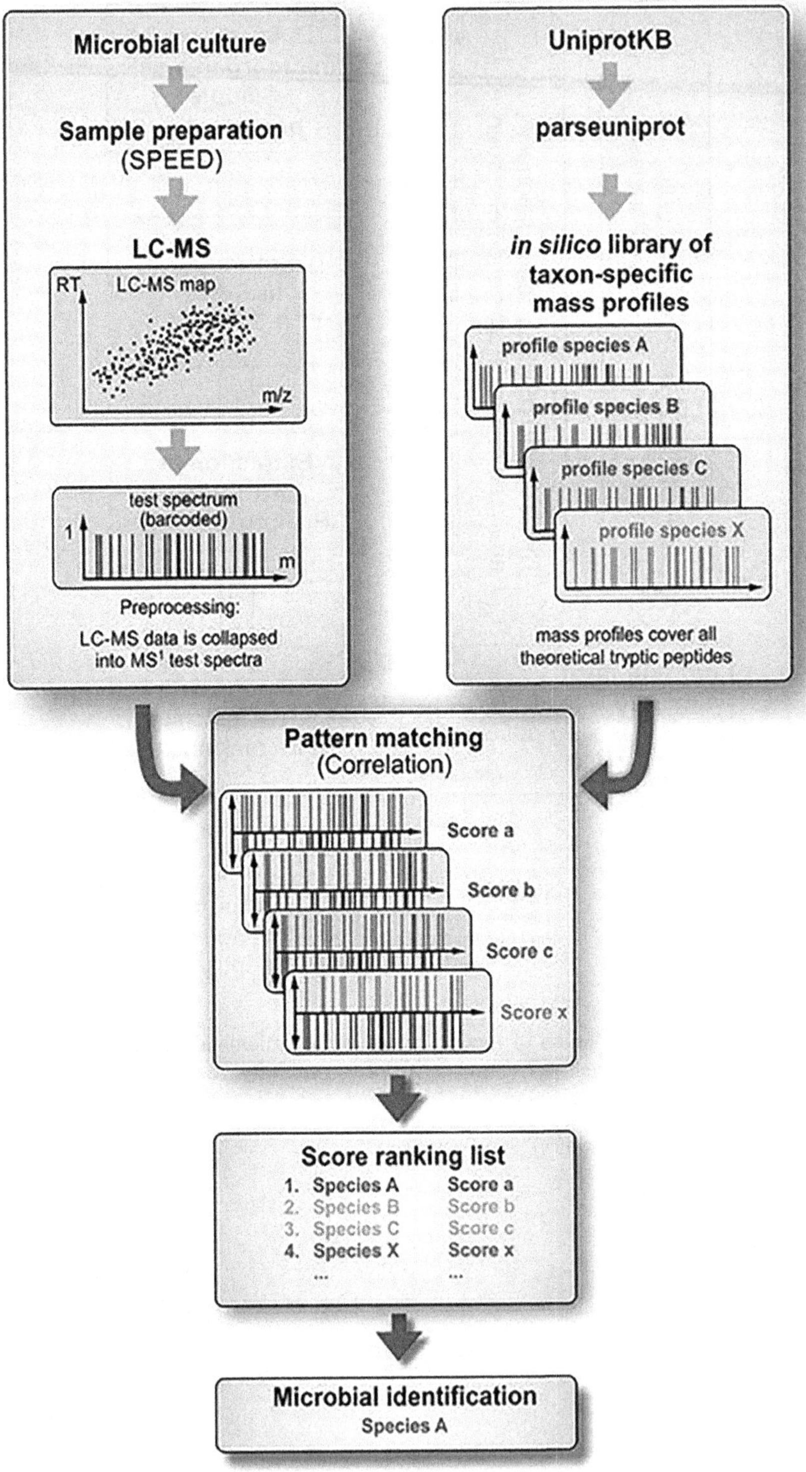

FIGURE 17.4 The workflow of LC-MS[1] data analysis to identify pathogenic microbes in food matrix. (Adapted from Lasch et al. 2020, Creative Commons Attribution 4.0 International.)

17.3 DETECTION OF FOODBORNE PATHOGENS USING MATRIX-ASSISTED LASER DESORPTION/IONIZATION (MALDI)-MS

Microbial contamination of foodstuffs is one of the most challenging aspects to protect humans from several foodborne illnesses (Stoev 2013). These foodstuffs are affected either at the source or during manufacturing (Pei et al. 2015; Grace et al. 2020). There have been several microbial foodborne outbreaks resulting in mortalities, and economic loss, which include MRSA detected in human and animal milk samples (Abolghait et al. 2020; Algammal et al. 2020a,b), the various *Enterobacteriaceae* isolates obtained from dairy and meat products (Ahmed and Shimamoto 2014; Makharita et al. 2020), *Pseudomonas aeruginosa* strains that are MDR were inoculated from fish farms (Saleh et al. 2014; Algammal et al. 2020c), *Salmonella enterica* contamination of food causing typhoid fever and so on. Therefore, this urges for a rapid detection method of such microorganisms in foodstuffs.

The emergence of MALDI-TOF-MS has transformed the foodborne bacterial identification that is essential for product quality (Pavlovic et al. 2013). It is a chemotaxonomic method that depends upon suitable biomarkers to analyze the similarities and differences. Therefore, it provides reliable data within a short time with low sample volume and reagent costs. However, the availability of reference data for comparison is necessary (Ryzhov and Fenselau 2001). For promising results, the eminence of the available reference data, the algorithm used for identification and protein fingerprinting references are essential. The biomarker assignment is typically based on selecting peaks with high intensity that refers to the ribosomal proteins. It is also possible to delineate closely related species using attuned software tools (Zeller-Péronnet et al. 2013). This is beneficial in improvising veterinary diagnostics, human clinical samples which include bacteria (Seng et al. 2009), fungi (Cassagne et al. 2011), and also environmental isolates classification (Decristophoris et al. 2011; Hijazin et al. 2012; Pavlovic et al. 2013) (Figure 17.5).

Several factors are involved in the identification process include the method used and its performance, cost-effective criteria, and facility acquired at the organization. The reliability and reproducibility of the method opted must be suitable for the identification of a broad spectrum of microorganisms, thereby providing a considerable amount of taxonomical resolution. Rapid analysis to ensure food safety with increased throughput reduces the cost of labor, providing reproducibility, and provides a platform for implementation (Pavlovic et al. 2013) (Figure 17.6).

The applicability of MALDI-TOF-MS in food microbiology has been encouraging by producing reference fingerprints. It has been successfully employed in milk and pork meat microbial spoilage (Nicolaou et al. 2012). The flexibility, reliability, and rapid analyses have made their

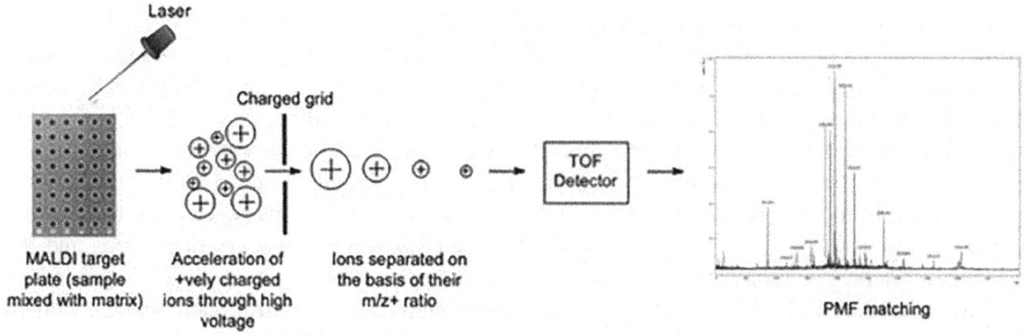

FIGURE 17.5 An overview of MALDI-TOF MS. (Adapted from Singhal et al. 2015, Creative Commons Attribution 4.0 International.)

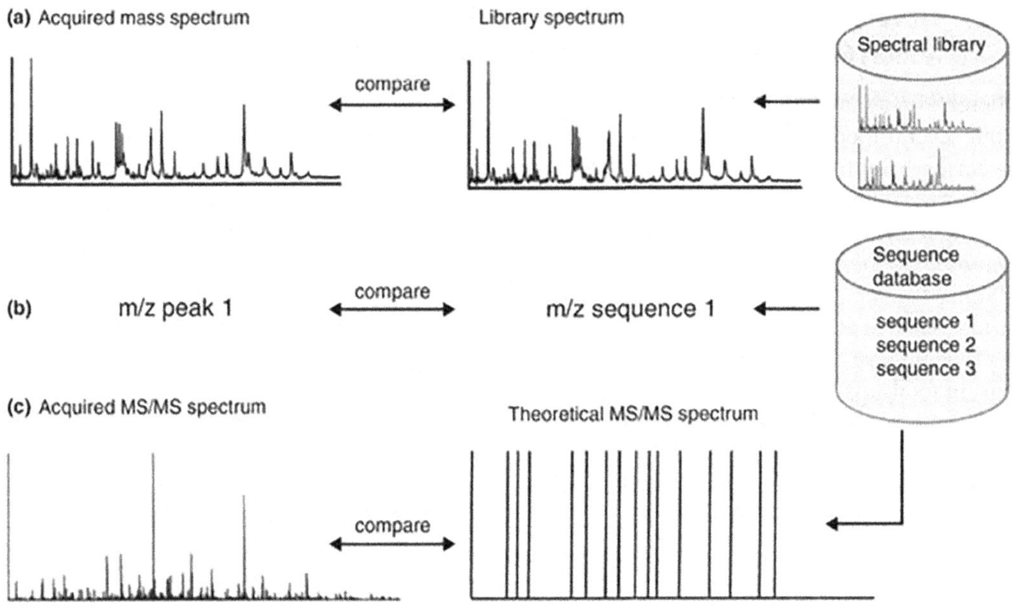

FIGURE 17.6 The scheme of data analysis in MALDI-TOF MS for detecting foodborne pathogens. (Reproduced with permission from de Koster and Brul 2016, © Elsevier.)

place in several food microbial laboratories. To successfully implement this technique, a thorough acquaintance of mass spectrometric-based microbial identification is essential, and replacing it with prevailing processes is a prerequisite. It is possible to quantify the microbial spoilage directly in food processing, thereby providing a platform to ease microbial food analytics (Pavlovic et al. 2013) (Figure 17.7).

17.4 ELECTROSPRAY IONIZATION-MASS SPECTROMETRY (ESI-MS) FOR IDENTIFICATION OF FOOD-SPOILAGE MICROORGANISMS

Because of speed, limited preparation of samples, and a high dynamic range, MS has been established as a valuable instrument in species classification (Demirev and Fenselau 2008; Yates et al. 2009; Ho and Reddy 2010, 2011). Most interestingly, MS enables a broad bacterial range to be detected, and they don't need any antibodies or DNA primers. Currently, the MS (MALDI) and MS (ESI) are two main ionization techniques for the analysis of biomolecules. MALDI and ESI research will achieve high-mass spectral profiles for proteins isolated from microbes such as bacteria. Those patterns of proteins can be built as markers for the classification of species (Everley et al. 2009; Giebel et al. 2010; Sauer and Kliem 2010). Another type of MS, also called Intact Cell MS, has also been established to distinguish bacteria without pretreatment (Edwards-Jones et al. 2000; Walker et al. 2002). The majority of the published research on microbial analysis through direct MS is currently focusing on the MALDI techniques' speed and simplicity. Alternatively, approaches were developed based on bottom-up digesting isolated proteins into peptides with particular proteases. Before the ESI-MS examination, the digests are isolated by LC. Improved instrumental effectiveness allows the sequencing of thousands of peptides in an individual sprint, leading to the discovery of proteins and the subsequent characterization of the proteins (VerBerkmoes et al. 2004; Dworzanski and Snyder 2005; Demirev and Fenselau 2008). In proteomic research, selected reaction control, such as the identification of disease markers, was used. By tracking several transformations for a given peptide obtained even from chromatographical coelution, the system will classify a single peptide. This procedure can be very helpful for the identification of bacterial markers in food samples.

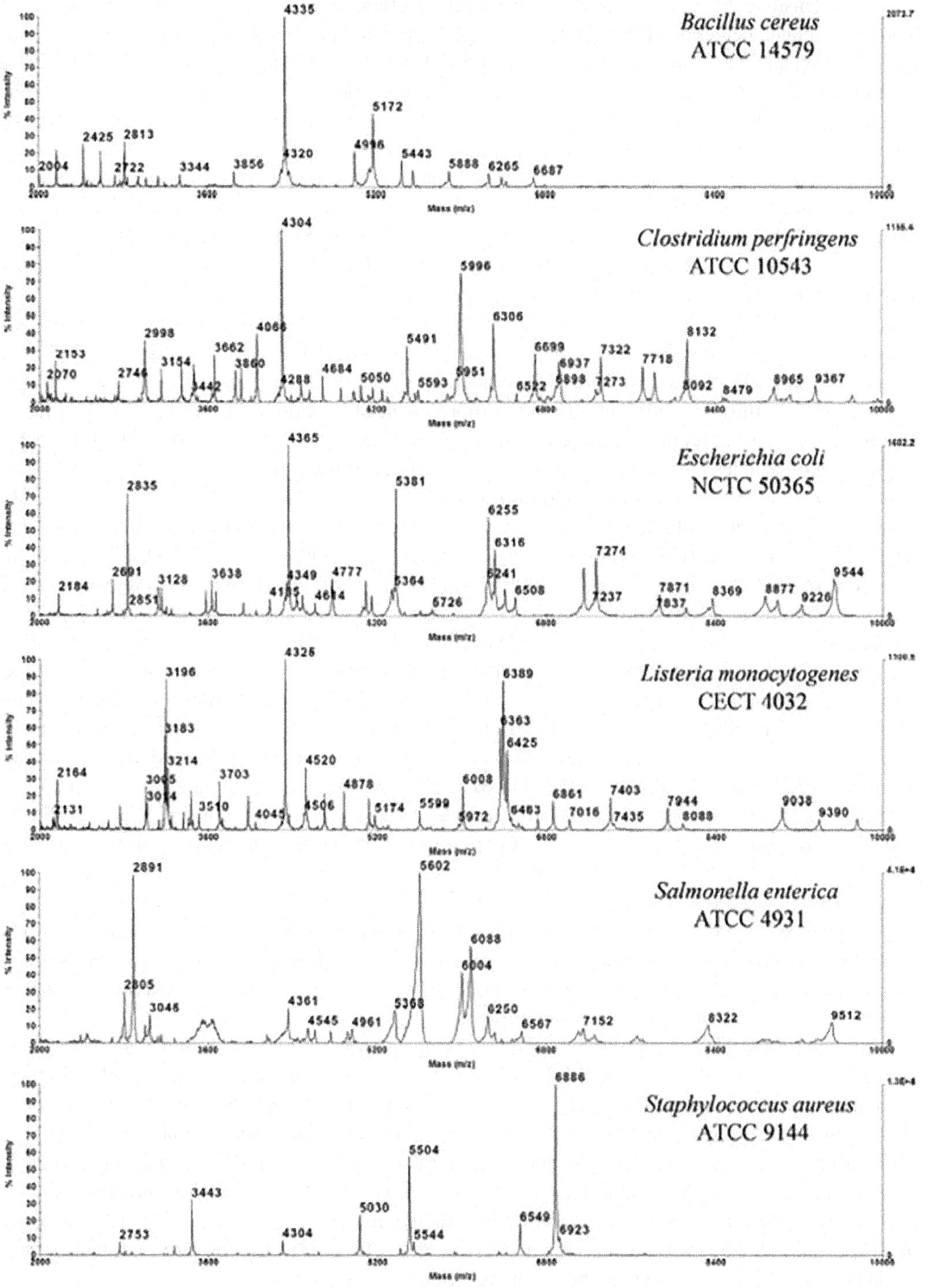

FIGURE 17.7 Representation of MALDI-TOF MS profiles of selected foodborne pathogens. (Reproduced with permission from Böhme et al. 2016a,b, © Elsevier.)

Highly sensitive and convenient methods based on nanoparticles have been used without MS in food for pathogen detection (Inbaraj and Chen 2016). Due to their high surface area-to-volume ratio and increased adequate binding sites, working particles with 10–100 nm diameters could efficiently combine with target cells (Soukka et al., 2003). Nanoparticle coating antibodies are a successful development for food-spoilage/pathogen-specific identification. Zhao et al. (2004) used antibody-conjugated nanomaterials in beef samples to mark the bacterial cells. Cell particle complexes have been observed using a flow cytometer, which reduces the overall analysis time to a maximum of 20 minutes and the detection limit to 1 CFU/g. Kim et al. (2007) used an impedimetric biosensor antibody coated to detect bacteria such as *Salmonella enteritidis*. Spiking quantum points into a cell buffer increased the detection limit to 104 CFU/mL from 106 CFU/mL. In addition, 105 CFU/mL of cells could be detected in milk by the biosensor.

Although MS methods apply to differentiating microbes, most studies cover medical applications, and comparatively few have attention on food spoiling/pathogens analyzing (Ochoa and Harrington 2005; Böhme et al. 2010; Jang and Kim 2018). Rich food matrix (e.g., proteins, lipids, salts) could minimize vulnerability in microorganism MS recognition. The enrichment step is, therefore, necessary to reduce interference and provide an adequate number of bacteria. For further analysis, nanoparticles have magnetic properties that could be used in the concentration of bacteria, simplifying purification procedures, and eliminating interference from complicated matrices. The successful identification of bacteria named *L. monocytogenes* at a concentration of 10 CFU/mL has been achieved on magnetic beads by covering oligonucleotides for selecting the pDNA target of the pathogens from milk (Amagliani et al. 2006). Additionally, Yang et al. (2007) combined RT-PCR with immunomagnetic separation to calculate bacteria detection limits. *L. monocytogenes* have been captured and observed in a milk sample at 226 CFU/0.5 mL by using magnetic nanoparticles. In addition, Yang et al. (2007) merged immunomagnetic isolation with RT-PCR to approximate bacteria detection limits. These techniques, however, include complex antibodies and pose problems deciding whether cells are alive or dead. Non-toxic zirconium hydroxide has been used by Chen et al. (2012) to capture bacteria in pudding and milk. After 5 hours of cultivation, several characteristic signals of isolated proteins could be found in MALDI-TOF-MS spectra. In non-fat dried milk, the solution had 33 CFU/mL as the detection threshold for *Enterococcus faecalis*. Chen et al. (2008) studied the trapping affinities between Fe_3O_4@TiO_2 and bacteria. Five bacterial strains based on possible biomarkers have been distinguished.

ESI-MS was used to classify bacteria rather than MALDI-MS. The model species *E. coli* and *Staphylococcus aureus* are two widespread bacteria that lead to recurrent global outbreaks. The bacterial cells have been concentrated with magnetic zirconia particles. Nano-LC-ESI-MS and library scanning have studied proteotypic peptides. The bacterial species were classified as bacterial markers by the identification of proteotypic peptides.

Although ESI-MS is less commonly used for direct microbial detection, Xiang et al. (2000) have characterized the microbes by global ESI-MS/MS studies of cellular lysates. Direct ESI-MS analysis of entire bacterial cells without prior separation was documented by Vaidyanathan et al. (2002). The recently developed DESI-MS allows the direct examination of condensed phase samples with electrosprayed solvent drops. In this method, several species of bacteria – including *Bordetella bronchiseptica*, *E. coli*, *Enterococcus* sp., *S. aureus*, *Bacillus subtilis*, *Bacillus thuringiensis*, and *Salmonella typhimurium* – are differentiated by their mass spectral DESI profile. Distinguished DESI spectra, in the 50–500 Da mass range, were obtained by positive and negative ion modes from whole bacteria; analytical time maybe 2 minutes (Song et al. 2009).

17.5 PCR-COUPLED MS FOR FOOD AUTHENTICATION

PCR assays are promising tools for food analysis because of their higher sensitivity and specificity. However, PCR-based methods are relatively useful to detect microbial strains, but they can't be opted for classification, particularly when microbial genomic sequences are unknown. PCR-coupled MS combination can be even more robust for identifying microbial species; in certain cases, this approach yields further details that usually can't be gained from either assay alone (Sampath et al. 2007a,b). ESI-based identification of PCR amplicons and MALDI-based resequencing provides comprehensive genomic data that are greatly helpful in the fast screening of microbes (Sauer and Kliem 2010). ESI-MS is suitable for the detection of large and multiple-charged PCR fragments (Mayr et al. 2005).

Consequently, the accurate molecular weight of PCR amplicons, obtained with ultra-high-throughput ESI-MS, has been manipulated to estimate their nucleotide compositions (Ecker et al. 2005). The mass spectrum of nucleic acid fragments is mostly aligned with theoretical information present in the genomic sequence database to recognize the species. LC-ESI-MS has also been effectively used to detect bacterial strains' PCR products (Mayr et al. 2005). Moreover, ion-pair reversed-phase with monolithic HPLC columns is successfully used to purify the PCR fragments rapidly and efficiently because it is completely automated with high throughput (Oberacher et al. 2000; Berger et al. 2002). PCR-coupled MS provides an ideal limit of detection and can also identify non-culturable microbes. However, primer designing requires absolute knowledge of genomic markers, and sample processing steps might need intensive labor.

Restriction digestion patterns are yielded from PCR fragments incubated with particular restriction enzymes. These reduced size digestive PCR amplicons also support ionization analysis and mass examination. MALDI-MS has also been successfully used to differentiate the microbial taxa by profiling restriction fragments of nucleic acids (Taranenko et al. 2002). Studies have utilized LC-MS to identify the targeted proteins that can be sequenced and reverse-engineered into PCR oligonucleotides peculiar to desired thermal tolerance and phenotypic trait (Williams et al. 2005). As the genome sequence of *E. sakazakii* was not available at that time, this technique provided the independent and unique means to recognize genetic differences present in closely related species without the requirement of any prior genome sequencing data.

17.6 GAS CHROMATOGRAPHY (GC)-MS IN FOOD MICROBE ANALYSIS

The other chromatographic technique with mass spectrometry (GC-MS) was also used to identify and detect other bacterial metabolites such as sugars, fatty acids, amino acids, nucleosides, organic acids, and secondary metabolites. This technique was used to identify metabolites for *L. monocytogenes*, *S. enterica*, and *E. coli* O157:H7 from contaminated beef samples (Jadhav et al. 2018). The main drawback of GC-MS is that it demands that the analytes are in a volatile form. As several metabolites are non-volatile, time-consuming derivatization steps are required (Rockwood et al. 2018). In the food industry, GC-MS is increasingly getting used in metabolite profiling to characterize foodborne microbes. GC-MS can analyze only those compounds that have a maximum 1,000 Da molecular weight. GC-MS often requires conversion of non-volatile analytes to volatile through derivatization before analysis. GC-MS/MS approach can improve the specificity and detection limit for carbohydrate biomarkers in complex samples. Distinctively, this system involves the transition monitoring of precursor ion to product ion in various selected reaction monitoring modes (Ho and Reddy 2010).

17.7 CAPILLARY ELECTROPHORESIS-MS FOR FOODBORNE PATHOGEN DETECTION

Analysis of foodborne pathogens was a time-consuming and labor-intensive task due to traditional techniques leading to a delayed detection system (Nugen and Baeumner 2008; Byrne et al. 2009; Pedrero et al. 2009). Characterizing pathogenic and non-pathogenic foodborne bacteria requires a rapid, sensitive, and specific detection system (Roda et al., 2012). CE was generally used in the detection of foodborne pathogens, in particular, in combination with new analytical techniques (García-Cañas et al. 2004; Hyytiä-Trees et al. 2007; Oh et al. 2008). A recent methodology that involved CE-MS is found to be promising in the detection of foodborne pathogens. Several studies implemented a CE-MS method for the classification of different pathogens. Determination of

Patulin, a type of mycotoxin frequently found in apple derivatives, was attempted using in-line molecularly imprinted polymer solid-phase extraction (MISPE) capillary zone electrophoresis-tandem mass spectrometry (CZE-MS/MS) method. The involvement of MISPE-CZE-MS/MS remarkably enhanced the degree of automatization, improved sensitivity of CE, and increased the laboratory throughput. The methodology was used in the presence of 5-(hydroxymethyl)furfural, which was known to be a main interfering agent in such a matrix. This method yielded a noticeable peak area of 14.9% and a migration time of 1.6% (Moreno-González et al. 2021). However, CE-MS was also used to detect pathogens such as *P. aeruginosa*, *Staphylococcus aureus*, and *Staphylococcus epidermidis* in the bacterial mixture (Hu et al. 2005). Adopting emerging CE-MS method in food industries could improve the detection of foodborne pathogens.

17.8 CONCLUSION

Food production and safety are considered to be essential aspects to ensure the overall quality of food products. Food contamination incidents often cause serious health problems all over the world. Rapid detection and analysis of food ingredients and maintaining optimal safety measures for food production are necessary. Technological advances are rigorously driving novel strategies for the authenticity assessment of foods. LC-MS-based food analysis represents a powerful technology that could drive improvements in pathogen identification. This method is easy to use and computationally efficient. The taxonomic resolution of the method is promising, but improvements, such as the application of feature selection methods, well-curated databases, are needed in a reliable, effective, and useful manner. PCR-based methods proved to be a reliable, fast, and highly sensitive technique for detecting food allergens. The PCR techniques can certainly reduce the increasing demands of consumers concerning allergens, genetically modified organisms, and adulterations. CE may be used to screen and quantify targeted genes in a single run with high reproducibility and observed to be an effective alternative DNA analysis tool for the detection of food contaminates. Overall, the application of advanced analytical methods in food analysis is crucial, and there are many advantages over conventional culture-based and biochemical-mediated methods.

REFERENCES

Abolghait, S. K., Fathi, A. G., Youssef, F. M. and Algammal, A. M. 2020. Methicillin-resistant *Staphylococcus aureus* (MRSA) isolated from chicken meat and giblets often produces staphylococcal enterotoxin B (SEB) in non-refrigerated raw chicken livers. *Int J Food Microbiol.* 328:108669.

Ahmed, A. M. and Shimamoto, T. 2014. Isolation and molecular characterization of *Salmonella enterica*, *Escherichia coli* O157:H7 and *Shigella* spp. from meat and dairy products in Egypt. *Int J Food Microbiol.* 168–169:57–62.

Algammal, A. M., Enany, M. E., El-Tarabili, R. M., Ghobashy, M. O. I. and Helmy, YA 2020b. Prevalence, antimicrobial resistance profiles, virulence and enterotoxins determinant genes of MRSA isolated from subclinical bovine mastitis in Egypt. *Pathogens* 9(5):362.

Algammal, A. M., Hetta, H. F., Elkelish, A., et al. 2020a. Methicillin-resistant *Staphylococcus aureus* (MRSA): one health perspective approach to the bacterium epidemiology, virulence factors, antibiotic-resistance, and zoonotic impact. *Infect Drug Resist.* 13:3255–3265.

Algammal, A. M., Mabrok, M., Sivaramasamy, E., et al. 2020c. Emerging MDR-*Pseudomonas aeruginosa* in fish commonly harbor oprL and toxA virulence genes and bla(TEM), bla(CTX-M), and tetA antibiotic-resistance genes. *Sci Rep.* 10(1):15961.

Amagliani, G., Omiccioli, E., Del Campo, A., Bruce, I. J., Brandi, G. and Magnani, M. 2006. Development of a magnetic capture hybridization-PCR assay for *Listeria monocytogenes* direct detection in milk samples. *J Appl Microbiol.* 100(2):375–383.

Asuming-Bediako, N., Parry-Hanson Kunadu, A., Abraham, S., et al. 2019. *Campylobacter* at the human–food interface: the African perspective. *Pathogens* 8:87.

Bahadir, E. B. and Sezginturk, M. K. 2017. Biosensor technologies for analyses of food contaminants. In: *Nanobiosensors*, ed. A. M. Grumezescu, 289–337. Cambridge, MA: Academic Press.

Beauchemin, D. 2006. Inductively coupled plasma mass spectrometry. *Anal Chem.* 78:4111–4135.

Berger, B., Holzl, G., Oberacher, H., et al. 2002. Single nucleotide polymorphism genotyping by on-line liquid chromatography-mass spectrometry in forensic science of the y-chromosomal locus m9. *J Chromatogr B* 782:89–97.

Böhme, C.-M., Monica Carrera, K., Caamaño-Antelo, J. M., et al. 2016a. Bacterial identification by LC-ESI-IT-MS/MS. In *Microbes in the spotlight: recent progress in the understanding of beneficial and harmful microorganisms*, ed. A. Méndez-Vilas, 160–164. Boca Raton, FL: BrownWalker Press.

Böhme, K., Antelo, S. C., Fernández-No, I. C., Quintela-Baluja, M., Barros-Velázquez, J., Cañas, B. and Calo-Mata, P. 2016b. Detection of foodborne pathogens using MALDI-TOF mass spectrometry. In *Antimicrobial Food Packaging*, ed. J. Barros-Velázquez, 203–214. Cambridge, MA: Academic Press. https://doi.org/10.1016/B978-0-12-800723-5.00015-2.

Böhme, K., Fernández-No, I. C., Barros-Velázquez, J. et al. 2012. SpectraBank: an open access tool for rapid microbial identification by MALDI-TOF MS fingerprinting. *Electrophoresis* 33:2138–2142.

Böhme, K., Fernández-No, I. C., Barros-Velázquez, J., Gallardo, J. M., Calo-Mata, P. and Canas, B. 2010. Species differentiation of seafood spoilage and pathogenic gram-negative bacteria by MALDI-TOF mass fingerprinting. *J Proteome Res.* 9(6):3169–3183.

Byrne, B., Stack, E., Gilmartin, N. and O'Kennedy, R. 2009. Antibody-based sensors: principles, problems and potential for detection of pathogens and associated toxins. *Sensors* 9:4407–4445. https://doi.org/10.3390/s90604407.

Cassagne, C., Ranque, S., Normand, A. C., et al. 2011. Mould routine identification in the clinical laboratory by matrix-assisted laser desorption ionization time-of flight mass spectrometry. *PLoS One* 6:28425.

CDC. 2016. *Foodborne Outbreak Online Database (FOOD tool).* Atlanta, GA: Centers for Disease Control and Prevention.

Chen, C. T., Reddy, P. M., Ma, Y. R. and Ho, Y. P. 2012. Mass spectrometric identification of pathogens in foods using a zirconium hydroxide immobilization approach. *Int J Mass Spectrom.* 312:45–52.

Chen, W. J., Tsai, P. J. and Chen, Y. C. 2008. Functional nanoparticle-based proteomic strategies for characterization of pathogenic bacteria. *Analytical Chemistry* 80(24):9612–21.

Davis, C. 2014. Enumeration of probiotic strains: review of culture dependent and alternative techniques to quantify viable bacteria. *J Microbiol Methods* 103:9–17.

De Koster, C. and Brul, S. 2016. MALDI-TOF MS identification and tracking of food spoilers and foodborne pathogens. *Curr Opin Food Sci.* 10:76–84.

Decristophoris, P., Fasola, A., Benagli, C., Tonolla, M. and Petrini, O. 2011. Identification of *Staphylococcus intermedius* group by MALDI-TOF MS. *Syst Appl Microbiol.* 34:45–51.

Demirev, P. A. and Fenselau, C. 2008. Mass spectrometry for rapid characterization of microorganisms. *Annu Rev Anal Chem.* 1:71–93.

Dworzanski, J. P. and Snyder, A. P. 2005. Classification and identification of bacteria using mass spectrometry-based proteomics. *Expert Rev Proteomics* 2(6):863–878.

Dworzanski, J., Dickinson, D. N., Deshpande, S. V., et al. 2010. Discrimination and phylogenomic classification of *Bacillus anthracis-cereus-thuringiensis* strains, based on LC-MS/MS analysis of whole cell protein digests. *Anal Chem.* 82:145–155.

Ecker, D. J., Sampath, R., Blyn, L. B., et al. 2005. Rapid identification and strain-typing of respiratory pathogens for epidemic surveillance. *Proc Natl Acad Sci USA.* 102:8012–8017.

Edwards-Jones, V., Claydon, M. A., Evason, D. J., Walker, J., Fox, A. J. and Gordon, D. B. 2000. Rapid discrimination between methicillin-sensitive and methicillin-resistant *Staphylococcus aureus* by intact cell mass spectrometry. *J Med Microbiol.* 49(3):295–300.

Elbehiry, A., Marzouk, E., Hamada, M., et al. 2017. Application of MALDI-TOF MS fingerprinting as a quick tool for identification and clustering of foodborne pathogens isolated from food products. *New Microbiol.* 40(4):278–296.

Everley, R. A., Mott, T. M., Toney, D. M. and Croley, T. R. 2009. Characterization of *Clostridium* species utilizing liquid chromatography/mass spectrometry of intact proteins. *J Microbiol Methods* 77(2):152–158.

Fall, B., Lo, C. I., Samb-Ba, B., et al. 2015. The ongoing revolution of MALDI-TOF mass spectrometry for microbiology reaches tropical Africa. *Am J Trop Med Hyg.* 92(3):641–647.

FAO. 2016. *Drivers, Dynamics and Epidemiology of Antimicrobial Resistance in Animal Production.* Rome: Food and Agriculture Organization.

Feng, B., Shi, H., Xu, F., et al. 2020. FTIR assisted MALDI-TOF MS for the identification and typing of bacteria. *Anal. Chim. Acta* 1111: 75–82.

García-Cañas, V., González, R. and Cifuentes, A. 2004. The combined use of molecular techniques and capillary electrophoresis in food analysis. *TrAC Trend Anal Chem.* 23:637–643 https://doi.org/10.1016/j.trac.2004.07.005.

Giebel, R., Worden, C., Rust, S. M., Kleinheinz, G. T., Robbins, M. and Sandrin, T. R. 2010. Microbial fingerprinting using matrix-assisted laser desorption ionization time-of-flight mass spectrometry (MALDI-TOF MS): applications and challenges. *Adv Appl Microbiol.* 71:149–184.

Grace, D., Wu, F. and Havelaar, A. H. 2020. MILK symposium review: foodborne diseases from milk and milk products in developing countries – review of causes and health and economic implications. *J Dairy Sci.* 103(11): 9715–9729.

Gwida, M. and El-Gohary, A. 2015. Prevalence and characterization of antibiotic resistance food borne pathogens isolated from locally produced chicken raw meat and their handlers. *J Dairy Vet Anim Res.* 2(6):238–244.

Hameed, S., Xie, L. and Ying, Y. 2018. Conventional and emerging detection techniques for pathogenic bacteria in food science: a review. *Trends Food Sci Technol.* 81:61–73.

Hijazin, M., Hassan, A. A., Alber, J., et al. 2012. Evaluation of matrix-assisted laser desorption ionization-time of flight mass spectrometry (MALDI-TOF MS) for species identification of bacteria of genera *Arcanobacterium* and *Trueperella*. *Vet Microbiol.* 157:243–245.

Ho, Y. P. and Reddy, P. M. 2010. Identification of pathogens by mass spectrometry. *Clin Chem.* 56(4):525–536. https://doi.org/10.1373/clinchem.2009.138867.

Ho, Y. P. and Reddy, P. M. 2011. Advances in mass spectrometry for the identification of pathogens. *Mass Spectrom Rev.* 30(6):1203–1224.

Hu, A., Tsai, P.-J. and Ho, Y.-P. 2005. Identification of microbial mixtures by capillary electrophoresis/selective tandem mass spectrometry. *Anal Chem.* 77:1488–1495. https://doi.org/10.1021/ac0484427.

Huber, I., Pavlovic, M., Maggipinto, M., et al. 2018. Interlaboratory proficiency test using MALDI-TOF MS for identification of food-associated bacteria. *Food Anal Methods* 11:1068–1075.

Hyytiä-Trees, E. K., Cooper, K., Ribot, E. M. and Gerner-Smidt, P. 2007. Recent developments and future prospects in subtyping of foodborne bacterial pathogens. *Future Microbiol.* 2:175–185. https://doi.org/10.2217/17460913.2.2.175.

Inbaraj, B. S. and Chen, B. H. 2016. Nanomaterial-based sensors for detection of foodborne bacterial pathogens and toxins as well as pork adulteration in meat products. *J Food Drug Anal.* 24(1):15–28.

Islam, M. A., Parveen, S., Rahman, M., et al. 2019. Occurrence and characterization of methicillin resistant *Staphylococcus aureus* in processed raw foods and ready-to-eat foods in an urban setting of a developing country. *Front Microbiol.* 10:503.

Jabbour, R. E., Deshpande, S. V., Wade, M. M., et al. 2010. Double-blind characterization of nongenome-sequenced bacteria by mass spectrometry-based proteomics. *Appl Environ Microbiol.* 76:3637–3644.

Jadhav, S. R., Shah, R. M., Karpe, A. V., et al. 2018. Detection of foodborne pathogens using proteomics and metabolomics-based approaches. *Front Microbiol.* 9:3132.

Jadhav, S. R. 2019. Identification of putative biomarkers specific to foodborne pathogens using metabolomics. In *Foodborne Bacterial Pathogens*, ed. A. Bridier, 149–64. Totowa, NJ: Humana Press.

Jang, K. S. and Kim, Y. H. 2018. Rapid and robust MALDI-TOF MS techniques for microbial identification: a brief overview of their diverse applications. *J Microbiol.* 56(4):209–216.

Jeng, K., Charlotte, A. G., Lawrence, B. B., et al. 2012. Comparative analysis of two broad-range PCR assays for pathogen detection in positive-blood-culture bottles: PCR–high-resolution melting analysis versus PCR-mass spectrometry. *J Clin Microbiol.* 50(10):3287–3292.

Kailasa, S. K., Koduru, J. R., Park, T. J., et al. 2019. Progress of electrospray ionization and rapid evaporative ionization mass spectrometric techniques for the broad-range identification of microorganisms. *The Analyst* 144(4):1073–103.

Karlsson, R., Gonzales-Siles, L., Boulund, F., et al. 2015. Proteotyping: proteomic characterization, classification and identification of microorganisms – a prospectus. *Syst Appl Microbiol.* 38(4):246–257.

Kim, G., Om, A. S. and Mun, J. H. 2007. Nanoparticle enhanced impedimetric biosensor for detection of foodborne pathogens. In *Journal of Physics: Conference Series*, Vol. 61(1), 1–112. Bristol: IOP Publishing.

Lasch, P., Schneider, A., Blumenscheit, C. and Doellinger, J. 2020. Identification of microorganisms by liquid chromatography-mass spectrometry (LC-MS[1]) and *in silico* peptide mass libraries. *Mol Cell Proteomics* 19(12):2125–2139. https://doi.org/10.1074/mcp.TIR120.002061.

Lee, L.-H. and Raghunath, P. 2018. Editorial: vibrionaceae diversity, multidrug resistance and management. *Front. Microbiol.* 9:563.

Li, H. and Zhu, J. 2017. Targeted metabolic profiling rapidly differentiates *Escherichia coli* and *Staphylococcus aureus* at species and strain level. *Rapid Commun Mass Spectrom.* 31:1669–1676.

Lucia, G., Olivier, P. and Jean, A. 2019. Pathogen proteotyping: a rapidly developing application of mass spectrometry to address clinical concerns. *Clin Mass Spectrom.* 14:9–17.

Makharita, R. R., El-Kholy, I., Hetta, H. F., Abdelaziz, M. H., Hagagy, F. I., Ahmed, A. A. and Algammal, A. M. 2020. Antibiogram and genetic characterization of carbapenem-resistant gram-negative pathogens incriminated in healthcare-associated infections. *Infect Drug Resist.* 13:3991–4002.

Mayr, B. M., Kobold, U., Moczko, M., et al. 2005. Identification of bacteria by polymerase chain reaction followed by liquid chromatography-mass spectrometry. *Anal Chem.* 77:4563–4570.

Moothoo-Padayachie, A., Kandappa, H. R., Krishna, S. B. N., et al. 2013. Biotyping *Saccharomyces cerevisiae* strains using matrix-assisted laser desorption/ionization time-of-flight mass spectrometry (MALDI-TOF MS). *Eur Food Res Technol.* 236:351–364.

Moreno-González, D., Jáč, P., Riasová, P. and Nováková, L. 2021. In-line molecularly imprinted polymer solid phase extraction-capillary electrophoresis coupled with tandem mass spectrometry for the determination of patulin in apple-based food. *Food Chem.* 334:127607. https://doi.org/10.1016/j.foodchem.2020.127607.

Moriconi, M., Acke, E., Petrelli, D., et al. 2017. Multiplex PCR based identification of *Streptococcus canis*, *Streptococcus zooepidemicus* and *Streptococcus dysgalactiae* subspecies from dogs. *Comp Immunol Microbiol Infect Dis.* 50:48–53.

Nagaraj, S., Ramlal, S., Kingston, J., et al. 2016. Development of IgY based sandwich ELISA for the detection of staphylococcal enterotoxin G (SEG), an EGC toxin. *Int J Food Microbiol.* 237:136–141.

Nicolaou, N., Xu, Y. and Goodacre, R. 2012. Detection and quantification of bacterial spoilage in milk and pork meat using MALDI-TOF-MS and multivariate analysis. *Anal Chem.* 84:5951–5958.

Nugen, S. R. and Baeumner, A. J. 2008. Trends and opportunities in food pathogen detection. *Anal Bioanal Chem.* 391:451. https://doi.org/10.1007/s00216-008-1886-2.

Oberacher, H., Krajete, A., Parson, W. and Huber, C. G. 2000. Preparation and evaluation of packed capillary columns for the separation of nucleic acids by ion-pair reversed-phase high-performance liquid chromatography. *J Chromatogr A* 893:23–35.

Ochoa, M. L. and Harrington, P. B. 2005. Immunomagnetic Isolation of enterohemorrhagic *Escherichia coli* O157: H7 from ground beef and identification by matrix-assisted laser desorption/ionization time-of-flight mass spectrometry and database searches. *Anal Chem.* 77(16):5258–5267.

Oh, M.-H., Park, Y. S., Paek, S.-H., Kim, H.-Y., Jung, G. Y. and Oh, S. 2008. A rapid and sensitive method for detecting foodborne pathogens by capillary electrophoresis-based single-strand conformation polymorphism. *Food Control* 19:1100–1104. https://doi.org/10.1016/j.foodcont.2007.11.009.

Pavlovic, M., Huber, I., Konrad, R. and Busch, U. 2013. Application of MALDI-TOF MS for the identification of food borne bacteria. *The Open Microbiol J.* 7:135–141.

Pedrero, M., Campuzano, S. and Pingarrón, J. M. 2009. Electroanalytical sensors and devices for multiplexed detection of foodborne pathogen microorganisms. *Sensors* 9:5503–5520. https://doi.org/10.3390/s90705503.

Pei, X., Li, N., Guo, Y., et al. 2015. Microbiological food safety surveillance in China. *Int J Environ Res Public Health.* 12(9):10662–10670.

Perreten, V., Endimiani, A., Thomann, A., et al. 2013. Evaluation of PCR electrospray-ionization mass spectrometry for rapid molecular diagnosis of bovine mastitis. *J Dairy Sci.* 96(6):3611–3620.

Petr, J., Ryparova, O., Ranc, V., et al. 2009. Assessment of CE for the identification of microorganisms. *Electrophoresis* 30:444–449.

Picó, Y. 2015. Mass spectrometry in food quality and safety: an overview of the current status. In *Comprehensive Analytical Chemistry*, Vol. 68, 3–76. Amsterdam: Elsevier. https://doi.org/10.1016/B978-0-444-63340-8.00001-7.

Pierce, S. E., Bell, R. L., Hellberg, R. S. and Cheng, C.-M. 2012. Detection and identification of *Salmonella enterica*, *Escherichia coli*, and *Shigella* spp. via PCR-ESI-MS: isolate testing and analysis of food samples. *Appl Environ Microbiol.* 78(23): 8403–8411.

Pinu, F. R. 2016. Early detection of food pathogens and food spoilage microorganisms: application of metabolomics. *Trends Food Sci. Technol.* 54(Suppl. C):213–215.

Quéro, L., Girard, V., Pawtowski, A., et al. 2019. Development and application of MALDI-TOF MS for identification of food spoilage fungi. *Food Microbiol.* 81:76–88.

Řezanka, T., Matoulková, D., Benada, O., et al. 2015. Lipidomics as an important key for the identification of beer-spoilage bacteria. *Lett Appl Microbiol.* 60(6): 536–543.

Rockwood, A. L., Kushnir, M. M. and Clarke, N. J. 2018. Mass spectrometry. In *Principles and Applications of Clinical Mass Spectrometry*, eds. N. Rifai, A. R. Horvath, C. T. Wittwer and A. Hoofnagle, 33–65. Amsterdam: Elsevier.

Roda, A., Mirasoli, M., Roda, B., Bonvicini, F., Colliva, C. and Reschiglian, P. 2012. Recent developments in rapid multiplexed bioanalytical methods for foodborne pathogenic bacteria detection. *Microchim Acta* 178:7–28. https://doi.org/10.1007/s00604-012-0824-3.

Ryzhov, V. and Fenselau, C. 2001. Characterization of the protein subset desorbed by MALDI from whole bacterial cells. *Anal Chem.* 73:746–50.

Saleh, F. O., Ahmed, H. A., Khairy, R. M. and Abdelwahab, S. F. 2014. Increased quinolone resistance among typhoid *Salmonella* isolated from Egyptian patients. *J Infect Dev Ctries.* 8(5):661–665.

Sampath, R., Hall, T. A., Massire, C. and Li, F. 2007a. Rapid identification of emerging infectious agents using PCR and electrospray ionization mass spectrometry. *Ann NY Acad Sci.* 1102:109–120.

Sampath, R., Russell, K. L., Massire, C., et al. 2007b. Global surveillance of emerging influenza virus genotypes by mass spectrometry. *PLoS One* 2:e489.

Sauer, S. and Kliem, M. 2010. Mass spectrometry tools for the classification and identification of bacteria. *Nat Rev Microbiol.* 8:74–82.

Seng, P., Drancourt, M., Gouriet, F., et al. 2009. Ongoing revolution in bacteriology: routine identification of bacteria by matrix-assisted laser desorption ionization time-of-flight mass spectrometry. *Clin Infect Dis.* 49:543–551.

Singhal, N., Kumar, M., Kanaujia, P. K. and Virdi, J. S. 2015. MALDI-TOF mass spectrometry: an emerging technology for microbial identification and diagnosis. *Front Microbiol.* 6:791. https://doi.org/10.3389/fmicb.2015.00791.

Song, Y., Talaty, N., Datsenko, K., Wanner, B. L. and Cooks, R. G. 2009. In vivo recognition of *Bacillus subtilis* by desorption electrospray ionization mass spectrometry (DESI-MS). *Analyst* 134(5):838–841.

Soukka, T., Antonen, K., Härmä, H., Pelkkikangas, A. M., Huhtinen, P. and Lövgren, T. 2003. Highly sensitive immunoassay of free prostate-specific antigen in serum using europium (III) nanoparticle label technology. *Clinica Chimica Acta* 328(1–2):45–58.

Stoev, S. D. 2013. Food safety and increasing hazard of mycotoxin occurrence in foods and feeds. *Crit Rev Food Sci Nutr.* 53(9):887–901.

Taranenko, N. I., Hurt, R., Zhou, J. Z., et al. 2002. Laser desorption mass spectrometry for microbial DNA analysis. *J Microbiol Methods* 48:101–106.

Vaidyanathan, S., Kell, D. B. and Goodacre, R. 2002. Flow-injection electrospray ionization mass spectrometry of crude cell extracts for high-throughput bacterial identification. *J Am Soc Mass Spectrom.* 13(2):118–128.

VerBerkmoes, N. C., Connelly, H. M., Pan, C. and Hettich, R. L. 2004. Mass spectrometric approaches for characterizing bacterial proteomes. *Expert Rev Proteomics* 1(4):433–447.

Walker, J., Fox, A. J., Edwards-Jones, V. and Gordon, D. B. 2002. Intact cell mass spectrometry (ICMS) used to type methicillin-resistant *Staphylococcus aureus*: media effects and inter-laboratory reproducibility. *J Microbiol Methods* 48(2–3):117–126.

Welker, M., van Belkum, A., Girard, V., et al. 2019. An update on the routine application of MALDI-TOF MS in clinical microbiology. *Expert Rev Proteomics* 8:695–710.

WHO. 2014. *Antimicrobial Resistance: Global Report on Surveillance.* Geneva: World Health Organization.

Williams, T. L., Monday, S. R., Edelson-Mammel., et al. 2005. A top-down proteomics approach for differentiating thermal resistant strains of *Enterobacter sakazakii. Proteomics* 5:4161–4169.

Xiang, F., Anderson, G. A., Veenstra, T. D., Lipton, M. S. and Smith, R. D. 2000. Characterization of microorganisms and biomarker development from global ESI-MS/MS analyses of cell lysates. *Anal Chem.* 72(11):2475–2481.

Yang, H., Qu, L., Wimbrow, A. N., Jiang, X. and Sun, Y. 2007. Rapid detection of *Listeria monocytogenes* by nanoparticle-based immunomagnetic separation and real-time PCR. *Int J Food Microbiol.* 118(2):132–138.

Yates, J. R., Ruse, C. I. and Nakorchevsky, A. 2009. Proteomics by mass spectrometry: approaches, advances, and applications. *Annu Rev Biomed Eng.* 11:49–79.

Zeller-Péronnet, V., Brockmann, E., Pavlovic, M., Timke, M., Busch, U. and Huber, I. 2013. Potential of MALDI-TOF MS for strain discrimination within the species *Leuconostoc mesenteroides* and *Leuconostoc pseudomesenteroides. J Für Verbraucherschutz und Lebensmittelsicherheit* 803:205–214.

Zhao, X., Hilliard, L. R., Mechery, S. J., Wang, Y., Bagwe, R. P., Jin, S. and Tan, W. 2004. A rapid bioassay for single bacterial cell quantitation using bioconjugated nanoparticles. *Proc Natl Acad Sci USA.* 101(42):15027–15032.

18 Biogenic Amines Analysis by Mass Spectrometry

Sadaf Jamal Gilani
Princess Noura Bint Abdulrahman University

Chandra Kala and Mohammad Taleuzzaman
Maulana Azad University

Syed Sarim Imam and Sultan Alsheri
King Saud University

Mohammad Asif
Lachoo Memorial College of Science and Technology

CONTENTS

18.1 Introduction 347
18.2 Extraction and Derivatization of BAs from the Sample 348
18.3 Mass Spectrometry 349
18.4 Liquid Chromatography-Mass Spectrometry (LC-MS) 350
18.5 Gas Chromatography-Mass Spectrometry 351
18.6 Capillary Electrophoresis-Mass Spectroscopy 351
18.7 Ion-Exchange Chromatography Coupled with Mass Spectroscopy 352
18.8 Conclusion 353
Acknowledgments 353
References 353

18.1 INTRODUCTION

Preparation, processing, and preservation are common steps that food goes through before being consumed [1]. Biogenic amines (BAs) are basic, polar, or semi-polar compounds having low molecular weight present in various foods such as alcoholic beverages, cheese, fish, and meat products. Microbial decarboxylation of amino acids (AAs) in food attributed to its improper storage can result in BAs' formation [2]. When consumed in high concentrations, these compounds may impose toxicological and detrimental effects on human health. Therefore, these concerns mandate the monitoring and accurate determination of BAs in food samples [3]. Histamine (HA) is one of the significant BAs found in spoiled food and is generated by histidine's decarboxylation. Therefore, these BAs can act as an indicator to determine food freshness or spoilage, and their quantification in foods helps assess food quality and their associated health risks [4].

For BAs determination in foods, different techniques such as paper chromatography, thin-layer chromatography, gas–liquid chromatography, and high-performance liquid chromatography (HPLC) have been employed. However, these methods are tedious, time-consuming, and expensive. Nevertheless, mass spectrometry (MS) coupled with gas chromatography (GC), liquid chromatography (LC), and capillary electrophoresis (CE) have been applied [5,6] and are reported to

DOI: 10.1201/9781003091226-21

be reliable and accurate detection techniques but are expensive and require technical skill, making them less prevalent in developing countries [5]. Liquid chromatography coupled with tandem mass spectroscopy (LC-MS/MS) can be directly coupled with automated sample cleanup, facilitating the reproducible and efficient handling of samples to be analyzed [7].

Some of the most common BAs reported in food are putrescine (PUT), cadaverine (CAD), spermine, and spermidine with an aliphatic structure; tyramine (TA), tryptamine, and β-phenylethylamine (β-PE) with an aromatic structure and HA with a heterocyclic structure [8]. These compounds are reported to exert various toxicological effects. All foods containing free AAs or protein are expected to produce BAs or are exposed to conditions that facilitate microbial or biochemical activity [9].

BAs exhibit their precursors' characteristics, which are considered when analytical methods are being developed for their determination. BAs act as a nitrogen source and are precursors of many biochemical substances such as hormones, alkaloids, proteins, and nucleic acids.

They are known to modulate several biological processes such as body temperature, alteration of blood pressure, nutrition intake, etc., in the organism. HA and TA are confirmed to be the most toxic BAs by The European Food Safety Authority, specifically relevant for food safety. Some amines are reported to increase HA toxicity. BAs are present in a wide range of products such as fish, fish products, meat, meat products, beer, wine, cheese, milk, dairy products, various beverages, condiments, fruits, vegetables, vinegar, tea, chocolate, and coffee). Thus, the development of new analytical methods or improving the current techniques for rapidity and reliability is of great interest [3]. Among many other methods, LC-MS/MS represents wider anticipation for development in the future [10].

This chapter reviews various popular analytical techniques coupled with mass spectroscopy such as LC-MS, GC-MS, ion-exchange chromatography coupled with mass spectroscopy (IC-MS), and CE-MS to determine, quantify, and detect BAs in food to furnish a reference for further studies.

18.2 EXTRACTION AND DERIVATIZATION OF BAs FROM THE SAMPLE

BAs are compounds of high polarity without intrinsic property. Moreover, these are present in low concentrations and more interfering compounds in complex matrices, making their determination and quantification difficult. The matrices containing BAs are subjected to extraction and derivatization to facilitate their rapid and efficient separation, determination, and quantification. Extraction and derivatization not only remove interferents but also concentrate the analytes [11].

Extraction of BAs can be carried out with water or in an acidic medium with hydrochloric acid (HCl), perchloric acid ($HClO_4$), trichloroacetic acid (TCA), or with organic solvents such as methanol, acetone, acetonitrile-$HClO_4$, or dichloromethane-$HClO_4$ [12]. Perchloric acid has demonstrated a 30% higher BAs extraction than chloric acid, sulfosalicylic acid, and TCA [13,14]. For instance, TCA (5%) and perchloric acid are recommended to extract BAs from fish and meat; 0.1 M HCl is considered best for amine extraction from cheese. Methanol:0.4 N HCl demonstrated good recovery for HA, putrescine, and cadaverine. A combination of organic solvents and acid (75% ethanol–0.4 N HCl) resulted in the complete recovery of HA from fresh, frozen, and canned fish tissue [15]. Besides, efficient sample preparation is required to lower the analyte loss, increase the recoveries and eliminate the sample complexity [16].

After primary extraction, pre-concentration can be done using solid-phase microextraction (SPME), liquid-phase microextraction (LPME), and electromembrane extraction, molecularly imprinted-SPME), dispersive liquid–liquid microextraction (DLLME), vortex-assisted surfactant-enhanced emulsification liquid–liquid microextraction (HF-LPME), salting-out assisted liquid–liquid extraction, and cloud point extraction [3].

For different types of analytes from samples, the procedure of extraction varies. For instance, the solvent choice varies for aliphatic amines, heterocyclic amines, and catecholamine. For methylxanthines, the extraction was facilitated at high temperatures or by soxhlet extraction [17].

As mentioned above, BAs lack chromophore or fluorophore groups in their structure and are highly polar compounds; their separation and detection are difficult. Therefore, a pre-column or post-column derivatization step is required to change these analytes into derivatives capable of being separated and detected, leading to improved sensitivity and selectivity [18,19]. Most commonly used derivatization agents are dansyl chloride (DNS-Cl), dabsyl chloride benzoyl chloride, 4-chloro-3,5-dinitrobenzotrifluoride, *o*-phthalaldehyde (OPA), diethyl ethoxymethylenemalonate, 6-aminoquinolyl-*N*-hydroxysuccinimidyl carbamate (AQC), 9H-fluoroen-9-ylmethyl chloroformate, tosyl chloride, phenylisothiocyanate (PITC), acetaldehyde, propionic anhydride with pyridine, 4-(*N*, *N*-dimethylaminosulfonyl)-7-(3-isothiocyanatopyrrolidin-1-yl)-2,1,3-benzoxadiaz-ole, perfluoroheptanoic acid, [20] and Ninhydrin [21].

For instance, isocyanates react with aliphatic and aromatic amines to form *N,N′*-disubstituted ureas. PITC reacted with 20 proteogenic AAs and 22 physiological amino compounds, forming phenylthiocarbamyl derivatives of acids effectively separated by reversed-phase LC (RP-LC) in 60 minutes [22]. A large number of fluorescence labeling reagents are also used for the derivatization of BAs. OPA reacts with primary amino compounds within 2 minutes at room temperature and is effectively separated with RP-LC [22].

Although MS can detect native AAs, pre-column derivatization procedures are frequently employed for better separation and high sensitivity. AQC is commonly used as a pre-column fluorescence derivatization reagent for BAs. AQC reacts with amines forming aminoquinoline-labeled compounds, separated by RP-LC and detected by electrospray ionization (ESI)-MS [22,23].

Unfortunately, there are certain disadvantages associated with the derivatization of analytes, such as derivative instability, interferences caused by reagent used, and the time-consuming process. Therefore, there is significant interest in developing and using techniques that allow BAs analysis without the need for derivatization [18].

18.3 MASS SPECTROMETRY

MS is widely used to identify and quantify various biogenic compounds. When coupled with different chromatography techniques, it can quantify various biogenic compounds at subnanomolar levels with great precision. MS has been used for analyzing diverse trace amines such as PE, phenylethanolamine *meta-* and *para-*tyramine (m-TA and p-TA) *ortho-*, *meta-* and *para-*octopamine, *meta-* and *para-*synephrine (*m*-SYN and+*p*-SYN), tryptamine, the catecholamines, dopamine, noradrenaline, adrenaline (epinephrine), 5-hydroxytryptamine, HA, the cycloalkyl amine, piperidine and the quaternary amines, choline and acetylcholine, *p*-tyrosine, and tryptophan, among many others. MS principally has three components contained in a high-vacuum system: an ion source in which molecules are transformed into a gaseous ion, an analyzer, and an ion detector.[24]. The gaseous ions are transferred into the mass analyzer, which sorts the ions according to their mass-to-charge ratio (*m/z*). An ion detector produces electrical signals that are converted into mass spectra. The detected ion may be of original molecules, their fragments, or another species generated during the process of ionization [25]. MS can be used to get a qualitative result by utilizing an appropriate internal standard. The mass analysis is based on the effect of electric field on the ion (time-of-flight [TOF]); electric and radiofrequency on the ion (quadrupole mass spectrometer); electric and magnetic field on the ion (magnetic sector mass spectrometer); electric, magnetic, and radiofrequency on the ion (Fourier transform [FT] mass spectrometer) [24].

MS detection combined with electrostatic separation methods has been an essential analytical tool over the last decades. ESI, in particular, is of great interest when used for the hyphenation of MS with liquid separation mode. Underivatized amines and easily protonizable amino groups are well suited to be detected using ESI-MS. It is highly selective and allows the evaluation of overlapping or even co-migrating signals. Recently, the use of multistage MS detectors has been increased, which provides meaningful information for structure elucidation [18].

18.4 LIQUID CHROMATOGRAPHY-MASS SPECTROMETRY (LC-MS)

The LC-MS system is one of the most powerful techniques used to determine different compounds in complex biological matrices with high quantitative accuracy [22].

When coupled with ultraviolet or fluorescence detection for BAs determination, the derivatization step is required to improve the BAs' separation and detection by transforming them into less polar products. Among all, the most effective method of derivatization used by many countries for BAs determination in protein-rich samples is the OPA method. In this method, OPA is condensed along with aliphatic thiol and the amine. Nevertheless, it has some disadvantages (uncertainty) in peak identification in complex matrices due to similar retention times and unstable reaction products [26].

Recently LC coupled with MS/MS has been employed for BAs determination without applying the derivatization step. However, a sample cleanup step was required to avoid inaccurate quantification due to interfering substances present in the food [4]. LC effectively separates the organic components from matrices, and MS gives information on molecular weight, concentration, and structure of organics and gives improved peak identification capabilities [10,26]. It is widely used to analyze mixed organic compounds of strong polarity and low volatility. Mass spectroscopy is based on ionization, and two types of ion sources with ESI and atmospheric pressure chemical ionization (APCI) are used. Li et al. used ultra-performance liquid chromatography (UPLC) with ESI tandem quadrupole simultaneously to determine 10 BAs in a thymopeptide injection [10].

LC-MS/MS has several advantages for BAs measurements, such as high sensitivity and specificity, minimal sample preparation, a wide concentration range, detection of both free and conjugated forms of BAs, and high throughput. It is attributed to MS's ability to identify and quantify compounds based on precursor ion transition to product ions. After the positive and negative ionization of compounds, selective reaction monitoring (SRM) or multiple reaction monitoring transition chromatograms are extracted. ESI in positive mode is preferred for highly polar BAs due to its ability to quantify low- and high-molecular-weight compounds with precision and accuracy. Although negative mode was also used for vanillylmandelic acid (VMA), homovanillic acid (HVA) and 3,4-dihydroxyphenylacetic acid (DOPAC) were mainly used to favor polar compounds. For quantification of BAs of low or medium polarity, APCI is also preferred [20].

Mass analyzers include triple quadrupole (QQQ), ion trap (IT), TOF, and other types. QQQ is found to be the most sensitive and quantitatively reproducible. Sagratini et al. used HPLC-MS/MS with a QQQ analyzer to analyze eight BAs in fish tissues [10]. LC-TOF-MS has been established as an important technique for routine food wholesomeness analysis [27]. As TOF relies on the TOF in linear space and has a wide mass range. Jia et al. simultaneously determined seven BAs in food samples by using LC-Q-TOF-MS methods [10]. The majority of analysis by LC-MS/MS relies on the highly sensitive and SRM mode of QQQ-MS/MS [27].

RP-LC with a volatile acid modifier coupled with MS/MS detection is generally used to determine BAs. It allows the rapid and simultaneous determination of AAs with high sensitivity [22].

HPLC coupled with various detection systems is one of the most standard techniques for BAs analysis. It involves pre-column and post-column derivatization, evaporative light scattering detector and ultraviolet (UV) or fluorescence (FL) detection, and more often with MS detector. BAs separation is generally performed with an alkyl chain column or hydrophilic interaction liquid chromatography (HILC) columns. Each method involves two steps: (i) extraction of BAs from food matrices with cleanup with solid-phase extraction or matrix solid-phase dispersion or DLLME and (ii) derivatization of appropriate compounds [3,14], which reduces organic solvent volume and increases enrichment factor [28].

Since 2004, UPLC has been used, which was more advantageous than HPLC. UPLC technique significantly decreased the separation time of BAs when a short column (5 cm) packed with smaller particles (<2 μm) and high flow rates were employed. In a study, BAs and 23 AAs were separated from cheese, beer, and sausage samples within 25–30 minutes with limits of detection (LODs) of 5–20 μg/L using UPLC/Q-TQFMS. In another study, UPLC-MS/MS (ESI+) method was used to determine nine BAs in rice wine samples with 21 minutes of run time with LODs of 0.1–4.6 μg/L.

If HPLC is coupled with an MS detector, the derivatization step can be left out. MS or MS/MS detectors provide considerable sensitivity and specific structural information for derivatized amines, making them the most efficient detection tools for low concentration metabolites. Additionally, MS/MS is used where isotopically labeled BAs are employed as internal standards (internal standard lowers the matrix interference in complex matrices). Here, the limits of quantification (LOQs) were at 50 ng/kg level. In another study, LC-MS/MS was employed to determine BAs-3,5-Bia-(trifluoromethyl)phenylisothiocyanate LODs ranged from 2.0 to 4.3 ng/L [3].

The HPLC-APCI-MS technique was employed in a study to determine the BAs present in ready-to-eat baby food. During the study, the pre-column derivatization with DNS-Cl was done. The two most toxic BAs with HA and TA were quantified at concentration levels as low as 2 ng/g. Simultaneous analysis of cadaverine, putrescine, spermine, and spermidine was also done simultaneously with low LODs [14].

18.5 GAS CHROMATOGRAPHY-MASS SPECTROMETRY

GC-MS and GC-MS/MS are commonly used techniques in clinical chemistry and allow the identification of analytes with even minute structural differences because of high resolution and high sensitivity. GC analysis requires a very low sample volume (1–15 μL). However, it is not used as a routine tool for BAs quantification clinically due to Bas' non-volatile and thermolabile structure. Therefore, before employing the sample for GC separation, a derivatization step is required to produce volatile and stable BAs adducts that fit ionization techniques. Proper derivatization is needed to make a single and stable derivative that can be subjected to an appropriate detection mode as multiple derivatives' formation hinders the analytes' identification and quantification. However, there are scanty procedures available in the literature for the same. Moreover, the sample preparation for extraction and derivatization procedures are time-consuming and labor-intensive [20].

The procedure involves an injection of the analyte solution into the GC apparatus. It gets evaporated and enters a capillary column where analyte separation occurs as carrier gas flows through the capillary. At the end of the capillary, analytes are detected via MS. Electron impact (EI) ionization and chemical ionization are ionization methods used for gas- and vapor-phase analysis [29,30]. The method of choice for amine analysis is the EI ionization mode at 70 eV and a quadrupole mass analyzer. In quantitative analysis, the selective ion monitoring (SIM) mode has been used.

In a study, 22 amines were quantified in port and grape wine using GC-MS. The MS parameters were ion source at 230°C, EI ionization at 70 eV, MS quadrupole at 150°C, and SIM as the MS operating mode. The method was found to be sensitive, with high separation, low LOD, and linearity. Huang et al. and Mohammadi et al. employed GC-MS to determine PUT, CAD, HA, and TA in fish samples and cheese samples [29].

Di- and polyamines and aromatic amines with heptafluorobutyric anhydride derivatization were determined in port wine and grape juice using GC-MS methods. Also, HA was quantified as pentafluorobenzyl-HA in the same matrices. Similarly, primary alkylamines were determined in wines with pentafluorobenzaldehyde derivatization using GC-MS [31].

Neofotistos et al. used a DLLME method simultaneously with a derivatization process to determine 18 BAs in beer samples using GC-MS with LODs of 0.3–2.9 μg/L. Recently, Huang et al. determined BAs in fish samples employing SPME coupled with GC-MS with LODs of 2.98–45.3 μg/kg [3]. Vanda et al. used GC-MS to quantify volatile and non-volatile BAs in port wine after derivatization with isobutyl chloroformate [32].

18.6 CAPILLARY ELECTROPHORESIS-MASS SPECTROSCOPY

Capillary electrophoresis (CE) is a popular method for quality control, food characterization, monitoring of food processing, or spoilage detection in food. It has been reported to analyze BAs in fish, wine, beer, vegetables, milk, cheese, and meat [33]. CE has been broadly utilized for routine analysis owing to its high resolving power, rapid separation speed, small sample volume, and flexibility in choosing separation modes. Nevertheless, sample introduction in nanoliters lowers CE sensitivity.

When CE is coupled with MS, the detection limit is improved significantly as MS is an analytical technique with high resolution and sensitivity to obtain structural information of an unknown sample. However, attention is to be paid to interfacing the technique used and the separation methodology selected. The popular interfaces used in CE-MS systems are ESI, inductively coupled plasma, matrix-assisted laser desorption ionization (with an off-line mode), and the utilization of microfluidic devices. Successfully employed CE separation modes with MS are capillary gel electrophoresis, capillary zone electrophoresis (CZE), capillary isoelectric focusing, capillary isotachophoresis, capillary electrochromatography, and micellar electrokinetic chromatography. Recently, all types of mass analyzers have been used with CE detection namely IT-MS, TOF MS, FT-ion cyclotron resonance mass spectrometer, quadrupole and triple quadrupole mass spectrometer (Q/QQQ), and Orbitrap MS [34].

Nevertheless, CE's concentration detection limit is greater than that of ion chromatography, and HPLC is attributed to the small quantity of injected samples and shorter optical path length. Additionally, CE's reproducibility is also less than HPLC and IC due to the electroosmotic flow (EOF) by solute adsorption on capillary walls. Under physiological conditions, BAs exist in protonated forms. Hence, their separation is done under acidic conditions with small EOF. Despite that, BAs separation is problematic with CZE due to low hydrophilic property, similar dissociation constant (K_a) values, and adsorption on the capillary wall [35].

Daniel et al. used the CE-MS/MS method to determine BAs in commercial samples of beer and wines. The separation was done in polyvinyl alcohol-coated silica capillary to suppress EOF. Separation efficiency and detection were enhanced using ESI-MS/MS with QQQ as a mass analyzer. The method determined and quantified putrescine, spermine, spermidine, cadaverine, HA tryptamine, TA, PE, and urocanic acid with LOD 1–2 g/L and LOQ ranging from 3 to 8 g/L. The method presented good precision and sensitivity and was simple, with the requirement of a sample in a small quantity and reagent in low concentration [32,36].

Santos et al. separated and determined nine BAs with HA, cadaverine, ethanolamine, isopropylamine, isoamylamine, pentaylamine, PE, and TA from red and white wine by using the CE-ESI-MS method [37].

Simo et al. used CE-IT-MS and CE-TOF-MS to analyze BAs in wines. They also compared these two methods in terms of LOD, the total number of BAs detected, and their quantification in various wine samples. The accuracy of quantification of CE-MS was confirmed by comparing its result with those obtained from standard HPLC procedures. Both the methods determined BAs in wines, but the results of CE-TOF-MS were better than CE-IT-MS as the former one allowed the determination of a higher number of BAs with higher sensitivity, and its sensitivity was comparable to that of standard HPLC-FD. Nevertheless, CE-TOF-MS was advantageous over HPLC-FD in terms of a fivefold faster time of analysis without the requirement of derivatization [33].

Woźniakiewicz et al. used extraction-free CE-MS and direct immersion-SPME combined with gas chromatography-mass spectrometry (DI-SPME-GC-MS) to determine BAs in wine and fruit wine. CE-MS sample preparation was easier and faster than GC-MS, but the LOD values for GC-MS (3.2 ng/mL) were lower than in CE-MS (25 mg/mL). However, the recovery percentage of BAs for both methods was similar [38].

18.7 ION-EXCHANGE CHROMATOGRAPHY COUPLED WITH MASS SPECTROSCOPY

IC-MS is a combination of conventional ion-exchange chromatography with MS. Components separation with IC is attributed to the ionic interaction between functional groups on a resin-based stationary phase and analytes present in the mobile phase. Elution occurs by exchanging analyte with mobile-phase ions typically of higher ion strength (hydroxide ions for anion exchange and protons for cation exchange). IC is made compatible with MS by situating an ion suppressor between the electrospray ion source of MS and the ion chromatography system. When chromatographic eluents pass through the suppressor, electrochemical conversion occurs, i.e., in anion-exchange mode, hydroxide ions are converted to water. In cation exchange, protons are converted to water, proving an electrospray-compatible chromatographic eluent that can be analyzed by MS directly [39].

Saccani et al. used cation-exchange chromatography and suppressed conductivity coupled with mass spectroscopy to determine BAs from processed meat products. They used methanesulfonic acid to extract the amines from the muscle tissue without any requirement of additional derivatization or sample cleanup. Separation of BAs was done on an IonPac CS17 cation-exchange column with gradient elution with a mass spectrometer and suppressed conductivity as the detector. The method separated cadaverine, putrescine, HA, agmatine, phenethylamine, and spermidine in processed meat products with high detection sensitivity with MS [40]. Methanesulfonic acid used in the procedure can form a solid deposit in the ion source of the MS detector, hindering the analysis, which necessitated the use of a suppressor. However, the use of suppressor removed TA which might be a problem when determining the BAs in cheese matrices. Scavnicer et al. employed ion chromatography coupled with MS/MS detection in a non-suppressed mode to determine BAs, including TA which is crucial in spoilage in cheese matrices. The BAs were extracted with water and separated using a cation-exchange column with gradient elution using a mixture of formic acid (1.00 M) and deionized water as eluents. MS/MS with multiple monitoring modes was used for detection. The technique separated trimethylamine, putrescine, cadaverine, HA, 2-PE, spermine, spermidine, tryptamine, agmatine, and TA with trimethylamine, TA, and cadaverine showing the best recoveries [41].

18.8 CONCLUSION

BAs content in various foods has been studied extensively over the last few decades. Comprehensive research has been carried out to develop and validate methods that can substantially determine and quantify these compounds without much complexity. However, many hurdles need consideration while adopting BAs determination and quantification practices, such as their polar nature, absence of chromophore or fluorophore in aliphatic amines, present in low concentration and complex matrices. Many techniques such as LC, GC, CE, and IC have been employed coupled with various detectors such as UV, electrochemical, FL, etc. However, proper pre- or post-column derivatization of the BAs exists. Haplessly, derivatization is a time-consuming process and may form unstable derivatives. However, when such methods with LC, GC, CE, and IC are coupled with an MS detector, the derivatization step can be phased out. The reliability, accuracy, and sensitivity of the method are enhanced. Nevertheless, there is an urgent need to develop techniques that can determine and quantify BAs without the need for derivatization, with low LODs and LOQs and less time-consuming, and with LODs and LOQs.

ACKNOWLEDGMENTS

The authors are thankful to the library of Process Noura bint Abdulrahman University, Riyadh, Kingdom of Saudi Arabia.

REFERENCES

1. Almeida, C., Fernandes, J.O., & Cunha, S.C. A novel dispersive liquid-liquid microextraction (DLLME) gas chromatography mass spectrometry (GC-MS) method for the determination of eighteen biogenic amines in beer. *Food Control*, 2012, 25, 380–388.
2. Ozogul, Y., & Ozogul, F. Chapter 1: Biogenic amines formation, toxicity, regulations in food, in *Biogenic Amines in Food: Analysis, Occurrence and Toxicity*, Saad, B. and Tofalo, R. Ed., 2019, Royal Society of Chemistry, London, pp. 1–17.

3. Antonios-Dionysios, G.N., Aristeidis, S.T., Danezis, G.P., & Proestos, C. *Emerging Trends in Biogenic Amines Analysis, Biogenic Amines, Charalampos Proestos*, IntechOpen, 2019, DOI: 10.5772/intechopen.81274.
4. Ochi, N. Simultaneous determination of eight underivatized biogenic amines in salted mackerel fillet by ion-pair solid-phase extraction and volatile ion-pair reversed-phase liquid chromatography-tandem mass spectrometry. *Journal of Chromatography A*, 2019, 1601, 115–120.
5. Yoon, H., Park, J.H., Choi, A., Hwang, H.J., & Mah, J.H. Validation of an HPLC analytical method for determination of biogenic amines in agricultural products and monitoring of biogenic amines in Korean fermented agricultural products. *Toxicology Research*, 2015, 31, 299–305.
6. Nalazek-Rudnicka, K., & Wasik, A. Development and validation of an LC-MS/MS method for the determination of biogenic amines in wines and beers. *Monatshefte für Chemie*, 2017, 148(9), 1685–1696.
7. de Jong, W.H., de Vries, E.G., Kema, I.P. Current status and future developments of LC-MS/MS in clinical chemistry for quantification of biogenic amines. *Clinical Biochemistry*, 2011, 44(1), 95–103.
8. Durak-Dados, A., Michalski, M., & Osek, J. Histamine and other biogenic amines in food. *Journal of Veterinary Research*, 2020, 64(2), 281–288.
9. Kelly, M.T., Blaise, A., & Larroque, M. Rapid automated high performance liquid chromatography method for simultaneous determination of amino acids and biogenic amines in wine, fruit and honey. *Journal of Chromatography A*, 2010, 1217(47), 7385–7392.
10. Zhang, Y.J., Zhang, Y., Zhou, Y., Li, G.H., Yang, W.Z., & Feng, X.S. A review of pretreatment and analytical methods of biogenic amines in food and biological samples since 2010. *Journal of Chromatography A*, 2019, 1605, 360–361.
11. Kabir, A., Mocan, A., Piccolantonio, S., Sperandio, E., Ulusoy, H. I., & Locatelli, M. Analysis of amines. *Recent Advances in Natural Products Analysis*, 2020, 569–591.
12. Moret, S., & Conte, L.S. High-performance liquid chromatographic evaluation of biogenic amines in foods an analysis of different methods of sample preparation in relation to food characteristics. *Journal of Chromatography A*, 1996, 729 (1–2), 363–369.
13. Munir, M. A., Badri, H.K., The importrance of derivatizing in chromatography applications for biogenic amine detetion in food and beverages. *Journal of Analytical Methods in Chemistry*, 2020, DOI: 10.1155/2020/5814389.
14. Czajkowska-Mysłek, A., & Leszczyńska, J. Liquid chromatography-single-quadrupole mass spectrometry as a responsive tool for determination of biogenic amines in ready-to-eat baby foods. *Chromatographia*, 2018, 81(6), 901–910.
15. Zarei, M., Fazlara, A., Najafzadeh, H., & Karahroodi, F.Z. Efficiency of different extraction solvents on recovery of histamine from fresh, frozen and canned fish. *Journal of Food Quality and Hazards Control*, 2014, 1, 72–76.
16. Han, S.Y., Hao, L.L., Shi, X., Niu, J.M., & Zhang, B. Development and application of a new QuEChERS method in UHPLC-QqQ-MS/MS to detect seven biogenic amines in Chinese wines. *Foods*. 2019, 8(11), 552.
17. Baranowska, I., & Płonka, J. Simultaneous determination of biogenic amines and methylxanthines in foodstuff – sample preparation with HPLC-DAD-FL analysis. *Food Analytical Methods*, 2015, 8, 963–972.
18. Klampfl, C.W. Determination of underivatized amines by capillary electrophoresis and capillary electrochromatography. *Quantitation of Amino Acids and Amines by Chromatography – Methods and Protocols*, 2005, 5, 525–558.
19. Plakidi, E.S., Maragou, N.C., Dasenaki, M.E., Megoulas, N.C., Koupparis, M.A., & Thomaidis, N.S. Liquid chromatographic determination of biogenic amines in fish based on pyrene sulfonyl chloride pre-column derivatization. *Foods*, 2020, 9(5), 609.
20. Plenis, A., Olędzka, I., Kowalski, P., Miękus, N., & Bączek, T. (2019). Recent trends in the quantification of biogenic amines in biofluids as biomarkers of various disorders: a review. *Journal of Clinical Medicine*, 2019, 8(5), 640.
21. Ho, P.C. HPLC of amines as 9-fluorenyimethyl chloroformate derivatives. *Journal of Chromatography Library*, 2005, 70, 471–501.
22. Koga, R., Miyoshi, Y., Todoroki, K., & Hamase, K. Amino acid and bioamine separations, in *Liquid Chromatography*, Riekkola, M.L., Ed., 2017, Elsevier, Amsterdam, pp. 87–106.
23. Busto, O., Guasch, J., & Borrull, F. Determination of biogenic amines in wine after precolumn derivatization with 6-aminoquinolyl-N-hydroxysuccinimidyl carbamate. *Journal of Chromatography A*, 1996, 737(2), 205–213.

24. Durden, D.A. High resolution and metastable mass spectrometry of biogenic amines and metabolites. in *Amines and Their Metabolites*. Neuromethods, vol 2. Boulton, A.A., Baker, G.B., Baker, J.M. Eds., 1985, Humana Press, 325. https://doi.org/10.1385/0-89603-076-8.
25. Urban P.L. Quantitative mass spectrometry: an overview. *Philosophical Transactions of the Royal Society Series A, Mathematical, Physical, and Engineering Sciences*, 2016, 374–379.
26. Bomke, S., Seiwert, B., Dudek, L., Effkemann, S., & Karst, U. Determination of biogenic amines in food samples using derivatization followed by liquid chromatography/mass spectrometry. *Analytical and Bioanalytical Chemistry*, 2008, 393(1), 247–256.
27. Malik, A.K., Blasco, C., & Picó, Y. Liquid chromatography-mass spectrometry in food safety. *Journal of chromatography A*, 2010, 1217(25), 4018–4040.
28. Kamankesh, M., Mohammadi, A., Mollahosseini, A., & Seidi, S. Application of a novel electromembrane extraction and microextraction method followed by gas chromatography-mass spectrometry to determine biogenic amines in canned fish. *Analytical Methods*, 2019, 11, 1898–1907.
29. Jubele, A. Chromatographic determination of amines in food samples. M.Sc. thesis University of Helsinki Department of Chemistry Laboratory of Analytical Chemistry, 2018.
30. Fernandes, J.O., Judas, I.C., Oliveira, M.B. et al. A GC-MS method for quantitation of histamine and other biogenic amines in beer. *Chromatographia*, 2001, 53, S327–S331.
31. Cunha, S.C., Faria, M.A., & Fernandes, J.O. Gas chromatography-mass spectrometry assessment of amines in Port wine and grape juice after fast chloroformate extraction/derivatization. *Journal of Agricultural and Food Chemistry*, 2011, 59(16), 8742–8753.
32. Miranda, A., Leça, J.M., Pereira, V., & Marques, J.C. Analytical methodologies for the determination of biogenic amines in wines: an overview of the recent trends. *Journal of Analytical, Bioanalytical and Separation Techniques*, 2017, 2(1), 52–57.
33. Simo, C., Moreno-Arribas, M.V., & Cifuentes, A. Ion-trap versus time-of-flight mass spectrometry coupled to capillary electrophoresis to analyze biogenic amines in wine. *Journal of Chromatography A*, 2008, 1195(1–2), 150–156.
34. Jiang, Y., He, M.-Y., Zhang, W.-J., Luo, P., Guo, D., Fang, X., & Xu, W. Recent advances of capillary electrophoresis-mass spectrometry instrumentation and methodology. *Chinese Chemical Letters*. 2017, 28, 1640–1652.
35. Chiu, T.-C., Lin, Y.-W., Huang, Y.-F., Chang, H.-T. Analysis of biologically active amines by CE. *Electrophoresis*, 2006, 27(23), 4792–4807.
36. Daniel, D., dos Santos, V.B., Vidal, D.T.R., & do Lago, C.L. Determination of biogenic amines in beer and wine by capillary electrophoresis–tandem mass spectrometry. *Journal of Chromatography A*, 2015, 1416, 121–128.
37. Santos, B., Simonet, B.M., Ríos, A., & Valcárcel, M. Direct automatic determination of biogenic amines in wine by flow injection-capillary electrophoresis-mass spectrometry. *Electrophoresis*, 2004, 25(20), 3427.
38. Wozniakiewicz, M., Wozniakiewicz, A., Nowak, P.M., Kłodzińska, E.; Namieśnik, J.; Płotka-Wasylka, J. CE-MS and GC-MS as "green" and complementary methods for the analysis of biogenic amines in wine. *Food Analytical Methods*, 2018, 11, 2614–2627.
39. Walsby-Tickle, J., Gannon, J., Hvinden, I. et al. Anion-exchange chromatography mass spectrometry provides extensive coverage of primary metabolic pathways revealing altered metabolism in IDH1 mutant cells. *Communications Biology*, 2020, 3, 247.
40. Saccani, G., Tanzi, E., Pastore, P., Cavalli, S., & Rey, M. Determination of biogenic amines in fresh and processed meat by suppressed ion chromatography-mass spectrometry using a cation-exchange column. *Journal of Chromatography A*, 2005, 1082(1), 43–50.
41. Ščavničar, A., Rogelj, I., Kočar, D., Köse, S., & Pompe, M. Determination of biogenic amines in cheese by ion chromatography with tandem mass spectrometry detection. *Journal of AOAC International*. 2018, 101, DOI: 10.5740/jaoacint.16-0006.

19 Mass Spectrometry of Food Preservatives

Emmanouil D. Tsochatzis
Aarhus University

CONTENTS

19.1 Introduction 357
19.2 Mass Spectrometry 358
19.3 Types of Food Preservatives 359
19.3.1 Natural Food Preservatives 359
19.4 Chemical Food Preservatives 360
19.4.1 Sulfur Dioxide and Sulfites 360
19.4.2 Nitrite and Nitrate Salts 361
19.4.3 Sorbic Acid and Esters 363
19.4.4 Benzoic Acid and Esters 367
19.4.5 p-Hydroxybenzoate Alkyl Esters 367
19.4.6 Antibiotics 371
19.5 Conclusions 373
References 373

19.1 INTRODUCTION

A wide range of physical, chemical, enzymatic, and microbial reactions can cause quality loss and food spoilage. Food preservatives are being used to prevent spoilage during storage, distribution, retailing, and consumption. These compounds are being intentionally added to prevent deteriorative reactions and thus protect the quality and extend the shelf life of foods and beverages. Preservatives are used in controlled amounts and usually at low levels, ranging from parts per million (ppm) to 1%–3% by weight.

Preservatives are added to food to fight spoilage caused by bacteria, molds, fungus, and yeast. Preservatives can keep food fresher for longer periods, extending its shelf life. Food preservatives are also used to slow or prevent changes in color, flavor, or texture and delay rancidity.

The use of preservatives in food products is strictly studied, regulated, and monitored by the U.S. Food and Drug Administration (FDA) and Directorate-General for Health and Food Safety (DG SANTE). Federal regulations require evidence that food additives are safe for their intended use. Preservatives in foods are subject to an ongoing safety review by the FDA as scientific understanding, and testing methods continue to improve. Hence, the FDA defines an antimicrobial agent as a substance used to preserve food by preventing or retard the growth of pathogenic and deteriorative microorganisms and subsequent spoilage. Most available antimicrobial agents do not fully inactivate microbial populations but rather work as growth inhibitors providing inhospitable environments for microorganisms preventing their development [1].

A "chemical preservative" may be used in non-standardized foods if the preservative is a suitable substance that is not a food additive as this term is defined in Section 201(s) of the Federal Food,

DOI: 10.1201/9781003091226-22

Drug, and Cosmetic Act; or if a food additive, it is used in conformity with the regulations of 21 CFR 172 and with good manufacturing practices. Furthermore, the use of chemical preservatives in foods is either a "Generally recognized as safe (GRAS)" for such use, or it is a food additive covered by food additive regulations prescribing conditions of safe use; not used in such a way as to conceal damage or inferiority or to make the food appear better or of greater value than it is; properly declared on the label of the food in which used [2].

Codex Alimentarius defines food additives as "any substance that its intentional addition of which to a food aiming for a technological (including organoleptic) purpose in the manufacture, processing, preparation treatment, packing, packaging, transport or holding of such food results, or may be reasonably expected to result, in it or its byproducts becoming a component of the food or otherwise affecting the characteristics of such foods" [3–5].

According to regulation EC No. 1333/2008, food additives are defined as substances that are not normally consumed as the food itself but are added to food intentionally for a technological purpose described in this regulation, such as the preservation of food [6]. In Annex I of this EU regulation, the functional classes of food additives in foods and food additives and food enzymes are defining "preservatives," as the substances which prolong the shelf life of foods by protecting them against deterioration caused by microorganisms and/or which protect against the growth of pathogenic microorganisms [6].

19.2 MASS SPECTROMETRY

Mass spectrometry (MS) is an emerging and powerful analytical tool for studying foods, which has an important role in the analysis of foods. Its major strength lies in measuring complex food mixtures, ranging from low molar masses (M) (i.e., sorbic acid) molecules to larger molar masses (M), such as peptides (i.e., Nisin). MS-based methods provide the capability for detecting various food components, including nutrients (e.g., amino acids, vitamins, carbohydrates, fatty acids, minerals, and proteins), additives (e.g., colorants, aromas, and preservatives), and contaminants (e.g., pesticides, plasticizers, and polycyclic aromatic hydrocarbons).

Recent advances in MS technology have focused on the development of (i) ionization source with high ionization efficiency and minimal in-source fragmentation; (ii) high-resolution mass spectrometry (HRMS) for accurate mass measurement of complex food matrices, identification and structural elucidation of chemical compounds; (iii) hybrid mass analyzer to elucidate the molecular structure through a characteristic fragmentation pattern; and (iv) chromatographic techniques that minimize signal suppression and reduce peak overlap for accurate quantitation. These achievements greatly expand the capability of MS-based approaches for food analysis [7].

In liquid chromatography (LC), the use of multiple reaction monitoring (MRM) or single reaction monitoring (SRM) is preferred. However, apart from these technologies, also the application of HRMS is of raising interest, despite the higher cost, such as quadrupole time-of-flight (TOF) MS or Orbitrap MS (Thermo®). The advantages of these technologies are the evaluation of molecular ions (*m*/*z*) or fragments with very high accuracy and precision, based on high resolution or resolving power. The latter is defined by a ratio (equation 19.1), for a single peak made up of singly charged ions at mass *m* in a mass spectrum; it may be expressed as the ratio:

$$R = \frac{m}{\Delta m} \tag{19.1}$$

m is the mass of the singly charged ion in a mass spectrum and Δm is the width of the peak at a height, which is a specified fraction of the maximum peak height (for example, 50%), or full width at half maximum (FWHM). Subsequently, this facilitates proper identification and elucidation of chemical molecules and fragmentation patterns.

19.3 TYPES OF FOOD PRESERVATIVES

Many of the existing food preservatives are multifunctional, serving to help extend different features as additives. Thus, they can act as preservatives by extending food's shelf life by inhibiting oxidation, browning, or enzymatic reactions or ensuring microbially and consumer's safety by bacterial inhibition of microorganisms or impacting and minimizing pathogenic bacteria [8].

Food preservatives may come in two main forms. They can be originated either from natural sources (natural food preservatives) or from synthetically produced/added (chemical food preservatives) sources. However, at an EU level, all food additives, either natural or not, shall be assessed for their potential risks to humans [6].

Several factors have to be considered when selecting a food preservative for a specific food product, including the type of microorganism present (pathogen or deteriorative), pH, composition, and the product's shelf life requirements (time), potential use, and final application. It is also important to consider possible effects by altering the product's organoleptic properties such as flavor or color [8,9]. In this chapter, we will focus mainly on the second group of chemical food preservatives that present a particular interest from a scientific and risk perspective.

19.3.1 NATURAL FOOD PRESERVATIVES

The use of natural antimicrobial compounds in food has gained great attention and popularity in the last years, especially from the consumers and the food industry, driven by the potential negative impact of synthetic chemical preservatives on health. In addition, the benefits of natural preservatives have motivated researchers to develop natural food products. Natural compounds with antimicrobial activity can be obtained from different sources, such as plants, animals, bacteria, fungi, and algae. Nisin, natamycin, and essential oils are some natural preservatives used in food processing. The antibiotics nisin and natamycin are both obtained from bacteria, while essential oils are plant-derived.

Herbs and spices have certain functional characteristics related to antimicrobial effects. Research in this area has proven the microbiological efficacy of many plant-derived compounds. Essential oils are considered as one of the most promising natural alternatives to synthetic preservatives. They consist of aromatic and volatile oily extracts obtained from different plant tissues, including flowers, buds, roots, bark, and leaves. Essential oils from thyme, clove, balm, ginger, oregano, rosemary, basil, coriander, citric fruits, marjoram, and lemongrass have shown a greater potential to be used as antibacterial and antifungal agents. Given that essential oils have been extensively used for centuries in food products, they are classified as GRAS and present a very low risk of microorganism resistance development. Despite their approval for use in the United States and in Europe, currently, few commercially available essential oils are marketed as food preservatives; nonetheless, their usage is expected to rise with the trending demand for natural and minimally processed foods.

Phenylpropenes and terpenoids constitute the most antimicrobial compounds found in essential oils, the most active being thymol, carvacrol, and eugenol. Molecular targets and action mechanisms are not yet fully understood, although many essential oil constituents have been proposed to act on the cell membrane. Aldehydes, such as cinnamaldehyde, and other nonphenolic substances may be responsible for important synergic activity as suggested by the greater antimicrobial activity of crude essential oils compared to blends of their major individual components.

Concentrations needed to achieve a significant antibacterial effect range between 0.5 and 20 mL/g in foods and about 0.1–10 mL/mL in fruit and vegetable washing solutions. Microbial inhibition by essential oils has greatly focused on dairy and meat products where they may also impart flavor to the final product. Special considerations have to be taken when using essential oils as preservative agents. Given the hydrophobicity of many essential oil components, application in foods with a high-fat content may impair antimicrobial activity. As a result, much higher inhibitory concentrations are typically required to reproduce in food the effects observed *in vitro*. This has been reported from twofold in semi-skimmed milk to 100-fold in soft cheese. In certain food products, the required inhibitory concentrations can strongly affect the product's flavor profile and even render it unpalatable [8].

19.4 CHEMICAL FOOD PRESERVATIVES

Since ancient years acids are the most effective and widely used preservatives, such as acetic or propionic acid. However, nowadays, additional organic acids are being used effectively and efficiently as food preservatives, either in their acidic form or in their weak lipophilic form (salts). However, except for salts, other types of antimicrobial/preservatives are being used, such as sulfur dioxide (SO_2) and sulfites, nitrites and nitrates, and *p*-hydroxybenzoic acid esters (parabens), or antibiotics. There are no antimicrobial preservatives with sufficient activity at varied pH values, except parabens. Emerging concerns on the extensive use of these types of preservatives and their effects on human health, and their inefficacy in pH levels closer to neutrality, have generated a market demand for natural alternatives.

19.4.1 Sulfur Dioxide and Sulfites

SO_2 and its derivatives have been extensively used in food products as food preservatives. They are added to inhibit non-enzymatic browning, enzymatic reactions, and control and inhibit the growth of microorganisms [10]. Approved forms of sulfite additives apart from SO_2 are also sodium sulfite (Na_2SO_3) (SO_3^-, HSO_3^-), sodium bisulfate, sodium metabisulfite ($Na_2S_2O_5$), potassium metabisulfite ($K_2S_2O_5$), calcium sulfite, calcium bisulfate, and potassium bisulfate. The concentration of sulfite additives is evaluated in terms of SO_2 equivalents since sulfite salts release molecular SO_2. The concentration in foods is highly dependent on the maximum approved levels for specific products. SO_2 concentrations range from 10 ppm in wines to a maximum of 2,000 ppm in dried fruits [10]. Sulfites are a family of food preservatives that commonly include sulfur dioxide (SO_2), sodium sulfite (Na_2SO_3), sodium metabisulfite ($Na_2S_2O_5$), potassium metabisulfite $K_2S_2O_5$, sodium bisulfite ($NaHSO_3$), and potassium bisulfite ($KHSO_3$). Sulfite is a challenging molecule to analyze in its highly unstable free form, leading to implementing a chemical derivatization step before its analysis [11–14].

Currently, three main methods exist (pre-treatment) for the determination of sulfites in foods, which are: (i) direct determination of sulfite (SO_3^{2-}) in the tested sample solution, (ii) the oxidation of sulfite to sulfate (SO_4^{2-}), and (iii) the derivatization of sulfite [13]. Hence, the latter approach is the one that is preferred the most to be applied in hyphenated techniques, combined with MS. Recent advances in this field are sulfite ion reaction with formaldehyde, which forms a stable compound, hydroxy methylsulfonate (HMS). The latter is more stable than its free form (Figure 19.1). However, any derivatization procedure that requires additional and more extensive sample preparation shall not be underestimated.

The liquid chromatography-tandem mass spectrometry (LC-MS/MS) was operating in negative electrospray ionization mode (ESI–) with curtain gas set to 35 arbitrary units (au), the collisionally activated dissociation gas was run at medium, an ion spray voltage of −1200 V was used, the source temperature was 550°C, gas 1 pressure was 70 au, and gas 2 pressure was 40 au. The MS/

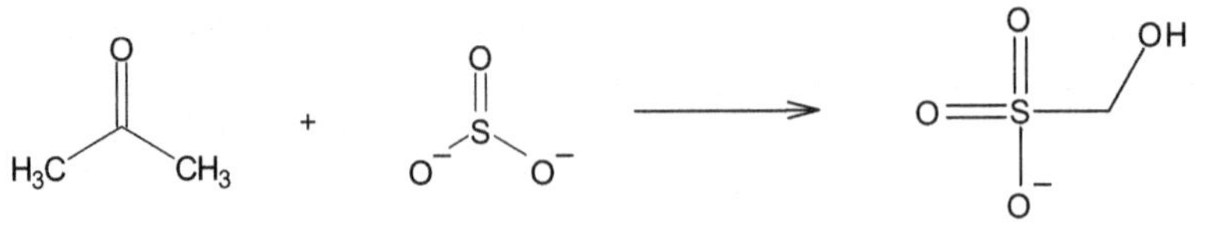

FIGURE 19.1 Chemical reaction of formaldehyde with sulfite ion for the formation of HMS (hydroxy methylsulfonate). (Adapted by Robbins et al. [11].)

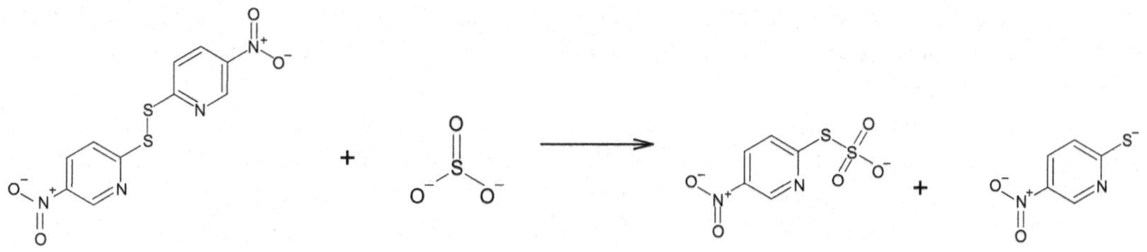

FIGURE 19.2 Chemical reaction of DTNP with sulfite ion. (Adapted by Yang et al. [13].)

MS data were acquired using the MRM mode (unscheduled) with a unit resolution of both Q1 and Q3, a dwell time of 170 ms, declustering potential (DP) of −25.0 V, entrance potential (EP) of −4.0 V, collision energy (CE) of −15.0 V, and collision cell exit potential of −6.0 V [11]. The monitored transition for HMS was set from 111 amu (Q1) to 81 amu (Q3). This transition represents the molecular ion of HMS fragmented into the methylene group attached to the sulfur and the hydroxy group (Figure 19.1). The HMS reaction and quantification results can be transformed into ppm of SO_2, following stoichiometrical calculation of SO_3^{2-} to SO_2. The LC-MS/MS method, based on the above-mentioned formation of HMS, has been officially registered to the FDA compendium of analytical laboratory methods (Method Nr. C-004.02) [14]. It has been applied to be compared with standard titrimetric methods [15] and within an inter-laboratory comparison [12], highlighting robustness and stability.

Another methodology for the stoichiometric formation of novel dithiol-compound ($RSSO_3^-$) and thiol (RS^+) substances is based on derivatization (pre-column) with 2,2′-dithiobis(5-nitropyridine) (DTNP) (Figure 19.2).

The generated thio-compounds are also being analyzed in negative ESI mode. The transition of sulfite–DTNP is 310 amu, and the main MS products are the sulfite-disulfuro fragment (*m/z* 235) and the nitrosulfuro thiol (*m/z* 155) [13].

Another method for analyzing sulfite is reported for dry fruits and vegetables, applying liquid chromatography inductively coupled plasma mass spectrometry (LC-ICP-MS), following *in situ* derivatization of the sulfite to HMS [16], to retain stability throughout the analysis. The LC-ICP-MS methodology focuses on the determination of some metal and metalloid species due to its easy connection to LC. The LC conditions were based on an ion-exchange column, an isocratic run of NH_4NO_3, and formaldehyde. During ICP-MS, the radiofrequency (RF) power was set to 1,400 W. Analog stage voltage, pulse stage voltage, and lens voltage were 1,670, 1,055, and 6.80 V, respectively. Plasma and auxiliary argon flow rates were 15 and 1.1 L/min, respectively. Dwell time was set at 500 ms. In this methodology, the sulfite is detected as $^{32}S^{16}O^+$ by ICP-MS in dynamic reaction cell O_2 gas mode. However, the high cost per analysis shall be taken into consideration [16].

In another methodology, the determination of the natural abundance sulfur isotope ratios in foods containing sulfite preservatives was reported. There are four stable isotopes of sulfur found in nature with atomic masses of 32, 33, 34, and 36. The major isotopes of interest are ^{32}S (95.0%) and ^{34}S (4.215% mean natural abundance). The natural abundance range for ^{34}S is reported to be from 4.05% to 4.45% [17].

19.4.2 Nitrite and Nitrate Salts

Salts of nitrite (NO_2^-) and nitrate (NO_3^-) are being used in curing mixtures as food additives, aiming to develop, fix the appearance of food products, especially meat, as well as to inhibit the growth of microorganisms [10]. Nitrites (NO^{2-}) are nitrogen-oxygen chemical units combined with several organic and inorganic compounds. Nitrites are one of the most talked-about ions in the meat industry due to their applications in curing and preserving various processed meat and poultry products. In the meat industry, they are usually mixed with meat binders or cure ingredients and are added to various meat products as sodium or potassium salt. Thus, the addition of sodium nitrite has led to the production of meat with better quality and safety as it extended storage life and improved the desirable color/taste [18].

However, it is proven that nitrite is formed during the reduction of nitrate of bacteria in the saliva and under a gastric acid condition in reaction with a secondary amine to form the carcinogenic nitrosamines, noxious compounds present in processed meats. The formation of these *N*-nitrous type carcinogens (nitrosamines) is among the most studied toxic compounds in these products [19–21]. It is also suggested that nitrite is 10 times more toxic than the nitrate and its potential as a human carcinogen (Group 2A) by the International Agency for Research on Cancer [22].

A ultra-performance liquid chromatography–mass spectrometry (UPLC-MS) method is reported for the quantitative determination of nitrite to food products (meat), based on UPLC-MS/MS. The monitoring conditions were also optimized. The parameters optimized were capillary voltage 3.0 kV, cone voltage 40 V, source temperatures 120°C, and desolvation temperature 300°C. The cone gas flow and the desolvation gas flow were set at 60 and 600 L/h, with the MS range being set from 30 to 70 amu. The parent ion of *m/z* 46 amu corresponds to the NO_2^- ion [18].

A similar UPLC-MS method is reported for the quantitative determination of nitrite to food products. The mass spectrometer was functioned at negative ionization mode to generate ions at *m/z* 62 for nitrate. The monitoring conditions were also optimized. The parameters optimized were capillary voltage 3.0 kV, cone voltage 40 V, source temperatures 120°C, and desolvation temperature 300°C. The cone gas flow and the desolvation gas flow were set at 60 and 600 L/h, with the MS range being set from *m/z* 50 to 100 amu. It was highlighted that analysis in the negative ionization mode presented much higher ions abundance compared to positive ionization. Most abundant ions were selected from the full scan (FS), with *m/z* 62 corresponding to the NO_3^- ion. Selected ion reaction was used for higher detection sensitivity [23].

Also, recently, the analysis of NO_3^- in food products, as a part of a multi-analyte method for additives quantitative determination, was reported. The analysis was performed in a heated electrospray interface in negative ion mode at a spray voltage of –2,500 V on a single-stage Exactive instrument (Thermoa®, Bremen, Germany). The sheath gas flow (nitrogen) was set to 50 units, the aux gas flow to 10 arbitrary units, the sweep gas flow to 0 arbitrary units, applying a capillary temperature of 160°C and a heater temperature of 250°C. The spray voltage was set to –37.5 V, the tube lens voltage to –125 V, and the skimmer voltage to –36 V, without applying quadrupole. Explicitly for nitrate, the ESI ion was the e^+, with a monoisotopic mass (*m/z*) of 61.988, while the resolving power was 25,000 FWHM, posing a maximum injection time of 50 ms. A limitation and drawback is the monitoring of NO_2^-. The utilized HRMS mass resolving power in combination with narrow mass windows produced a significantly higher selectivity than the obtainable selectivity when using conventional conductivity-based ion chromatography (IC) or ultraviolet detection, where due to the low mass of this ion (*m/z* 46), it falls below the mass range of the specific instrument (Orbitrap; *m/z* 50) [24].

However, the interesting point and novelty of this work, concerning the MS detection of nNO_3–, was the protonation approach, by applying dibutyl amine (DBA) in the presence of 1,1,1,3,3,3-hexafluoro-2-propanol (HFIP). DBA acts as an ion-pairing agent, while a weak acid but volatile compound can be added to the eluent to ensure the protonation, as it is HFIP. This ion pair agent is added to the mobile phase, and the reaction is being supported *in situ* during the chromatographic process (courier ion). Furthermore, in the case of HFIP, it increases relatively by 3.1 times the signal compared to formic acid, which is a very common protonating agent in UPLC-MS(MS) mobile phases [24].

Furthermore, an IC-MS with a single-stage quadrupole detector installed in series with the conductivity detector was used to confirm the identity of both NO_3^- and NO_2^- anions. The MS instrument was operated in the negative ESI– mode, having a cone voltage of 50 V, probe temperature of 400°C, mass range from 20 to 500 amu (*m/z*), and operating at single-ion monitoring (SIM). The dwell time was 0.2 second and nebulizing gas (N_2), 75 psi, while the molecular ions $[M-H]^-$ of nitrate were detected at the respective *m/z* (amu). Although IC-MS seems to be a good solution, it

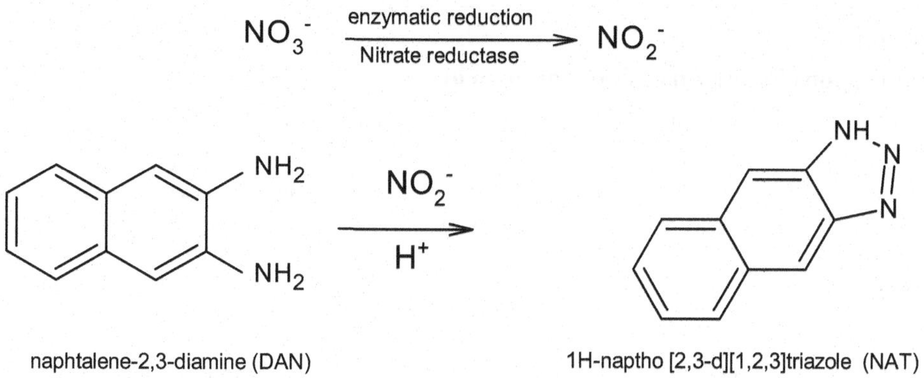

FIGURE 19.3 Reduction of nitrate to NO_2^- and chemical reaction of DAN with nitrite for the formation of NAT. (Adapted by Akyüz and Ata 2009 [26].)

has certain drawbacks as the presence of interferences close to the nitrate peak, mainly coming from a phosphate-containing compound (sugar-phosphate), which is a metabolite resulting from the glycolysis pathway. The second drawback is that the NO_2^- and NO_3^- quantification in meat products can interfere with several matrix components, such as nitrogen-containing compounds, resulting from biochemical proteolysis and the presence of chloride ions (NaCl salt) [25].

Another GC-MS method involved derivatization with 2,3-diaminonaphthalene (DAN), enzymatic reduction of NO_3^- to NO_2^-, extraction with toluene, and chromatographic analyses by using GC-MS in SIM (Figure 19.3). Nitrite and nitrates were extracted from solid food samples. The GC-MS/MS instrument was operating in the electron impact (EI) ionization and SIM mode focusing on *m/z* 114, 141, and 169 amu. Scan range (*m/z*) was set at 50–250 amu and electron energy at 70 eV for EI. The method's limits of detection (LODs) are lower but not significantly lower to compensate for the intensive and laborious sample preparation that is required, either for the derivatization but also for the reduction of NO_3^- to NO_2^- [26].

19.4.3 Sorbic Acid and Esters

Sorbic acid ($C_6H_8O_2$) is considered a very efficient food preservative in preventing or inhibiting mold growth, while it contributes very little to the flavor. It is generally applied by direct incorporation, on surface coatings, or into packaging [10]. Therefore, the addition of $C_6H_8O_2$ can prolong shelf life, while it has become increasingly important in controlling the growth of microbes [27]. Organic acids, such as $C_6H_8O_2$, can occur naturally in raw food materials or are produced during the food fermentation process, or they can also be found in fruits and vegetables (prunes and cranberries) [28].

There are several analytical strategies to analyze $C_6H_8O_2$, applying different hyphenated techniques, mainly LC or gas chromatography (GC), combined with MS. For the latter, different applications are reported either by using single-quadrupole MS, MS/MS, or HRMS (Table 19.1).

For GC-MS and EI ionization, single-quadrupole MS is preferred, operating at SIM. In this case, electron ionization, set at 70 eV, is used in various products (Table 19.1). A typical fragmentation scheme of $C_6H_8O_2$ through EI is presented in Figure 19.4, where typical abundant fragments denote the removal of the hydroxyl group of the carboxy group and the formation of the respective unsaturated aliphatic alcohol (*m/z* 97).

Regarding analysis with and MS, the use of MRM or SRM is preferred, while also HRMS technologies, such as quadrupole TOF-MS or Orbitrap MS (Thermo®) are applied. However, $C_6H_8O_2$ (so as benzoic acid) doesn't show good fragmentation at medium fragmenting voltages, and thus, a higher fragmenting voltage of 130 V is needed [30]. These technologies' advantages are the evaluation of molecular ions (*m/z*) or fragments with very high accuracy and precision. Subsequently, this

TABLE 19.1
MS Applications for the Analysis of Sorbic Acid

Analyte	Hyphenated Technique	MS		MS Conditions (Ionization)		Matrix	References
Sorbic acid	LC	HRMS	Orbitrap	ESI (−)	−H	Food products	[24]
Sorbic acid and its potassium and calcium salts	LC	HRMS	QTOF-MS	ESI (+)	+H	Study of the degradation products of food preservatives	[29]
Benzoic, sorbic, and propionic acid	LC	MS	QQQ	ESI (−)	−H	Vegetables	[28]
Sorbic acid	LC	HRMS	QTOF	ESI (−)	−H	Beverages	[30]
Sorbic acid	LC	LC	QQQ	ESI (−)	−H	Biological samples	[31,32]
Sorbic acid and potassium salts	LC	MS	QQQ	ESI	N/a	Food and beverages	[33]
Sorbic acid	GC	MS	NS	EI		beverages, vinegar, sauces, and drinks	[34]
Sorbic acid and salts	LC	QTRAP	MS	ESI (+)	+H	Meat and fish product	[19]
Sorbic acid	LC	QTRAP	MS	ESI (+)	+H	Cheeses and cream	[35]
Sorbic acid	GC	Q	MS	EI	EI	Food samples	[36]
Sorbic acid	GC	Q	MS	EI		Food samples	[37]
Sorbic acid	GC	MS	Q	EI		Fishery products	[38]
Sorbic acid	LC	MS	QQQ	ESI	ESI	Cheeses	[39]
Sorbic acid	LC	MS	QQQ	ESI (+)	+H	Food product	[40]

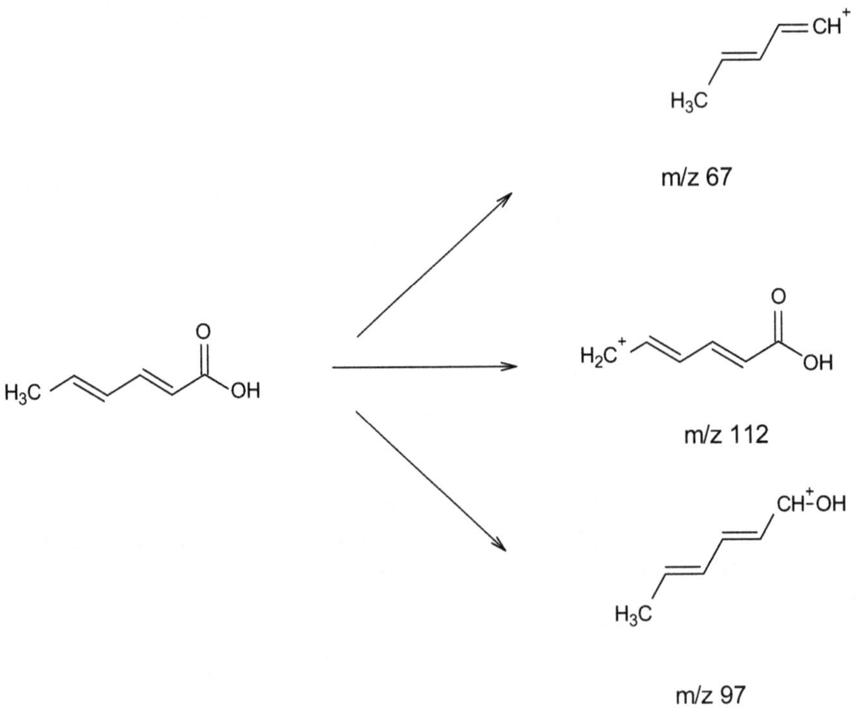

FIGURE 19.4 Typical GC-MS EI fragments of sorbic acid.

facilitates proper identification and elucidation of chemical molecules and fragmentation patterns. Moreover, for all the ESI modes mentioned before, both polarities are used to analyze $C_6H_8O_2$, either in the negative (–) or in the positive (+) mode. The intensity of the molecular ion in the ESI (–) is significantly higher than the respective ESI (+). A typical total ion chromatogram and fragmentation pattern (triple quadrupole) are given in Figures 19.5 and 19.6.

Horiyama et al. reported experiments for analyzing $C_6H_8O_2$ in negative ion mode, using ESI and atmospheric pressure chemical ionization (APCI) in negative polarity. The main peak in both ionizations was the deprotonated molecule $[M-H]^-$ at *m/z* 111 (Figure 19.6). However, ESI (–) generates different fragments than the respective APCI (Figure 19.6). The peak intensity at *m/z* 111 in the ESI (–) was higher than the respective *m/z* 113 in the ESI (+) mode. As $C_6H_8O_2$ includes a carbonyl group, it is not a good proton acceptor, rather a good proton donor compound, which explains the higher intensity in the negative mode [31].

Furthermore, for MS/MS, both ESI and APCI interfaces (Figure 19.6) have respective parent and product ions for the protonated or the deprotonated molecular ion. Thus, in the case of positive ionization, the protonated molecular ion ($[M+H]^+$) has a typical parent ion of *m/z* 113. In contrast, in the deprotonated form ($[M-H]^-$), it has a typical parent ion of *m/z* 111. Their major ions

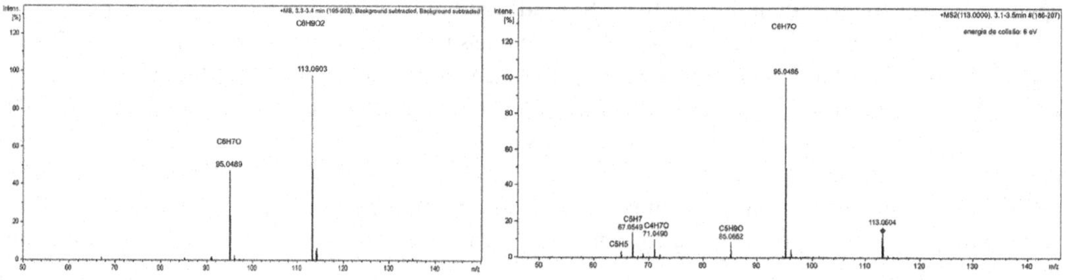

FIGURE 19.5 LC-MS extracted ion chromatogram (EIC) for sorbic acid using ESI positive (+) ionization (*m/z* 113.0604). (From de Jesus et al. [29].)

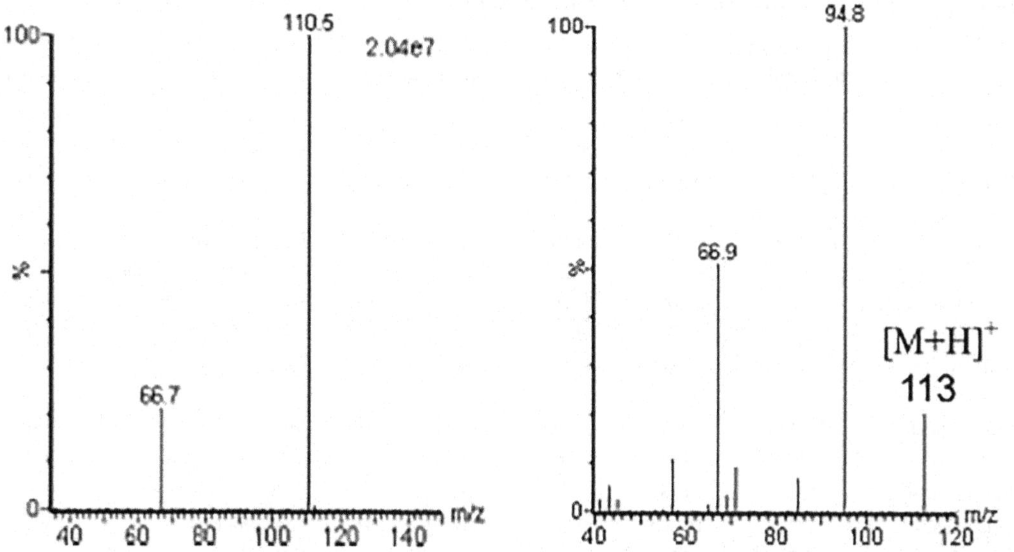

FIGURE 19.6 Typical negative ESI and APCI mass spectra and fragmentation pattern of sorbic acid. (Horiyama et al. [31].)

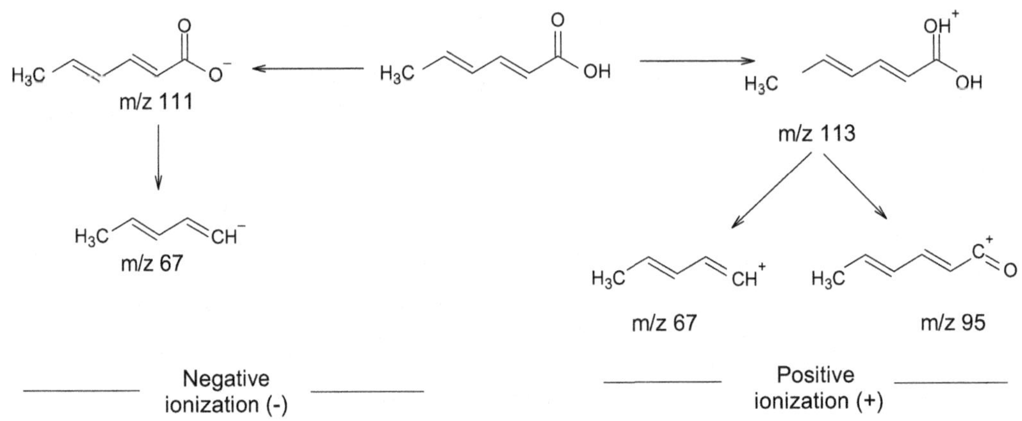

FIGURE 19.7 ESI (+) and ESI (–) MRM transition fragments for the analysis of sorbic acid. (Adapted from Horiyama et al. [31,32].)

appeared at *m/z* 67, and 95 in the positive ESI mode, and only *m/z* 67 in the negative ion mode ESI and APCI production mass spectra. Therefore, the mass transition patterns, *m/z* 113 → *m/z* 67 and 95, are regularly selected to monitor sorbic acid in the positive ion mode, while the transition from *m/z* 111 to 67 is preferred in case of the negative ESI mode for MRM analyses under ESI and APCI mode (Figure 19.7). The mass transition patterns of *m/z* 111 and 67 indicate the $[M-H]^-$ breakdown to $[M\text{-}CO_2H]^-$ and from *m/z* 113 to 95 and/or 67 refer to $[M+H]^+$, $[M-OH]^+$ and $[M\text{-}CO_2H]^+$ [31,32]. Nevertheless, the relative abundance of the main fragment ions (at *m/z* 67) can be low, and significant interfering peaks coming from mass reference solution can be presented, depending on the food matrix [30].

TABLE 19.2
MS Applications for the Analysis of Benzoic Acid

Analyte	Hyphenated Technique	MS		MS Conditions (Ionization)		Matrix	References
Benzoic acid	LC	MS	QQQ	ESI (+)	+H	Vegetables	[28]
Benzoic acid	LC	HRMS	Orbitrap	ESI (–)	–H	Food products	[24]
benzoic acid and salts	LC	QTRAP	MS	ESI (+)	+H	Meat and fish product	[19]
Benzoic acid	TD-GC/MS	Q	MS	EI		Beverages, vinegar, aqueous sauces, drinks	[34]
Benzoic acid	GC	Q	MS	EI	E–	Food and beverages	[36]
Benzoic	GC	Q	MS	EI		Food samples	[37]
Sodium benzoate	LC	MS	QQQ	ESI	n/a	Food samples	[33]
benzoic acid	FIA-MS	MS	QQQ	MS/MS	–H	MRM	[43]
benzoic acid	HS.GC-MS	Trap	Q	EI		SIM	[43]
Benzoic acid	GC	MS	Q	EI		Fishery products	[38]
Benzoic acid	LC	HRMS	TOF	ESI (–)	–H	Food products	[30]
Benzoic	KC	MS	QQQ	ESI (+)	+H	SRM	[39]
Benzoic acid	GC	MS	Q	EI		SIM (70=eV)	[42]

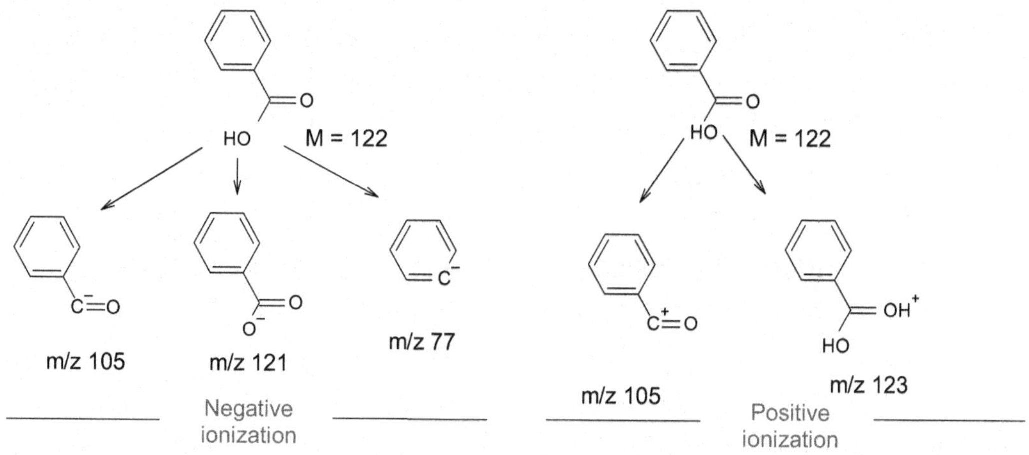

FIGURE 19.8 ESI (+) and ESI (–) MRM transition fragments for the analysis of benzoic acid.

19.4.4 Benzoic Acid and Esters

Benzoic acid, as also in sorbic and propionic acid, occurs naturally in raw food materials or is produced. The World Health Organization [41] stated that benzoic acid is produced in many plants as an intermediate in the formation of other compounds and can be found at significantly high concentrations in berries and also in animals [28].

The analysis of benzoic acid involves both GC or LC techniques, combined with MS (see Table 19.2). In the case of GC, in certain cases, derivatization is needed. An example is a silylation with MSTFA to achieve lower detection limits and facilitate selectivity by using MS [42]. For LC-MS analysis of benzoic acid, as in $C_6H_8O_2$, the existence of a hydroxy group (–OH) facilitates the deprotonation. Sometimes that justifies the negative ESI mode of the deprotonated molecule $[M-H]^-$ [31,32]. However, LC in combination with HRMS, benzoic acid shows no good fragmentation at medium voltages, and thus higher fragmenting voltage (130 V) is needed [30].

Regarding the different ionization strategies, using ESI, either in positive or negative, specific fragments and transitions are used (Figure 19.8). Hence, in the case of negative ionization, the molecular ion (M) is being fragmented to benzaldehyde (*m/z* 105) with low CE (10–20 V). Also, an additional transition is used from *m/z* 123 to 51 using high CE of 50–60 V, indicating an opening of the benzene ring, however, without identifying the fragments' structure [19].

F negative ESI, the fragments are presenting higher abundance what makes this the preferred technique. Thus, the deprotonated M is fragmented to benzene (*m/z* 77) or the respective benzaldehyde (*m/z* 105) using low collision energies (10–20 V) and typical 3–4 kV cone voltages [30,43].

19.4.5 p-Hydroxybenzoate Alkyl Esters

Esters of 4-hydroxybenzoic acid (4-hydroxybenzoate), most commonly known as parabens, are being used as antimicrobial preservatives in cosmetics and pharmaceuticals. Parabens occur naturally in foods having long-chain esters of 4-hydroxybenzoic acid, high antimicrobial activity, and low water solubility [44]. Parabens are also known to possess certain estrogenic and genotoxic activities, and there is a great interest in their analysis of foods [44,45]. Because of their widespread use as preservatives in various personal care products, cosmetics, pharmaceuticals, and foods, parabens may be introduced to humans via many different environmental sources (including water, soil, sediment and sludge, air and dust, and biota) [44]. Typical molecules that have been regulated [6] to be used as food preservatives are methyl 4-hydroxybenzoate, ethyl 4-hydroxybenzoate, propyl 4-hydroxybenzoate, and butyl 4-hydroxybenzoate (Figure 19.9).

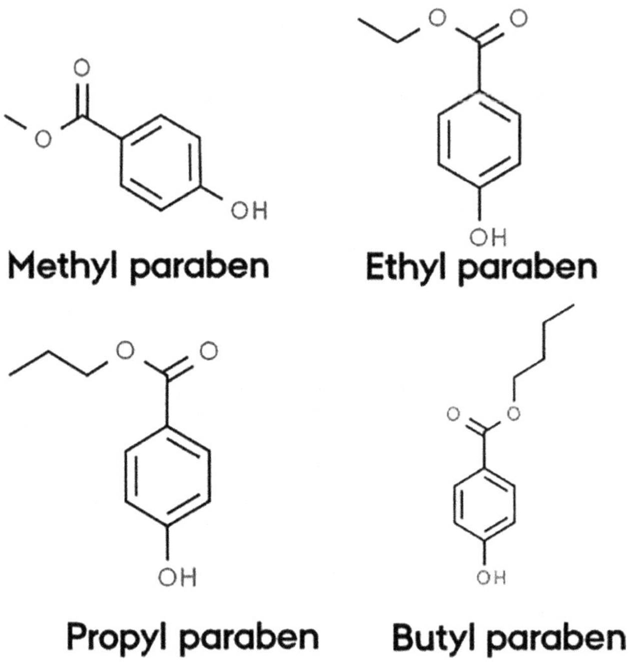

FIGURE 19.9 Chemical structures of regulated 4-hydroxybenzoate to be used as food preservatives [6].

Different applications using MS for the analysis of 4-hydroxybenzoate in various food products are reported, as given in Table 19.3.

GC methods using MS or MS/MS as detectors usually require a relatively large amount of sample and a derivatization step due to some analytes' relatively low volatility. The commonly used derivatization agents were *N*-methyl-*N*-(tert-butyl dimethylsilyl) trifluoroacetamide, a mixture of *N*,*O*-bis(trimethylsilyl)trifluoroacetamide and trimethylchlorosilane (MSTFA/1%TMCS), and acetic anhydride [59]. However, the use of polar GC-analytical columns facilitates the separation and analysis of these compounds to avoid the time-consuming step of derivatization [60]. On the other hand, LC is one of the most extensively used methods for the determination of parabens. Among all LC methods to determine parabens, C18 and C8 columns with different diameters, lengths, and particle sizes were commonly used.

The mass spectrometer was optimized to achieve maximum sensitivity in the quadrupoles (Table 19.4). According to Commission Decision 2002/657/EC [61], a precursor ion and 2 second-generation ions must meet the criteria for identifying analytes by LC-MS/MS. Thus, these types of methods are using the more intense fragments to quantitate the analytes, and the second fragments are used only for the identification and confirmation of these target analytes. Hence, typical MRM transitions for quantitation and quantification of 4-hydroxybenzoate are given in Table 19.4.

In the case of GC analysis, typical fragments of parabens, no matter what the alkyl chain is placed on the *para*-position, are *m/z* 137, referring to *p*-hydroxybenzoic acid, *m/z* 121 to *p*-hydroxybenzaldehyde, and *m/z* 93 to benzyl alcohol (Figure 19.10).

Apart from GC, LC combined with MS is also a technique of interest that can be applied either to positive or negative ESI, following MRM, to increase selectivity and sensitivity. For the latter, and based on the directive 2002/657/EC, two transitions are required for quantitation and quantification. For both ESI (+) and ESI (–), specific transitions are shown in Table 19.4. However, although both ionizations have been applied, the vast majority of MS applications refer to negative ESI, which has

TABLE 19.3
MS Applications for the Analysis of 4-Hydroxybenzoate Food Preservatives in Food Matrices

Analyte[a]	Hyphenated Technique	MS		MS Conditions (Ionization)		Matrix	References
MP, EP, PP, BP	LC	MS	QTRAP	ESI	–H	Vegetables	[46]
MP, EP, PP, isoPP	LC	MS	QTRAP	ESI	–H	Meat and fish products	[19]
MP, EP, PP, BP	TD-GC/MS	MS	Q	EI		beverages, vinegar, sauces	[34]
MP, EP, PP	UPLC	MS	QQQ	ESI	+H	Beverages	[47]
MP	GC	MS	QQQ	EI	EI	Water	[48]
MP, EP, PP, BP, BenzP, PhP, PP, IsoBP, PentP	LC	MS	QQQ	ESI	–H	Water	[49]
MP, EP, PP, BP	MALDI	MS	TOF	MALDI[b]	+H	Pharmaceuticals	[44]
MP, EP, PP, BP, isoPP, isoBP	HPLC	MS	QQQ	ESI	–H	food samples	[50]
MP, EP, PP, BP	GC	MS	Q	EI	EI	Vegetables	[51]
MP, EP, PP, BP	GC	MS	Q	EI	EI	Vegetables	[52]
MP, EP, PP, BP, BenzP	LC	MS	HR-orbitrap	ESI	–H	Fish and fish products	[53]
MP, EP, PP, BP	GC	MS	Q-ion trap	MS	EI	Seafood	[54,55]
MP, EP, PP, BP	GC	MS	Q	MS	PCI[c]	Food samples	[56]
MP, EP, PP, BP	LC	MS	Q-trap (LIT)	MS	–H	Seafood	[57]
MP, EP, PP, BP	GC	MS	Q	MS	EI	Food samples	[36]
MP, EP, PP, BP	GC	Q	MS	EI		Food samples	[37]
MP, EP, PP, BP, IsopP, IsoBP, HeptP	LC	HRMS	TOF	ESI+	+H	Beverages	[30]
MP, EP, PP, BP, IsopP, IsoBP, HeXP, HeptP	LC	MS	QQQ	ESI	–H	Spices	[58]

[a] Methyl paraben (MP), ethyl paraben (EP), propyl paraben (PP), butyl paraben (BP), isopropyl paraben (IsopP), Isobutyl paraben (IsoBP), hexyl paraben (HexP), heptyl paraben (HeptB), Benzyl paraben (BenzP), phenyl paraben (PhP), pentyl paraben (PentP).
[b] MALDI = matrix-assisted laser desorption ionization.
[c] PCI = positive chemical ionization.

been proven to be more, selective, and with acceptable sensitivity. Furthermore, hydroxyl groups (–OH) are known as acting as a good proton acceptor, rather than a good proton donor compound, which explains the higher intensity in the negative mode of the deprotonated molecule $[M-H]^-$.

Furthermore, HRMS has been reported for the analysis of 4-hydroxybenzoate in fish and fish products, using negative ESI orbitrap® [53] and in beverages applying a positive ESI-TOF [30]. Although fragmentation patterns are possible to be recognized in both cases, the high mass accuracy permits the identification even with EIC, or fragmentations can be used to overcome potential interferences. Thus, in the case of the Orbitrap, MS operated in FS, and data-independent analysis mode, which permitted to obtain full MS scan data followed by MS/MS acquisition spectra required for a confirmatory response for parabens, while resolving power was 140.000 FWHM, followed by a data-independent scan [53]. For LC-TOF-MS analysis of beverages, the capillary voltage was set at +4,000 V and a 130 V fragment *r* voltage [30].

Besides, a simple matrix-assisted, laser desorption ionization (MALDI)-TOF-MS method to screen for parabens in pharmaceutical products has been reported. A simple dilution method for

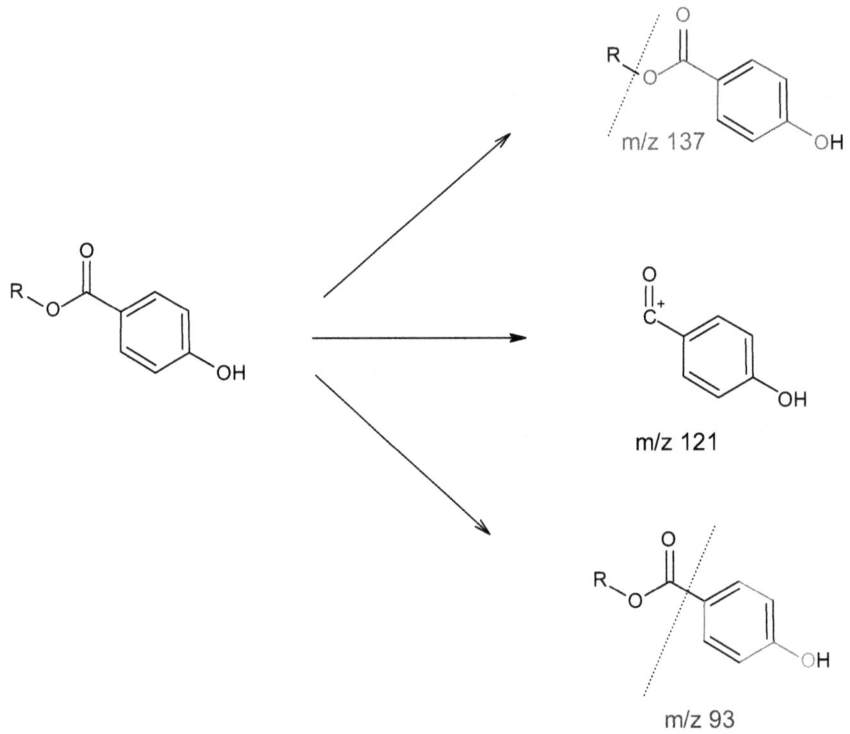

FIGURE 19.10 Typical and abundant fragments for GC-EI-MS analysis and identification of 4-hydroxybenzoate.

TABLE 19.4
Typical MRM Transition for Quantitation and Quantification of Parabens with ESI (–) and ESI (+) Mode

	MRM Transitions			MS Operation	
Analyte	**Parent Ion (*m/z*)**	**Quantification Ion (*m/z*)**	**Quantitation Ion (*m/z*)**	**CE (V)**	**Declustering Potential (DP/V)**
Negative ESI (–)					
Methyl paraben	151.0	91.6	135.9	20/15	35/35
Ethyl paraben	165.11	92.2	136.7	19/13	27/27
Isopropyl paraben	179.12	92.7	137.0	21/15	27/27
Butyl paraben	193.21	92.0	136.3	23/17	31/31
Benzyl paraben	227.16	91.9	135.9	13/15	27/23
Phenyl paraben	213.16	92.9	64.9	15/53	23/23
Propyl paraben	179.1	91.8	137.1	13/20	30/30
Isobutyl paraben	194.04	92.7	137.0	21/15	27/27
Pentyl paraben	207.21	92.0	136.4	25/17	33/33
Positive ESI (+)					
Methyl paraben	151.0	92.0	135.9	18/13	34/34 V
Ethyl paraben	166.0	92.1	136.9	20/14	34/34 V
Propyl paraben	179.1	92	136.2	24/16	40/40 V
Isopropyl paraben	179.1	92	136.2	24/16	40/40 V

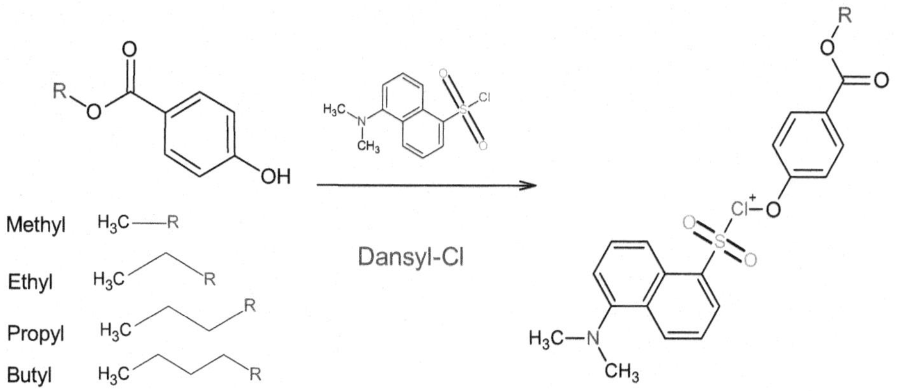

FIGURE 19.11 Derivatization reaction (dansylation) of parabens for analysis with MALDI-TOF. (Adapted from Lee et al. [44].)

sample preparation and applying chemical derivatization to label eight parabens to increase detection sensitivity, using Dansyl-Cl (5-(dimethylamino)naphthalene-1-sulfonyl chloride) as the reagent to label parabens resulted in the best signal for paraben analysis. These results have highlighted that derivatization (dansylation, Figure 19.11) is suitable for the analysis of phenol-containing compounds, such as parabens by MS combined with MALDI interface [44].

Typical mass spectra in MALDI-TOF-MS of the paraben derivatives under optimal derivatization conditions are based on the protonated form ($[M+H]^+$), and the signals at *m/z* 386.2, 400.2, 405.2, 414.2, and 428.2, representing Dansyl-MP, Dansyl-EP, Dansyl-PP, and Dansyl-BP, respectively. The MS/MS fragmentation revealed, as expected, common 4-hydroxybenzoate' fragments, such as *m/z* 170.12, 354.06, and 372.14, based on the respective chemical structures (see also Figure 19.11) [44].

19.4.6 Antibiotics

Antibiotics are agents that pose an antimicrobial activity to a range of microorganisms (molds and bacteria). They exhibit a selective and targeted mode of action, and they are being approved to be used as food preservatives. However, there is an increasing concern for using antibiotics to cause the formation of resistant organisms and strains [10].

Nisin is a peptide consisting of 34 amino acids, including the unusual 2-amino butyric acid (Abu), dehydrobutyrine (Dhb), dehydroalanine (Dha), and the thioether amino acids lanthionine (Ala-S-Ala) and β-methyl lanthionine (Abu-S-Ala). The antimicrobial effect of Nisin A ($C_{143}H_{230}N_{42}O_{37}S_7$, M=838.8923) was discovered between the late 1920s and early 1930s. In 1991, another natural nisin variant was found to be produced by the *L. lactis* strain NIZO 22186, which was called nisin Z ($C_{141}H_{229}N_{41}O_{38}S_7$, M=833.1396). It was shown that the structure of nisin Z differs from nisin only at position 27, where histidine is replaced by asparagine [62]. Nisin was then renamed as nisin A. Natamycin (also known as pimaricin), was first discovered in 1955, and it has a molecular structure of $C_{33}H_{47}NO_{13}$ and a molar mass of 666.3120. It is a macrolide polyene antifungal agent produced during fermentation by the bacteria *Streptomyces natalensis*, *Streptomyces chmanovgensis*, and *Streptomyces gilvosporeus* [62]. The chemical structures of Nisin A and natamycin are given in Figure 19.12.

For the analysis of either nisin or natamycin, LC technology combined with MS is applied due to these molecules' high molecular mass. The MS/MS technology is preferred within this framework, using specifying transitions for both molecules (Table 19.5). Furthermore, HRMS can also be applied by using high-resolution power and mass accuracy to elucidate and selectively identify while providing high sensitivity for their analysis, as reported in Table 19.5. For Nisin A, the parent

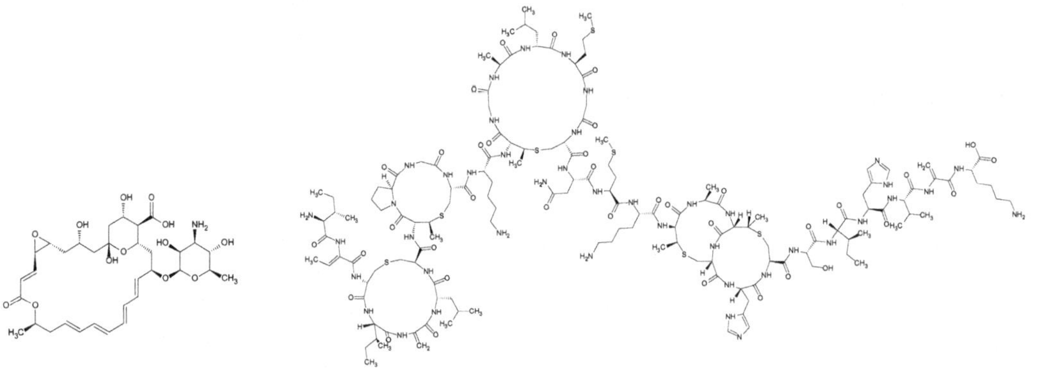

Molecular Formula: $C_{33}H_{47}NO_{13}$
Natamycin

Molecular Formula: $C_{143}H_{230}N_{42}O_{37}S_7$
Nisin A

FIGURE 19.12 Chemical structures of Nisin A and Natamycin.

ion of *m/z* 671.6 is selected, and the transitions of *m/z* 811 (quantification) and *m/z* 649 (quantitation). For nisin Z, the parent ion is *m/z* 667.2, and the transitions are *m/z* 805.4 (quantification) and *m/z* 739.1 (quantitation). MS operating conditions are normally with a cone voltage of 4,000 V, in the ESI (+) mode, CE of 15–25 eV, and DPs of 56 V (Table 19.5). Nisin is characterized by one specific transition of the $[\text{nisin}+5H]^{5+}$ molecule to a higher *m/z* to increase the method's selectivity [63]. The latter has already been implemented to an ISO standard to analyze nisin in cheese [64].

Recently, an advanced and novel method has been developed and validated using isotopic dilution. Here, nisin A's determination was based on the usage of labeled peptide sequence MSTKDFNLDLVSVSKKDSGASP(R) (without thioether bridges) as an internal standard, improving the MS detection of nisin A significantly. The labeled peptide has an average molar mass of 2491.3 Da, while the MRM was defined by *m/z* 499.5 (precursor ion), *m/z* 103.9 (product ion), dwell time of 200 ms, a DP of 56 V, and CE of 22 eV [65].

TABLE 19.5
MS Applications for the Analysis of Nisin A, Nisin Z, and Natamycin in Food Matrices

Analyte	Hyphenated Technique	MS		MS Conditions (Ionization)		Matrix	References
Nisin	LC	QTRAP	MS	ESI	+H	Meat and fish product	[19]
Nisin, natamycin	LC	QTRAP	MS	ESI	+H	Cheeses and cream	[35]
Natamycin, Nisin	LC	MS	QQQ	ESI	+H	Cheeses	[39]
Natamycin	LC	MS	QQQ	ESI	+H	Food products	[40]
Nisin A, Nisin Z, Natamycin	LC	MS	HRMS (Orbitrap)	ESI	+H	Food products	[62]
Nisin A, Nisin Z	LC	MS	QQQ	ESI	+H	Food products	
Nisin A, Nisin Z	LC	MS	QQQ	ESI	+H	Cow milk	[66]
Nisin A	LC	MS	QQQ	ESI	+H	cheese, mascarpone, processed cheese, and ripened cheese	[65]
Nisin A, Nisin Z	LC	MS	QQQ	ESI	+H	Cheeses	[63]

TABLE 19.6
Typical MRM Transition for Quantitation and Quantification of Nisin A, Nisin Z, and Natamycin with ESI (+)

	MRM Transitions			MS Operation		
Analyte	Parent Ion (*m/z*)	Quantification Ion (*m/z*)	Quantitation Ion (*m/z*)	CE (V)	Declustering Potential (DP/V)	References
Natamycin	666.2	503.2	485.2	17/23	56	[19]
Nisin	671.1	811.3	649.3	23/23	56	
Natamycin	666.2	503.2	485.2	17/21	56	[35]
Nisin	671.24	811.3	649.3	23/23	56	
Natamycin	666	629	633/647	n/a	n/a	[39]
Nisin	567	373	431			
Natamycin	666.2	503.3	485	20/20	15/15	[40]
Nisin A	671.6	810.9	790.3	10/10	n.a	[66]
Nisin Z	667.0	804.9	738.4	10/10	n.a	
Nisin A	671.2	811.3	649.3	23/23	56/56	[65]
ISTD (labeled peptide)	499.4	103.9	103.9	22	56	
Nisin A	671.7	811.2	790.5	25/25	101/101	[63]
Nisin Z	667.2	805.4	739.1	25/25	70/70	

A standard was scanned using positive ion ESI to detect the $[M+4H]^{4+}$ and $[M+5H]^{5+}$ molecule ions. Afterward, scans with different collision energies were performed to find the most abundant fragment ions. When multiply charged molecules are analyzed, the transitions to a higher *m/z* ratio are highly specific. Therefore, the three most intense fragment ions with a higher *m/z* ratio than the precursor ions were chosen to develop the SRM methods for the different analytes. The applied potentials of the mass spectrometer and the ESI source's adjustment were optimized for the SRM transitions. Thus, three highly specific transitions from the precursor ions $[M+4H]^{4+}$ or $[M+5H]^{5+}$ (Table 19.6) to higher *m/z* ratios were identified [63]. Typical transitions are reported in different studies are shown in Table 19.6.

19.5 CONCLUSIONS

Food preservatives are basic additives in processed foods. They serve the purpose of extending the shelf life of products and preventing deterioration and spoilage. This has led to the extensive application of chemical or synthetic preservatives, while governments and research centers have become increasingly concerned with their possible effects on human health. Thus, there is an increasing demand for their quantification, where analytical needs lead to the implementation of hyphenated techniques together with MS, MS/MS, or even HRMS. The main goals are to lower the LODs in combination with the selectivity that MS can certainly provide. There is an existing discussion about the safety and potential risks resulting from the use of chemical food preservatives, so we now have a public demand for "more natural" preservative technologies and natural or plant-derived antimicrobial agents, such as herbs and spices.

REFERENCES

1. U.S. Food and Drug Administration, Overview of Food Ingredients, Additives & Colors, U.S. Food and Drug Administration (2010). https://www.fda.gov/food/food-ingredients-packaging/overview-food-ingredients-additives-colors (accessed November 15, 2020).

2. U.S. Food and Drug Administration, CFR 172 – Code of Federal Regulations Title 21. Chapter I – Food and Drug Administration Department of Health and Human Services Subchapter B – Food for human consumption (continued). Part 172 food additives permitted for direct addition to food for human consumption (2019). https://www.accessdata.fda.gov/scripts/cdrh/cfdocs/cfcfr/CFRSearch.cfm?CFRPart=172 (accessed November 15, 2020).
3. Codex Alimentarius, Codex General Standard for Food Additives (GSFA) Online Database, Codex General Standard for Food Additives (GSFA) Online Database (2020). http://www.fao.org/fao-who-codexalimentarius/sh-proxy/en/?lnk=1&url=https%253A%252F%252Fworkspace.fao.org%252Fsites%252Fcodex%252FStandards%252FCXS%2B192-1995%252FCXS_192e.pdf (accessed October 11, 2020).
4. K. Ntrallou, H. Gika, E. Tsochatzis, Analytical and sample preparation techniques for the determination of food colorants in food matrices, *Foods*. 9 (2020) 58. https://doi.org/10.3390/foods9010058.
5. M. Carocho, P. Morales, I.C.F.R. Ferreira, Natural food additives: Quo vadis?, *Trends in Food Science & Technology*. 45 (2015) 284–295. https://doi.org/10.1016/j.tifs.2015.06.007.
6. European Union, Regulation (EC) No. 1333/2008 of the European Parliament and of the Council on food additives (2008). https://eur-lex.europa.eu/legal-content/EN/TXT/PDF/?uri=CELEX:32008R1333&from=EN (accessed November 12, 2020).
7. G. Cao, K. Li, J. Guo, M. Lu, Y. Hong, Z. Cai, Mass spectrometry for analysis of changes during food storage and processing, *Journal of Agricultural and Food Chemistry*. 68 (2020) 6956–6966. https://doi.org/10.1021/acs.jafc.0c02587.
8. R. García-García, S.S. Searle, Preservatives: Food use, in: *Encyclopedia of Food and Health*, Elsevier, 2016: pp. 505–509. https://doi.org/10.1016/B978-0-12-384947-2.00568-7.
9. B. Caballero, P. Finglas, F. Toldra, *Encyclopedia of Food and Health*, 1st Edition, Academic Press, Cambridge, MA, 2015.
10. S. Damodaran, K.L. Parkin, *Fennema's Food Chemistry*, 5th Edition, CRC Press, Taylor & Francis Group, Boca Raton, FL, 2017.
11. K.S. Robbins, R. Shah, S. MacMahon, L.S. de Jager, Development of a liquid chromatography–tandem mass spectrometry method for the determination of sulfite in food, *Journal of Agricultural and Food Chemistry*. 63 (2015) 5126–5132. https://doi.org/10.1021/jf505525z.
12. K.S. Carlos, L.S. de Jager, Determination of sulfite in food by liquid chromatography tandem mass spectrometry: Collaborative study, *Journal of AOAC International*. 100 (2017) 1785–1794. https://doi.org/10.5740/jaoacint.17-0033.
13. K. Yang, C. Zhou, Z. Yang, L. Yu, M. Cai, C. Wu, P. Sun, Establishing a method of HPLC involving precolumn derivatization by 2,2′-dithiobis (5-nitropyridine) to determine the sulfites in shrimps in comparison with ion chromatography, *Food Science and Nutrition*. 7 (2019) 2151–2158. https://doi.org/10.1002/fsn3.1060.
14. U.S. Food and Drug Administration, Method C-004.02: Determination of sulfites in food using liquid chromatography-tandem mass spectrometry (LC-MS/MS) (2020). https://www.fda.gov/media/114411/download (accessed October 15, 2020).
15. K.S. Carlos, M. Treblin, L.S. de Jager, Comparison and optimization of three commercial methods with an LC–MS/MS method for the determination of sulfites in food and beverages, *Food Chemistry*. 286 (2019) 537–540. https://doi.org/10.1016/j.foodchem.2019.02.042.
16. Z. Ni, F. Tang, Y. Liu, D. Shen, R. Mo, rapid determination of low-level sulfite in dry vegetables and fruits by LC-ICP-MS, *FSTR*. 21 (2015) 1–6. https://doi.org/10.3136/fstr.21.1.
17. S.D. Kelly, M.J. Scotter, R. Macarthur, L. Castle, M.J. Dennis, Survey of stable sulfur isotope ratios ($^{34}S/^{32}$) of sulfite and sulfate in foods, *Food Additives and Contaminants*. 19 (2002) 1003–1009. https://doi.org/10.1080/02652030210153587.
18. M.R. Siddiqui, S.M. Wabaidur, M.A. Khan, Z.A. ALOthman, M.Z.A. Rafiquee, A.A. Alqadami, A rapid and sensitive evaluation of nitrite content in Saudi Arabian processed meat and poultry using a novel ultra performance liquid chromatography–mass spectrometry method, *Journal of Food Science and Technology*. 55 (2018) 198–204. https://doi.org/10.1007/s13197-017-2908-x.
19. L. Molognoni, H. Daguer, L.A. de Sá Ploêncio, J. De Dea Lindner, A multi-purpose tool for food inspection: Simultaneous determination of various classes of preservatives and biogenic amines in meat and fish products by LC-MS, *Talanta*. 178 (2018) 1053–1066. https://doi.org/10.1016/j.talanta.2017.08.081.
20. L. Molognoni, H. Daguer, R.B. Hoff, J. Rodrigues, A.C. Joussef, J. De Dea Lindner, Assessing the mutagens ethylnitrolic acid and 2-methyl-1,4-dinitro-pyrrole in meat products: Sample preparation and simultaneous analysis by LC–MS/MS, *Journal of Chromatography A*. 1609 (2020) 460512. https://doi.org/10.1016/j.chroma.2019.460512.

21. L. Molognoni, G.E. Motta, H. Daguer, J. De Dea Lindner, Microbial biotransformation of N-nitro-, C-nitro-, and C-nitrous-type mutagens by *Lactobacillus delbrueckii* subsp. bulgaricus in meat products, *Food and Chemical Toxicology*. 136 (2020) 110964. https://doi.org/10.1016/j.fct.2019.110964.
22. S. Chamandust, M.R. Mehrasebi, K. Kamali, R. Solgi, J. Taran, F. Nazari, M.-J. Hosseini, Simultaneous determination of nitrite and nitrate in milk samples by ion chromatography method and estimation of dietary intake, *International Journal of Food Properties*. 19 (2016) 1983–1993. https://doi.org/10.1080/10942912.2015.1091007.
23. M.R. Siddiqui, S.M. Wabaidur, Z.A. ALOthman, M.Z.A. Rafiquee, Rapid and sensitive method for analysis of nitrate in meat samples using ultra performance liquid chromatography–mass spectrometry, *Spectrochimica Acta Part A: Molecular and Biomolecular Spectroscopy*. 151 (2015) 861–866. https://doi.org/10.1016/j.saa.2015.07.028.
24. A. Kaufmann, M. Widmer, K. Maden, P. Butcher, S. Walker, Analysis of a variety of inorganic and organic additives in food products by ion-pairing liquid chromatography coupled to high-resolution mass spectrometry, *Analytical and Bioanalytical Chemistry*. 410 (2018) 5629–5640. https://doi.org/10.1007/s00216-018-0904-2.
25. G. Saccani, E. Tanzi, S. Cavalli, J. Rohrer, Determination of nitrite, nitrate, and glucose-6-phosphate in muscle tissues and cured meat by IC/MS, *Journal of AOAC International*. 89 (2006) 712–719. https://doi.org/10.1093/jaoac/89.3.712.
26. M. Akyüz, Ş. Ata, Determination of low level nitrite and nitrate in biological, food and environmental samples by gas chromatography–mass spectrometry and liquid chromatography with fluorescence detection, *Talanta*. 79 (2009) 900–904. https://doi.org/10.1016/j.talanta.2009.05.016.
27. C. Xu, J. Liu, C. Feng, H. Lv, S. Lv, D. Ge, K. Zhu, Investigation of benzoic acid and sorbic acid in snack foods in Jilin province, China, *International Journal of Food Properties*. 22 (2019) 670–677. https://doi.org/10.1080/10942912.2019.1599011.
28. S.S. Yun, J. Kim, S.J. Lee, J.S. So, M.Y. Lee, G. Lee, H.S. Lim, M. Kim, Naturally occurring benzoic, sorbic, and propionic acid in vegetables, *Food Additives & Contaminants: Part B*. 12 (2019) 167–174. https://doi.org/10.1080/19393210.2019.1579760.
29. J.H.F. de Jesus, I.M. Szilágyi, G. Regdon, E.T.G. Cavalheiro, Thermal behavior of food preservative sorbic acid and its derivates, *Food Chemistry*. 337 (2021) 127770. https://doi.org/10.1016/j.foodchem.2020.127770.
30. X.Q. Li, F. Zhang, Y.Y. Sun, W. Yong, X.G. Chu, Y.Y. Fang, J. Zweigenbaum, Accurate screening for synthetic preservatives in beverage using high performance liquid chromatography with time-of-flight mass spectrometry, *Analytica Chimica Acta*. 608 (2008) 165–177. https://doi.org/10.1016/j.aca.2007.12.010.
31. S. Horiyama, C. Honda, K. Suwa, Y. Umemoto, Y. Okada, M. Semma, A. Ichikawa, M. Takayama, Sensitive and simple analysis of sorbic acid using liquid chromatography with electrospray ionization tandem mass spectrometry, *Chemical and Pharmaceutical Bulletin*. 56 (2008) 578–581. https://doi.org/10.1248/cpb.56.578.
32. S. Horiyama, C. Honda, K. Suwa, Y. Okada, M. Semma, A. Ichikawa, M. Takayama, Negative and positive ion mode LC/MS/MS for simple, sensitive analysis of sorbic acid, *Chemical and Pharmaceutical Bulletin*. 58 (2010) 106–109. https://doi.org/10.1248/cpb.58.106.
33. A.C. Gören, G. Bilsel, A. Şimşek, M. Bilsel, F. Akçadağ, K. Topal, H. Ozgen, HPLC and LC–MS/MS methods for determination of sodium benzoate and potassium sorbate in food and beverages: Performances of local accredited laboratories via proficiency tests in Turkey, *Food Chemistry*. 175 (2015) 273–279. https://doi.org/10.1016/j.foodchem.2014.11.094.
34. N. Ochiai, K. Sasamoto, M. Takino, S. Yamashita, S. Daishima, A.C. Heiden, A. Hoffmann, Simultaneous determination of preservatives in beverages, vinegar, aqueous sauces, and quasi-drug drinks by stir-bar sorptive extraction (SBSE) and thermal desorption GC–MS, *Analytical and Bioanalytical Chemistry*. 373 (2002) 56–63. https://doi.org/10.1007/s00216-002-1257-3.
35. L. Molognoni, L.A. de Sá Ploêncio, A.C. Valese, J. De Dea Lindner, H. Daguer, A simple and fast method for the inspection of preservatives in cheeses and cream by liquid chromatography–electrospray tandem mass spectrometry, *Talanta*. 147 (2016) 370–382. https://doi.org/10.1016/j.talanta.2015.10.008.
36. J. Yang, D. Li, C. Sun, Simultaneous determination of eleven preservatives in foods using ultrasound-assisted emulsification micro-extraction coupled with gas chromatography-mass spectrometry, *Analytical Methods*. 4 (2012) 3436. https://doi.org/10.1039/c2ay25406a.
37. M. Ding, W. Liu, J. Peng, X. Liu, Y. Tang, Simultaneous determination of seven preservatives in food by dispersive liquid-liquid microextraction coupled with gas chromatography-mass spectrometry, *Food Chemistry*. 269 (2018) 187–192. https://doi.org/10.1016/j.foodchem.2018.07.002.

38. D.-B. Kim, G.-J. Jang, M. Yoo, G. Lee, S.S. Yun, H.S. Lim, M. Kim, S. Lee, Sorbic, benzoic and propionic acids in fishery products: A survey of the South Korean market, *Food Additives & Contaminants: Part A*. 35 (2018) 1071–1077. https://doi.org/10.1080/19440049.2018.1447692.
39. F. Fuselli, C. Guarino, A. La Mantia, L. Longo, A. Faberi, R.M. Marianella, Multi-detection of preservatives in cheeses by liquid chromatography–tandem mass spectrometry, *Journal of Chromatography B*. 906 (2012) 9–18. https://doi.org/10.1016/j.jchromb.2012.07.035.
40. L. Molognoni, A.C. Valese, A. Lorenzetti, H. Daguer, J. De Dea Lindner, Development of a LC–MS/MS method for the simultaneous determination of sorbic acid, natamycin and tylosin in Dulce de leche, *Food Chemistry*. 211 (2016) 748–756. https://doi.org/10.1016/j.foodchem.2016.05.105.
41. World Health Organization (WHO), Concise International Chemical Assessment Document 26. Benzoic acid and sodium benzoate (2005). https://www.who.int/ipcs/publications/cicad/cicad26_rev_1.pdf (accessed October 15, 2020).
42. C. Mak, Y.-L. Wong, C. Mok, S. Choi, An accurate analytical method for the determination of benzoic acid in curry paste using isotope dilution gas chromatography-mass spectrometry (GC-MS), *Analytical Methods*. 4 (2012) 3674. https://doi.org/10.1039/c2ay25517k.
43. L. Junqueira de Carvalho, E. Cristina Pires do Rego, B. Carius Garrido, Quantification of benzoic acid in beverages: The evaluation and validation of direct measurement techniques using mass spectrometry, *Analytical Methods*. 8 (2016) 2955–2960. https://doi.org/10.1039/C5AY03168K.
44. Y.-H. Lee, Y.-C. Lin, C.-H. Feng, W.-L. Tseng, C.-Y. Lu, A derivatization-enhanced detection strategy in mass spectrometry: Analysis of 4-hydroxybenzoates and their metabolites after keratinocytes are exposed to UV radiation, *Scientific Reports*. 7 (2017) 39907. https://doi.org/10.1038/srep39907.
45. P.D. Darbre, P.W. Harvey, Parabens can enable hallmarks and characteristics of cancer in human breast epithelial cells: A review of the literature with reference to new exposure data and regulatory status: Parabens and breast cancer, *Journal of Applied Toxicology*. 34 (2014) 925–938. https://doi.org/10.1002/jat.3027.
46. X. Zhou, S. Cao, X. Li, B. Tang, X. Ding, C. Xi, J. Hu, Z. Chen, Simultaneous determination of 18 preservative residues in vegetables by ultra high performance liquid chromatography coupled with triple quadrupole/linear ion trap mass spectrometry using a dispersive-SPE procedure, *Journal of Chromatography B*. 989 (2015) 21–26. https://doi.org/10.1016/j.jchromb.2015.02.030.
47. Q. Yin, Y. Zhu, Y. Yang, Dispersive liquid–liquid microextraction followed by magnetic solid-phase extraction for determination of four parabens in beverage samples by ultra-performance liquid chromatography tandem mass spectrometry, *Food Analytical Methods*. 11 (2018) 797–807. https://doi.org/10.1007/s12161-017-1051-7.
48. J. Regueiro, E. Becerril, C. Garcia-Jares, M. Llompart, Trace analysis of parabens, triclosan and related chlorophenols in water by headspace solid-phase microextraction with in situ derivatization and gas chromatography–tandem mass spectrometry, *Journal of Chromatography A*. 1216 (2009) 4693–4702. https://doi.org/10.1016/j.chroma.2009.04.025.
49. A.V. Marta-Sanchez, S.S. Caldas, A. Schneider, S.M.V.S. Cardoso, E.G. Primel, Trace analysis of parabens preservatives in drinking water treatment sludge, treated, and mineral water samples, *Environmental Science and Pollution Research*. 25 (2018) 14460–14470. https://doi.org/10.1007/s11356-018-1583-4.
50. S. Cao, Z. Liu, L. Zhang, C. Xi, X. Li, G. Wang, R. Yuan, Z. Mu, Development of an HPLC–MS/MS method for the simultaneous analysis of six kinds of parabens in food, *Analytical Methods*. 5 (2013) 1016–1023. https://doi.org/10.1039/C2AY26283E.
51. Y. Chen, S. Cao, L. Zhang, C. Xi, Z. Chen, Modified quechers combination with magnetic solid-phase extraction for the determination of 16 preservatives by gas chromatography–mass spectrometry, *Food Analytical Methods*. 10 (2017) 587–595. https://doi.org/10.1007/s12161-016-0616-1.
52. Y. Chen, S. Cao, L. Zhang, C. Xi, X. Li, Z. Chen, G. Wang, Preparation of size-controlled magnetite nanoparticles with a graphene and polymeric ionic liquid coating for the quick, easy, cheap, effective, rugged and safe extraction of preservatives from vegetables, *Journal of Chromatography A*. 1448 (2016) 9–19. https://doi.org/10.1016/j.chroma.2016.04.045.
53. L.M. Chiesa, R. Pavlovic, S. Panseri, F. Arioli, Evaluation of parabens and their metabolites in fish and fish products: A comprehensive analytical approach using LC-HRMS, *Food Additives & Contaminants: Part A*. 35 (2018) 2400–2413. https://doi.org/10.1080/19440049.2018.1544721.
54. R. Djatmika, C.-C. Hsieh, J.-M. Chen, W.-H. Ding, Determination of paraben preservatives in seafood using matrix solid-phase dispersion and on-line acetylation gas chromatography–mass spectrometry, *Journal of Chromatography B*. 1036–1037 (2016) 93–99. https://doi.org/10.1016/j.jchromb.2016.10.005.

55. R. Djatmika, W.-H. Ding, H. Sulistyarti, Determination of parabens by injection-port derivatization coupled with gas-chromatography-mass spectrometry and matrix solid phase dispersion, *IOP Conference Series: Materials Science and Engineering.* 299 (2018) 012005. https://doi.org/10.1088/1757-899X/299/1/012005.
56. R. Jain, M.K.R. Mudiam, A. Chauhan, R. Ch, R.C. Murthy, H.A. Khan, Simultaneous derivatisation and preconcentration of parabens in food and other matrices by isobutyl chloroformate and dispersive liquid–liquid microextraction followed by gas chromatographic analysis, *Food Chemistry.* 141 (2013) 436–443. https://doi.org/10.1016/j.foodchem.2013.03.012.
57. C. Han, B. Xia, X. Chen, J. Shen, Q. Miao, Y. Shen, Determination of four paraben-type preservatives and three benzophenone-type ultraviolet light filters in seafoods by LC-QqLIT-MS/MS, *Food Chemistry.* 194 (2016) 1199–1207. https://doi.org/10.1016/j.foodchem.2015.08.093.
58. J. Lv, L. Wang, X. Hu, Z. Tai, Y. Yang, Rapid determination of 10 parabens in spices by high performance liquid chromatography-mass spectrometry, *Analytical Letters.* 45 (2012) 1960–1970. https://doi.org/10.1080/00032719.2012.680089.
59. C. Piao, L. Chen, Y. Wang, A review of the extraction and chromatographic determination methods for the analysis of parabens, *Journal of Chromatography B.* 969 (2014) 139–148. https://doi.org/10.1016/j.jchromb.2014.08.015.
60. E.D. Tsochatzis, J. Alberto Lopes, E. Hoekstra, H. Emons, Development and validation of a multi-analyte GC-MS method for the determination of 84 substances from plastic food contact materials, *Analytical and Bioanalytical Chemistry.* 412 (2020) 5419–5434. https://doi.org/10.1007/s00216-020-02758-7.
61. European Commission, Commission Decision 2002/657/EC implementing Council Directive 96/23/EC concerning the performance of analytical methods and the interpretation of results (2002). https://eur-lex.europa.eu/legal-content/EN/TXT/PDF/?uri=CELEX:32002D0657&from=EN.
62. A.L.L. Duchateau, W.B. van Scheppingen, Stability study of a nisin/natamycin blend by LC-MS, *Food Chemistry.* 266 (2018) 240–244. https://doi.org/10.1016/j.foodchem.2018.05.121.
63. N. Schneider, K. Werkmeister, M. Pischetsrieder, Analysis of nisin A, nisin Z and their degradation products by LCMS/MS, *Food Chemistry.* 127 (2011) 847–854. https://doi.org/10.1016/j.foodchem.2011.01.023.
64. International Standards Organization (ISO), ISO/TS 27106:2009, Cheese – Determination of nisin A content by LC-MS and LC-MS-MS (2009). https://www.iso.org/standard/43998.html.
65. C.W. Lim, K.Y. Lai, W.T. Ho, S.H. Chan, Isotopic dilution assay development of nisin A in cream cheese, mascarpone, processed cheese and ripened cheese by LC-MS/MS method, *Food Chemistry.* 292 (2019) 58–65. https://doi.org/10.1016/j.foodchem.2019.04.040.
66. K.Y. Ko, S.R. Park, C.A. Lee, M. Kim, Analysis method for determination of nisin A and nisin Z in cow milk by using liquid chromatography-tandem mass spectrometry, *Journal of Dairy Science.* 98 (2015) 1435–1442. https://doi.org/10.3168/jds.2014-8452.

20 Mass Spectrometry for Mycotoxin Detection in Food and Food Byproducts

Devarajan Thangadurai
Karnatak University

Saher Islam
Lahore College for Women University

Afshan Shafi
MNS University of Agriculture

Namita Bedi
Amity University

Jeyabalan Sangeetha
Central University of Kerala

Adil Hussain
Abdul Wali Khan University

Ravichandra Hospet and Jarnain Naik
Karnatak University

Muhammad Zaki Khan
MNS University of Agriculture

Anand Torvi
Jain University

Muniswamy David
Karnatak University

Zaira Zaman Chowdhury
University of Malaya

Amjad Iqbal
Abdul Wali Khan University Mardan

DOI: 10.1201/9781003091226-23

CONTENTS

20.1 Introduction 380
20.2 Application of Liquid Chromatography–Tandem Mass Spectrometry (LC–MS/MS) in Mycotoxins Analysis 383
20.3 Determination of Mycotoxins with LC–High-Resolution Mass Spectrometry (HRMS) 393
20.4 LC-Electrospray Ionization (ESI)–MS/MS for Mycotoxins 393
20.5 Identification of Mycotoxins with High-Performance Liquid Chromatography (HPLC)–MS/MS 394
20.6 Gas Chromatography (GC)–MS for the Sensitive Estimation of Mycotoxins 395
20.7 Conclusion 396
References 396

20.1 INTRODUCTION

Fungi are ubiquitously distributed in the environment and cause spoilage of food and foodborne diseases (Adejumo and Adejoro 2014). These fungi directly cause mycosis or produce mycotoxicoses by secondary metabolites known as mycotoxins. They cause harmful effects in livestock, domestic animals, and humans (Ashiq 2015; Jeswal and Kumar 2015). Mycotoxins as potent contaminants of food lead to a remarkable reduction in yield and quality of agriculture grains (Moretti et al. 2017). The fungal genera mainly producing mycotoxins are represented by *Aspergillus, Penicillium, Fusarium, Trichoderma, Trichothecium*, and *Alternaria* (Richard 2007; Ashiq 2015) (Table 20.1).

TABLE 20.1
Fungal Species Producing Mycotoxins that Contaminate Food Commodities

Fungal Species	Mycotoxin	Food Commodity	Maximum Permitted Levels[a] (μg/kg)
Aspergillus parasiticus	Aflatoxins B1, B2, G1, G2	Maize, wheat, rice, sorghum, groundnuts, tree nuts, and figs	B1: 0.10–12 B1 + B2 + G1 + G2: 4–15
Aspergillus flavus	Aflatoxins B1 y B2	Idem	
Metabolite of aflatoxin B1 in mammals	Aflatoxin M1	Milk and milk products	0.025–0.050
Fusarium sporotrichioides	Toxins T-2 and HT-2	Cereals and cereal products	T-2 + HT-2: 15–2,000[b]
Fusarium graminearum	Deoxynivalenol zearalenone	Cereals and cereal products	DON: 200–1,750 ZEN: 20–400
Fusarium moniliforme (*F. verticillioides*)	Fumonisins B1, B2	Maize, maize products, sorghum, and asparagus	B1 + B2: 200–4,000
Penicillium verrucosum, Aspergillus ochraceus	Ochratoxin A	Cereals, wine, fruits, coffee, and spices	0.50–80
Penicillium expansum	Patulin	Apples, apple juice, and apple products	25–50
Aspergillus, Penicillium, and *Monascus*	Citrinin	Cereals, red rice, fruits, and cheese	Nr
Aspergillus versicolor	Sterigmatocystin	Cereals, coffee, ham, pepper, and cheese	Nr
Hypocreales (*Claviceps purpúrea*), Eurotiales	Ergot alkaloids	Cereals	Nr

Source: Adapted from Arroyo-Manzanares et al. (2014), Creative Commons Attribution 3.0 International.
nr: not regulated.
[a] Maximum permitted level range in EU.
[b] Recommended level.

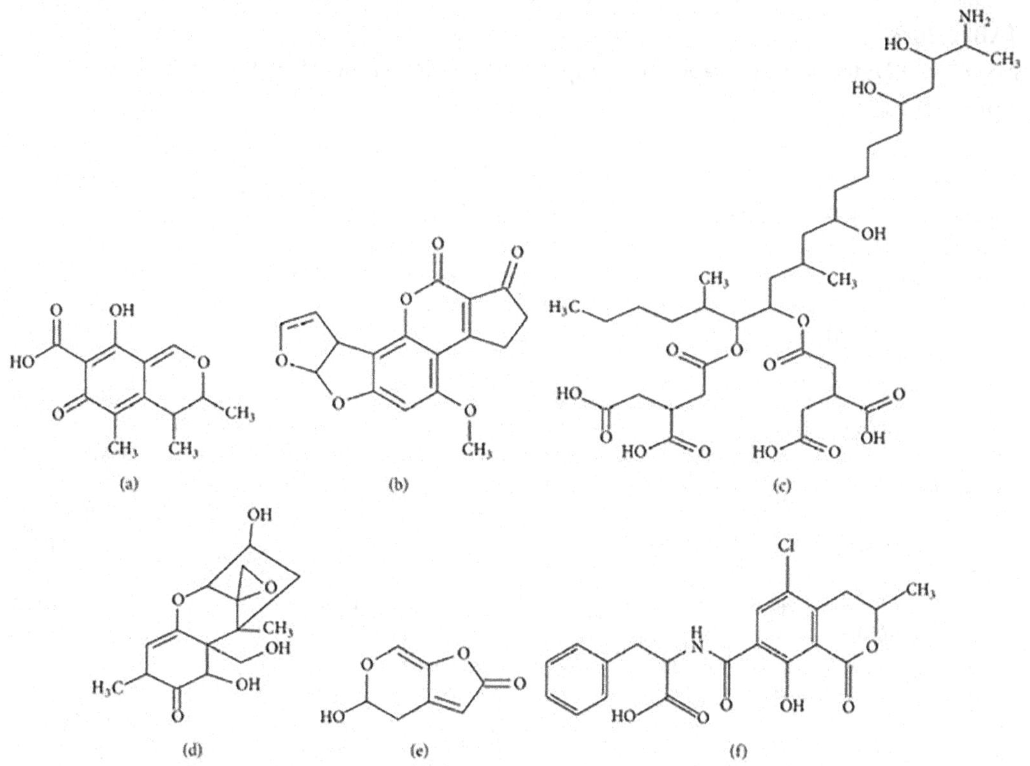

FIGURE 20.1 Structures of mycotoxins: (a) citrinin, (b) aflatoxin B1, (c) fumonisin B1, (d) deoxynivalenol, (e) patulin, and (f) ochratoxin A. (Adapted from Arroyo-Manzanares et al. (2014), Creative Commons Attribution 3.0 International.)

The most significant mycotoxins in terms of food safety are fumonisins (FUMs), trichothecenes, zearalenone (ZEA), aflatoxins (AFLAs), ergot alkaloids, and ochratoxin A (OTA), which have mutagenic, carcinogenic, cytotoxic, neurotoxic, teratogenic, nephrotoxic, estrogenic, dermotoxic, and immunotoxic effects (Gámiz-Gracia et al. 2020) (Figure 20.1). Mycotoxins in food and feed have detrimental effects and impose a challenge for public health (Pereira et al. 2014). For mycotoxins detection in food, many countries implemented specific limits for mycotoxins and have given directives for the maximum permitted levels for mycotoxins (European Commission 2006, 2007; Juan et al. 2012).

Due to varied mycotoxins in agriculture products, there is a need for new approaches capable of simultaneously identifying several mycotoxins. In the past and till now, various analytical methods are used to assess the low doses of mycotoxins, which require toxin extraction from the matrix using chromatographic methods (Krska et al. 2008). The selection of an analytical procedure primarily depends on mycotoxins' nature (Soares et al. 2018). Chromatographic techniques are mainly used for the identification of mycotoxins, especially in cereals. The use of expensive tools, specialized technicians, and laborious sample preparation steps are some factors that limit the extensive application of chromatography. Liquid chromatography and tandem mass spectrometry (LC–MS/MS) is the primary technique for analyzing many mycotoxins. Because of its high accuracy and sensitivity, this technique has determined diverse mycotoxins in the same matrix.

During the past years, LC–MS/MS has shown great potential in determining traces of residues that have low molecular weight (Anumol et al. 2017). LC–MS facilitates the simplified sample preparation steps by QuEChERS (quick, easy, cheap, effective, rugged, and safe) from complex matrices saving time and minimizing the overall cost (Shephard 2016; Zhao et al. 2017). MS detectors

TABLE 20.2
Mycotoxin Detection in Grape and Grape Products Based on Mass Spectrometry Approaches

Mycotoxin(s)	Commodity	Extraction/Clean-Up Procedure	Analytical Method
Eighteen mycotoxins, including AFUMs, AOH, DON, FUMs, HT-2, NIV, OTA, T-2, ZEN, etc.	Grape, grape juice, wine	LLE	UHPLS-MS/MS
AOH, AME, TEN, TeA	Apple, grape, and blueberry juice	QuEChERS/d-SPE	HPLC-ESI-MS/MS
ALS, ALT, AME, AOH, TEN, TeA	Grapes	QuEChERS	UHPLC-MS/MS
AFUMs, FB_{1-2}, trichothecenes	Grape	-	HPLC–MS/MS
Apicidin, beauvericin, enniatins, etc.	Dried grapes	Ultrasound-assisted SLE	HPLC–MS/MS
AFUMs	Wine (red and white)	LLE	DART-HRMS
AFUMs, ergocornine, FB_1, OTA, ZEN,	Red wine	SPE (polymeric-type cartridges)	HPLC-ESI-TOF-MS
Thirty-six mycotoxins include AFUMs, DAS, DON, FUMs, HT-2, NIV, OTA, T-2, etc.	Wine	QuEChERS	UPLC-MS/MS
Thirty-three mycotoxins include AFUMs, AOH, DAS, DON, FUMs, HT-2, NIV, OTA, T-2, etc.	Raisins	LLE	HPLC-ESI-MS/MS
Fourteen mycotoxins, include AFUMs, AOH, DON, FUMs, HT-2, OTA, PAT, T-2, ZEN, etc.	Wine (red and white)	SPE	UHPLS-MS/MS
Twenty mycotoxins, include AFUMs, DAS, DON, FUMs, HT-2, OTA, T-2, ZEN, etc.	Grape, wine	QuEChERS/various d-SPE and sorbent mixtures	UHPLS-MS/MS
Alternaria spp. toxins (AA-III, AAL TB_1, and TB_2, ALS, ALT, isoALT, AME, AOH, ATL, ATX-I, ATX-II, STTX-III, TEN, TeA)	Currant juice, grape juice, wine (red and white)	Diatomaceous earth SPE	HPLC-(ESI, APCI, APPI)-MS/MS

Source: Adapted from Kizis et al. (2021) with modifications, Creative Commons Attribution 4.0 International.

applied for mycotoxin examination include ion trap, triple quadrupole (QQQ), time-of-flight (TOF), orbital ion trap (OIT), and hybrid structures by coupling two different kinds of analyzers (Hird et al. 2014). Many quantitative studies have been published to determine mycotoxins in cereals, sugarcane, coffee beverages, animal-derived foods, fish, etc., using LC–MS/MS (Chen et al. 2013; Jeyakumar et al. 2018; Malachová et al. 2018; Deng et al. 2020). The main limitations of LC–MS methods include matrix influences and isobaric interference to determine foodstuff contaminants (Malachová et al. 2018) (Table 20.2).

To increase the mass resolution in complex matrices, there is a need to decrease the signal-to-noise ratio using high-resolution mass spectrometry (HRMS; Lacina et al. 2012). HRMS analyzers function in full scan mode and therefore provide target, post-target, and non-target analyses in a single run without consuming time and energy by optimizing multiple reaction monitoring settings for each analyte (Murray et al. 2013). High-performance liquid chromatography (HPLC) is an advanced technique with high resolution (Guo et al. 2019). HPLC combined with fluorescence detector is used to detect ochratoxin, with several advantages such as high sensitivity and reliability in a single round (Lai et al. 2015), without the presence of the chromophore. However, for other mycotoxins such as fumonisins, derivatization is needed as a mandatory step because they don't contain chromophores in their chemical structure (Ma et al. 2012). Electrospray ionization mass spectrometry (ESI-MS) is highly suited for the identification of mycotoxins, particularly for less volatile toxins. Bi et al. (2018) reported the application of ultra-performance liquid chromatography (UPLC)–ESI-MS/MS to determine 11 of the most crucial mycotoxins (AFG2, AFG1, AFB2, AFB1, citrinin (CIT), FB1, FB2, HT-2, T-2, ZON, and OTA) in 10 batches of *Panax ginseng* samples.

Applications of biosensors in the food industry may reduce a load of mycotoxins in food products by providing potential benefits, including a fast, accurate, cost-effective sample analysis, stability, reproducibility, and on-site examination of samples (Pirinçci et al. 2018). There are many reports of detecting mycotoxins from peanut, corn, cereal, red wine, milk, and juice using biosensors (Gu et al. 2019; Agriopoulou et al. 2020). Hence for fast and rapid mycotoxin analysis, immunoassays and biosensors are preferred. LC–MS is still holding a remarkable place as a reliable analytical method for multiple mycotoxins in a food matrix due to its sensitivity and accuracy.

20.2 APPLICATION OF LIQUID CHROMATOGRAPHY–TANDEM MASS SPECTROMETRY (LC–MS/MS) IN MYCOTOXINS ANALYSIS

Various chromatographic methods with selective detection and accuracy have been discovered to detect and quantify mycotoxins in food samples. The chromatographic techniques include gas chromatography (GC), thin-layer chromatography, liquid chromatography combined with mass spectrometry (LC–MS), and HPLC. LC is fast, simple, cheap, and excellent for the initial screening of mycotoxins but has limitations in terms of automation. Likewise, GC, combined with a flame ionization detector (FID) or electron capture detector or coupled with a mass spectrometer detector, can only detect volatile mycotoxins; therefore, its commercial use is minimal. On the other hand, HPLC combined with detectors such as diode array detector, ultraviolet detector, MS detector, and fluorescence detector; have a broader application in the food toxicology field. HPLC with MS (LC–MS/MS) is one of the best options to detect various mycotoxins, including OTA, AFT, ZEN, DON, patulin (PAT), CIT, and fumonisins (FUM) in complex food matrices.

The fundamental principle of MS/MS is mainly based on the selection and separation of precursor ions and quantification of the mass-to-charge ratio (*m/z*) ratio of product ions. MS/MS can be carried out in terms of space or time, where QQQ is used in space, and quadrupole-time-of-flight (QqTOF) and Orbitrap hybrid is used in time.

Orbitrap hybrid is designed from trap ions that consist of either 3-dimensional quadrupole ion traps, linear ion traps (LIT), or Fourier transform ion cyclotron resonance. The food samples can be examined for ZEA and trichothecenes (TCTs) using the QqQ LC–MS/MS technique, atmospheric pressure chemical ionization, and ESI interfaces. ESI–QqQ–MS in all polarity modes is best for identifying three FUM, TCTs, alpha-zearalenol, and ZEA. Also, LC–ESI–QqQ–MS/MS in the positive-ion mode may be operated for the simultaneous determination of diverse mycotoxins.

The ESI in positive mode usually comprises HPLC or UHPLC coupled with a mass spectrometer (Figures 20.2 and 20.3). However, QqQ and QLIT mass spectrometers are standard techniques for identifying targeted mycotoxins present in food products. Quite recently, a QqQ/QLIT method has been introduced for the identification of 300 different mycotoxins along with their metabolites. Improvement in sensitivity and high scan speed led to further QTOF and Orbitrap MS/MS techniques. The resolving power of TOF and Orbitrap mass spectrometers is in the range of 10,000–100,000 and 140,000–240,000 *m/z*. This shows their sensitivity and detection power to identify and quantify the mycotoxins present in complex food matrices. A Q-OIT hybrid technique for mycotoxins detection in food specimens provides a high-performance quadrupole range for precursor ions along with remarkable resolution mass detection. LC–MS/MS is regarded as a top priority method for detecting various mycotoxins in food matrices, but due to changes in different properties of these mycotoxins, improved LC–MS/MS methods and sample preparation techniques are required. The new methods should be simple, fast, cheap, robust, and accurate in detecting various mycotoxins in complex food matrices in a single run (Tables 20.3 and 20.4).

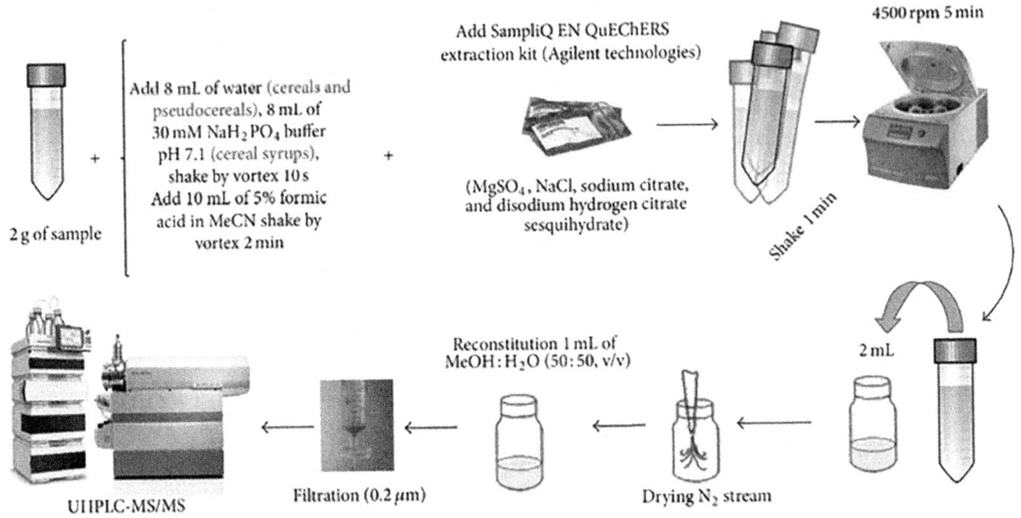

FIGURE 20.2 Sample preparation for mycotoxin detection in cereals and pseudocereals. (Adapted from Arroyo-Manzanares et al. (2014), Creative Commons Attribution 3.0 International.)

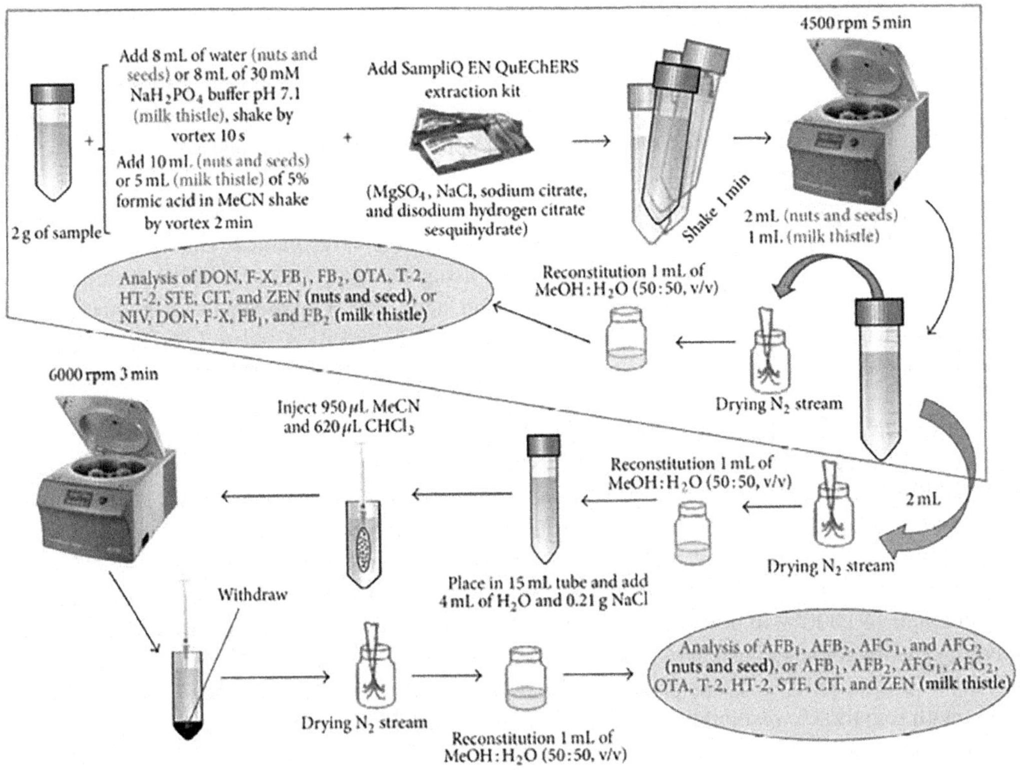

FIGURE 20.3 Sample preparation for mycotoxin detection in nuts, seeds, and milk thistle. (Adapted from Arroyo-Manzanares et al. (2014), Creative Commons Attribution 3.0 International.)

TABLE 20.3
Selected LC/MS-MS Studies in Mycotoxin Analysis in Foods

Mycotoxin	Sample	Extraction Solution	Extraction Method	Clean-Up	LOD	LOQ
AFUMs	Walnut kernel	Methanol–water (70:30, *v/v*)	Sonicating	Self-made amino-function nanometer Fe_3O_4 magnetic polymer SPE	0.004–0.013 μg/kg	0.012–0.042 μg/kg
AFUMs, OTA *Fusarium* mycotoxins	Cereals and derived products	Methanol–water (60:40, *v/v*)	Blending	IAC	1 μg/kg for AFUMs and OTA 5–30 μg/kg for *Fusarium* toxins	Nd
5 *Alternaria* mycotoxins, CIT	Tomato and tomato juice	Methanol 2,4-dinitrophenylhydrazine	Vortex	SPE cartridge	1–20 μg/kg	2–50 μg/kg
4 *Alternaria* mycotoxins	Wheat kernel	Acetonitrile–water–methanol (45:45:10, *v/v/v*)	Sonicating	SPE cartridge	0.04–1.3 μg/kg	0.1–4.2
AFUMs, FB1, FB2, DON, OTA, ZEA	Brown rice	Acetonitrile with 10% (*v/v*) acetic acid	Vortex	QuEChERS	1.4–25 μg/kg	4.1–75 μg/kg
15 mycotoxins	Cow milk	Acetonitrile (2% formic acid)	Shaking	Sodium acetate	0.02–10.14 ng/mL	Nd
16 mycotoxins	Vegetable oils	85% Acetonitrile	Shaking	QuEChERS	0.04–2.9 ng/g	Nd
11 mycotoxins	Infant cereals	Acetonitrile/water/formic acid, (80:19.9:0.1, *v/v/v*)	Shaking	Nd	0.01–10.0 ng/g	0.05–50 ng/g
12 *Fusarium* mycotoxins	Beer	Acetonitrile/water (70:30, *v/v*) Acetonitrile/water (84:16, *v/v*)	Vortex	SPE cartridge	0.05–6.9 μg/L	0.15–20 μg/L
AFB1 OTA FB1 DON T2 HT-2 ZEA	Cereal-based samples	Acetonitrile–water–acetic acid (79:20:1, *v/v/v*)	Shaking	Nd	0.06–0.13 μg/kg for AFB1 0.4–0.8 μg/kg for OTA 8–16 μg/kg for FB1 20 μg/kg for DON 4–8 μg/kg for T–2 20 μg/kg for HT–2 1.6–3.2 μg/kg for ZEA	Nd
13 mycotoxins	Cereal grains	Methanol 80%, containing 0.5% acetic acid	Shaking	IAC	0.1–18.1 ng/g	0.4–54.8 ng/g
20 mycotoxins	Soybean Paste	Methanol–water (60:40, *v/v*) and PBS	Blending	IAC	0.06–4.68 μg/kg	0.17–13.24
6 *Alternaria* toxins	Grapes	Acetonitrile and dispersive solid-phase extraction	Shaking	QuEChERS	0.03–0.21 μg/kg	0.09–0.48 μg/kg
AFUMs, ZEA, α-ZOL	Vegetable oils	Acetonitrile	Shaking	QuEChERS	Nd	0.5 μg/kg for AFUMs 1 μg/kg for ZEA and α-ZOL

Source: Adapted from Agriopoulou et al. (2020) with modifications, Creative Commons Attribution 4.0 International.
α-ZOL: α-zearalenol; CIT: citrinin; Nd: not described.

TABLE 20.4
Overview on Advanced LC–MS- and LC–MS/MS-Based Methods to Study the Multiple Mycotoxins in Cereals and Cereal-Based Food

Matrix	Analytes	Extraction	Clean-Up Step	LC Conditions	Chromatographic Column	Sensitivity (μg/kg) LOD	Sensitivity (μg/kg) LOQ
Maize	NIV	CH_3CN-H_2O (84 :16, v/v)	Clean-up columns, MycoSep® #226 and #227 from Romer Labs®	(i) Eluent A H_2O-CH_3OH (80:20, v/v), eluent B H_2O-CH_3OH (10:90, v/v), both containing 5 mM NH_4CH_3COO (ii) Gradient: 0.5 min 0% eluent B, linear gradient to 100% eluent B to 4.5 min, 100% eluent B to 7 min, 7.1 min 0% eluent B, re-equilibration 3 min, total run 10 min	Thermo Electronaquasil® RP-18 column	3.7	18.3
	DON					0.8	2.7
	FUS-X					1.6	5.6
	ADONs					0.9	3.5
	3ADON					3.8	13.4
	DAS					0.3	1.1
	HT2					1.0	3.5
	T2					0.3	0.8
	ZON/ZAN					0.9	3.2
Wheat and maize	18 trichothecene mycotoxins (i) Type-A trichothecenes (T-2, HT-2, T-2 triol, T-2 tetraol, DAS, MAS, NEO, DacVOL, and VOL) (ii) Type-B trichothecenes (DON, 3-acDON, 15-acDON, NIV, and FUSX) (iii) Type-D trichothecenes (SG, SH, RA, and VA)	Dilute and shoot	MycoSep®-226	(i) Eluent A CH_3OH-H_2O-CH_3COOH (10:89:1, v/v/v), eluent B CH_3OH·H_2O-CH_3COOH (97:2:1, v/v/v), both containing 5 mM NH_4CH_3COO- (ii) Gradient 2 min at 100% eluent A, linear increase to 100% eluent B within 12 min, held at 100% eluent B for 3 min, re-equilibration at 100% eluent A for 4 min, total run 19 min	Gemini® C18 column, 150×4.6 mm i.d., 5 μm particle size, equipped with a C18 4×3 mm i.d. security guard cartridge (all from Phenomenex, Torrance, CA, US)	[0.064–0.41] [0.092–0.70] [0.20–0.38]	[0.22–1.4] [0.31–2.2] [0.67–1.3]
Maize, durum wheat, corn flakes, and maize crackers	DON	2-step extraction: (1) H_2O; (2) CH_3OH. Evaporation and redissolution in PBS before IAC	AOFZDT2TM column at 1–2 drops per second; the column was then washed with 20 mL	(i) Eluent A H_2O, eluent B CH_3OH, both containing 0.5% CH_3COOH and 1 mM NH_4CH_3COO- (ii) Gradient 3 min at 20% eluent B, jump to 40% eluent B, linear increase to 63% eluent B within 35 min, 63% eluent B for 11 min, re-equilibration at 20% eluent B for 10 min, total run 59 min	Gemini1 C18 column (150 mm, 2 mm, 5 mm particles; Phenomenex, Torrance, CA, USA), preceded by a Gemini C18 guard column (4 mm, 2 mm, 5 mm particles)	4.2	-
	AFG_2					0.8	
	AFG_1					0.4	
	AFB_2					0.3	
	AFB_1					0.6	
	FB_1					1.1	
	FB_2					0.4	
	OTA					0.6	
	HT-2					1.9	
	T-2					1.5	
	ZEA					0.7	

(Continued)

TABLE 20.4 (*Continued*)
Overview on Advanced LC–MS- and LC–MS/MS-Based Methods to Study the Multiple Mycotoxins in Cereals and Cereal-Based Food

Matrix	Analytes	Extraction	Clean-Up Step	LC Conditions	Chromatographic Column	Sensitivity (μg/kg) LOD	LOQ
Various foods and feed	ATB_1	CH_3CN-H_2O (84:16, v/v)	Mycosep #226 and #228 Aflazon+ multifunctional cartridges	(i) Eluent A ESI+ 10 mM NH_4CH_3COO-, ESI– 0.1% 0.1% (v/v) aqueous NH_3, eluent B CH_3OH (ii) Gradient initially 20% eluent B, linear increase from 5.5% to 85% eluent B, 100% eluent B within 0.3 min, re-equilibration for 2 min at 20% eluent B, total run 10 min	UPLCBEH C18 column (1.7 μm, 100 mm × 2.1 mm i.d., Waters)	0.003	0.01
	ATB_2					0.003	0.01
	ATG_1					0.003	0.01
	ATG_2					0.006	0.02
	ATM_1					0.003	0.01
	T-2					0.060	0.20
	HT-2					0.006	0.02
	VCG					0.006	0.02
	CTN					0.101	0.35
	OTA					0.064	0.21
	3-ADON					0.182	0.60
	15-ADON					0.182	0.60
	FX					0.152	0.50
	ZON					0.091	0.30
	NIV					0.182	0.60
	DON					0.212	0.70
Rice, corn, wheat, rye, oat, barley, infant cereals, soya, and corn gluten	$AflB_1$	QuEChERS	n-hexane (5 mL) under agitation and centrifugation	(i) Eluent A 0.15% (v/v) HCOOH+ 10 mM NH_4HCOO-, eluent B 0.05% HCOOH (v/v) in CH_3OH (ii) Gradient: 0% eluent B at 1 min, linear increase to 100% eluent B until 15 min, 100% eluent B for 5 min, re-equilibration at 0% eluent B for 5 min, total run 25 min	ZorbaxBonus-RP column 150 mm 2.1 mm i.d., 3.5 μm, equipped with a ZorbaxRB C8 guard column 12.5 mm, 2.1 mm i.d., 5 μm (both from Agilent Technologies, Geneva, Switzerland)	-	[1–10]
	$AflB_2$						[1–10]
	$AflG_1$						[1–10]
	$AflG_2$						[1–10]
	DON						[50–250]
	NIV						[100–500]
	15-AcDON						[50–250]
	FUSX						[25–125]
	NEO						[25–125]
	HT-2						[25–125]
	T-2						[25–125]
	FB_1						[5–25]
	FB_2						[50–250]
	ZON						[50–250]
	OTA						[50–250]
							[0.50–2.5]

(*Continued*)

TABLE 20.4 (*Continued*)
Overview on Advanced LC–MS- and LC–MS/MS-Based Methods to Study the Multiple Mycotoxins in Cereals and Cereal-Based Food

Matrix	Analytes	Extraction	Clean-Up Step	LC Conditions	Chromatographic Column	Sensitivity (μg/kg) LOD	LOQ
Barley	3ADON, 15ADON, DON, DON-3-Glc, FUSX, NIV, HT2, T2, DAS, NEO, AFUMs, OTA, FUMs, STER, ZEN, penitrem A, BEA, Alternaria toxins, ergot alkaloids	QuEChERS (2 g sample, 10 mL 0.1% HCOOH in H_2O, 3 min shaking, 10 mL CH_3CN, 3 min shaking, 4 g $MgSO_4$, 1 g NaCl, shaking)	C18-SPE clean-up procedure was performed with Oasis HLB cartridges (150 mg) from Waters (Milford, MA, USA)	(i) Eluent A H_2O with 5 mM NH_4HCOO- and 0.1% HCOOH, eluent B CH_3OH (ii) Gradient: start with 5% eluent B, increase to 50% eluent B in 6 min, increase to 95% eluent B within 4 min, keep until 15 min of the run, re-equilibration at 5% eluent B for 3 min	Acquity UPLC HSS T3 analytical column (100 mm, 2.1 mm i.d., 1.8 mm; Waters, Milford, MA, USA)	-	-
Barley and barley products	T-2, HT-2, T-2 triol, T-2 tetraol, DAS, NEO, DON, NIV, 3-AcDON, 15-AcDON, FUSX	Acetonitrile:water (84/16)	A MycoSep® column (no. 226, CoringSystem Diagnostix, Germany)	A binary linear gradient was applied, which consisted of eluent A (methanol + 5 mmol/L ammonium formate) and eluent B (water + 5 mmol/L ammonium formate) with a total flow rate of 0.4 mL/min: 0 min 95% B, 11 min 95% B, 22 min 35% B, 26 min 35% B, 27 min 95% B, 35 min 95% B	Synergi™ polar-RP® 150×2 mm, 4 μm (Phenomenex, Aschaffenburg, Germany)	[0.02–2.25]	-
Cereal and cereal products	NIV	Extraction with matrix solid-phase dispersion (MSPD) method	—	The gradient that started at 100% A (5 mM ammonium formate in water) and 0% B (5 mM ammonium formate in methanol) increased linearly to 100% B in 10 min, followed by a linear decrease to 80% B in 5 min, then to 70% B in 10 min. Afterward, the initial conditions were maintained for 5 min.	GeminiNX C18 (150 mm, 4.6 mm ID, 5 μm particle size) analytical column supplied by Phenomenex (Barcelona, Spain), preceded by a guard column C18 (4 mm, 2 mm ID.)	-	85.24
	DON						31.25
	AFB_1						0.25
	AFB_2						1.50
	AFG_1						0.25
	AFG_2						0.75
	DAS						5.00
	FB_1						83.33
	FB_2						83.75
	HT-2						35.5
	T-2						12.50
	OTA						3.00
	BEA						1.00

(*Continued*)

TABLE 20.4 (*Continued*)
Overview on Advanced LC–MS- and LC–MS/MS-Based Methods to Study the Multiple Mycotoxins in Cereals and Cereal-Based Food

Matrix	Analytes	Extraction	Clean-Up Step	LC Conditions	Chromatographic Column	Sensitivity (µg/kg) LOD	LOQ
Maize, wheat, rye, oat, oat flakes, and flours (maize, wheat, rye, and oat)	MON	Acetonitrile/water (84/16)	1M hydrochloric acid from *n* exchanger material (SAX). The SAX column (Bond Elut-SAX, 500 mg, 3 mL) (Agilent Technologies, Böblingen, Germany)	Solvent A: 1% formic acid in methanol, and solvent B: 1% formic acid in water. The detection was set to 260 nm. An isocratic run at 20% A was performed for 10 min at a flow rate of 250 µL/min. An isocratic run at 95% A was used for 10 min	A 150 mm×2.1 mm i.d., 5 µm, Synchronis HILIC with a 10 mm×2.1 mm i.d. guard column (Thermo Scientific, Dreieich, Germany), a 150 mm×2.1 mm i.d., 3.5 µm	0.7	2.5
Maize	FMs	Water/methanol (30/70)	Sep-Pak C18 cartridges	Gradient elution was performed using bidistilled water (eluent A) and acetonitrile (eluent B), both acidified with 0.2% formic acid: initial condition at 100% A, 0–5 min isocratic step, 5–30 min linear gradient to 100% B, 30–35 min isocratic step, 35–36 min linear gradient to 100% A and re-equilibration step at 100% A for 14 min (total analysis time: 50 min)	A 250×2.1 mm i.d., 5 mm, XTerra C18; the flow rate was 0.2 mL/min	20	-
Cereals and cereal products (wheat, wheat-based noodles, rice, rice-based noodles, and corn)	NEO DAS T-2 HT-2 DON NIV 15-ac-DON FUSX	QuEChERS method	—	Mobile phase A consists of 1% acetic acid and 5 mM ammonium acetate in water, and mobile phase B consisted of 1% acetic acid and 5 mM ammonium acetate in methanol. The gradient was changed to 80% mobile phase B over 10 min and then maintained for 3 min. After 13 min of run time, the gradient was returned to 30% mobile phase B over 1 min	ZORBAX Eclipse XBD-C18, 2.1 mm, 100 mm, 1.8 mm (PN 928700-902) column	0.02 0.02 0.05 0.02 0.045 0.05 0.02 0.04	-

(*Continued*)

TABLE 20.4 (*Continued*)
Overview on Advanced LC–MS- and LC–MS/MS-Based Methods to Study the Multiple Mycotoxins in Cereals and Cereal-Based Food

Matrix	Analytes	Extraction	Clean-Up Step	LC Conditions	Chromatographic Column	Sensitivity (μg/kg) LOD	LOQ
Cereals (rice, wheat, oat, barley, and maize)	Acetonitrile/water/acetic acid (79/20/1)	AFUMs, OTA, ZEN, DON, FB_1, FB_2, T-2, HT-2	—	Different proportions of mobile phase consisted of methanol or acetonitrile and acetic acid (0%–1%), different flow rates (0.2–0.3 mL/min)	A column, 150 mm, 4.6, 3 μm particle size C18 columns (Thermo Scientific, CA, USA)	[10^{-5}–0.02]	[2×10^{-5}–0.04]
Wheat	NIV	Acetonitrile/water (84/16)	A SecurityGuard™ cartridge C18 (4.0 3.0 mm i.d. 5 μm).	Mobile phase A consisted of an $H_2O/CH_3OH/CH_3COOH$ mixture (89:10:1, v/v/v) containing 5 mM ammonium acetate, while mobile phase B: $H_2O/CH_3OH/CH_3COOH$ mixture (2:97:1, v/v/v) containing 5 mM ammonium acetate. The following gradient was applied: initial condition 55% B; 0–3 min, 70% B; 3–8 min, 100% B; 8–11 min constant at 100% B; 11–13 min returning to the initial conditions and maintain during 2 min 55% B.	Phenomenex (Castel Maggiore, Italy) Gemini C18 (150 mm, 2.0 mm, i.d. 5 μm particle size, 110A)	5.5	15
	DON					1	10
	FUSX					5.5	20
	15-AcDON					2	10
	3-AcDON					2	10
	DAS					2	10
	NEO					5.5	20
	HT-2					2	6
	T-2					3	8
	ZEN					1.5	5
	α-ZOL					1.5	5
	β-ZOL					2	2.5
Maize and other cereals (sorghum, millet, rice, sesame, wheat, infant food, cuscus, cornflakes, and cookies)	AFB_1, AFB_2, AFG_1, AFG_2, AFM_1, FB_1, FB_2, FB_3, OTA, DON, NIV, ZEN, MON, CIT, ENA, ENA1, ENB, ENB1, ENB2, BEA, STC	Acetonitrile/water/acetic acid (79/20/1)	—	Both eluents contained 5 mM ammonium acetate and were composed of methanol/water/acetic acid 10:89:1 (v/v/v; eluent A) or 97:2:1 (eluent B), respectively. After an initial time of 2 min at 100% A, the proportion of B was increased linearly to 100% within 12 min, followed by a hold-time of 3 min at 100% B and 4 min column re-equilibration at 100% A	A Gemini C18 column, 150×4.6 mm, 5 μm particle size, equipped with a C18 4×3 mm guard cartridge, all from Phenomenex (Torrance, CA, USA).	[0.005–250]	-

(*Continued*)

TABLE 20.4 (*Continued*)
Overview on Advanced LC–MS- and LC–MS/MS-Based Methods to Study the Multiple Mycotoxins in Cereals and Cereal-Based Food

Matrix	Analytes	Extraction	Clean-Up Step	LC Conditions	Chromatographic Column	Sensitivity (μg/kg) LOD	LOQ
Breakfast and infant cereals	FB_1, FB_2, FB_3	Acetonitrile/water (85/15)	A C18 security guard cartridge (4 mm×2 mm i.d., 5 μm), both Phenomenex (Madrid, Spain)	The sing gradient elution with water as mobile phase A and methanol as mobile phase B, both containing 0.5% formic acid. After an isocratic step of 65% B for 3 min, it was gradually increased to 95% B in 4 min and held constant for 3 min. Afterward, the initial conditions were maintained for 10 min.	A Luna C18 analytical column (150 mm×4.6 mm i.d., 5 μm) Phenomenex (Madrid, Spain)	-	-
Barley, maize breakfast cereals, and peanuts	AFB_1	CH_3CN-H_2O-CH_3COOH (79.5:20:0.5, v/v/v). Evaporation and redissolution in PBS before IAC		(i) Eluents A H_2O, eluent B CH_3OH, both containing 5 mM $NH_4CH_3COO^-$ (ii) Gradient 5% eluent B increased to 50% eluent B in 1 min, linear increase to 100% eluent B within 6 min, 100% eluent B to 8 min, at 8.1 min initial conditions 5% eluent B, re-equilibration at 5% eluent B for 2 min, total run 10 min		0.05	0.1
	AFB_2					0.05	0.1
	AFG_1					0.05	0.1
	AFG_2					0.05	0.1
	FB_1					5	10
	FB_2					5	10
	FB_3					5	10
	DON					1	5
	ZEN					1	5
	HT2					0.5	1
	T2					0.5	1
	OTA					0.1	0.25
Maize and maize-beer	69 mycotoxins	Extraction with many solvents (ACN/water/glacial acetic acid 79:20:1, v/v/v)	—	(i) Two eluting solvents (eluent A and eluent B) that each contained 5 mM ammonium acetate were prepared using MeOH/water/glacial acetic acid (10:89:1, v/v/v) (eluent A) and (97:2:1, v/v/v) (eluent B) (ii) After an initial time of 2 min at 100% eluent A, the proportion of eluent B was increased linearly to 50% within 2–5 min and to 100% within 5–14 min, followed by a holding time of 4 min at 100% eluent B and 2.5 min column re-equilibration at 100% eluent A	A Gemini C18 column (Phenomenex, Torrance, CA, US).	[0.05–0.14]	[3–41]

(*Continued*)

TABLE 20.4 (*Continued*)
Overview on Advanced LC–MS- and LC–MS/MS-Based Methods to Study the Multiple Mycotoxins in Cereals and Cereal-Based Food

Matrix	Analytes	Extraction	Clean-Up Step	LC Conditions	Chromatographic Column	Sensitivity (μg/kg)	
						LOD	LOQ
Maize and cereal-based products	AFB_1	Raw extract	—	(i) Eluent A H_2O–HCOOH (99.9:0.1, v/v), eluent B CH_3OH–HCOOH (99.9:0.1, v/v) both containing 5 mM NH_4HCOO- (ii) Gradient: 0.5 min at 30% eluent B, linear increase to 100% eluent B in 7.5 min, hold at 100% eluent B for 1.5 min, at 9.6 min back to 30% eluent B, re-equilibration at 30% eluent B for 2 min, total run 11.5 min	A ZORBAX RRHD Eclipse Plus C18 (100×2.1 mm, 1.8 μm) column from Agilent Technologies	0.04	0.1
	AFB_2					0.04	0.1
	AFG_1					0.2	0.1
	AFG_2					0.1	0.4
	DON					3.4	1
	FB_1					1.4	4.3
	FB_2					1.3	3.9
	HT2					0.8	2.5
	OTA					0.1	0.4
	T2					0.1	0.2
	ZEN					1.2	2.9
Barley, malt, oat, wheat, and maize	DON	CH_3CN-H_2O (84:16, v/v),	—	(i) Eluent A H_2O-HCOOH (99.9:0.1, v/v), eluent B CH_3OH-HCOOH (99.9:0.1, v/v); gradient ESI– 2 min at 10% eluent B, linear increase to 99% eluent B in 6 min, hold at 99% eluent B for 7.5 min, for 2 min back to 10% eluent B, re-equilibration at 10% eluent B for 9.5 min, total run 25 min; ESI+ 2 min at 10% eluent B, linear increase to 87% eluent B in 6 min, hold at 87% eluent B for 7 min, increase to 100% eluent B in 5 min, hold at 100% eluent B for 3.5 min, for 2 min back to 10% eluent B, re-equilibration at 10% eluent B for 9.5 min.	A Shimadzu LC-20A Prominence system (Shimadzu, Kyoto, Japan) using a Hydrosphere RP-C18 column (150×3.0 mm^2, S-3 μm, 12 nm, YMC Europe GmbH, Dinslaken, Germany).	0.9	2.6
	3-ADON					1.7	13.5
	15-ADON					4.6	13.5
	HT2					0.1	0.2
	T2					0.2	0.5
	BEA					0.3	0.7
	FUSX					5	3.1
	NIV					5	2.9
	ZEA					0.5	1.5

Source: Adapted from Smaoui et al. (2020) with modifications, Creative Commons Attribution 4.0 International.
n.d.: not determined.

20.3 DETERMINATION OF MYCOTOXINS WITH LC–HIGH-RESOLUTION MASS SPECTROMETRY (HRMS)

Rapid Alert System for Food and Feed (RASFF) has notified the mycotoxins as one of the top 10 food hazards reported mainly in nuts and cereals (RASFF 2021). Such a drastic state makes it crucial to develop sensitive analytical protocols for accurate mycotoxins detection in various food matrices. In this view, LC–MS/MS is a powerful technique for the unique quantification of different analytes. In addition, the powerful application of HRMS includes the revealing of novel mycotoxins types, identification of untargeted complexes, proficiency in recording complete scan spectra by way of measuring the correct masses of compounds, a high level of structural validation, and retrospective evaluation data that provides an option to investigate those elements that have not been considered during spectral acquisition (Natalia et al. 2021).

Narváez et al. (2020), who have developed an LC–HRMS tool to identify mycotoxins released by *Cannabis sativa* fungi, recently analyzed the scarcely studied food matrix named Cannabidiol nutritional supplements (Narváez et al. 2020). They identified six unique *Fusarium* mycotoxins in seven samples; the most prevailing ones were ZEN with 60% abundance and ENN B1 with 30% abundance.

Another subject seeking great attention is the biological monitoring of body fluids as a complementary approach to assess the health risk. Although metabolic pathways are still to be investigated comprehensively for all types of mycotoxins, most existing LC–MS tools used to evaluate personal exposure focus mainly on identifying parent analytes. In this view, Slobodchikova et al. (2019) accomplished human microsomal *in vitro* incubations for 17 mycotoxins to characterize all the released metabolites by LC–HRMS. These results made them develop an extensive LC–MS archive. Moreover, this method may also investigate 188 additional analytes, including 100 different metabolites isolated for the first time. Such contributions make the LC–HRMS method one of the most comprehensive tools for biomonitoring of mycotoxins and metabolites.

20.4 LC-ELECTROSPRAY IONIZATION (ESI)–MS/MS FOR MYCOTOXINS

Mycotoxins are a category of poisonous, fungal chemicals (molds) produced in various food commodities, including cereals, nuts, spices, berries, oilseeds, or coffee. These toxins can be created before and even after harvesting when the environment is favorable for fungal development. Mycotoxins are beyond doubt one of the most severe threats to food safety exacerbated by global climate change. Fungi's capacity to generate mycotoxins depends largely on precipitation, temperature, plant stress, and relative humidity (Paterson and Lima 2010). For example, fungi that produce aflatoxins (in particular, *Aspergillus* species like *A. parasiticus* and *A. flavus*), which are generally common in (humid) tropical and subtropical regions, are likely to be in areas such as eastern and southern Europe or the United States where temperatures above 30°C (that temperature is the optimal one for aflatoxin production) due to these conditions they spread rapidly (Battilani et al. 2016).

Any of the few hundred mycotoxins found so far threaten food safety. Maximum level (ML) in foodstuffs for mycotoxins have been developed globally to protect public health. MLs for aflatoxins (AFLA), OTA, fumonisins (FUMs), ZEN and deoxynivalenol (DON) (EC 1881:2006) (EC 1881:2006) (European Commission 2006) and recently suggestive amounts of T-2 and HT 2 toxins (EC 165/2013) (European Commission 2013) in a wide variety of food crops were set out in EU regulations (often considered the most stringent). Lower MLs food for babies and young children has also been developed (Smith et al. 2016).

LC–MS/MS is widely used in various food matrices for quantitative and qualitative determination of mycotoxins (Fan et al., 2016; Paoloni et al., 2018). ESI is the most common type of liquid molecular ionization and is consistent with most chromatographic separation systems. Many factors such as the flow rate of eluent, analyte form, solvent method, the voltage applied, and gas drying are responsible for ionization efficacy. The validation and development of the LC–MS/MS method, quantitative/qualitative for optimal selectivity and sensitivity, is always time-intensive and complicated (Sargent 2013). So, to improve separation efficiency, it is necessary to optimize factors that affect peak zone and retention time (Tahboub 2014). System architecture and optimization typically occur using the standard univariate approach in LC–MS/MS (Szerkus et al. 2016). The influence of one parameter is explored in this method by holding the other variables constant and changing the value of one variable at a time. The time-consuming univariate optimization technique (Low et al. 2012), unable to measure many (and not economic) components at a time (Low et al. 2009), and interactions with variables that can influence the source of ions are not evaluated. These problems can be solved by a multivariate optimization technique (Dos Santos Depoi et al. 2012). The multivariate optimization approach effectively explores and has successfully optimized LC–MS/MS analyses (Rodrigues et al. 2014; Karapinar et al. 2016).

Many methods of pre-treatment, such as solid-phase extraction (SPE) analysis, solid–liquid extraction, and liquid–liquid extraction in food samples, have been introduced to reduce matrix results. The QuEChERS is the most common sample pre-treatment method for mycotoxins detection and other food-species pollutants analysis (González-Jartín et al. 2019).

Some studies use the chemometric strategy to optimize the LC–MS/MS system to improve ionization with a shorter retention time (analytical time) and the highest peak area. The chemometric approach was taken to assess significant factors using Plackett–Burman design (PBD). In the meantime, the Box–Behnken architecture was introduced to optimize major factors defined by the PBD process. Compared to the time-consuming traditional univariate optimization technique, the LC–MS/MS chemometric optimization approach has better benefits and has become a practical alternative. The LC–MS/MS method was paired with the QuEChERS-Dispersive SPE (low, no, and high fat) techniques for detecting the phenomenon in different industrial food samples of multi-mycotoxins (OTA, AFB1, AFG1, AFB2, and AFG2). The determination of mycotoxin content allowed a better view of the occurrence of mycotoxins in different food matrices (Rejczak and Tuzimski 2015).

20.5 IDENTIFICATION OF MYCOTOXINS WITH HIGH-PERFORMANCE LIQUID CHROMATOGRAPHY (HPLC)–MS/MS

Mycotoxins are secondary metabolites produced by fungi that contaminate agricultural commodities that include coffee, nuts, spices, tea, peanut, corn, and grains. They are contaminated throughout the food supply chain of agricultural production, including harvesting, storage, food processing, and transportation. Several mycotoxins such as aflatoxin, ochratoxin A (OTA), ZEN, and T-2 toxin occur in agricultural products and are proven to be carcinogenic, teratogenic, and mutagenic (Zhang et al. 2016; Pakshir et al. 2020). Lately, HPLC has become the vital method for the investigation of mycotoxins in food as well as in biological samples. HPLC and a mass spectrometer are sensitive, specific, and selective (Songsermsakul and Razzazi-Fazeli 2008). Determination of mycotoxins depends on HPLC by employing an adsorbent based on the chemical and physical structure of the mycotoxin. Depending upon their polarity, normal and reversed-phase columns were used for the separation and purification. Large-scale preparative columns are used to prepare mycotoxin standards, and sample pre-treatment was done in small mini columns. Most of the procedures for mycotoxin detection by HPLC are similar. Commonly used methods are fluorescence detector or ultraviolet spectrometer that relies on a chromophore present in the molecules. Numerous mycotoxins such as aflatoxin, OTA have natural fluorescence that HPLC-fluorescence detection (FD) can easily detect, but some do not have chromophores (Turner et al. 2009). The analysis of multi-mycotoxins requires high sensitivity HPLC–MS/MS systems. As this instrumentation is usually expensive, time for analysis is also limited, and solvent consumption must be reduced. To overcome these problems, a micro-HPLC–MS/MS technique was industrialized for real-time quantification of 26 mycotoxins in 42 maize samples from the field trials of South Africa. Preparation of samples includes extraction and dilute and shoot approach without additional clean-up of grains. It is challenging to develop a fast and sensitive method that analyses mycotoxins from highly polar to non-polar components because several aspects have to be considered. While using micro-HPLC–MS/MS, the time of measurement cycle becomes a crucial parameter (Hickert et al. 2015).

The analysis and determination of the concentration of OTA in total of 480 retail spices products were done by applying extraction, OchraTestTM immunoaffinity column clean-up, and then HPLC-FD. Chili, cinnamon, aniseed, and cumin samples marketed in China showed a higher incidence of OTA in the range of 0.41–30.73 μg/kg, 0.55–2.02 μg/kg, 0.41–0.98 μg/kg, and 0.31–3.25 μg/kg, respectively (Zhao et al. 2014). The determination of 18 mycotoxins in European beers was studied using SPE procedure with HPLC–QQQ–MS/MS and ultra-high-resolution mass analyzer (Rubert et al. 2013). HPLC–MS/MS technique has been used for the analysis of mycotoxins released by *Fusarium* species such as trichothecenes B (deoxynivalenol, nivalenol, 3- and 15-acetyldeoxynivalenol, and fusarenon X) and macrocyclic lactones (α and β zearalenol and ZEN) in maize (Cavaliere et al. 2005).

20.6 GAS CHROMATOGRAPHY (GC)–MS FOR THE SENSITIVE ESTIMATION OF MYCOTOXINS

About 25% of the global agricultural output has been polluted by mycotoxins, causing economic losses to food, feed, and grain industries (FAO 1995; Cazzaniga et al. 2001). Toxicity caused by mycotoxin exposure, which exists at very low concentrations in food and feedstuffs, necessitates the use of sensitive and accurate sampling, identification, and further analysis. Mycotoxin detection and food contamination surveillance using GC–MS have become more rapid, sensitive, specific, and reproducible (Krska et al. 2008; Medina et al. 2021). When mycotoxins and their derivatives are volatiles, GC–MS may be used to determine them in a sensitive/accurate manner. In practice, the GC–MS method may not be appropriate for routine analysis, but it can be used for validation (Soleas et al., 2001).

Several safe and accurate analytical approaches have been evaluated and validated for the multiple determinations of mycotoxins in human food and animal feedstuffs (Shanakhat et al. 2018). There are plenty of established and validated GC–MS methods for detecting mycotoxins in cereals such as HT-2 toxin, DON, nivalenol, and T2 toxin (Eskola et al. 2001; Krska et al. 2001; Pettersson and Langseth 2002; Eke et al. 2004; Melchert and Pabel 2004; Kralj Cigić and Prosen 2009; Pascale 2009; Pereira et al. 2014; Kirinčič et al. 2015). Clin Elut columns were used to extract trichothecenes directly from sorghum, which were then cleaned on Florisil Sep-Pak cartridges and determined by GC–MS after hydrolysis (Black et al. 1987). Fused silica capillary GC–MS was used to detect aflatoxin B. When an FID was used, aflatoxins B1, B2, G1, and G2 could be distinguished with limit of quantification (LOQ) of 1–2 ng, and the limits of detection (LOD) were slightly lower. OTA was transformed to *O*-methyl OTA, which was detected using capillary GC–MS (Scott 1993). Also, Soleas et al. (2001) utilized GC with mass-selective detection for monitoring OTA in wines and beers. The identification and quantification limits were 0.1 and 2 g/L, respectively, while recovery and imprecision were 69%–75% and 9%–11.1%, respectively.

Tanaka et al. (2000) prepared trimethylsilyl derivatives of seven trichothecene mycotoxins (3-acetyldeoxynivalenol, diacetoxyscirpenol, DON, fusarenon X, neosolaniol, nivalenol, and T-2 toxin) and ZEN in cereals using SPE clean-up with Florisil cartridge columns. Using GC–MS, Olsson et al. (2002) determined the existence of OTA and DON for determining grain quality. Nielsen and Thrane (2001) extracted on a micro-scale. They derivatized *Fusarium* cultures to produce 15-acetoxyscirpenol, 15-acetyl-DON, 3-acetyl-DON, 4,15-diacetoxyscirpenol, 4-acetoxyscirpentriol, acetyl T-2 toxin, DON, fusarenon X, HT-2 toxin, iso-neosolaniol, neosolaniol, NIV, scirpentriol, T-2 tetraol, T-2 toxin, and T-2 triol using pentafluoropropionic anhydride before detection by GC–MS/MS. In respect to multiple mycotoxin detection by GC–MS in semolina and corn grits, A and B type trichothecenes (4,15-diacetoxy-scirpenol, DON, nivalenol, and T2-toxin) were derivatized using a clean-up cartridge containing alumina, ammonium sulfate, C18 with *N,N*-dimethyl-trimethylsilylcarbamate, celite, and charcoal. GC–FID and GC with mass selective detection (GC-MSD) determined these with an LOD from 0.30 to 0.47 mg/kg and 0.05 to 0.35 mg/kg, respectively (Eke et al.

2004). Schollenberger et al. (2007) have determined 13 trichothecenes in fermented products, infant formula, roasted soy, textured soy protein, tofu, and whole beans through GC–MS.

A novel GC–MS approach was developed to detect T-2 and HT-2 toxin analogs using ^{13}C isotope labeling (Brodacz 2004, 2007, 2008; Breidbach et al. 2007). In another study, GC–MS was used to evaluate the presence of mycotoxins like DON, ergosterol, and OTA in food grains (Turner et al., 2009). Also, GC–MS has been operated to detect a minute concentration of PAT present in fruit products and quince by injecting the PAT labeled with $^{13}C_{5-7}$ as an internal standard with a LOD and LOQ of 0.4 and 1.6 μg/kg, respectively (Cunha et al. 2009; Fernández-Cruz et al. 2010). A new heart-cutting GC–MS was introduced for the simultaneous detection and quantification of 15-acetyldeoxynivalenol, deoxynivalenol, Fusarenon X, nivalenol, and ZEN in cereals and flours (Cunha and Fernandes 2010). GC–MS was optimized and validated in popcorn (unpopped and popped) to determine the presence of several mycotoxins such as 15-acetyl-deoxynivalenol, deoxynivalenol, fusarenon X, nivalenol, and ZEN (Ferreira et al. 2012).

A GC–MS technique was also developed for the sensitive and precise quantification of mycotoxins such as PAT, trichothecenes, and ZEA in grain products (Rodríguez-Carrasco et al. 2012, 2014). GC–MS/MS detection and validation of PAT, trichothecenes, and ZEN were carried out in milled grains by Rodríguez-Carrasco et al. (2014). Moreover, volatile mycotoxins such as PAT and TCTC have been detected using GC coupled with MS detector (Pereira et al. 2014; Alshannaq and Yu 2017) via derivatization by silylation, acylation, and bromination (Köppen et al. 2010). A QuEChERS method is used for a multiple mycotoxin determination (12 A and B type trichothecenes) from the cereal-based infant formula by GC–MS (Pereira et al. 2015). McMaster et al. (2019) quantified the presence of deoxynivalenol in Sorghum by GC–MS. Although GC–MS-based mycotoxin detection provides several advantages, viz. structure, sensitivity, multiple determination, and confirmation, this approach is expensive and has problematic derivatization and poor reproducibility (Singh and Mehta 2020).

20.7 CONCLUSION

Food and food byproducts are usually contaminated by various toxins producing fungus isolates. Consequently, many potential techniques such as GC–MS, LC–MS/MS, and LC–ESI-MS/MS are continuously playing a key role in identifying mycotoxins. Samples with smaller amounts can be analyzed much faster than conventional methods. Compared to other detection methods, LC–MS/MS method is proved to be the best and more suitable method. Extraction protocol and easy clean-up and reduced matrix effects are among the few extremely important features for developing, optimizing, and advancing analytical methods. The high sensitivity and selectivity of LC–MS/MS instruments have made significant contributions to the exact quantification of multi-mycotoxins. A fully validated qualitative and quantitative LC–MS/MS and LC–HRMS were observed as essential approaches to determine mycotoxins and their metabolites in food samples.

REFERENCES

Adejumo, T. O. and Adejoro, D. O. 2014. Incidence of aflatoxins, fumonisins, trichothecenes and ochratoxins in Nigerian foods and possible intervention strategies. *Food Science and Quality Management* 31:127–46.

Agriopoulou, S., Stamatelopoulou, E. and Varzakas, T. 2020. Advances in analysis and detection of major mycotoxins in foods. *Foods* 9:518. doi:10.3390/foods9040518.

Alshannaq A. and Yu J. -H. 2017. Occurrence, toxicity, and analysis of major mycotoxins in food. *International Journal of Environmental Research and Public Health* 14(6):632. doi:10.3390/ijerph14060632.

Anumol, T., Lehotay, S. J., Stevens, J., et al. 2017. Comparison of veterinary drug residue results in animal tissues by ultrahigh-performance liquid chromatography coupled to triple quadrupole or quadrupole-time-of-flight tandem mass spectrometry after different sample preparation methods, including use of a commercial lipid removal product. *Analytical and Bioanalytical Chemistry* 409(10):2639–53.

Arroyo-Manzanares, N., Huertas-Pérez, J. F., García-Campaña, A. M. and Gámiz-Gracia, L. 2014. Mycotoxin analysis: New proposals for sample treatment. *Advances in Chemistry* 2014:547506. doi:10.1155/2014/547506.

Ashiq, S. 2015. Natural occurrence of mycotoxins in food and feed: Pakistan perspective. *Comprehensive Reviews in Food Science and Food Safety* 14:159–75.

Battilani, P., Toscano, P., Van der Fels-Klerx, H. J., Moretti, A., Leggieri, M. C., Brera, C. and Robinson, T. 2016. Aflatoxin B 1 contamination in maize in Europe increases due to climate change. *Scientific Reports* 6(1):1–7.

Bi, B., Bao, J., Xi, G., et al. 2018. Determination of multiple mycotoxin residues in *Panax ginseng* using simultaneous UPLC-ESI-MS/MS. *Journal of Food Safety* 38(4):12458.

Black, R. M., Clarke, R. J. and Read R. W. 1987. Detection of trace levels of trichothecene mycotoxins in environmental residues and foodstuffs using gas chromatography with mass spectrometric or electron-capture detection. *Journal of Chromatography A* 388(2):365–78.

Breidbach, A., Povilaityte, V., Mischke, C., Doncheva, I. and Stroka, J. 2007. T-2 and HT-2 by GC/MS for official food control – a validated method. *Poster anl. 29. Mykotoxin-Workshop* 14.-16.05.2007, Fellbach, Germany.

Brodacz, W. 2004. Entwicklung optimierter GC-Trennungen in der Mykotoxinanalytik. *LaborPraxis* 6:26–8.

Brodacz, W. 2007. Isotopenverdünnungsanalytik von Mykotoxinen mit GC/MSD. *LaborPraxis* 9:46–9.

Brodacz, W. 2008. Stabilisotopenverdünnungs-GC/MS: Hochtoxischen Mykotoxinen auf der Spur. *Chemiereport* 1:32–3.

Cavaliere, C., D'Ascenzo, G., Foglia, P., Pastorini, E., Samperi, R. and Lagana, A. 2005. Determination of type B trichothecenes and macrocyclic lactone mycotoxins in field contaminated maize. *Food Chemistry* 92(3):559–68.

Cazzaniga, D., Basilico, J., Gonzalez, R., Torres, R. and Greef, D. 2001. Mycotoxins inactivation by extrusion cooking of corn flour. *Letters in Applied Microbiology* 33(2):144–7.

Chen, N., Cao, X., Tao, Y., et al. 2013. Development of a liquid chromatography–tandem mass spectrometry with ultrasound-assisted extraction and auto solid-phase clean-up method for the determination of *Fusarium* toxins in animal derived foods. *Journal of Chromatography* 1311:21–9.

Cunha, S. C. and Fernandes, J. O. 2010. Development and validation of a method based on a QuEChERS procedure and heart-cutting GC-MS for determination of five mycotoxins in cereal products. *Journal of Separation Science* 33:600–9. doi:10.1002/jssc.200900695.

Cunha, S. C., Faria, M. A. and Fernandes, J. O. 2009. Determination of patulin in apple and quince products by GC–MS using$^{13}C_{5-7}$ patulin as internal standard. *Food Chemistry* 115(1):352–9. doi:10.1016/j.foodchem.2008.11.074.

Deng, Y., Wang, Y., Deng, Q., et al. 2020. Simultaneous quantification of aflatoxin B1, T-2 toxin, Ochratoxin A and deoxynivalenol in dried seafood products by LC-MS/MS. *Toxins* 12(8):488.

Dos Santos Depoi, F., Bentlin, F. R., Ferrão, M. F. and Pozebon, D. 2012. Multivariate optimization for cloud point extraction and determination of lanthanides. *Analytical Methods* 4(9):2809–14.

Eke, Z., Kende, A. and Torkos, K. 2004. Simultaneous detection of A and B trichothecenes by gas chromatography with flame ionization or mass-selective detection. *Microchemical Journal* 78(2):211–6.

Eskola, M., Parikka, P. and Rizzo, A. 2001. Trichothecenes, ochratoxin A and zearalenone contamination and *Fusarium* infection in Finnish cereal samples in 1998. *Food Additives & Contaminants* 18(8):707–18.

European Commission. 2007. Commission Regulation (EC) No 1126/2007 of 28 September 2007 amending Regulation (EC) No 1881/2006 setting maximum levels for certain contaminants in foodstuffs as regards *Fusarium* toxins in maize and maize products.

European Commission. 2006. Commission Regulation (EC) No 1881/2006 of 19 December 2006 setting maximum levels for certain contaminants in foodstuffs. *Office Journal of the European Union* L364:5–24.

European Commission. 2013. 2013/165/EU Commission Recommendation of 27 March 2013 on the presence of T-2 and HT-2 toxin in cereals and cereal products Text with EEA relevance. *Office Journal of the European Union* L91:12–5.

Fan, Z. B., Jin, B., Fan, P., Guo, K., Zhao, Z. & Han, Z. 2016. Development and validation of an ultra-high performance liquid chromatography-tandem mass spectrometry method for simultaneous determination of four type B trichothecenes and masked deoxynivalenol in various feed products. *Molecules* 21:747.

FAO. 1995. *Worldwide regulations for mycotoxins.* Food and Agriculture Organization.

Fernández-Cruz, M. L., Mansilla, M. L. and Tadeo, J. L. 2010. Mycotoxins in fruits and their processed products: Analysis, occurrence and health implications. *Journal of Advanced Research* 1(2):113–22. doi:10.1016/j.jare.2010.03.002.

Ferreira, I., Fernandes, J. O. and Cunha, S. C. 2012. Optimization and validation of a method based in a QuEChERS procedure and gas chromatography–mass spectrometry for the determination of multi-mycotoxins in popcorn. *Food Control* 27(1):188–93. doi:10.1016/j.foodcont.2012.03.014.

Gámiz-Gracia, L., García-Campaña, A. M. and Arroyo-Manzanares, N. 2020. Application of LC–MS/MS in the mycotoxins studies. *Toxins* 12:272. doi:10.3390/toxins12040272.

González-Jartín, J. M., Alfonso, A., Rodríguez, I., Sainz, M. J., Vieytes, M. R. and Botana, L. M. 2019. A QuEChERS based extraction procedure coupled to UPLC-MS/MS detection for mycotoxins analysis in beer. *Food Chemistry* 275:703–10.

Gu, Y., Wang, Y., Wu, X., et al. 2019. Quartz crystal microbalance sensor based on covalent organic framework composite and molecularly imprinted polymer of poly(o-aminothiophenol) with gold nanoparticles for the determination of aflatoxin B1. *Sensors & Actuators, B: Chemical* 291:293–7.

Guo, W., Fan, K., Nie, D., et al. 2019. Development of a QuEChERS-Based UHPLC-MS/MS method for simultaneous determination of six *Alternaria* toxins in grapes. *Toxins* 11:87.

Hickert, S., Gerding, J., Ncube, E., Hübner, F., Flett, B., Cramer, B. and Humpf, H. U. 2015. A new approach using micro HPLC-MS/MS for multi-mycotoxin analysis in maize samples. *Mycotoxin Research* 31(2):109–15.

Hird, S. J., Lau, B. P. -Y., Schuhmacher, R., et al. 2014. Liquid chromatography-mass spectrometry for the determination of chemical contaminants in food. *Trends in Analytical Chemistry* 59:59–72.

Jeswal, P. and Kumar, D. 2015. Mycobiota and natural incidence of aflatoxins, ochratoxin A, and citrinin in Indian spices confirmed by LC-MS/MS. *International Journal of Microbiology* 2015:242486.

Jeyakumar, J. M. J., Zhang, M. and Thiruvengadam, M. 2018. Determination of mycotoxins by HPLC, LC-ESI-MS/MS, and MALDI-TOF MS in *Fusarium* species-infected sugarcane. *Microbial Pathogenesis* 123:98–110.

Juan, C., Ritieni, A. and Mañes, J. 2012. Determination of trichothecenes and zearalenones in grain 1134 cereal, flour and bread by liquid chromatography tandem mass spectrometry. *Food Chemistry* 134(4):2389–97.

Karapinar, I., Ertaş, F. N., Şahintürk, B., Aftafa, C. and Kiliç, E. 2016. LC-MS/MS signal enhancement for estrogenic hormones in water samples using experimental design. *RSC Advances* 6(45):39188–97.

Kirinčič, S., Škrjanc, B., Kos, N., Kozolc, B., Pirnat, N. and Tavčar-Kalcher, G. 2015. Mycotoxins in cereals and cereal products in Slovenia – Official control of foods in the years 2008–2012. *Food Control* 50:157–65. doi:10.1016/j.foodcont.2014.08.034.

Kizis, D., Vichou, A. -E. and Natskoulis, P. I. 2021. Recent advances in mycotoxin analysis and detection of mycotoxigenic fungi in grapes and derived products. *Sustainability* 13:2537. doi:10.3390/su13052537.

Köppen, R., Koch, M., Siegel, D., et al. 2010. Determination of mycotoxins in foods: Current state of analytical methods and limitations. *Applied Microbiology and Biotechnology* 86:1595–612. doi:10.1007/s00253-010-2535-1.

Kralj Cigić, I. and Prosen, H. 2009. An overview of conventional and emerging analytical methods for the determination of mycotoxins. *International Journal of Molecular Sciences* 10(1):62–115. doi:10.3390/ijms10010062.

Krska, R., Schubert-Ullrich, P., Molinelli, A., Sulyok, M., MacDonald, S. and Crews, C. 2008. Mycotoxin analysis: An update. *Food Additives and Contaminants: Part A* 25(2):152–63. doi:10.1080/02652030701765723.

Krska, R., Baumgartner, S. and Josephs, R. 2001. The state-of-the-art in the analysis of type-A and -B trichothecene mycotoxins in cereals. *Fresenius Journal of Analytical Chemistry* 371(3):285–99.

Lacina, O., Zachariasova, M., Urbanova, J., et al. 2012. Critical assessment of extraction methods for the simultaneous determination of pesticide residues and mycotoxins in fruits, cereals, spices and oil seeds employing ultra-high performance liquid chromatography-tandem mass spectrometry. *Journal of Chromatography A.* 1262:8–18.

Lai, X., Liu, R., Ruan, C., et al. 2015. Occurrence of aflatoxins and ochratoxin A in rice samples from six provinces in China. *Food Control* 50:401–4.

Low, K. H., Zain, S. M. and Abas, M. R. 2012. Evaluation of microwave-assisted digestion condition for the determination of metals in fish samples by inductively coupled plasma mass spectrometry using experimental designs. *International Journal of Environmental Analytical Chemistry* 92(10):1161–75.

Low, K. H., Zain, S. M., Abas, M. R. and Ali Mohd, M. 2009. Evaluation of closed vessel microwave digestion of fish muscle with various solvent combinations using fractional factorial design. *ASM Science Journal* 3(1):71–6.

Ma, L., Xu, W., He, X., et al. 2012. Determination of fumonisins B1 and B2 in Chinese rice wine by HPLC using AQC precolumn derivatisation. *Journal of the Science and Food Agriculture* 93:1128–33.

Malachová, A., Stránská, M., Václavíková, M., et al. 2018. Advanced LC–MS-based methods to study the co-occurrence and metabolization of multiple mycotoxins in cereals and cereal-based food. *Analytical and Bioanalytical Chemistry* 410:801–25.

McMaster, N., Acharya, B., Harich, K., et al. 2019. Quantification of the mycotoxin deoxynivalenol (DON) in sorghum using GC-MS and a Stable Isotope Dilution Assay (SIDA). *Food Analytical Methods* 12:2334–43. doi:10.1007/s12161-019-01588-3.

Medina, D. A. V., Borsatto, J. V. B., Maciel, E. V. S. and Lanças, F. M. 2021. Current role of modern chromatography and mass spectrometry in the analysis of mycotoxins in food. *TrAC Trends in Analytical Chemistry* 135:116156. doi:10.1016/j.trac.2020.116156.

Melchert, H. U. and Pabel, E. 2004. Reliable identification and quantification of trichothecenes and other mycotoxins by electron impact and chemical ionization-gas chromatography-mass spectrometry, using an ion-trap system in the multiple mass spectrometry mode: Candidate reference method for complex matrices. *Journal of Chromatography A* 1056:195–9.

Moretti, A., Logrieco, A.F. and Susca, A. 2017. Mycotoxins: An underhand food problem. In *Mycotoxigenic Fungi: Methods and Protocols, Methods in Molecular Biology*, Vol. 1542, eds. A. Moretti, A. Susca, pp. 154–96. Berlin/Heidelberg, Germany: Springer.

Murray, K. K., Boyd, R. K., Eberlin, M. N., et al. 2013. Definitions of terms related to mass spectrometry (IUPAC recommendation 2013). *Pure and Applied Chemistry* 85(7):1515–609.

Narváez, A., Rodríguez-Carrasco, Y., Castaldo, L., Izzo, L. and Ritieni, A. 2020. Ultra-high-performance liquid chromatography coupled with quadrupole Orbitrap high-resolution mass spectrometry for multi-residue analysis of mycotoxins and pesticides in botanical nutraceuticals. *Toxins* 12:114.

Natalia, A. -M., Natalia, C., Ignacio, L. -G., Manuel, H. -C. and Pilar, V. 2021. High-resolution mass spectrometry for the determination of mycotoxins in biological samples. A review. *Microchemical Journal* 166:106197.

Nielsen, K. F. and Thrane, U. 2001. Fast methods for screening of trichothecenes in fungal cultures using gas chromatography–tandem mass spectrometry. *Journal of Chromatography A* 929(1–2): 75–87. doi:10.1016/S0021-9673(01)01174-8.

Olsson, J., Borjesson, T., Lundstedt, T. and Schnurer, J. 2002. Detection and quantification of ochratoxin A and deoxynivalenol in barley grains by GC–MS and electronic nose. *International Journal of Food and Microbiology* 72:203–14.

Pakshir, K., Mirshekari, Z., Nouraei, H., Zareshahrabadi, Z., Zomorodian, K., Khodadadi, H. and Hadaegh, A. 2020. Mycotoxins detection and fungal contamination in black and green tea by HPLC-based method. *Journal of Toxicology* 2020:2456210. doi:10.1155/2020/2456210.

Paoloni, A., Solfrizzo, M., Bibi, R. and Pecorelli, I. 2018. Development and validation of LC-MS/MS method for the determination of Ochratoxin A and its metabolite Ochratoxin α in poultry tissues and eggs. *Journal of Environmental Science and Health, Part B* 53(5):327–333.

Pascale, M. N. 2009. Detection methods for mycotoxins in cereal grains and cereal products. *Proceedings of Natural Sciences, Matica Srpska Novi Sad* 117:15–25. doi:10.2298/ZMSPN0917015P.

Paterson, R. R. M. and Lima, N. 2010. *Toxicology of mycotoxins*. In *Molecular, Clinical and Environmental Toxicology: Volume 2: Clinical Toxicology*, ed. A. Luch, pp. 31–63. Basel, Switerland: Birkhäuser.

Pereira, V. L., Fernandes, J. O. and Cunha, S. C. 2014. Mycotoxins in cereals and related foodstuffs: A review on occurrence and recent methods of analysis. *Trends in Food Science and Technology* 36(2):96–136. doi:10.1016/j.tifs.2014.01.005.

Pereira, V. L., Fernandes, J. O. and Cunha, S. C. 2015. Comparative assessment of three clean-up procedures after QuEChERS extraction for determination of trichothecenes (type A and type B) in processed cereal-based baby foods by GC–MS. *Food Chemistry* 182:143–9. doi:10.1016/j.foodchem.2015.01.047.

Pettersson, H. and Langseth, W. 2002. *Intercomparison of trichothecene analysis and feasibility to produce certified calibrants and reference material, Final report I.* Method studies. BCR Information, Project Report EUR 20285/1, 1–82. European Committee for Standardization.

Pirinçci, S., Ertekin, Ö. and Laguna, D. E. 2018. Label-Free QCM immunosensor for the detection of ochratoxin A. *Sensors* 18:1161.

RASFF. 2021. *Rapid Alert System for Food and Feed.* https://ec.europa.eu/food/safety/rasff_en (accessed on 26 May 2021).

Rejczak, T. and Tuzimski, T. 2015. A review of recent developments and trends in the QuEChERS sample preparation approach. *Open Chemistry* 13:980–1010.

Richard, J. L. 2007. Some major mycotoxins and their mycotoxicosis - An overview. *International Journal of Food Microbiology* 119:3–10.

Rodrigues, K. L. T., Sanson, A. L., de Vasconcelos Quaresma, A., de Paiva Gomes, R., da Silva, G. A. and Afonso, R. J. D. C. F. 2014. Chemometric approach to optimize the operational parameters of ESI for the determination of contaminants of emerging concern in aqueous matrices by LC-IT-TOF-HRMS. *Microchemical Journal* 117:242–9.

Rodríguez-Carrasco, Y., Moltó, J.C., Berrada, H. and Mañes, J. 2014. A survey of trichothecenes, zearalenone and patulin in milled grain-based products using GC–MS/MS. *Food Chemistry* 146:212–9. doi:10.1016/j.foodchem.2013.09.053.

Rodríguez-Carrasco, Y., Berrada, H., Font, G. and Mañes, J. 2012. Multi-mycotoxin analysis in wheat semolina using an acetonitrile-based extraction procedure and gas chromatography-tandem mass spectrometry. *Journal of Chromatography A* 1270:28–40. doi:10.1016/j.chroma.2012.10.061.

Rubert, J., Soler, C., Marín, R., James, K.J. and Mañes, J. 2013. Mass spectrometry strategies for mycotoxins analysis in European beers. *Food Control* 30(1):122–8.

Sargent, M. 2013. *Guide to Achieving Reliable Quantitative LC-MS Measurements*. Middlesex, UK: RSC Analytical Methods Committee.

Schollenberger, M., Muller, H. M., Rufle, M., Terry-Jara, H., Suchy, S., Plank, S. and Drochner, W. 2007. Natural occurrence of Fusarium toxins in soy food marketed in Germany. *International Journal of Food Microbiology* 113(2):142–6.

Scott, P. 1993. Recent developments in analysis for mycotoxins in foodstuffs. *Trends in Analytical Chemistry* 12:373–81.

Shanakhat, H., Sorrentino, A., Raiola, A., Romano, A., Masi, P. and Cavella, S. 2018. Current methods for mycotoxins analysis and innovative strategies for their reduction in cereals: An overview. *Journal of the Science of Food and Agriculture* 98:4003–13. doi:10.1002/jsfa.8933.

Shephard, G. S. 2016. Current status of mycotoxin analysis: A critical review. *Journal of AOAC International* 99(4):842–8.

Singh, J. and Mehta, A. 2020. Rapid and sensitive detection of mycotoxins by advanced and emerging analytical methods: A review. *Food Science & Nutritions* 8:2183–204. doi:10.1002/fsn3.1474.

Slobodchikova, I., Sivakumar, R., Rahman, M. S. and Vuckovic, D. 2019. Characterization of Phase I and glucuronide Phase II metabolites of 17 mycotoxins using liquid chromatography-High-resolution mass spectrometry. *Toxins* 11:433.

Smaoui, S., Braïek, O. B. and Hlima, H. B. 2020. Mycotoxins analysis in cereals and related foodstuffs by liquid chromatography-tandem mass spectrometry techniques. *Journal of Food Quality* 2020:8888117. doi:10.1155/2020/8888117.

Smith, M. C., Madec, S., Coton, E. and Hymery, N. 2016. Natural co-occurrence of mycotoxins in foods and feeds and their in vitro combined toxicological effects. *Toxins* 8(4):94.

Soares, R. R. G., Ricelli, A., Fanelli, C., et al. 2018. Advances, challenges and opportunities for point-of-need screening of mycotoxins in foods and feeds. *Analyst* 143:1015–35.

Soleas, G. J., Yan, J. and Goldberg, D. M. 2001. Assay of ochratoxin A in wine and beer by high-pressure liquid chromatography photodiode array and gas chromatography mass selective detection. *Journal of Agricultural and Food Chemistry* 49(6):2733–40.

Songsermsakul, P. and Razzazi-Fazeli, E. 2008. A review of recent trends in applications of liquid chromatography-mass spectrometry for determination of mycotoxins. *Journal of Liquid Chromatography and Related Technologies* 31(11–12):1641–86.

Szerkus, B., Mpanga, A., Kaliszan, R. and Siluk, D. 2016. Optimization of the electrospray ionization source with the use of the design of experiments approach for the LC–MS-MS determination of selected metabolites in human urine. *Spectroscopy* 14:8–16.

Tahboub, Y. R. 2014. Chromatographic behavior of co-eluted plasma compounds and effect on screening of drugs by APCI-LC–MS (/MS): Applications to selected cardiovascular drugs. *Journal of Pharmaceutical Analysis* 4(6):384–91.

Tanaka, T., Yoneda, A., Inoue, S., Sugiura, Y. and Ueno, Y. 2000. Simultaneous determination of trichothecene mycotoxins and zearalenone in cereals by gas chromatography-mass spectrometry. *Journal of Chromatography A* 882:23–8.

Turner, N. W., Subrahmanyam, S. and Piletsky, S. A. 2009. Analytical methods for determination of mycotoxins: A review. *Analytica Chimica Acta* 632(2):168–80. doi:10.1016/j.aca.2008.11.010.

Zhang, Z., Hu, X., Zhang, Q. and Li, P. 2016. Determination for multiple mycotoxins in agricultural products using HPLC–MS/MS via a multiple antibody immunoaffinity column. *Journal of Chromatography B. Analytical Technologies in the Biomedical and Life Science* 1021:145–152.

Zhao, X., Yuan, Y., Zhang, X. and Yue, T. 2014. Identification of ochratoxin A in Chinese spices using HPLC fluorescent detectors with immunoaffinity column clean-up. *Food Control* 46:332–7.

Zhao, Y., Huang, J., Ma, L., et al. 2017. Development and validation of a simple and fast method for simultaneous determination of aflatoxin B1 and sterigmatocystin in grains. *Food Chemistry* 221:11–7.

Section 4

MS Analysis in Food Authentication

21 Mass Spectrometry in Food Authentication

Javed Ahamad and Subasini Uthirapathy
Tishk International University

Javed Ahmad
Najran University

CONTENTS

21.1 Introduction 403
21.2 Separation Techniques Coupled with Mass Spectrometry 404
21.3 Application of Mass Spectrometry in Food Authentication 405
 21.3.1 Honey 405
 21.3.2 Wine 406
 21.3.3 Milk and Dairy Products 407
 21.3.4 Cheese 408
 21.3.5 Cereals 408
 21.3.6 Edible Oils 409
 21.3.7 Meat and Meat Products 409
21.4 Conclusion 409
References 410

21.1 INTRODUCTION

Testing the quality of food products is of utmost importance for assuring their nutritional values and safety. There are several advanced analytical methods available to provide reliable qualitative and quantitative data to ensure food products' quality and safety. They provide data in compliance with criteria set globally. Food materials are made up of complex organic compounds and have extensive chemical variabilities; hence they pose analytical challenges for their correct analysis (Stefanoa et al., 2012; Ahamad et al., 2020a). Mass spectrometry (MS) has emerged as a powerful analytical tool for studying food products, pharmaceuticals, and health supplements (Wanga et al., 2013). *Foodomics* is a new discipline that studies food and nutritional products with the help of modern analytical tools, including MS, to ensure the quality and safety of food and nutritional products, and it includes proteomics and metabolomics. Proteomics, is currently being used as a powerful tool in food authentication, as it has been used for the authentication of food items such as cereals, wine, honey, chocolates, and other commercial products (Stefanoa et al., 2012; Ahamad et al., 2020b).

MS is a powerful analytical technique for correct and reproducible analysis of food and nutritional products. MS-based studies offer excellent analysis of food products by biomarker analysis, molecular profiling, and authentication. The major advantage of MS-based studies is that a wide range of products ranging from small molecules to very large proteins such as amino acids, proteins, vitamins, fatty acids, minerals present, etc. present in food products including honey, cereals, milk, dairy products, meat, and its products may be analyzed

DOI: 10.1201/9781003091226-25

(Herrero et al., 2012). It is also regularly used to detect food allergens present in food products such as cereals, seafood, peanut, eggs, and soybean, etc., and so preventing food allergies in susceptible individuals.

Recent advancements in separation techniques and hyphenation with mass spectrometry such as liquid chromatography (LC-MS/MS), gas chromatography (GC-MS), and in ionization techniques in mass spectrometry such as electrospray ionization (ESI-MS-MS), matrix-assisted laser desorption/ionization-time-of-flight (MALDI-TOF-MS), surface-enhanced laser desorption/ionization (SELDI-TOF-MS), desorption electrospray ionization (DESI-MS), direct analysis in real-time mass spectrometry (DART-MS), extractive electrospray ionization (EESI-MS), and tandem-MS approaches makes MS-based methods as the first choice for authentication of food products (Lu et al., 2018; Ibáñez et al., 2015; Wang et al., 2013).

Separation techniques such as LC, GC, and capillary electrophoresis (CE) reduce peak overlap, minimize ion suppression effects, and enhance the analysis of low-abundance food constituents such as amino acids and proteins. Thus, MS emerged as the most important and specific analytical tool for ensuring quality and safety of food products. This chapter comprehensively summarizes applications in assuring the quality and safety of food products such as honey, milk, dairy products, wine, edible oils, cereals, etc.

21.2 SEPARATION TECHNIQUES COUPLED WITH MASS SPECTROMETRY

LC, GC, and CE are the most suitable and frequently used techniques for the separation of analytes before analysis by MS. These separation techniques significantly reduce peak overlap, minimizes ion suppression effects, and allow greater sensitivity for analysis of low-abundance food constitutes. LC is most frequently used to separate non-volatile and polar components present in food products and for the identification and quantification in conjunction with MS. The advancement in LC, such as ultra-high pressure liquid chromatography (UPLC) and short chromatographic columns with smaller particles of stationary phase, improves separation efficiency, and reduces the time of analysis. Melough et al. (2017) developed a UPLC-MS method to identify and quantify furocoumarins in 29 food items. Klimczak and Gliszczyńska-Świgło (2015) performed a UPLC and HPLC comparative study to quantify vitamin C in fruit beverages. This study showed that the UPLC method was faster (R_t 6 minutes), more sensitive, and required less mobile phase than HPLC (R_t 15 minutes).

GC is the preferred method applied for analyzing volatile, non-polar, and low-molecular weight organic compounds such as essential oils, edible oils, and low-molecular weight proteins and amino acids, aromas, and plasticizers. GC methods are frequently used to separate essential oils components. Milkovska-Stamenova et al. (2015) used a GC-MS method for separating and characterizing 22 carbohydrates such as fructose, glucose, and myo-inositol in food products using the pre-column derivation method. GC-MS is also applied to analyze glycosylated and non-glycosylated polyphenols present in apples, pears, etc., using the injection port derivatization method (Marsol-Vall et al., 2016). Vismeh et al. (2018) developed a method for detection and analysis of *N*-acetylated compounds in milk, beef, and coffee samples using derivatization of acetamide with 9-xanthydrol.

Ion mobility spectrometry (IMS) is a non-chromatographic separation method frequently used to analyze food materials. In this technique, the separation of food components is based on charge, size, and shapes through a drift cell (Hernández-Mesa et al., 2018). Mittermeier et al. (2020) developed an LC-ESI-IMS method to quantify sensometabolites in chanterelles and other mushrooms. The study results showed the presence of octadecadien-12-ynoic acids.

Arroyo-Manzanares et al. (2018) developed a GC-IMS method to study Iberian ham's volatile compounds. The major target in this study was octan-2-one and trans-2-octenal and their proton-bound dimers. The study aimed to prevent labeling fraud.

21.3 APPLICATION OF MASS SPECTROMETRY IN FOOD AUTHENTICATION

21.3.1 Honey

Honey has been consumed since ancient times, as it has health benefits and nutritional value and is used as an adjuvant in pharmaceutical formulations. Honey can be categorized as a luxury product, typically perceived as a high-quality and appreciated product because of its desirable flavor and taste due to characteristics that are essentially or exclusively related to the specific area or local climate and flora. Traditionally, honey is commonly related to its healing properties and is used to cure wounds and burns and treat colds and sore throats. Several studies on honey proved its medicinal effects such as hepatoprotective, antibacterial, antihypertensive, hypoglycemic, gastroprotective, antifungal, anti-inflammatory, and antioxidant (Frans et al., 2001; Akbulut et al., 2009; Gomes et al., 2010; Erejuwa et al., 2012). The main components of honey are sugars, vitamins, minerals, volatile matters, amino acids, proteins, and enzymes (Schievano et al., 2013). As honey is a widely consumed natural product as a medicinal and nutritional agent, its authenticity is a major concern (Soares et al., 2015). Geographical and botanical sources have been the key issues related to honey authenticity, but other substances such as sugar syrups are also an important issue. Adulteration practices, such as overfeeding of bees with sucrose or other sugars, pre-maturity harvesting, and overuse of veterinary drugs, are still being carried out by some producers around the world to respond to the demands of a competitive market (Bogdanov et al., 2004; Sahinler et al., 2004; Guler et al., 2007).

Today, MS-based protein identification methods have become a common platform in proteomics (Delahunty and Yates, 2005; Rauh, 2012; Matros et al., 2011). Several factors, such as protein abundance, size, hydrophobicity, and other electrophoretic properties, limit the application of proteomic approaches to honey samples (Bhushan et al., 2007). For the authentication of honey, matrix-assisted laser desorption/ionization time-of-flight mass spectrometry (MALDI-TOF-MS) has proven to be a powerful tool for proteome identification and characterization (Peiren et al., 2005). MALDI-TOF-MS became the first-choice method in protein, and amino acids identification and characterization because of its advantages such as high sensitivity, high speed of analysis, robustness, and the prevalence of singly charged ions (Schmelzer et al., 2004; Matysiak et al., 2011; Thomas, 2001; Gonnet et al., 2003). The commonly used matrix solution for the ultraviolet-absorbing matrix system is cyano-4-hydroxycinnamic acid in a solution containing 30%–70% acetonitrile and 0.1% trifluoroacetic acid at a concentration of 5–20 g/L (Peiren et al., 2005). The matrix solution is combined with the digested protein to characterize honey protein and then directly spotted on the MALDI target plate. The sample is dried with a gentle air stream (approximately 35°C). The mass spectra are then registered in a reflection or linear positive-ion mode by collecting data from 200 to 1,000 laser shots for each spectrum with a mass accuracy between 50 and 75 ppm (Matysiak et al., 2011).

Matysiak et al. (2011) used cyano-4-hydroxycinnamic acid in acetonitrile containing 0.1% trifluoroacetic acid and sinapinic acid as a matrix system for characterization of honey bee venom. The study shows that the sinapinic acid matrix system helps identify peptides compared to the cyano-4-hydroxycinnamic acid matrix system. Corbella and Cozzolino (2005) used MALDI-TOF and principal component analysis to determine the geographical origin of honey from Hawaiian bees based on the proteome. Honey samples from Hawaii origin (n = 16) were used as the references for building a database library of protein mass spectra and 38 commercial samples, including 15 labeled Hawaii origin. The study results show that the four non-Hawaiian commercial samples yielded a correlation coefficient in the same range as the Hawaii-origin ones. Garcia-Alvarez et al. (2000) classified honey varieties (e.g., chestnut, acacia, sunflower, eucalyptus, and orange) using a combination of combinatorial hexapeptide ligand libraries enrichment and 2-DE analysis. The study results show that the method was found unsuccessful since no plant proteins were identified.

Kivrak (2015) analyzed free amino acids using UPLC-ESI-MS/MS of 58 unifloral kinds of honey of 17 different floras from different regions across Turkey (e.g., cedar, eucalyptus, vitex, carob, clover, honeydew, sunflower, citrus, heather, thyme, flower, chestnut, sideritis, acacia, lavender, cotton, and other kinds of honey). The study results show that phenylalanine, proline, tyrosine, and isoleucine as major amino acids are present in Turkish kinds of honey. Chua et al. (2015) identified honey proteins from *Acacia*, *Tualang*, and *Gelam* from Malaysia. The authors used the dialysis method and ammonium sulfate, and sodium tungstate precipitation extraction methods to isolate the honey proteins. The isolated proteins were analyzed by MS. The study shows that the honey contained major royal jelly proteins (MRJPs) such as MRJP-1, MRJP-2, MRJP-5, and MRJP-7. MRJP-1 was the most abundant protein, particularly in *Acacia* samples.

Pirini et al. (1992) applied capillary-GC to identify and characterize free amino acids in six honey samples of different botanical sources (e.g., acacia, citrus fruit, chestnut-tree, rhododendron, rosemary, and lime-tree). The study shows that arginine, tryptophan, and cysteine are characteristic of a particular kind of honey. Gilbert et al. (1981) used GC to determine 17 free amino acids in 45 samples of honey obtained from the UK, Australia, Argentina, and Canada. The study shows good discrimination between the samples from Australia, Argentina, and Canada, showing the potential of applying MS-based techniques in assessing the quality and safety of honey samples.

21.3.2 Wine

Wine is a complex matrix. The most common wine ingredients are water, ethanol, glycerol, sugars, organic acids, phenolic compounds, volatile components, and inorganic compounds, which are present in a wide concentrations range (Ribereau-Gayon et al., 2006). Wine also contains a low amount of amino acids; the most abundant amino acids are proline, histamine, tyramine, phenethylamine, alanine, glutamic acid, glutamine, and arginine, and γ-aminobutyric acid. The amino acid profile is an important parameter in the authentication and traceability of wine. The presence of amino acids is paramount in wine production because they act as a source of nitrogen for yeast during fermentation and influence its aromatic composition (Lehtonen, 1996). Amino acids in wine are originated from various sources such as raw grapes, fermentation products from live or dead yeast, and are produced by enzymatic degradation of grape proteins. The chemical composition of wine is greatly influenced by many factors, including grape variety, climate, vitivinicultural practices, geographical location, vintage, yeast strains, and fermentation processes (Ribereau-Gayon et al., 2006).

Wine authenticity is a very important issue in ensuring quality, safety, and efficacy. The purity of wine has been thoroughly investigated; however, due to its complex chemical composition and availability worldwide, wine is an easily adulterated commodity (Ahamad et al., 2018). The development of various protocols for analysis was initiated by many parameters influencing the quality of the wine. The components of wine are strictly regulated to prevent bribery and health threats by foreign organizations or government agencies. Wine contains amino acids that can be used as a maker/parameter for determining wine variety, geographical origin, and year of production. Therefore, amino acids are used as a marker for authentication by oenologists and food chemists (Huang and Ough, 1991). MS-based approaches emerge as the methods of choice for characterization, identification, and authentication of amino acids and proteins in wine. The isolation and identification of individual wine pigments is a tedious process and generally requires preliminary fractionation/purification steps before any analytical procedure such as MS or nuclear magnetic resonance spectroscopy. Recently, LC-MS emerged as the analytical method to study grape extracts and wine (Wang et al., 2003). The presence of procyanidins in wine protein fractions has been studied using LC/ESI-MS. Both heat-induced and natural hazes isolated from white wines revealed procyanidins' presence (Waters et al., 1995). MALDI-TOF and SELDI-TOF-MS have been used to determine the protein and peptide fingerprints of eight wines from Chardonnay, Sauvignon Blanc, and Muscat of Alexandria grape varieties. The study found 15 protein peaks with masses between 7 and 86 kDa, and protein with 21.3–kDa mass was found as a major protein. Other significant peaks with masses 7.2, 9.1, 13.1, and 22.2 kDa were found (Weiss et al., 1998). Kwon (2004) studied soluble proteins in wine by nano-high-performance liquid chromatography/tandem mass spectrometry. The study identified 80 peptides corresponding to 20 proteins unambiguously from Sauvignon Blanc wine. Five proteins are derived from grapes, 12 from yeast, 2 from bacteria, and 1 from fungi.

De Person et al. (2004) performed an LC/ESI-MS study to determine low-molecular-weight peptides in champagne wine. The champagne wine samples were ultra-filtrated with a molecular weight 1,000 cutoff membrane. Analyses were performed with a triple quadrupole mass spectrometer in positive-ion mode using dipeptide Phe-Arg as an internal standard. The authors reported nine dipeptides and one tripeptide. MALDI-MS technique has been used for the analysis of different red wine samples to check their vintage. The authors selected 20 wine samples. The study shows that 2,5-dihydroxybenzoic acid is far superior to all the matrices tested. More than 80 molecular species were detected in different wines. The predominant amino acids found in wine samples are tyramine, tryptamine, arginine, and tyrosine (Carpentieri et al., 2007).

Nunes-Miranda et al. (2012) utilized MALDI-MS for the analysis of wine for its classification. Ten different algorithms have been assessed as classification tools using the experimental data collected after analyzing 14 types of wine. They demonstrated that α-Cyano with a sample to matrix ratio of 1:0.75 is the best matrix. The study also suggested that to classify the wines correctly, profiling a minimum of five bottles per type of wine is required, with a minimum of three MALDI spot replicates for each bottle. Resetar et al. (2016) developed a MALDI-TOF-MS method to differentiate Istrian Malvazija wine and four other Croatian white wines (*viz* Zlahtina, Grasevina, Chardonnay, and Sauvignon Blanc). α-Cyano-4-hydroxycinnamic acid as a MALDI matrix in a solution mixed with the unmodified white wine sample in a volume ratio of 1:1 has been applied at ambient conditions. Mass spectrometric data showed that this was an efficient way to differentiate white wines of Croatian origin.

GC coupled with MS can be a useful alternative to LC-MS for the analysis of amino acids in wine samples (Duncan and Poljak, 1998). For rapid profiling and screening of 17 different wines for protein and non-protein amino acid contents, GC-MS has been used. The GC profiles analysis indicates that the tool is potentially useful for wine brand classification (Kim et al., 1996). In another study, GC with chiral stationary phase was used for the determination of *D*-amino acids and the separation of enantiomers in wines, using a capillary column coated with dimethyl(oxo)silane anchored to (S)-(−)-t-Leu-t-butyl-amide (Abe et al., 2002).

21.3.3 Milk and Dairy Products

Mill and dairy products make important parts of the human diet. Milk is considered a complete diet because of its nutritional value. Milk and dairy products also play a very important role in infant nutrition and are considered a valuable product economically. In general, milk proteins are classified as caseins, whey proteins, and milk fat globule membrane proteins (Swaisgood, 1992). Milk proteins are complex and are present in a complex matrix. Hence they pose an analytical challenge for their correct analysis, and milk proteins are the major target compounds for analysis (Cunsolo et al., 2011). Mass spectrometry, coupled with chromatographic separation techniques such as HPLC, UPLC, and GC, has been extensively applied. Hyphenated MS-based approaches are mostly applied for the authentication of milk and milk products. Amino acids, proteins, carbohydrates, and lipids are the main target in milk analysis (Casado et al., 2009; Guy and Fenaille, 2006).

The objectives are to explain the relationship between the composition of proteins and nutritional and technical properties (Cunsolo et al., 2005); to improve traceability methods for milk (Czerwenka et al., 2006); and to identify adulterated dairy products (Muller et al., 2008). Milk is also a big source of food allergens. In identifying and characterizing allergenic proteins, MS-based approaches are effective (Weber et al., 2006; D'Amato et al., 2010). Donkey milk is hypoallergic, and for cows' milk protein allergy, it represents a special safe 'natural' remedy in most cases. Proteomic research shows that no major variations in composition are demonstrated by cow and donkey milk proteomes. Compared to cow's milk, donkey milk's low allergenic property seems to be probably related to a remarkable difference in the primary structure of their proteins (Cunsolo et al., 2011).

EESI-MS is useful for analyzing melamine in raw milk, wheat gluten extracts, and milk powder (Li et al., 2009). Similarly, direct examination of melamine in raw milk has been done by nano-EESI-MS (Zhu et al., 2009). Also, novel membrane electrospray ionization mass spectrometry integrated molecular imprinting technology has shown excellent performance for direct ciprofloxacin study in fresh milk with a quantification limit of 1 ng/mL (Lin et al., 2017).

In the dairy industry, antibiotics such as sulfonamides are extensively used. Nevertheless, their presence in daily milk can pose a serious public health risk and contribute to the spread of bacterial antibiotic resistance (Lin et al., 2017). For the quantitative analysis of trace fluoroquinolones in raw milk samples, solid-phase extraction based on magnetic molecularly imprinted polymers coupled with iEESI-MS was designed to display a limit of detection (LOD) lower than 0.03 μg/L and a high analytical speed (about 4 minutes per sample) using the developed process (Zhang et al., 2017).

Ambient mass spectrometry (AMS) techniques may be used for the direct detection of antibiotic compounds based on good analysis results. Laser diode thermal desorption with atmospheric pressure chemical ionization mass spectrometry (LDTD-APCI-MS/MS) was used to identify the presence of seven sulfonamide residues in milk, with a quantity limit of 2–14 μg/L (Segura et al., 2010). Busman et al. (2015), by direct examination in real-time mass spectrometry, determined aflatoxin M1 (AFM1) in milk (DART-MS).

21.3.4 Cheese

Cheese is a dairy product derived from milk and produced by coagulation of milk protein (casein). The quality and safety of cheese are efficiently assessed by MS. In the evaluation of cheese, the most concerning things like mycotoxins derived from milk or microorganisms used during the dairy process (such as aflatoxin M) can adversely affect cheese safety. For the quality control of cheese, peptides, and proteins are major focused targets. Proteins and peptides are complex materials. For their identification and characterization, LC-based separation is the first step, and then in the second step, MS-based identification and characterization (Sforza et al., 2006; Piraino et al., 2007). Peptides are used as a marker to control the production of cheese, and also used to differentiate between different brands of cheese (Sforza et al., 2004), establishing the authenticity of the ingredients; discriminating between cheese produced by cow milk, sheep milk (Sforza et al., 2008), or water buffalo milk (Somma et al., 2008); and to differentiate the aging time (Masotti et al., 2010).

21.3.5 Cereals

Cereals, for example, wheat, rice, oat, and barley, make major components of the human diet and are the main agricultural products. Food product quality is based upon their nutritional values that largely depend upon their protein pattern. There are several types of cereal proteins, such as prolamins (e.g., gliadins and glutenins). The protein pattern of cereals forms the basis for its analysis. LC and GC, coupled with mass spectrometry, have been considered the method of choice to analyze cereal proteins (Agrawal and Rakwal, 2011; Finnie and Svensson, 2009). LC-MS-based analysis of cereal proteins has been used for many purposes, including identification, characterization, traceability, and allergen detection (Muccilli et al., 2011). Prolamins have been considered important in the analysis of cereal proteins because they are a key factor in bread making and the major determinant of gluten properties (Colgrave et al., 2012). Specific amino acid sequences of prolamins with a long repeating motif and few tryptic cleavage sites contribute to a pool of peptides that make MS/MS characteristics unfavorable to prolamins (Muccilli et al., 2010). That is why we need special analytical tools required for the analysis of cereal proteins. Some alternative enzymes (such as chymotrypsin) are required for analysis of prolamins before MS/MS analysis (Mamone et al., 2005). For improving the identification and characterization of cereal proteins, hyphenated techniques are applied, such as high-throughput sequential identification steps in combination with error-tolerant

Basic Local Alignment Search Tool based on database-dependent scanning and database independent *de novo* sequencing (Carpentier et al., 2008).

21.3.6 Edible Oils

Edible oils obtained from olive, mustard, soybean, sunflower, palm, and rapeseed constitute major parts. They are used for cooking food items, and together they represent about 80% of the edible oil market. Olive oil is considered the healthiest among edible oil, and it is the most consumed oil in the Mediterranean region. The olive and its oil have several health benefits: antioxidant, antidiabetic, anticancer, anti-obese, neuroprotective, or cardioprotective (Ahamad et al., 2019). Olive oil only represents 2% of the global oil market, but scientifically it has been studied often and is closely regulated. Due to its beneficial role in human health and high price, adulteration of olive oil has become one of the biggest sources of agricultural fraud worldwide. For the analysis of olive oils and other edible oils, GC-MS and LC-MS are considered the best suitable methods. The olive oil procured from (Erbil) Iraq, has been studied by GC-MS. The major components are oleic acid, 3-(octadecyloxy) propyl ester (19.34%), arachidonic acid (11.25%), oleic acid (6.07%), docosahexaenoic acid (DHA) (9.50%), pentadecanoic acid (5.53%), palmitic acid (3.86%), and linoleic acid (3.13%) (Ahamad et al., 2020c). Several studies have been reported for the determination of the chemical composition of olive oil by GC-MS, HPLC-MS, and HPLC-MS/MS (Nagy et al., 2005; Agozzino et al., 2010; Chen et al., 2011; Zarrouk et al., 2009; Zarrouk et al., 2009).

21.3.7 Meat and Meat Products

Meat and meat products make an important part of the human diet, and their quality depends upon protein content. The major source of meat includes buffalo, cow, goat, sheep, and pigs. The quality of meat and meat products is evaluated by determining its underivatized biogenic amines and toxicological issues with the help of MS coupled with cation-exchange chromatography. With cation-exchange chromatography-MS in hams, major biogenic amines such as cadaverine, putrescine, histamine, agmatine, phenethylamine, and spermidine were determined (Saccani et al., 2005). LC-ESI-MS is used to characterize non-volatile components, such as peptides in meat and its products. For peptide characterization in hams, HPLC-ESI-MS techniques are applied. Proteomic analysis of hams based on MS shows that unique protein patterns are linked to technological treatment, enabling the identification of meat processing protein biomarkers and quality (Pioselli et al., 2011). Electronic nose techniques based on MS include identifying volatile components of meat and its products and have also been used to identify suppliers and product quality (Llobet et al., 2007).

Table 21.1 summarizes the application of MS in the analysis and authentication of different food products.

21.4 CONCLUSION

Proteomics, a discipline under foodomics, is currently being used as a powerful tool in food authentication. It has been used to authenticate food items such as cereals, wine, honey, meat, and dairy products. MS-based studies made it simpler, easier, and feasible for protein detection and characterization of samples containing low protein abundance. Identification and structural characterization of food components by MS are now efficiently applied. Advancements in MS-based techniques such as MALDI-TOF-MS, HRMS SELDI-TOF-MS, DESI-MS, DART-MS, EESI-MS, and tandem-MS make a step forward in ensuring the safety and efficacy of food products. The findings of several studies suggest that MS-based approaches can be successfully applied to authenticate food products, thus ensuring the quality and safety of food products and providing methods to detect adulteration and deceptive practices.

TABLE 21.1
Application of Mass Spectrometry in the Analysis of Different Food Products

Food Products	Target Compound	Analytical Method	Reference
Mango, papaya, strawberries and vegetables	Pesticide	DESI-MS	Gerbig et al. (2017); Garcia-Reyes et al. (2009)
Wheat	Strobilurin, fungicides	DART-MS	Schurek et al. (2008)
Garlic	Trisulfane s-oxide	DART-MS	Block et al. (2010)
Tea and coffee	Caffeine	DART-MS	Danhelova et al. (2012)
Propolis	Flavonoids	DART-MS	Chernetsova et al. (2012)
Orange juice and cow milk		DART-MS	Martins et al. (2018)
Meat and meat products	Peptides	LC-MS	Montowska et al. (2014; 2015)
Cabbage, lettuce	Atrazine, diuron	LC-MS	Pereira et al. (2016)
Water, orange juice, and beer	Lead	EESI-MS	Liu et al. (2012)
Milk and milk products	Melanin	EESI-MS	Zhu et al. (2009)
	Melanin	DART-MS	Kiyota et al. (2017)
	Aflatoxin M1	nanoDESI-MS	Hartmanova et al. (2010)
	Ciprofloxacin	MESI-MS	Lin et al. (2017)
	Sulfonamides	LDTD-APCI-MS/MS	Segura et al. (2010)
Wine	Pesticides	LTP-MS	Beneito-Cambra et al. (2015)
Wine, raw milk	Lysozyme, melanin	EESI-MS	Zhou et al. (2012)
Soft drinks	Nicotinamide and caffeine	nanoEESI-MS	Li et al. (2009)
Honey	Volatile matters and chloramphenicol	ND-EESI-MS	Huang et al. (2014)
Honey	Chloramphenicol	LDTD-APCI-MS/MS	Blachon et al. (2013)
Olive oil	Oleic acid and other oil components	GC-MS	Ahamad et al. (2020b, c)

REFERENCES

Abe, I., Minami, H., Nakao, Y., Nakahara, T. (2002). N-pivaloyl methyl ester derivatives of amino acids for separation of enantiomers by chiral-phase capillary gas chromatography. *J. Sep. Sci.*, 25: 661–664.

Agozzino, P., Avellone, G., Bongiorno, D., Ceraulo, L., Indelicato, S., Indelicato, S., Vekey, K. (2010). Determination of the cultivar and aging of Sicilian olive oils using HPLCMS and linear discriminant analysis. *J. Mass Spectrom.*, 45: 989–995.

Agrawal, G.K., and Rakwal, R. (2011). Rice proteomics: a move toward expanded proteome coverage to comparative and functional proteomics uncovers the mysteries of rice and plant biology. *Proteomics*, 11: 1630–1649.

Ahamad, J., Ahmad, J., Ameen, M.S.M., Anwer, E.T., Mir, S.R., Ameeduzzafa. (2020b). Chapter 12: Proteomics in authentication of honey (Edited by Leo M.L. Nollet, Semih Ötleş), *Proteomics for Food Authentication*. CRS Press, Taylor & Francis Group, Boca Raton, FL.

Ahamad, J., Ahmad, J., Shahzad, N., Mohsin, N. (2018). *Authentication and traceability of wine. Fingerprinting Techniques in Food Authentication and Traceability*, vol. 1. CRS Press, Taylor & Francis Group, Boca Raton, FL.

Ahamad, J., Mir, S.R., Kaskoos, R.A., Ahmad, J. (2020a). Chapter 11: Proteomics in authentication of wine (Edited by Leo M.L. Nollet, Semih Ötleş), *Proteomics for Food Authentication*. CRS Press, Taylor & Francis Group, Boca Raton, FL.

Ahamad, J., Toufeeq, I., Khan, M.A., Ameen, M.S.M., Anwer, E.T., Uthirapathy, S., Mir, S.R., Ahmad J. (2019). Oleuropein: a natural antioxidant molecule in the treatment of metabolic syndrome. *Phytotherapy Res.*, 33(12): 3112–3128.

Ahamad, J., Uthirapathy, S., Ameen, M.S.M., Anwer, E.T., Mir, S.R. (2020c). Chemical composition and *in-vitro* antidiabetic effects of *Olea europaea* Linn. (Olive). *Current Bioactive Comp.*, 16(8): 1157–1163.

Akbulut, M., Ozcan, M.M., Coklar, H. (2009). Evaluation of antioxidant activity, phenolics, mineral contents and some physicochemical properties of several pine honeys collected from Western Anatolia. *Int. J. Food Sci. Nutr.*, 60(7): 577–589.

Arroyo-Manzanares, N., Martín-Gómez, A., Jurado-Campos, N., Garrido-Delgado, R., Arce, C., Arce, L. (2018). Target vs spectral fingerprint data analysis of Iberian ham samples for avoiding labelling fraud using headspace–gas chromatography–ion mobility spectrometry. *Food Chem.*, 246: 65–73.

Beneito-Cambra, M., Pérez-Ortega, P., Molina-Díaz, A., García-Reyes, J.F. (2015). Rapid determination of multiclass fungicides in wine by low-temperature plasma (LTP) ambient ionization mass spectrometry. *Anal. Methods*, 7: 7345–7351.

Bhushan, D., Pandey, A., Choudhary, M.K., Datta, A., Chakraborty, S., Chakraborty, N. (2007). Comparative proteomics analysis of differentially expressed proteins in chickpea extracellular matrix during dehydration stress. *Mol. Cell Proteomics*, 6: 1868–1884.

Blachon, G., Picard, P., Tremblay, P., Demers, S., Paquin, R., Babin, Y., Fayad, P.B. (2013). Rapid determination of chloramphenicol in honey by laser diode thermal desorption using atmospheric pressure chemical ionization-tandem mass spectrometry. *J. AOAC Int.*, 96: 676–679.

Block, E., Dane, A.J., Thomas, S., Cody, R.B. (2010). Applications of direct analysis in real time mass spectrometry (DART-MS) in Allium chemistry. 2-propenesulfenic and 2-propenesulfinic acids, diallyl trisulfane S-oxide, and other reactive sulfur compounds from crushed garlic and other Alliums. *J. Agric. Food Chem.*, 58: 4617–4625.

Bogdanov, S., Ruoff, K., Persano Oddo, L. (2004). Physico-chemical methods for the characterization of unifloral honeys: a review. *Apidologie*, 35(Suppl 1): S4–S17.

Busman, M., Bobell, J.R., Maragos, C.M. (2015). Determination of the aflatoxin M1 (AFM1) from milk by direct analysis in real time-mass spectrometry (DART-MS). *Food Control.*, 47: 592–598.

Carpentier, S.C., Panis, B., Vertommen, A., Swennen, R., Sergeant, K., et al. (2008). Proteome analysis of non-model plants: a challenging but powerful approach. *Mass Spectrom. Rev.*, 27: 354–377.

Carpentieri, A., Marino, G., Amoresano, A. (2007). Rapid fingerprinting of red wines by MALDI mass spectrometry. *Anal. Bioanal. Chem.*, 389: 969–982.

Casado, B., Affolter, M., Kussmann, M. (2009). OMICS-rooted studies of milk proteins, oligosaccharides and lipids. *J. Proteomics*, 73: 196–208.

Chen, H.L., Angiuli, M., Ferrari, C., Tombari, E., Salvetti, G., Bramanti, E. (2011). Tocopherol speciation as first screening for the assessment of extra virgin olive oil quality by reversed-phase high-performance liquid chromatography/fluorescence detector. *Food Chem.*, 125: 1423–1429.

Chernetsova, E.S., Bromirski, M., Scheibner, O., Morlock, G.E. (2012). DART-Orbitrap MS: a novel mass spectrometric approach for the identification of phenolic compounds in propolis. *Anal. Bioanal. Chem.*, 403: 2859–2867.

Chua, L.S., Lee, J.Y., and Chan, G.F. (2015). Characterization of the Proteins in Honey. *Anal. Lett.*, 48(4): 697–709.

Colgrave, M.L., Goswami, H., Howitt, C.A., Tanner, G.J. (2012). What is in a beer? Proteomic characterization and relative quantification of hordein (gluten) in beer. *J. Proteome Res.*, 11: 386–396.

Corbella, E., and Cozzolino, D. (2005). The use of visible and near infrared spectroscopy to classify the floral origin of honey samples produced in Uruguay. *J. Near Infrared Spectrosc.*, 13: 63.

Cunsolo, V., Galliano, F., Muccilli, V., Saletti, R., Marletta, D., Bordonaro, S., Foti, S. (2005). Detection and characterization by high-performance liquid chromatography and mass spectrometry of a goat β-casein associated with a CSN2 null allele. *Rapid Commun. Mass Spectrom.*, 19: 2943–2249.

Cunsolo, V., Muccilli, V., Fasoli, E., Saletti, R., Righetti, P.G., Foti, S. (2011). Poppea's bath liquor: the secret proteome of she-donkey's milk. *J. Proteomics*, 74: 2083–2099.

Cunsolo, V., Muccilli, V., Saletti, R., Foti, S. (2011). Applications of mass spectrometry techniques in the investigation of milk proteome. *Eur. J. Mass Spectrom. (Chichester, Eng)*, 17: 305–320.

Czerwenka, C., Maier, I., Pittner, F., Lindner, W. (2006). Investigation of the lactosylation of whey proteins by liquid chromatography-mass spectrometry. *J. Agric. Food Chem.*, 54: 8874–8882.

D'Amato, A., Kravchuk A.V., Bachi A., Righetti P.G. (2010). Noah's nectar: the proteome content of a glass of red wine. *J. Proteomics*, 73: 2370–2377.

Danhelova, H., Hradecky, J., Prinosilova, S., Cajka, T., Riddellova, K., Vaclavik, L., Hajslova, J. (2012). Rapid analysis of caffeine in various coffee samples employing direct analysis in real-time ionization-high-resolution mass spectrometry. *Anal. Bioanal. Chem.*, 403: 2883–2889.

De Person, M., Sevestre, A., Chaimbault, P., Perrot, L., Duchiron, F., Elfakir, C. (2004). Characterization of low-molecular weight peptides in champagne wine by liquid chromatography/tandem mass spectrometry. *Anal. Chim. Acta*, 520: 149–158.

Delahunty, C., and Yates, J.R. (2005). Protein identification using 2D-LC-MS/MS. *Methods*, 35: 248–255.

Duncan, M.W., and Poljak, A. (1998). Amino acid analysis of peptides and proteins on the femtomole scale by gas chromatography/mass spectrometry. *Anal. Chem.*, 70: 890–896.

Erejuwa, O.O., Sulaiman, S.A., Ab Wahab, M.S. (2012). Honey: a novel antioxidant. *Molecules*, 17(4): 4400–4423.

Finnie, C., and Svensson, B. (2009). Barley seeds proteomics from spots to structures. *J. Proteomics*, 72: 315–334.

Frans, T., Sias, G., Itzhak, G. (2001). The antifungal action of three South African honeys on Candida albicans. *Apidologie*, 32(4): 371–379.

García-Reyes, J.F., Jackson, A.U., Molina-Dı́az, A., Cooks, R.G. (2009). Desorption electrospray ionization mass spectrometry for trace analysis of agrochemicals in food. *Anal Chem.*, 81: 820–829.

Garcia-Alvarez, M., Huidobro, J.F., Hermida, M., Rodriguez-Otero, J.L., Major, J.L. (2000). Major components of honey analysis by near-infrared transflectance spectroscopy. *J. Agric. Food Chem.*, 48: 5154.

Gerbig, S., Stern, G., Brunn, H.E., During, R.A., Spengler, B., Schulz, S. (2017). Method development towards qualitative and semi-quantitative analysis of multiple pesticides from food surfaces and extracts by desorption electrospray ionization mass spectrometry as a preselective tool for food control. *Anal. Bioanal. Chem.*, 409: 2107–2117.

Gilbert, J., Shepherd, M.J., Wallwork M.A., and Harris, R.G. (1981). Determination of the geographical origin of honeys by multivariate analysis of gas chromatographic data on their free amino acid content. *J. Apicultural Res.*, 20(2): 125–135.

Gomes, S., Dias, L.G., Moreira, L.L., Rodrigues, P., Estevinho, L. (2010). Physicochemical, microbiological and antimicrobial properties of commercial honeys from Portugal. *Food Chem. Toxicol.*, 48(2): 544–548.

Gonnet, F., Lemaitre, G., Waksman, G., Tortajada, J. (2003). MALDI/MS peptide mass fingerprinting for proteome analysis: identification of hydrophobic proteins attached to eucaryote keratinocyte cytoplasmic membrane using different matrices in concert. *Proteome Sci.*, 1: 1–7.

Guler, A., Bakan, A., Nisbet, C., Yavuz, O. (2007). Determination of important biochemical properties of honey to discriminate pure and adulterated honey with sucrose (*Saccharum officinarum* L.) syrup. *Food Chem.*, 105(3): 1119–1125.

Guy, P.A., and Fenaille, F. (2006). Contribution of mass spectrometry to assess quality of milk-based products. *Mass Spectrom. Rev.*, 25: 290–326.

Hartmanova, L., Ranc, V., Papouskova, B., Bednar, P., Havlicek, V., Lemr, K. (2010). Fast profiling of anthocyanins in wine by desorption nano-electrospray ionization mass spectrometry. *J. Chromatogr. A.*, 1217: 4223–4228.

Hernández-Mesa, M., Le Bizec, B., Monteau, F., García-Campaña, A.M., Dervilly-Pinel, G. (2018). Collision Cross Section (CCS) database: an additional measure to characterize steroids. *Anal. Chem.*, 90: 4616–4625.

Herrero, M., Simó, C., García-Cañas, V., Ibáñez, E., Cifuentes, A. (2012). Foodomics: MS-based strategies in modern food science and nutrition. *Mass Spectrom. Rev.*, 31: 49–69.

Huang, X.Y., Fang, X.W., Zhang, X., Dai, X.M., Guo, X.L., Chen, H.W., Luo, L.P. (2014). Direct detection of chloramphenicol in honey by neutral desorption-extractive electrospray ionization mass spectrometry. *Anal. Bioanal. Chem.*, 406: 7705–7714.

Huang, Z., and Ough, C.S. (1991). Amino acid profiles of commercial grape juices and wines. *Am. J. Enol. Vitic.*, 42: 261–267.

Ibáñez, C., Simó, C., García-Cañas, V., Acunha, T., Cifuentes, A. (2015). The role of direct high-resolution mass spectrometry in foodomics. *Anal. Bioanal. Chem.*, 407: 6275–6287.

Kim, K.R., Kim, J.H., Cheong, E., Jeong, C. (1996). Gas chromatographic amino acid profiling of wine samples for pattern recognition Nuclear Magnetic Resonance Spectroscopy (NMR). *J. Chromatogr. A*, 722(1–2): 303–309.

Kivrak, I. (2015). Free amino acid profiles of 17 Turkish unifloral honeys. *J. Liquid Chromatogr. Related Technol.*, 38: 855–862.

Kiyota, K., Kawatsu, K., Sakata, J., Yoshimitsu, M., Akutsu, K., Satsuki-Murakami, T., et al., (2017). Development of monoclonal antibody-based ELISA for the quantification of orange allergen Cit s 2 in fresh and processed oranges. *Food Chem.*, 232: 43–48.

Klimczak, I., Gliszczyńska-Świgło, A. (2015). Comparison of UPLC and HPLC methods for determination of vitamin C. *Food Chem.*, 175: 100–105.

Kwon, S.W. (2004). Profiling of soluble proteins in wine by nano-high-performance liquid chromatography/tandem mass spectrometry. *J. Agric. Food Chem.*, 52: 7258–7263.

Lehtonen, P. (1996). Determination of amines and amino acids in wine -A review. *Am. J. Enol. Vitic.*, 47: 127–133.
Li, M., Hu, B., Li, J.Q., Chen, R., Zhang, X., Chen, H.W. (2009). Extractive electrospray ionization mass spectrometry toward in situ analysis without sample pretreatment. *Anal Chem.*, 81: 7724–7731.
Lin, H., Zhang, J., Chen, H.J., Wang, J.M., Sun, W.C., Zhang, X., et al., (2017). Effect of temperature on sulfonamide antibiotics degradation, and on antibiotic resistance determinants and hosts in animal manures. *Sci. Total Environ.*, 607–608: 725–732.
Liu, C.X., Zhang, X.L., Xiao, S.J., Jia, B., Cui, S.S., Shi, J.B., et al., (2012). Detection of trace levels of lead in aqueous liquids using extractive electrospray ionization tandem mass spectrometry. *Talanta*, 98: 79–85.
Llobet, E., Gualdron, O., Vinaixa, M., El-Barbri, N.J. et al. (2007). Efficient feature selection for mass spectrometry based "electronic nose" applications. *Chemometrics, Intell. Lab. Syst.*, 85: 253–261.
Lu, H., Zhang, H., Chingin, K., Xiong, J., Fang, X., Chen, H. (2018). Ambient Mass Spectrometry for Food Science and Industry. *Trends in Anal. Chem.*, 107: 99–115.
Mamone, G., Addeo, F., Chianese, L., Di Luccia, A., De Martino, A., Nappo, A., et al. (2005). Characterization of wheat gliadin proteins by combined two-dimensional gel electrophoresis and tandem mass spectrometry. *Proteomics*, 5: 2859–2865.
Marsol-Vall, A., Balcells, M., Eras, J., Canela-Garayoa, R. (2016). Injection-port derivatization coupled to GC–MS/MS for the analysis of glycosylated and non-glycosylated polyphenols in fruit samples. *Food Chem.*, 204, 210–217.
Martins, C., Assuncao, R., Cunha, S.C., Fernandes, J.O., Jager, A., Petta, T., Oliveira, C.A., Alvito, P. (2018). Assessment of multiple mycotoxins in breakfast cereals available in the Portuguese market. *Food Chem.*, 239: 132–140.
Masotti, F., Hogenboom, J.A., Rosi, V., De Noni, I., Pellegrino, L. (2010). Proteolysis indices related to cheese ripening and typicalness in PDO Grana Padano cheese. *Int. Dairy J.*, 20: 352–359.
Matros, A., Kaspar, S., Witzel, K., Mock, H.P. (2011). Recent progress in liquid chromatography-based separation and label-free quantitative plant proteomics. *Phytochemistry*, 72: 963–974.
Matysiak, J., Schmelzer, C.E.H., Neubert, R.H.H., Kokot, Z.J. (2011). Characterization of honeybee venom by MALDI-TOF and nanoESI-QqTOF mass spectrometry. *J. Pharm. Biomed. Anal.*, 54: 273–278.
Melough, M.M., Lee, S.G., Cho, E., Kim, K., Provatas, A.A., Perkins, C., et al., (2017). Identification and quantitation of furocoumarins in popularly consumed foods in the US Using QuEChERS extraction coupled with UPLC-MS/MS analysis. *J. Agric. Food Chem.*, 65: 5049–5055.
Milkovska-Stamenova, S., Schmidt, R., Frolov, A., Birkemeyer, C. (2015). GC-MS method for the quantitation of carbohydrate intermediates in glycation systems. *J. Agric. Food Chem.*, 63: 5911–5919.
Mittermeier, V.K., Pauly, K., Dunkel, A., Hofmann, T. (2020). Ion Mobility based LC-MS-Quantitation of Taste Enhancing Octadecadien-12-ynoic Acids in Mushrooms. *J. Agric. Food Chem.*, 68: 5741–5751.
Montowska, M., Alexander, M.R., Tucker, G.A., Barrett, D.A. (2014). Rapid detection of peptide markers for authentication purposes in raw and cooked meat using ambient liquid extraction surface analysis mass spectrometry. *Anal. Chem.*, 86: 10257–10265.
Montowska, M., Alexander, M.R., Tucker, G.A., Barrett, D.A. (2015). Authentication of processed meat products by peptidomic analysis using rapid ambient mass spectrometry. *Food Chem.*, 187: 297–304.
Muccilli, V., Cunsolo, V., Saletti, R., Foti, S., Margiotta, B., Scossa, F., Masci, S., Lafiandra, D. (2010). Characterization of a specific class of typical low molecular weight glutenin subunits of durum wheat by a proteomic approach. *J. Cereal Sci.*, 51: 134–139.
Muccilli, V., Lo Bianco, M., Cunsolo, V., Saletti, R., Gallo, G., Foti, S. (2011). High molecular weight glutenin subunits in some durum wheat cultivars investigated by means of mass spectrometric techniques. *J. Agric. Food Chem.*, 59: 12226–12237.
Muller, L., Bartak, P., Bednar, P., Frysova, I., Sevcik, J., Lemr, K. (2008). Thirty years of capillary electrophoresis in food analysis laboratories: potential applications. *Electrophoresis*, 29(10): 2088–2093.
Nagy, K., Bongiorno, D., Avellone, G., Agozzino, P., Ceraulo, L., Vekey, K. High-performance liquid chromatography-mass spectrometry-based chemometric characterization of olive oils. *J. Chromatogr. A*, 1078 (2005) 90–107.
Nunes-Miranda, J.D., Santos, H.M., Reboiro-Jato, M., Fdez-Riverola, F., Igrejas, G., Lodeiro, C., Capelo, J.L. (2012). Direct matrix assisted laser desorption ionization mass spectrometry based analysis of wine as a powerful tool for classification purposes. *Talanta*, 91: 72–76.
Peiren, N., Vanrobaeys, F., de Graaf, D.C., Devreese, B., Beeumen, J.V., Jacobs, F.J. (2005). The protein composition of honeybee venom reconsidered by a proteomic approach. *Biochim. Biophys. Acta.*, 1752: 1–5.
Pereira, I., Rodrigues, S.R.M., de Carvalho, T.C., Carvalho, V.V., Lobón, G.S., Bassane, J.F.P., et al., (2016). Rapid screening of agrochemicals by paper spray ionization and leaf spray mass spectrometry: which technique is more appropriate? *Anal. Methods*, 8: 6023–6029.

Pioselli, B., Paredi, G., Mozzarelli, A. (2011). Proteome analysis of pork meat in the production of cooked ham. *Mol. Biosyst.*, 7: 2252–2260.

Piraino, P., Upadhyay, V.K., Rossano, R., Riccio, P., Parente, E., Kelly, A.L., McSweeney, P.L.H. (2007). Use of mass spectrometry to characterize proteolysis in cheese. *Food Chem.*, 101: 964–972.

Pirini, Annalisa, Conte, Lanfranco S., Francioso, Ornella, Lercker, Giovanni. (March 1992). Capillary gas chromato-graphic determination of free amino acids in honey as a means of discrimination between different botanical sources. *J High Resolut Chromatogr.*, 15(3): 165–170.

Rauh, M. (2012). LC-MS/MS for protein and peptide quantification in clinical chemistry. *J. Chromatogr. B.*, 883–884: 59–67.

Resetar, D., Marchetti-Deschmann, M., Allmaier, G., Katalinic, J.P., Pavelic, S.K., (2016). Matrix assisted laser desorption ionization mass spectrometry linear time-of-flight method for white wine fingerprinting and classification. *Food Control*, 64: 157–164.

Ribereau-Gayon P., Dubourdie D., Doneche B., Lovaud A. (2006). *Handbook of Enology.* John Wiley & Sons, Ltd ISBN: 0-470-01034-7.

Saccani, G., Tanzi, E., Pastore, P., Cavalli, S., Rey, A. Determination of biogenic amines in fresh and processed meat by suppressed ion chromatography-mass spectrometry using a cation-exchange column. *J. Chromatogr. A*, 1082 (2005) 43–50.

Sahinler, N., Sahinler, S., Gul, A. (2004). Biochemical composition of honeys produced in Turkey. *J. Apicult. Res.*, 43(2): 53–56.

Schievano, E., Morelato, E., Facchin, C., Mammi, S. (2013). Characterization of markers of botanical origin and other compounds extracted from unifloral honeys. *J. Agric. Food Chem.*, 61(8): 1747–1755.

Schmelzer, C.E., Schops, R., Ulbrich-Hofmann, R., Neubert, R.H., Raith, K. (2004). Mass spectrometric characterization of peptides derived by peptic cleavage of bovine β-casein. *J. Chromatogr. A*, 1055: 87–92.

Schurek, J., Vaclavik, L., (Dick) Hooijerink, H., Lacina, O., Poustka, J., Sharman, M., et al., (2008). Control of strobilurin fungicides in wheat using direct analysis in real time accurate time-of-flight and desorption electrospray ionization linear ion trap mass spectrometry. *Anal Chem.*, 80: 9567–9575.

Segura, P.A., Tremblay, P., Picard, P., Gagnon, C., Sauve, S. (2010). High-throughput quantitation of seven sulfonamide residues in dairy milk using laser diode thermal desorption-negative mode atmospheric pressure chemical ionization tandem mass spectrometry. *J. Agric. Food Chem.*, 58: 1442–1446.

Sforza, S., Aquino, G., Cavatorta, V., Galaverna, G., Mucchetti, G., Dossena, A., Marchelli, R. (2008). Proteolytic oligopeptides as molecular markers for the presence of cows' milk in fresh cheeses derived from sheep milk. *Int. Dairy J.*, 18: 1072–1080.

Sforza, S., Ferroni, L., Galaverna, G., Dossena, A., Marchelli, R. (2006). Effect of extended aging of parma dry-cured ham on the content of oligopeptides and free amino acids. *J. Agric. Food Chem.*, 54(25): 9422–9429.

Sforza, S., Galaverna, G., Neviani, E., Pinelli, C., Dossena, A., Marchelli, R. (2004). Study of the oligopeptide fraction in grana padano and parmigiano-reggiano cheeses by liquid chromatography-electrospray ionisation mass spectrometry. *Eur. J. Mass Spectrom.*, 10(3): 421–427.

Soares, S., Amaral, J.S., Oliveira, M.B.P.P., Mafra, I. (2015). Improving DNA isolation from honey for the botanical origin identification. *Food Control*, 48(0): 130–136.

Somma, A., Ferranti, P., Addeo, F., Mauriello, R., Chianese, L. (2008). Peptidomic approach based on combined capillary isoelectric focusing and mass spectrometry for the characterization of the plasmin primary products from bovine and water buffalo β-casein. *J. Chromatogr. A*, 1192: 294–300.

Stefanoa, V.D., Avellonea, G., Bongiornoa, D., Cunsolob, V., Muccilli, V., et al., (2012). Applications of liquid chromatography–mass spectrometry for food analysis. *J. Chromatogr. A*, 1259: 74–85.

Swaisgood H.E. (1992). Chemistry of the caseins (Edited by F.P. Fox), *Advanced Dairy Chemistry.* Elsevier, London, p. 63.

Vismeh, R., Haddad, D., Moore, J., Nielson, C., Bals, B., Campbell, T., et al., (2018). Exposure assessment of acetamide in milk, beef, and coffee using xanthydrol derivatization and gas chromatography/mass spectrometry. *J. Agric. Food Chem.*, 66: 298–305.

Wang, H., Race, E.J., Shrikhande, A.J. (2003). Anthocyanin Transformation in Cabernet Sauvignon Wine during Aging. *J. Agric. Food Chem.*, 51: 7989–7994.

Wang, X., Wang, S., Cai, Z. (2013). The latest developments and applications of mass spectrometry in food-safety and quality analysis. *TrAC-Trend Anal. Chem.*, 52: 170–185.

Wanga, X., Wang, S., Cai, Z. (2013). The latest developments and applications of mass spectrometry in food-safety and quality analysis. *Trends in Anal. Chem.*, 52: 170–185.

Waters, E.J., Peng, Z., Pocock, K.F., Williams, P.J. (1995). Proteins in white wine, I: procyanidin occurrence in soluble proteins and insoluble protein hazes and its relationship to protein instability. *Aust. J. Grape Wine Res.*, 1: 86–93.

Weber, D., Raymond, P., Ben-Rejeb, S., Lau, B. (2006). Development of a liquid chromatography-tandem mass spectrometry method using capillary liquid chromatography and nano electrospray ionization-quardipole time-of-flight hybrid mass spectrometer for the detection of milk allergens. *J. Agric. Food Chem.*, 54: 1604–1710.

Weiss, K.C., Yip, T.T., Hutchens, T.W., Bisson, L.F. (1998). Rapid and sensitive fingerprinting of wine proteins by matrix-assisted laser desorption/ionisation time-of-flight (MALDI-TOF) mass spectrometry. *Am. J. Enol. Vitic.*, 49(3): 231–239.

Zarrouk, W., Carrasco-Pancorbo, A., Zarrouk, M., Segura-Carretero, A., Fernandez-Gutierrez, A. (2009). Multi-component analysis (sterols, tocopherols and triterpenic dialcohols) of the unsaponifiable fraction of vegetable oils by liquid chromatography-atmospheric pressure chemical ionization-ion trap mass spectrometry. *Talanta*, 80: 924–934.

Zhang, H., Kou, W., Bibi, A., Jia, Q., Su, R., Chen, H.W., Huang, K.K. (2017). Internal extractive electrospray ionization mass spectrometry for quantitative determination of fluoroquinolones captured by magnetic molecularly imprinted polymers from raw milk. *Sci. Rep.*, 7: 14714.

Zhou, Z.Q., Jiang, J., Li, M., Zhao, Z.F., Fu, J. (2012). Fast screening of chicken egg lysozyme in white wine products by extractive electrospray ionization mass spectrometry. *Chem. Res. Chinese U.*, 28: 200–203.

Zhu, L., Gamez, G., Chen, H.W., Chingin, K., Zenobi, R. (2009). Rapid detection of melamine in untreated milk and wheat gluten by ultrasound-assisted extractive electrospray ionization mass spectrometry (EESI-MS). *Chem. Commun. (Camb.)*, 5: 559–561.

Section 5

Emerging Fields

22 Challenges and Trends of Mass Spectrometry for Food Analysis

Robert Winkler
CINVESTAV Irapuato

CONTENTS

22.1 Introduction 419
22.2 Contaminants of Emerging Concern (CECs) 419
22.2.1 Microbial Toxins 420
22.2.2 Chemical Contamination 420
22.2.3 Mineral Oil 420
22.2.4 Microplastics 421
22.2.5 Metals from Electronic Waste and High-Technology Products 421
22.3 Technological Advances in Mass Spectrometry 421
22.3.1 Mass Spectrometry Fingerprinting 421
22.3.2 Ambient Ionization Mass Spectrometry (AIMS) Methods 422
22.3.3 Portable and Miniature Devices 422
22.3.4 3D-Printing 423
22.4 Mass Spectrometry Data Processing 423
22.4.1 Compound Identification and Structure Elucidation 423
22.4.2 Data Mining 424
22.4.3 Open-Source Software and Databases 424
22.5 Conclusion and Perspectives 425
References 425

22.1 INTRODUCTION

"Food quality" is a continuously developing concept. Various chapters in this book already deal with new topics such as the nutraceutical value and authenticity of the food. The first section of this chapter gives an overview of contaminants of emerging concern (CECs). CECs are a moving target because improved analytical methods and discoveries from related research – e.g., toxicology and ecology – frequently reveal new possible hazards to food consumers. The next section shows technical innovations in mass spectrometry (MS). Some of them are still in an experimental stage, such as 3D-printed devices, whereas others, e.g., some miniature MS analyzers, are already usable for routine testing of food products. Finally, new MS data processing strategies are introduced.

22.2 CONTAMINANTS OF EMERGING CONCERN (CECs)

For many compounds of known origin and toxicity, official limits have been defined, and their presence in food is monitored using approved methods. Newly discovered compounds, which might represent risks to the consumers, are studied as so-called "contaminants of emerging concern (CECs)."

DOI: 10.1201/9781003091226-27

The targeted and untargeted screening of food samples with MS-based methods, combined with suitable data processing/interpretation, is one of the most powerful tools for the detection of food contaminants [1]. Currently, multiclass approaches enable the detection of up to 1,000 compounds using a combination of liquid chromatography (LC) and tandem mass spectrometry (MS/MS) or high-resolution mass spectrometry [2].

22.2.1 Microbial Toxins

Different forms of shellfish poisoning – amnestic, diarrheal, neurotoxic, and paralytic – are caused by diverse chemical substances. More than 100 phytoplankton and bacterioplankton species produce hazardous compounds, which accumulate in filter-feeding mollusks [3,4]. Outbreaks are often localized with a vast variability of toxin profiles and related biological potential [5,6]. Therefore, LC–MS methods facilitate the monitoring of shellfish toxins [7,8].

Various fungal and bacterial toxins are routinely determined in food, using official standard methods. However, untargeted analyses with state-of-the-art MS strategies can reveal surprising microbial contamination of food materials and novel toxins. Testing maize-*fufu* from 50 households in a village in Cameroon led to the detection of 74 microbial toxins, including well-known mycotoxins such as aflatoxins, fumonisins, patulin, trichothecenes, and zearalenone derivatives. Besides less common fungal metabolites, such as the putatively identified nigragilin, three bacterial compounds, including the *Bacillus cereus* toxin cereulide and the plant toxin xanthotoxin, were found [9]. As another example, 14 *Alternaria* toxins, including the highly genotoxic altertoxin II, were proven to occur in naturally infected apples [10]. Both studies used LC–MS/MS techniques for the identification and quantification of compounds.

22.2.2 Chemical Contamination

Undesired materials can contaminate food products during production and handling steps and may cause serious health problems to their consumers [11]. Chemical CECs include metals, pharmaceutical drugs, personal care products, artificial sweeteners, pesticides, endocrine disruptors, polycyclic aromatic hydrocarbons, and brominated flame retardants [12,13].

Polystyrene (PS) is a popular material for food packing, especially for "take-away," although the styrene monomer is suspected to be a human carcinogen. In addition to the monomer, dimers and trimers are present as residues in PS products. The migration of PS into food is low, but monitoring all PS-derived compounds' daily uptake is required for a proper risk assessment [14].

Notably, the bioavailability of CECs might change during food processing. For example, steaming reduced the bioavailability of polybrominated diphenyl ether flame retardants, the pharmaceutical drug venlafaxine, and the ultraviolet-filter octocrylene in seafood [15].

Using GC–MS, seven commercial Chlorella and Spirulina samples showed no detectable levels of 31 tested CECs. However, since microalgae have increasing importance as a food supplement, further research is needed to evaluate their safety [16].

Limited resources of freshwater promote the use of treated water for irrigation. GC–MS/MS allowed the detection of residual CECs in vegetables from local markets in Spain, mainly organophosphates and bisphenol A [17]. Although the daily uptake of CECs from vegetables irrigated with recycled water was predicted to be below critical levels, investigating the fate of CECs in water recycling is recommendable [18]. Advanced oxidation processes could degrade CECs in aquatic systems and their entry into the food chain [13].

The following sections discuss some special classes of chemical contaminants.

22.2.3 Mineral Oil

Mineral oil hydrocarbons (MOHs) can migrate from treated jute bags into raw materials such as nuts, coffee, cocoa, and rice [19]. Other primary sources of food contamination are lubricant and "non-lubricant" additives to various product-related materials (rubber and plastics) [20]. Even the direct – accidental or fraudulent – admixture of (waste or "white") MOHs to food products is possible, and their use for "beautifying" produces, as well as the MOH content in packing made from recycled paper [21,22].

MOHs are divided into mineral oil saturated hydrocarbons and mineral oil aromatic hydrocarbons [23]. These two fractions display a different toxicological behavior [20].

The standard method for MOH analysis is liquid chromatography coupled with gas chromatography, with flame ionization detection (LC–GC/FID). However, more comprehensive studies on the chemical composition of oil mixture need MS, with prior separation by one- or two-dimensional GC [24,25].

22.2.4 Microplastics

Microplastics (MPs) are either directly released from polymers, such as the debris of car tires [26], or result from degradation of larger pieces. Especially the pollution of water bodies with plastics represents a huge environmental problem. A recent study found up to 0.3 mg polymers per gram of seafood tissue [27]. Lately, also MPs in freshwater have attracted attention [28]. Furthermore, processing and packing can contaminate the food products with MPs [29].

The possible impact of MPs on human health still needs further investigation, but their adverse effect on marine wildlife is evident. Therefore, more research for evaluating the risk of seafood consumption containing MPs is indicated [30].

22.2.5 Metals from Electronic Waste and High-Technology Products

Fifteen lanthanide group elements, yttrium, and scandium are summarized as the "high-technology rare earth elements (REEs)," which are used in electronics, high-tech devices, and even daily use products. REEs are released during their life cycle and pollute soil and aquatic systems. Entering the food chain, REEs may have various adverse health effects on humans, including kidney, lung, bone, and neurological disorders, as well as male sterility. Further studies on the environmental behavior of individual REEs and their toxicological effects are needed. One method to track REEs in environmental and biological samples is by inductively coupled plasma MS [31,32].

22.3 TECHNOLOGICAL ADVANCES IN MASS SPECTROMETRY

22.3.1 Mass Spectrometry Fingerprinting

Mass spectrometry fingerprinting (MSF) is the acquisition of spectra without pre-separation of the samples. The analytical performance of MSF methods is lower compared with chromatography-coupled MS, especially for highly complex mixtures. Nevertheless, MSF data are informative about the chemical composition of materials, and the analysis time is significantly shorter. Therefore, MSF methods can be used for the high-throughput screening of food samples [33–35].

The most commonly used methods for MSF are direct liquid injection-electrospray ionization (DLI-ESI) MS [36,37], matrix-assisted laser desorption/ionization time-of-flight (MALDI-ToF) MS [38,39], and direct analysis in real-time (DART) MS [40,41].

In addition to these established techniques, ambient ionization mass spectrometry (AIMS) methods are becoming interesting for routine MSF applications. MSF methods are often combined with advanced statistics and data mining (more information on AIMS and data mining is given below).

DLI-ESI-MS is very versatile and available in most MS laboratories. Liquid samples can be directly injected into an ESI analyzer and enable, for example, rapidly testing the authenticity of liquors such as tequila [42] and the characterization of olive oils [36]. Solid food products need to be extracted with a suitable solvent, such as mixtures of water, methanol, and formic acid. DLI-ESI MSF of extracts was used, for example for the classification of coffee products [43] and for testing the authenticity of organic tomato [44].

For MALDI-ToF MS, the analytes need to be co-crystallized with a matrix substance, making the sample preparation more difficult. In return, lipid profiles in complex food materials such as dairy and coffee products can be determined [39,45].

DART-MS can be directly applied to most sample types and is robust against matrix effects such as metal adducts and ion suppression [46]. Therefore, DART-MS enables the rapid and semi-quantitative detection of food contaminants such as pesticides [47].

22.3.2 Ambient Ionization Mass Spectrometry (AIMS) Methods

The era of AIMS started with the introduction of desorption electrospray ionization (DESI) [48]. The great advantage of AIMS methods is the possibility to analyze sample materials in their native state without, or with very simple, pre-treatment. Therefore, dozens of AIMS techniques have been reported, sometimes with only slight differences and confusing acronyms [49,50]. Several of them have been adopted for food and environmental analyses; Ref. [51] gives an overview of the current state-of-the-art.

Since AIMS is mainly used for fingerprinting, the accurate quantification of compounds of interest is difficult. However, AIMS can be used for the high-throughput screening of food products. Suspicious samples can then be analyzed with conventional methods. Since the field is very broad, only a few examples are given here.

The most established AIMS method for food analysis is DART, introduced in 2005 [40] because commercial ion sources for different mass spectrometers are readily available. Applications include the detection of melamine in powdered milk [52], for screening food packing additives [53], for monitoring tea fermentation [54], and for testing the origin of food such as beer and honey [46,55,56].

For desorption electrospray ionization (DESI) [48], also multiple applications were reported [57], such as the detection of contaminants in foodstuff at trace levels, food forensics, etc. [58,59]. However, DESI-MS requires solvents for operation, which could be seen as a disadvantage compared with plasma-based methods such as DART, developing "green" and portable strategies.

Low-temperature plasma (LTP) ionization was used for diverse food analysis studies. Direct LTP-MS of fruit peels and extracts could detect agrochemicals of different chemical classes in picogram levels [60] and is suitable for the rapid classification of Agave spirits [42] and coffee products [43].

LTP and paper-spray (PS) ionization MS are efficient olive oil analysis methods and provide complementary information. LTP-MS can detect free fatty acids, phenolics, and volatiles, directly from untreated olive oils, allowing their fast quality assessment [61]. PS-MS reported more compounds with higher mass, such as triacylglycerides. Adding $AgNO_3$ to the PS samples led to the formation of Ag+ adducts, which further improved the spectra quality and enabled the analysis of unsaturated compounds such as squalene [62].

Besides PS ionization, various methods generate an electrospray by applying a high voltage to a support material with a suitable solvent, such as leaf spray [63], (wooden) tip spray [64], and thin-layer chromatography spray [65]. Those methods provide simple sampling and sample logistics, require low volumes of solvents, and are compatible with later high-throughput analytics.

22.3.3 Portable and Miniature Devices

High analytical performance is often desirable in food analysis. However, for the *in situ* screening of products, e.g., in customs or for checking the identity of raw materials, the portability of mass spectrometers is more important than excellent resolution. Besides, miniature mass spectrometers (MMSs) are less expensive than full-size instruments and suitable for certain routine analyses. As well, MMSs require less bench space and shave lower running costs. Most MMSs are ion trap analyzers, such as the "Mini" series of Purdue University [66]. Various research groups developed different types of ion trap-based MMSs, and some of them were successfully commercialized [67]. Other mass spectrometers that were miniaturized used also ToF [68], magnetic sector [69], and quadrupole/triple quadrupole analyzers [70,71].

Such miniature devices are even more useful if combined with ambient ionization sources [67]. Coupling an LTP probe with a portable mass analyzer allowed the sensitive detection of melamine in complex food matrices such as fish, milk, and milk powder. The method is suitable to detect the fraudulent and dangerous admixture of melamine to food, which is sometimes practiced to simulate a higher protein content [72]. Agrochemicals such as diphenylamine and thiabendazole were detected in fruits with a handheld mass spectrometer, using LTP and PS ionization [73].

A "brick" mass spectrometer with photo-ionization could classify fruits based on their volatile compounds' profiles [74].

Coupling solid-phase microextraction with an MMS, using extraction nano-ESI enabled the direct analysis of chemical contaminants in infant drinks [75]. DLI-ESI with an MMS analyzer was suitable to detect food adulterants such as illegal chilly dyes and a plant growth regulator in watermelon [76].

As well, the real-time authentication of milk, fish, and coffee is possible, using data classification software [77].

Altogether, there is considerable potential for portable and miniature devices in food analysis, especially if combined with ambient ionization and data mining methods.

22.3.4 3D-Printing

3D printing is a disruptive technology in the development of analytical methods. The fast creation of custom devices and prototypes enables numerous applications in sample handling [78], automation [79], separation [80,81], and detection [82]. Combining different new concepts results in innovative analytical workflows. For example, 3D-printed cones can be used for sampling and the following analysis by ambient ESI-MS [83]. In the last years, 3D-printed components were also used to construct ion mobility spectrometry (IMS) and MS systems. Often, the capabilities of existing MS devices can be extended with simple 3D-printed gadgets. Besides, novel functional units can be integrated, such as special ion sources and ion manipulation components. Some recent developments of 3D-printing in IMS and MS are reviewed in Ref. [84].

22.4 MASS SPECTROMETRY DATA PROCESSING

22.4.1 Compound Identification and Structure Elucidation

Matching mass spectra with reference databases is a standard method for the identification of compounds. Gas chromatography coupled with electron impact ionization mass spectrometry (GC-EI-MS), with a subsequent spectral search using the National Institute of Standards and Technology (NIST) [85] database, is generally accepted as the "gold standard" for the analysis of small molecules. High-resolution mass spectra also enable the *de novo* generation of sum formulas and even structure elucidation.

The active MS community develops software tools and databases for compound identification and structure elucidation, which can be freely used for food quality analyses. For example, the MS data analysis suite MZmine (http://mzmine.github.io, [86,87]) can directly send MS queries to different publicly available databases:

Meta-databases of compounds and MS spectra

- PubChem, https://pubchem.ncbi.nlm.nih.gov [88]
- ChemSpider, http://www.chemspider.com
- MassBank.eu, https://massbank.eu [89]

Metabolic pathway databases

- KEGG, https://www.kegg.jp [90]
- MetaCyc, https://metacyc.org [91]

Specialized databases

- HMDB human metabolome database, https://hmdb.ca [92]
- YMDB yeast metabolome database, http://www.ymdb.ca [93]
- LIPID MAPS, https://www.lipidmaps.org [94]

Fragmentation data from untargeted LC–MS/MS experiments can be analyzed on the Global Natural Products Social Molecular Networking (GNPS, https://gnps.ucsd.edu) server. The platform collects community data and generates molecular networks from submitted datasets [95].

SIRIUS+CSI:FingerID provides state-of-the-art functions for structure elucidation from high-resolution MS/MS spectra (https://bio.informatik.uni-jena.de/software/sirius/). The framework uses machine-learning algorithms for predicting structural features and takes into account isotope patterns and fragmentation trees [96–99].

22.4.2 Data Mining

Conventional statistics methods are limited for the evaluation of massive datasets, such as untargeted MS experiments. Data mining methods use machine learning and artificial intelligence to excavate important information and create predictive models from "big data."

The statistical computing language R (https://www.r-project.org) [100] is frequently used for MS data processing workflows [101]. Identified features in tabular format can be put into the R package Rattle, which provides a graphical interface for data mining [102,103].

Usually, data mining models build, for example, with the Random Forest or Support Vector Machine algorithms display much lower classification errors than unsupervised methods such as Principle Component Analysis and Hierarchical Cluster Analysis [104].

The possible applications of machine learning and artificial intelligence in MS data processing are almost infinite and include noise filtering [105], compound identification [99], biomarker search [106], and sample classification [107].

It is important to note that such data mining models are unbiased by the researchers' knowledge and support the discovery of unexpected findings [108].

22.4.3 Open-Source Software and Databases

MS devices are usually sold with software for instrument control and data analysis. Having such an integrated system facilitates user training and the fast-starting of productive work. However, a fully commercial informatics infrastructure also has several drawbacks:

1. The programs of different vendors are often not compatible with each other, which make the set-up of data processing workflows for laboratories with machines from diverse providers complicated.
2. Software updates can be very costly (thousands of USD for a single program).
3. Changing later to an alternative data processing platform can be difficult if proprietary data formats are used ("Locked-in syndrome").
4. Correcting software errors or implementing new features is impossible.

Therefore, various academic research groups are developing open-source software for MS data analysis. Some of the open MS programs have similar or better functions than commercial ones [109]. Several data analysis platforms are also available in the following two versions: a free community edition and a professional edition. For example, KNIME (https://www.knime.com/) and LabKey (https://www.labkey.com/) offer such hybrid licensing models, which bring together the advantages of open and commercial software.

For the sharing and long-term use of MS data, raw files should be converted into community formats such as mzML, which was defined by the Proteomics Standards Initiative of the Human Proteome Organization (HUPO-PSI) (http://www.psidev.info) [110]. With the ProteoWizard (http://proteowizard.sourceforge.net), most raw MS files can be read and exported to HUPO standard files [111].

It is also good practice to archive project MS data in public repositories. Data of small molecules can be submitted, for example, to the MassBank.eu (https://massbank.eu) [89] and GNPS [95]. Any scientific data also can be freely stored on the Zenodo (https://zenodo.org) server [112] and become citable with a Digital Object Identifier.

22.5 CONCLUSION AND PERSPECTIVES

There is a dynamic development of MS methods and strategies for food analysis. Ambient ionization and miniature analyzers enable the *in situ* screening of food products, which translates to a faster evaluation of food quality and the early detection of possible threats to the consumers. Computer algorithms assist food chemists in the structure analysis of emerging contaminants and the interpretation of untargeted MS data.

Now, the instrumental and software innovations have to be integrated into the current quality assurance systems. It is important to note that expert knowledge and conventional methods should not be replaced but complemented by those novel technologies. Before their routine use in evaluating human food, such methods need to be thoroughly tested and validated.

REFERENCES

1. Tsagkaris AS, Nelis JLD, Ross GMS, et al (2019) Critical assessment of recent trends related to screening and confirmatory analytical methods for selected food contaminants and allergens. *TrAC Trends in Analytical Chemistry* 121:115688. doi:10.1016/j.trac.2019.115688.
2. Steiner D, Malachová A, Sulyok M, Krska R (2020) Challenges and future directions in LC-MS-based multiclass method development for the quantification of food contaminants. *Analytical and Bioanalytical Chemistry*. doi:10.1007/s00216-020-03015-7.
3. Farabegoli F, Blanco L, Rodríguez LP, et al (2018) Phycotoxins in Marine Shellfish: Origin, Occurrence and Effects on Humans. *Marine Drugs* 16:188. doi:10.3390/md16060188.
4. Munday R, Reeve J (2013) Risk assessment of shellfish toxins. *Toxins* 5:2109–2137. doi:10.3390/toxins5112109.
5. Goya AB, Tarnovius S, Hatfield RG, et al (2020) Paralytic shellfish toxins and associated toxin profiles in bivalve mollusc shellfish from Argentina. *Harmful Algae* 99:101910. doi:10.1016/j.hal. 2020.101910.
6. Harwood DT, Selwood AI, van Ginkel R, et al (2014) Paralytic shellfish toxins, including deoxydecarbamoyl-STX, in wild-caught Tasmanian abalone (Haliotis rubra). *Toxicon* 90:213–225. doi:10.1016/j.toxicon.2014.08.058.
7. Riccardi C, Buiarelli F, Di Filippo P, et al (2018) Liquid chromatography–tandem mass spectrometry method for the screening of eight paralytic shellfish poisoning toxins, Domoic acid, 13-Desmethyl Spirolide C, Palytoxin and Okadaic Acid in seawater. *Chromatographia* 81:277–288. doi:10.1007/s10337-017-3440-x.
8. Yang L, Singh A, Lankford SK, et al (2020) A rapid method for the detection of diarrhetic shellfish toxins and azaspiracid shellfish toxins in washington state shellfish by liquid chromatography tandem mass spectrometry. *Journal of AOAC International* 103:792–799. doi:10.1093/jaoacint/qsaa009.
9. Abia WA, Warth B, Ezekiel CN, et al (2017) Uncommon toxic microbial metabolite patterns in traditionally home-processed maize dish (fufu) consumed in rural Cameroon. *Food and Chemical Toxicology* 107:10–19. doi:10.1016/j.fct.2017.06.011.
10. Puntscher H, Marko D, Warth B (2020) First determination of the highly genotoxic fungal contaminant altertoxin II in a naturally infested apple sample. *Emerging Contaminants* 6:82–86. doi:10.1016/j.emcon.2020.01.002.
11. Rather IA, Koh WY, Paek WK, Lim J (2017) The sources of chemical contaminants in food and their health implications. *Front Pharmacol* 8. doi:10.3389/fphar.2017.00830.
12. Vandermeersch G, Lourenço HM, Alvarez-Muñoz D, et al (2015) Environmental contaminants of emerging concern in seafood – European database on contaminant levels. *Environmental Research* 143:29–45. doi:10.1016/j. envres.2015.06.011.

13. Salimi M, Esrafili A, Gholami M, et al (2017) Contaminants of emerging concern: A review of new approach in AOP technologies. *Environmental Monitoring and Assessment* 189:414. doi:10.1007/s10661-017-6097-x.
14. Genualdi S, Nyman P, Begley T (2014) Updated evaluation of the migration of styrene monomer and oligomers from polystyrene food contact materials to foods and food simulants. *Food Additives & Contaminants Part A: Chemistry, Analysis, Control, Exposure & Risk Assessment* 31:723–733. doi:10.1080/19440049.2013.878040.
15. Alves RN, Maulvault AL, Barbosa VL, et al (2017) Preliminary assessment on the bioaccessibility of contaminants of emerging concern in raw and cooked seafood. *Food and Chemical Toxicology* 104:69–78. doi:10.1016/j.fct.2017.01.029.
16. Martín-Girela I, Albero B, Tiwari BK, et al (2020) Screening of Contaminants of Emerging Concern in Microalgae Food Supplements. *Separations* 7:28. doi:10.3390/separations7020028.
17. Albero B, Sánchez-Brunete C, Miguel E, Tadeo JL (2017) Application of matrix solid-phase dispersion followed by GC–MS/MS to the analysis of emerging contaminants in vegetables. *Food Chemistry* 217:660–667. doi:10.1016/j.foodchem.2016.09.017.
18. González García M, Fernández-López C, Polesel F, Trapp S (2019) Predicting the uptake of emerging organic contaminants in vegetables irrigated with treated wastewater – Implications for food safety assessment. *Environmental Research* 172:175–181. doi:10.1016/j.envres.2019.02.011.
19. Grob K, Lanfranchi M, Egli J, Artho A (1991) Determination of food contamination by mineral oil from jute sacks using coupled LC-GC. *Journal of Association of Official Analytical Chemists* 74:506–512. doi:10.1093/jaoac/74.3.506.
20. Pirow R, Blume A, Hellwig N, et al (2019) Mineral oil in food, cosmetic products, and in products regulated by other legislations. *Critical Reviews in Toxicology* 49:742–789. doi:10.1080/10408444.2019.1694862.
21. Biedermann M, Grob K (2009) How "white" was the mineral oil in the contaminated Ukrainian sunflower oils? *European Journal of Lipid Science and Technology* 111:313–319. doi:10.1002/ejlt.200900007.
22. Grob K (2018) Mineral oil hydrocarbons in food: A review. *Food Additives & Contaminants: Part A* 35:1845–1860. doi:10.1080/19440049.2018.1488185.
23. Biedermann M, Fiselier K, Grob K (2009) Aromatic hydrocarbons of mineral oil origin in foods: Method for determining the total concentration and first results. *Journal of Agricultural and Food Chemistry* 57:8711–8721. doi:10.1021/jf901375e.
24. Spack LW, Leszczyk G, Varela J, et al (2017) Understanding the contamination of food with mineral oil: The need for a confirmatory analytical and procedural approach. *Food Additives & Contaminants: Part A* 34:1052–1071. doi:10.1080/19440049.2017.1306655.
25. Weber S, Schrag K, Mildau G, et al (2018) Analytical methods for the determination of mineral oil saturated hydrocarbons (MOSH) and mineral oil aromatic hydrocarbons (MOAH)—A short review. *Analytical Chemistry Insights* 13:1177390118777757. doi:10.1177/1177390118777757.
26. Sommer F, Dietze V, Baum A, et al (2018) Tire abrasion as a major source of microplastics in the environment. *Aerosol and Air Quality Research* 18:2014–2028. doi:10.4209/aaqr.2018.03.0099.
27. Ribeiro F, Okoffo ED, O'Brien JW, et al (2020) Quantitative analysis of selected plastics in high-commercial-value Australian seafood by pyrolysis gas chromatography mass spectrometry. *Environmental Science & Technology* 54:9408–9417. doi:10.1021/acs.est.0c02337.
28. Lambert S, Wagner M (2018) Microplastics are contaminants of emerging concern in freshwater environments: An overview. In: Wagner M, Lambert S (eds) *Freshwater Microplastics: Emerging Environmental Contaminants?* Springer International Publishing, Cham, pp 1–23.
29. Diaz-Basantes MF, Conesa JA, Fullana A (2020) Microplastics in honey, beer, milk and refreshments in Ecuador as emerging contaminants. *Sustainability* 12:5514. doi:10.3390/su12145514.
30. Barboza LGA, Dick Vethaak A, Lavorante BRBO, et al (2018) Marine microplastic debris: An emerging issue for food security, food safety and human health. *Marine Pollution Bulletin* 133:336–348. doi:10.1016/j.marpolbul.2018.05.047.
31. Gwenzi W, Mangori L, Danha C, et al (2018) Sources, behaviour, and environmental and human health risks of high-technology rare earth elements as emerging contaminants. *Science of the Total Environment* 636:299–313. doi:10.1016/j.scitotenv.2018.04.235.
32. Jensen H, Gaw S, Lehto NJ, et al (2018) The mobility and plant uptake of gallium and indium, two emerging contaminants associated with electronic waste and other sources. *Chemosphere* 209:675–684. doi:10.1016/j.chemosphere.2018.06.111.
33. Ellis DI, Brewster VL, Dunn WB, et al (2012) Fingerprinting food: Current technologies for the detection of food adulteration and contamination. *Chemical Society Reviews* 41:5706–5727. doi:10.1039/C2CS35138B.

34. Ren J-L, Zhang A-H, Kong L, Wang X-J (2018) Advances in mass spectrometry-based metabolomics for investigation of metabolites. *RSC Advances* 8:22335–22350. doi:10.1039/C8RA01574K.
35. Medina S, Pereira JA, Silva P, et al (2019) Food fingerprints – A valuable tool to monitor food authenticity and safety. *Food Chemistry* 278:144–162. doi:10.1016/j.foodchem.2018.11.046.
36. Goodacre R, Vaidyanathan S, Bianchi G, Kell DB (2002) Metabolic profiling using direct infusion electrospray ionisation mass spectrometry for the characterisation of olive oils. *Analyst* 127:1457–1462. doi:10.1039/B206037J.
37. Garcia-Flores M, Juarez-Colunga S, Garcia-Casarrubias A, et al (2015) Metabolic profiling of plant extracts using direct-injection electrospray ionization mass spectrometry allows for high-throughput phenotypic characterization according to genetic and environmental effects. *J Agric Food Chem* 63:1042–1052. doi:10.1021/jf504853w.
38. Karas M, Bachmann D, Bahr U, Hillenkamp F (1987) Matrix-assisted ultraviolet laser desorption of non-volatile compounds. *International Journal of Mass Spectrometry and Ion Processes* 78:53–68. doi:10.1016/0168-1176(87)87041-6.
39. Pazmiño-Arteaga JD, Chagolla A, Gallardo-Cabrera C, et al (2019) Screening for green coffee with sensorial defects due to aging during storage by MALDI-ToF mass fingerprinting. *Food Analytical Methods*. doi:10.1007/s12161-019-01485-9.
40. Cody RB, Laramée JA, Durst HD (2005) Versatile new ion source for the analysis of materials in open air under ambient conditions. *Analytical Chemistry* 77:2297–2302. doi:10.1021/ac050162j.
41. Guo C, Tang F, Chen J, et al (2015) Development of dielectric-barrier-discharge ionization. *Analytical and Bioanalytical Chemistry* 407:2345–2364. doi:10.1007/s00216-014-8281-y.
42. Martínez-Jarquín S, Moreno-Pedraza A, Cázarez-García D, Winkler R (2017) Automated chemical fingerprinting of Mexican spirits derived from: Agave (tequila and mezcal) using direct-injection electrospray ionisation (DIESI) and low-temperature plasma (LTP) mass spectrometry. *Analytical Methods* 9:5023–5028. doi:10.1039/c7ay00793k.
43. Gamboa-Becerra R, Montero-Vargas JM, Martínez-Jarquín S, et al (2017) Rapid classification of coffee products by data mining models from direct electrospray and plasma-based mass spectrometry analyses. *Food Analytical Methods* 10:1359–1368. doi:10.1007/s12161-016-0696-y.
44. García-Casarrubias A, Winkler R, Tiessen A (2019) Mass fingerprints of tomatoes fertilized with different nitrogen sources reveal potential biomarkers of organic farming. *Plant Foods for Human Nutrition* 74:247–254. doi:10.1007/s11130-019-00726-w.
45. Calvano CD, Zambonin CG (2013) MALDI-Q-TOF-MS ionization and fragmentation of phospholipids and neutral lipids of dairy interest using variable doping salts. *Advances in dairy Research* 1:1–8.
46. Guo T, Yong W, Jin Y, et al (2017) Applications of DART-MS for food quality and safety assurance in food supply chain. *Mass Spectrometry Reviews* 36:161–187. doi:10.1002/mas.21466.
47. Schurek J, Vaclavik L, Hooijerink H(, et al (2008) Control of strobilurin fungicides in wheat using direct analysis in real time accurate time-of-flight and desorption electrospray ionization linear ion trap mass spectrometry. *Analytical Chemistry* 80:9567–9575. doi:10.1021/ac8018137.
48. Takáts Z, Wiseman JM, Gologan B, Cooks RG (2004) Mass spectrometry sampling under ambient conditions with desorption electrospray ionization. *Science* 306:471–473. doi:10.1126/science.1104404.
49. Chen H, Gamez G, Zenobi R (2009) What can we learn from ambient ionization techniques? *Journal of the American Society for Mass Spectrometry* 20:1947–1963. doi:10.1016/j.jasms.2009.07.025.
50. Alberici RM, Simas RC, Sanvido GB, et al (2010) Ambient mass spectrometry: Bringing MS into the "real world". *Analytical and Bioanalytical Chemistry* 398:265–294. doi:10.1007/s00216-010-3808-3.
51. Nollet LML, Munjanja BK (2019) *Ambient Mass Spectroscopy Techniques in Food and the Environment*. CRC Press, Boca Raton, FL.
52. John Dane A, B. Cody R (2010) Selective ionization of melamine in powdered milk by using argon direct analysis in real time (DART) mass spectrometry. *Analyst* 135:696–699. doi:10.1039/B923561B.
53. Ackerman LK, Noonan GO, Begley TH (2009) Assessing direct analysis in real-time-mass spectrometry (DART-MS) for the rapid identification of additives in food packaging. *Food Additives & Contaminants: Part A* 26:1611–1618. doi:10.1080/02652030903232753.
54. Fraser K, Lane GA, Otter DE, et al (2013) Monitoring tea fermentation/manufacturing by direct analysis in real time (DART) mass spectrometry. *Food Chemistry* 141:2060–2065. doi:10.1016/j.foodchem.2013.05.054.
55. Cajka T, Riddellova K, Tomaniova M, Hajslova J (2011) Ambient mass spectrometry employing a DART ion source for metabolomic fingerprinting/profiling: A powerful tool for beer origin recognition. *Metabolomics* 7:500–508. doi:10.1007/s11306-010-0266-z.

56. Lippolis V, De Angelis E, Fiorino GM, et al (2020) Geographical origin discrimination of monofloral honeys by direct analysis in real time ionization-high resolution mass spectrometry (DART-HRMS). *Foods* 9. doi:10.3390/foods9091205.
57. Gentili A, Fanali S, Mainero Rocca L (2019) Desorption electrospray ionization mass spectrometry for food analysis. *TrAC Trends in Analytical Chemistry* 115:162–173. doi:10.1016/j.trac.2019.04.015.
58. García-Reyes JF, Jackson AU, Molina-Díaz A, Cooks RG (2009) Desorption electrospray ionization mass spectrometry for trace analysis of agrochemicals in food. *Analytical Chemistry* 81:820–829. doi:10.1021/ac802166v.
59. Nielen MWF, Hooijerink H, Zomer P, Mol JGJ (2011) Desorption electrospray ionization mass spectrometry in the analysis of chemical food contaminants in food. *TrAC Trends in Analytical Chemistry* 30:165–180. doi:10.1016/j.trac.2010.11.006.
60. Wiley JS, García-Reyes JF, Harper JD, et al (2010) Screening of agrochemicals in foodstuffs using low-temperature plasma (LTP) ambient ionization mass spectrometry. *Analyst* 135:971–979. doi:10.1039/b919493b.
61. García-Reyes JF, Mazzoti F, Harper JD, et al (2009) Direct olive oil analysis by low-temperature plasma (LTP) ambient ionization mass spectrometry. *Rapid Communications in Mass Spectrometry* 23:3057–3062. doi:10.1002/rcm.4220.
62. Lara-Ortega FJ, Beneito-Cambra M, Robles-Molina J, et al (2018) Direct olive oil analysis by mass spectrometry: A comparison of different ambient ionization methods. *Talanta* 180:168–175. doi:10.1016/j.talanta.2017.12.027.
63. Liu J, Wang H, Cooks RG, Ouyang Z (2011) Leaf spray: Direct chemical analysis of plant material and living plants by mass spectrometry. *Analytical Chemistry* 83:7608–7613. doi:10.1021/ac2020273.
64. Yang B-c, Wang F, Deng W, et al (2015) Wooden-tip electrospray ionization mass spectrometry for trace analysis of toxic and hazardous compounds in food samples. *Analytical Methods* 7:5886–5890. doi:10.1039/C5AY00917K.
65. Himmelsbach M, Waser M, Klampfl CW (2014) Thin layer chromatography–spray mass spectrometry: A method for easy identification of synthesis products and UV filters from TLC aluminum foils. *Analytical and Bioanalytical Chemistry* 406:3647–3656. doi:10.1007/s00216-014-7639-5.
66. Snyder DT, Pulliam CJ, Ouyang Z, Cooks RG (2016) Miniature and fieldable mass spectrometers: Recent advances. *Analytical Chemistry* 88:2–29. doi:10.1021/acs.analchem.5b03070.
67. Guo X-Y, Huang X-M, Zhai J-F, et al (2019) Research advances in ambient ionization and miniature mass spectrometry. *Chinese Journal of Analytical Chemistry* 47:335–346. doi:10.1016/S1872-2040(19)61145-X.
68. White AJ, Blamire MG, Corlett CA, et al (1998) Development of a portable time-of-flight membrane inlet mass spectrometer for environmental analysis. *Review of Scientific Instruments* 69:565–571. doi:10.1063/1.1148695.
69. Sinha MP, Tomassian AD (1991) Development of a miniaturized, light-weight magnetic sector for a field-portable mass spectrograph. *Review of Scientific Instruments* 62:2618–2620. doi:10.1063/1.1142240.
70. Malcolm A, Wright S, Syms RRA, et al (2010) Miniature mass spectrometer systems based on a microengineered quadrupole filter. *Analytical Chemistry* 82:1751–1758. doi:10.1021/ac902349k.
71. Wright S, Malcolm A, Wright C, et al (2015) A microelectromechanical systems-enabled, miniature triple quadrupole mass spectrometer. *Analytical Chemistry* 87:3115–3122. doi:10.1021/acs.analchem.5b00311.
72. Huang G, Xu W, Visbal-Onufrak MA, et al (2010) Direct analysis of melamine in complex matrices using a handheld mass spectrometer. *Analyst* 135:705–711. doi:10.1039/b923427f.
73. Soparawalla S, Tadjimukhamedov FK, Wiley JS, et al (2011) In situ analysis of agrochemical residues on fruit using ambient ionization on a handheld mass spectrometer. *The Analyst*. doi:10.1039/c1an15493a.
74. Meng X, Zhang X, Zhai Y, Xu W (2018) Mini 2000: A robust miniature mass spectrometer with continuous atmospheric pressure interface. *Instruments* 2:2. doi:10.3390/instruments2010002.
75. Guo X, Bai H, Ma X, et al (2020) Online coupling of an electrochemically fabricated solid-phase microextraction probe and a miniature mass spectrometer for enrichment and analysis of chemical contaminants in infant drinks. *Analytica Chimica Acta* 1098:66–74. doi:10.1016/j.aca.2019.11.021.
76. Meng X, Zhai Y, Yuan W, et al (2020) Ambient ionization coupled with a miniature mass spectrometer for rapid identification of unauthorized adulterants in food. *Journal of Food Composition and Analysis* 85:103333. doi:10.1016/j.jfca.2019.103333.
77. Gerbig S, Neese S, Penner A, et al (2017) Real-time food authentication using a miniature mass spectrometer. *Analytical Chemistry* 89:10717–10725. doi:10.1021/acs.analchem.7b01689.

78. Barthels F, Barthels U, Schwickert M, Schirmeister T (2020) FINDUS: An open-source 3D printable liquid-handling workstation for laboratory automation in life sciences. *SLAS Technology: Translating Life Sciences Innovation* 25:190–199. doi:10.1177/2472630319877374.
79. Sosnowski, P, Hopfgartner, G (2020) Application of 3D printed tools for customized open port probe-electrospray mass spectrometry. *Talanta* 215:120894. doi:10.1016/j.talanta.2020.120894.
80. Kalsoom U, Nesterenko PN, Paull B (2018) Current and future impact of 3D printing on the separation sciences. *TrAC Trends in Analytical Chemistry* 105:492–502. doi:10.1016/j.trac.2018.06.006.
81. Li F, Ceballos MR, Balavandy SK, et al (2020) 3D Printing in analytical sample preparation. *Journal of Separation Science* 43:1854–1866. doi:10.1002/jssc.202000035.
82. Xu Y, Wu X, Guo X, et al (2017) The boom in 3D-printed sensor technology. *Sensors* 17:1166. doi:10.3390/s17051166.
83. Brown HM, Fedick PW (2021) Rapid, low-cost, and in-situ analysis of per- and polyfluoroalkyl substances in soils and sediments by ambient 3D-printed cone spray ionization mass spectrometry. *Chemosphere* 129708. doi:10.1016/j.chemosphere.2021.129708.
84. Guillén-Alonso H, Rosas-Román I, Winkler R (2021) The emerging role of 3D-printing in ion mobility spectrometry and mass spectrometry. *Analytical Methods*. doi:10.1039/D0AY02290J.
85. National Institute of Standards and Technology (2020) NIST20: Updates to the NIST tandem and electron ionization spectral libraries.
86. Pluskal T, Castillo S, Villar-Briones A, Oresic M (2010) MZmine 2: Modular framework for processing, visualizing, and analyzing mass spectrometry-based molecular profile data. BMC Bioinformatics 11:395. doi:10.1186/1471-2105-11-395.
87. Pluskal T, Korf A, Smirnov A, et al (2020) Chapter 7: Metabolomics Data Analysis Using MZmine. In: *Processing Metabolomics and Proteomics Data with Open Software*. pp. 232–254.
88. Kim S, Chen J, Cheng T, et al (2021) PubChem in 2021: New data content and improved web interfaces. *Nucleic Acids Research* 49:D1388–D1395. doi:10. 1093/nar/gkaa971.
89. Horai H, Arita M, Kanaya S, et al (2010) MassBank: A public repository for sharing mass spectral data for life sciences. *Journal of Mass Spectrometry* 45:703–714. doi:10.1002/jms.1777.
90. Kanehisa M, Goto S, Sato Y, et al (2014) Data, information, knowledge and principle: Back to metabolism in KEGG. *Nucleic Acids Research* 42:D199–D205. doi:10.1093/nar/gkt1076.
91. Caspi R, Billington R, Fulcher CA, et al (2018) The MetaCyc database of metabolic pathways and enzymes. *Nucleic Acids Research* 46:D633–D639. doi:10.1093/nar/gkx935.
92. Wishart DS, Feunang YD, Marcu A, et al (2018) HMDB 4.0: The human metabolome database for 2018. *Nucleic Acids Research* 46:D608–D617. https://doi.org/10.1093/nar/gkx1089
93. Ramirez-Gaona M, Marcu A, Pon A, et al (2017) YMDB 2.0: A significantly expanded version of the yeast metabolome database. *Nucleic Acids Research* 45:D440–D445. doi:10.1093/nar/gkw1058.
94. Fahy E, Sud M, Cotter D, Subramaniam S (2007) LIPID MAPS online tools for lipid research. *Nucleic Acids Research* 35:W606–W612. doi:10.1093/nar/gkm324.
95. Wang M, Carver JJ, Phelan VV, et al (2016) Sharing and community curation of mass spectrometry data with Global Natural Products Social Molecular Networking. *Nature Biotechnology* 34:828–837. doi:10.1038/nbt.3597.
96. Böcker S, Letzel MC, Lipták Z, Pervukhin A (2009) SIRIUS: Decomposing isotope patterns for metabolite identification. *Bioinformatics* 25:218–224. doi:10.1093/bioinformatics/btn603.
97. Dührkop K, Fleischauer M, Ludwig M, et al (2019) SIRIUS 4: A rapid tool for turning tandem mass spectra into metabolite structure information. *Nature Methods* 16:299. doi:10.1038/s41592-019-0344-8.
98. Dührkop K, Böcker S (2015) Fragmentation trees reloaded. In: Przytycka TM (ed) *Research in Computational Molecular Biology*. Springer International Publishing, Cham, pp. 65–79.
99. Dührkop K, Shen H, Meusel M, et al (2015) Searching molecular structure databases with tandem mass spectra using CSI:FingerID. *PNAS* 112:12580–12585. doi:10.1073/pnas.1509788112.
100. R Core Team (2018) *R: A Language and Environment for Statistical Computing*. R Foundation for Statistical Computing, Vienna, Austria.
101. Palmblad M, Ovando-Vázquez C (2020) Chapter 15: Proteomic workflows with R/R markdown. In: *Processing Metabolomics and Proteomics Data with Open Software*. pp 371–380.
102. Williams G (2011) *Data Mining with Rattle and R: The Art of Excavating Data for Knowledge Discovery (Use R!)*, 1st ed. Springer Science+Business Media, New York.
103. Williams GJ (2009) Rattle: A data mining GUI for R. *The R Journal* 1:45–55.
104. Winkler R (2015) An evolving computational platform for biological mass spectrometry: Workflows, statistics and data mining with MASSyPup64. *Peer Journal* 3:1–34. doi:10.7717/peerj.1401.

105. Ovchinnikova K, Kovalev V, Stuart L, Alexandrov T (2020) OffsampleAI: Artificial intelligence approach to recognize off-sample mass spectrometry images. *BMC Bioinformatics* 21:129. doi:10.1186/s12859-020-3425-x.
106. Thomas A, Tourassi GD, Elmaghraby AS, et al (2006) Data mining in proteomic mass spectrometry. *Clinical Proteomics* 2:13–32. doi:10.1385/CP:2: 1: 13.
107. Creydt M, Fischer M (2020) Food phenotyping: Recording and processing of non-targeted liquid chromatography mass spectrometry data for verifying food authenticity. *Molecules* 25:3972. doi:10.3390/molecules25173972.
108. Winkler R (2016) Popper and the Omics. *Frontiers in Plant Science* 7:1–3. doi:10.3389/fpls.2016.00195.
109. Winkler R (2020) *Processing Metabolomics and Proteomics Data with Open Software: A Practical Guide*, 1st edition. Royal Society of Chemistry, Cambridge, UK.
110. Martens L, Chambers M, Sturm M, et al (2011) mzML - a community standard for mass spectrometry data. *Molecular Cell Proteomics* 10:R110.000133. doi:10.1074/mcp.R110.000133.
111. Chambers MC, Maclean B, Burke R, et al (2012) A cross-platform toolkit for mass spectrometry and proteomics. *Nature Biotechnology* 30:918–920. doi:10.1038/nbt.2377.
112. European Organization For Nuclear Research, OpenAIRE (2013) Zenodo. CERN.

Index

Note: **Bold** page numbers refer to tables and *italic* page numbers refer to figures.

Aarhus Protocol 280
abiotic hazards 50–51
Acaryochloris marina 122
acetone, mass spectra of 11, *11*
acetonitrile (AcN) 70, 71, 76, 143, 256, 286, 405
acetylene (C_2H_2) 10
acid hydrolysis 69, 76, 125, 219, 299
ACQUITY UPLCTM system 303
acylation 77, 396
acyl carrier protein (ACP) 69
adulterants 35, 316, 317, 320
adulteration 316
  of dairy products 317–319
  in functional foods 317
  in health foods 317
  in herbal medicines 317
  in honey 320
  in meat and meat products 320–321
  non-targeted screening with LC/HR-MS 321, *322*
  in oils 319–320
  in protein-containing foods 316–317
adzuki beans, origin testing 298
AEDA *see* aroma extract dilution analysis (AEDA)
Aeromonas hydrophila 52, 330
aflatoxins (AFLA) 194, **380**, 381, 393–395, 408, 420
aldoses 87, 88, 99
Alexandrium minutum 50
Alexandrium tamarense 50
3-alkyl-2-methoxypyrazines 146
alloys 233–235
α-amylase 68, 69, 71, 75, 306
α-Gal syndrome 109–110
α-spinasterol 191
altertoxin II 420
ambient desorption ionization 34
ambient ionization mass spectrometry (AIMS) 421, 422
ambient mass spectrometry (AMS) 408
American Oil Chemists' Society (AOCS) 31
amino acids (AAs) 13, 48, 114, 150–153, 196, 197, 347, 371, 404–408, 406
aminoalditols 92
4-amino-biphenyl 216
4-aminocarminic acid (4-ACA) *123*, 124, 125, **126**
ammonia 6, 142, 150
ammonium formate (AF) 76, 299
analytes, elution of 183
Anemonia sulcata 305
angiotensin-converting enzyme (ACE) 197
anthocyanins 119, *120*, 184
  fingerprinting analysis 303
  HPLC-ESI-MS analysis 299
  LC separation for 304
anthraquinones 119, *120*
antibiotic-resistance genes/proteins (ARG/Ps) 51
antibiotic resistance, in aquaculture 51–53
antibiotics 371–373
  in aquaculture 51
  natamycin 359, 371
  nisin 359, 371
  tetracyclines 52, 249
  veterinary drugs 247
antioxidants 51, 68, 184, 191, 197, 236
APCI *see* atmospheric-pressure chemical ionization (APCI)
Aphanizomenon flos-aquae 122
Aphanizomenon gracile 122
API *see* atmospheric pressure ionization (API)
APPI *see* atmospheric pressure photoionization (APPI)
apple *(Malus domestica)* 188
aquaculture 43
  antibiotic resistance in 51–53
  dietary management in 47
  fish welfare and stress response in 47–49
  food safety in 49–51
    abiotic hazards 50–51
    biotic hazards 49–50
  proteomics in 43–44
aqueous two-phase extraction 122
aroma 137, 138
  and GC-MS 140–148
  intensity of gluten free bread 166
  mint 160
  release, nose-space analysis 159, *159*, 160, *161*
  wine aroma persistence, *in vivo* PTR-ToF-MS analysis 160
aroma extract dilution analysis (AEDA) 145, 147, 150
o-arsanilic acid 306
arsenic (As), speciation analysis 304–307
Arthrospira platensis 121
artificial dyes 119
artificial intelligence 424
ascorbic acid *see* vitamin C
atmospheric-pressure chemical ionization (APCI) 34, 67, 139, 218
atmospheric pressure gas chromatography (APGC) 236, 240
atmospheric pressure ionization (API) 34, 40, 218, 222
atmospheric pressure photoionization (APPI) 19, 34, 67, 144
atmospheric solid analysis probe (ASAP)-MS 166
authenticity 33, 140, 166, 167, 194, 295–297, 303, 405, 406, 421
Automatic Mass spectral Deconvolution and Identification System (AMDIS) 148
Averrhoa carambola 188
azaphilones 119–121, *120*, 125–132

bacterial exopolysaccharides (BEPs) 199
baijiu, origin testing 297
base peak 9

beefy, meaty peptide (BMP) 151
beer
  biogenic amines in 348, 351, 352
  bitterness, investigations on 154–156
  ochratoxin A monitoring in 395
  origin testing 296
benzene (C_6H_6) 10
benzoic acid 367, **367**, *368*
benzoic acid and esters 367, **367**, *368*
β-glucan 47
β-lactam antibiotics, detection of 251
β-lactoglobulin **114**, 167, 319
β-phenylethylamine (β-PE) 348
β-sitosterol 191, 192, 195
betalains 119, *120*
bioactive peptides, analysis of 196–197
biogenic amines (BAs) analysis
  CE-MS for 351–352
  derivatization 348–349
  extraction 348–349
  GC-MS for 351
  IC-MS for 352–353
  LC-MS for 350–351
biomaterials 236
biosensors
  limitations 112
  protein 111–112
biotic hazards 49–50
biotin *see* vitamin B7
bisphenol A (BPA) 233, 235, 264, 420
bisphenol A diglycidyl ether (BADGE) 235
bis-pyrazolone 93
bitter peptides 151–152
bitter taste receptors 149, 152
BLAST 45
bond numbers 95, *96*
borodeuteride reduction 88
Box–Behnken architecture 394
broccoli, deuterium-labeled 78
bromobenzene
  chemical structure of 12, *12*
  mass spectra 12, *12*
bulk oils, lipids analysis in 33–37, **36–37**

CA *see* carminic acid (CA)
cadaverine (CAD) 348
capillary electrophoresis (CE) 19, 47, 347, 351, 404
capillary electrophoresis coupled with inductively coupled plasma mass spectrometry (CE–ICP-MS)
  arsenic species, quantification of 306
  lead, speciation analysis 308
capillary electrophoresis-mass spectroscopy (CE-MS)
  for biogenic amine analysis 351–352
  for foodborne pathogen detection 341–342
Captiva ND-Lipids filters 259, 261
caramels 119, *120*
carbohydrates 33
  anomeric configurations, determination of
    gas-phase IR spectrometry 101–102
    ion mobility spectrometry 101–102
  classification 87
  isomeric carbohydrates, separation of 99–101, *101*
  methylation analysis 88–92, *89–91*
  qualitative and quantitative HPLC-MS analysis 93, *94*, *95*
  small polysaccharides and oligosaccharide mixtures, profiling of *92*, 92–93
  tandem mass spectrometry
    de novo determination of absolute configuration of monosaccharides 97–99, *97–100*
    structure elucidation of oligosaccharides and glycoconjugates 94–97, *96*
carbon-Se-carbon (C-Se-C) 308
carmidole solution, chromatogram of 13
carmine 122–123, *123*
carminic acid (CA) 120, 122–125, **126**
carotenoids 68, 71, **72**, 119, *120*, 192
catechin 153, 241
CE *see* capillary electrophoresis (CE)
CECs *see* contaminants of emerging concern (CECs)
cellulose 198, 199, 235
CE-MS *see* capillary electrophoresis-mass spectroscopy (CE-MS)
Centroceras clavulatum 122
ceramics 235
cereals
  authentication of 408–409
  cereal-based food, mycotoxins detection in *384*, **386–392**
cereulide 420
cesium antimony (CsSb) 8
Cheddar cheese 151, 162
cheese, authentication of 408
Chemical Abstract Service (CAS) 14, 132
chemical contamination 165, 420
chemical food preservatives 360–373
  antibiotics 371–373
  benzoic acid and esters 367, **367**, *368*
  p-hydroxybenzoate alkyl esters 367–371
  nitrite and nitrate salts 361–363
  sorbic acid and esters 363–366
  sulfur dioxide and sulfites 360–361
chemical ionisation mass spectrometry (CIMS) 139, 163
chemical ionization (CI) 6, **6**, 67, 142, 351
chemical speciation 292, 304
Chinese goji berries 192
Chlorella 71, 420
chlorophylls 119, *120*
chlortetracycline (CTC) 52, 251
Chorda filum 92
chromium (Cr), speciation analysis 308
chymotrypsin 113, 122, 408
CID *see* collision-induced dissociation (CID)
cinnamaldehyde 359
cis-isomers 13
CODEX Alimentarius Commission 282, 358
coenzyme A (CoA) 69
collision energy (CE) 124, 260, 361
collision-induced dissociation (CID) 8, 23, 96, 141, 199, 236, 303
colorants 119, *120*
complex lipids 189
compound structure, determination of 11, *11*
conjugated linoleic acid (CLA) 33, 195
contaminants of emerging concern (CECs)
  chemical contamination 420
  metals from electronic waste 421

microbial toxins 420
microplastics 421
mineral oil 420–421
rare earth elements 421
contamination
chemical 420
of food with pesticides 214
microbial 337
persistent organic pollutants 283
correspondence analysis (CA) 295
cortisol 48
coumestans 192, 193
countercurrent chromatography (CCC) 77
counter-propagation artificial neural networks (CPANN) 296
C-PCs *see* phycocyanins (C-PCs)
Crassostrea angulata 50
Crataegus species 188
cross-contamination 110, 214, 282
curcuminoids 119, *120*
cured meats 195
cyanocobalamin *see* vitamin B12
cyano-4-hydroxycinnamic acid 405
cyclotron frequency 8

Dactylopius coccus 122
dairy products
adulteration of 317–319
authentication of 407–408
lipids analysis in 37–40, **38–39**
Dansyl-Cl (5-(dimethylamino)naphthalene-1-sulfonyl chloride) 371
data-dependent analysis (DDA) 21, 22, *22*
data-independent acquisition (DIA) 22–23, *23*, 321
data mining 424
data processing 423–425
compound identification 423–424
data mining 424
open-source software and databases 424–425
structure elucidation 423–424
DDA *see* data-dependent analysis (DDA)
deconvolution 93, 141, 148
dehydroascorbic acids 69
de novo sequencing
fish allergens 44
monosaccharides, absolute configuration of 97–99, *97–100*
phycocyanin analysis 121
deoxynivalenol (DON) *381*, 393, 395, 396
derivatization
of biogenic amines 348–349
of carbohydrates 93, *94*, *95*
of parabens 371, *371*
desorption chemical ionization (DCI) 150
desorption electrospray ionization (DESI)-MS 34, *34*, 35, 139, 166, 241, 242, 340, 404, 409, 422
deuteromethylation 90
DIA *see* data-independent acquisition (DIA)
2, 3-diaminonaphthalene (DAN) 363, *363*
dibutyl amine (DBA) 362
dichlorodiphenyltrichloroethane (DDT) 280
Diels–Alder reaction 9
dietary management, in aquaculture 47
2, 6-dihydroxyacetophenone (DHAP) 184
2, 5-dihydroxybenzoic acid (DHB) 92, 188, 407
3, 4-dihydroxyphenylacetic acid (DOPAC) 350
dimethylarsinic acid (DMA) 304–307, **312**
DIMS *see* direct-injection mass spectrometry (DIMS)
direct analysis in real-time (DART) MS 34, 35, *35*, 139, 166, 296, 421
direct immersion-SPME combined with gas chromatography-mass spectrometry (DI-SPME-GC-MS) 352
direct-injection mass spectrometry (DIMS) 139, 163, 164, 166
direct liquid injection-electrospray ionization (DLI-ESI) MS 421
Directorate-General for Health and Food Safety (DG SANTE) 357
discriminant partial least squares (DPLS) 294
dispersive liquid–liquid microextraction 67
dispersive solid-phase extraction (d-SPE) 214, 216
distal cholesterol biosynthesis (DCB) 192
distillers dried grains and solubles, origin testing 296–297
2, 2′-dithiobis(5-nitropyridine) (DTNP) 361, *361*
Djulis (*Chenopodium formosanum* Koidz.) hull 192
DNA-based methods, detection of food allergens 112
DNA replication 112
docosahexaenoic acid (DHA) 371, 409
Domon-Costello nomenclature 94–95, 97
doxycycline 52
drift time 101
drug metabolism 14
durum wheat, origin testing 298–299
dynamic headspace (DHS) sampling 138
dynamic headspace–thermal desorption–GC-MS (DHS-TD-GC-MS) 141
dynodes 8

E. coli 330–332, 340, 341
edible oils
arsenic, speciation analysis of 307
authentication of 409
lipids analysis in 33–37, **36–37**
Edwardsiella tarda 52
egg (*Gallus gallus*), allergenic protein identification 113, **114**
electrochemical biosensors 112
electron capture detector 286, 383
electron-capture dissociation (ECD) 97
electron detachment dissociation (EDD) 97
electronic nose (e-nose) 139, 293, 409
electronic pneumatic control (EPC) 147
electron impact (EI) ionization 4, 6, **6**, 67, 85, 214, 351
electron multiplier 8
electron transfer dissociation (ETD) 97
electroosmotic flow (EOF) 352
electrospray ionization (ESI) **6**, 7, 34, 47, 70, 85, 257
electrospray ionization-collision-induced dissociation mass spectrometry to collision-induced dissociation (ESI-CID-MS) 199
electrospray ionization-collision-induced dissociation-tandem mass spectrometry (ESI-CID-MS/MS) 199
electrospray ionization-mass spectrometry (ESI-MS) 331, 338–340
Enterococcus faecalis 340

enzymatic digestion 68, 125
enzymatic hydrolysis 68, 69, 76, 197
enzymatic proteolysis 124
enzyme-linked immunosorbent assays (ELISA)
  food allergens, detection of 111, **111**
  limitations 111
equol 193
essential oils 6, 359, 404
ethanolamine 142, 349, 352
ethanol, mass spectrum of 10, *11*
ethylene bisdithiocarbamate (EBDC) 219
ethylenediaminetetraacetic acid (EDTA) 261–263
ethylene oxide 143
ethylene thiourea (ETU) 219
EU Commission Regulation 33, 122, 231, 240
European Food Safety Authority (EFSA) 121
Euterpe oleracea 303
exposome 256
extracted ion chromatograms (EICs) 21, *23*, 23–24, 273
extraction
  acetonitrile 286
  biogenic amines 348–349
  fat-soluble vitamins 67–68
  lipids 33
  Monascus pigments 127
  water-soluble vitamins 68–69
  xenobiotics 258
extractive electrospray ionization (EESI-MS), melamine analysis in raw milk 408
extra virgin olive oil (EVOO) 33, **36**

falcarindiol 300
Faraday cup mass detector 8
fast-atom bombardment (FAB) 6, **6**, 86
fat-soluble vitamins (FSVs)
  extraction 67–68
  gas chromatography 77–78
  liquid chromatography 70–71, **72**
  simultaneous determination of **72**, 75–77
fatty acid methyl esters (FAME) 33
fatty acids (FAs) 31, 39, 52, 183, 189, 194, 195, 303
FCMs *see* food contact materials (FCMs)
Federal Food, Drug, and Cosmetic Act 357–358
fermentation 127, 150, 151, 166, 195, 406
FFAs *see* free fatty acids (FFAs)
fish welfare and stress response, in aquaculture 47–49
fixed ligand kinetic method (FLKM) 97–98, *97–99*
flame ionization detector (FID) 33, 66, 86, 383
flavanol-3-glycosides 157
flavan-3-ols epicatechin 153
flavour 137
  global analyses of 162–167
  perception 137, 138, 158
  release and perception, *in vivo* analysis
    non-volatiles 161–162
    volatiles 158–161
flavour dilution (FD) 145
flavouromics 162–167
flaxseed oil 319
fluorescence detector (FLD) 70, 382, 383, 394
fluoroquinolones (FQs) 247, 251, 261–262, 408
folate *see* vitamin B9
folic acid *see* vitamin B9
food additives, defined 358
food adulteration 194
food allergens
  detection
    DNA-based methods 112
    enzyme-linked immunosorbent assays 111, **111**
    protein biosensors 111–112
    using MS 113–114, **114**, *114*, *115*
  food allergy 109–110
food allergy
  allergens 109–110
  management 110
  prevalence of 110
Food and Drug Administration (FDA) 31, 121, 357
food authentication
  cereals 408–409
  cheese 408
  edible oils 409
  honey 405–406
  meat and meat products 409, **410**
  milk and dairy products 407–408
  PCR-coupled MS for 331, 340–341
  separation techniques coupled with mass spectrometry 404
  wine 406–407
foodborne diseases 330
foodborne pathogen (FBP), detection of
  CE-MS for 341–342
  ESI-MS for 331, 338–340
  GC-MS for 341
  LC-MS for 331, 332, *335*, *336*
  MALDI-TOF-MS for 330–331, 337–338, *337–339*
  PCR-coupled MS 340–341
food contact materials (FCMs)
  ambient ionisation–accurate MS 241–242
  biomaterials 236
  cellulose 235
  ceramics 235
  chemical migration of 231
  DESI-HRMS 242
  DESI-MS 241
  GC-MS 236–240
  glass 235
  ICP-MS 241
  inks and photoinitiators 236
  intentionally added substances 232
  LC-MS/MS 240–241
  LESA-nano-ESI-MS 241
  metals and alloys 233–235
  migration limits for 231–232
  MS analysis for various materials **237–240**
  non-intentionally added substances 232
  paper and cardboard 233
  plastics 233
  rubber 235
  types and legislations 232, *232*
  UHPLC 241
  wood and fibre 236
food contamination 214
food fortification 66
food-labeling regulations 110
food lipids 33
food microbe analysis, GC-MS in 341
food pigments 119, *120*
  azaphilones 120–121, 125–132

carminic acid 120, 122–125
phycocyanins 119, 121–122
food preservatives
chemical 360–373
antibiotics 371–373
benzoic acid and esters 367, **367**, *368*
p-hydroxybenzoate alkyl esters 367–371
nitrite and nitrate salts 361–363
sorbic acid and esters 363–366
sulfur dioxide and sulfites 360–361
natural 359–360
food quality 419
food safety, in aquaculture 49–51
abiotic hazards 50–51
biotic hazards 49–50
food-spoilage microorganisms, identification of 331, 338–340
formic acid (FA) 298
Fourier transform (FT)-ion cyclotron resistance mass analyzer 8
Fourier transform ion cyclotron resonance mass spectrometry (FT-ICR-MS), in phenolics compounds analysis 183
FQs *see* fluoroquinolones (FQs)
fractionation 138, 140, 149, 167, 406
fragmentation rules 9
fragment ion peak 9
fragment ions 9, 11, 13, 19, 21, 99, 140–143, 318, 373
free fatty acids (FFAs) 31, 33, 34
free induction decay (FID) 8
FSVs *see* fat-soluble vitamins (FSVs)
full width at half maximum (FWHM) 67, 260, 318, 321, 358, 369
fumonisins (FUMs) 381–383, 393, 420
functional foods, adulteration in 317
fungi polysaccharides 198–199
fuzzy optimal associative memory (FOAM) 301
fuzzy rule-building expert system (FuRES) 301
FWHM *see* full width at half maximum (FWHM)

Galderia phlegrea (GpPC) 122
gallium phosphide (GaP) 8
γ-glutamyl dipeptides 150
γ-glutamyl peptides 157
γ-glutamyltransferase 151
gas chromatography (GC) 66, 404
fat-soluble vitamins 77–78
retention index 77
water-soluble vitamins 77–78
gas chromatography coupled with electron impact ionization mass spectrometry (GC-EI-MS)
aroma and 140–142
4-hydroxybenzoate, identification of 368, 370
sorbic acid 363, *364*
gas chromatography-electrospray ionization mass spectrometry (GC-ESI-MS) 194
gas chromatography-high-resolution mass spectrometry (GC-HRMS) 147–148
gas chromatography-ion mobility spectrometry (GC-IMS) 147–148
gas chromatography-mass spectrometry (GC-MS) 138
aroma and
databanks and software 148
GC-CI-MS 142–144
GC-EI-MS 140–142
GC-HRMS 147–148
GC-IMS 147–148
GC-O-MS 145–146
MDGC- and GC×GC-MS 144–145
quantification 146–147
biogenic amine analysis 351
cured meats 195
food contact materials 236–240
food microbe analysis 341
mycotoxins, sensitive estimation of 395–396
persistent organic pollutants analysis 282–283, **284**
pesticide analysis 214–217, **215**, *217*
phospholipids 191
phytoestrogens, analysis of 193
phytosterols 191
tocopherols 191
gas chromatography–olfactometry (GC-O) 138
gas chromatography-quadrupole mass spectrometry (GC-QMS) 192
gas chromatography-tandem mass spectrometry (GC-MS/MS)
multiresidue pesticide analysis 212–213, *217*
persistent organic pollutants analysis 283–284
phytoestrogens, analysis of 193
gas chromatography with time-of-flight mass spectrometry (GC-Q-TOF-MS) 78
adzuki beans, origin testing 298
pesticides determination in edible vegetable oils 194
gas-phase ions 4, 331
gas-phase IR spectrometry, carbohydrate analysis 101–102
GC *see* gas chromatography (GC)
GC-CI-MS, aroma and 142–144
GC-EI-Orbitrap 147
GC×GC-MS, aroma and 144–145
GC-O-MS, aroma and 145–146
gelatine, origin testing 297
gel-free approach 44
gel permeation chromatography (GPC) 95
generally recognized as safe (GRAS) 358
genistein 193
Gibbs' free energy 98
Girard's T reagent 93
glass 235
glycerolipids 32, *32*
glycerophospholipids 32, *32*
glycosylation points determination, in oligo-and polysaccharides 88–92, *89–91*
G-protein-coupled (G-PC) transmembrane heterodimer protein 149
guanosine monophosphate (5′GMP) 98

harmful algal blooms (HABs) 50
Hazard Analysis Critical Control Points (HACCP) 49
hazards
abiotic 50–51
biotic 49–50
headspace autosampler 293
headspace solid-phase microextraction (HS-SPME)
honey, origin testing 294
organotin, determination of 308
health foods, adulteration in 317
heat sublimation, determination of 13
heat vaporization, determination of 13

herbal medicines, adulteration in 317
heterocyclic amines (HAs) assay, LC-MS for 222
1, 1,1, 3,3, 3-hexafluoro-2-propanol (HFIP) 362
high-density polyethene (HDPE) **234**
higher-energy collisional dissociation (HCD) 96
high-performance liquid chromatography-electrospray ionisation mass spectrometry (HPLC-ESI-MS)
  anthocyanins identification 299
  peptide characterization in hams 409
high-performance liquid chromatography-mass spectrometry (HPLC-MS)
  identification of mycotoxins 394–395
  qualitative and quantitative analysis of derivatized carbohydrates 93, *94*, *95*
  taste and 148–158
high-performance size exclusion chromatography coupled with multi-angle laser light scattering and refractive index detector (HPSEC-MALLS-RID) 198
high-resolution extracted ion chromatograms (HR-XICs) 21–24
high-resolution liquid chromatography combined with time-of-flight mass spectrometry (HRLC-TOF-MS) 250
high-resolution mass spectrometry (HRMS) 70–71, 79, 147
  POPs detection *286*, 287
  veterinary drugs 248
histamine (HA) 347
homovanillic acid (HVA) 350
honey
  adulterants 320
  adulteration in 320
  authentication of 405–406
  origin testing 293–295
HPLC-APCI-MS technique, biogenic amine analysis 351
HPLC fingerprinting 302
HR-XICs *see* high-resolution extracted ion chromatograms (HR-XICs)
Human Metabolome Database 260
human serum, vitamin K1 determination in 78
humulone 154, *156*
hybrid triple quadrupole-linear ion trap 67
hydrophilic interaction chromatography and tandem mass spectrometry (HILIC-MS/MS) 316
hydrophilic interaction chromatography-evaporative light scattering detection-electrospray time-of-flight mass spectrometry (HILIC-ELSD-ESI-TOF-MS) 198
hydrophilic interactions 183
hydrophobic interactions 183
4-hydroxybenzaldehyde 184, 368
4-hydroxybenzoate *see* parabens
4-hydroxybenzoic acid *see* parabens
hydroxy methylsulfonate (HMS), formation of 360, *360*
4-hydroxyphenethyl alcohol (4-HPEA) 184
hypertrophy 48

Ichthyophthirius multifiliis 50
identification points (IPs) 219
immunoglobulin E (IgE) 109, 110, 113
immunoproteomics 115
inductively coupled plasma mass spectrometry (ICP-MS)
  arsenic determination 304, 306
  chemical speciation 304
  food contact materials 241
  honey, origin testing 295
  truffles, authentication of 297–298
  wine, origin testing 295–296
inductively coupled plasma quadrupole mass spectrometry (ICPQ-MS) 303
inks 236
intentionally added substances (IASs) 232
internal residue loss (IRL) 97
International Olive Council 33
International Union of Pure and Applied Chemists 31
ion deflection 5
ion-exchange chromatography coupled with mass spectroscopy (IC-MS)
  biogenic amine analysis 352–353
  nitrites, determination of 362–363
ionisation energy (IE) 141
ionization potential, measurement of 13
ionization techniques 5–7, **6**
  desorption phase
    electrospray ionization **6**, 7
    fast-atom bombardment 6, **6**
    matrix-assisted laser desorption ionization **6**, 6–7
  gas phase
    chemical ionization 6, **6**
    electron impact 6, **6**
ionizing agent 6
ion mobility-mass spectrometry (IM-MS) 101
ion mobility spectrometry (IMS), carbohydrate analysis
  anomeric configurations 101–102
  isomeric carbohydrates, separation of 99–101, *101*
ion–molecule reactions, determination of 13
ion peaks 9
ions, trajectory of 7
ion traps (ITs) 67, 141, 350
IRMS *see* isotope ratio mass spectrometry (IRMS)
IR multiphoton dissociation (IRMPD) 96–97
irradiation 6–7
isobaric tags for relative and absolute quantitation (iTRAQ) 45
isobutane 142
isobutene 6
3-isobutyl-2-methoxypyrazine 146
isocyanates 349
isoflavones 192, 193
isotope abundance assessment 12
isotope ratio, determination of 13
isotope ratio mass spectrometry (IRMS) 140, 296–297
isoxanthohumol M 156, *156*

kale (*Brassica oleracea* L.) 188
kefir milk 197
ketoses 87, 99
6-keto-sitostanol 191
kinetic energy 140, 142, 296
kokumi effect 157

laser diode thermal desorption with atmospheric pressure chemical ionization mass spectrometry (LDTD-APCI-MS/MS) 408
L-asparagine 99
L-aspartic acid 98
LC *see* liquid chromatography (LC)
LDA *see* linear discriminant analysis (LDA)
lead, speciation analysis 308
Lewis acids 90

lignans 192
limits of detection (LODs) 93, 218, 219, 317
limits of quantitation (LOQs) 249
linear discriminant analysis (LDA) 293–295
linear ion traps (LIT) 383
linkage analysis
  nitrogen-containing carbohydrates 90
  partially methylated alditol acetates 88, *89*
linoleic acid 195
lipases 68
lipids
  classifications *32*, 32–33
  extraction 33
  phospholipids 32
  triacylglycerols 32
lipid-accumulating microorganisms 196
lipid analysis 189–196
  animal products
    cured meats 195
    fermented 195
    marine lipids 195–196
  in bulk oil/edible oils 33–37, **36–37**
  in dairy food products 37–40, **38–39**
  microbial and insects 196
  plant-based
    characterization of 190–194
    phospholipids 190–192
    phytoestrogens 192–194
    phytosterols 190–192
    quality control 194
    tocopherols *190*, 190–192
Lipid Metabolites and Pathways Strategy (LIPID) 189
liquid chromatography (LC) 66, 70, 404
  fat-soluble vitamins 70–71, **72**
  normal-phase 70
  reverse-phase 70
  water-soluble vitamins 71–75, **73–74**, *75*
liquid chromatography coupled with atmospheric pressure chemical ionization (LC–APCI–MS) 188
liquid chromatography coupled with diode array detection and electrospray ionization (LC–DAD–ESI-MS) 188
liquid chromatography coupled with diode array detection and electrospray ionization tandem mass spectrometry (LC–DAD–ESI–MS/MS) 188
liquid chromatography-electrospray ionization-mass spectrometry (LC-ESI-MS)
  low-molecular-weight peptides in champagne wine 407
  meat and meat products, authentication of 409
liquid chromatography-electrospray ionization quadrupole time-of-flight tandem mass spectrometry (LC-ESI-Q-TOF-MS/MS) 48, 184
liquid chromatography-electrospray ionization-tandem mass spectrometry (LC-ESI-MS/MS)
  conjugated linoleic acid detection 195
  mycotoxins detection 393–394
liquid chromatography-high-resolution mass spectrometry (LC-HRMS) 393
liquid chromatography inductively coupled plasma mass spectrometry (LC-ICP-MS) 361
liquid chromatography-mass spectrometry (LC-MS)
  for biogenic amine analysis 350–351
  carminic acid analysis in food 125, ***126***
  extracted ion chromatogram for sorbic acid 365, *365*
  food pathogens, identification of 331, 332, *335*, *336*
  heterocyclic amines assay 222
  persistent organic pollutants analysis *286*, 286–287
  pesticide analysis 217–222, *218*, **220–221**
  phytoestrogens, analysis of 194
liquid chromatography-tandem mass spectrometry (LC-MS/MS) 44
  adulteration in meat and meat products 320–321
  biogenic amine analysis 350
  cereals and cereal-based food, mycotoxins detection in 383, *384*, **386–392**
  food contact materials 240–241
  multi-residue pesticides, analysis of 222, **223–226**
  phytoestrogens, analysis of 193
  sulfites determination in foods 360–361
  veterinary drugs 249
liquid chromatography–triple quadrupole tandem mass spectrometry (LC-QQQ-MS/MS) 251
liquid extraction surface analysis nanoelectrospray mass spectrometry (LESA-nano-ESI-MS) 241
liquid-liquid extraction (LLE) 67, 71, 79, 256
liquid-phase microextraction (LPME) 67, 68
LLE *see* liquid-liquid extraction (LLE)
low-density polyethene (LDPE) **234**, 236
low-temperature plasma (LTP) ionization 422
L-serine 99
lupulone 154, *156*

machine learning 424
magnesium sulfate ($MgSO_4$) 257
major royal jelly proteins (MRJPs) 406
marine polysaccharides 199
mass analyzers 7–8, 113
  Fourier transform–ion cyclotron resistance 8
  functions 7
  orbitrap 8
  quadrupole 7–8
  tandem 8
  time-of-flight 7
mass detectors 8
MasSpec Pen technology 166
mass-selective detection 93
mass spectrometry (MS)
  analyzers in *334*
  application 9–14, *10*
    qualitative 10–11
    qualitative and quantitative 14
    quantitative 12–14
  data processing 423–425
    compound identification 423–424
    data mining 424
    open-source software and databases 424–425
    structure elucidation 423–424
  fragmentation rules 9
  instrumentation 5–8
    ionization techniques 5–7, **6**
    mass analyzers 7–8
    mass detectors 8
  interpretation 9
  ionization sources *333*
  peak types 9
  principles 4–5
mass spectrometry fingerprinting (MSF) 421–422
matrix-assisted laser desorption ionization (MALDI) **6**, 6–7, 14, 34, 47, 85

matrix-assisted laser desorption/ionization Fourier transform ion cyclotron resonance mass spectrometry (MALDI-FT-ICR-MS) 35
matrix-assisted laser desorption/ionization mass spectrometry (MALDI-MS)
  azaphilone, identification of 127
  phycocyanobilin, characterization and identification of 121
matrix-assisted laser desorption/ionization time-of-flight mass spectrometry (MALDI-TOF-MS) 421
  foodborne pathogens, detection of 330–331, 337–338, *337–339*
  honey, authentication of 405
  parabens, analysis of 369–370, *371*
  phenolics compounds analysis 188
  wines, authentication of 406
maximum residue levels (MRLs) 213
meat and meat products
  adulteration in 320–321
  authentication of 409, **410**
metabolomics 139
Metab R 148
metals 233–235
metastable ion peak 9
methane 6, 142
methanesulfonic acid 353
methanol (MeOH) 256, 261
methicillin-resistant *Staphylococcus aureus* (MRSA) 331, 337
method detection limits (MDL) 316
methylated glycans 93
methylation analysis 88–92, *89–91*
N-methyl-*N*-(trimethylsilyl)trifluoroacetamide (MSTFA) 78
microbes 330
microbial contamination, of food 337
microbial toxins 420
microplastics (MPs) 421
microwave-assisted extraction 67
microwave-induced plasma atomic emission (MIP-AED) 308
migration 231
milk
  allergenic proteins, identification of 113, **114**
  authentication of 407–408
  lipids, detection of 37–39, **38–39**
  melamine analysis in 408
  sulfonamide identification in 408
  veterinary antibiotic residues, determination of 251
mineral oil 420–421
mineral oil hydrocarbons (MOHs) 420–421
miniature mass spectrometers (MMSs) 422
minocycline residues, detection of 251
molecular formula, determination of 10
molecular ion peak 9
molecularly imprinted polymer solid-phase extraction-capillary zone electrophoresis tandem mass spectrometry (MISPE-CZE-MS/MS) method 342
molecular weight, determination of 10
Monascus pigments 125, *127*
  extraction 127
  fragmentation patterns of 132
Mongolian goji berries 192
monomethylarsonic acid (MMA) 306
monosaccharides, *de novo* determination of absolute configuration 97–99, *97–100*
monosodium glutamate (MSG) 149, 150
mozambioside 154, *155*
MRM *see* multiple-reaction monitoring (MRM)
MSF *see* mass spectrometry fingerprinting (MSF)
multicharged ion peak 9
multidimensional GC (MDGC), aroma and 144–145
multidimensional mass spectrometry 248
multidrug-resistant (MDR) foodborne pathogens 330
multi-elemental speciation 309, **310–315**
multilayer perceptron (MLP) 295
multiple-reaction monitoring (MRM) 21, 249, 317, 350, 358
multi-residue pesticides, LC-MS/MS for analysis of 222, **223–226**
multivariate analysis of variance (MANOVA) 293
multivariate statistical analysis 294
muscle atrophy 48
mushrooms 198
mycotoxins
  detection
    in cereals and pseudocereals *384*
    in grape and grape products 382, **382**
    LC–MS/MS 383, **385–392**
    in nuts, seeds, and milk thistle *384*
  determination with LC–HRMS 393
  fungal species 380, **380**
  GC–MS for sensitive estimation of 395–396
  identification with HPLC–MS/MS 394–395
  LC–ESI-MS/MS for 393–394
  structures of *381*
Mytilus galloprovincialis 50

N-acetyl hexosamines 90
nano high-pressure liquid chromatography-electrospray ionization-OrbiTrap-tandem mass spectrometry (nano-HPLC-ESI-OrbiTrap-MS/MS) 121–122
nanoparticle pollution 51
2-naphthylamine 216
natamycin 359, 371, **372**, *372*
National Health Interview Survey 110
National Institute of Standards and Technology (NIST) 423
natural colors 119
natural food preservatives 359–360
N-deglycosylated protein 95
negative chemical ionization (NCI) 143, 214
negative electron-capture dissociation (NECD) 97
negative electron transfer dissociation (NETD) 97
negative ion peak 9
niacin *see* vitamin B3
nicotinic acid *see* vitamin B3
nisin 359, 371, **372**, *372*
NIST Hybrid Search 148
nitrate salts 361–363
nitrites (NO_2–) 361–363
nitrosamines 362
N-methyl-*N*-(tert-butyl dimethylsilyl) trifluoroacetamide 368
non-intentionally added substances (NIASs) 232, 233
non-polar solvents 256
non-targeted screening, with LC/HR-MS 321, *322*
non-volatile compounds, *in vivo* flavour release analysis 161–162

nose-space analysis *159*, 160, 166
nuclear magnetic resonance (NMR) spectroscopy 127
nutraceuticals 181–182
nutraceutical value, evaluation of
bioactive peptides 196–197
lipids 189–196
phenolic compounds 183–188, **185–187**
polysaccharides 198–199
sample preparation 182–183

Oasis hydrophilic–lipophilic balance (HLB) 257, 259, 261–262
ochratoxin 382
3, 5-*O*-dicaffeoyl-*epi*-δ-quinide
HPLC-MS/MS analysis 154, *155*
LC-MS/MS spectrum 154, *154*
in roast coffee beverage 153, *153*
odour activity values (OAV) 147
odourants 139, 148
oil, adulteration in 319–320
oil crop, pesticides detection in 213, **215**, *217*
oleaginous microorganisms *see* lipid-accumulating microorganisms
olfactometry 145
oligosaccharides
cleavages of 94, 95, *96*
fragmentation of 94
olive oil, origin testing 292–293
one-dimensional (1D) gas chromatography 144
o-phthalaldehyde (OPA) 349, 350
optical biosensors 112
orbitrap mass analyzers 8
organically *vs.*onventionally grown foodstuffs 299–304
organic brown rice syrup (OBRS) 304
organic products, analysis of 302–304
organochlorine pesticides (OCPs) 216, 280, 282
organophosphorus insecticides 217
organotin, speciation analysis 308–309
origin testing
adzuki beans 298
baijiu 297
beer 296
distillers dried grains and solubles 296–297
durum wheat 298–299
gelatine 297
honey 293–295
olive oil 292–293
truffles 297–298
wine 295–296
oxidation, of Met 122
oxidative degradation 88
oxygen adical antioxidant capacity (ORAC) 299
oxytetracycline (OXY) 52

palmitic acid 37
pantothenic acid 69; *see also* vitamin B5
paper and cardboard 233
paper-spray (PS) ionization 422
parabens 367, *368*, **369**, **370**
PARAFAC2-based Deconvolution and Identification System (PARADISe) 148
PARAllel FACtor analysis 2 (PARAFAC2) 148
parallel reaction monitoring (PRM) 45, 50
partial least squares-discriminant analysis (PLS-DA) 164, 293
partially methylated alditol acetates (PMAAs)
EI mass spectra 88, *91*
fragmentation pathways for molecular ions 88, *90*
linkage analysis 88, *89*
partial molecular formula, determination of 10
patulin 420
PCR-coupled MS, for food authentication 331, 340–341
peanuts, allergenic protein identification 113, **114**
Pearson VII universal kernel (PUK) 294
Pegasus 4D instrument 300
Penicillium 125
pentadecanoic acid 195
peptides 44, 150
perchloric acid ($HClO_4$) 348
perfluorinated acids (PFCAs) 286
perfluorinated compounds (PFCs) 286
perfluorinated sulfonates (PFASs) 286
perfluorooctane sulfonate 280
persistent organic pollutants (POPs)
analytical approaches **285**
GC-MS analysis 282–283, **284**
LC-MS analysis *286*, 286–287
source identification using GC-MS/MS 283–284
types 280
pesticide analysis 213
GC-MS 214–217, **215**, *217*
LC–MS 217–222, *218*, **220–221**
LC–MS/MS 213, 222, **223–226**
phenolic compounds, analysis of 183–188, **185–187**
phenoxy herbicides 217
1-phenyl-3-methyl-2-pyrazoline-5-one (PMP) 93
phenylpropenes 359
phospholipids, determination of 190–192
photoinitiators 236
photomultipliers 8
phycocyanins (C-PCs) 119, *120*, 121–122
phycocyanobilin (PCB) 121
p-hydroxybenzoic acid 184
phytochemicals 183
analysis 14
unknown, structural elucidation of 14
phytoestrogens, determination of 192–194
phytosterols, determination of 190–192
pimaricin *see* natamycin
Plackett–Burman design (PBD) 394
plant-based lipids
characterization of 190–194
phospholipids 190–192
phytoestrogens 192–194
phytosterols 190–192
quality control 194
tocopherols *190*, 190–192
plant polysaccharides 198
plastic food-packaging materials 233, **234**
plastics 233
polyacetylenes quantification, in carrot roots 300
polyaromatic hydrocarbons (PAHs) 280
polybrominated diphenyl ethers (PBDEs) 280
polychlorinated biphenyls (PCBs) 280
polychlorinated dibenzofurans (PCDFs) 280
polychlorinated dibenzo-*p*-dioxins (PCDDs) 280
polyethene terephthalate (PETE) **234**

polyketides (PK) 32, *32*
polymerase chain reaction (PCR) 112
polymethoxylated flavones (PMFs) 153
polypropylene (PP) **234**
polysaccharides
  bacterial exopolysaccharides 199
  fungi 198–199
  marine 199
  plant 198
  small, profiling of *92*, 92–93
polystyrene (PS) **234**, 420
polytetrafluoroethylene (PTFE) 35
polyunsaturated fatty acids (PUFAs) 33
polyvinyl chloride (PVC) **234**
POPs *see* persistent organic pollutants (POPs)
Porphyrophora polonica 122
portable and miniature devices 422–423
positive ionisation (PCI) 142, 143
prenol lipids 32, *32*
pressurized liquid extraction (PLE) 67, 79, 124
primary, secondary amine (PSA) 216
principal component analysis (PCA) 164, *165*, 293
a priori 86
proanthocyanidins 152
procyanidins 153
progenesis 44
prolamins 408
protein-containing foods, adulteration in 316–317
proteins 33, 47
  as allergens 110
  biosensors in detection of food allergens 111–112
  identification of
    bottom-up method 113
    top-down method 113
    workflow for 114, *114*
proteolytic-peptide analysis 332
proteomics 44, 115, 403
  in aquaculture 43–44
  farming conditions on food quality and safety in aquaculture products 44–46, *46*
    abiotic hazards 50–51
    antibiotic resistance 51–53
    biotic hazards 49–50
    dietary management 47
    fish welfare and stress response 47–49
  mass analyzers in 113
  workflow 44–46, *45*
proton-transfer chemical-ionization mass spectroscopy 121
proton transfer reaction (PTR) 139
pseudocyanocobalamin 75
Pseudomonas aeruginosa 337
putrescine (PUT) 348
pyridoxine *see* vitamin B6
Pyropia yezoensis 196

Q-OIT hybrid technique, for mycotoxins detection 383
QqQ/QLIT method, for mycotoxins detection 383
quadrupole filters 218
quadrupole ion traps 218
quadrupole mass analyzer 7–8
quadrupole time-of-flight (Q-TOF) 67, 183, 188, 218
quercetin-3-glucoside 300
Quick, Easy, Cheap, Effective, Rugged and Safe (QuEChERS) method 213, 214, 216, 394

radiofrequency (RF) 7, 141, 361
raisins (dried grapes) 188
random forest (RF) 295
Rapid Alert System for Food and Feed (RASFF) 393
rapid resolution liquid chromatography-tandem mass spectrometry (RRLC-ESI-MS/MS) 76
rare earth elements (REEs) 421
rearrangement ion peak 9
reductive amination 93
regenerated cellulose films 235
regulations
  food additive 358
  food-labeling 110
relative standard deviation (RSD) 261, 264, 266
reprogramming proteomics 53
resistome 51, 53
retention index 77, 145
Rhodiola imbricata 76
riboflavin *see* vitamin B2
rice, GC×GC-ToF-MS analysis 144
rubber 235

saccharolipids (SL) 32, *32*
Salmonella enterica 337
Salmonella enteritidis 340
saponification 68, 78
saturated fatty acids 33
Schiff base 92, 93, 167
secoisolariciresinol 193
secondary ions 6
selected ion flow tube mass spectrometry (SIFT-MS) 293
selective ion monitoring (SIM) 20, *20*, 77, 138, 146, 214, 351
selective reaction monitoring (SRM) 21, *21*, 66, 113, 317, 350
selenium (Se), speciation analysis 307–308
sensometabolomics 151
sensomics 138, 149, 152, 157
sensoproteomics 151
sequential windowed acquisition of all theoretical fragment ion mass spectra (SWATH) 45, 51, 53, 157
shotgun proteomics 44
sialic acid 93
silicone rubber 235
silver nanoparticles (AgNPs) 51
silver (Ag), speciation analysis 309
silylation 77
SIM *see* selective ion monitoring (SIM)
simple lipids 189
simultaneous determination
  of fat-soluble vitamins (FSVs) **72**, 75–77
  of water-soluble vitamins **73–74**, 75–77
sinapinic acid 405
single mass spectrometry, veterinary drugs 248
single particle (SP)-ICP-MS, AgNPs quantification 309
size-exclusion chromatography (SEC) 92
sodium chloride (NaCl) 257
sodium cyanoborohydride 93
sodium dodecyl sulphatepolyacrylamide gel electrophoresis (SDS-PAGE) 47
soft independent modelling of class analogies (SIMCAs) 294
solid-phase extraction (SPE) 67, 68, 71, 79, 182–183, 257

solid-phase extraction–liquid chromatography/electrospray ionization–tandem mass spectrometry (SPE-LC-ESI-MS/MS) 251
solid-phase microextraction (SPME) 236
sorbic acid ($C_6H_8O_2$) 363–366, **364**
GC-MS EI fragments of 363, *364*
LC-MS extracted ion chromatogram for 365, *365*
MRM transition fragments 366, *366*
negative ESI and APCI mass spectra 365, *365*
soxhlet extraction 67, 348
soybean, allergenic protein identification 113, **114**
SPE *see* solid-phase extraction (SPE)
speciation
arsenic 304–307
chemical 304
chromium 308
lead 308
multi-elemental 309, **310–315**
organotin 308–309
selenium 307–308
silver 309
specific migration limit (SML) 231
spermidine 348
spermine 348
sphingolipids 32, *32*
Spirulina 420
SRM *see* selective reaction monitoring (SRM)
stable-isotope dilution assay (SIDA) 138
standard reference material (SRM) 305
Staphylococcus aureus 49, 340
sterol lipids 32, *32*
Stockholm Convention 280, 282
Streptomyces chmanovgensis 371
Streptomyces gilvosporeus 371
Streptomyces natalensis 371
strong anion exchange (SAX) cartridges 69
strong aroma-type baijiu (SAB) 297
structural elucidation 423–424
of azaphilone 127
of marine polysaccharides 199
of oligosaccharides and glycoconjugates by tandem mass spectrometry 94–97, *96*
of unknown phytochemicals 14
sulfites 360–361
sulfur dioxide (SO_2) 360–361
supercritical fluid chromatography (SFC) 71
supercritical fluid extraction (SFE) 67, 124
support vector machines (SVM) 294
surface plasmon resonance (SPR) 112
symmetrization, of methylated carbohydrates 88
synthetic compounds, sensory evaluation of 150

tagging
of reducing carbohydrate 93, *94*
of reducing disaccharide by hydrazone formation 93, *95*
TAGs *see* triacylglycerols (TAGs)
Talaromyces 125
tandem mass analyzers 8
tandem mass spectrometry (MS/S)
carbohydrate analysis
de novo determination of absolute configuration of monosaccharides 97–99, *97–100*
structure elucidation of oligosaccharides and glycoconjugates 94–97, *96*
veterinary drugs 248
tandem mass tag (TMT) 45
targeted analysis
features 19, *20*
multiple reaction monitoring 21
m/z ratio of ions 20
selected ion monitoring 20, *20*
selected reaction monitoring 21, *21*
targeted acquisition using MS 20–21
using MS/MS 21
workflow 24, *24*
targeted proteomics 45, 50, 51
taste activity values (TAV) 149
taste, and HPLC-MS 148–158
taste dilution analysis (TDA) 150, 151
temporal check-all-that-apply (TCATA) 139
temporal dominance of sensations (TDS) 139
terpenoids 359
tetracyclines 52
tetramethylsilane 142
thermodynamics 13
thiamine *see* vitamin B1
3D printing 423
time intensity (TI) 138
time-of-flight (ToF) mass analyzer 7, 142
time-of-flight mass spectrometry (TOF-MS) 184, 196
titanium dioxide nanotubes (TDNTs) 308
tocopherols *190*, 190–192
TonB protein 50
total anthocyanin content (TAC) 299
total phenolic content (TPC) 299
trans-fatty acids 33
trans-isomers 13
trapped ion mobility spectrometry (TIMS) 101
triacylglycerols (TAGs) 31–34, 189, 194
trichloroacetic acid (TCA) 348
trichothecenes (TCTs) 383, 420
triethyl lead chloride (TEL) 308
trifluoroacetic acid 405
triglycerides 68
trimeric ion complex 99, *100*
trimethyl lead chloride (TML) 308
trimethylsilylation 77
triple quadrupole (QQQ) 66–67, 141, 350
tripyrrole bilin (TPB) 122
tropomyosin 47, 50
truffles, origin testing 297–298
trypsin 44, 122
tryptamine 348
Tuber indicum 298
Tuber magnatum 298
Tuber melanosporum 298
turbo ion-spray source–mass spectrometry (TIS-MS) 286
two-dimensional (2D) chromatography 142
two-dimensional gel electrophoresis (2-DE) 44, 47, 48
two-dimensional LC×LC-DAD-APCI-MS method 71
tyramine (TA) 348

ultra-high-performance liquid chromatography (UHPLC) 70
ultra-high-performance liquid chromatography coupled with quadrupole Orbitrap high-resolution mass spectrometry (UHPLC-Q-Orbitrap-HRMS) 252

ultra-high-performance liquid chromatography coupled with triple quadrupole mass spectrometry (UHPLC-QQQ-MS) 194, 241
ultra high-performance liquid chromatography–high-resolution mass spectrometry (UHPLC-HRMS) 197
ultra-high-performance liquid chromatography-quadrupole time-of-flight (UHPLC-QToF) 125
ultra-high-performance liquid chromatography-tandem mass spectrometry (UHPLC-MS/MS)
  phytoestrogens, analysis of 193
  veterinary drug residues, detection of 251–252
ultra-performance liquid chromatography coupled with quadrupole time-of-flight tandem mass spectrometry (UPLC–Q-TOF-MS/MS) 194
ultra-performance liquid chromatography–mass spectrometry (UPLC-MS) 362
ultra-performance liquid chromatography (UPLC)-MS/MS
  bitter peptides, identification of 152
  cocoa analysis 153
  veterinary antibiotic residues in raw milk, determination of 251
ultrasonic-assisted enzymatic extraction (UAEE) 304, 305
ultrasonic liquid processor (ULP) 300
ultraviolet photodissociation 97
undermethylation 88
unknown compounds, identification of 14
untargeted analysis
  data-dependent analysis 21, 22, *22*
  data-independent acquisition 22–23, *23*
  features 19, *20*
  workflow 24, *25*
uronic acids 88
US Food Labelling and Consumer Protection Act 110

vanillylmandelic acid (VMA) 350
vapor pressure (VP) 13
vegetable oils 33, 34, 47, 195
veterinary drugs
  antibiotics 247
  classification 247, **248**
  high-resolution mass spectrometry 248
  multidimensional mass spectrometry 248
  residues, determination in food samples **249–250**
  single mass spectrometry 248
Vibrio mimicus 49
virgin olive oil (VOO) 33, 34, **36**
vitamers 61
vitamins
  dietary allowances of 61
  fat-soluble 61
    extraction 67–68
    gas chromatography 77–78
    liquid chromatography 70–71, **72**
    simultaneous determination of 75–77
  food fortification with 66
  functions 61, **62–65**
  instability 66
  sources 61, **62–65**
  structure 61, **62–65**
  water-soluble 61
    extraction 68–69
    gas chromatography 77–78
    liquid chromatography 71–75, **73–74**, *75*
    simultaneous determination of 75–77
vitamin A **64**
vitamin B1 **62**, 69, 71, 77
vitamin B2 **62**, 69, 71
vitamin B3 **62**, 69, 71
vitamin B5 **62**, 71, 77
vitamin B6 **63**, 69, 71
vitamin B7 **63**, 69
vitamin B9 **63**, 69, 71, 75
vitamin B12 **64**, 69, 71
vitamin C **64**, 69, 77
vitamin D **64**
vitamin E **64**, 77
vitamin K **64**, 68
volatile organic compounds (VOCs) 137, 139, 141, 162, 164, 236
volatiles compounds, *in vivo* flavour release analysis 158–161
volatilomics 139–140
VOO *see* virgin olive oil (VOO)

water-soluble extracts (WSE) 150–152
water-soluble vitamins (WSVs)
  extraction 68–69
  gas chromatography 77–78
  liquid chromatography 71–75, **73–74**, *75*
  simultaneous determination of **73–74**, 75–77
wheat, allergenic protein identification 113, **114**
wine
  authentication of 406–407
  origin testing 295–296
wood and fibre 236
World Allergy Organization 109
World Health Organization (WHO) 282, 306, 367

xanthotoxin 420
xenobiotics analysis, in milk by UHPLC-HRMS/MS
  absolute and apparent recovery *265*, 265–266
  additives, influence of 262–264, **263**
  clean-up 259
  commercial and breast milk samples
    suspect analysis 270, *270*, **271–272**
    target analysis 267, **268–269**
  instrumental and procedural repeatability **266**, 266–267
  limits of identification (LOIs) 267
  limits of quantification (LOQs) 264
  linearity ranges and determination coefficients 264
  milk samples 257
  optimization
    EDTA 261–262
    Fluoroquinolones 261–262
    Oasis HLB 261–262
    protein precipitation 261
  reagents 257
  suspect screening 260
  target analysis 260
  UHPLC-qOrbitrap 259–260
  validation method 260–261
  xenobiotics, extraction of 258

zearalenone 420

Printed in the United States
by Baker & Taylor Publisher Services